T0237625

ASTRONOMY AND ASTROPHYSICS LIBRARY

Series Editors: Martin A. Barstow, University of Leicester, Leicester, UK
Andreas Burkert, University Observatory Munich, Munich, Germany
Athena Coustenis, Paris-Meudon Observatory, Meudon, France
Roberto Gilmozzi, European Southern Observatory (ESO), Garching, Germany
Georges Meynet, Geneva Observatory, Versoix, Switzerland
Shin Mineshige, Department of Astronomy, Kyoto, Japan
Ian Robson, The UK Astronomy Technology Centre, Edinburgh, UK
Peter Schneider, Argelander-Institut für Astronomie, Bonn, Germany
Steven N. Shore, Università di Pisa, PISA, Italy
Virginia Trimble, University of California, Irvine, USA
Derek Ward-Thompson, University of Central Lancashire, Preston, UK

Shoji Kato • Jun Fukue

Fundamentals of Astrophysical Fluid Dynamics

Hydrodynamics, Magnetohydrodynamics, and Radiation Hydrodynamics

Springer

Shoji Kato
Kyoto University
Kyoto, Japan

Jun Fukue
Division of Science Education
Osaka Kyoiku University
Kashiwara, Osaka, Japan

ISSN 0941-7834 ISSN 2196-9698 (electronic)
Astronomy and Astrophysics Library
ISBN 978-981-15-4176-6 ISBN 978-981-15-4174-2 (eBook)
https://doi.org/10.1007/978-981-15-4174-2

© Springer Nature Singapore Pte Ltd. 2020, corrected publication 2023
This work is subject to copyright. All rights are reserved by the Publisher, whether the whole or part of the material is concerned, specifically the rights of translation, reprinting, reuse of illustrations, recitation, broadcasting, reproduction on microfilms or in any other physical way, and transmission or information storage and retrieval, electronic adaptation, computer software, or by similar or dissimilar methodology now known or hereafter developed.
The use of general descriptive names, registered names, trademarks, service marks, etc. in this publication does not imply, even in the absence of a specific statement, that such names are exempt from the relevant protective laws and regulations and therefore free for general use.
The publisher, the authors, and the editors are safe to assume that the advice and information in this book are believed to be true and accurate at the date of publication. Neither the publisher nor the authors or the editors give a warranty, expressed or implied, with respect to the material contained herein or for any errors or omissions that may have been made. The publisher remains neutral with regard to jurisdictional claims in published maps and institutional affiliations.

This Springer imprint is published by the registered company Springer Nature Singapore Pte Ltd.
The registered company address is: 152 Beach Road, #21-01/04 Gateway East, Singapore 189721, Singapore

Preface

Astrophysical fluid dynamics is one of the important branches of astrophysics. In astrophysics, we apply all kinds of physics to investigate and understand the structure and evolution of the Universe. Because information from the Universe were primarily brought to us through optical observations, physics of radiative transfer was first developed in astrophysics. Since then, various branches of physics including nuclear physics, high energy physics, and general relativity have been applied and developed in astrophysics. The application of fluid dynamics to astrophysics is, however, relatively late, compared with the above-mentioned fields of physics, since full-blown studies on dynamics of astrophysical objects began after elucidation of their structure.

In the present advanced stages of astrophysics, however, studies on astrophysical objects from the dynamical viewpoints are important. Since many astrophysical objects are gaseous systems subject to magnetic and radiation fields, their studies from the viewpoints of magnetized fluids and radiative fluids are important. The purpose of this textbook is to provide the basics of astrophysical hydrodynamics, magnetohydrodynamics, and radiation hydrodynamics to advanced undergraduates and above. At the same time, we hope that the materials presented in this textbook are helpful for researchers to organize their knowledge on astrophysical fluid dynamics.

The present book consists of three parts. In Part I, we describe the fundamental and important processes in astrophysical fluid dynamics, and, in Part II, some important magnetohydrodynamical processes are presented. In Part III, starting from the basic radiative transfer, we describe radiative fluid dynamics.

The original version of the book has been revised. A correction to this book can be found at https://doi.org/10.1007/978-981-15-4174-2_27

Finally, we appreciate many colleagues we met in our research activities. Without collaborations and arguments with them, the present book could not be written. We thank the Springer Nature for publishing this textbook of astrophysical fluid dynamics.

Kyoto, Japan
Osaka, Japan
2 February 2020

Shoji Kato
Jun Fukue

Contents

Part II Magnetohydrodynamical Phenomena in Astrophysical Objects

List of Symbols

Symbols	Meanings
$(+,-,-,-)$	Space-time signature
Greek suffix $\alpha, \beta, \gamma, \cdots$	1, 2, 3
Latin suffix $i, j, k, \cdots$	0, 1, 2, 3
(x, y, z)	Cartesian coordinates
(r, φ, z)	Cylindrical coordinates
(r, θ, φ)	Spherical coordinates
$\boldsymbol{A}$	Magnetic vector potential defined by $\boldsymbol{B} = \text{curl}\,\boldsymbol{A}$
$\boldsymbol{A}, A_i$	Schwarzschild discriminant vector
A	albedo
$\mathcal{B}$	Boltzmann number
$\boldsymbol{B} = (B_r, B_\varphi, B_z)$	Magnetic flux density
$\boldsymbol{B}_\mathrm{p}$	Poloidal magnetic field
$\boldsymbol{B}_\mathrm{t}$	Toroidal magnetic field
$B = \sigma_\mathrm{SB} T^4/\pi,\ B_\nu$	(Frequency-integrated) blackbody intensity
C_P	Specific heat of constant pressure
C_V	Specific heat of constant volume
$\mathcal{D}$	Denominator
D	Distance
$\mathcal{E}$	Total energy of particle
$\boldsymbol{E}, E^i$	Electric field
E	Energy
E, E_ν	(Frequency-integrated) radiation energy density
E_b	Binding energy of unit mass
E_co	Radiation energy density in comoving frame
$\mathcal{F}$	Energy flux
$F^{\mu\nu}$	Electromagnetic field tensor

Symbols	Meanings
$\boldsymbol{F}$, F^i	Radiative flux vector
F, F_ν	(Frequency-integrated) radiative flux
$F_{\rm co}$	Radiative flux in comoving frame
G	Gravitational constant
$G(r)$	Torque at radius r
H, H_ν	Eddington flux
H	Disk half-thickness
$H = U + p/\rho$	Enthalpy
H	Magnetic helicity
H	Scale height
$H_{\rm F}$	Fluid helicity
$H_{\rm c}$	Cross helicity
$\Im$	Imaginary part
I, I_ν	(Frequency-integrated) specific intensity
J, J_ν	Mean intensity
K, K_ν	K integral
K	Thermal conductivity
$K_0, K_1, \cdots$	Modified Bessel functions
$K_{\rm rad} = c^2 \chi_{\rm rad}$	Radiative conductivity
L	Luminosity
L	Size of systems
L	Specific angular momentum
$L_{\rm E}$	Eddington luminosity
$L_{\rm adv}$	Advective luminosity
$L_{\rm d}$	Disk luminosity
$L_{\rm diff}$	Diffusive luminosity
$L_{\rm mb}$	Specific angular momentum of marginally bound circular orbit
$L_{\rm ms}$	Specific angular momentum of marginally stable circular orbit
$L_\odot$	Solar luminosity
$\mathcal{M}$	Mach number
$\mathcal{M}$	Magnetic energy
M^{ij}	Energy-momentum tensor for magnetic field
M_{ij}	Maxwell stress tensor (Chap. 11)
M	Mass
$M_\odot$	Solar mass
$\dot{M}$	Mass-flow rate
$\dot{M}_{\rm crit} = L_{\rm E}/c^2$	Eddington accretion rate
$\mathcal{N}$	Numerator
$\boldsymbol{N}$, N^i	Viscous force
N	Brunt–Väisälä frequency
N	(Collisional) number

Symbols	Meanings
P^{ij}, P_ν^{ij}	(Frequency-integrated) radiation stress tensor
P	Period
$P = aT^4/3$	Radiation pressure
$P_{\rm co}$	Radiation pressure in comoving frame
Q	Toomre's Q value
$Q^- = \int q^- dz$	Vertically integrated cooling rate
$Q^-_{\rm adv} = \int q^-_{\rm adv} dz$	Vertically integrated advective cooling rate
$Q^-_{\rm rad} = \int q^-_{\rm rad} dz$	Vertically integrated radiative cooling rate
$\Re$	Real part
$\mathcal{R}$	Gas constant
$\mathcal{R} = v\ell/\nu$	Reynolds number
$\mathcal{R}_{\rm m} = v\ell/\eta$	Magnetic Reynolds number
$R^{\mu\nu}$	Energy-momentum tensor for radiation
$R_{\rm crit}$	Critical radius
$R_{\rm ph}$	Photospheric radius
$R_{\rm trap}$	Photon trapping radius
$R_\odot$	Solar radius
S, S_ν	Source function
T^{ij}, T_{ij}	Energy-momentum tensor
$T_{ij} = \int t_{ij} dz$	Vertically integrated viscous stress tensor
T	Temperature
T_0	Unperturbed temperature
T_1	Perturbed temperature
$T_{\rm b}$	Brightness temperature
$T_{\rm c}$	Color temperature
$T_{\rm e}$	Electron temperature
$T_{\rm eff}$	Effective temperature
$T_{\rm i}$	Ion temperature
$T_{\rm rad}$	Radiation temperature
$T_{\rm vir}$	Virial temperature
$\boldsymbol{U}$	Mean flow
U	Internal energy per unit mass
V	Volume
$V_{\rm A}$	Alfvén speed
$W(r)$	Dilution factor
X	Hydrogen abundance
Y	Helium abundance
Z	Metal abundance
$Z_{\rm i}$	Atomic number

Symbols	Meanings
$a = 4\sigma/c$	Radiation constant
a_d	Dust radius
a_*	Spin parameter
$\boldsymbol{b} = (b_r, b_\varphi, b_z)$	Perturbed magnetic flux density
b	Impact parameter
c	Speed of light
c_A	Alfvén speed
c_T	Isothermal sound speed
c_g	Group velocity
c_rad	Radiation sound speed
c_s	Adiabatic sound speed
d	Distance
e	Elementary charge
e	Internal energy per unit volume
f^{ij}	Eddington tensor
$\boldsymbol{f}_\mathrm{rad}$, f_rad	Radiative force
$f(\boldsymbol{r}, \boldsymbol{v}, t)$	Distribution function
$f \propto I_\nu/\nu^3$	Invariant phase space density of photons
f, f^i	Radiation flux (vector)
f, f_ν	Eddington factor
g_{ij}	Space-time metric
$\boldsymbol{g}$, g	Gravitational acceleration
h	Planck constant
$\boldsymbol{j}$	Electric current
j, $j_\nu = 4\pi\epsilon_\nu$	(Frequency-mean) mass emissivity
k^μ	Four-momentum of photon
k, $\boldsymbol{k}$	Wave number (vector)
k_B	Boltzmann constant
k_R, k_I	Real, imaginary parts of wave number
$\boldsymbol{l}$, l^i	Direction cosine
l	Mean free path
l	Mixing length
$\ell(= r^2\Omega)$	Specific angular momentum
ℓ, ℓ_ν	Mean free path
$\ell_* = \sqrt{N}\ell_\nu$	Effective mean path
M	Mass
$M_\odot$	Solar mass
$m = M/M_\odot$	Normalized mass
m	Number of nodes in azimuthal direction
m	Particle mass
m_H	Mass of hydrogen atom

Symbols	Meanings
$m_{\rm e}$	Electron mass
$m_{\rm i}$	Ion mass
$m_{\rm p}$	Proton mass
$\dot{m} = \dot{M}/\dot{M}_{\rm crit}$	Normalized accretion rate
n	Number density
$n_{\rm e}$	Electron number density
$n_{\rm i}$	Ion number density
$n_{\rm p}$	Proton number density
$n_\nu \propto I_\nu/\nu^3$	Photon occupation number
$p = (\mathcal{R}/\bar{\mu})\rho T$	Gas pressure
p_0	Unperturbed pressure
p_1	Eulerian perturbation of pressure
$p_{\rm gas}$	Gas pressure
$p_{\rm mag}$	Magnetic pressure
$p_{\rm rad}$	Radiation pressure
$p_{\rm tot}$	Total pressure
$q(\tau)$	Hopf function
q^+	Heating rate per unit volume
q^-	Cooling rate per unit volume
$\boldsymbol{r}$	Position vector
r	Radius
$r_{\rm IL}$	Radius of inner Lindblad resonance
$r_{\rm OL}$	Radius of outer Lindblad resonance
$r_{\rm S} = 2GM/c^2$	Schwarzschild radius
$r_{\rm c}$	Corotation radius
$r_{\rm gy}$	Gyration radius
$r_{\rm in}$	Inner radius of accretion disks
$r_{\rm mb}$	Marginally bound circular radius
$r_{\rm ms}$	Marginally stable circular radius
s	Specific entropy
s	Path length along light-ray
t	Time
$t^{\mu\nu}$	Viscous stress tensor
t_{ij}	Viscous stress tensor
$t_{\rm cool}$	Cooling timescale
$t_{\rm dyn}$	Dynamical timescale
$t_{\rm ff}$	Freefall timescale
$t_{\rm th}$	Thermal timescale
$\boldsymbol{u}$	Relative velocity
u^μ	Four-velocity of matter
u, u_ν	Radiation energy density
$\boldsymbol{u} = (u_r, u_\varphi, u_z)$	Perturbation of velocity

Symbols	Meanings
$\boldsymbol{v} = (v_r, v_\varphi, v_z)$	Velocity vector
$\boldsymbol{v}_1$	Perturbed velocity vector
x^μ	Space-time coordinate
z	Redshift
$\Gamma = L/L_\mathrm{E}$	Normalized luminosity, Eddington parameter
$\Gamma_1, \Gamma_2, \Gamma_3$	Generalized ratios of specific heats
$\Gamma_\mathrm{d} = L_\mathrm{d}/L_\mathrm{E}$	Normalized disk luminosity
$\Delta\nu_\mathrm{D}$	Doppler width
$\Lambda_\nu = 1/\sqrt{3\varepsilon_\nu}$	Thermalization length
$\Pi = \int p dz$	Vertically integrated pressure
Π_e	Vertically integrated electron pressure
Π_{ij}	Vertically integrated viscous stress tensor
$\Sigma = \int \rho dz$	Surface density
Ω	Angular speed of rotation
Ω	Solid angle
$\Omega_\perp$	Vertical epicyclic frequency
Ω_c	Angular speed at corotation radius r_c
Ω_K	Keplerian angular speed
α	Viscosity parameter
$\alpha_\nu = \kappa_\nu \rho$	Absorption coefficient [cm^{-1}]
$\beta = v/c$	Normalized velocity
$\beta = p_\mathrm{gas}/(p_\mathrm{rad} + p_\mathrm{gas})$	Ratio of gas to total pressure
$\beta_\mathrm{ad} = g/C_\mathrm{p}$	Adiabatic temperature gradient
$\beta_\mathrm{mag} = p_\mathrm{gas}/p_\mathrm{mag}$	Plasma beta
$\beta_\nu = \sigma_\nu \rho$	Scattering coefficient [cm^{-1}]
γ	Specific heat ratio, adiabatic index
$\gamma = (1 - v^2/c^2)^{-1/2}$	Lorentz factor
$\gamma_{ij} = -g_{ij}$	Three-dimensional part of space-time metric
δ	Lagrangian variation
δ_{ij}	Kronecker's delta
ϵ	Heat generation rate by subatomic processes
$\epsilon_\nu = j_\nu/4\pi = \eta_\nu/\rho$	Mass emissivity
ε	Relativistic internal energy
$\varepsilon, \varepsilon_\nu = \kappa_\nu/(\kappa_\nu + \sigma_\nu)$	Photon destruction probability
ζ (= curl $\boldsymbol{v}/2$)	Vorticity
ζ	Bulk viscosity
ζ	Vorticity
$\boldsymbol{\eta}$	Lagrangian displacement vector
$\hat{\boldsymbol{\eta}}$	Displacement vector defined by $\boldsymbol{\eta} = \Re[\hat{\boldsymbol{\eta}}\exp(\mathrm{i}\omega t)]$
η	Dynamical viscosity
η	Energy-conversion efficiency

Symbols	Meanings
η	Magnetic diffusivity
$\eta_\nu = \epsilon_\nu \rho = j_\nu \rho / 4\pi$	Volume emissivity
$\eta_{\rm rad}$	Radiative viscosity
θ	Polar angle
$\theta_{\rm e} = k_{\rm B} T_{\rm e} / m_{\rm e} c^2$	Normalized electron temperature
κ	Epicyclic frequency
κ, κ_ν	Absorption opacity
$\kappa_{\rm R}$	Rosseland mean opacity
$\kappa_{\rm es}$	Electron scattering opacity
$\kappa_{\rm ff}, \kappa_\nu^{\rm ff}$	Free-free opacity
$\kappa_{\rm th} = K / \rho C_{\rm p}$	Thermometric conductivity (Chaps. 8 and 16)
λ	Eigenvalue
λ	Wavelength
$\lambda_{\rm C}$	Compton wavelength
$\lambda_{\rm D}$	Debye length
λ_ν	Flux limiter
$\mu = \cos\theta$	Direction cosine
$\bar{\mu}$	Mean molecular weight
ν	Frequency
$\nu = \eta / \rho$	Kinematic viscosity
$\boldsymbol{\xi}$	Lagrangian displacement vector
$\hat{\boldsymbol{\xi}}$	Displacement vector defined by $\boldsymbol{\xi} = \Re[\hat{\boldsymbol{\xi}} \exp(i\omega t]$
$\breve{\boldsymbol{\xi}}$	Displacement vector defined by $\boldsymbol{\xi} = \Re[\breve{\boldsymbol{\xi}} \exp(i\omega t - im\varphi)]$
ξ	Dimensionless entropy gradient
$\varpi, \varpi_\nu = 1 - \varepsilon_\nu$	Single-scattering albedo
$\rho = mn$	Density
ρ_0	Unperturbed density
ρ_1	Eulerian perturbation of density
σ	Charge density
σ	Electric conductivity (Chap. 10)
σ, σ_ν	Scattering opacity
$\sigma_{\rm SB} = ac/4$	Stefan–Boltzmann constant
$\sigma_{\rm T}$	Thomson scattering cross section
τ_{ij}	Viscous tensor
τ, τ_ν	Optical depth
$\tau_{\rm S}$	Sobolev optical depth
$\tau_{\rm corr}$	Correlation timescale
$\tau_{\rm d}$	Dynamical time
$\tau_{\rm dif}$	Diffusion timescale
$\tau_{\rm dyn}$	Dynamical timescale
$\tau_{\rm gyr}$	Gyration timescale

Symbols	Meanings
$\tau_{\rm r}$, $\tau_{\rm rel}$	Relaxation timescale
$\tau_* = \sqrt{(\tau_{\rm es} + \tau_{\rm ff})\tau_{\rm ff}}$	Effective optical depth
ϕ	Gravitational potential
ϕ_ν	Scattering redistribution function
φ	Azimuthal angle
χ	Ionization degree
χ, $\chi_\nu = \alpha_\nu + \beta_\nu$	Extinction coefficient
$\psi < 0$	Gravitational potential
ω	Angular frequency of waves
$\omega_{\rm D}$	Angular frequency of tidal waves
$\omega_{\rm R}$, $\omega_{\rm I}$	Real, imaginary parts of angular frequency of waves
$\omega_{\rm i}$	Imaginary part of frequency
$\omega_{\rm i}$	Ion cyclotron frequency
$\omega_{\rm e}$	Electron cyclotron frequency

Chapter 1
Introduction

In this textbook we study dynamical phenomena in astrophysical objects from fluid dynamical and related viewpoints. In this chapter, as the first step for the study, we consider temperature–density diagram, and discuss what astrophysical objects can be regarded as fluid dynamical systems in this diagram. To clarify the conditions under which astrophysical systems can be regarded as fluid dynamical ones, we examine Debye length, mean free path, gyration radius. We see that the domain on the diagram where the fluid dynamical treatments are possible is rather broad, and many astrophysical objects are diverged on the domain, from low to high temperature and from low to high density. Since the domain where hydrodynamical treatments are possible is wide, in some situations magnetic fields, radiation fields, and sometimes general relativity are important to determine the fluid dynamical behaviors.

1.1 Objects Treated in Astrophysics

Astronomy is the science to investigate the structure and the evolution of the Universe in order to deepen our knowledge and understanding on the Universe. Astrophysics is a part of astronomy to make such investigation from the viewpoints of physics. Objects of the investigation cover all ingredients of the Universe, from stars and interstellar gases to galaxies, clusters of galaxies, and the Universe itself.

To know objects outside ourselves, observation is necessary. In the case of astronomy, in particular, we cannot touch or handle celestial objects except for those in the Solar system, and thus observation is essential. Before the World War II, our ways for observation of astrophysical objects are only through visual light. Nowadays, however, we can get information of the Universe through all wave range of electromagnetic waves, i.e., radio waves to γ-rays. In addition to electromagnetic

© Springer Nature Singapore Pte Ltd. 2020, corrected publication 2023

S. Kato, J. Fukue, *Fundamentals of Astrophysical Fluid Dynamics*,
Astronomy and Astrophysics Library,
https://doi.org/10.1007/978-981-15-4174-2_1

waves, cosmic rays and neutrinos have been observed from the Universe. Quite recently, furthermore, gravitational waves have been detected.

To know physical states of observed objects from information obtained, we need a priori what physical states present us what information. In this sense, the branch of astrophysics developed first was astrophysical spectroscopy and the theory of radiative transfer. Subsequently, the theory of stellar structure was developed. As observing wave ranges and observational techniques extend, astrophysics has been widened into various branches.

Research work in astrophysics is broadly divided into two steps. The first one is the step to know the physical states of observed objects from observed quantities (observations and analyses). The next is the step to examine how the observed physical states are realized (theory). This book is concerned with the second stage, and the purpose is to review the dynamical processes which are needed to understand the physical states of observed objects. Particularly, we concern in this book with fluid dynamical processes in astrophysical objects. Important ingredients in fluid dynamical systems are gases, photons, magnetic fields, and high energy particles. They interact with each other and their energies are exchanged among them. In this introduction we try to have a rough picture concerning physical states of the gases of astrophysical objects. For example, gases in astrophysical objects are characterized by their large scale. Due to this, the astrophysical fluid systems have various different characteristics from those on the Earth. We briefly emphasize important characteristics of astrophysical fluid systems in following sections.

1.2 Gaseous States of Astrophysical Objects on Temperature–Density Diagram

In this section we first examine on which domain of the temperature-density diagram gaseous systems can be regarded as an ensemble of free particles. Next, on which parts of the domain effects of electron degeneracy, magnetic fields, radiation fields, and general relativity are of importance in consideration of gaseous motions are briefly mentioned.

1.2.1 Condition for Ensemble of Free Particles

Let us consider fully ionized gases. In particular, we consider hydrogen gases in order to avoid unnecessary complications. If mean kinetic energy of a particle is sufficiently larger than mean interaction energy of the particle with surrounding particles, the particle can be regarded as a free particle. This condition is written as

$$\frac{3}{2}k_{\rm B}T \gg \frac{e^2}{r}, \tag{1.1}$$

where T is temperature of the gas, k_{B} is the Boltzmann constant, e is the elementary charge, and r is the mean distance among particles. Since the number density n of particles and r are related by $4\pi nr^3/3 = 1$, the above relation can be written as

$$T \gg 1.8 \times 10^{-3} n^{1/3}. \tag{1.2}$$

Condition (1.2) is realized in many astrophysical gases unless extreme situations are considered. For example, in the central part of the Sun its density n is more than $100\,\mathrm{g\,cm^{-3}}$, i.e., $n > 6 \times 10^{25}\,\mathrm{cm^{-3}}$. This density is rather high in the earth, but condition (1.2) is realized at the center of the Sun, since the temperature is as high as 10^7 K. Before presenting the domain where condition (1.1) or (1.2) is realized on temperature–density diagram, we shall rewrite the condition (1.1) in an elegant form which represents more deeply the meaning of an ensemble of free particles.

1.2.1.1 Debye Shielding Length λ_{D}

The gases concerned in astrophysics are generally ionized (i.e., plasma). Ionized gases consist of electrons and ions. Even if they are neutral as the whole, the behaviors of ionized gases are different from gases consisting of neutral particles. There are two main differences. One is that the particles of different charge interact with a long range force (Coulomb force). The second is the presence of plasma oscillations in the ensemble of the same charged particles. The former characteristic is discussed here in relation to the condition of an ensemble of free particles, i.e., inequality (1.1).

Here, we pay our attention on an ion. Around this ion, many electrons path through by thermal motions. Since an attractive force acts between ion and electron, electrons cluster statistically around the ion, although they pass through the ion. By this clustering of electrons, the Coulomb force of the ion cannot reach far outside beyond a certain distance.

Let us consider quantitatively this shielding effect by electrons. If there is no shielding effect, the potential of the ion, say $\phi(r)$, is given by $\phi(r) = e/r$. If the shielding effect acts, the potential decreases more sharply with increase of r. This sharp decrease of $\phi(r)$ is described by the Poisson equation:

$$\nabla^2\phi = -4\pi\sigma(r), \tag{1.3}$$

where $\sigma(r)$ is the distribution of charge around the ion. In the state where there is no clustering, ion and electron densities are balanced so that the gas is neutral. The number density of electrons in such mean neutral state is denoted by n_{e}. Then, by the clustering around an ion, the number density of electron around the ion, n, increases from n_{e} to

$$n(r) = n_{\mathrm{e}}\exp(e\phi/k_{\mathrm{B}}T), \tag{1.4}$$

and the distribution of charge around the ion, $\sigma(r)$, is

$$\sigma(r) = -e(n - n_e). \tag{1.5}$$

By assuming that $e\phi/k_BT$ is smaller than unity, we expand $\exp(e\phi/k_BT)$. Then, from Eqs. (1.3)–(1.5) we have

$$\nabla^2\phi = \frac{4\pi e^2 n_e}{k_BT}\phi. \tag{1.6}$$

By adopting the boundary condition that at the far outside ϕ vanishes, we have as the solution of Eq. (1.6)

$$\phi \propto \frac{\exp(-\kappa r)}{r}, \tag{1.7}$$

where

$$\kappa = \left(\frac{4\pi e^2 n_e}{k_BT}\right)^{1/2}. \tag{1.8}$$

Equation (1.7) shows that by the screening effect of electrons the ion potential decreases with r in the form of the Yukawa potential. The characteristic length, λ_D, of the screening can be regarded as $1/\kappa$. This length is called *Debye shielding distance* or *Debye screening radius*:

$$\lambda_D = \left(\frac{k_BT}{4\pi e^2 n_e}\right)^{1/2}. \tag{1.9}$$

If Debye length is used, condition (1.1) is written in an elegant form of

$$n\lambda_D^3 \gg 1. \tag{1.10}$$

This inequality means that in gases which can be regarded as an ensemble of free particles many electrons are in the Debye shielding radius. That is, to screen the ion potential many electrons are needed. In other words, in the gases consisting of almost free particles, one collision between ion and electron is inefficient to screen the ion potential. In Fig. 1.1 of temperature–density diagram, the line of $n\lambda_D^3 = 1$ is shown. The left-hand side of this line is the domain where the gas can be regarded as free particles. This figure shows that in a broad domain in the temperature–density diagram gases can be regarded as ensembles of free particles.

For some astrophysical objects, typical values of number density n, temperature T, and a few physical quantities calculated from them are shown in Table 1.1. We see that the value of $n\lambda_D^3$ is much larger than unity in many objects. That is, we can treat many astrophysical objects as gases of free particles. In such gases, the

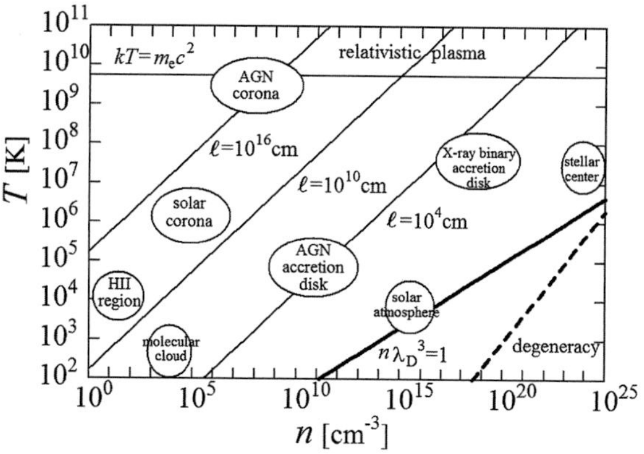

Fig. 1.1 Classification of gaseous states on temperature–density diagram. The left-hand side of $n\lambda_D^3 = 1$ is the region where gases can be regarded as ensemble of free particles. The lines on which mean free path is constant are also shown. Typical positions of some astrophysical objects in this diagram are shown by circles

Table 1.1 Some parameter values representing gaseous state of astrophysical objects

	Number density $n(\mathrm{cm}^{-3})$	Temperature $T(\mathrm{K})$	$n\lambda_D^3$	Mean free path ℓ (cm)	Typical size (cm)
Solar atmosphere	10^{17}	$6 \cdot 10^3$	0.2	10^{-4}	10^7
Solar corona	10^6	10^6	10^8	$4 \cdot 10^{11}$	10^{10}
Interstellar gases					
H II region	1	10^4	10^8	$4 \cdot 10^{13}$	10^{18}
H I region	10^2	10^2	10^4	$4 \cdot 10^7$	10^{18}
Molecular cloud	10^4	10^2	10^3	$4 \cdot 10^5$	10^{18}
Accretion disks					
X-ray stars	10^{20}	10^8	10^4	$4 \cdot 10$	10^4 (thickness)
Active galactic nuclei	10^{18}	10^5	3	$4 \cdot 10^{-3}$	10^9 (thickness)

velocity distribution of the particles is the Maxwell one, when we are interested in systems whose linear size is much larger than the mean free path of the particles (the mean free path is described later). What we treat in this textbook is such objects. In some particular situations, however, the distribution of particles deviates from the Maxwell one. Such cases are outside of our concern in this textbook. One of such cases is extremely high density gases where the electron gas degenerates. One more typical example is the case where the gases system is subject to strong magnetic fields, and electrons gyrate around magnetic fields before collisions. The domains of the temperature–density diagram of these two cases are discussed later in Sects. 1.2.2 and 1.2.4.

The condition of free particles, i.e., Eq. (1.10), implies that the mean free path of electrons, ℓ, is much longer than the Debye length. (If not, i.e., if a single electron can effectively screen the potential of an ion, this means that the orbit of an electron is appreciably changed in the radius of the Debye length.) In the next subsection, we show that the mean free path of electrons is really much longer than the Debye length in ionized gases consisting of free particles.

1.2.1.2 Mean Free Path ℓ

In addition to Debye length, mean free path of electrons is one of important quantities describing characteristic lengths in plasma. Furthermore, the mean free path is deeply related to the Debye length.

Here, we consider fully ionized gases. An ion is taken to have a charge of Ze. An electron moving with velocity v can be regarded as collided with an ion if the electron approaches to the ion so close that the potential energy, Ze^2/r, between them becomes equal to the kinetic energy of the electron. That is, the collision radius r is determined by $Ze^2/r = m_e v^2/2$. Then, the collision cross section, σ, is given by

$$\sigma \sim r^2 \sim \left(\frac{Ze^2}{m_e v^2}\right)^2, \tag{1.11}$$

where m_e is the mass of electron.

By using the collision cross section obtained above, we can determine the *mean free path* ℓ of electrons from the relation $n_i \ell \sigma = 1$, where n_i is the number density of ions. The kinetic energy of electron, $m_e v^2$, is roughly $k_B T$. Consequently, we have

$$\ell \sim \frac{1}{n_i}\left(\frac{k_B T}{Ze^2}\right)^2. \tag{1.12}$$

By using $Zn_i = n_e$ and inserting numerical values, we have

$$\ell \sim 3.5 \times 10^5 \frac{T^2}{Zn_e} \text{ cm}. \tag{1.13}$$

The mean free path, ℓ, for various objects are shown in Table 1.1. Furthermore, the T-n relation for which $\ell =$ const. is also shown in Fig. 1.1 with $Z = 1$.

It is noted here that if we use the Debye radius defined by Eq. (1.9) we see that ℓ and λ_D are related by

$$\ell \sim \lambda_D (n\lambda_D^3). \tag{1.14}$$

In gases of almost free particles, $n\lambda_D^3 \gg 1$ and we have $\ell \gg \lambda_D$ as expected.

1.2.2 *Condition of Degenerate Gas*

In extremely high density gases, the velocity distribution of free particles deviates from the Maxwell one by degeneracy. Since such gases are outside of our interest in this textbook, we shall mention the domain of such gases on the temperature–density diagram.

Here, we derive the condition of degeneracy of electron gases. It is noticed that the degeneracy of proton gases is not considered here, since it occurs at much higher density than that where the degeneracy of electrons occurs. Let us pack electrons successively from the lowest energy level to higher level in phase space, and let the highest energy level occupied by the last electron be E_F (Fermi energy). Then, the condition that the degeneracy of electron gases becomes important in velocity distribution is found to be

$$E_F \gg k_B T. \tag{1.15}$$

Next, we consider a relation between E_F and number density of electron, n_e. In the completely degenerated state, the momentum of the electron at the highest level, p_F, is related to n_e by

$$n_e = 8\pi p_F^3/3h^3, \tag{1.16}$$

since in a phase space element of h^3 two electrons can be packed, where h is the Planck constant. In the case of non-relativistic gases, p_F and E_F are related by $E_F = p_F^2/2m_e$, where m_e is the rest mass of electron. Hence, we have

$$E_F = \frac{h^2}{8m_e}\left(\frac{3}{\pi}\right)^{2/3} n_e^{2/3}. \tag{1.17}$$

Then, from Eqs. (1.17) and (1.15) we have the condition of degeneracy in the form:

$$T \lesssim 4.2 \times 10^{-11} n_e^{2/3}. \tag{1.18}$$

The line showing the critical state of degeneracy where $T = 4.2 \times 10^{-11} n_e^{2/3}$ is shown in Fig. 1.1. The right-hand side of the line is the domain of the degeneracy.

As Eq. (1.17) shows, Fermi energy E_F increases with increase of n_e. Hence, in very high density gases, E_F becomes larger than the rest mass energy, $m_e c^2$, of an electron. By using Eq. (1.17), we have this condition of $E_F > m_e c^2$ (relativistic degeneracy) as

$$n_e \gtrsim \frac{2^{9/2}\pi}{3}\left(\frac{m_e c}{h}\right)^3 = 1.7 \times 10^{30}\text{cm}^{-3}. \tag{1.19}$$

In the case of relativistic degeneracy, using the relation $E_F = p_F c$, we have the relation between E_F and n_e as

$$E_F = \left(\frac{3}{8\pi}\right)^{1/3} hc n_e^{1/3}. \tag{1.20}$$

Hence, the condition of relativistic degeneracy is found to be

$$T \lesssim 0.71 n_e^{1/3} \mathrm{K} \tag{1.21}$$

as is shown by substituting Eq. (1.20) into Eq. (1.15). The line showing the critical condition of the relativistic degeneracy is $T = 0.71 n_e^{1/3}$. The relativistic degeneracy occurs for high density [see inequality (1.19)]. Hence, the curve of relativistic degeneracy does not appear in the domain shown in Fig. 1.1.

1.2.3 Relativistic Gases

There is another important restriction concerning the gases which we treat in this textbook. This restriction is at extremely high temperature. In the state where temperature of gases is very high, various relativistic effects become important on behaviors of gases. Such high temperature can be regarded to be

$$k_B T \gtrsim m_e c^2. \tag{1.22}$$

In the high temperature gas where its kinetic energy becomes comparable with the electron rest mass energy, the collisions between electron and photon become inelastic (e.g., Compton scattering). Furthermore, the interactions bring about generation and annihilation of electron, positron, and photon.

In the high energy region, interactions between electrons and protons are weak, and radiation processes are mainly related to electron gases. Hence, electron and proton temperatures are not equal in general (two-temperature gas). If we focus our attention on electron gases, the condition of relativistic gases is $k_B T_e \gtrsim m_e c^2$, where T_e is the electron temperature. By substituting numerical values, we can write this condition as

$$T_e \gtrsim 5.9 \times 10^9 \mathrm{K}. \tag{1.23}$$

The line of $T = 5.9 \times 10^9$ K is shown in Fig. 1.1.

Summing up the above considerations in three subsections, the gases which we are mainly concerned with in this book are those consisting of non-relativistic and non-generated free particles. Figure 1.1 shows that such gases are in a broad domain on the temperature–density diagram. Typical positions of various astrophysical objects in this domain are shown in Fig. 1.1.

1.2.4 Gyration of Electrons around Magnetic Fields

In the above subsections, we have shown that in a wide domain in the temperature–density diagram, the gas can be considered to be an ensemble of free particles. Here, we examine how much this conclusion is affected or not when the gaseous system is subject to strong magnetic fields. In magnetized systems, ionized particles gyrate around magnetic field lines. Hence, their distribution deviates from the isotropic Maxwell one.

Let us consider the *gyration radius* r_{gy} of an electron rotating around magnetic field $\boldsymbol{B}$ with circular velocity v. The balance between the centrifugal force v^2/r_{gy} and the magnetic force $(e/c)vB$ toward the gyration center gives $r_{gy} = cv/eB$. Hence, the gyration time t_{gy} is estimated by $t_{gy} \sim 2\pi r_{gr}/v \sim 2\pi c/(eB)$. On the other hand, the collision time t_{coll} by particle mutual collisions is ℓ/v. If the mean free path ℓ is estimated by Eq. (1.12), we have

$$\frac{t_{gy}}{t_{coll}} \sim \frac{2\pi c e^3}{m_e^{1/2} k_B^{3/2}} \frac{n}{BT^{3/2}}, \tag{1.24}$$

where $n_e = n_i = n$ and $Z = 1$ have been adopted. By substituting numerical values, we have

$$\frac{t_{gy}}{t_{coll}} \sim 4.4 \times 10^{20} \frac{n}{BT^{3/2}}. \tag{1.25}$$

This result shows that in the domain of the temperature–density diagram of our interest, t_{gy}/t_{coll} is much larger than unity, unless rather low density and highly magnetized gases are considered. In other words, we do not need to consider deviation from the Maxwell distribution of electron gas in normal situations.

1.3 Characteristics of Physical States of Astrophysical Objects

Our main interest is dynamical behaviors of gaseous systems in the Universe. We saw in Sect. 1.2 that the gaseous systems can be regarded as an ensemble of free particles in many cases. These gaseous systems are, however, quite different from those on the Earth. Most important differences are related to their scale, mass, and rotation. Furthermore, the ingredients of the fluid systems are not ionized gases alone. Magnetic fields, radiation, and high energy particles are involved and they have strong interactions with the gases. In these interactions, magnetic fields, radiation, and high energy particles are not simple supplementary ingredients, but their presence is the origins of various important activities of the fluid systems.

1.3.1 Gravitational Force and Self-Gravity

First, we emphasize importance of scale and mass of astrophysical objects. This is not simply a quantitative difference from small scale objects on the earth, but brings about qualitative differences.

There are four fundamental forces in the Universe. The gravitational force is the most weak one among them, but is a long range force. The electrostatic force is also a long range force, but the force cannot affect distant objects, since the force is screened by oppositely charged particles, as mentioned in Sect. 1.2. Different from this, there is no shielding in gravitational force. Hence, as we consider a larger scale object the effects of gravitational force become stronger.

Here, we demonstrate a rough image how the effects of self-gravity are important in astrophysical objects. As a measure we compare the gravitational energy of a gaseous system with the internal thermal energy of the gas. Let us consider, for simplicity, a uniform spherical gas of radius r with temperature T and density ρ. Assuming that the gas is fully ionized hydrogen gas, we find that the internal energy of the gas is $3k_{\rm B}T(M/m_{\rm H})$, where $M(= 4\pi\rho r^3/3)$ is the mass of the gas as the whole and $m_{\rm H}$ is the mass of hydrogen. The gravitational energy of the gas is on the order of GM^2/r. Hence, the condition that the gravitational energy becomes larger than the internal thermal energy is $M > (3k_{\rm B}T/m_{\rm H}G)r$. Thus, this condition is written as

$$\frac{M}{M_\odot} > 1.3 \times 10^{-7} T \frac{r}{R_\odot}, \tag{1.26}$$

where $M_\odot$ and $R_\odot$ are solar mass and solar radius, respectively. Rough positions of typical astrophysical objects on the $M - r$ diagram are shown in Fig. 1.2. As this figure shows, various astrophysical objects are in the state where thermal energy and gravitational energy are roughly balanced, depending on their temperature. That is, many astrophysical objects are constructed by the balance between the attractive force of self-gravity and the repulsive force of pressure.

One of important points to be noticed here is that the gravitational energy of a gas system with mass M increases with M^2, while the thermal energy of mass M is proportional to M. This nonlinearity of gravitational energy has important effects on thermodynamical equilibrium of gravitating systems, and leads to gravothermal instability (see Sect. 7.5).

In large scale objects their dynamical equilibria are realized by the presence of forces against the gravitational force. One of important forces against the gravitational force is the pressure force, which leads to many nearly spherical objects, as shown in Fig. 1.2. Another important force which leads astrophysical objects to dynamical equilibrium states against gravitational force is centrifugal force due to rotation of the objects. There are many astrophysical objects where dynamical equilibria are realized by the balance between gravitational force and centrifugal one. Typical examples of such systems are proto-planetary disks and

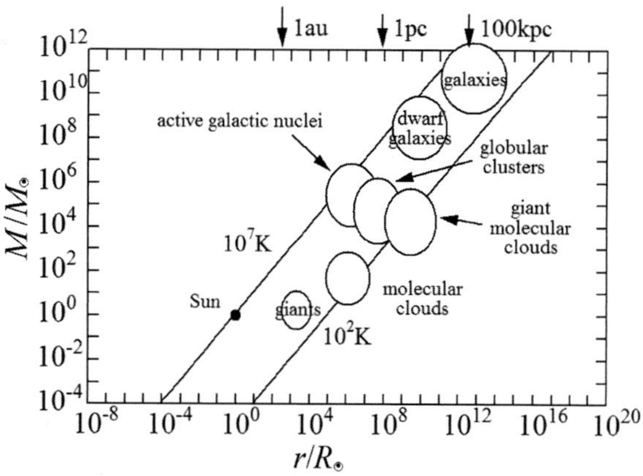

Fig. 1.2 Mass-radius relation of self-gravitating gaseous systems. Positions of typical astrophysical objects in this diagram are also shown for some typical temperatures

accretion disks. In this sense, studies of dynamics of disk systems are of importance in astrophysics, as well as those of spherical systems.

1.3.2 Magnetic Fields

One of measures whether magnetic fields have important effects on configurations of astrophysical objects is the ratio of magnetic energy to gaseous internal energy. As will be found in Part II, the magnetic energy per unit volume is $B^2/8\pi$, where B is the magnetic flux density. The thermal energy in unit volume is $3k_\mathrm{B}nT$ in the case of fully ionized hydrogen gas, where n is the number density of hydrogen atom. Hence, the condition that magnetic energy overcomes thermal energy is $B \geq (24\pi k_\mathrm{B}nT)^{1/2}$. By inserting numerical values, we can write this condition in the form

$$B \gtrsim 1.0 \times 10^{-7}(nT)^{1/2}\,\mathrm{G}, \tag{1.27}$$

or

$$T \lesssim 10^{14}\frac{(B/\mathrm{gauss})^2}{n}. \tag{1.28}$$

In many astrophysical objects the above condition is not satisfied. That is, magnetic fields in normal astrophysical objects are not so strong as those required by the above condition. This is related to the fact that the system is unstable if the magnetic energy is as strong as the internal energy of the system (Virial theorem, see Sect. 15.1).

Real astrophysical objects are affected by much weaker magnetic fields. A measure will be the ratio of Alfvén speed to acoustic speed. In magnetized gases, information propagate at Alfvén speed (see, e.g., Chap. 13) as well as acoustic

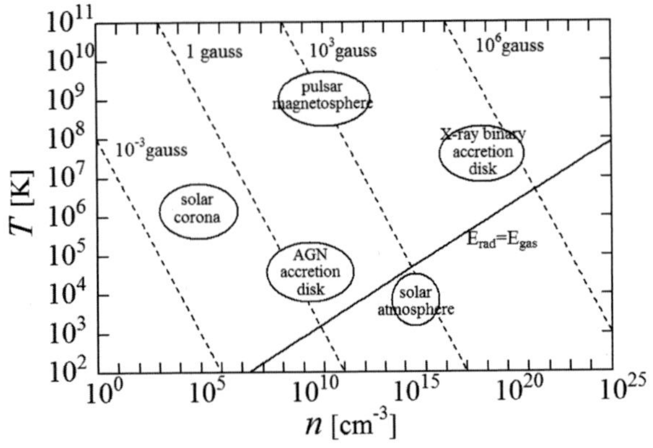

Fig. 1.3 Temperature-density relation for $c_A^2 = c_s^2$ for four cases of $B[\text{gauss}] = 10^{-3}$, 1, 10^3, and 10^6. For each case, the left side of the line labeled by the value of B is the domain of $c_A > c_s$ for that value of B. Typical positions of some astrophysical objects in this diagram are shown. The line of radiation energy density ($E_{rad} \equiv aT^4$) is equal to gas thermal energy density ($E_{gas} \equiv 3nk_BT$) is also shown

speed. Hence, a parameter to measure the importance of magnetic fields is the ratio of Alfvén speed, c_A, to acoustic speed, c_s, where $c_A^2 = B^2/(8\pi n m_H)$ for fully ionized gas, and $c_s^2 = \gamma p_{gas}/(2nm_H)$, γ being the ratio of specific heats. By substituting numerical values, we can write the condition of $c_A^2 \gtrsim c_s^2$ as

$$B \gtrsim 1.0 \times 10^{-4}\left(\frac{n}{10^{15}\text{cm}^{-3}}\right)T^{1/2}\,\text{G}, \tag{1.29}$$

or

$$T \lesssim 10^8\left(\frac{B^2}{\text{gauss}}\right)\left(\frac{n}{10^{15}\text{cm}^{-3}}\right)^{-2}. \tag{1.30}$$

For four cases of $B = 10^{-3}$ G, 1 G, 10^3 G, and 10^6 G the $T - n$ relation for $c_A^2 = c_s^2$ is shown in the temperature–density diagram in Fig. 1.3. In this figure, below a line labeled by a value of B, we have $c_A > c_s$ for that labeled B. Rough positions of some astrophysical objects in this diagram are also shown in Fig. 1.3. This result shows that magnetic fields have appreciable effects on dynamical behavior of many astrophysical objects.

1.3.3 Radiation Fields

For simplicity, let us consider a gaseous system in thermodynamical equilibrium with temperature T. In such a system there is isotropic radiation field, and its energy density, $E_{\rm rad}$, is given by aT^4, where a is the radiation constant (see Part III). One of the indicators to measure the importance of radiation is thus the ratio of aT^4 to the thermal energy density of the gas, $E_{\rm gas}$.[1] The latter is $3nk_{\rm B}T$ in fully ionized hydrogen gas. The condition of $aT^4 > 3nk_{\rm B}T$ is thus written as $T > (3k_{\rm B}/a)^{1/3}n^{1/3}$ and by using numerical values we can write it as

$$T \gtrsim 0.38\,n^{1/3}. \tag{1.31}$$

The $T - n$ relation for $aT^4/3nk_{\rm B}T = 1$ is shown also in Fig. 1.3. On the domain of the left-hand side of the line, radiation energy (aT^4) is higher than gas energy ($3nk_{\rm B}T$). The line of $aT^4/3nk_{\rm B}T = 1$ in Fig. 1.3 shows that the cases where radiation energy density becomes as high as thermal energy density is rather special. It should be emphasized, however, that $aT^4/3nk_{\rm B}T = 1$ is one of measures of importance of radiation fields. Radiation comes to us usually from optically thin parts of gases. Hence, it is also important to examine how radiation fields decouple from matter in optically thin media, and thus how they deviate from the Plank distribution, since radiation is the most important measure to know the physical states of the our outside world.

In summary, many astrophysical objects are ionized gaseous systems consisting of free particles. Important ingredients in the systems are self-gravity, magnetic field, and radiation. The purpose of this textbook is to overview the essence of dynamical processes in such gaseous systems. We separate this textbook into three parts. Part I is devoted to hydrodynamical phenomena where magnetic fields and radiation fields are not main players, but in Part II we focus our attention to phenomena where the presence of magnetic fields is important sources of phenomena, and finally in Part III phenomena associated with radiation are studied.

[1]This comparison is qualitatively the same with the comparison between radiation pressure and gas pressure.

Part I
Hydrodynamical Phenomena in Astrophysical Objects

Chapter 2
Basic Equations for Hydrodynamics

Hydrodynamical equations are usually derived from the Boltzmann equation. In this chapter, however, assuming continuous hydrodynamical systems from the beginning, we summarize the basic set of hydrodynamical equations. There are two ways to describe hydrodynamical motions. One is the Eulerian description and the other is the Lagrangian one. Each one has merits and demerits. In this chapter both descriptions are presented, and basic characteristics of fluid motions described by these descriptions are summarized. A derivation from the Boltzmann equation is made in Chap. 10 as a part of derivation of hydromagnetic fluid equations.

2.1 Condition for Fluid Approximations

In Chap. 1 we have derived $n_e\lambda_D^3 \gg 1$ as the condition for ionized gases being regarded as an ensemble of free particles. Furthermore, we showed that in this case of $n_e\lambda_D^3 \gg 1$, the mean free path of ions, ℓ, is much longer than the Debye length. The subject to be concerned here is in what cases such gases can be regarded as fluid. It will be obvious that if the scale under consideration, L, is longer than ℓ, the system can be regarded as fluids. The lines of ℓ =const. on the temperature–density diagram have been shown in Fig. 1.1. This figure shows that in large scale astrophysical phenomena in gaseous systems the condition $L \gg \ell$ is sufficiently satisfied.

Let us describe the above conditions of fluids from the viewpoint of timescales. Three timescales are defined: $\tau_{\rm corr} = \lambda_D/c_s$, $\tau_r = \ell/c_s$, and $\tau_d = L/c_s$, where c_s is the acoustic speed. The timescale $\tau_{\rm corr}$ is the time for ions passing through the Debye radius, and represents the timescale during which correlation among particles are important. The timescale τ_r is the time by which the distribution of particles approaches to the Maxwellian, i.e., the timescale by which the temperature is defined. Finally, τ_d is the timescale of global dynamical phenomena. For gases to

© Springer Nature Singapore Pte Ltd. 2020, corrected publication 2023

S. Kato, J. Fukue, *Fundamentals of Astrophysical Fluid Dynamics*,
Astronomy and Astrophysics Library,
https://doi.org/10.1007/978-981-15-4174-2_2

be regarded as fluids, timescale τ_d in consideration needs to be much longer than τ_r. That is, the condition is written as

$$\tau_\mathrm{corr} \ll \tau_\mathrm{r} \ll \tau_\mathrm{d}. \tag{2.1}$$

It should be noted here that there are cases where inequalities (2.1) are not realized. Such systems are collisionless ones. Stellar systems such as galaxies and globular clusters belong to the collisionless systems. Even in the case of ionized gases there are cases where the condition is not satisfied, i.e., collisionless plasma. Of course, such systems are outside of our present interest.

2.2 Eulerian Description of Hydrodynamical Equations

There are two ways to describe fluid motions: Eulerian and Lagrangian descriptions. The latter is better to argue some basic properties of fluid motions, but the former is convenient to study practical problems. First we present Eulerian equations without detailed derivation, since they are well-known and described in many excellent textbooks. The basic equations are those describing the time variations of velocity $\boldsymbol{v}(\boldsymbol{r}, t)$, density $\rho(\boldsymbol{r}, t)$, pressure $p(\boldsymbol{r}, t)$, and temperature $T(\boldsymbol{r}, t)$ of fluid elements, where $\boldsymbol{r}$ is the coordinates representing position.

(a) **Equation of Continuity**
The time variation of density is given by the equation of continuity, which is

$$\frac{d\rho}{dt} = \left(\frac{\partial}{\partial t} + \boldsymbol{v}\cdot\nabla\right)\rho = -\rho\,\mathrm{div}\,\boldsymbol{v}, \tag{2.2}$$

where d/dt is the Lagrangian time derivative, and related to the Eulerian time derivative $\partial/\partial t$ by

$$\frac{d}{dt} = \frac{\partial}{\partial t} + (\boldsymbol{v}\cdot\nabla). \tag{2.3}$$

(b) **Equation of Motions**
The time variation of $\boldsymbol{v}$ is described by the equation of motion, which is

$$\rho\frac{d\boldsymbol{v}}{dt} = -\rho\nabla\psi - \nabla p + \rho\boldsymbol{N}. \tag{2.4}$$

On the right-hand side of Eq. (2.4), ψ is the gravitational potential, which consists of two parts due to an external one, $\psi_\mathrm{ext}(\boldsymbol{r}, t)$, and self-gravity, $\psi_\mathrm{self}(\boldsymbol{r}, t)$, i.e.,

$$\psi = \psi_\mathrm{ext}(\boldsymbol{r}, t) + \psi_\mathrm{self}(\boldsymbol{r}, t), \tag{2.5}$$

and the latter one is related to density distribution through the Poisson equation:

$$\nabla^2 \psi_{\rm self} = 4\pi G\rho. \tag{2.6}$$

If we consider optically thick gases, the pressure consists of radiation pressure $p_{\rm rad}$ in addition to gas pressure $p_{\rm gas}$, i.e., $p = p_{\rm gas} + p_{\rm rad}$. The final term, $\rho \boldsymbol{N}$, on the right-hand side of Eq. (2.4) represents forces other than gravitational and pressure ones, including magnetic force, viscous force, and so on. They are not shown here explicitly. These terms will be considered when we need them.

(c) **Conservation of Thermal Energy**
Next, we need an equation describing time variation of pressure. This is obtained from the first law of thermodynamics. The law gives

$$\rho\frac{dU}{dt} + p\,{\rm div}\,\boldsymbol{v} = \rho\epsilon - {\rm div}\,\mathcal{F} + \Phi, \tag{2.7}$$

where U is the internal energy of fluids (gas and radiation), and given by

$$\rho U = \frac{1}{\gamma - 1}p_{\rm gas} + 3p_{\rm rad}, \tag{2.8}$$

where γ is the ratio of specific heats. The right-hand side of Eq. (2.7) represents the rate of heat generation per unit volume, which consists of the rate of heat generation by subatomic processes, $\rho\epsilon$, and the rate of thermal energy loss by divergence of thermal flux, $\mathcal{F}$, and the heat generation by viscous processes, Φ.

(d) **Equation of State**
Finally, we need a relation among p, ρ, and T, which is the equation of state and given by

$$p = p_{\rm gas} + p_{\rm rad} = \frac{k_{\rm B}}{\bar{\mu} m_{\rm H}}\rho T + \frac{1}{3}aT^4 = \frac{\mathcal{R}}{\bar{\mu}}\rho T + \frac{1}{3}aT^4, \tag{2.9}$$

where $\bar{\mu}$ is the mean molecular weight, $k_{\rm B}$ the Boltzmann constant, and $\mathcal{R}$ the gas constant.

2.3 Comments on Energy Conservation

We have various kinds of energy, e.g., kinetic, thermal, gravitational, and total energies. We summarize here relations among these energies.

2.3.1 Mechanical Energy and Its Conservation

First, kinetic energy per unit mass, $v^2/2$, is considered. The equation of motion, Eq. (2.4), can be written as

$$\frac{d}{dt}\left(\frac{1}{2}v^2\right) = \boldsymbol{v}\cdot\left(\boldsymbol{F} - \frac{1}{\rho}\nabla p\right), \tag{2.10}$$

where forces other than the pressure one have been summarized, for simplicity, by $\boldsymbol{F}$. Equation (2.10) shows the well-known fact in dynamics. That is, the rate of increase of kinetic energy of an object with unit mass is equal to the rate of work done on the object by forces.

2.3.2 Conservation of Total Energy

As is shown in Fig. 2.1, we consider a volume element A in fluid. The element is assumed to have density $\rho(\boldsymbol{r})$, pressure $p(\boldsymbol{r})$, internal energy per unit mass $U(\boldsymbol{r})$, and to move with velocity $\boldsymbol{v}(\boldsymbol{r})$.

We assume that this fluid element is subject to an external force $\boldsymbol{F}(\boldsymbol{r})$ per unit volume, and on the surface S of the element there are energy exchange with the external fluid. The energy flux through the surface is assumed to be $\mathcal{F}$ per unit surface. It is noted that there is no mass exchange through the surface by its definition of the surface. It is further assumed that the fluid element has direct subatomic energy exchange in the medium. The rate of the exchange is denoted by $\epsilon(\boldsymbol{r})$ per unit mass. Then, the conservation of total energy in the element is written as

$$\frac{d}{dt}\iiint_{\mathrm{A}}\rho\left(U + \frac{1}{2}v^2\right)dV = \iiint_{\mathrm{A}}\rho\boldsymbol{v}\cdot\boldsymbol{F}dV - \iint_{\mathrm{S}}\mathcal{F}\cdot d\boldsymbol{S} - \iint_{\mathrm{S}}p\boldsymbol{v}\cdot d\boldsymbol{S} + \iiint_{\mathrm{A}}\rho\epsilon dV, \tag{2.11}$$

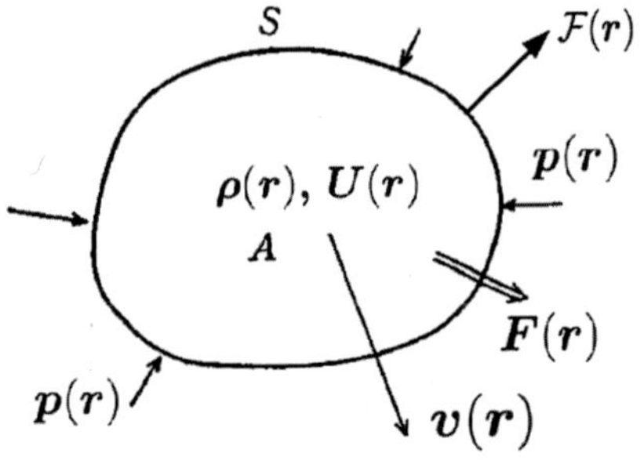

Fig. 2.1 Relation between fluid element A and surrounding medium through boundary S

where dV is a volume element and $d\boldsymbol{S}$ is a surface element. The third term on the right-hand side of the above equation represents the rate of energy decrease by moving against the pressure at the surface.

Mass element ρdV can be commuted with $1/dt$. The surface integrals can be changed to volume integrals by use of the Gauss formula. Furthermore, the volume element in Eq. (2.11) is arbitrary. Consequently, from Eq. (2.11) we have

$$\rho\frac{d}{dt}\left(U+\frac{1}{2}v^2\right)=\rho\boldsymbol{F}\cdot\boldsymbol{v}-\mathrm{div}\mathcal{F}-\mathrm{div}(p\boldsymbol{v})+\rho\epsilon. \tag{2.12}$$

This equation can be written by use of the equation of continuity:

$$\frac{\partial}{\partial t}\left[\rho\left(U+\frac{1}{2}v^2\right)\right]+\mathrm{div}\left[\rho\left(H+\frac{1}{2}v^2\right)\boldsymbol{v}+\mathcal{F}\right]=\rho\boldsymbol{F}\cdot\boldsymbol{v}+\rho\epsilon, \tag{2.13}$$

where H is enthalpy defined by $H\equiv U+p/\rho$.

If the force $\boldsymbol{F}$ is gravitational one, it is related to gravitational potential ψ by

$$\rho\boldsymbol{F}=-\rho\,\mathrm{grad}\psi. \tag{2.14}$$

In the case where the gravitational force comes from a steady external one, we have

$$-\rho\boldsymbol{F}\cdot\boldsymbol{v}=\mathrm{div}(\rho\boldsymbol{v}\psi)+\psi\frac{\partial\rho}{\partial t}=\frac{\partial}{\partial t}(\rho\psi)+\mathrm{div}(\rho\boldsymbol{v}\psi) \tag{2.15}$$

and when $\epsilon=0$, Eq. (2.13) is summarized in a conservation form as

$$\frac{\partial}{\partial t}\left[\rho\left(U+\psi+\frac{1}{2}v^2\right)\right]+\mathrm{div}\left[\rho\left(H+\psi+\frac{1}{2}v^2\right)\boldsymbol{v}+\mathcal{F}\right]=0. \tag{2.16}$$

In the case where the force acting on the gases, $\boldsymbol{F}$, is magnetic force, we have (see Chap. 10)

$$\rho\boldsymbol{F}=\frac{1}{c}\boldsymbol{j}\times\boldsymbol{B}, \tag{2.17}$$

where $\boldsymbol{j}$ and $\boldsymbol{B}$ are, respectively, electric current and magnetic field. The conservation form extending Eq. (2.13) to this case will be presented in Chap. 11.

2.3.3 Conservation of Thermal Energy

Equation of thermal energy conservation is obtained by substituting the equation of kinetic energy conservation from that of total energy conservation. That is, substituting Eq. (2.10) from Eq. (2.12), we have

$$\rho \frac{dU}{dt} + p \,\mathrm{div}\, \boldsymbol{v} = -\mathrm{div}\mathcal{F} + \rho\epsilon. \tag{2.18}$$

This is equivalent with Eq. (2.7). If entropy S is introduced, the above equation is written as

$$\rho T \frac{dS}{dt} = \rho \frac{dU}{dt} + p \,\mathrm{div}\, \boldsymbol{v} = -\mathrm{div}\mathcal{F} + \rho\epsilon. \tag{2.19}$$

2.4 Bernoulli Theorem and Vorticity Conservation

Here, we mention two astrophysically important characteristics of barotropic flows. They are the Bernoulli theorem and conservation of *fluid helicity*. *Barotropic gases* are defined by those where pressure p and density ρ are directly related as $p = p(\rho)$. In this case the pressure force per unit mass, $-(1/\rho)\nabla p$, can be written by a potential force as $-\nabla H$, where H is given by

$$H = \frac{\gamma}{\gamma - 1}\frac{p}{\rho} \tag{2.20}$$

in the case of polytropic gas of $p \propto \rho^\gamma$.

In barotropic gases the equation of motion for inviscid flows is written in the form:

$$\frac{\partial \boldsymbol{v}}{\partial t} - \boldsymbol{v} \times \mathrm{curl}\, \boldsymbol{v} = -\nabla\left(\frac{1}{2}\boldsymbol{v}^2 + H + \psi\right), \tag{2.21}$$

where ψ is the gravitational potential.

2.4.1 Bernoulli Theorem

Starting from Eq. (2.21), we show first the Bernoulli theorem. In the case of steady flows, the above equation is reduced to

$$\boldsymbol{v} \times \mathrm{curl}\, \boldsymbol{v} = \nabla\left(\frac{1}{2}\boldsymbol{v}^2 + H + \psi\right). \tag{2.22}$$

Since the left-hand side of this equation is perpendicular to the flow direction, the right-hand side vanishes along the flow, i.e., along stream line we have

$$\frac{1}{2}\boldsymbol{v}^2 + H + \psi = \text{const.} \tag{2.23}$$

The constant depends on each stream line. This is called the *Bernoulli theorem.*

In a particular case where the system in consideration is axisymmetric, the right-hand side of Eq. (2.22) vanishes along poloidal component of the flow. Hence, we have

$$\frac{1}{2}\boldsymbol{v}_p^2 + \frac{L^2}{2r^2} + H + \psi = \text{const} \tag{2.24}$$

along the poloidal component of the flow, where $\boldsymbol{v}$ has been decomposed into poloidal component, $\boldsymbol{v}_\mathrm{p}$, and toroidal one, $\boldsymbol{v}_\mathrm{t}$ as $\boldsymbol{v} = \boldsymbol{v}_\mathrm{p} + \boldsymbol{v}_\mathrm{t} = \boldsymbol{v}_\mathrm{p} + (L/r)\boldsymbol{i}_\varphi$, L being specific angular momentum of rotation and $\boldsymbol{i}_\varphi$ being the unit vector in the azimuthal direction.

2.4.2 *Helmholtz Theorem and Helicity Conservation*

Taking curl of Eq. (2.21) we have

$$\frac{\partial \boldsymbol{\zeta}}{\partial t} = \text{curl}(\boldsymbol{v} \times \boldsymbol{\zeta}), \tag{2.25}$$

where $\boldsymbol{\zeta}$ is *vorticity* defined by $\boldsymbol{\zeta} \equiv (1/2)\text{curl}\,\boldsymbol{v}$. This equation shows how vorticity evolves with time.

An important characteristic shown by this equation is that the vorticity through a closed surface which moves with flow is time-independent. To show this, let us introduce an arbitrary closed curve, $\boldsymbol{l}$, in fluid, and consider a surface, $\boldsymbol{S}$, bounded by this curve. The purpose here is to examine how the vorticity through the surface, i.e., $\int \boldsymbol{\zeta} \cdot d\boldsymbol{S}$, changes with time. Because the boundary curve is taken to move with fluid, we have

$$\frac{d}{dt}\iint_\mathrm{S} \boldsymbol{\zeta} \cdot d\boldsymbol{S} = \iint_\mathrm{S} \frac{\partial \boldsymbol{\zeta}}{\partial t} \cdot d\boldsymbol{S} + \int_l \boldsymbol{\zeta} \cdot (\boldsymbol{v} \times d\boldsymbol{l}), \tag{2.26}$$

where a line element of the boundary is denoted by $d\boldsymbol{l}$ and a surface element by $d\boldsymbol{S}$. Because $\boldsymbol{\zeta} \cdot (\boldsymbol{v} \times d\boldsymbol{l}) = -(\boldsymbol{v} \times \boldsymbol{\zeta}) \cdot d\boldsymbol{l}$, the right-hand side of Eq. (2.26) is written by changing the line integration into surface integration by use of the Stokes theorem as

$$\iint_\mathrm{S} \left[\frac{\partial \boldsymbol{\zeta}}{\partial t} - \text{curl}\,(\boldsymbol{v} \times \boldsymbol{\zeta})\right] \cdot d\boldsymbol{S}. \tag{2.27}$$

Because the time evolution of $\boldsymbol{\zeta}$ is governed by Eq. (2.25), the above expression (2.27) vanishes, and we have

$$\frac{d}{dt}\iint_{\mathrm{S}} \boldsymbol{\zeta}\cdot d\boldsymbol{S} = 0. \tag{2.28}$$

This equation shows that the vorticity passing through a surface in a fluid is unchanged if the boundary of the surface moves with fluid. This is called the *Helmholtz theorem* concerning vorticity conservation. This theorem implies that in turbulent flows stream lines are tangled, but vorticity is conserved in the sense given by Eq. (2.28), if the fluid is barotropic and inviscid.

There is another measure for topological structure of flows. This is the *fluid helicity* defined by

$$H_{\mathrm{F}} \equiv \iiint_{\mathrm{V}} (\boldsymbol{v}\cdot\boldsymbol{\zeta}) dV. \tag{2.29}$$

Now, helicity density, h_{F}, is defined by $h_{\mathrm{F}} \equiv \boldsymbol{v}\cdot\boldsymbol{\zeta}$. Then, the use of Eqs. (2.21) and (2.25) leads to[1]

$$\begin{aligned}\frac{\partial h_{\mathrm{F}}}{\partial t} &= \frac{\partial}{\partial t}(\boldsymbol{v}\cdot\boldsymbol{\zeta}) = \boldsymbol{v}\cdot \mathrm{curl}(\boldsymbol{v}\times\boldsymbol{\zeta}) - \mathrm{div}\left[\left(\frac{1}{2}v^2 + H + \psi\right)\boldsymbol{\zeta}\right]. \\ &= -\mathrm{div}\left[\boldsymbol{v}\times(\boldsymbol{v}\times\boldsymbol{\zeta}) + \left(\frac{1}{2}v^2 + H + \psi\right)\boldsymbol{\zeta}\right],\end{aligned} \tag{2.30}$$

which is further written as[2]

$$\frac{\partial h_{\mathrm{F}}}{\partial t} + \mathrm{div}\left[h_{\mathrm{F}}\boldsymbol{v} + \left(-\frac{1}{2}v^2 + H + \psi\right)\boldsymbol{\zeta}\right] = 0. \tag{2.31}$$

It should be emphasized that the time evolution of helicity density has a form of conservation. In this sense it is said that in addition to mass, momentum, and energy, the helicity is the fourth conserved quantity in fluid in the case where the gas is barotropic.

[1] In deriving the last equality in Eq. (2.30), a formula of vector analyses:

$$\mathrm{div}(\boldsymbol{A}\times\boldsymbol{B}) = \boldsymbol{B}\cdot\mathrm{curl}\boldsymbol{A} - \boldsymbol{A}\cdot\mathrm{curl}\boldsymbol{B}$$

is used, where $\boldsymbol{A}$ and $\boldsymbol{B}$ are arbitrary vectors.

[2] A formula of vector analyses:

$$\boldsymbol{A}\times(\boldsymbol{B}\times\boldsymbol{C}) = \boldsymbol{B}(\boldsymbol{A}\boldsymbol{C}) - \boldsymbol{C}(\boldsymbol{A}\boldsymbol{B})$$

is used.

If we consider a volume whose boundary surface moves with fluid, the time variation of fluid helicity is given by

$$\begin{aligned}\frac{dH_{\rm F}}{dt} = \frac{d}{dt}\iiint_{\rm V} h_{\rm F} dV &= \iiint \frac{\partial h_{\rm F}}{\partial t} dV + \iint (h_{\rm F}\boldsymbol{v}) \cdot d\boldsymbol{S} \\ &= \iiint \left[\frac{\partial h_{\rm F}}{\partial t} + {\rm div}(h_{\rm F}\boldsymbol{v})\right] dV \\ &= -\iint \left[\left(-\frac{1}{2}v^2 + H + \psi\right)\boldsymbol{\zeta}\right]\boldsymbol{S}. \end{aligned} \tag{2.32}$$

In deriving the third and fourth equalities, the Gauss theorem has been used to write the surface integral to the volume integral. This equation shows that the fluid helicity is conserved if there is no input or output of $(-v^2/2 + H + \psi)$ through the surface.

The concepts corresponding to vorticity conservation and helicity conservation in MHD systems are the frozen-in of magnetic fields to fluid and the magnetic helicity conservation, respectively. They are of importance in magnetohydrodynamics (see Chap. 11).

2.5 Lagrangian Description of Displacement Vector $\boldsymbol{\xi}$

In Lagrangian formulation we take our attention on a fluid element, and examine how the element moves and how physical quantities associated with the element change with time. Let us denote the displacement vector of a fluid element from unperturbed flow by $\boldsymbol{\xi}(\boldsymbol{r}, t)$, where $\boldsymbol{r}(t)$ represents the position of the fluid element in the unperturbed flow. Our purpose here is to derive an equation describing time evolution of $\boldsymbol{\xi}$.

We start from the Eulerian equation (2.4), and consider its Lagrangian variation:

$$\delta\left(\frac{d\boldsymbol{v}}{dt}\right) = \delta\left(-\nabla\psi - \frac{1}{\rho}\nabla p + \boldsymbol{N}\right), \tag{2.33}$$

where δ represents the Lagrangian variation. The Lagrangian variation of a quantity, $Q(\boldsymbol{r}, t)$, i.e., δQ, is defined by[3]

$$\delta Q = Q[\boldsymbol{r} + \boldsymbol{\xi}(\boldsymbol{r}), t] - Q_0(\boldsymbol{r}, t), \tag{2.34}$$

[3]The Eulerian variation of a quantity, $Q(\boldsymbol{r}, t)$, i.e., Q_1, is defined by

$$Q_1(\boldsymbol{r}, t) = Q(\boldsymbol{r}, t) - Q_0(\boldsymbol{r}, t).$$

where $Q[\boldsymbol{r}+\boldsymbol{\xi}(\boldsymbol{r}), t]$ is the value of Q along the perturbed flow, and $Q_0(\boldsymbol{r}, t)$ is the value of Q in the unperturbed flow.

Here, we should emphasize the presence of an important characteristic of the Lagrangian operator δ. This is that for arbitrary quantity $\boldsymbol{Q}$ we have (Lynden-Bell and Ostriker 1967)

$$\delta\left(\frac{d\boldsymbol{Q}}{dt}\right)=\frac{d_0}{dt}(\delta\boldsymbol{Q}). \tag{2.35}$$

Here, d_0/dt is the time derivative along the unperturbed flow, $\boldsymbol{v}_0$, and defined by

$$\frac{d_0}{dt}=\frac{\partial}{\partial t}+\boldsymbol{v}_0\cdot\nabla. \tag{2.36}$$

By adopting Eq. (2.35), we have (Lynden-Bell and Ostriker 1967)

$$\delta\left(\frac{d\boldsymbol{v}}{dt}\right)=\frac{d_0}{dt}\delta\boldsymbol{v}=\frac{d_0^2\boldsymbol{\xi}}{dt^2}. \tag{2.37}$$

Then, from Eq. (2.33) we see that the hydrodynamical equation describing displacement vector, $\boldsymbol{\xi}$, is written as (Lynden-Bell and Ostriker 1967)

$$\frac{d_0^2\boldsymbol{\xi}}{dt^2}=\delta\left(-\nabla\psi-\frac{1}{\rho}\nabla p+\boldsymbol{N}\right). \tag{2.38}$$

This equation is the basic equation for Lagrangian description of dynamical flows, and is valid for arbitrary magnitude of perturbations, not restricted only to small amplitude perturbations.

Before proceeding to arguments on equations expressed in terms of Lagrangian displacement vector $\boldsymbol{\xi}$, it will be instructive to mention a direct relation between Eulerian velocity perturbation $\boldsymbol{v}_1$ and Lagrangian displacement vector $\boldsymbol{\xi}$. Let us express fluid velocity of perturbed flow by $\boldsymbol{V}(\boldsymbol{r}, t)$, i.e., $\boldsymbol{V}(\boldsymbol{r}, t) = \boldsymbol{v}_0(\boldsymbol{r}, t) + \boldsymbol{v}_1(\boldsymbol{r}, t)$, where $\boldsymbol{v}_0$ is the unperturbed flow at $\boldsymbol{r}$ and $\boldsymbol{v}_1$ is the Eulerian perturbation of the flow. Then, by the definition of the displacement vector $\boldsymbol{\xi}$, we have (Lynden-Bell and Ostriker 1967)

$$\frac{d_0\boldsymbol{\xi}}{dt}=\boldsymbol{V}(\boldsymbol{r}+\boldsymbol{\xi}, t)-\boldsymbol{V}_0(\boldsymbol{r}, t). \tag{2.39}$$

This gives in the first order of approximation with respect to perturbations:

$$\frac{d_0\boldsymbol{\xi}}{dt}=\frac{\partial\boldsymbol{\xi}}{\partial t}+(\boldsymbol{v}_0\cdot\nabla)\boldsymbol{\xi}=\boldsymbol{v}_1(\boldsymbol{r}, t)+(\boldsymbol{\xi}\cdot\nabla)\boldsymbol{v}_0. \tag{2.40}$$

2.5.1 Equation Describing Small-Amplitude Perturbations in Terms of Displacement Vector $\boldsymbol{\xi}$

Equation (2.38) is general. That is, no restriction on the magnitude of displacement vector $\boldsymbol{\xi}$. In many studies of wave motions and stability, however, an equation linearized with respect to $\boldsymbol{\xi}$ and expressed explicitly in terms of $\boldsymbol{\xi}$ alone is useful. Here, such an equation is derived.

For simplicity, we restrict here our attention to cases where the gravitational potential is external (nonself-gravitating system).[4] Then, for small amplitude perturbations, we have

$$\delta(\nabla\psi) = \xi_j \frac{\partial}{\partial r_j}(\nabla\psi_0), \tag{2.41}$$

since $\psi_1 = 0$.[5] Expressing $\delta(\nabla p/\rho)$ in terms of $\boldsymbol{\xi}$ is somewhat complicated. First, in the first order with respect to perturbations we have

$$\begin{aligned}\delta\left(\frac{1}{\rho}\nabla p\right) &= -\frac{\delta\rho}{\rho_0^2}\nabla p_0 + \frac{1}{\rho_0}\left[\nabla p_1 + \xi_j \frac{\partial}{\partial r_j}\nabla p_0\right] \\ &= -\frac{\delta\rho}{\rho_0^2}\nabla p_0 + \frac{1}{\rho_0}\left[\nabla(\delta p) - \nabla\left(\xi_j \frac{\partial}{\partial r_j} p_0\right) + \xi_j \frac{\partial}{\partial r_j}\nabla p_0\right].\end{aligned} \tag{2.42}$$

The Lagrangian pressure variation, δp, is related to the Lagrangian density variation, $\delta\rho$, in the case of small amplitude adiabatic perturbations by

$$\frac{\delta p}{p_0} = \Gamma_1 \frac{\delta\rho}{\rho_0}, \tag{2.43}$$

where Γ_1 is a generalized adiabatic exponent, and equal to γ in the case of no radiation pressure (see Sect. 19.6 for a detailed expression for Γ_1). Furthermore, the Lagrangian density variation, $\delta\rho$, is related to $\boldsymbol{\xi}$ in the case of small amplitude perturbations by

$$\delta\rho + \rho_0 \mathrm{div}\boldsymbol{\xi} = 0. \tag{2.44}$$

Substitution of Eqs. (2.43) and (2.44) into Eqs. (2.42) gives an expression for $\delta(\rho^{-1}\nabla p)$ in terms of $\boldsymbol{\xi}$, which is

$$\rho_0 \delta\left(\frac{1}{\rho}\nabla p\right) = \nabla\left[(1-\Gamma_1) p_0 \mathrm{div}\boldsymbol{\xi}\right] - p_0 \nabla \mathrm{div}\boldsymbol{\xi} + (\boldsymbol{\xi}\cdot\nabla)\nabla p_0. \tag{2.45}$$

[4]General cases of self-gravitating systems have been examined by Lynden-Bell and Ostriker (1967).

[5]The summation abbreviation is used when the same Latin subscript appears in a term twice. The summation is performed over three-dimensional coordinate space.

Consequently, in the case of $\boldsymbol{N} = 0$ Eq. (2.38) is written in terms of $\boldsymbol{\xi}$ as

$$\rho_0 \frac{\partial^2}{\partial t^2}\boldsymbol{\xi} + 2\rho_0(\boldsymbol{v}_0 \cdot \nabla)\frac{\partial}{\partial t}\boldsymbol{\xi} + \mathcal{L}(\boldsymbol{\xi}) = 0, \tag{2.46}$$

where $\mathcal{L}(\boldsymbol{\xi})$ is a linear operator with respect to $\boldsymbol{\xi}$, and is given by (Lynden-Bell and Ostriker 1967)

$$\mathcal{L}(\boldsymbol{\xi}) = \rho_0(\boldsymbol{v}_0 \cdot \nabla)(\boldsymbol{v}_0 \cdot \nabla)\boldsymbol{\xi} + \rho_0(\boldsymbol{\xi} \cdot \nabla)\nabla\psi_0$$
$$+ \nabla\left[(1 - \Gamma_1)p_0\mathrm{div}\boldsymbol{\xi}\right] - p_0\nabla\mathrm{div}\boldsymbol{\xi} - \nabla\left[(\boldsymbol{\xi} \cdot \nabla)p_0\right] + (\boldsymbol{\xi} \cdot \nabla)\nabla p_0. \tag{2.47}$$

An extension of Eq. (2.46) to quasi-nonlinear cases is important in studying turbulent excitation of stellar non-radial oscillations (Sect. 5.3) and wave-wave resonant excitation of oscillations in deformed systems (Sect. 6.7). Hence, an extension of Eq. (2.46) to quasi-nonlinear cases is presented in Appendix A.

2.5.2 *Hermitian Character of Operator $\mathcal{L}(\boldsymbol{\xi})$*

The Lagrangian description of perturbations is useful in deriving various important concepts in isolated systems. Here we present three issues: (1) the operator $\mathcal{L}$ is Hermitian, (2) in the case of time-periodic perturbations in isolated systems, we have a concept of wave energy which is constant during the motions, (3) time-periodic normal modes of perturbations are orthogonal each other if there is no internal motion in the unperturbed state. These characteristics hold even in MHD systems, since what we use to derive them is the Hermitian character of operator $\mathcal{L}(\boldsymbol{\xi})$ and this holds even in MHD perturbations (see Sect. 15.2).

First we notice that the operator $\mathcal{L}$ is Hermitian (self-adjoint operator) in the following sense, i.e.,

$$\int \boldsymbol{\eta} \cdot \mathcal{L}(\boldsymbol{\xi})dV = \int \boldsymbol{\xi} \cdot \mathcal{L}(\boldsymbol{\eta})dV, \tag{2.48}$$

where the integrations are performed over the whole volume of the system, and $\boldsymbol{\eta}$ and $\boldsymbol{\xi}$ are any non-singular functions defined in the unperturbed volume and having continuous first derivatives everywhere. This can be shown by performing the integration by part (for details in more general cases of self-gravitating systems, see Lynden-Bell and Ostriker 1967).

2.5.3 Wave Energy and Its Conservation

Let us consider a time-periodic adiabatic perturbation in isolated systems (oscillations or waves). For such perturbations we can introduce a concept of wave energy and can show that it is conserved during the motions.

To do so, we assume that a weakly growing artificial force is acting on an isolated system, which is $\delta \boldsymbol{f}$ per unit mass. At the initial epock ($t = -\infty$) when the force began to work on a perturbation, the perturbation had negligible amplitude. After that, by the effects of the force, the perturbation grows. The work done on the perturbation between $t = -\infty$ and t should be regarded as the wave energy, E, of the perturbation at t. In this sense, the wave energy of perturbation, E, can be written as

$$E(t) = \int \rho_0 \bigg[\int_{-\infty}^{t} \frac{\partial \boldsymbol{\xi}}{\partial t} \cdot \delta \boldsymbol{f} dt \bigg] dV. \tag{2.49}$$

Here, the volume integration is performed over the whole volume where the system exists.

Since $\delta \boldsymbol{f}$ is an external force acting on unit mass, the integrand of Eq. (2.49) can be expressed as

$$\frac{\partial \boldsymbol{\xi}}{\partial t} \bigg[\rho_0 \frac{\partial^2 \boldsymbol{\xi}}{\partial t^2} + 2\rho_0 (\boldsymbol{v}_0 \cdot \nabla) \frac{\partial \boldsymbol{\xi}}{\partial t} + \mathcal{L}(\boldsymbol{\xi}) \bigg]. \tag{2.50}$$

Integration by part with respect to t under the assumption of no perturbation at $t = -\infty$ gives

$$\int_{-\infty}^{t} \rho_0 \frac{\partial \boldsymbol{\xi}}{\partial t} \frac{\partial^2 \boldsymbol{\xi}}{\partial t^2} dt = \frac{1}{2} \rho_0 \bigg(\frac{\partial \boldsymbol{\xi}}{\partial t} \bigg)^2. \tag{2.51}$$

The volume integration of $2\rho_0 (\partial \boldsymbol{\xi}/\partial t)(\boldsymbol{v}_0 \cdot \nabla)(\partial \boldsymbol{\xi}/\partial t)$ vanishes if the integration is performed by part under consideration of vanishing of the surface integrations and $\nabla(\rho_0 \boldsymbol{v}_0) = 0$ (the unperturbed state is steady). Next, considering that the operator $\mathcal{L}$ is Hermitian, we have

$$\int_{-\infty}^{t} dt \int \rho_0 \frac{\partial \boldsymbol{\xi}}{\partial t} \mathcal{L}(\boldsymbol{\xi}) dV = \frac{1}{2} \int_{-\infty}^{t} dt \int \rho_0 \frac{\partial}{\partial t} \bigg[\boldsymbol{\xi} \cdot \mathcal{L}(\boldsymbol{\xi}) \bigg] dv = \frac{1}{2} \int \rho_0 \boldsymbol{\xi} \cdot \mathcal{L}(\boldsymbol{\xi}) dV. \tag{2.52}$$

Combining the above results, we have from Eq. (2.49)

$$E = \frac{1}{2} \int \rho_0 \bigg[\bigg(\frac{\partial \boldsymbol{\xi}}{\partial t} \bigg)^2 + \boldsymbol{\xi} \cdot \mathcal{L}(\boldsymbol{\xi}) \bigg] dV. \tag{2.53}$$

This can be regarded as wave energy of perturbations. The first term in the brackets of Eq. (2.53) can be regarded as kinetic energy, while the second term as potential energy.

Conservation of wave energy, E, can be derived directly from Eq. (2.53) by taking its time derivative. That is, we have

$$\begin{aligned}\frac{\partial E}{\partial t} &= \frac{1}{2}\int \rho_0\left[2\frac{\partial \boldsymbol{\xi}}{\partial t}\frac{\partial^2 \boldsymbol{\xi}}{\partial t^2} + \frac{\partial \boldsymbol{\xi}}{\partial t}\cdot \mathcal{L}(\boldsymbol{\xi}) + \boldsymbol{\xi}\cdot \mathcal{L}\left(\frac{\partial \boldsymbol{\xi}}{\partial t}\right)\right]dV \\ &= \int \rho_0 \frac{\partial \boldsymbol{\xi}}{\partial t}\left(\frac{\partial^2 \boldsymbol{\xi}}{\partial t} + \mathcal{L}(\boldsymbol{\xi})\right)dV = -2\int \rho_0 \frac{\partial \boldsymbol{\xi}}{\partial t}\left[(\boldsymbol{v}_0\cdot\nabla)\frac{\partial \boldsymbol{\xi}}{\partial t}\right]dV = 0.\end{aligned} \tag{2.54}$$

In deriving the second relation the Hermitian of $\mathcal{L}$ has been used, and in the last relation integration by part has been adopted.

2.5.4 Orthogonality of Normal Modes

Next, a less general issue is considered. The solutions of Eq. (2.46) with some relevant boundary conditions consist of a set of time-periodic motions. Each solution, $\boldsymbol{\xi}_\alpha(\boldsymbol{r}, t)$, is written as

$$\boldsymbol{\xi}_\alpha(\boldsymbol{r}, t) = \Re[\hat{\boldsymbol{\xi}}_\alpha \exp(i\omega_\alpha t)] \qquad (\alpha = 1, 2, 3\ldots), \tag{2.55}$$

and satisfies

$$-\omega_\alpha^2 \rho_0 \hat{\boldsymbol{\xi}}_\alpha + 2i\omega_\alpha \rho_0 (\boldsymbol{v}_0\cdot\nabla)\hat{\boldsymbol{\xi}}_\alpha + \mathcal{L}(\hat{\boldsymbol{\xi}}_\alpha) = 0, \tag{2.56}$$

where ω_α is real. The subscript α attached to $\hat{\boldsymbol{\xi}}$ is to distinguish oscillation modes (not to represent a component of vector).

Now, Eq. (2.56) is multiplied by $\hat{\boldsymbol{\xi}}^*_\beta(\boldsymbol{r}, t)$ and integrated over the whole volume, where the superscript $*$ denotes the complex conjugate and $\beta \neq \alpha$. The volume integration of $\rho_0\hat{\boldsymbol{\xi}}^*_\beta(\boldsymbol{r}, t)\hat{\boldsymbol{\xi}}_\alpha(\boldsymbol{r}, t)$ over the whole volume is written hereafter as $\langle\rho_0\hat{\boldsymbol{\xi}}^*_\beta\hat{\boldsymbol{\xi}}_\alpha\rangle$. Then, we have

$$-\omega_\alpha^2\langle\rho_0\hat{\boldsymbol{\xi}}^*_\beta\hat{\boldsymbol{\xi}}_\alpha\rangle + 2i\omega_\alpha\langle\rho_0\hat{\boldsymbol{\xi}}^*_\beta(\boldsymbol{v}_0\cdot\nabla)\hat{\boldsymbol{\xi}}_\alpha\rangle + \langle\hat{\boldsymbol{\xi}}^*_\beta\cdot\mathcal{L}(\hat{\boldsymbol{\xi}}_\alpha)\rangle = 0. \tag{2.57}$$

Similarly, the linear wave equation with respect to $\hat{\boldsymbol{\xi}}^*_\beta$ is integrated over the whole volume after being multiplied by $\hat{\boldsymbol{\xi}}_\alpha$ to lead

$$-\omega_\alpha^2\langle\rho_0\hat{\boldsymbol{\xi}}_\alpha\hat{\boldsymbol{\xi}}^*_\beta\rangle - 2i\omega_\beta\langle\rho_0\hat{\boldsymbol{\xi}}_\alpha(\boldsymbol{v}_0\cdot\nabla)\hat{\boldsymbol{\xi}}^*_\beta\rangle + \langle\hat{\boldsymbol{\xi}}_\alpha\cdot\mathcal{L}(\hat{\boldsymbol{\xi}}^*_\beta)\rangle = 0. \tag{2.58}$$

Since the operator $\mathcal{L}$ is Hermitian, we have the relation:

$$\langle \hat{\boldsymbol{\xi}}_\alpha \cdot \mathcal{L}(\hat{\boldsymbol{\xi}}^*_\beta)\rangle = \langle [\mathcal{L}(\hat{\boldsymbol{\xi}}^*_\alpha)]^* \cdot \hat{\boldsymbol{\xi}}^*_\beta\rangle = \langle \mathcal{L}(\hat{\boldsymbol{\xi}}_\alpha) \cdot \hat{\boldsymbol{\xi}}^*_\beta\rangle. \tag{2.59}$$

Hence, the difference of the above two equations [Eqs. (2.57) and (2.58)] gives, when $\omega_\beta \neq \omega_\alpha$,

$$(\omega_\alpha + \omega_\beta)\langle \rho_0 \hat{\boldsymbol{\xi}}_\alpha \hat{\boldsymbol{\xi}}^*_\beta\rangle + 2i\Big\langle \rho_0 \hat{\boldsymbol{\xi}}_\alpha [(\boldsymbol{v}_0 \cdot \nabla)\hat{\boldsymbol{\xi}}^*_\beta]\Big\rangle = 0 \tag{2.60}$$

or

$$(\omega_\alpha + \omega_\beta)\langle \rho_0 \hat{\boldsymbol{\xi}}_\alpha \hat{\boldsymbol{\xi}}^*_\beta\rangle - 2i\Big\langle \rho_0 \hat{\boldsymbol{\xi}}^*_\beta [(\boldsymbol{v}_0 \cdot \nabla)\hat{\boldsymbol{\xi}}_\alpha]\Big\rangle = 0. \tag{2.61}$$

In deriving these equalities, we have used an integration by part, assuming that ρ_0 vanishes on the surface of the system.

In the special cases where unperturbed systems have no motion, i.e., $\boldsymbol{v}_0 = 0$, the eigenfunctions of normal modes are orthogonal, i.e.,

$$\int \rho_0 \hat{\boldsymbol{\xi}}_\alpha \hat{\boldsymbol{\xi}}^*_\beta \, dV = 0 \qquad \text{for} \quad \alpha \neq \beta. \tag{2.62}$$

Reference

Lynden-Bell, D., Ostriker, J.P.: Mon. Not. Roy. Astron. Soc. **136**, 293 (1967)

Chapter 3
Astrophysical Fluid Flows

In Chap. 2 we have derived the basic equations of astrophysical hydrodynamics, and summarized various relations or characteristics of hydrodynamical motions. In this chapter we consider steady/stationary flows, and review some typical astrophysical applications; steady spherical accretion and wind, viscous accretion flows, and meridional circulation in stars. Finally, as a powerful method to study astrophysical flows, the similarity method is briefly mentioned.

3.1 Parker Wind and Bondi Accretion

We first consider the simplest case: spherically symmetric steady accretion and wind under the gravitational field around a point mass. Spherical accretion onto a gravitating body was first studied by Bondi (1952) and is often called *Bondi accretion*, which is now believed to be quite important in various astrophysical situations. Spherical outflow, on the other hand, was examined by Parker (1958) in relation to *solar wind*, the existence of solar and stellar winds is now well established. See, e.g., Holzer and Axford (1970) for the fundamentals of spherical transonic flows.

Some basic characteristics of winds and flows in subject to magnetic fields are briefly discussed in Chap. 12.

© Springer Nature Singapore Pte Ltd. 2020, corrected publication 2023
S. Kato, J. Fukue, *Fundamentals of Astrophysical Fluid Dynamics*,
Astronomy and Astrophysics Library,
https://doi.org/10.1007/978-981-15-4174-2_3

3.1.1 Basic Equations for Steady Spherical Flow

Let us consider a spherically symmetric flow around an object of mass M. The flow is supposed to be steady and one-dimensional in the radial (r) direction. The flow is further assumed to be inviscid and adiabatic (or isothermal), and magnetic and radiation fields are ignored.

Under the Newtonian approximation, the continuity equation and the equation of motion are, respectively,

$$\frac{1}{4\pi r^2}\frac{d}{dr}\left(4\pi r^2 \rho v\right) = 0, \tag{3.1}$$

$$v\frac{dv}{dr} = -\frac{1}{\rho}\frac{dp}{dr} - \frac{GM}{r^2}, \tag{3.2}$$

where v is the flow velocity (positive for wind and negative for accretion), ρ the density, and p the pressure. The polytropic relation is assumed,

$$p = K\rho^\gamma, \tag{3.3}$$

where K and γ are constants.

Integrating the above equations yields the mass conservation and the Bernoulli equation:

$$4\pi r^2 \rho v = \dot{M}, \tag{3.4}$$

$$\frac{1}{2}v^2 + \frac{\gamma}{\gamma - 1}\frac{p}{\rho} - \frac{GM}{r} = E, \tag{3.5}$$

where $\dot{M}$ is the mass-loss rate for wind or mass-accretion rate for accretion (which is constant in the present case), and E is the *Bernoulli constant.*[1,2]

[1] In the isothermal case ($\gamma = 1$), the Bernoulli equation becomes

$$\frac{1}{2}v^2 + \frac{p}{\rho}\ln\rho - \frac{GM}{r} = E.$$

[2] In the case of the solar wind, which is almost isothermal at temperature $T_0 = 10^6$ K, $\bar{\mu} = 0.5$, and $\gamma = 1.05$, at the wind base, where $v \sim 0$ and $r \sim R_\odot$, the Bernoulli constant is about $E = \gamma/(\gamma - 1) * 2\mathcal{R}T_0 - GM_\odot/R_\odot \sim 1.6 \times 10^{15}$ erg.

3.1.2 Sound Speed and Critical Points

Let us introduce the *adiabatic sound speed* c_s, which is defined by $c_s^2 \equiv (dp/d\rho)_s$ ($= \gamma p/\rho$ for a polytropic gas),[3] and rewrite the basic equations as[4]

$$4\pi r^2 c_s^{\frac{2}{\gamma-1}} v = (K\gamma)^{\frac{1}{\gamma-1}} \dot{M}, \tag{3.6}$$

$$\frac{1}{2}v^2 + \frac{1}{\gamma-1}c_s^2 - \frac{GM}{r} = E. \tag{3.7}$$

From the logarithmic differentiation of Eq. (3.4) we have $2/r + d\rho/(\rho dr) + dv/(vdr) = 0$, and by eliminating $d\rho/dr$ from Eq. (3.2) we finally obtain, after some manipulation, the *wind equation* as[5]

$$(v^2 - c_s^2)\frac{1}{v}\frac{dv}{dr} = \frac{2}{r}c_s^2 - \frac{GM}{r^2}. \tag{3.8}$$

Here, the sound speed is expressed as

$$c_s^2 = (\gamma - 1)\left(E + \frac{GM}{r} - \frac{1}{2}v^2\right). \tag{3.9}$$

In the above Eq. (3.8), the term $(v^2 - c_s^2)$ on the left-hand side becomes zero at the *transonic point* (*critical point*), where the flow velocity equals the sound speed. As long as the velocity gradient, dv/dr, is finite at the transonic point, the right-hand side of Eq. (3.8) must vanish simultaneously at the critical point. This condition (the *regularity condition*) gives the locations of critical points r_c as follows (the subscript "c" is for "critical").

In the adiabatic case, inserting r_c, v_c, c_c [$\equiv c_s(r_c)$] into continuity and Bernoulli equations (3.6) and (3.7), and considering the regularity condition, $v_c = c_c$ and $r_c = GM/2c_c^2$, we have

$$(K\gamma)^{1/(\gamma-1)}\dot{M} = 4\pi r_c^2 |v_c|^{(\gamma+1)/(\gamma-1)}, \tag{3.10}$$

---

[3]In the case of the solar wind, where $T \sim 10^6$ K and $\gamma \sim 1$, the sound speed is about $c_s \sim \sqrt{2\gamma \mathcal{R} T} \sim 130\,\mathrm{km\,s^{-1}}$.

[4]In the isothermal case, the Bernoulli equation is rewritten as

$$\frac{1}{2}v^2 + c_T{}^2 \ln \rho - \frac{GM}{r} = E,$$

where $c_T{}^2$ ($\equiv p/\rho$) is the *isothermal sound speed* and constant.

[5]In the isothermal case, c_s is simply replaced by constant c_T.

$$v_c^2 = \frac{2(\gamma - 1)}{5 - 3\gamma} E. \tag{3.11}$$

These give the relations between the quantities at the critical points and the flow parameters. Furthermore, the critical radius r_c is expressed in terms of γ and E as[6]

$$r_c = \frac{GM}{2c_c^2} = \frac{GM}{2v_c^2} = \frac{(5 - 3\gamma)GM}{4(\gamma - 1)E}. \tag{3.12}$$

Moreover, in order for the steady transonic solution to exist, E must be positive. Hence, the condition

$$1 < \gamma < 5/3 \tag{3.13}$$

should be satisfied in the case of spherically symmetric adiabatic flows.

3.1.3 Solution Behavior Near to Center or at Infinity

Before considering the transonic solutions, we examine the asymptotic behavior of the present spherical flow near to the center or at sufficiently large radius.

1. Near to the center of spherical wind ($r \to R_\odot; r \ll r_c$)
   In the case of outflow like solar wind, the flow speed is sufficiently small ($v \to 0$) near to the center ($r \ll r_c$). In this case, we safely drop the kinetic energy in the Bernoulli equation, and we have

$$c_s^2 \to (\gamma - 1)\left(E + \frac{GM}{r}\right) \sim (\gamma - 1)\frac{GM}{r}, \quad r \ll r_c. \tag{3.14}$$

   This is just the static spherical atmosphere around the central point mass. Namely, the structure of the spherical wind in the central subsonic region is similar to that of the static spherical atmosphere.
   Substituting this asymptotic solution into the continuity equation (3.6), we have the behavior of the velocity field near to the center; $v \propto r^{(3-2\gamma)/(\gamma-1)}$ ($r \to 0$).
2. At infinity of spherical wind ($r \to \infty; r \gg r_c$)
   On the other hand, at far from the center ($r \to \infty$), the flow density and sound speed as well as gravity become sufficiently small, compared with the kinetic term. In this case, from the Bernoulli equation, the flow speed approaches constant at infinity:

$$v \to v_\infty = \sqrt{2E}, \quad r \to \infty. \tag{3.15}$$

[6] In the case of the solar wind, $r_c = GM/(2c_c^2) \sim GM_\odot/(2c_s^2) \sim 5.73R_\odot$.

This asymptotic value is called *terminal velocity*.

Substituting this result into the continuity equation (3.6), we have the behavior of the sound speed at far from the center; $c_s \propto r^{-(\gamma-1)}$ $(r \to \infty)$.

3. At infinity of spherical accretion ($r \to \infty$; $r \gg r_c$)

In the case of accretion flow, the flow infall speed approaches 0 at infinity, but the sound speed does not, whereas the gravitational potential vanishes at infinity. In this case, from the Bernoulli equation, the sound speed approaches constant at infinity:

$$c_s \to c_\infty = \sqrt{(\gamma - 1)E}, \qquad r \to \infty. \tag{3.16}$$

This is the sound speed of interstellar gas at rest at infinity. This is also the boundary condition at infinity.

Substituting this result into the continuity equation (3.6), we have the behavior of the flow velocity at far from the center; $v \propto r^{-2}$ $(r \to \infty)$.

4. Near to the center of spherical accretion ($r \to 0$; $r \ll r_c$)

On the other hand, the accretion flow is supersonic near to the center ($r \ll r_c$), and the gravitational potential also diverses at the center. In this case, we drop the enthalpy term in the Bernoulli equation, and we have

$$v \to \sqrt{2\left(E + \frac{GM}{r}\right)} \sim \sqrt{\frac{2GM}{r}}, \qquad r \to 0. \tag{3.17}$$

This is just the velocity field of the simple freefall motion. Namely, the spherical accretion in the central supersonic region is quite similar to the simple freefall without the effect of gas pressure.

Inserting this result into the continuity equation (3.6), we have the behavior of the sound speed near to the center; $c_s \propto r^{-3(\gamma-1)/4}$ $(r \to 0)$.

3.1.4 Mach Number and Singular Point Analysis

Let us next introduce the *Mach number* $\mathcal{M}$, which is defined by $\mathcal{M} \equiv v/c_s$, and derive the wind equation on $\mathcal{M}$.

In the adiabatic case, we easily derive

$$\frac{d\mathcal{M}}{dr} = \frac{\mathcal{N}}{\mathcal{D}}, \tag{3.18}$$

where the denominator $\mathcal{D}$ and numerator $\mathcal{N}$ are, respectively,

$$\mathcal{D} = \mathcal{M}^2 - 1, \tag{3.19}$$

$$\mathcal{N} = \mathcal{M}\left(\frac{\gamma - 1}{2}\mathcal{M}^2 + 1\right)\left[\frac{2}{r} - \frac{\gamma + 1}{2(\gamma - 1)}\frac{1}{E + \dfrac{GM}{r}}\frac{GM}{r^2}\right], \tag{3.20}$$

after some manipulations.[7]

In order to examine the behavior of solutions near to the critical points, we linearize the wind equation in the vicinity of the critical points. That is, we expand the denominator and numerator as

$$\mathcal{D}(r, \mathcal{M}) \sim \mathcal{D}|_\mathrm{c} + \left.\frac{\partial \mathcal{D}}{\partial r}\right|_\mathrm{c} dr + \left.\frac{\partial \mathcal{D}}{\partial \mathcal{M}}\right|_\mathrm{c} d\mathcal{M}, \tag{3.21}$$

$$\mathcal{N}(r, \mathcal{M}) \sim \mathcal{N}|_\mathrm{c} + \left.\frac{\partial \mathcal{N}}{\partial r}\right|_\mathrm{c} dr + \left.\frac{\partial \mathcal{N}}{\partial \mathcal{M}}\right|_\mathrm{c} d\mathcal{M}. \tag{3.22}$$

Here, $|_\mathrm{c}$ means the quantities evaluated at the critical points (i.e., $\mathcal{D}|_\mathrm{c} = 0 = \mathcal{N}|_\mathrm{c}$ trivially). Other coefficients are

$$\lambda_{11} \equiv \left.\frac{\partial \mathcal{D}}{\partial r}\right|_\mathrm{c} = 0, \tag{3.23}$$

$$\lambda_{12} \equiv \left.\frac{\partial \mathcal{D}}{\partial \mathcal{M}}\right|_\mathrm{c} = 2, \tag{3.24}$$

$$\lambda_{21} \equiv \left.\frac{\partial \mathcal{N}}{\partial r}\right|_\mathrm{c} = \frac{5 - 3\gamma}{r_\mathrm{c}^2}, \tag{3.25}$$

$$\lambda_{22} \equiv \left.\frac{\partial \mathcal{N}}{\partial \mathcal{M}}\right|_\mathrm{c} = 0 \tag{3.26}$$

for the present adiabatic case.[8]

Using these coefficients, λ_{ij}, the wind equation (3.18) can be expressed near to the critical point in a general form as

$$\frac{d\mathcal{M}}{dr} = \frac{\lambda_{21}dr + \lambda_{22}d\mathcal{M}}{\lambda_{11}dr + \lambda_{12}d\mathcal{M}} = \frac{\lambda_{21} + \lambda_{22}\dfrac{d\mathcal{M}}{dr}}{\lambda_{11} + \lambda_{12}\dfrac{d\mathcal{M}}{dr}}. \tag{3.27}$$

[7] In the isothermal case, the numerator is expressed as

$$\mathcal{N} = \mathcal{M}\left(\frac{2}{r} - \frac{GM}{c_\mathrm{T}{}^2 r^2}\right).$$

[8] In the isothermal case, $\lambda_{21} = 2/r_\mathrm{c}^2$.

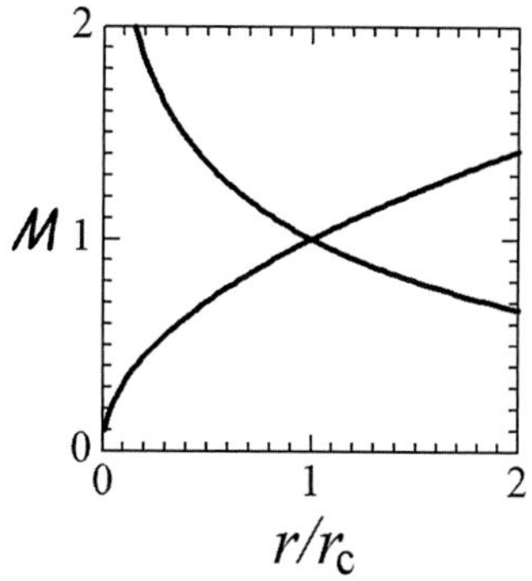

Fig. 3.1 Transonic solutions in the vicinity of critical points. One corresponds to the Bondi spherical accretion, and the other to the Parker solar wind

Hence, we have a quadratic equation on $d\mathcal{M}/dr$,

$$\lambda_{12}\left(\frac{d\mathcal{M}}{dr}\right)^2 + (\lambda_{11} - \lambda_{22})\frac{d\mathcal{M}}{dr} - \lambda_{21} = 0, \tag{3.28}$$

which has two roots:

$$\frac{d\mathcal{M}}{dr} = \frac{-\lambda_{11} + \lambda_{22} \pm \sqrt{(\lambda_{11} - \lambda_{22})^2 + 4\lambda_{12}\lambda_{21}}}{2\lambda_{12}}. \tag{3.29}$$

In the present case of adiabatic flows[9] with neither viscosity nor other dissipative processes, Eq. (3.29) simply reduces to

$$\frac{d\mathcal{M}}{dr} = \pm\sqrt{\frac{5 - 3\gamma}{2r_c^2}}. \tag{3.30}$$

Since $d\mathcal{M}/dr$ is just the slope of solutions on the $(r, \mathcal{M})$-diagram, there are two slopes near to the critical points if transonic solutions exist (Fig. 3.1). One corresponds to the Bondi spherical accretion, and the other to the Parker solar wind.

3.1.5 A General Argument on Topology around Critical Point

In general there are four types of critical points (*topologies*): *saddle*, *center*, *node*, and *spiral* (Fig. 3.2). These types/topologies are determined by the eigenvalue of the matrix λ_{ij}:

$$\Lambda = \begin{bmatrix} \lambda_{11} & \lambda_{12} \\ \lambda_{21} & \lambda_{22} \end{bmatrix}, \tag{3.31}$$

[9]In the isothermal case, $d\mathcal{M}/dr = \pm 1/r_c$.

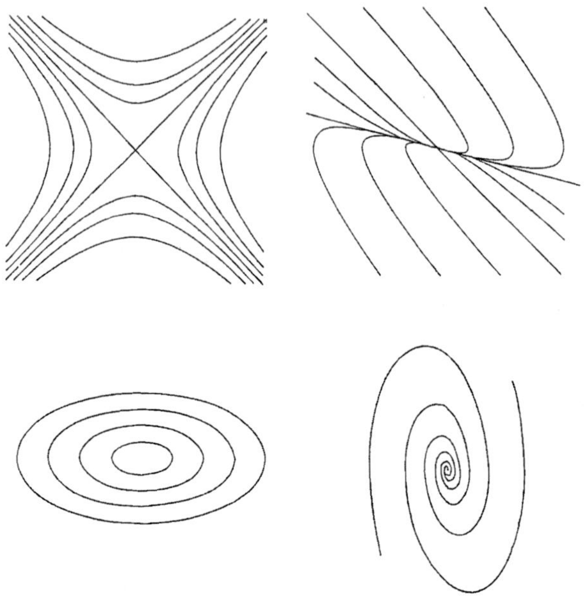

Fig. 3.2 Types (topologies) of critical points. From top-left to bottom-right, types are saddle, node, center, and spiral

whose trace and determinant are, respectively,

$$\mathrm{tr}\Lambda = \lambda_{11} + \lambda_{22}, \tag{3.32}$$

$$|\Lambda| = \begin{vmatrix} \lambda_{11} & \lambda_{12} \\ \lambda_{21} & \lambda_{22} \end{vmatrix} = \lambda_{11}\lambda_{22} - \lambda_{12}\lambda_{21}. \tag{3.33}$$

Then, the characteristic equation of the matrix Λ is expressed by

$$\lambda^2 - (\mathrm{tr}\Lambda)\lambda + |\Lambda| = 0, \tag{3.34}$$

and the roots of this equation are given as

$$\lambda = \frac{1}{2}\left[\mathrm{tr}\Lambda \pm \sqrt{(\mathrm{tr}\Lambda)^2 - 4|\Lambda|}\right], \tag{3.35}$$

where $Det = (\mathrm{tr}\Lambda)^2 - 4|\Lambda|$ is the discriminant.

The signature and properties of the roots are related to the types/topologies of the critical points as shown in Table 3.1.

The above general results concerning the relation between the root of the characteristic equation and the type of critical points can be derived from the following considerations. For this purpose, let us generalize the case of solar wind.

Table 3.1 Roots and types of critical points

Conditions	Roots	Types of critical points
$\lvert\Lambda\rvert < 0$	Two real roots with opposite sign	Saddle
$\lvert\Lambda\rvert > 0$ and $Det > 0$	Two real roots with same sign	Node
$\lvert\Lambda\rvert > 0$ and $Det = 0$	Double root	Node
$\lvert\Lambda\rvert > 0$ and $\mathrm{tr}\Lambda = 0$	Pure imaginary roots	Center
$\lvert\Lambda\rvert > 0$ and $Det < 0$	Complex roots	Spiral

We denote the difference of $\mathcal{M}$ from unity by y, i.e., $y = \mathcal{M} - 1$, and the difference of r from r_c by x, i.e., $x = r - r_c$. Then, by introducing a parametric variable s we can express a generalized form of Eq. (3.27) by a set of equations as

$$\frac{dx}{ds} = \lambda_{11}x + \lambda_{12}y \tag{3.36}$$

and

$$\frac{dy}{ds} = \lambda_{21}x + \lambda_{22}y. \tag{3.37}$$

Let us denote the two solutions of Eq. (3.35) by λ_1 and λ_2. Then, the solutions of the set of the above two equations are

$$x = C_1\exp(\lambda_1 s) + C_2\exp(\lambda_2 s), \tag{3.38}$$

$$y = D_1\exp(\lambda_1 s) + D_2\exp(\lambda_2 s), \tag{3.39}$$

where D_1 and D_2 are related to C_1 and C_2 by

$$D_1 = \frac{1}{\lambda_{12}}(\lambda_1 - \lambda_{11})C_1 \quad \text{and} \quad D_2 = \frac{1}{\lambda_{12}}(\lambda_2 - \lambda_{11})C_2. \tag{3.40}$$

Free parameters of integration constants are C_1 and C_2.

First, we consider a simpler case where λ_1 and λ_2 are real with the same sign. In this case, by eliminating the term of $\exp(\lambda_1 s)$ from Eqs. (3.38) and (3.39) we have

$$y - \frac{D_1}{C_1}x = \left(D_2 - \frac{C_2}{C_1}D_1\right)\exp(\lambda_2 s). \tag{3.41}$$

Similarly, eliminating $\exp(\lambda_2 s)$ from the same set of equations, we have

$$y - \frac{D_2}{C_2}x = \left(D_1 - \frac{C_1}{C_2}D_2\right)\exp(\lambda_1 s). \tag{3.42}$$

By eliminating the exponential terms with parametric variable s from the above two equations, we have

$$y - \frac{D_1}{C_1}x = \frac{D_2 - (C_2/C_1)D_1}{[D_1 - (C_1/C_2)D_2]^{\lambda_2/\lambda_1}}\left(y - \frac{D_2}{C_2}x\right)^{\lambda_2/\lambda_1}, \tag{3.43}$$

where

$$D_2 - \frac{C_2}{C_1}D_1 = \frac{C_2}{\lambda_{12}}(\lambda_2 - \lambda_1) \quad \text{and} \quad D_1 - \frac{C_1}{C_2}D_2 = \frac{C_1}{\lambda_{12}}(\lambda_1 - \lambda_2). \tag{3.44}$$

Here, we assume $\lambda_2/\lambda_1 > 1$ without loss of generality. Then, Eq. (3.43) shows that curves pass through the critical point are a family of curves which touch to the line of $y = D_1/C_1x$ and an exceptional curve of $y = (D_2/C_2)x$, which crosses the line of $y = D_1/C_1x$ at the critical point. This means that the topology around the critical point is nodal (see Table 3.1).

Next, we consider the case where λ_1 and λ_2 are real with opposite sign, i.e., the case of saddle-type critical point. A general argument on this case is complicated. Hence, we shall demonstrate only a special case of $\lambda_{11} = \lambda_{22} = 0$, which is the case of solar wind. In this case we have $\lambda_1 = -\lambda_2 \equiv \lambda$, and $\lambda = (-\lambda_{12}\lambda_{21})^{1/2}$. The set of parametric expressions for x and y are then

$$x = C_1\exp(\lambda s) + C_2\exp(-\lambda s), \tag{3.45}$$

$$y = C_1\frac{\lambda}{\lambda_{12}}\exp(\lambda s) - C_2\frac{\lambda}{\lambda_{12}}\exp(-\lambda s). \tag{3.46}$$

From the above two equations we can easily eliminate the parametric variable s to lead to

$$\frac{x^2}{\lambda_{12}} - \frac{y^2}{\lambda_{21}} = 2C_1C_2\frac{\lambda_{12} + \lambda_{21}}{\lambda_{12}\lambda_{21}}, \tag{3.47}$$

where $\lambda^2 = \lambda_{12}\lambda_{21}$ has been used. This equation presents a set of hyperbolic curves, and the lines passing through the critical point are $y = \pm(\lambda_{21}/\lambda_{12})^{1/2}x$. That is, the topology is saddle. By similar ways we can show the presence of critical points of elliptical and spiral types, which are, however, omitted here.

In the case of a simple solar wind, the type of critical points is saddle, since the characteristic equation has two real roots with opposite sign. When the centrifugal force and/or the relativistic effect are included, the center-type critical points appear. The diffusion processes, such as viscosity, conduction, and radiative diffusion, often cause nodal-type critical points.

3.2 Accretion Disks

Accretion disks are one of important ingredients in the Universe. Activities of various active astrophysical objects are attributed to accretion disks which are formed around such compact objects as white dwarfs, neutron stars, stellar black holes, and massive black holes at galactic centers. The protoplanetary systems are also important astrophysical objects with accretion disks. The spiral disk galaxies themselves are also considered to be disk systems as the whole, although they consist mainly of stars, not gaseous matter. Since the pioneering work by Shakura and Sunyaev (1973) the structure of gaseous disks has been extensively studied for various parameter cases (see, for example, review books by Frank et al. 1992, 2002; Kato et al. 1998, 2008). Here, we present fundamental properties of accretion disks and briefly describe some typical disk models.

3.2.1 Standard Accretion Disk Model

Although many variety of disk models are possible, we present first the classical standard model by Shakura and Sunyaev (1973), which is an optically thick, radiation-cooling-dominated disk. The model is called *standard model* or α *model* by a parameter α being introduced in description of viscous stress tensor.

1. **Nonself-Gravity of Disks**
   Disks are subject to the gravitational potential of a central object. In protoplanetary disks as well as galactic disks, self-gravity of the disk matter has important effects on structure and dynamics of disks, in addition to that of the central object. There are two criteria measuring the importance.

   To examine this issue, let us consider a geometrically thin disks surrounding around a central object of mass, M, with nearly Keplerian rotation. Cylindrical coordinates (r, φ, z) whose center is at the disk center (the position of the central star) and the z-axis is the rotation axis are introduced. A point $P(r, \varphi, z)$ which is slightly above the disk equatorial plane, i.e., $z \ll r$, is considered. Then, the vertical component of gravitational force working at $P(r, \varphi, z)$ by the central object is $-(GM/R^2)(z/R) \sim -(GM/r^3)z$, which acts towards the equator, R^2 being $R^2 = r^2 + z^2$. On the other hand, the force acting towards the equator by the self-gravity of the disk material is $\sim -4\pi G\rho z$, where ρ is the density in the disk and G the gravitational constant.[10] The effects of self-gravity on disk structure are negligible if $4\pi G\rho < (GM/r^3)$. Because surface density, Σ, of the

[10]If the gravitational potential due to self-gravity is written as $\psi(r, z)$, the Poisson equation is $\nabla^2\psi = 4\pi G\rho$. This gives $-\partial\psi/\partial z \sim -4\pi G\rho z$.

disk is related to ρ through disk thickness, H, by $\Sigma \sim 2\rho H$, the above inequality leads to

$$\Sigma < \frac{M}{2\pi r^2}\frac{H}{r}. \tag{3.48}$$

Roughly speaking, this inequality implies that the disk mass ($\sim \pi r^2 \Sigma$) needs to be smaller than the mass of the central source by H/r. This condition is satisfied in many disks. Hereafter, we restrict our attention to disks which satisfy this condition.

Even if the above inequality is satisfied, the effects of self-gravity are not always negligible, if behaviors of perturbations on the disks are considered. As will be described in Chap. 4, if the disk surface density Σ is higher than a critical value, the disk is gravitationally unstable to local perturbations. The criterion for this instability is characterized by a parameter Q called *Toomre's Q value*, which is (see also Sect. 4.5)

$$Q = \frac{\kappa c_s}{\pi G \Sigma}, \tag{3.49}$$

where c_s is the acoustic speed in disks, κ is the epicyclic frequency. This Q-parameter needs to be larger than unity for local stability (see Sect. 4.5). Hereafter, we consider the disks with $Q > 1$.

2. **Vertical Hydrostatic Balance**
In nonself-gravitating disks their vertical thickness is determined by the force balance between pressure force and the gravitational force of the central source, which is written as

$$-\frac{\partial p}{\partial z} = -\rho g_z, \tag{3.50}$$

where g_z is given by

$$g_z \sim -\frac{GMz}{r^3}, \tag{3.51}$$

M being the mass of the central source. Equation (3.50) gives, under the one-zone approximation, $p/H = -\rho g_z(H)$, where $g_z(H)$ is the gravitational acceleration at the surface of the disk ($z = H$), i.e., $g_z(H) \sim -(GM/r^2)(H/r) = -\Omega_K^2 H$, where Ω_K is the angular velocity of the Keplerian rotation. Hence, we can write the vertical hydrostatic balance as

$$\Omega_K^2 H^2 = \frac{p}{\rho} = c_s^2 \quad \text{or} \quad H = \frac{c_s}{\Omega_K}, \tag{3.52}$$

where c_s is the speed of sound in the disk. The disk is geometrically thin in the sense that H/r is on the order of c_s over rotational speed.

3. **Angular Momentum Transport by Viscous Force**
The fact that disks are thin in the vertical direction implies that the pressure force (including radiation and magnetic forces if exist) in the radial direction is much smaller than the radial gravitational force by the factor of $(H/r)^2$. Thus, in the lowest order of approximations the force balance in the radial direction is realized between the gravitational force of the central object and the centrifugal force of disk rotation. That is, the angular velocity of disk rotation, $\Omega(r)$, is Keplerian:

$$\Omega(r) = \Omega_{\rm K}(r) = \left(\frac{GM}{r^3}\right)^{1/2}. \tag{3.53}$$

Since this rotation is differential and the rotational velocity is faster in the inner region, angular momentum is transported outwards by viscous force. The outward flux of angular momentum is given by $-r^2\nu\rho d\Omega/dr$, where ν is kinematic viscosity defined by μ/ρ, μ being viscosity. The divergence of this angular momentum flux is positive, implying that the gas at r loses angular momentum and falls towards the disk center. Hence, in steady disks, angular momentum conservation requires that to compensate this angular momentum loss at r, the disk gas must be supplied from the outer part (accretion). By considering cylindrical surface at radius r with thickness $2H$, we can express the angular momentum balance as

$$-\frac{\partial}{\partial r}\left[2Hr\left(r^2\nu\rho\frac{d\Omega}{dr}\right)\right] = 2Hr\rho v\frac{d}{dr}(r^2\Omega), \tag{3.54}$$

where $v > 0$ is the accretion speed in the radial direction and is related to mass-accretion rate, $\dot{M}(>0)$, by

$$\dot{M} = 2\pi r\Sigma v, \tag{3.55}$$

where $\Sigma(r) = 2H\rho$ is the surface density.

The use of $\dot{M}$ reduces Eq. (3.54) to the form:

$$\dot{M}\left[\frac{d}{dr}(r^2\Omega)\right] = -2\pi\frac{d}{dr}\left(r^3\nu\Sigma\frac{d\Omega}{dr}\right). \tag{3.56}$$

Since $\dot{M}(>0)$ is constant in steady disks, the above equation can be integrated to lead to

$$\frac{\dot{M}}{2\pi}(\ell - \ell_{\rm in}) = -r^3\nu\Sigma\frac{d\Omega}{dr}, \tag{3.57}$$

where $\ell = \sqrt{GMr}$ and $\ell_{\rm in} = \sqrt{GMr_{\rm in}}$ is the specific angular momentum at the inner edge, $r_{\rm in}$, of disks. It has been assumed that there is no stress at the inner edge of disks.

4. **Viscosity Prescription**
The viscosity in the accretion disks plays two important roles. One is transport of angular momentum as mentioned above. The other is heating of the disk plasma, which is an important ingredient in considering heat balance in disks. The viscosity will come from (magnetic) turbulence [magneto-rotational instability (Balbus and Hawley 1991; Hawley and Balbus 1991)]. Since the disks are rotating, the major component of the viscous stress tensor is the $r\varphi$-one, say $t_{r\varphi}$. In the conventional standard model (Shakura and Sunyaev 1973), $t_{r\varphi}$ is prescribed by introducing a free parameter α as

$$t_{r\varphi} = \rho\nu r\frac{d\Omega}{dr} = -\alpha p \quad \text{or} \quad T_{r\varphi} = \nu\Sigma r\frac{d\Omega}{dr} = -\alpha\Pi, \tag{3.58}$$

where α is called *alpha-parameter* and taken to be a dimensionless quantity smaller than unity, and $T_{r\varphi}$ and Π are, respectively, the vertically integrated stress tensor and pressure.

5. **Energy Balance**
Among many possible heating processes, the major one is that due to viscous shear of differential rotation. Retaining the term alone, we write the heating rate per unit volume, $q^+_{\rm vis}$, as $t_{r\varphi}rd\Omega/dr$. The vertical integrated heating rate per unit surface, $Q^+_{\rm vis}$, is thus

$$Q^+_{\rm vis} = \int_{-\infty}^{\infty}\nu\rho\left(r\frac{d\Omega}{dr}\right)^2 dz = \frac{9}{4}\nu\Sigma\Omega^2 = -\frac{3}{2}T_{r\varphi}\Omega, \tag{3.59}$$

where Eq. (3.58) has been used.

Now, we assume that the disks are optically thick in the vertical direction, and the radiative cooling occurs from the surface. The radiative flux, $F(z)$, in the z-direction at height z is

$$F(z) = -\left(\frac{4acT^3}{3\bar{\kappa}_{\rm R}\rho}\right)_z\frac{\partial T}{\partial z}, \tag{3.60}$$

where a is the radiation constant ($\sigma_{\rm SB} = ac/4$), and $\bar{\kappa}_{\rm R}$ is the *Rosseland mean opacity* (see Sect. 20.3). Hence, we can express the cooling rate at unit surface, $Q^-_{\rm rad}$, in terms of the temperature $T_{\rm c}$ at $z = 0$ as

$$Q^-_{\rm rad} = 2\frac{4acT_{\rm c}^4}{3\tau} = \frac{32\sigma_{\rm SB}T_{\rm c}^4}{3\tau}, \tag{3.61}$$

where τ is the optical depth in the vertical direction:

$$\tau = \bar{\kappa}_{\rm R}\rho H = \frac{\bar{\kappa}_{\rm R}\Sigma}{2} \gg 1. \tag{3.62}$$

The local energy balance at each radius is, therefore, given by

$$Q^+_{\rm vis} = Q^-_{\rm rad}. \tag{3.63}$$

It is noted that for this expression to be valid, the cooling timescale needs to be much shorter than the matter advection timescale. Otherwise, the gas heated by viscosity falls inwards before cooled down by radiation (see Sect. 3.3.2).

6. **Opacity**
In high-temperature disks with $T \geq 10^4$K, the opacity $\bar{\kappa}_{\rm R}$ comes from electron scattering, $\kappa_{\rm es}$, and free-free absorption, $\kappa_{\rm ff}$:

$$\bar{\kappa}_{\rm R} = \kappa_{\rm es} + \kappa_{\rm ff} = \kappa_{\rm es} + \kappa_0 \rho T^{-3.5}, \tag{3.64}$$

where $\kappa_{\rm es} = 0.4\,{\rm cm^2\,g^{-1}}$ and $\kappa_0 = 6.4 \times 10^{22}\,{\rm cm^2\,g^{-1}}$ for pure hydrogen plasmas.

7. **Equation of State**
In optically thick gases pressure p consists of gas pressure $p_{\rm gas}$ and radiation pressure $p_{\rm rad}$ as

$$p = p_{\rm gas} + p_{\rm rad} = \frac{2k_{\rm B}}{m_{\rm H}}\rho T_{\rm c} + \frac{aT^4}{3}, \tag{3.65}$$

where $m_{\rm H}$ is the mass of a hydrogen atom and $k_{\rm B}$ is the Boltzmann constant. The vertical integration of Eq. (3.65) is

$$\Pi = \Pi_{\rm gas} + \Pi_{\rm rad} = \frac{2k_{\rm B}}{m_{\rm H}}\Sigma T_{\rm c} + \frac{aT_{\rm c}^4}{3}\cdot 2H. \tag{3.66}$$

In summary, the basic set of equations describing steady axisymmetric non-magnetized, optically thick, geometrically thin, and radiation-cooling-dominated disks are Eqs. (3.53) (radial force balance), (3.57) (azimuthal force balance or angular momentum balance), (3.52) (vertical force balance), (3.55) (continuity), (3.63) (energy balance), (3.66) (equation of state), with subsidiary relations describing the $r\varphi$-component of stress tensor [Eq. (3.58)], and an expression for opacity [Eq. (3.64)]. That is, the basic equations are six for six quantities, Ω, Σ, $T_{\rm c}$, Π, H, and v, with three parameters M (mass of the central object), $\dot{M}$ (mass-accretion rate), and α (viscosity parameter). These basic equations are all algebraic, and thus we can solve these equations to derive H, Σ, Π, $T_{\rm c}$, and v as functions of r with parameters α, M, and $\dot{M}$. The results show that the disks can be separated into three regions:

- The inner region where the density and temperature are high, and $p \sim p_{\rm rad}$, and $\bar{\kappa}_{\rm R} \sim \kappa_{\rm es}$.
- The middle region where the density and temperature are modest, and $p \sim p_{\rm gas}$ and $\bar{\kappa}_{\rm R} \sim \kappa_{\rm es}$.

- The outer region where the density and temperature are low, and $p \sim p_{\rm gas}$ and $\bar{\kappa}_{\rm R} \sim \kappa_{\rm ff}$.

Details on the above-mentioned standard disk are shown, for example, in Sect. 3.2.3 of Kato et al. (2008).

3.3 Slim Disks and Advection-Dominated Accretion Flows

In Sect. 3.2.1 we have described the steady standard disk model. This disk model, however, cannot explain all observational evidence related to accretion phenomena in compact sources. For example, some observational evidence requires higher temperature and less luminous disks, and some other requires much luminous disks. Here, related to the latter, *slim disk* models by Abramowicz et al. (1988), and, related to the former, *advection-dominated accretion flows* (ADAFs) by Ichimaru (1977), Narayan and Yi (1995), and Abramowicz et al. (1996) are briefly described. In addition to these models, magnetized disk models have been proposed as transition ones between ADAF and standard disks (a series of works started from Machida et al. 2006).

3.3.1 Slim Disks

In the above-mentioned standard disks, thermal energy generated by viscosity is assumed to be radiated away at the same radius from the disks in the vertical direction. If vertical optical depth of disks becomes high by high density, however, the thermal energy generated at radius r cannot be radiated away from disk surface at the same radius r. That is, the thermal energy generated is transported inwards before radiated away there (slim disks by Abramowicz et al. 1988). The energy balance is then realized between $Q^+_{\rm vis}$ and advective cooling $Q^-_{\rm adv}$ which is neglected in the standard model.

Let us now write energy equation in the form:

$$\rho T(\boldsymbol{v}\cdot\nabla)s = \rho(\boldsymbol{v}\cdot\nabla)e + p\,{\rm div}\,\boldsymbol{v} = q^+ - q^-, \tag{3.67}$$

or

$${\rm div}\left[\rho\left(e+\frac{p}{\rho}\right)\right] - (\boldsymbol{v}\cdot\nabla)p = q^+ - q^-, \tag{3.68}$$

where s is the specific entrophy, e is the internal energy per unit mass, and q^+ and q^- are, respectively, the heating and cooling rates per unit volume. The left-hand side of Eq. (3.68) represents the advective cooling. After vertically integrating equation (3.68), we find after some manipulations that the vertically integrated equation of energy balance is

$$Q_{\rm adv}^{-} = Q_{\rm vis}^{+} - Q_{\rm rad}^{-}, \tag{3.69}$$

where

$$Q_{\rm adv}^{-} = \frac{\Pi}{\Sigma}\frac{\dot{M}}{2\pi r^2}\xi, \tag{3.70}$$

and ξ is a dimensionless quantity of the order of unity (see, e.g., Kato et al. 2008).

Let us now compare $Q_{\rm adv}^{-}$ with $Q_{\rm rad}^{-}$. To do so, we approximate the angular momentum balance [Eq. (3.57)] as

$$\Pi \sim \frac{\Omega_{\rm K}\dot{M}}{2\pi\alpha}, \tag{3.71}$$

where the alpha-model given by Eq. (3.58) has been adopted and $\ell_{\rm in}$ is neglected compared with ℓ. Then, considering that in the high-temperature optically thick disks the radiation pressure dominates over the gas pressure, we can write the cooling rate $Q_{\rm rad}^{-}$ in the form

$$Q_{\rm rad}^{-} = \frac{8acT_{\rm c}^4}{3\bar{\kappa}_{\rm R}\rho_0 H} = \frac{8c\Pi}{\kappa_{\rm es}\Sigma H} \sim \frac{8c}{(2\pi)^{1/2}\kappa_{\rm es}}\Omega_{\rm K}^{3/2}\frac{\dot{M}^{1/2}}{\alpha^{1/2}\Sigma^{1/2}}. \tag{3.72}$$

In deriving the last relation, $H^2\Omega_{\rm K}^2 = \Pi/\Sigma$ has been used in addition to Eq. (3.71). On the other hand, the vertically integrated advective cooling can be expressed as

$$Q_{\rm adv}^{-} = \frac{\Pi}{\Sigma}\frac{\dot{M}}{2\pi r^2}\xi, \sim \frac{\xi}{4\pi^2}\left(\frac{\Omega_{\rm K}}{r^2}\right)\frac{\dot{M}^2}{\alpha\Sigma}. \tag{3.73}$$

The above two equations express $Q_{\rm rad}^{-}$ and $Q_{\rm adv}^{-}$ as functions of $\dot{M}$ for fixed r and Σ. For comparison, the viscous heating $Q_{\rm vis}^{+}$ [Eq. (3.59)] is also expressed here as a function of $\dot{M}$ for fixed r and Σ, which is

$$Q_{\rm vis}^{+} = rT_{r\varphi}\frac{d\Omega}{dr} = -\alpha r\frac{d\Omega}{dr}\Pi \sim \frac{3}{4\pi}\Omega_{\rm K}^2\dot{M}. \tag{3.74}$$

To understand how distinct branches of solution appear, we fix r and Σ and illustrate the $\dot{M}$-dependence of $Q_{\rm vis}^{+}$, $Q_{\rm adv}^{-}$, and $Q_{\rm rad}^{-}$ in Fig. 3.3. When Σ and r are fixed, $Q_{\rm vis}^{+}$ is proportional to $\dot{M}$, while $Q_{\rm rad}^{-}$ is proportional to $\dot{M}^{1/2}$. Hence, we have $Q_{\rm rad}^{-} = Q_{\rm vis}^{+}$ for a value of $\dot{M}$, which correspond to the standard model. If $\dot{M}$ is higher than this $\dot{M}$, we have another $\dot{M}$ satisfying the balance between cooling and heating, since $Q_{\rm adv}^{-}$ increases with $\dot{M}^2$ (see Fig. 3.3). This implies that there is another steady disk model, which is the slim disk.

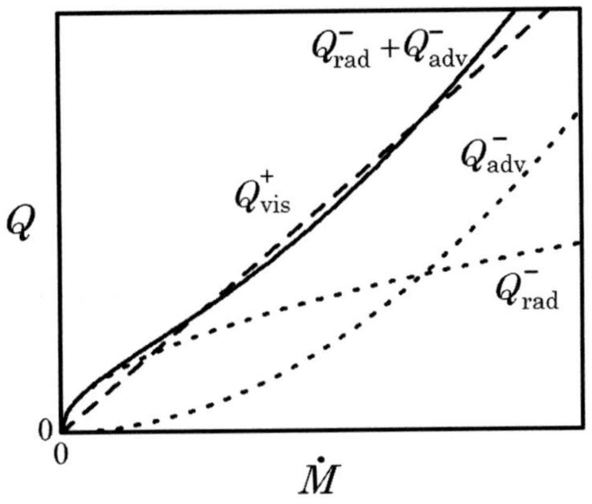

Fig. 3.3 Schematic picture showing the $\dot{M}$-dependence of Q^+_{vis}, Q^-_{adv}, and Q^-_{rad} in optically thick disks. (After Kato et al. 2008, *Black-Hole Accretion Disks—Towards a New Paradigm*)

3.3.2 Advection-Dominated Accretion Flows (ADAFs)

Next, let us consider optically thin, low-density disks. In this case radiative cooling has very complicated parameter dependences. Here, we consider only bremsstrahlung cooling by nonrelativistic electrons, neglecting relativistic bremsstrahlung as well as synchrotron and Compton coolings:

$$Q^-_{rad} = \epsilon_{br}\rho^2 T^{1/2} H \sim \frac{\epsilon_{br}}{4}\left(\frac{k_B}{\mu m_H}\right)^{-1/2} \Omega_K \Sigma^2, \tag{3.75}$$

where $\epsilon_{br} = 1.24 \times 10^{21}$ erg s^{-1} cm^{-2}. Expressions for Q^-_{adv} and Q^+_{vis} are unchanged from those given by Eqs. (3.73) and (3.74), respectively. In the present case, Q^-_{adv} increases with the square of $\dot{M}$. Hence, if Σ is below some critical value Σ_{max}, the curve of $Q^-(= Q^-_{adv} + Q^-_{rad})$ crosses two times the curve of Q^+_{vis}, as sketched in Fig. 3.4. That is, there are two thermal equilibrium states for a given surface density. One is for a smaller $\dot{M}$, and the other is for a larger $\dot{M}$. In the former solution of a smaller $\dot{M}$, Q^+_{vis} is roughly balanced by Q^-_{rad}. This is the *radiation-cooling-dominated disk* found by Shapiro et al. (1976).[11] The other solution has a larger $\dot{M}$, and Q^+_{vis} is balanced by Q^-_{adv}. This is an *advection-dominated accretion flow* (ADAF).

3.3.3 Thermal Equilibrium Curves

In Sects. 3.3.1 and 3.3.2 we have shown that there are a few disk models by the difference of disk's vertical optical depth and the related cooling processes.

[11]This disk model is thermally unstable, and will not be realized.

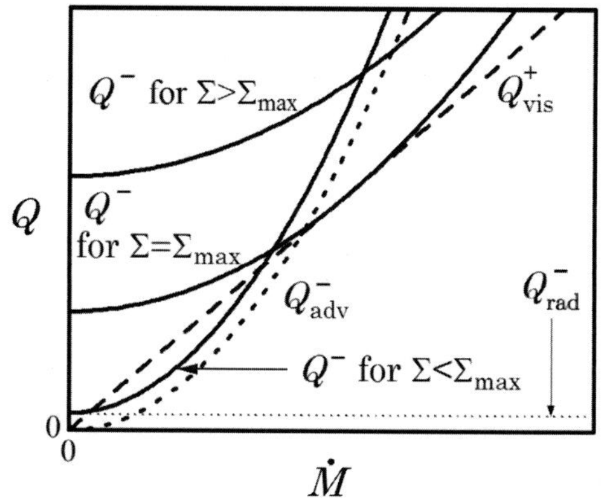

Fig. 3.4 Schematic picture showing how Q^+_{vis}, Q^-_{rad}, Q^-_{adv}, and $Q^-(\equiv Q^-_{rad}+Q^-_{adv})$ depend on $\dot{M}$ when Σ and r are fixed for optically thin disks. In the cases of $\Sigma \geq \Sigma_{max}$, only the total cooling rate, Q^- are plotted, while the values of the individual cooling rates, Q^-_{rad} and Q^-_{adv}, are shown solely for a case of $\Sigma < \Sigma_{max}$. In the case of $\Sigma < \Sigma_{max}$, Q^-, which is a solid curve, crosses the curve of Q^+_{vis} at two points. When $\Sigma = \Sigma_{max}$, the curve of Q^- touches that of Q^+_{vis} only at one point, while it does not contact with the curve of Q_{vis} when $\Sigma > \Sigma_{max}$. (After Kato et al. 2008, *Black-Hole Accretion Disks—Towards a New Paradigm*)

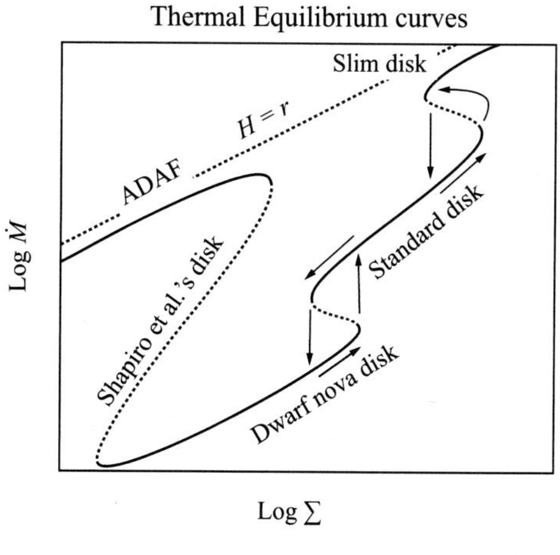

Fig. 3.5 Schematic diagram showing thermal equilibrium curve on $\dot{M} - \Sigma$ plane. The radius r and parameters α, M are fixed. The dotted parts are thermally unstable branches. (After Kato et al. 2008, *Black-Hole Accretion Disks—Towards a New Paradigm*)

The presence of various disk models can be visualized by showing the thermal equilibrium sequence of disks on $\dot{M} - \Sigma$ diagram, fixing r and other parameters such as α, M, and $\dot{M}$. The sequence is schematically shown in Fig. 3.5.

In this figure the dotted parts of the curves represent thermally unstable ones (for detailed arguments on thermal stability see, e.g., Kato et al. 2008). It is noted that the equilibrium sequence extends below the branch of the standard disks toward the lower $\dot{M}$ region. In the region of lower $\dot{M}$ another types of cooling processes appear due to recombinations of helium and hydrogen. The equilibrium sequence in this region corresponds to disks of dwarf novae, although we do not describe them in this section (see, for details, chap. 5 of Kato et al. 2008).

It is also noted that if effects of global magnetic fields are taken into account, we have another type equilibrium branch, which is *magnetic-fields-supported disks*, and they bridge between ADAFs and the standard disks, corresponding to the intermediate state of hard-soft transition of X-ray star disks (Machida et al. 2006; Oda et al. 2007, and so on).

3.4 Meridional Circulation in Stars

In single stars with neither magnetic fields nor rotation, their constitution is spherically symmetric and can be static without any internal motion. This will be obvious, but in rotating stars this is not the case.

Let us consider a cylindrically rotating star. The angular velocity of rotation is taken to be $\Omega(r)$, where r is the distance from a rotating axis. Then, the hydrostatic balance of the star is

$$-\frac{1}{\rho}\nabla p - \nabla\psi + r\Omega^2(r)\boldsymbol{i}_r = 0, \tag{3.76}$$

where $\boldsymbol{i}_r$ is the unit vector in the radial direction in cylindrical coordinates (r, φ, z). The gravitational potential ψ is governed by the Poisson equation:

$$\nabla^2\psi = 4\pi G\rho. \tag{3.77}$$

By introducing the effective gravitational potential Ψ defined by

$$\Psi(r, z) = \psi - \int_0^r r\Omega^2(r)dr. \tag{3.78}$$

Equation (3.76) is reduced to

$$-\frac{1}{\rho}\nabla p - \nabla\Psi = 0. \tag{3.79}$$

Operating curl to this equation, we have

$$\text{curl}\ p \times \text{curl}\ \rho = 0. \tag{3.80}$$

This equation shows that p needs to be a function of ρ, i.e., the gas is *barotropic*. The above results imply that the equi-pressure (isobaric), equi-density, and isothermal surfaces coincide with the effective gravitational potential Ψ, i.e.,

$$p = p(\Psi),\ \rho = \rho(\Psi),\ T = T(\Psi). \tag{3.81}$$

It is noticed that different from the case of non-rotating stars, the contour of a constant $\Psi(r, z)$ is oblate due to rotation.

Next, let us consider heat balance in the stellar interior. In the region of radiative equilibrium in optically thick medium, the radiative heat flux, $\boldsymbol{F}$, is given by (Chap. 20)

$$\boldsymbol{F} = -\frac{4ac}{3}\frac{T^3}{\bar{\kappa}_{\rm R}\rho}\nabla T, \tag{3.82}$$

where a is the radiation constant and $\kappa_{\rm R}$ is the Rossland mean opacity. In the region where heat is wholly transported by radiation alone, div $\boldsymbol{F} = 0$ is required for the star have no motion except for a steady rotation. This condition is written as

$$\text{div}\,\boldsymbol{F} = \frac{d}{d\Psi}\biggl(-\frac{4ac}{3}\frac{T^3}{\kappa\rho}\frac{dT}{d\Psi}\biggr)|\nabla\Psi|^2 + \biggl(-\frac{4ac}{3}\frac{T^3}{\kappa\rho}\frac{dT}{d\Psi}\biggr)\nabla^2\Psi = 0, \tag{3.83}$$

where

$$\nabla^2\Psi = 4\pi G\rho - \frac{1}{r}\frac{d}{dr}(r^2\Omega^2). \tag{3.84}$$

Equation (3.83) cannot be satisfied in general in purely rotating stars. Let us first consider the case of uniform rotation. In this case, all terms in Eq. (3.83) except for $|\nabla\Psi|^2$ are functions of Ψ alone. On the other hand, $|\nabla\Psi|^2$ is not constant on the surface of Ψ = const., since by the effects of rotation the surface of constant Ψ is oblate and $|\nabla\Psi|^2$ on the same equi-potential surface is different in equatorial and polar directions. In the case where $\Omega(r)$ is not constant, $d(r^2\Omega^2)/rdr$ is constant on cylindrical surface, and this surface is different from both surfaces of Ψ = const. and of $|\nabla\Psi|^2$ = const.. That is, Eq. (3.83) cannot be satisfied in the whole space in stars. This result is known as *von Zeipel's paradox* (or *Von Zeipel's theorem*).

This paradox suggests that the stellar rotation is generally not cylindrical and $\Omega = \Omega(r, z)$, or there is meridional circulation in rotating stars. The former means $\partial\Omega(r, z)/\partial z \neq 0$, but as will be shown in Sect. 7.4 such rotation is thermally unstable in long thermal timescale (the Goldreich–Shubert–Fricke instability). In summary, rotating stars will generally have meridional circulation. Such circulation is considered to be important in evolution of stars, since it brings about element mixing and some effects on dynamo processes. Recent helioseismology, for example, shows the presence of meridional motions in solar convection zone. Similar situations are also expected in disks.

3.5 Self-Similar Spherical Accretion

As a powerful tool for studying astrophysical flows, the self-similar treatment has been applied in various situations, including a point explosion such as supernovae (Sedov 1959), spherical flows around a central object (e.g., Sakashita 1974;

Sakashita and Yokosawa 1974; Cheng 1977; Fukue 1984; Tsuribe et al. 1995), a gravitational collapse of a self-gravitating cloud (e.g., Larson 1969; Penston 1969; Shu 1977; Hunter 1977; Whitworth and Summers 1985; Lynden-Bell and Lemos 1988), a gravitational collapse of a globular cluster (e.g., Hénon 1961, 1965; Larson 1970), as well as a two-dimensional rotating flow of a self-gravitating cloud (e.g., Hayashi et al. 1982; Toomre 1982), a hydrodynamical wind from an accretion disk (e.g., Clarke and Alexander 2016), a magnetohydrodynamical wind from an accretion disk (e.g., Blandford and Payne 1982; Li et al. 1992; Ostriker 1997), a radiation hydrodynamical flow (e.g., Lucy 2005; Falize et al. 2011; Fukue 2018), a relativistic blast wave (Blandford and McKee 1976; Sari 1997; Nakayama and Shigeyama 2005; Hendrick 2014).

Although similarity solutions have been widely found in many astrophysical flows, there are restrictions and limitations for the self-similar treatment. In time-dependent flows, in order to rescale time and space, the numbers of physical scales are restricted. For example, in the case of an energy-driven explosion with constant energy E into the surrounding medium with uniform density ρ, the radius r varies with time t as $r \propto (E/\rho)^{1/5} t^{2/5}$ (Sedov 1959). In the case of a momentum-driven outflow with constant luminosity L, another relation is imposed. In the case of a time-dependent flow around a central object of constant mass M, the gravitational field introduces a new timescale, the freefall time, into the problem, and therefore, the reference radius must vary as $r \propto (GM)^{1/3} t^{2/3}$ in order for the self-similarity to be maintained (e.g., Sakashita 1974; Sakashita and Yokosawa 1974; Cheng 1977; Fukue 1984). If, however, the central mass varies with time, this restriction is removed. Since magnetic field is scale free, the magnetohydrodynamical flows are also easy to construct various similarity solutions. Although radiation field itself is also scale free, the radiation hydrodynamical self-similar flows are limited due to the existence of the coupling constant between radiation and matter, i.e., opacity. However, there are a few studies on the self-similar treatments of radiation hydrodynamical flows (see Falize et al. 2011 and reference therein). Finally, the self-similar treatments of relativistic flows are restrictive, since there appear several physical quantities, such as the speed of light and the Schwarzschild radius.

In this section, in order to demonstrate such a useful self-similar treatment, we present a simple time-dependent spherical accretion onto a central gravitating body.

3.5.1 Self-Similar Treatments

Let us suppose a time-dependent spherical accretion in the radial (r) direction onto a central gravitating body with mass M, surrounded by a uniform medium with density ρ_0. In order to focus our attention on the self-similar behavior, we ignore the gas pressure (cf. Fukue 1984 for the case with gas pressure).

The continuity equation for the time-dependent spherical flow is

$$\frac{\partial \rho}{\partial t} + \frac{1}{r^2}\frac{\partial}{\partial r}(r^2 \rho v) = 0, \tag{3.85}$$

where ρ is the gas density, v the flow velocity. The equation of motion is

$$\frac{\partial v}{\partial t} + v\frac{\partial v}{\partial r} = -\frac{GM}{r^2}, \tag{3.86}$$

where the gas pressure is ignored, as already stated.

By the dimensional considerations on the constant GM in the present system, we have $[GM] = \mathrm{L}^3\,\mathrm{T}^{-2}$, or $[(GM)^{1/3}t^{2/3}] = \mathrm{L}$. Hence, we can introduce the non-dimensional similarity coordinate ξ as

$$\xi \equiv (GM)^{-1/3} t^{-\delta} r, \quad \delta = 2/3, \tag{3.87}$$

and the similarity variables as

$$v = -(GM)^{1/3} t^{\delta - 1} V(\xi), \tag{3.88}$$

$$\rho = \rho_0 D(\xi), \tag{3.89}$$

where the similarity variables, $V(\xi)$ and $D(\xi)$, are functions of only ξ.[12] In order to explicitly express *accretion*, the minus sign is attached in the velocity transformation. In addition, time t runs from infinity to zero for accretion.

In terms of the similarity coordinate ξ, the partial differentials are transformed as

$$\frac{\partial}{\partial t} = \frac{\partial \xi}{\partial t}\frac{d}{d\xi} = -\delta \frac{\xi}{t}\frac{d}{d\xi}, \tag{3.90}$$

$$\frac{\partial}{\partial r} = \frac{\partial \xi}{\partial r}\frac{d}{d\xi} = \frac{\xi}{r}\frac{d}{d\xi} = (GM)^{-1/3} t^{-\delta}\frac{d}{d\xi}. \tag{3.91}$$

Using these similarity transformations, after some manipulations, we can transform the basic equations (3.85) and (3.86) to the following set of ordinary differential equations:

$$(V + \delta\xi)\frac{1}{D}\frac{dD}{d\xi} + \frac{dV}{d\xi} + \frac{2}{\xi}V = 0, \tag{3.92}$$

$$(V + \delta\xi)\frac{dV}{d\xi} = -(1 - \delta)V - \frac{1}{\xi^2}. \tag{3.93}$$

[12] In general, substituting $r = a(t)\xi$, $v = b(t)V(\xi)$, and $\rho = c(t)D(\xi)$ into the basic equations, we can determine the forms of the similarity transformation. Depending on the problems and situations, we may obtain non-power law type transformations, such as an exponential type.

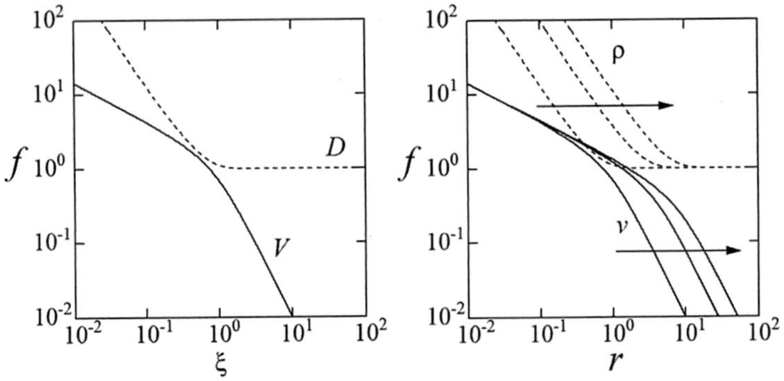

Fig. 3.6 Examples of self-similar solutions. Left: solutions in the similarity coordinate. A solid curve denotes V, and a dashed one expresses D. Right: time-dependent solutions in the real space. Solid curves denote υ, and dashed ones represent ρ

These transformed equations can be easily solved under the appropriate boundary conditions, since they are simultaneous ordinary differential equations.

3.5.2 Typical Solutions

As boundary conditions for the present accretion flow onto the central object surrounded by a uniform medium, we impose that the density D is constant at infinity. From Eq. (3.92), we then obtain an asymptotic solution on V at infinity as $V \sim -1/\xi^2$. Using these asymptotic solutions at infinity, we can easily integrate equations (3.92) and (3.93) toward the center.

Examples of self-similar solutions are shown in Fig. 3.6. In the left panel of Fig. 3.6 the infall velocity V (a solid curve) and density D (a dashed one) are plotted as a function of the similarity coordinate ξ, whereas in the right panel the time-dependent behaviors of the velocity υ (solid curves) and density ρ (dashed ones) in the real space are depicted.

In the present accretion flow onto the central body, from (3.93) the asymptotic solution of V in the vicinity of the center must become $V = -\sqrt{2/\xi}$; i.e., a freefall solution. From (3.92), on the other hand, the asymptotic solution of D toward the center is $D \propto \xi^{-3/2}$. Hence, in the real space, the central freefalling accretion region expands into the surrounding uniform medium, as time passes.

References

Abramowicz, M.A., Czerny, B., Lasota, J.-P., Szuszkiewicz, E.: Astrophys. J. **332**, 646 (1988)
Abramowicz, M.A., Chen, X., Kato, S., Lasota, J.-P., Regev, O.: Astrophys. J. **471**, 762 (1996)

Balbus, S.A., Hawley, J.F.: Astrophys. J. **376**, 214 (1991)
Blandford, R.D., McKee, C.F.: Phys. Fluids **19**,1130 (1976)
Blandford, R.D., Payne, D.G.: Mon. Not. R. Astron. Soc. **199**, 883 (1982)
Bondi, H.: Mon. Not. R. Astron. Soc. **112**, 195 (1952)
Cheng, A.F.: Astrophys. J. **213**, 537 (1977)
Clarke, C.J., Alexander, R.D.: Mon. Not. R. Astron. Soc. **460**, 3044 (2016)
Falize, E., Michaut, C., Bouquet, S.: Astrophys. J. **730**, 96 (2011)
Frank, J., King, A., Raine, D.: Accretion Power in Astrophysics, 2nd edn. Cambridge University Press, Cambridge (1992)
Frank, J., King, A., Raine, D.: Accretion Power in Astrophysics, 3rd edn. Cambridge University Press, Cambridge (2002)
Fukue, J.: Publ. Astron. Soc. Jpn. **36**, 87 (1984)
Fukue, J.: Publ. Astron. Soc. Jpn. **70**, 43 (2018)
Hawley, J.F., Balbus, S.A.: Astrophys. J. **376**, 223 (1991)
Hayashi, C., Narita, S., Miyama, S.: Prog. Theor. Phys. **68**, 1949 (1982)
Hénon, M.: AnAp **24**, 369 (1961)
Hénon, M.: AnAp **28**, 62 (1965)
Holzer, T.E., Axford, W.I.: ARA&Ap **8**, 31 (1970)
Hunter, C.: Astrophys. J. **218**, 834 (1977)
Ichimaru, S.: Astrophys. J. **214**, 840 (1977)
Kato, S., Fukue, J., Mineshige, S.: Black-Hole Accretion Disks. Kyoto University Press, Kyoto (1998)
Kato, S., Fukue, J., Mineshige, S.: Black-Hole Accretion Disks—Towards a New Paradigm. Kyoto University Press, Kyoto (2008)
Larson, R.B.: Mon. Not. R. Astron. Soc. **145**, 271 (1969)
Larson, R.B.: Mon. Not. R. Astron. Soc. **150**, 93 (1970)
Li, Z.-Y., Chiueh, T., Begelman, M.C.: Astrophys. J. **394**, 459 (1992)
Lucy, L.B.: Astron. Astrophys. **429**, 31 (2005)
Lynden-Bell, D., Lemos, J.P.S.: Mon. Not. R. Astron. Soc. **233**, 197 (1988)
Machida, M., Nakamura, K.E., Matsumoto, R.: Publ. Astron. Soc. Jpn. **58**, 193 (2006)
Nakayama, K., Shigeyama T.: Astrophys. J. **627**, 310 (2005)
Narayan, R., Yi, I.: Astrophys. J. **444**, 231 (1995)
Ostriker, E.C.: Astrophys. J. **486**, 291 (1997)
Oda, H., Machida, M., Nakamura, K.E., Matsumoto, R.: Publ. Astron. Soc. Jpn. **59**, 457 (2007)
Parker, E.N.: ApJ **132**, 175 (1958)
Penston, M.V.: Mon. Not. Roy. Astron. Soc. **144**, 425 (1969)
Sakashita, S.: Astrophys. Space Sci. **26**, 183 (1974)
Sakashita, S., Yokosawa, M.: Astrophys. Space Sci. **31**, 251 (1974)
Sedov, L.I.: Similarity and Dimensional Methods in Mechanics. Infosearch, London (1959)
Sari, R.: Astrophys. J. **489**, L37 (1997)
Shakura, N.I., Sunyaev, R.A.: Astron. Astrophys. **24**, 337 (1973)
Shapiro, S.L., Lightman, A.P., Eardley, D.H.: Astrophys. J. **204**, 187 (1976)
Shu, F.H.: Astrophys. J. **214**, 488 (1977)
Toomre, A.: Astrophys. J. **259**, 535 (1982)
Tsuribe, T., Umemura, M., Fukue, J.: Publ. Astron. Soc. Jpn. **47**, 73 (1995)
van Hendrick, E.: Mon. Not. R. Astron. Soc. **442**, 3495 (2014)
Whitworth, A., Summers, D.: Mon. Not. Roy. Astron. Soc. **214**, 1 (1985)

Chapter 4
Wave Phenomena in Astrophysical Objects

Astrophysical objects have many kinds of configurations due to subtle balances among such various forces as gravitational, pressure, rotational, and magnetic ones. Related to this, many kinds of restoring force work on fluid motions when the fluid is perturbed from equilibrium states, which leads to various kinds of wave motions (oscillations) and instabilities. In this chapter, we explore wave motions resulting from various restoring forces except for magnetic and radiation fields, including shock waves. Wave motions resulting from magnetic fields are described in Chap. 13, while those from radiation are given in Chap. 23.

4.1 Waves in Gravitationally Stratified Atmosphere

The effects of gravitational field on wave motions are twofold. One is as self-gravity, and the other is as external field. The former will be argued in Sect. 4.5 in relation to density waves in galactic disks, and also in Sect. 6.4 in relation to gravitational instability. Here, we consider wave motions in stratified atmospheres subject to a constant external gravitational field.

The direction of the external gravitational acceleration is taken in the direction anti-parallel to the z-direction, and the atmosphere is assumed to be uniform in the directions (x- and y-directions) perpendicular to the z-direction. The equation describing the hydrostatic structure in the z-direction is

$$-\frac{dp_0}{dz} = \rho_0 g, \tag{4.1}$$

where $p_0(z)[= k_B\rho_0(z)T_0(z)/\bar{\mu}m_H]$ is gas pressure, $\rho_0(z)$ is gas density, and $T_0(z)$ is temperature.

© Springer Nature Singapore Pte Ltd. 2020, corrected publication 2023

S. Kato, J. Fukue, *Fundamentals of Astrophysical Fluid Dynamics*,
Astronomy and Astrophysics Library,
https://doi.org/10.1007/978-981-15-4174-2_4

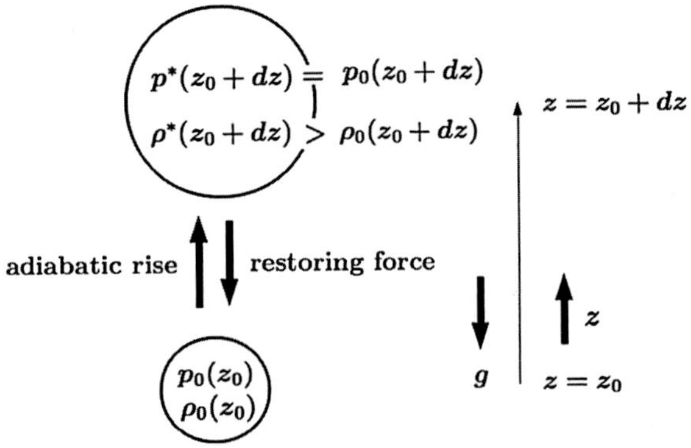

Fig. 4.1 Schematic picture showing adiabatic rise of a fluid element in stratified media. If density $\rho^*(z_0 + dz)$ in an adiabatically ascending fluid element is higher than that of the surrounding gas, $\rho_0(z_0 + dz)$, i.e., $\rho^*(z_0 + dz) > \rho_0(z_0 + dz)$, the element is returned to the original position, and makes a harmonic oscillation, whose frequency is called Brunt–Väisärä frequency. On the other hand, if $\rho^*(z_0 + dz) < \rho_0(z_0 + dz)$, the fluid element departs further from the original position by buoyancy. Such a medium is called convectively unstable (see Fig. 5.1 and Sect.5.1)

4.1.1 Brunt–Väisälä Frequency

Let us consider a fluid element of unit mass and move it from the equilibrium height, say z_0, to a height of $z_0 + dz$ as is shown in Fig. 4.1. The element has initially the same density $\rho_0(z_0)$ and the same pressure $p_0(z_0)$ as those of the surrounding gas at the height z_0. The displacement is made adiabatically, keeping the pressure balance with the surrounding media. Then, after the displacement of dz, the displaced fluid element has a density ρ^* given by $\rho_0 + (d\rho/dp)_{\rm ad} dp$, where the subscript ad denotes the adiabatic change, and the pressure change by the displacement is denoted by $dp[= p_0(z_0 + dz) - p_0(z_0)]$.

Compared with this, the density in the surrounding medium at the displaced height is $\rho_0 + (d\rho_0/dp_0)dp$, where $d\rho_0/dp_0$ represents the density-pressure relation in the unperturbed medium. Hence, the density of a rising element ($dz > 0$) is higher than that of the surrounding one by $\Delta\rho$, which is

$$\Delta\rho = \left(\frac{d\rho}{dp}\right)_{\rm ad} dp - \frac{d\rho_0}{dp_0} dp = -\frac{\rho_0}{p_0}\left[\left(\frac{d\ln\rho}{d\ln p}\right)_{\rm ad} - \left(\frac{d\ln\rho}{d\ln p}\right)_0\right]\rho_0 g dz, \tag{4.2}$$

where $dp_0 = -\rho_0 g dz$ [Eq. (4.1)] has been used. If $\Delta\rho$ is positive when $dz > 0$ (or, if $\Delta\rho$ is negative when $dz < 0$), the fluid element is returned to the original position by negative buoyancy (see Fig. 4.1). Since the restoring force per unit mass is $(\Delta\rho/\rho_0)g$, and proportional to dz, the displaced element makes a harmonic oscillation whose frequency N is given by

$$N^2 = \frac{\rho_0}{p_0}\left[\left(\frac{d\ln\rho}{d\ln p}\right)_0 - \left(\frac{d\ln\rho}{d\ln p}\right)_{\rm ad}\right] g^2. \tag{4.3}$$

This frequency N characterizes the degree of stratification of the media, and is called the *Brunt–Väisälä frequency*. In particular, in the isothermal atmosphere which we consider hereafter, we have

$$N = (\gamma - 1)^{1/2} \frac{g}{c_s}, \tag{4.4}$$

where γ is the ratio of specific heats, i.e., $(d\ln\rho/d\ln p)_{ad} = 1/\gamma$ and the acoustic speed c_s is related to ρ and p by $c_s^2 = \gamma p/\rho$.

It is noticed here that in stellar physics ∇ defined by $\nabla \equiv d\ln T/d\ln p$ is often used. If this notation is used, Eq. (4.3) can be expressed as[1]

$$N^2 = \frac{\rho_0}{p_0}(\nabla_{ad} - \nabla) g^2. \tag{4.5}$$

4.1.2 Acoustic and Gravity Waves

After the above preparations, we consider small amplitude perturbations on an isothermal atmosphere consisting of ideal gas. Since the temperature, T_0, is taken to be constant, Eq. (4.1) is easily integrated to give

$$\rho_0(z) = \rho_{00} \exp\left(-\frac{z}{H} \right), \tag{4.6}$$

where the integration constant, ρ_{00}, is the density at $z = 0$, and

$$H \equiv \frac{k_B T_0}{\mu m_H g} \tag{4.7}$$

represents the height where density changes by $1/e$ and is called *scale height*. As Eqs. (4.6) and (4.7) show, the isothermal atmosphere extends infinitely with exponentially decreasing density and pressure in the z-direction.

We consider small amplitude perturbations on the above isothermal atmosphere. Equations describing perturbations are given by

$$\frac{\partial \rho_1}{\partial t} + \mathrm{div}(\rho_0 \boldsymbol{v}) = 0, \tag{4.8}$$

$$\rho_0 \frac{\partial \boldsymbol{v}}{\partial t} = -\mathrm{grad} p_1 + \rho_1 \boldsymbol{g}, \tag{4.9}$$

[1]The condition of convective instability is often expressed as $\nabla_{ad} < \nabla$. This is the case where $N^2 < 0$. Convection is discussed in Chap. 5.

$$\rho_0 C_V \frac{\partial T_1}{\partial t} + p_0 \mathrm{div}\, \boldsymbol{v} = 0, \tag{4.10}$$

$$\frac{p_1}{p_0} = \frac{\rho_1}{\rho_0} + \frac{T_1}{T_0}, \tag{4.11}$$

where $\boldsymbol{g} = (0, 0, -g)$, C_V is the specific heat of constant volume, $\boldsymbol{v}$ is the perturbed velocity, and quantities with subscripts 0 and 1 represent, respectively, unperturbed and perturbed quantities. Other notations have their usual meanings. The above four equations are in turn equation of continuity, equation of motion, equation of thermal energy conservation, and equation of state.

Let us derive an equation expressed in terms of a variable by eliminating other variables from Eq. (4.8)–(4.11). We adopt here $p_1/\rho_0^{1/2}$ as the variable to lead to

$$\left[\frac{\partial^2}{\partial t^4} - \frac{\partial^2}{\partial t^2}\left(c_s^2\nabla^2 - \frac{c_s^2}{4H^2}\right) - (\gamma - 1)g^2\nabla_1^2\right]\left(\frac{p_1}{\rho_0^{1/2}}\right) = 0, \tag{4.12}$$

where ∇^2 is the Laplacian and ∇_1^2 is the Laplacian in the horizontal plane:

$$\nabla_1^2 \equiv \frac{\partial^2}{\partial x^2} + \frac{\partial^2}{\partial y^2}. \tag{4.13}$$

Equation (4.12) is a partial differential equation with constant coefficients. Hence, we can examine normal mode oscillations by taking $p_1/\rho_0^{1/2}$ as $p_1/\rho_0^{1/2} \propto \exp[i(\omega t - \boldsymbol{k}\cdot\boldsymbol{x})]$, where ω is the frequency and $\boldsymbol{k}$ is the wavenumber. Using this expression for $p_1/\rho_0^{1/2}$ we have a dispersion relation from Eq. (4.12):

$$\omega^4 - (c_s^2k^2 + \omega_s^2)\omega^2 + c_s^2\omega_g^2k_1^2 = 0, \tag{4.14}$$

where

$$\omega_s (= N_{ac}) = \frac{c_s}{2H} = \frac{\gamma}{2}\frac{g}{c_s}, \tag{4.15}$$

$$\omega_g (= N) = (\gamma - 1)^{1/2}\frac{g}{c_s}, \tag{4.16}$$

and k_1 is the wavenumber in the horizontal direction, i.e., $k_1^2 = k_x^2 + k_y^2$. As long as $\gamma < 2$, we have $\omega_s > \omega_g$. It is noted here that in many literatures instead of ω_s and ω_g, N_{ac} and N are used, respectively.

Here, it may be helpful to make a comment. Since we are considering a stratified medium, we must be careful about adoption of the dependent variable which can be taken to be proportional to $\exp[i(\omega t - \boldsymbol{k}\cdot\boldsymbol{x})]$. Since we are considering wave motions in a conservative medium, wave energy density does not change with

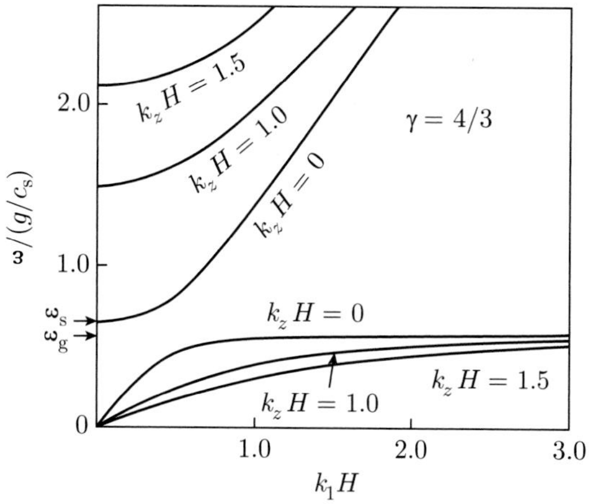

Fig. 4.2 Dispersion relation of acoustic and gravity waves in the isothermal atmosphere subject to constant external gravitational field. The wave in high frequency domain is acoustic wave, and that in low frequency one is gravity wave

propagation of waves. This means that in the normal mode analyses, dependent variables which guarantee the above requirement must be adopted. Such variables are $\rho_0^{1/2} v_z$, $p_1/\rho^{1/2}$, and so on. If we adopt inappropriate quantities, say v_z and p_1, in such a form as $\exp[i(\omega t - \boldsymbol{k} \cdot \boldsymbol{x})]$, then we have fictitious grow or damping of waves in results.

The dispersion relation (4.14) gives the relation between wavenumber $\boldsymbol{k}$ and wave frequency ω. The ω–k_1 relation is shown in Fig. 4.2 in the case of $\gamma = 4/3$, $k_z H$ being taken as a parameter. This figure shows that the oscillation region of waves is separated into two, i.e., $\omega > \omega_s$ and $\omega < \omega_g$. The waves in the former frequency region ($\omega > \omega_s$) are *acoustic waves* (p-mode), while the waves in lower frequency region ($\omega < \omega_g$) are *gravity waves* (g-mode).[2] Between ω_s and ω_g there is an evanescent region of waves.[3]

Let us consider why there are no acoustic waves whose frequency is smaller than ω_s. To examine this, we consider the wave which propagates in the vertical direction ($k_1 = 0$), since the effects of stratification of the medium appear most characteristically on this wave. In the limit of short wavelength in the vertical direction (i.e., k_z is large), the effects of stratification of the medium are negligible. As the wavelength in the vertical direction becomes longer, however, the pressure restoring force decreases, and the relative importance of the restoring force due to stratification of the medium increases. Even in the limit of $k_z = 0$, the latter restoring force remains, and the frequency of waves cannot become smaller than a critical frequency, which is ω_s. In the case of gravity waves, the restoring force acts on fluid element most efficiently when $k_1 = \infty$, but their frequencies cannot become higher than ω_g as discussed in the previous Sect. 4.1.1. As horizontal wavelength

[2]The waves resulting from metric variation in general relativity are called gravitational waves.

[3]The presence of the evanescent region is due to idealization of isothermal atmosphere. In general cases of non-isothermal atmosphere, there is no clear separation of frequency domains.

of the waves becomes longer, the efficiency of restoring force decreases, and the frequency decreases. In the limit of $k_1 = 0$, the gravity waves disappear.

A typical example of stratified atmospheres is the solar photosphere. Leighton et al. (1962) found the so-called 5-min-oscillations in solar photosphere. The solar atmosphere is by no means isothermal, and furthermore, observed oscillations are not local. In spite of them, Fig. 4.1 is helpful in approximately representing the wave nature on the solar photosphere. Nowadays, it is considered that the 5-min oscillations are standing (trapped) acoustic eigenmodes of the Sun with frequencies around the evanescent frequency region [see a monograph by Unno et al. 1989 for details]. Studies of 5-min-oscillations stimulated studies of nonradial oscillations of stars, leading to *helioseismology* and *asteroseismology*. Helioseismology and asteroseismology are the field of astrophysics where internal structures of the Sun and stars are examined by using their oscillations.

4.2 Radial and Nonradial Oscillations of Stars

Almost all stars are variables. Furthermore, the cause of their variability comes mainly from their interiors (pulsation of stars). Pulsation theory of stars has been developed since the early era of studies of stellar interiors. There are many excellent reviews on theory of stellar pulsations and nonradial oscillations (e.g., Cox 1980; Unno et al. 1979, 1989). Hence, in this section we restrict our attention only on classification of nonradial oscillations of stars, from the view point of restoring force.

4.2.1 Basic Equations Describing Nonradial Oscillations of Stars

The unperturbed stars are assumed to be spherically symmetric with no rotation. Let us denote displacement vector associated with oscillations by $\boldsymbol{\xi}$. Then, the basic hydrodynamical equations describing small amplitude adiabatic oscillations are

$$\rho_1 + \mathrm{div}(\rho_0 \boldsymbol{\xi}) = 0, \tag{4.17}$$

$$\rho_0 \frac{\partial^2 \boldsymbol{\xi}}{\partial t^2} = -\nabla p_1 + \frac{\rho_1}{\rho_0} \nabla p_0 - \rho_0 \nabla \psi_1, \tag{4.18}$$

$$p_1 + (\boldsymbol{\xi} \cdot \nabla) p_0 = \frac{\Gamma_1 p_0}{\rho_0} [\rho_1 + (\boldsymbol{\xi} \cdot \nabla) \rho_0], \tag{4.19}$$

$$\nabla^2 \psi_1 = 4\pi G \rho_1, \tag{4.20}$$

where the quantities with subscript 0 are unperturbed ones, while those with subscript 1 are small amplitude perturbed quantities. The above equations are, in turn, equation of continuity, equation of motion, adiabatic relation, and the Poisson equation, Γ_1 being the ratio of specific heats generalized by taking into account the effects of radiation pressure (see Sect. 19.6).[4]

Since the unperturbed stars are spherically symmetric, it is convenient to introduce spherical coordinates $(r, \theta\ \varphi)$ whose center is the center of the star. Then, when we consider small amplitude normal mode oscillations, they can be characterized by a single spherical surface harmonic function (e.g., Unno et al. 1989), $Y_l^m(\theta, \varphi)$:

$$Y_l^m(\theta, \varphi) = P_l^m(\cos\theta)\exp(im\varphi), \tag{4.21}$$

where $P_l^m(\cos\theta)$ is the associated Legendre function with argument $\cos\theta$, m being an integer satisfying $|m| \leq l$. For example, a scalar function such as p_1 can be expressed as

$$p_1(r, \theta, \varphi; t) = p_1(r)Y_l^m\exp(-i\omega t), \tag{4.22}$$

where ω is the frequency of normal mode oscillations. The displacement vector, $\boldsymbol{\xi}$, can be expressed in terms of two scalar functions, $\xi_n(r)$ and $\xi_t(r)$, as

$$\boldsymbol{\xi} = \left[\xi_n(r), \xi_t(r)\frac{\partial}{\partial\theta}, \xi_t(r)\frac{\partial}{\sin\theta\,\partial\varphi}\right]Y_l^m(\theta, \varphi)\exp(-i\omega t). \tag{4.23}$$

Hereafter, for example, we write $p_1(r)$ as p_1 by neglecting the argument r. Then, Eqs. (4.17)–(4.20) are reduced to

$$\rho_1 + \frac{1}{r^2}\frac{d}{dr}(r^2\rho_0\xi_n) - \frac{l(l+1)}{r}\rho_0\xi_t = 0, \tag{4.24}$$

$$-\rho_0\omega^2\xi_n = -\frac{dp_1}{dr} + \frac{\rho_1}{\rho_0}\frac{dp_0}{dr}, \tag{4.25}$$

$$-\rho_0\omega^2\xi_t = -\frac{p_1}{r} - \frac{\rho_0}{r}\psi_1, \tag{4.26}$$

$$p_1 + \xi_n\frac{dp_0}{dr} = \frac{\Gamma_1 p_0}{\rho_0}\left(\rho_1 + \xi_n\frac{d\rho_0}{dr}\right), \tag{4.27}$$

$$\frac{1}{r^2}\frac{d}{dr}\left(r^2\frac{d\psi_1}{dr}\right) - \frac{l(l+1)}{r^2}\psi_1 = 4\pi G\rho_1. \tag{4.28}$$

[4]The meaning of subscript 1 to Γ is different from that to other quantities.

Here, we introduce an approximation which is often used in studying nonradial oscillations of stars. Since the variation of gravitational potential, ψ_1, is generally small in studying nonradial oscillations of stars, we put $\psi_1 = 0$ (Cowling approximation) and omit the Poisson equation (4.28). Then, ξ_t is expressed by p_1. Furthermore, ρ_1 is expressed in terms of p_1 and ξ_n by using Eq. (4.27). By substituting these ξ_t and ρ_1 into Eqs. (4.24) and (4.25), we have a set of two equations concerning p_1 and ξ_n, which are

$$\frac{1}{r^2}\frac{d}{dr}(r^2\xi_n) - \frac{g}{c_s^2}\xi_n + \left(1 - \frac{L_l^2}{\omega^2}\right)\frac{p_1}{\rho_0 c_s^2} = 0, \tag{4.29}$$

$$\frac{1}{\rho_0}\frac{dp_1}{dr} + \frac{g}{c_s^2\rho_0}p_1 + (N^2 - \omega^2)\xi_n = 0, \tag{4.30}$$

where $g(r)$ is the gravitational acceleration at radius r, i.e., $g = -(1/\rho_0)dp_0/dr$, and L_l is *Lamb frequency* defined by

$$L_l^2(r) = \frac{l(l+1)}{r^2}c_s^2 \tag{4.31}$$

and N is the *Brunt–Väisälä frequency* defined by

$$N^2(r) = \left(\frac{1}{\Gamma_1}\frac{d\ln p_0}{dr} - \frac{d\ln\rho_0}{dr}\right)g, \tag{4.32}$$

and is the same as Eq. (4.3).

Following Unno et al. (1989), we examine the characteristics of waves described by the set of Eqs. (4.29) and (4.30). In order to obtain equations of the standard form, we introduce new variables, y and z defined, respectively, by

$$y = r^2\xi_n \exp\left(-\int_0^r \frac{g}{c_s^2}dr\right), \tag{4.33}$$

$$z = \frac{p_1}{\rho_0}\exp\left(-\int_0^r \frac{N^2}{g}dr\right). \tag{4.34}$$

By these transformations of variables, we can reduce the set of Eq. (4.29) and (4.30) to

$$\frac{dy}{dr} = h(r)\frac{r^2}{c_s^2}\left(\frac{L_l^2}{\omega^2} - 1\right)z, \tag{4.35}$$

$$\frac{dz}{dr} = \frac{1}{r^2 h(r)}(\omega^2 - N^2)y, \tag{4.36}$$

where $h(r)$ is defined by

$$h(r) = \exp\left[\int_0^r \left(\frac{N^2}{g} - \frac{g}{c_s^2}\right) dr\right] > 0. \tag{4.37}$$

After obtaining Eq. (4.35) and (4.36) of the standard form, we consider local perturbations with radial wavenumber k_r as

$$y(r),\ z(r) \propto \exp(ik_r r). \tag{4.38}$$

Substitution of $y(r)$ and $z(r)$ given by Eq. (4.38) into Eqs. (4.35) and (4.36) leads to a dispersion relation:

$$k_r^2 = \frac{(\omega^2 - L_l^2)(\omega^2 - N^2)}{\omega^2 c_s^2}. \tag{4.39}$$

By defining the wavenumber in the horizontal direction, k_h, by

$$k_h^2 = \frac{l(l+1)}{r^2} \tag{4.40}$$

we can rewrite Eq. (4.39) as

$$\omega^4 - (N^2 + k^2 c_s^2)\omega^2 + N^2 k_h^2 c_s^2 = 0, \tag{4.41}$$

where

$$k^2 = k_r^2 + k_h^2. \tag{4.42}$$

It is noticed that the dispersion relation (4.41) is similar to the dispersion relation (4.14) derived for wave motions in isothermal atmospheres. Dispersion relation (4.41) shows the presence of two wave modes, as in the case of dispersion relation (4.14). The high frequency mode corresponds to acoustic one (p-mode), and low frequency one is gravity mode (g-mode). In the case of $N^2 < 0$, however, the g-modes disappear and they changed to convection modes.[5] It is also noted that the frequencies of wave modes are independent of m, i.e., the frequencies for different m's are degenerated, but this degeneracy is split in rotating stars although it is not shown here.

The above arguments are based on local analyses. In actual situations the internal structure of stars is highly inhomogeneous. Hence, the regions where oscillations mainly present are different by modes. The p-mode oscillations usually have large

[5]In the case considered in Sect. 4.1.2, the change to the convection mode does not occur since the medium was assumed to be isothermal.

amplitudes in the envelope with higher frequency, while the g-modes are in the core region with lower frequency. There is a mode which extends through the whole region with frequency between p- and g-modes, which is called the *fundamental mode* (f-mode).

Recent development of helioseismology (the research field of studying the structure and dynamics of the Sun through its oscillations) makes clear the structure of solar internal structure. For example, the rotation is roughly separated into a rigidly rotating core and a differentially rotating convective envelope. The boundary layer, which is called *tachocline*, is known to be an important place for dynamo processes in the Sun. For details, consult the above-mentioned text books and related papers.

4.3 Waves in Rotating Spherical Shell

In Sects. 4.1 and 4.2 we have considered acoustic and gravity waves. Frequencies of these oscillation modes are modified or split if effects of rotation are taken into account as perturbation on oscillations. These are due to a restoring force of rotation. Different from them, we present here that rotation can introduce a different type of oscillation mode. This is an oscillation mode related to conservation of absolute vorticity in rotating spherical shell, and is known as the *Rossby wave* in the field of meteorology. It is noted that the restoring force due to rotation is also important in stability of systems as will be mentioned in Chaps. 6–8.

4.3.1 Rossby–Haurwitz Waves

In Sect. 4.1 we described the presence of acoustic waves and gravity waves. The acoustic waves are basically longitudinal waves resulting from the restoring force of expansion and compression of the medium, while gravity waves are transverse waves resulting from (negative) buoyancy force. In rotating medium, there is another important transverse wave mode, which is the *Rossby-type wave*. In order to demonstrate this situation, let us consider a medium cylindrically rotating around the z-axis (see Fig. 4.3). For simplicity, the gravitational force is taken in the negative z-direction. The gravity waves are transverse waves which oscillate in the direction parallel to the direction of gravitational force (z-direction) and propagate in the horizontal direction (say, the x-direction), while the Rossby waves are those which oscillate perpendicular to the direction of gravitational force.

In order to demonstrate the presence of restoring force resulting from rotation, let us remember the Helmholtz theorem. It is well-known that in inviscid barotropic gas

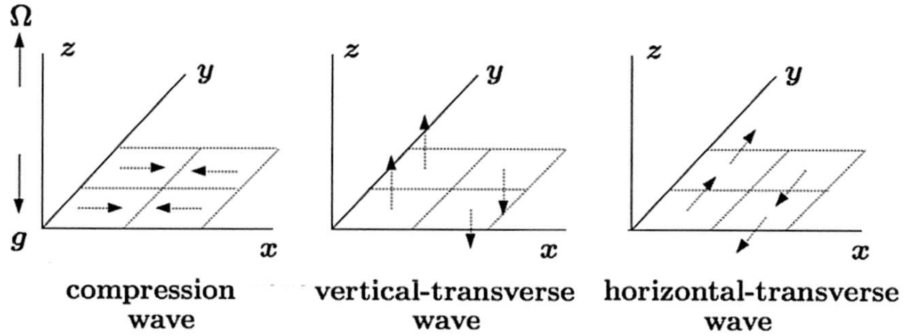

Fig. 4.3 Showing the presence of three types of wave motions in rotating stratified media. In this figure, the compression waves (longitudinal wave) correspond to acoustic waves, the vertical transverse wave to gravity waves, and the horizontal-transverse waves to Rossby waves

[i.e., $p = p(\rho)$] the vorticity, $\boldsymbol{\zeta}$ ($\equiv \mathrm{curl}\boldsymbol{v}/2$),[6] is frozen in fluid motions (Sect. 2.4), which is called *vorticity conservation* (the *Helmholtz theorem*) and expressed in the form:

$$\frac{\partial \boldsymbol{\zeta}}{\partial t} = \mathrm{curl}(\boldsymbol{v} \times \boldsymbol{\zeta}). \tag{4.43}$$

This equation shows that fluid motions are frozen to vorticity lines (analogous to the fact that in MHD flows fluid motions are frozen to magnetic field lines, which will be discussed in Chap. 11). In other words, fluid motions receive a restoring force when they cross vorticity lines, and a wave motion is induced. It is noticed that the horizontal-transverse wave shown in Fig. 4.3 has a component perpendicular to $\boldsymbol{\Omega}$.

In order to treat a concrete example well-known in meteorology, let us consider a gaseous system in a thin spherical shell which rotates with a constant angular velocity Ω. Here, we introduce non-rotating spherical coordinates (r,θ,φ), where the origin is at the center of the sphere and $\theta = 0$ is in the direction of the rotation axis. Our purpose here is to consider gas motions in this thin shell, assuming that the gas has perturbed from the pure rotating flow in the shell. The whole velocity of gas, including the rotation, is expressed by (v_r, v_θ, $r\Omega \sin\theta + v_\varphi$).

For simplicity, we assume that the gas is incompressible and the flows have no radial component since the shell is thin:

$$v_r = 0, \tag{4.44}$$

$$\frac{\partial}{\partial \theta}(v_r \sin\theta) + \frac{\partial}{\partial \varphi} v_\varphi = 0. \tag{4.45}$$

[6] In this subsection $\boldsymbol{\zeta}$ is used to denote vorticity. It should not be confused with the displacement vector.

In the present particular case where Eqs. (4.44) and (4.45) hold, the radial component of Eq. (4.43) leads to

$$\frac{d\zeta_r}{dt} = 0, \tag{4.46}$$

where d/dt is the Lagrangian time derivative defined by

$$\frac{d}{dt} = \left(\frac{\partial}{\partial t} + \Omega\frac{\partial}{\partial \varphi}\right) + v_\theta\frac{\partial}{r\partial\theta} + v_\varphi\frac{\partial}{r\sin\theta\partial\varphi}. \tag{4.47}$$

Equation (4.46) shows that the radial component of vorticity is conserved along the flow. In the fields of geophysics, ζ_r is called *absolute vorticity* and Eq. (4.46) is known as the conservation of absolute vorticity.

If we consider fluid motions in a rotating spherical shell, the absolute vorticity associated with rotation changes as the latitude of the fluid element changes. Equation (4.46), however, requires that the whole absolute vorticity (including the absolute vorticity associated with perturbation) is conserved. That is, the fluid element must move in the shell so that the absolute vorticity is conserved. This brings about an oscillatory motion of the fluid element. Let us more concretely consider the motions. Then, from the definition of absolute vorticity, we have

$$\zeta_r = \Omega\cos\theta + \frac{1}{2r\sin\theta}\left[\frac{\partial}{\partial\theta}(v_\varphi\sin\theta) - \frac{\partial v_\varphi}{\partial\varphi}\right]. \tag{4.48}$$

The equation of conservation of absolute vorticity is then written as, neglecting the nonlinear terms,

$$\left(\frac{\partial}{\partial t} + \Omega\frac{\partial}{\partial\varphi}\right)\left[\frac{\partial}{\partial\theta}(v_\varphi\sin\theta) - \frac{\partial v_\theta}{\partial\varphi}\right] - 2\Omega v_\theta\sin^2\theta = 0. \tag{4.49}$$

Eliminating v_φ from this equation by using the equation of continuity:

$$\frac{\partial}{\partial\theta}(v_\theta\sin\theta) + \frac{\partial}{\partial\varphi}v_\varphi = 0 \tag{4.50}$$

we have an equation with respect to v_θ. By assuming that perturbations are proportional to $\exp[i(\omega t - m\varphi)]$ (m is a positive integer), we can write the resulting equation as

$$\left[\frac{\partial}{\partial\mu}(1-\mu^2)\frac{\partial}{\partial\mu} - \frac{m^2}{1-\mu^2} - \frac{2m\Omega}{\omega - m\Omega}\right][v_\theta(1-\mu^2)^{1/2}] = 0, \tag{4.51}$$

where

$$\mu = \cos\theta. \tag{4.52}$$

Equation (4.51) is the differential equation for an associated Legendre function. From the condition that the solution needs to be regular at $\mu = -1$ and 1, we have for ν of $\nu \geq m$

$$-\frac{2m\Omega}{\omega - m\Omega} = \nu(\nu + 1). \tag{4.53}$$

That is, eigenfunction v_θ is

$$v_\theta \propto (1 - \mu^2)^{-1/2} P_\nu^m(\mu) \tag{4.54}$$

and eigenvalue is

$$\omega - m\Omega = -\frac{2m\Omega}{\nu(\nu + 1)}. \tag{4.55}$$

This wave is called the *Rossby–Haurwitz wave*. An important characteristics of this wave is that the frequency of oscillations seen from the rotating frame, i.e., $\omega - m\Omega$, is negative and propagates only in the direction opposite to the rotation.

This wave is of importance in meteorology, and related to the fact that the weather changes from the west.

As is shown above, the Rossby–Haurwitz wave can be seen directly only when acoustic waves and gravity waves are filtered out by introducing some approximations (incompressibility and nearly horizontal motions). In other words, in real situations in stars, the Rossby–Haurwitz waves are masked by more short timescale wave motions such as acoustic waves and gravity waves. Because of this, the Rossby–Haurwitz waves are not well-known in astrophysics unlike in meteorology and planetary science. In rotating stars, however, they will be of importance in considering long timescale phenomena.

In the remaining part of this section we consider local phenomena of this Rossby–Haurwitz wave by introducing the so-called the *β-plane approximation* which is widely used in meteorology. We assume that wave motions are uniform in the latitudinal direction, and reduce Eq. (4.49) to

$$\left(\frac{\partial}{\partial t} + \Omega\frac{\partial}{\partial \varphi}\right)\frac{\partial v_\theta}{r \sin\theta \partial\varphi} + \beta^* v_\theta = 0, \tag{4.56}$$

where

$$\beta^* = \frac{2\Omega}{r}\sin\theta. \tag{4.57}$$

Let us introduce local Cartesian coordinate system (x, y, z) as is shown in Fig. 4.4, where x-axis is in the direction of φ, y-axis is in the opposite direction of θ-direction. Then, Eq. (4.56) is written as

$$\frac{\partial^2 v_y}{\partial t \partial x} + v_0\frac{\partial^2 v_y}{\partial x^2} + \beta^* v_y = 0, \tag{4.58}$$

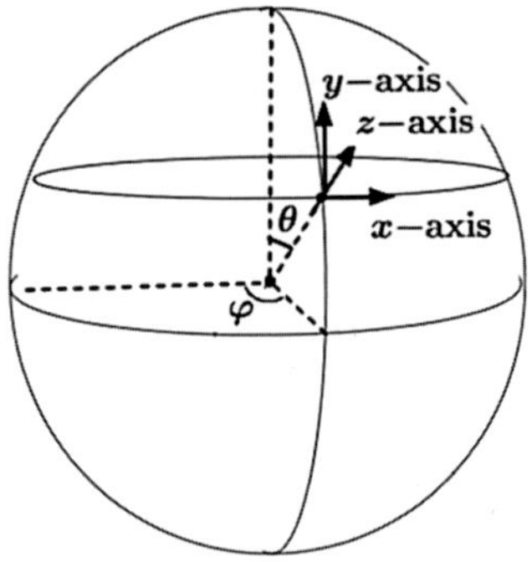

Fig. 4.4 The Cartesian coordinate system (x, y, z) in the β-plane approximation

where $v_y = -v_\theta$ is the latitudinal component of velocity, and $v_0 = r\Omega \sin\theta$ is the speed of rotation at latitude of $\pi/2 - \theta$. Equation (4.58) shows wave phenomena. To demonstrate this and to know the phase speed, v_y is written as

$$v_y = A \exp[i\alpha(x - ct)] \tag{4.59}$$

and substituted into Eq. (4.58). Then we have

$$c = v_0 - \frac{\beta^*}{\alpha^2}. \tag{4.60}$$

This is a simplified dispersion relation of Rossby waves. It is emphasized again that this wave propagates only in one direction. If we see this wave from the frame rotating with $v_0(= r\Omega \sin\theta)$, it propagates in the opposite direction to the rotation with a slow speed of β^*/α^2.

This wavy motion is related to the conservation of absolute vorticity. If a fluid element moves in the polar direction on the spherical surface, keeping its longitudinal coordinate, the absolute vorticity resulting from the rotation decreases. Hence, in order to keep absolute vorticity, the fluid element returns to its original latitude with smaller longitudinal position.

4.4 Waves in Nonself-Gravitating Disks

One of typical astrophysical ingredients in the Universe is disks. Spiral galaxies are popular as disk systems, but the main components in spiral galaxies are stars and they are collisionless systems. The most popular disk systems consisting of gases are protoplanetary systems and accretion disks surrounding a central gravitating body. In protoplanetary disks the effects of self-gravity are sometimes important. In accretion disks surrounding compact objects such as white dwarfs, neutron stars, and stellar black-holes, however, self-gravity of disks can be neglected in studying disk oscillations (rather, general relativistic effects are important). In Sect. 4.4 waves

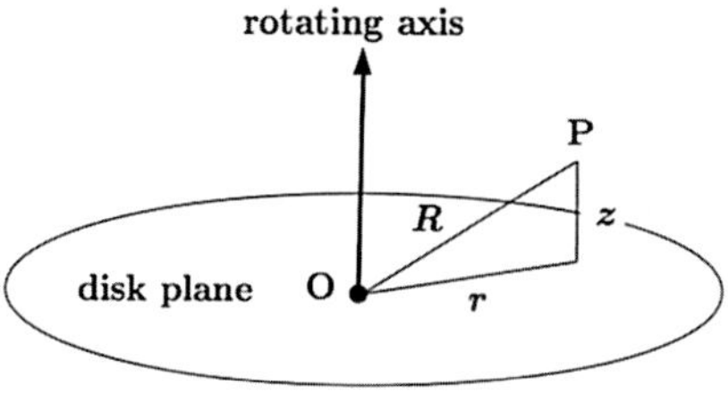

Fig. 4.5 Coordinates used in representing a point P in a disk system. Cylindrical coordinate of a point P is (r, φ, z), and $R^2 = r^2 + z^2$

in nonself-gravitating disks are considered, and the oscillations in self-gravitating ones will be described in Sect. 4.5. In the study of self-gravitating disks, for simplicity, we concentrate our attention only on oscillations with no node(s) in the vertical direction.

It is noted that the readers who are interested in details of oscillations in nonself-gravitating disks are recommended to see Kato (2001), Kato et al. (2008), and Kato (2016).

4.4.1 Unperturbed State

First, unperturbed disk structure is briefly described by focusing on the vertical structure. Cylindrical coordinates (r, φ, z) are introduced as is shown in Fig. 4.5, where the z-axis is perpendicular to the disk plane and its origin is at the disk center. Without constructing unperturbed equilibrium disk models in detail, we start from simplified disk models by introducing some approximations. That is, unperturbed disks are assumed to be axisymmetric and to rotate with a nearly Keplerian angular velocity, $\Omega(r)$, under the force balance among the gravitational force of central object, the centrifugal force due to disk rotation, and the pressure force acting supplementary to the balance.

The vertical disk structure is governed by the balance between the pressure force, $-\partial p_0/\rho_0\partial z$, which acts so as to expand disk in the vertical direction and the gravitational force, $-(GM/R^2)(z/R)$, which acts so as to return gases toward the equator, where M is the mass of the central point source and $R^2 = r^2 + z^2$. In disks which are sufficiently thin in the vertical direction, the latter gravitational force is approximately $-(GM/r^2)(z/r)$, which is written as $-\Omega_{\rm K}^2 z$ with use of the Keplerian angular velocity, $\Omega_{\rm K}(r)[= (GM/r^3)^{1/2}]$. In cases where there is a tidal force (the case of binary system) or the central star is a relativistic source with spin, however, the force acting towards the equator is not equal to $-\Omega_{\rm K}^2 z$. Considering such cases, we write the force as $-\Omega_\perp^2 z$, where $\Omega_\perp$ is the vertical epicyclic frequency and slightly different from $\Omega_{\rm K}$ (see Sect. 9.4). Keeping the difference between $\Omega_{\rm K}(r)$ and $\Omega_\perp(r)$, we write the vertical hydrodynamical equilibrium as

$$-\frac{\partial p_0}{\partial z} - \rho_0 \Omega_\perp^2 z = 0. \tag{4.61}$$

Furthermore, we introduce an approximation that the disk gases are isothermal in the vertical direction, i.e., $T_0 = T_0(r)$. Then, Eq. (4.61) can be integrated to give

$$\rho_0(r, z) = \rho_{00}(r)\exp\left(-\frac{z^2}{2H^2}\right), \tag{4.62}$$

where the disk half-thickness, $H(r)$, is related to the isothermal acoustic speed, $c_{\rm T}(r)(\equiv p_0/\rho_0)$, and the vertical epicyclic frequency, $\Omega_\perp$, by

$$H^2(r) = \frac{c_{\rm T}^2}{\Omega_\perp^2(r)}. \tag{4.63}$$

4.4.2 Small Amplitude Adiabatic Perturbations

We consider small amplitude adiabatic perturbations on the axisymmetric equilibrium disks described above. The perturbations are assumed to be proportional to $\exp[i(\omega t - m\varphi)]$, where ω and m are frequency and azimuthal wavenumber, respectively, of the perturbations. Then, hydrodynamical equations describing small amplitude perturbations are

$$i\tilde{\omega}\rho_1 + \frac{\partial}{r\partial r}(r\rho_0 v_r) - i\frac{m}{r}\rho_0 v_\varphi + \frac{\partial}{\partial z}(\rho_0 v_z) = 0, \tag{4.64}$$

$$i\tilde{\omega}v_r - 2\Omega v_\varphi = -\frac{\partial h_1}{\partial r}, \tag{4.65}$$

$$i\tilde{\omega}v_\varphi + \frac{\kappa^2}{2\Omega}v_r = i\frac{m}{r}h_1, \tag{4.66}$$

$$i\tilde{\omega}v_z = -\frac{\partial h_1}{\partial z}, \tag{4.67}$$

where h_1 and $\tilde{\omega}$ are defined, respectively, by

$$h_1 = \frac{p_1}{\rho_0} = c_{\rm T}^2\frac{\rho_1}{\rho_0}, \tag{4.68}$$

$$\tilde{\omega} = \omega - m\Omega, \tag{4.69}$$

and $\kappa(r)$ is the epicyclic frequency defined by

$$\kappa^2 = 2\Omega\left(2\Omega + r\frac{d\Omega}{dr}\right) = \frac{2\Omega}{r}\frac{d}{dr}(r^2\Omega). \tag{4.70}$$

Here, (v_r, v_φ, v_z), p_1, and ρ_1 are the Eulerian velocity, pressure, and density perturbations over the unperturbed ones, respectively. Equation (4.64) is the equation of continuity, Eqs. (4.65)–(4.67) are in turn the r-, φ-, and z- components of equation of motion. Equation (4.68) is the definition of h_1, and due to the assumption of adiabatic perturbations in barotropic disks, we have $-[(1/\rho_0)\nabla p]_1 = -\nabla h_1$, which has been adopted in Eqs. (4.65)–(4.67).

Elimination of v_φ from Eqs. (4.65) and (4.66) gives a relation between h_1 and v_r, which is

$$\frac{\partial h_1}{\partial r} - \frac{2m\Omega}{r\tilde{\omega}}h_1 = \frac{\tilde{\omega}^2 - \kappa^2}{i\tilde{\omega}}v_r. \tag{4.71}$$

Next, another relation between h_1 and v_r is derived by Eqs. (4.65) and (4.67) being substituted into Eq. (4.64). After some manipulations, we find

$$\begin{aligned}&\frac{1}{\rho_0}\frac{\partial}{\partial z}\left(\rho_0\frac{\partial h_1}{\partial z}\right)+\left(\frac{\tilde{\omega}^2}{c_s^2} - \frac{m^2}{r^2}\right)h_1 \\ &\quad = i\tilde{\omega}\frac{\partial v_r}{\partial r} + i\tilde{\omega}\left[\frac{\partial \ln(r\rho_0)}{\partial r} + \frac{m\kappa^2}{2r\tilde{\omega}\Omega}\right]v_r.\end{aligned} \tag{4.72}$$

Equations (4.71) and (4.72) are a set of two differential equations describing wave motions on disks in terms of v_r and h_1.

Eliminating v_r from Eqs. (4.71) and (4.72), we have a partial differential equation with respect to h_1, which is written explicitly in the form

$$\begin{aligned}&\frac{1}{\rho_0}\frac{\partial}{\partial z}\left(\rho_0\frac{\partial h_1}{\partial z}\right)+\left(\frac{\tilde{\omega}^2}{c_s^2} - \frac{m^2}{r^2}\right)h_1 + \tilde{\omega}\frac{\partial}{\partial r}\left[\frac{\tilde{\omega}}{\tilde{\omega}^2-\kappa^2}\left(\frac{\partial h_1}{\partial r} - \frac{2m\Omega}{r\tilde{\omega}}h_1\right)\right] \\ &\quad + \frac{\tilde{\omega}^2}{\tilde{\omega}^2-\kappa^2}\left[\frac{\partial \ln(r\rho_0)}{\partial r} + \frac{m\kappa^2}{2r\tilde{\omega}\Omega}\right]\left(\frac{\partial h_1}{\partial r} - \frac{2m\Omega}{r\tilde{\omega}}h_1\right) = 0.\end{aligned} \tag{4.73}$$

This equation is a second-order partial differential equation for $h_1(r, z)$. Solving this equation in a general way is difficult (see, e.g., Kato 2016). In the following we solve the equation by introducing approximations.

4.4.3 Eigenvalues and Eigenfunctions

For simplicity, the disk half-thickness, $H(r)$, is assumed to be independent of radius r. Then, changing the independent variable (r, z) to (r, η), where the variable, η, is defined by $\eta = z/H$, we can reduce Eq. (4.73) to

$$\left(\frac{\partial^2}{\partial \eta^2} - \eta\frac{\partial}{\partial \eta}\right)h_1 + H^2\mathcal{L}(h_1) = 0, \tag{4.74}$$

where

$$\mathcal{L}(h_1) = \left(\frac{\tilde{\omega}^2}{c_s^2} - \frac{m^2}{r^2}\right)h_1 + \tilde{\omega}\frac{\partial}{\partial r}\left[\frac{\tilde{\omega}}{\tilde{\omega}^2 - \kappa^2}\left(\frac{\partial h_1}{\partial r} - \frac{2m\Omega}{r\tilde{\omega}}h_1\right)\right]$$
$$+\frac{\tilde{\omega}^2}{\tilde{\omega}^2 - \kappa^2}\left[\frac{d\ln(r\rho_{00})}{dr} + \frac{m\kappa^2}{2r\tilde{\omega}\Omega}\right]\left(\frac{\partial h_1}{\partial r} - \frac{2m\Omega}{r\tilde{\omega}}h_1\right). \tag{4.75}$$

This equation can be solved by separating $h_1(r, \eta)$ as

$$h_1(r, \eta) = f(r)g(\eta). \tag{4.76}$$

This separation shows that Eq. (4.74) is reduced to two ordinary differential equations:

$$\frac{1}{g(\eta)}\left(\frac{d^2}{d\eta^2} - \eta\frac{d}{d\eta}\right)g(\eta) = -K, \tag{4.77}$$

and

$$\frac{1}{f(r)}H^2\mathcal{L}(f) = K, \tag{4.78}$$

where K is a separation constant, and should be determined later from boundary conditions.

First, Eq. (4.77) is solved. We impose the boundary condition that $\rho_0(\eta)g(\eta)$ does not become infinite at infinity of $\eta = \pm\infty$. This gives that K is zero or a positive integer, i.e., $K = n$ with $n = 0, 1, 2, 3, \ldots$, and the eigenfunction corresponding to $K = n$ is the Hermite polynomial of order n, i.e., $g(\eta) = \mathcal{H}_n(\eta)$ (Okazaki et al. 1987). The number n corresponds to node number of $g(\eta)$ in the vertical direction. Then, putting $K = n$ we can derive from Eq. (4.78)

$$\frac{\tilde{\omega}^2 - \kappa^2}{\tilde{\omega}}\frac{d}{dr}\left[\frac{\tilde{\omega}}{\tilde{\omega}^2 - \kappa^2}\left(\frac{df}{dr} - \frac{2m\Omega}{r\tilde{\omega}}f\right)\right]$$
$$+\left[\frac{d\ln(r\rho_{00})}{dr} + \frac{m\kappa^2}{2r\tilde{\omega}\Omega}\right]\left(\frac{df}{dr} - \frac{2m\Omega}{r\tilde{\omega}}f\right)$$
$$+\frac{(\tilde{\omega}^2 - \kappa^2)(\tilde{\omega}^2 - m^2c_s^2/r^2 - n\Omega_\perp^2)}{\tilde{\omega}^2c_s^2}f = 0. \tag{4.79}$$

Equation (4.79) is a second-order ordinary differential equation with respect to $f(r)$, and describes the radial behavior of $f(r)$.

By introducing a new variable defined by

$$\bar{f} = \left(\frac{r\rho_{00}}{\tilde{\omega}^2 - \kappa^2}\right)^{1/2} f, \tag{4.80}$$

we can reduce Eq. (4.79) to

$$\frac{d^2}{dr^2}\bar{f} + \bar{K}^2\bar{f} = 0, \tag{4.81}$$

where

$$\bar{K}^2 = \frac{1}{4}\left[\frac{d}{dr}\ln\left(\frac{r\rho_{00}}{\tilde{\omega}^2-\kappa^2}\right)\right]^2 - \frac{1}{2}\frac{d^2}{dr^2}\ln\left(\frac{r\rho_{00}}{\tilde{\omega}^2-\kappa^2}\right)$$
$$+\left[\frac{2m\Omega}{r\tilde{\omega}}\frac{d}{dr}\ln\left(\frac{\tilde{\omega}^2-\kappa^2}{\rho_{00}\Omega}\right) + \frac{\tilde{\omega}^2-\kappa^2}{c_s^2}\left(1-n\frac{\Omega_\perp^2}{\tilde{\omega}^2}\right) - \frac{m^2}{r^2}\right]. \tag{4.82}$$

If we consider oscillations whose radial wavelength is so short that radial variations of all unperturbed disk quantities (except for the radial variations of Ω and c_s) are neglected and d^2/dr^2 is written as $-k^2$, Eq. (4.81) is reduced to

$$k^2\tilde{\omega}^2c_s^2 = (\tilde{\omega}^2-\kappa^2)(\tilde{\omega}^2-n\Omega_\perp^2). \tag{4.83}$$

This equation is a simplified local dispersion relation of oscillations (Okazaki et al. 1987). In deriving the above dispersion relation, the term proportional to $1/\tilde{\omega}$ in Eq. (4.82) has been neglected. Effects of this term are, however, very important for waves which have non-negligible amplitude at the radius of $\tilde{\omega} = 0$ (*corotation radius*). This point will be discussed in the following subsection, and also in Sect. 6.6 concerning *corotation resonance*.

4.4.4 Classification of Oscillation Modes and Rossby Wave Mode

Dispersion relation (4.83) is formally analogous to Eq. (4.39) for stellar oscillations. Dispersion relation (4.83) is a quadratic equation with respect to $\tilde{\omega}^2$, implying that there are two wave modes. In the special case of $n = 0$, however, we have only one wave mode, and the dispersion relation is

$$\tilde{\omega}^2 = \kappa^2 + k^2c_s^2. \tag{4.84}$$

In this case of $n = 0$, oscillations are nearly horizontal motions: They have no node in the vertical direction. Restoring forces working on these oscillations are those of rotation and pressure in the radial direction. That is, in the limit of long wavelength where the pressure restoring force is weak, $\tilde{\omega}^2$ tends to κ^2 (inertial waves), while in the opposite limit of short wavelength where the pressure restoring force predominates over the restoring one due to rotation, $\tilde{\omega}^2$ tends to $k^2c_s^2$ (acoustic

Table 4.1 Classification of disk oscillation modes

	Node number in the vertical direction		
Frequency	$n = 0$	$n = 1$	$n \geq 2$
Higher $\tilde{\omega}^2$	p-mode (inertial-acoustic mode)	c-mode (corrugation mode)	vertical p-mode (vertical-acoustic mode)
Lower $\tilde{\omega}^2$	None	g-mode (gravity or r-mode)	g-mode (gravity or r-mode)
Complex $\tilde{\omega}$	Rossby mode		

waves). In this sense the mode in the case of $n = 0$ is called *inertial-acoustic mode* (see Table 4.1).

In the case of $n \geq 1$ we have two modes of $\tilde{\omega}^2$, say $\tilde{\omega}_1^2$, and $\tilde{\omega}_2^2$ with $\tilde{\omega}_1^2 < \tilde{\omega}_2^2$. Let us consider the case where $\Omega_\perp^2 > \kappa^2$ (this is realized in many cases).[7] Then, we have $\tilde{\omega}_1^2 < \kappa^2$ and $\tilde{\omega}_2^2 > n\Omega_\perp^2$. The oscillation modes of $\tilde{\omega}_1$ are conventionally called *g-mode* (or *r-mode*) (see Table 4.1), but we should be careful about the difference of this mode from the g-mode in stellar oscillations. In the latter case the appearance of the g-mode is related to $N^2 > 0$, where N is *Brunt–Väisälä frequency*. In the present disk oscillations, N^2 has been taken to be zero.

Concerning $\tilde{\omega}_2$ modes of $n \geq 1$, it will be relevant to consider separately the mode of $n = 1$ and those of $n \geq 2$. In the mode of $n = 1$, $h_1(r, \eta)$ has just one node in the vertical direction, but v_z has no node in the vertical direction.[8] This means that the disk plane makes up-down oscillation with a frequency close to $\Omega_\perp$. This frequency comes from the gravitational force which acts so as to return the fluid elements to the original equatorial plane. Furthermore, a careful examination shows that the motion is roughly incompressible (e.g., see Kato et al. 2008; Kato 2016). In this sense this mode of oscillation of $n = 1$ is sometimes called *corrugation mode* (see Table 4.1). In the oscillation modes of $n \geq 2$, $\tilde{\omega}_2^2$ is higher than $\Omega_\perp^2$, which means that the pressure restoring force acting on density variation increases the frequencies higher than $\Omega_\perp$. In this sense, the mode is called often *vertical p-mode*. The classification of the above-mentioned oscillation modes is summarized in Table 4.1.

4.4.4.1 Rossby-Type Mode

In deriving dispersion relation (4.83), the term proportional to $1/\tilde{\omega}$ in Eq. (4.82), i.e., $(2m\Omega/r)d\ln[(\tilde{\omega}^2 - \kappa^2)/(\rho_{00}\Omega)]/dr]\tilde{\omega}^{-1}$, has been neglected. This does not capture important effects of corotation resonance which exist at the radius of $\tilde{\omega} = 0$,

[7]Except for Keplerian disks with no temperature ($\Omega_\perp^2 = \Omega_K^2 = \kappa^2$), $\Omega_\perp^2 > \kappa^2$ holds in general relativistic disks and tidally deformed disks.

[8]In general, v_z has $n - 1$ node(s) in the vertical direction, when h_1 has n node(s) in the vertical direction.

when the wave amplitude is non-negligible near the corotation radius of $\tilde{\omega} = 0$. The term is important, since at the radius the phase velocity of waves is equal to the rotation speed of the disk, and thus a resonant interaction between the waves and disk rotation occurs, leading to instability or damping of waves. This is the case even when the coefficient of $1/\tilde{\omega}$ at the radius, i.e., $(2m\Omega/r)d\ln[(\kappa^2)/(\rho_{00}\Omega)]/dr$, vanishes (see Sect. 6.7). This issue of corotation resonance will be argued in Chap. 6.

Because of this resonance, the frequency of waves which pass the corotation radius becomes complex. In other words, $\bar{K}^2$ given by Eq. (4.82) is essentially complex and an additional wave propagation region and wave modes appears around the radius of the corotation resonance. The wave resulting from this appearance of wave propagation region around the corotation radius is called *Rossby-type waves* from an analogy of Rossby waves described in Sect. 4.3 (Lovelace et al. 1999).

This presence of a new type of oscillation mode could be understood if the term of $(2m\Omega/r\tilde{\omega})d\ln[\kappa^2/(\rho_{00}\Omega)]/dr$ in Eq. (4.82) is formally included in deriving a dispersion relation. In this case we have a dispersion relation which is of the fifth-order with respect to $\tilde{\omega}$, when d^2/dr^2 is written as $-k^2$ [compare with dispersion relation (4.83) of the fourth order one (the second order with respect to $\tilde{\omega}^2$)]. This shows the appearance of a wave mode which propagates only in one direction as Rossby waves described in Sect. 4.3.

Finally, it is noted that the quantity $\kappa^2/2\Omega$ is the z-component of vorticity, and thus $(\kappa^2/2\Omega)/\rho_{00}$ is a quantity called *vortensity* (specific vorticity). The sign of the radial gradient of vortensity at the corotation radius is an important quantity of the characteristics of wave amplification at the corotation resonance and of Rossby-type waves (see Sect. 6.6). In non-barotropic disks, however, the quantity corresponding to vortensity is found to be vortensity times entropy, which is shown in Sect. 6.6.3.

4.5 Self-Gravitating Disks and Toomre's Q Parameter

Spiral arms in disk galaxies are known to be wave patterns (called *density waves*). Density waves in disk galaxies are, however, different from those in the disks surrounding compact stars in two aspects. First, the materials consisting of the disks are stars, which are collisionless particles (not fluids). Second, density waves in spiral galaxies are self-gravitating. Keeping in mind the differences of collisionless particles and continuous fluids, we consider in this section mainly density waves in self-gravitating continuous fluids. Some differences between density waves in stellar systems and fluid ones will be briefly mentioned in the final part of this section. We emphasize here that our interest in this subsection is focused only on the $n = 0$ mode (*inertial p-mode*), since we are considering the density waves in galaxies. Hence, in order to make the analyses simpler, we treat vertically integrated quantities in this section, without considering vertical structure of disks.

4.5.1 Derivation of Dispersion Relation

Let us again introduce cylindrical coordinates (r, φ, z) whose origin is at the disk center and the z-axis is the axis of disk rotation. The unperturbed disk is assumed to be differentially rotating with angular velocity $\Omega(r)$. As mentioned before, we are interested in the fundamental p-mode ($n = 0$), and thus we treat here the vertically integrated or average quantities. That is, the vertically integrated density (surface density) and the vertically integrated pressure in the unperturbed disks are denoted by $\Sigma_0(r)$ and $P_0(r)$, respectively. Their variations by perturbations are denoted by Σ_1 and P_1, respectively. The small amplitude velocity perturbations (which are vertically averaged) superposed on rotation, $(0, r\Omega, 0)$, is written as $(v_r, v_\varphi, 0)$.

Then, equation of continuity and equations of motions describing perturbations are, respectively,

$$\left(\frac{\partial}{\partial t} + \Omega\frac{\partial}{\partial\varphi}\right)\Sigma_1 + \Sigma_0\left[\frac{1}{r}\frac{\partial}{\partial r}(rv_r) + \frac{\partial v_\varphi}{r\partial\varphi}\right] = 0, \tag{4.85}$$

$$\left(\frac{\partial}{\partial t} + \Omega\frac{\partial}{\partial\varphi}\right)v_r - 2\Omega v_\varphi = -\frac{1}{\Sigma_0}\frac{\partial P_1}{\partial r} + \frac{\Sigma_1}{\Sigma_0^2}\frac{dP_0}{dr} - \frac{\partial\psi_1}{\partial r}, \tag{4.86}$$

$$\left(\frac{\partial}{\partial t} + \Omega\frac{\partial}{\partial\varphi}\right)v_\varphi + \frac{\kappa^2}{2\Omega}v_r = -\frac{1}{\Sigma_0}\frac{\partial P_1}{r\partial\varphi} - \frac{\partial\psi_1}{r\partial\varphi}, \tag{4.87}$$

where $\kappa(r)$ is the epicyclic frequency defined by Eq. (4.70). Perturbations are assumed to be adiabatic and we adopt

$$P_1 = c_s^2\Sigma_1, \tag{4.88}$$

where c_s is an average of the acoustic speed in the vertical direction. Quantity, ψ_1, in equations of motion is a perturbation of gravitational potential. Since we are considering self-gravitating systems, ψ_1 is related to the Poisson equation through density variation:

$$\nabla^2\psi_1 = 4\pi G\Sigma_1\delta(z). \tag{4.89}$$

Since the disk is assumed to be geometrically thin, the delta function, $\delta(z)$, appears in the Poisson equation.

Hereafter, we consider tightly wound spiral waves with frequency ω and azimuthal wavenumber m (the number of arms). That is, perturbations are assumed to be proportional to

$$\exp[i(\omega t - kr - m\varphi)]. \tag{4.90}$$

The tightly wound pattern means $|kr| \gg 1$

Equation (4.90) implies that the locus where the phase of pattern is constant is given by $r = -(m/k)\varphi +$ const. This means that in the wave of $k > 0$ the arm extending outward runs in the direction opposite to that of rotation (trailing pattern) while in the wave of $k < 0$ the arm is forward to the rotation (leading pattern). In the case of tightly wound pattern, the second term on the right-hand side of Eq. (4.86) needs to be neglected compared with the first term. The observed number of arms in spiral galaxies is usually two. This implies that when we consider tightly wound waves, the whole terms of right-hand side of Eq. (4.87) should be also neglected. This is because if we compare Eqs. (4.86) and (4.87), we see that the terms on the left-hand sides are comparable, but those of right-hand side are smaller in Eq. (4.87) compared with those in Eq. (4.86) by $m/|k|(\ll 1)$ Hence, we reduce Eqs. (4.85)–(4.87) to

$$i(\omega - m\Omega)\Sigma_1 - ik\Sigma_0 v_r = 0, \tag{4.91}$$

$$i(\omega - m\Omega)v_r - 2\Omega v_\varphi = ik\frac{P_1}{\Sigma_0} + ik\psi_1, \tag{4.92}$$

$$i(\omega - m\Omega)v_\varphi + \frac{\kappa^2}{2\Omega}v_r = 0. \tag{4.93}$$

The next issue is to solve the Poisson equation. In the case of tightly wound waves, Lin and Shu (1964) solved the Poisson equation by using the WKBJ method. After some careful consideration related to the presence of the delta function in the Poisson equation, they obtained[9]

$$\psi_1 = -2\pi G\frac{\Sigma_1}{|k|}. \tag{4.94}$$

From Eqs. (4.91)–(4.94), and (4.88), we have finally a dispersion relation:

$$(\omega - m\Omega)^2 = \kappa^2 + c_s^2k^2 - 2\pi G\Sigma_0|k|. \tag{4.95}$$

It should be noted that in the case of nonself-gravitating disks, the last term on the right-hand side of Eq. (4.95) is absent, and the equation reduces to the dispersion relation of the inertial p-mode oscillations obtained in the previous subsection [see Eq. (4.84)].

Equation (4.95) shows that in the case where wavenumber k is sufficiently small, we have $(\omega - m\Omega)^2 \sim \kappa^2$ and the waves are inertial waves, while when k is sufficiently large we have $(\omega - m\Omega)^2 \sim c_s^2k^2$, which are acoustic waves. In the intermediate wavelength region, both restoring forces of rotation and pressure are weak and thus the effects of gravitational force become relatively important. This

[9]Equation (4.94) has the different sign from their expression since we adopt the opposite sign of k.

shows that there is a possibility of gravitational instability of $(\omega - m\Omega)^2 < 0$ in an intermediate wavelength region.[10] The condition that the instability does not occur even in the intermediate wavelength can be found by examining dispersion relation (4.95). The right-hand side of Eq. (4.95) becomes the minimum as a function of k at $|k| = \pi G \Sigma_0 / c_s^2$. The condition that we have still $(\omega - m\Omega)^2 > 0$ at the minimum is $c_s > \pi G \Sigma_0 / \kappa$. That is, if we define $c_{\rm min}$ by

$$c_{\rm min} = \frac{\pi G \Sigma_0}{\kappa}, \tag{4.96}$$

the condition that the disks do not become unstable is $c_s > c_{\rm min}$. This condition of stability is expressed by introducing a dimensionless quantity Q (*Toomre's Q value*) defined by[11]

$$Q \equiv \frac{c_s}{c_{\rm min}} = \frac{c_s \kappa}{\pi G \Sigma_0} \tag{4.97}$$

as

$$Q > 1. \tag{4.98}$$

In the limiting case of $c_s = c_{\rm min}$ or $Q = 1$, the disks are marginally stable at the wavenumber defined by

$$k_{\rm crit} = \frac{\pi G \Sigma_0}{c_{\rm min}^2} = \frac{\kappa^2}{\pi G \Sigma_0}. \tag{4.99}$$

Let us define dimensionless frequency, ω_*, and dimensionless wavenumber, k_*, by introducing κ and $\kappa / c_{\rm min}$ as the units of frequency and wavenumber, respectively:

$$\omega_* = \frac{\omega - m\Omega}{\kappa}, \quad k_* = \frac{k}{k_{\rm crit}} = \frac{k c_{\rm min}}{\kappa}. \tag{4.100}$$

Then, the dispersion relation (4.95) is written as

$$\omega_*^2 = 1 + k_*^2 \left(\frac{c_s}{c_{\rm min}} \right)^2 - 2|k_*|. \tag{4.101}$$

Equation (4.101) gives the relation between ω_* and k_* with a parameter $c_s / c_{\rm min}$. This relation between ω_* and k_* is shown in Fig. 4.6 with parameter $c_s / c_{\rm min}$. This figure shows that in disks with $c_s / c_{\rm min} < 1$, the disks become gravitationally

[10] For gravitational instability, see Sect. 6.4.

[11] In stellar disks the value corresponding to c_s is slightly different from c_s, and thus the expression for Q-value is different from that given here.

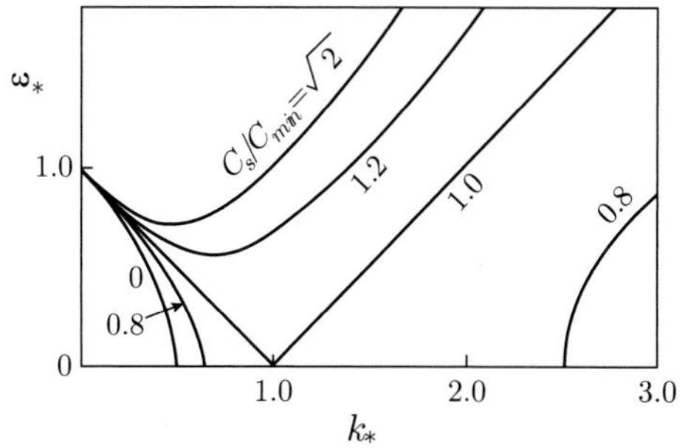

Fig. 4.6 Dispersion relation (4.101) of self-gravitating density waves on gaseous disks, where ω_* and k_* are dimensionless corotating frequency and dimensionless wavenumber, respectively, and defined by Eqs. (4.100). The parameter is disk temperature. It is noticed that in the case of $c_s/c_{\rm min} < 1$, perturbations become unstable in the intermediate wavelength by the effects of self-gravity. The dispersion relation in stellar disks (collisionless systems) is shown in Fig. 4.10 for comparison

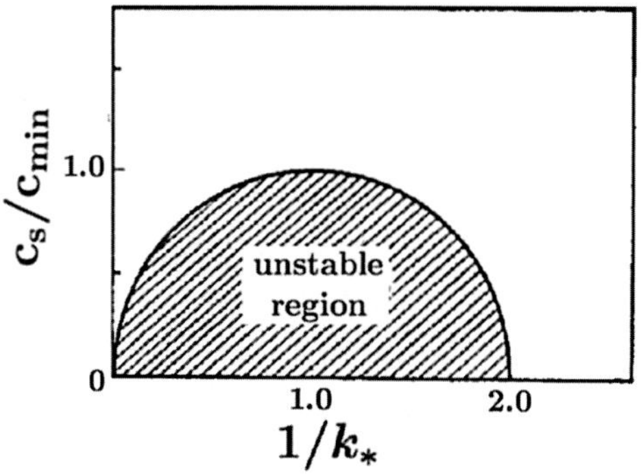

Fig. 4.7 Unstable region of perturbations. In the case where disk temperature is lower than a critical one, perturbations in a particular wavelength range become unstable. For comparison, the unstable region in stellar disks is shown in Fig. 4.9

unstable against perturbations in a certain wavelength range around $k_* = 1$. In order to show how the wavelength range where the disks become gravitationally unstable depends on the parameter $c_s/c_{\rm min}$, the unstable region on the $c_s^2/c_{\rm min}^2 - 1/k_*^2$ plane is shown in Fig. 4.7. Gravitational instability will be discussed in Sect. 6.4.

Finally, it is noted that the density waves have four types. In the case of $k > 0$, the solution of Eq. (4.95) with respect to k is

$$k = k_0\left[1 \pm \left(1 - Q^2(1 - \omega_*^2)\right)^{1/2}\right], \tag{4.102}$$

where k_0 is defined by

$$k_0 = \frac{\pi G \Sigma_0}{c_s^2} = \frac{\kappa^2}{c_s^2 k_{\rm crit}}. \tag{4.103}$$

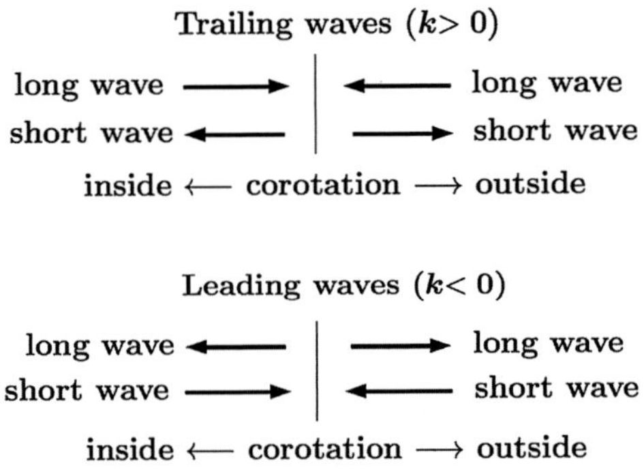

Fig. 4.8 Relations among trailing/leading waves, the direction of group velocity, and the position of corotation resonance. For example, long trailing waves ($k > 0$ but $k < k_0$) have positive group velocity inside the corotation radius, while they have negative group velocity outside the corotation radius. That is, long trailing waves propagate towards the corotation radius. In short trailing waves ($k > 0$ and $k > k_0$) they have negative group velocity inside the corotation radius, while they have positive group velocity outside the corotation radius. That is, short trailing waves depart from the corotation radius. In the case of leading waves situations are changed

In the case of $k < 0$, the first term in the large brackets in Eq. (4.102), i.e., 1, is changed to -1. That is, we have four types of solutions. Since wavy perturbations are taken to have the form of Eq. (4.90), $k > 0$ is a trailing wave, and $k < 0$ is a leading wave.

Since the group velocity of waves, $c_{\rm g}$, is defined by

$$c_{\rm g}^{-1} = \frac{\partial k}{\partial \omega}, \tag{4.104}$$

from Eq. (4.102) we have

$$c_{\rm g} = \frac{c_{\rm s}^2}{\kappa}(k \mp k_0)\frac{1}{\omega_*}, \tag{4.105}$$

where minus of $\mp$ is for $k > 0$ (*trailing wave*) and plus for $k < 0$ (*leading wave*). The angular velocity of rotation, $\Omega(r)$, generally decreases outwards in spiral galaxies, and thus ω_* is negative inside the corotation radius (the radius of $\omega_* = 0$), and is positive outside the radius. These results are summarized in Fig. 4.8.

4.5.2 *Collisionless Stellar Disks and Comparison with Gaseous Disks*

It is instructive to compare the above results on gaseous disks with those on collisionless stellar disks. A detailed derivation of dispersion relation in collisionless stellar disks is not the purpose of this subsection. Hence, we are satisfied with presenting the outline of the derivation and results.

The distribution function of stars in the unperturbed disk, $f_0(\boldsymbol{r}, \boldsymbol{v}, t)$, must satisfy the collisionless Boltzmann equation. As the simplest case, we assume that f_0 can be described by Oort's model (Oort 1928). Then, velocity distribution over a rotation $r\Omega(r)$ is expressed as

$$f_0(\boldsymbol{r}, \boldsymbol{v}, t) \propto \exp\left[-\left(\frac{u_r^2}{u_\mathrm{s}^2} + \frac{4\Omega^2}{\kappa^2}\frac{u_\varphi^2}{u_\mathrm{s}^2}\right)\right], \tag{4.106}$$

where

$$u_r = v_r, \quad u_\varphi = v_\varphi - r\Omega \tag{4.107}$$

and u_s^2 gives velocity dispersion of stars, which is an arbitrary parameter. In Oort's model the angular velocity of rotation, $\Omega(r)$, is required to have a particular form of one parameter family. It is noted that the velocity dispersion in the azimuthal direction is usually smaller than that of radial direction since $\kappa^2/4\Omega^2$ is smaller than unity. The quantity, u_s, corresponds to isothermal acoustic speed in gaseous disks, c_s, in the radial direction.

4.5.2.1 Dispersion Relation

On the above unperturbed distribution function, f_0, a small amplitude perturbation, f_1, which is $f_1 = \exp(ikr)f_1^*$, is superposed. The function, f_1, must satisfy the collisionless Boltzmann equation with respect to f_1. The equation of f_1 is integrated over velocity space in order to obtain a relation between the perturbation of surface density, Σ_1, which is $\exp(ikr)\Sigma_1^* = \exp(ikr)\int f_1^* du_r du_\varphi$, and the perturbation of gravitational potential, ψ_1, which is $\exp(ikr)\psi_1^*$. Combining this equation with a perturbation of the Poisson equation, which is another relation between Σ_1^* and ψ_1^*, we have a relation between ω and k (dispersion relation of perturbations). This procedure is complicated, but after lengthy analyses, we have a dispersion relation (e.g., Kato 1972):

$$\frac{|k|u_\mathrm{s}^2}{2\pi G\Sigma_0} - 1 = -\sum_{n=-\infty}^{n=\infty}\frac{\omega - m\Omega}{\omega - m\Omega + n\kappa}\exp(-\chi^2)I_n(\chi^2), \tag{4.108}$$

where $I_n(\chi^2)$ is the modified Bessel function of index n with argument χ^2, and χ^2 is defined by

$$\chi^2 = \frac{k^2u_\mathrm{s}^2}{\kappa^2}. \tag{4.109}$$

The dispersion relation for stellar disks was first obtained by Lin and Shu (1966), although the expression is different from Eq. (4.108). Their dispersion relation is rewritten by Toomre (1969) in the following form:

$$(\omega - m\Omega)^2 = \kappa^2 - 2\pi G\Sigma_0|k|\mathcal{D}(\chi^2), \tag{4.110}$$

where

$$\mathcal{D}(\chi^2) = \frac{2}{\chi^2}(1-\omega_*^2)\exp(-\chi^2)\sum_{n=1}^{\infty}\left(1-\frac{\omega_*^2}{n^2}\right)^{-1} I_n(\chi^2), \tag{4.111}$$

and

$$\omega_* = \frac{\omega - m\Omega}{\kappa}. \tag{4.112}$$

It is noted that if an identical equation:

$$\exp(\chi^2)\sum_{n=-\infty}^{\infty} I_n(\chi^2) = 1 \tag{4.113}$$

is used, we see that Eqs. (4.108) and (4.110) are the same. Before discussing the ω - k relation given by Eq. (4.110), we consider local stability of disks.

4.5.2.2 Local Stability Condition

Equation (4.108) is the dispersion relation for tightly wound spiral perturbation. If $\Im(\omega - m\Omega) < 0$, the disk is unstable against tightly wound perturbations. The dispersion relation (4.108) shows that the marginal state of stability occurs at $\omega - m\Omega = 0$. Hence, by putting $\omega - m\Omega = 0$ in Eq. (4.108), we obtain the condition of marginal stability:

$$\frac{|k|u_s^2}{2\pi G\Sigma_0} - 1 = -\exp(-\chi^2)I_0(\chi^2). \tag{4.114}$$

By introducing critical wavenumber, k_{crit}, and dimensionless wavenumber k_* defined, respectively, by [cf., Eqs. (4.99) and (4.100)]

$$k_{crit} = \frac{\kappa^2}{\pi G\Sigma_0}, \quad k_* = \frac{k}{k_{crit}}, \tag{4.115}$$

we can write Eq. (4.114) in the form:

$$\frac{\chi^2}{2k_*} - 1 = -\exp(-\chi^2)I_0(\chi^2). \tag{4.116}$$

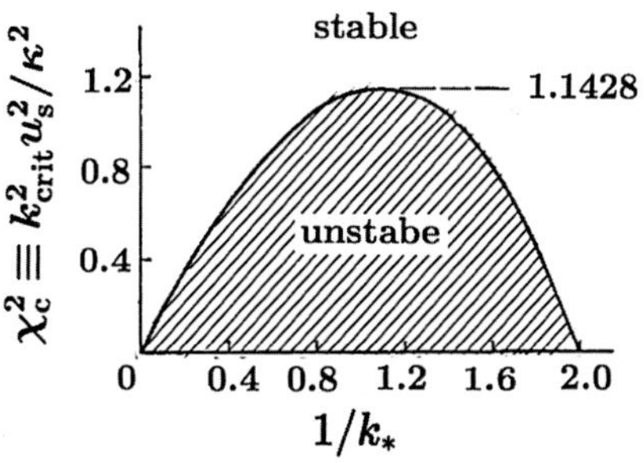

Fig. 4.9 Unstable region of perturbations in stellar disks. In the case where velocity dispersion of stars is lower than a critical one, perturbations in an intermediate wavelength range become unstable. The unstable region in the dimensionless diagram of u_s^2–$1/k_*$ is shown, which is similar to that in gaseous disks (see Fig. 4.7 and notice that χ_c corresponds to c_s/c_{min} in gaseous systems)

Equation (4.116) gives the relation between velocity dispersion, u_s^2, and wavenumber k, for marginal stability. This relation is shown in Fig. 4.9 by using dimensionless quantities $\chi_c^2(\equiv k_{crit}^2 u_s^2/\kappa^2)$ and k_*. It is noticed that χ_c^2 corresponds to c_s^2/c_{min}^2 in gaseous systems, since $\chi_c^2 = u_s^2(\kappa/\pi G\Sigma_0)^2$ and in gaseous systems we have $\pi G\Sigma_0/\kappa = c_{min}$. In this figure the hatched region is unstable. For perturbations with $\chi_c^2 < 1.1428$, the disks are stable for short and long wavelength perturbations (this means the perturbations become oscillations), but perturbations in the intermediate sizes are unstable. Long and short wavelengths perturbations are oscillatory due to the effects of rotation and pressure restoring force, respectively. Only perturbations of intermediate size are gravitationally unstable. To make the disks stable for whole sizes of perturbations, the velocity dispersion u_s^2 needs to be large in the sense that $\chi_c^2 > 1.1428$.

Let us now introduce $u_{s,min}$ defined by

$$u_{s,min} = (1.1428)^{1/2}\frac{\kappa}{k_{crit}} = (1.1428)^{1/2}\frac{\pi G\Sigma_0}{\kappa} \sim 1.069\frac{\pi G\Sigma_0}{\kappa}. \tag{4.117}$$

Then, the condition that disks are stable for all sizes of perturbations, i.e., $\chi_c^2 > 1.1428$, is expressed as $u_s > u_{s,min}$. As in the case of gaseous disks, if the condition is expressed as $Q \equiv u_s/u_{s,min} > 1$, we have

$$Q = \frac{u_s}{u_{s,min}} = 0.935\frac{\kappa u_s}{\pi G\Sigma_0}. \tag{4.118}$$

The quantity Q is called *Toomre's Q parameter* (Toomre 1964). This expression for Q is slightly different from that in the case of gaseous disks [see Eq. (4.97)].

It is noticed that the value of $u_{s,min}$ corresponds to $c_{s,min}$ given by Eq. (4.96) for gaseous disks. The value of Q for stellar disks, Eq. (4.118), and that for gaseous disks, Eq. (4.97), are also almost the same. The dispersion relation for perturbations in stellar disks, however, has an essential difference from that of gaseous disks. To

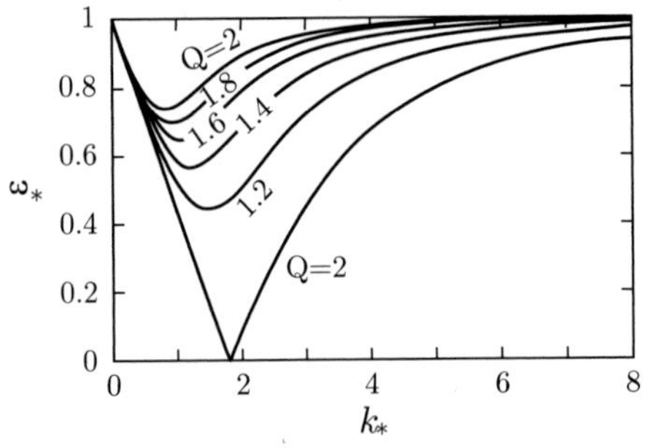

Fig. 4.10 Dispersion relation of density waves on stellar disks, showing ω_*–k_* relation. The dispersion relation is shown only to disks with $Q > 1$. The parameter Q is defined by $Q = u_s/u_{s,min}$. In disks with $Q < 1$ density waves become unstable to intermediate wavelength. A difference from the dispersion relation for gaseous disks should be noticed.In gaseous disks ω_* increases with increase of wavenumber k without limit (see Fig. 4.6). In the present case of stellar disks, however, ω_* is saturated to unity. This difference comes from the absence of pressure restoring force in stellar disks

demonstrate this difference, let us draw the frequency-wavenumber relation derived from the dispersion relation (4.108) in the $\omega_* - k_*$ diagram in the case of $Q > 1$. The results are shown in Fig. 4.10, which should be compared with Fig. 4.6 for gaseous disks. It is important to recognize that in the limit of short wavelength the frequency ω_* increases in the case of gaseous systems due to pressure restoring force, while in the present case of stellar disks the limit of ω_* is the epicyclic frequency, κ, and does not increase beyond κ, because of the absence of pressure restoring force in collisionless systems.

4.6 Shock Waves and Shock Conditions

Shock waves are often observed in daily life. For example, we know the explosive noise of a jet fighter and waves in front of bow of ships. They are due to objects moving faster than the velocity of waves. They are called *bow shocks*. In the case of sudden explosion of a volcano we can see a wave front propagating away from the mountain. Similarly, the explosion of supernovae generates waves propagating into the interstellar media. Such waves generated by sudden explosions are called *blast waves*. If chemical or nuclear reactions are associated with shocks, they are called *detonation waves*.

Even in small amplitude waves they grow to shock waves as they propagate, by the effects of nonlinearity of the waves. A well-known example of such growth is waves approaching to seaside in the sea. In the fields of astrophysics, it is well-known that the waves generated in the solar convection zone increases their amplitude as they propagate outwards to chromosphere and corona due to density decrease and finally grow to shocks.

4.6.1 Growth of Finite-Amplitude Acoustic Waves to Shock Waves

As the simplest example demonstrating the growth of finite-amplitude waves to shock waves, we consider here adiabatic acoustic waves in a static barotropic uniform medium. The waves, for simplicity, are assumed to be plane waves, and their propagation is in the direction of the x-axis. Then, the equation of motion and the equation of continuity are written, respectively, as

$$\frac{\partial v}{\partial t} + v\frac{\partial v}{\partial x} = -\frac{\partial p}{\rho \partial x}, \tag{4.119}$$

$$\frac{\partial \rho}{\partial t} + \frac{\partial}{\partial x}(\rho v) = 0, \tag{4.120}$$

where $v(x)$ is the velocity of wave motions in the x-direction.

Let us introduce ϖ defined by

$$\varpi = \int \frac{dp}{\rho}. \tag{4.121}$$

Then, changing dependent variables representing wave motions from (v, ρ) to (v, ϖ), from Eqs. (4.119) and (4.120), we have

$$\frac{\partial v}{\partial t} + v\frac{\partial v}{\partial x} = -\frac{\partial \varpi}{\partial x}, \tag{4.122}$$

$$\frac{\partial \varpi}{\partial t} + v\frac{\partial \varpi}{\partial x} = -c_{\rm s}^2\frac{\partial v}{\partial x}, \tag{4.123}$$

where $c_{\rm s}$ is the acoustic speed defined by $c_{\rm s} = (dp/d\rho)^{1/2}$. In order to change the above equations to a set of symmetric equations we introduce w defined by

$$w = \int \frac{d\varpi}{c_{\rm s}} = \int \left(\frac{dp}{d\rho}\right)^{1/2}\frac{d\rho}{\rho}. \tag{4.124}$$

Then we have

$$\frac{\partial v}{\partial t} + v\frac{\partial v}{\partial x} = -c_{\rm s}\frac{\partial w}{\partial x}, \tag{4.125}$$

$$\frac{\partial w}{\partial t} + v\frac{\partial w}{\partial x} = -c_{\rm s}\frac{\partial v}{\partial x}. \tag{4.126}$$

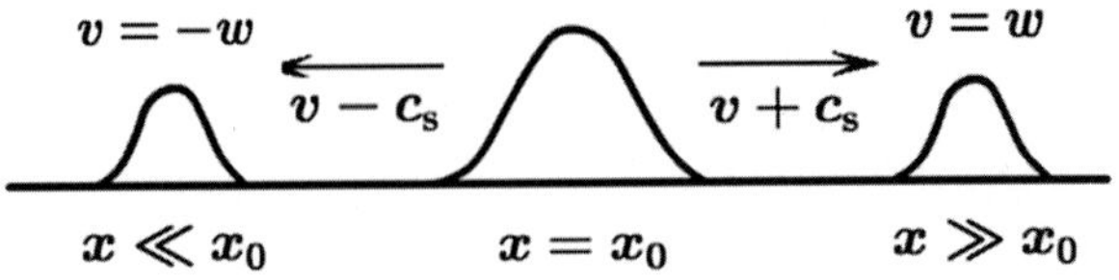

Fig. 4.11 Separation of a nonlinear perturbation into two acoustic waves propagating into opposite directions

The sum and difference of the above two Eqs. (4.125) and (4.126) give, respectively,

$$\left[\frac{\partial}{\partial t} + (v + c_{\rm s})\frac{\partial}{\partial x}\right](w + v) = 0 \tag{4.127}$$

and

$$\left[\frac{\partial}{\partial t} + (v - c_{\rm s})\frac{\partial}{\partial x}\right](w - v) = 0. \tag{4.128}$$

Let us now consider behaviors of $w + v$ and $w - v$ on the $x - t$ plane. Along the curve defined by $dx/dt = v + c_{\rm s}$ (this curve is called characteristic curve C_+) and along the curve defined by $dx/dt = v - c_{\rm s}$ (this curve is called characteristic curve C_-), $w + v$ and $w - v$ are constant, respectively. In other words, $w + v$ propagates along the characteristic curve C_+, while $w - v$ does along the characteristic curve C_-, respectively.

Now, let us consider a perturbation which was localized around a point x_0 at a time $t = 0$. The perturbation is separated into two parts which propagate along the characteristic curve C_+ in the direction of $x > x_0$, and those which propagate along the characteristic curve of C_- in the direction $x < x_0$ (see Fig. 4.11). After a sufficiently long time, the part of $w - v \neq 0$ is propagated away along the characteristic curve C_-, and thus in the perturbations propagating in the region of $x \gg x_0$, we can take $w = v$. That is, we have[12]

$$v = \int \frac{c_{\rm s}}{\rho} d\rho. \tag{4.129}$$

Similarly, in the region of $x \ll x_0$ we have

$$v = -\int \frac{c_{\rm s}}{\rho} d\rho. \tag{4.130}$$

[12]This equation is an equation generalizing the relation $v = c_{\rm s}\rho_1/\rho_0$ which is obtained from the set of Eqs. (4.119) and (4.120) for small amplitude outgoing plane acoustic waves.

Here, we consider how the wave form changes with time as the wave propagates outward in the domain of $x \gg x_0$. To do so, following Landau and Lifshitz (1959), we examine the sign of $d(v + c_s)/d\rho$. From equation (4.129) we have

$$\frac{d}{d\rho}(v + c_s) = \frac{1}{\rho}\frac{d}{d\rho}(\rho c_s). \tag{4.131}$$

By introducing the specific volume, V, by $V \equiv 1/\rho$ we can rewrite ρc_s as

$$\rho c_s = \rho\left(\frac{dp}{d\rho}\right)_s^{1/2} = \left(-\frac{dp}{dV}\right)_s^{1/2} = \left(-\frac{dV}{dp}\right)_s^{-1/2}, \tag{4.132}$$

where we have attached the subscript s in order to emphasize that the adiabatic relations are considered. Combining the above two relations, we have

$$\frac{1}{\rho}\frac{d}{d\rho}(\rho c_s) = \frac{c_s^2}{\rho}\frac{\partial}{\partial p}\left[\left(-\frac{\partial V}{\partial p}\right)_s^{-1/2}\right] = \frac{1}{2}\rho^2 c_s^5\left(\frac{\partial^2 V}{\partial p^2}\right)_s > 0. \tag{4.133}$$

That is, we see that $d(v + c_s)/d\rho > 0$, since $(\partial^2 V/\partial p^2)_s > 0$.

The above results show that the region where density is higher propagates outwards faster. On the contrary, the region where density is lower propagates outwards slower. Hence, as shown in Fig. 4.12, the wave front becomes sharper as the wave propagates outwards. In fluids, density cannot become multi-valued function. This means that the wave changes to a shock wave as is shown in Fig. 4.12. After becoming a shock wave, the wave decreases its amplitude, since the wave region where amplitude is lower becomes to the front. This is related to dissipation of shocks as will be discussed in the next subsection.

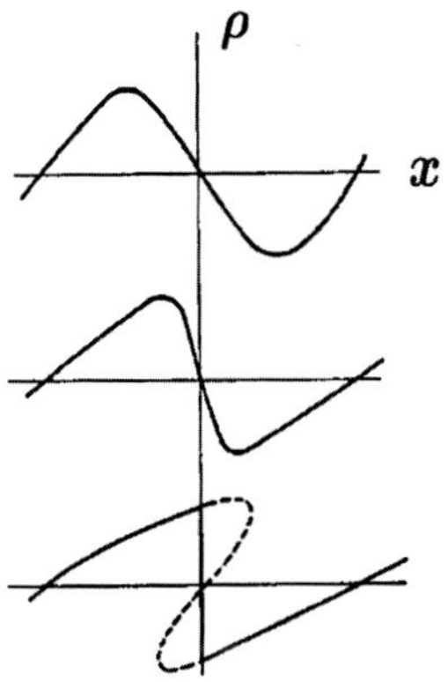

Fig. 4.12 Development of nonlinear acoustic wave pattern into shock

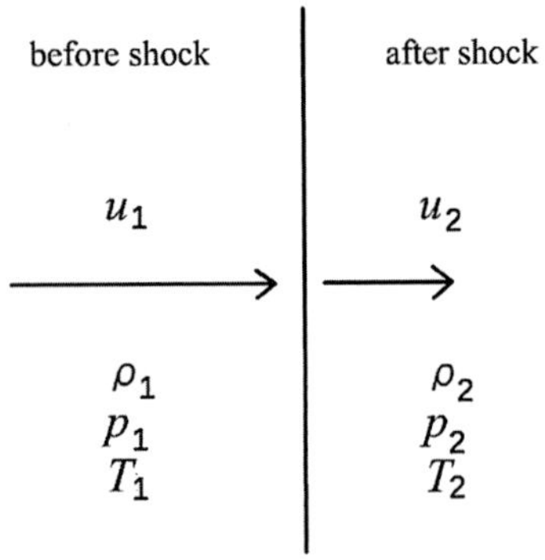

Fig. 4.13 Physical quantities before and after the shock front

4.6.2 Condition of Discontinuity at Shock Front

At shock fronts physical quantities become discontinuous. Without loss of generality, we can take the front to be static, and the flow velocity relative to the front, $\boldsymbol{u}$, is taken to be perpendicular to the front. Furthermore, as shown in Fig. 4.13, quantities in front of the shock are represented by attaching subscript 1, and those after the front by attaching subscript 2.

Since we consider the cases where the flow is adiabatic, equations describing gaseous motions are written in the form of $\partial A/\partial t + \mathrm{div}\boldsymbol{B} = 0$. The continuity condition at a discontinuous plane is found to be $[B_n] = 0$ by integrating the equation in a narrow region sandwiching the front, where B_n is the normal component of $\boldsymbol{B}$ to the shock front. Here and hereafter, the difference between the values of any quantity on the two sides of the discontinuous surface is denoted by enclosing it in square brackets. By considering this, from the equation of continuity and the equation of motion we have, respectively,[13]

$$[\rho u] = 0, \tag{4.134}$$

$$[p + \rho u^2] = 0. \tag{4.135}$$

By using adiabatic approximation, we have from energy equation [e.g., Eq. (2.16)]

$$\left[\frac{1}{2}\rho u^2 + H\right] = 0, \tag{4.136}$$

where H is enthalpy and defined by $H = U + p/\rho$ by using internal energy per unit mass U.

[13]The subscript n is neglected hereafter, since the flow has no transverse component.

The solutions of Eqs. (4.134)–(4.136) are separated into two cases, where the mass flow through the discontinuous surface, j ($\equiv \rho_1 u_1 = \rho_2 u_2$), is zero or not. The former case is called *contact discontinuity*. This case is outside of our present interest. We thus consider only the case of $j \neq 0$. From Eq. (4.134) we have

$$u_1 = \frac{j}{\rho_1}, \quad u_2 = \frac{j}{\rho_2}. \tag{4.137}$$

Substitution of this relation into Eq. (4.135) leads to

$$j^2 = \frac{p_2 - p_1}{1/\rho_1 - 1/\rho_2}. \tag{4.138}$$

Finally, substituting Eqs. (4.137) and (4.138) into Eq. (4.136) we have

$$U_1 - U_2 + \frac{1}{2}\left(\frac{1}{\rho_1} - \frac{1}{\rho_2}\right)(p_1 + p_2) = 0. \tag{4.139}$$

This equation is called the *Rankine–Hugoniot relation.*

The internal energy, U, is generally a function of ρ and p. Hence, Eq. (4.139) can be regarded as a relation between (ρ_1, p_1) and (ρ_2, p_2). Let us now consider the p–$1/\rho$ plane, and represent the position before shock discontinuity on the plane by P_1 (see Fig. 4.14). Then, the point P_2 behind the shock discontinuity is on the curve which passes P_1 and satisfying relation (4.139). This curve is called the *shock adiabat*. If mass flux j is specified, the gradient of the curve connecting points P_1 and P_2 is $-j^2$. By this relation, the position of point P_2 on the p–$1/\rho$ plane is determined.

As the above arguments show, if two independent quantities such as ρ_1 and p_1 are given with a parameter j, the physical quantities behind the shock discontinuity are all determined. In this sense, solutions are one parameter family. As the parameter, we take usually the Mach number $\mathcal{M}_1 (= u_1/c_s)$ or the pressure ratio p_2/p_1.

As long as we see Eqs. (4.134)–(4.136), the cases where the gases are compressed behind shock ($p_2/p_1 > 1$) and the cases of expansion ($p_2/p_1 < 1$) are both allowed. As is shown below, however, entropy is required to increase at the shock, and thus the latter case is excluded.

Fig. 4.14 Position of the shock front on p–$1/\rho$ plane. Point P_1 represents the position before the shock front and P_2 does the position after the shock

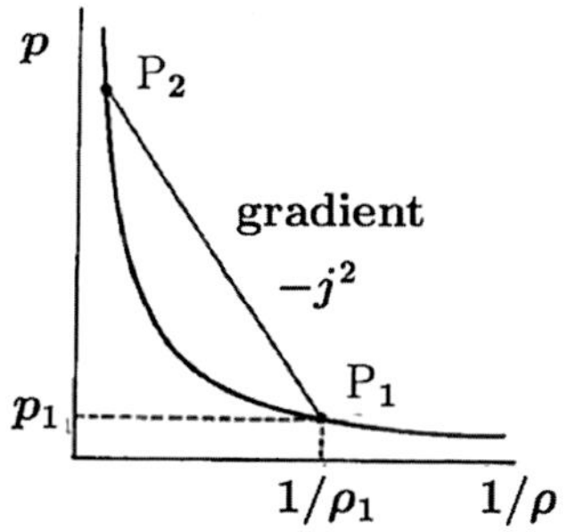

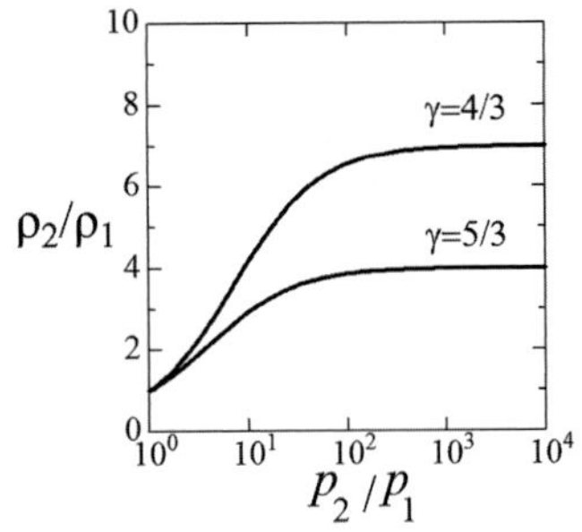

Fig. 4.15 Density discontinuity at shock front as a function of pressure discontinuity. Two cases of $\gamma = 4/3$ and $\gamma = 5/3$ are shown

4.6.3 *Strength of Shock in Case of Ideal Gases*

Considering ideal gases, we summarize how physical quantities change at shock front. First, the case where p_2/p_1 is adopted as a parameter of shock strength is considered. In the ideal gases we have $U = (\gamma - 1)^{-1} p/\rho$, where γ is the ratio of the specific heats. Hence, Rankine–Hugoniot relation (4.139) gives

$$\frac{\rho_2}{\rho_1} = \frac{(\gamma - 1) + (\gamma + 1)(p_2/p_1)}{(\gamma + 1) + (\gamma - 1)(p_2/p_1)}. \tag{4.140}$$

The ratio ρ_2/ρ_1 is shown as a function of p_2/p_1 in Fig. 4.15 for two cases of $\gamma = 4/3$ and $\gamma = 5/3$. It should be noticed that even in a strong shock where the ratio p_2/p_1 is large, the compression rate at shock front, ρ_2/ρ_1, is limited at a finite value, depending on the value of γ. That is, we have

$$\left(\frac{\rho_2}{\rho_1}\right)_{\max} = \frac{\gamma + 1}{\gamma - 1} = \begin{cases} 4 & ; \gamma = 5/3 \\ 7 & ; \gamma = 4/3 \\ \infty & ; \gamma = 1. \end{cases} \tag{4.141}$$

It should be noted that, different from ρ_2/ρ_1, the ratios of pressure, p_2/p_1, and temperature, T_2/T_1, can increase infinitely at the shock front. To demonstrate this, it is convenient to adopt the Mach number of preshock flow, $\mathcal{M}_1 (\equiv u_1/c_s)$, as a shock parameter, because it can become infinite in the limit of strong shock. We have then

$$\frac{\rho_2}{\rho_1} = \frac{(\gamma + 1)\mathcal{M}_1^2}{(\gamma - 1)\mathcal{M}_1^2 + 2}, \tag{4.142}$$

$$\frac{p_2}{p_1} = \frac{2\gamma\mathcal{M}_1^2}{\gamma + 1} - \frac{\gamma - 1}{\gamma + 1}, \tag{4.143}$$

$$\frac{T_2}{T_1} = \frac{[2\gamma\mathcal{M}_1^2 - (\gamma - 1)][(\gamma - 1)\mathcal{M}_1^2 + 2]}{(\gamma + 1)^2\mathcal{M}_1^2}, \tag{4.144}$$

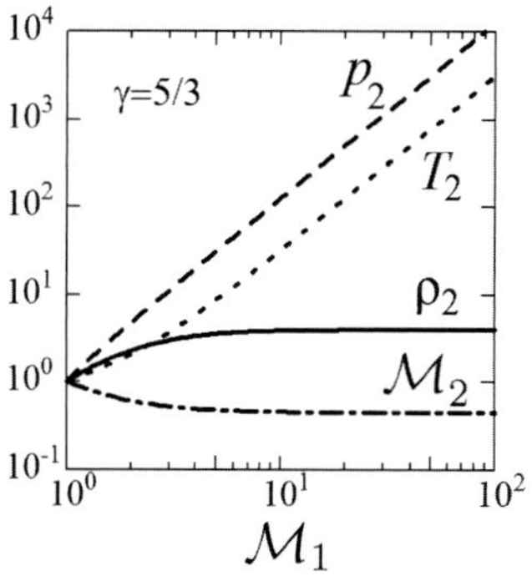

Fig. 4.16 Postshock quantities as functions of the Mach number of preshock flows, i.e., $\mathcal{M}_1 = u_1/c_s$, c_s being acoustic speed in the preshock region. All quantities are normalized by preshock quantities

$$\mathcal{M}_2^2 (\equiv u_2^2/(c_s^2)_2) = \frac{2 + (\gamma - 1)\mathcal{M}_1^2}{2\gamma\mathcal{M}_1^2 - (\gamma - 1)}. \tag{4.145}$$

The above results are summarized in Fig. 4.16 for $\gamma = 5/3$.

Finally, we should comment on *isothermal shock* ($\gamma = 1$). Spatial changes of such fluid quantities as density are realized by collisions of particles. Hence, the shock discontinuity has at least a thickness of the order of mean free path of fluid particles. On the other hand, there are many cases where thermal energy is transported by photons which have longer mean free path than fluid particles have, especially in optically thin media. In such cases, there is practically no temperature change at the shock front. In the framework of the present treatment such cases can be taken into account by taking $\gamma = 1$ (isothermal shock). In isothermal shocks in optically thin media, fluids are strongly compressed. Remember that the maximum value of ρ_2/ρ_1 is $(\gamma + 1)/(\gamma - 1)$. Because of this, propagation of shock front in the interstellar media is one of possible causes of star formation.

4.6.4 Entropy Increase by Shock Discontinuity

Entropy increase at a shock front is considered. Even if diffusion processes are weak, a finite amount of entropy increase at the shock front is present. This is because if diffusion processes are weak, the discontinuity of shocks becomes sharp and gradients of fluid quantities at the shock become large. This leads to independence of entropy increase at the shock front.

Let us now estimate the amount of entropy increase at a shock. Since the entropy increase is small, the adiabatic relations obtained in the previous subsection are valid, as the zeroth-order solutions, to estimate the entropy increase. By defining the entropy increase by ΔS, we see that it is given by

$$\Delta S = S_2 - S_1 = \ln\left(\frac{p_2}{p_1}\right) - \gamma \ln\left(\frac{\rho_2}{\rho_1}\right). \tag{4.146}$$

In the case of ideal gases this is written as [see Eq. (4.140)]

$$\Delta S = \ln\frac{p_2}{p_1} - \gamma \ln\frac{(\gamma-1)p_1 + (\gamma+1)p_2}{(\gamma+1)p_1 + (\gamma-1)p_2}. \tag{4.147}$$

This is a positive quantity in the case of $p_2 > p_1$. In a weak shock where ε defined by $p_2 = p_1(1+\varepsilon)$ is small (i.e., $\varepsilon \ll 1$), Eq. (4.147) is reduced to

$$\Delta S = \frac{\gamma^2 - 1}{12\gamma^2}\varepsilon^3. \tag{4.148}$$

That is, the entropy increase at a shock is a small quantity of the order of ε^3.

Finally, we show how entropy increase is realized at shock. Expressing thermal conductivity by K and two viscosity coefficients by η and ζ, we write energy equation in the form (e.g., Landau and Lifshitz 1959):

$$\rho T\frac{dS}{dt} = \mathrm{div}(K\,\mathrm{grad}T) + \Phi + \zeta(\mathrm{div}\boldsymbol{u})^2, \tag{4.149}$$

where

$$\Phi = \frac{\eta}{2}\left(\frac{\partial u_i}{\partial x_k} + \frac{\partial u_k}{\partial x_i} - \frac{2}{3}\delta_{ij}\frac{\partial u_m}{\partial x_m}\right)^2. \tag{4.150}$$

Since

$$\rho\frac{dS}{dt} = \frac{\partial}{\partial t}(\rho S) + \mathrm{div}(\rho S\boldsymbol{u}), \tag{4.151}$$

by dividing Eq. (4.149) by T, we have

$$j\Delta S = \int\frac{K(\mathrm{grad}T)^2}{T^2}dz + \int\frac{1}{T}[\Phi + \zeta(\mathrm{div}\boldsymbol{u})^2]dz, \tag{4.152}$$

where ΔS is the entropy change at the front. In deriving the first term, we have performed integration of $T^{-1}\mathrm{div}(K\,\mathrm{grad}T)$ by part, and the surface integral has been taken to be zero. The right-hand side of Eq. (4.152) is positive, which means $\Delta S > 0$.

4.7 Propagation of Blast Waves and Sedov–Taylor Similarity Solution

Blast waves emanated by sudden energy release by a point source are of importance in studying the time behavior of supernova explosion. There are two issues. What is the explosion speed of the blast waves and what is the structure behind the front? These issues can be studied by using the similarity solutions developed by Sedov (1946, 1959) and Tayor (1950), if the explosion is strong.

In this section we focus our attention mainly on propagation of the front of blast waves, assuming spherically symmetric geometry. The gas density in the preshock region is taken to be ρ_1, which is assumed to be constant. Then, the mass, $M(t)$, swept by the shock is $M(t) \sim \rho_1 R^3(t)$, where $R(t)$ is the radius of the shock at time t. In strong shocks the gas velocity behind the shock, u_2, and that of the preshock region, u_1, are comparable unless the ratio of specific heats is not close to unity [see Eq. (4.141) and $\rho_1 u_1 = \rho_2 u_2$]. The gas velocity in the postshock region is thus close to u_1, which is the expansion velocity of the shock front, i.e., $u_1 = dR/dt \sim R/t$. Hence, the kinetic energy of the gas behind the shock, $E_{\rm kin}$, is roughly

$$E_{\rm kin} \sim M u_1^2 \sim M\left(\frac{R}{t}\right)^2 \sim \rho_1 \frac{R^5}{t^2}. \tag{4.153}$$

The thermal energy in the postshock region, $E_{\rm thermal}$, is roughly of the order of

$$E_{\rm thermal} \sim p_2 R^3, \tag{4.154}$$

where p_2 is the pressure in the postshock region. The ratio of p_2/p_1 is given by (4.143) and in the limit of strong shock we have $p_2/p_1 = [2\gamma/(\gamma+1)]\mathcal{M}^2$, where p_1 is the gas pressure in the preshock region and $\mathcal{M}$ is the Mach number of the shock flow (the subscript 1 to $\mathcal{M}$ has been omitted here). Hence, we have

$$E_{\rm thermal} \sim p_1 \mathcal{M}^2 R^3 \sim \rho_1 u_1^2 R^3 \sim E_{\rm kin}. \tag{4.155}$$

Two energies, $E_{\rm kin}$ and $E_{\rm thermal}$, are comparable and the total energy, $E = E_{\rm kin} + E_{\rm thermal}$, will be constant if the flow is adiabatic. Hence, we have

$$\frac{R^5}{Et^2/\rho_1} \sim \text{const.} \tag{4.156}$$

This shows that the radius of the blast wave propagating in the uniform media with constant energy expands as

$$R(t) \propto t^{2/5} \tag{4.157}$$

and the expansion velocity, $v_{\rm s}$, is

$$v_{\rm s} = \frac{dR(t)}{dt} \propto t^{-3/5}. \tag{4.158}$$

The above arguments are made, assuming that the density in the preshock region is constant. When a blast wave propagates inside a star, the density in the preshock region may be a function of radius. An extension of the above argument to such cases can be made easily. An extension to such cases is, however, made here by using a slightly different way in order to demonstrate a powerful method of dimensional analysis.

The density in the preshock region is now assumed to decrease outwards as $\rho_1(r) \propto r^{-\alpha}$, where α is a constant. Then, physical quantities which are constant in the present problem are E and $\tilde{\rho}_1 \equiv \rho_1 r^{\alpha}$. The dimensions of these quantities are

$$[\tilde{\rho}_1] = \frac{\mathrm{M}}{\mathrm{L}^{3-\alpha}}, \quad [E] = \frac{\mathrm{ML}^2}{\mathrm{T}^2}, \quad [t] = \mathrm{T}. \tag{4.159}$$

Hence, a dimensional analysis shows that the length scale L which can be derived from the above quantities is

$$\left[\left(\frac{Et^2}{\tilde{\rho}_1}\right)^{1/(5-\alpha)}\right] = \mathrm{L}. \tag{4.160}$$

This suggests that we can define a similarity variable ξ as

$$\xi \equiv r\left(\frac{Et^2}{\tilde{\rho}_1}\right)^{1/(\alpha-5)}. \tag{4.161}$$

For the shock position we can have

$$R(t) = \xi_{\mathrm{s}}\left(\frac{Et^2}{\tilde{\rho}_1}\right)^{1/(5-\alpha)} \propto t^{2/(5-\alpha)}, \tag{4.162}$$

where ξ_{s} is the value of ξ at the shock radius. As expected, the expansion is accelerated when the wave propagates in a medium where density decreases outwards ($\alpha > 0$).

The structure behind the shock front can be examined by using similarity variable ξ. That is, we assume that all dynamical quantities behind the shock front are functions only of ξ. Then, partial differential equations describing gas motions (i.e., equation of continuity, equation of motion, and adiabatic relation) are reduced to ordinary differential equations with respect to ξ. These equations are a set of coupled nonlinear equations, but it is not difficult in principle to solve them numerically with relevant boundary conditions. Details of them will be beyond the scope of this book. For recent studies in this direction in relation to supernovae explosions, see Masuyama et al. (2016).

References

Cox, J.P.: Theory of Stellar Pulsation Colorado, University, Boulder (1980)
Kato, S.: Publ. Astron. Soc. Jpn. **24**, 61 (1972)
Kato, S.: Publ. Astron. Soc. Jpn. **53**, 1 (2001)

Kato, S., Fukue, J., Mineshige, S.: Black-Hole Accretion Disks—Towards a New Paradigm. Kyoto University Press, Kyoto (2008)
Kato, S.: Oscillation of Disks. Astrophysics and Space Science Library vol. 437. Springer, Tokyo (2016)
Landau, L.D., Lifshitz, E.M.: Fluid Mechanics. Pergamon Press, Oxford (1959)
Leighton, R.B., Noyes, R.W., Simon, G.W.: Astrophys. J. **135**, 474 (1962)
Lin, C.C., Shu, F.H.: Astrophys. J. **140**, 646 (1964)
Lin, C.C., Shu, F.H.: Proc. Nat. Acad. Sci. **55**, 229 (1966)
Lovelace, R.V.E. Li, H., Colgate, S.A., Nelson, A.F.: Astrophys. J. **513**, 805 (1999)
Masuyama, M., Shigeyama, T., Tsuboki, Y.: Publ. Astron. Soc. Jpn. **68**, 22 (2016)
Okazaki, A.T., Kato, S., Fukue, J.: Publ. Astron. Soc. Jpn. **39**, 457 (1987)
Oort, J.H.: Bull. Astron. Inst. Neth. **4**, 269 (1928)
Sedov, L.I.: Prikl. Mat. Mekh. **10**, 241 (1946)
Sedov, L.I.: Similarity and Dimensional Methods in Mechanics. Academic, New York (1959)
Tayor, G.I.: Proc. Roy. Soc. Lond. A **201**, 175 (1950)
Toomre, A.: Astrophys. J. **139**, 1217 (1964)
Toomre, A.: Astrophys. J. **158**, 899 (1969)
Unno, W., Osaki, Y., Ando, H., Shibahashi, H.: Nonradial Oscillations of Stars. University of Tokyo Press, Tokyo (1979)
Unno, W., Osaki, Y., Ando, H., Saio, H., Shibahashi, H.: Nonradial Oscillations of Stars, 2nd edn. University of Tokyo Press, Tokyo (1989)

Chapter 5
Convection and Related Topics

Convection and turbulence are one of the important fluid dynamical subjects in astrophysics. For example, the efficiency of energy transport by turbulent convection is an important issue in constructing models of stellar interiors. This is especially true in late type and red giant stars, because outer convective zones extend deeply and energy transport by convection in these regions is important to determine the stellar radii. In cases of accretion disks the efficiency of angular momentum transport by turbulent convection is important in determining disk structures. In general, turbulent convection has strong interactions with rotation and magnetic fields. Hence, in various stages of astrophysical phenomena convection and turbulence are important. Our quantitative understanding of turbulent convection is, however, limited, because turbulence is highly nonlinear phenomena. We present here a few basic topics related to convection.

5.1 Condition of Convective Instability

Like in Sect. 4.1, we consider a plane-parallel stratified medium in subject to a constant gravitational acceleration directed downward. Temperature and density decrease upwards so that the resulting pressure force acts upwards and balances the gravitational force. Taking the direction of the z-axis upwards, we can describe the hydrostatic balance by

$$-\frac{dp_0(z)}{dz}=\rho_0(z)g, \tag{5.1}$$

where $g>0$ is the gravitational acceleration.

As in Sect. 4.1, we consider slow adiabatic rise of a fluid element. At the initial time when the rise begins, the element is assumed to be at $z=z_0$ and have the

© Springer Nature Singapore Pte Ltd. 2020, corrected publication 2023

S. Kato, J. Fukue, *Fundamentals of Astrophysical Fluid Dynamics*,
Astronomy and Astrophysics Library,
https://doi.org/10.1007/978-981-15-4174-2_5

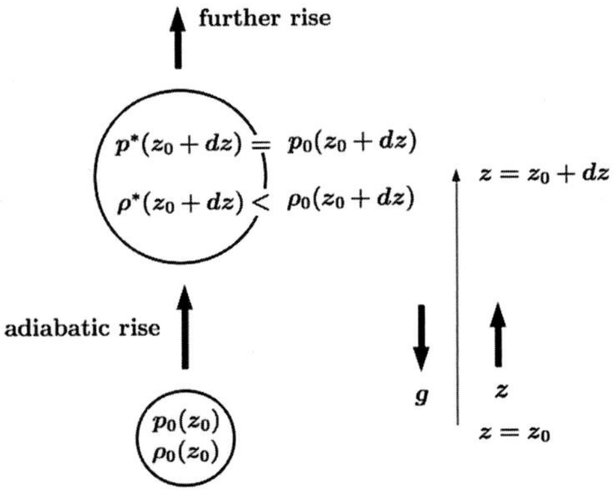

Fig. 5.1 Schematic picture showing an adiabatic rise of a fluid element in a stratified medium where gravitational force is working downward, and gas pressure and density decrease upward. A fluid element rising upward, keeping pressure balance with the surroundings, expands. If the expanded element has lower density compared with the surroundings, it rises further by buoyancy. This is convective instability. If the element has higher density, it is returned to the original position, and makes a harmonic oscillation (see Fig. 4.1), whose frequency is Brunt-Väisälä frequency (see Sect. 4.1)

same density, ρ_0, pressure, p_0, and temperature, T_0, as those of the surrounding fluid (see Fig. 5.1). Since the rise is slow, the fluid element expands, keeping the pressure balance with the surrounding media. At $z = z_0 + dz$ the density in the element, $\rho^*(z_0 + dz)$, is $\rho_0(z_0) + (d\rho/dp)_{\rm ad} dp$, where the subscript ad means the adiabatic change and dp is the pressure difference at two heights of $z = z_0$ and $z = z_0 + dz$. On the other hand, the density in the surrounding medium, $\rho_0(z_0 + dz)$, is $\rho_0(z_0) + (d\rho_0/dp_0)dp$. Hence, if

$$\left(\frac{d\rho}{dp}\right)_{\rm ad} > \left(\frac{d\rho_0}{dp_0}\right), \tag{5.2}$$

the element has a lower density compared with that of the surrounding medium when the element rises upward ($dp < 0$), and the element rises further by buoyancy. This means that the condition of convective instability is

$$\left(\frac{d\ln\rho}{d\ln p}\right)_{\rm ad} > \frac{d\ln\rho_0}{d\ln p_0}. \tag{5.3}$$

In the case where $p \propto \rho T$ and the adiabatic exponent is γ (the ratio of specific heats), the above instability condition is written as

$$\frac{d\ln T_0}{d\ln p_0} > \frac{\gamma - 1}{\gamma}. \tag{5.4}$$

In the field of solar and stellar physics, it is custom to use ∇ and $\nabla_{\rm as}$ defined by

$$\nabla \equiv \frac{d\ln T_0}{d\ln p_0}, \quad \nabla_{\rm ad} \equiv \left(\frac{d\ln T}{d\ln p}\right)_{\rm ad}. \tag{5.5}$$

Then, the above condition of convective instability is

$$\nabla > \nabla_{\rm ad} \equiv \frac{\gamma - 1}{\gamma}. \tag{5.6}$$

To proceed further, we consider more general situations. We remember that in the interior of stars the mean molecular weight, μ, is not always uniform. There are cases where the region of nuclear burning shrinks with time in mass coordinate. In such cases non-uniformity of mean molecular weight appears inside stars. Such situations actually occur in an evolutional stage of massive stars. In massive stars, furthermore, radiation pressure cannot be neglected compared with gas pressure: We need to take into account radiation pressure, $p_{\rm rad}$, in addition to gas pressure, $p_{\rm gas}$, in the equation of state as

$$p = p_{\rm gas} + p_{\rm rad}, \tag{5.7}$$

where

$$p_{\rm gas} = \frac{k_{\rm B}}{\mu m_{\rm H}}\rho T \equiv \beta p, \quad p_{\rm rad} = \frac{1}{3}aT^4 \equiv (1-\beta)p, \tag{5.8}$$

where β is defined by $\beta = p_{\rm gas}/p$. Then, the condition of convective instability, (5.6), is generalized as, after some manipulations (see Sect. 8.3),

$$\nabla > \nabla_{\rm ad} + \frac{\beta}{4-3\beta}\frac{d\ln\mu}{d\ln p}, \qquad \text{(Ledoux criterion)} \tag{5.9}$$

with

$$\nabla_{\rm ad} = \frac{\Gamma_2 - 1}{\Gamma_2}, \tag{5.10}$$

where Γ_2 is an adiabatic exponent in the case where radiation pressure is taken into account (see Sect. 19.6).

Equation (5.9) is the condition of convective instability in the medium with inhomogeneity of mean molecular weight (Ledoux criterion). In general, $d\ln\mu/d\ln p > 0$, and thus the Ledoux criterion is severer than the Schwarzschild criterion:

$$\nabla > \nabla_{\rm ad} \qquad \text{(Schwarzschild criterion)}. \tag{5.11}$$

If effects of thermal conduction is taken into account, overstable motions (*semi-convention*) can occur, even if the Ledoux criterion is not satisfied, as long as the Schwarzschild criterion is satisfied. The issue of semi-convection will be argued later in Sect. 8.3.

5.1.1 Boussinesq Approximation and Criterion of Convective Instability

The Boussinesq approximation is a useful tool to study convection and related motions by filtering out the acoustic mode. In studying stability criteria of rotating gas systems (Solberg-Høiland and Goldreich–Schubert–Fricke criteria) in Chaps. 7 and 8, we also use the Boussinesq approximation. Hence, the essence of the approximation is described here, and the criterion of convective instability is derived by using this approximation.

The *Boussinesq approximation* consists of three parts. (1) Since we are interested in slow motions, acoustic motions are filtered out by introducing approximation of incompressibility, i.e., $\mathrm{div}\,\boldsymbol{v} = 0$ is adopted as equation of continuity. (2) The density variation associated with buoyancy force, however, needs to be taken into account in equation of motion, since we are interested in the phenomena related to buoyancy. (3) Since slow motions which occur under the pressure balance with the surrounding medium are considered, Eulerian pressure variation is neglected in evaluating the buoyancy force and in the equation of state. The terms of pressure variation in the equation of motion are, however, taken into account, since they are necessary to describe the change of direction of motions. On detailed arguments on validity and limitation of the Boussinesq approximation, see Spiegel and Veronis (1960).

Let us now consider a plane-parallel gas system heated from below. The system is uniform in the x–y directions but is stratified in the z-direction under the balance between pressure and gravitational forces [see Eq. (5.1)]. For simplicity, radiation pressure is neglected and no distinction between γ, Γ_1, and Γ_2 are considered. Furthermore, no spatial variation of mean molecular weight is also considered.

In such a plane-parallel system small amplitude perturbations are superposed. Since we use the Boussinesq approximation, the equation of continuity is approximated as

$$\mathrm{div}\boldsymbol{v} = 0, \tag{5.12}$$

where $\boldsymbol{v} = (v_x, v_y, v_z)$ is velocity perturbation over the equilibrium state. The equation of motion is written as $d\boldsymbol{v}/dt = -(1/\rho)\nabla p - \nabla\psi$, where gravitational potential, $\psi(z)$, is time-independent in the present problem. Hence, for small amplitude perturbations, we have

$$\frac{\partial v_x}{\partial t} = -\frac{1}{\rho_0}\frac{\partial p_1}{\partial x}, \tag{5.13}$$

$$\frac{\partial v_y}{\partial t} = -\frac{1}{\rho_0}\frac{\partial p_1}{\partial y}, \tag{5.14}$$

$$\frac{\partial v_z}{\partial t} = -\frac{1}{\rho_0}\frac{\partial p_1}{\partial z} - \frac{\rho_1}{\rho_0}g, \tag{5.15}$$

where the subscript 1 to density ρ and pressure p denote the Eulerian perturbations over the unperturbed density ρ_0 and pressure p_0, respectively. The energy equation is now written as $dp/dt - c_s^2 d\rho/dt = 0$. Since Eulerian pressure variation is neglected (Boussinesq approximation), the perturbed part of this equation is written as $-c_s^2\partial\rho_1/\partial t + v_z(\partial p_0/\partial z - c_s^2\partial\rho_0/\partial z) = 0$. By using ∇ and $\nabla_{\rm ad}$ defined by Eq. (5.5), the above energy equation is written as

$$c_s^2\frac{\partial\rho_1}{\partial t} + \gamma(\nabla - \nabla_{\rm ad})\rho_0 g v_z = 0. \tag{5.16}$$

Let us take the derivative of Eq. (5.13) with respect to x and the derivative of Eq. (5.14) with respect to y. Summing the resulting two equations, using Eq. (5.12), we have

$$\frac{\partial}{\partial t}\Delta v_z = -\frac{1}{\rho_0}g\Delta_1^2\rho_1, \tag{5.17}$$

where $\Delta_1 = \partial^2/\partial x^2 + \partial^2/\partial y^2$, and $\Delta = \Delta_1 + \partial^2/\partial z^2$. Eliminating ρ_1 from Eqs. (5.16) and (5.17), we have

$$\frac{\partial^2}{\partial t^2}\Delta v_z = \gamma\frac{g^2}{c_s^2}(\nabla - \nabla_{\rm ad})\Delta_1 v_z. \tag{5.18}$$

This result shows that perturbations grow when $(\nabla - \nabla_{\rm ad}) > 0$ and the growth rate, n, is given by

$$n^2 = \gamma\frac{g^2}{c_s^2}(\nabla - \nabla_{\rm ad})\frac{k_1^2}{k^2}, \tag{5.19}$$

where k_1 is the wavenumber of perturbations in the horizontal direction, i.e., $k_1^2 = k_x^2 + k_y^2$, and $k^2 = k_1^2 + k_z^2$, k_z being the wavenumber of perturbations in the z-direction.

In the above derivation of criterion of convective instability, adiabatic displacement of a convective element has been assumed. If the movement of convective element is slow, energy exchange between the element and the surrounding media occurs. Hence, we may have a question whether the above assumption of adiabatic motion is relevant in deriving the convective instability. There is, however, no problem in adopting the above approximation. The energy exchange between the fluid element and the surrounding media certainly occurs, and reduces the temperature difference between a convective element and the surrounding medium. This is, however, a passive process and may decrease the growth rate, but brings about no change of the criterion. The situation is the same when viscosity is present without thermal energy exchange. If thermal conduction and viscosity are present simultaneously, however, the onset of convection is slightly affected. That is, for the onset of convection, a numerical quantity called the *Rayleigh number* needs to become larger than a certain critical value. The Rayleigh number, say $\mathcal{R}$, is defined by

$$\mathcal{R} = \frac{\alpha(\beta - \beta_{\rm ad})gh^4}{\kappa\nu}, \tag{5.20}$$

where

$$\alpha \equiv \frac{1}{T_0}, \quad \beta \equiv -\frac{dT_0}{dz}, \quad \beta_{\rm ad} \equiv -\left(\frac{dT_0}{dz}\right)_{\rm ad} \tag{5.21}$$

and h is the thickness of convection layer (β defined here should not be confused to $p_{\rm gas}/p$). Furthermore, κ and ν in Eq. (5.20) are, respectively, thermometric conductivity defined by $\kappa = K/\rho_0 C_{\rm p}$, and kinematic viscosity defined by $\nu = \mu/\rho_0$, where K, $C_{\rm p}$, and μ are conductivity, specific heat with constant pressure, and viscosity, respectively. The critical Rayleigh number, $\mathcal{R}_{\rm c}$, is about 10^3, depending on the boundary conditions (Pellew and Southwell 1940).

In non-conductive and inviscid media, the onset of convection occurs for vertically elongated cells as soon as the criterion (5.11) is satisfied. In conductive and viscous media, however, the onset occurs for nearly hexagonal cells after the Rayleigh number of the media exceeds the critical Rayleigh number. Convective cells at such a stage are called *Benard cells.* At $\mathcal{R} \sim \mathcal{R}_{\rm c}$, the convection pattern is nearly regular and spectrum of convection is discrete. If Rayleigh number increases further beyond $\mathcal{R}_{\rm c}$, convective eddies become semi-regular, quasi-steady and continuous components appear. If Rayleigh number of the medium increases further, discrete components of convections are masked below the continuous components, and finally convection tends to turbulence. The thermal convection is often quoted as a typical example of pattern formations in non-equilibrium open systems.

It is noted that in astrophysical objects thermal conduction is not molecular one, but that due to radiation processes (see Sect. 19.5.2), which is much larger than molecular one. However, the Rayleigh number of astrophysical objects is much larger than the critical one, $\mathcal{R}_{\rm c}$, because very large scale phenomena are considered, i.e., h is very large. Due to this, the convection in stellar interiors is highly turbulent, i.e., $\mathcal{R} \gg \mathcal{R}_{\rm c}$.

In solar photosphere, however, we see granulations which are similar to Benard cells. In the uppermost part of the solar convection zone, the degree of convective instability is high due to partial ionization of hydrogen gas (i.e., $\nabla_{\rm ad}$ is close to unity). Granulations might be Benard cells restricted in this highly unstable layer. In the solar surface we also observe giant cells (i.e., supergranulation) (Leighton et al. 1962). These giant cells might be convective cells extending over the whole convection zone under the effects of turbulent viscosity resulting from smaller turbulent convection.

Finally, we should mention briefly a limitation of the Boussinesq approximation. We have adopted the Boussinesq approximation in studying the onset of convection. The approximation is certainly valid in examining the condition of onset of monotonically growing convection. This approximation is, however, not always relevant in studying the criterion of convection of oscillatory growth (overstable convection) (Chaps. 7 and 8). For example, in the case where the medium has strong magnetic fields, the instability sets in for oscillatory motions (overstability). In this case, stability is examined for oscillatory motions, and thus compressibility of the medium cannot be neglected in studying stability criterion. In other words, in such cases adoption of the Boussinesq approximation is inadequate (e.g., Kato 1966).

5.2 Mixing Length Theory and Beyond

As mentioned before, energy transport by turbulent convection is of importance to construct stellar models, especially those of outer convection zones. Because of incompleteness of our understanding on turbulence, the *mixing length theory* developed in the 1950s is still used for simplicity. Here, we introduce the essence of the theory. The theory is based on the assumption that a turbulent convective element born in turbulent media is transported upwards by buoyancy and is destroyed after propagating a *mixing length* ℓ to be mixed to the surrounding medium. By this process physical quantities such as thermal energy is transported. The mixing length theory stands on the works of Biermann (1951) and Böhm-Vitense (1958).

Let us take the z-axis in the direction opposite to a constant gravitational acceleration, and the medium is taken to be homogeneous in the perpendicular direction to the z-axis. The medium is in a turbulent state by being heated from the below and thermal energy is transported upwards by turbulent convection. The energy flux transported upwards by turbulent convection, $\pi F_{\rm c}(z)$, can be expressed as

$$\pi F_{\rm c}(z) = \langle \rho C_{\rm p} \Delta T(\boldsymbol{x}, z) v(\boldsymbol{x}, z) \rangle, \tag{5.22}$$

where v is the z-component of fluid velocity of a turbulent element, and ΔT is the temperature deviation of the element from the horizontally averaged temperature (the direction on the horizontal plane is shown by $\boldsymbol{x}$). Furthermore, $\langle\ \rangle$ represents the average in the $\boldsymbol{x}$-plane. Equation (5.22) will be understood if we consider that the excess thermal energy of a fluid element (including the energy of expansion or contraction) compared with that of the surrounding media is $\rho C_{\mathrm{p}} \Delta T$, where C_{p} is the specific heat for constant pressure.

In the mixing length theory, Eq. (5.22) is approximated as

$$\pi F_{\mathrm{c}} = \rho C_{\mathrm{p}} \overline{\Delta T}\, \overline{(v^2)}^{1/2}, \tag{5.23}$$

where $\overline{\Delta T}$ represents the mean temperature difference between a convection element and the surrounding media during birth and destruction of the element, i.e., during the time of propagation of mixing length ℓ. $\overline{(v^2)}^{1/2}$ also has a similar meaning. The temperature difference ΔT which is realized during the time interval between birth and rise of a distance h is

$$\Delta T = h\left\{\left(\frac{dT}{dz}\right)' - \left(\frac{dT}{dz}\right)\right\} = \frac{h}{H} T(\nabla - \nabla'), \tag{5.24}$$

where $'$ represents the value which a convective element has, and H is the distance where pressure changes $1/e$ (scale height) and is $H = p/\rho g$. As the mean value of ΔT, we adopt ΔT at $h = \ell/2$, i.e.,

$$\overline{\Delta T} = \frac{\ell}{2H} T(\nabla - \nabla'). \tag{5.25}$$

Next, let us consider about v. When a convective element moves, keeping pressure balance with the surrounding media, the buoyancy force acting per unit mass is $(\Delta T/T)g$. By integrating this over the movement of h, we see that the kinetic energy of the convective element, $v^2/2$, is found to be $v^2 = gh^2(\nabla - \nabla')/H$. If we adopt as the mean value of v^2 the value at $h = \ell/2$, we have

$$\overline{v^2} = \frac{g\ell^2}{4H}(\nabla - \nabla'). \tag{5.26}$$

By substituting Eqs. (5.25) and (5.26) into Eq. (5.23), we have

$$\pi F_{\mathrm{c}} = \rho C_{\mathrm{p}} \frac{g^{1/2}\ell^2}{4H^{3/2}} T(\nabla - \nabla')^{3/2}. \tag{5.27}$$

Next, we need to know the value of ∇'. The most rough approximation is $\nabla' = \nabla_{\mathrm{ad}}$, since if energy exchange between the convective element and surrounding media is neglected we have $\nabla' = \nabla_{\mathrm{ad}}$. In the outer convection zone of stars, density is low and thus internal energy of convective element per unit volume is low. Because of this, energy exchange between a convective element and surrounding media by radiation cannot be neglected. That is, $\nabla > \nabla' > \nabla_{\mathrm{ad}}$ is realized, and in the convection zone of stars we have in general

$$\nabla_{\rm rad} > \nabla > \nabla' > \nabla_{\rm ad}, \tag{5.28}$$

where $\nabla_{\rm rad}$ is the value of $d\ln T/d\ln p$ when energy is transported by radiation alone. An evaluation of the value of ∇' in the case where radiation effects are considered is omitted here. For details, see Vitense (1953), Böhm-Vitense (1958), and Unsöld (1955).

If ∇' is expressed in terms of physical quantities of the surrounding medium, the energy flux due to convection is estimated by use of Eq. (5.27). It is noted that the value of ∇ is determined if a model of convection zone is obtained. A model of the convection zone is obtained by imposing that the sum of convective energy transport described above and radiative energy transport is constant.

It is mentioned that in constructing the convection cores of stars, we do not need to worry about the value of mixing length, since the internal energy of convective cell is large due to high density and thus convection can transport energy efficiently. Hence, we can take the temperature gradient in the convective cores is adiabatic with good accuracy, i.e., $\nabla = \nabla_{\rm ad}$.

Finally, let us comment on the mixing length ℓ. It is a parameter to be determined by introducing an additional considerations or by comparing stellar models obtained with a given ℓ with observed stellar models. The mixing length theory can be said to be a one-mode, one-eddy, one-length scale model. Turbulence in high Reynolds number objects such as astrophysical ones, however, could not be fully represented by a one-eddy model, since the spectrum of turbulence is widely extended. Canuto and Mazzitelli (1991), for example, constructed an alternative model where turbulent spectrum is taken into account. Their theory, however, still have a free parameter. It is desirable to construct a theory with no parameter. Such an attempt was made by Pasetto et al. (2014). They developed the first theory of stellar convection that is fully self-consistent and scale-free, and applied the theory to obtain stellar models.

5.3 Generation of Oscillations by Turbulence

Excitation of oscillations by turbulence is of importance in various aspects in astrophysics. For example, high temperature of solar and stellar corona is attributed to dissipation (thermalization) of waves which are generated below the photosphere and propagated to the corona. Thus, it is of importance to know what kinds of waves are generated below the photosphere and how much their amount is. Furthermore, many stars are now known to be subject to non-radial oscillations, and they are one of very powerful tools to examine the internal structure of the Sun and stars (*helioseismology* and *asteroseismolog*). The main source of excitation of these oscillations is considered to be turbulence (Goldreich and Keeley 1977), although κ-mechanism (Ando and Osaki 1976) will be partially important (see Sect. 8.1 for the *κ-mechanism*, and see Unno et al 1989 for thermal excitation processes of stellar oscillations in general). In this subsection, we present the basic theory of wave generation from turbulence by Lighthill (1952), and briefly mention the basic parts of Goldreich and Keeley (1977).

5.3.1 *Lighthill Theory of Excitation of Acoustic Oscillations by Turbulence*

Lighthill (1952) examined the efficiency of acoustic wave generation from turbulence in order to explore the noise from jet engine of airplanes. This theory is applied to estimate the wave generation from turbulence in stellar convection zone.

First, let us briefly consider about the efficiency of generation of acoustic waves in homogeneous media. As Table 5.1 shows, the most efficient process of acoustic wave generation in fluid is a time-periodic mass input to the fluid. Its typical example is siren. The next is drum, in which there is no mass input to the fluid, but waves are generated by time variation of momentum input to the fluid. Compared with them, wave generation by turbulence has less efficiency, which is proportional to $\mathcal{M}^5$, where $\mathcal{M}$ is the Mach number of turbulent velocity.

In the following, wave generation from turbulence in homogeneous media is considered. Equations describing motions in the media are

$$\frac{\partial \rho}{\partial t} + \frac{\partial}{\partial x_i}(\rho v_i) = 0, \tag{5.29}$$

and

$$\frac{\partial \rho v_i}{\partial t} + \frac{\partial}{\partial x_j}(\rho v_i v_j) = -\frac{\partial p}{\partial x_i}. \tag{5.30}$$

By eliminating ρv_i from the above two equations, we have

$$\left(\frac{\partial^2}{\partial t^2} - c_\mathrm{s}^2 \nabla^2\right)\rho = \frac{\partial T_{ij}}{\partial x_i \partial x_j}, \tag{5.31}$$

Table 5.1 Generation processes of acoustic waves and their efficiency

Type of radiation	Source of generation	Efficiency	Example
Monopole	Variation of mass	$\propto \mathcal{M}$	Siren
Dipole	Variation of momentum	$\propto \mathcal{M}^3$	Drum
Quadrupole	Variation of momentum flow	$\propto \mathcal{M}^5$	Turbulence

$\mathcal{M}$ is the Mach number of velocity fluctuation

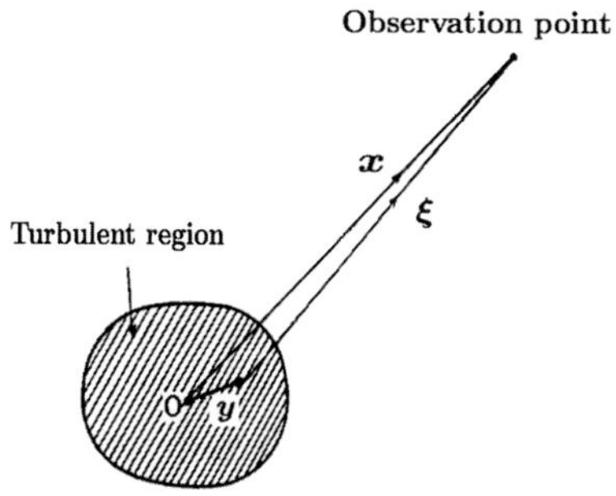

Fig. 5.2 Showing the relation between turbulent region and observation point

where

$$T_{ij} = \rho v_i v_j + (p - c_s^2 \rho)\delta_{ij}, \tag{5.32}$$

and c_s is the acoustic speed. The left-hand side of Eq. (5.31) shows that density variation propagates as acoustic waves, and the right-hand side of the equation can be regarded as the source term. To examine the generation of acoustic waves by the source term, we solve the wave equation (5.31) by using the retarded potential as

$$\rho(\boldsymbol{x}, t) - \rho_0 = \frac{1}{4\pi c_s^2} \int \frac{1}{\xi} \left[\frac{\partial}{\partial y_i \partial y_j} T_{ij}(\boldsymbol{y}, t') \right]_{t'=t-\xi/c_s} d\boldsymbol{y}, \tag{5.33}$$

where the volume integration is performed over the whole turbulent region, ρ_0 is the gas density in the state where there is no turbulence, and $\xi = |\boldsymbol{x} - \boldsymbol{y}|$ (see Fig. 5.2). By performing the integration with respect to $\boldsymbol{y}$ by part and replacing the derivative with respect to $\boldsymbol{y}$ to that with respect to $\boldsymbol{x}$, we have

$$\rho(\boldsymbol{x}, t) - \rho_0 = \frac{1}{4\pi c_s^2} \frac{\partial}{\partial x_i \partial x_j} \int \frac{1}{\xi} T_{ij}\left(\boldsymbol{y}, t - \frac{\xi}{c_s}\right) d\boldsymbol{y}. \tag{5.34}$$

The presence of $\partial^2/\partial x_i \partial x_j$ in Eq. (5.34) represents that the radiation is quadratic.

We are interested in density $\rho(\boldsymbol{x}, t)$ far outside the turbulent region. Hence, in Eq. (5.34) the derivatives are operated only to T_{ij}, and not operated to ξ. $1/\xi$ is written as $1/x$. Then, Eq. (5.34) is reduced to

$$\rho(\boldsymbol{x}, t) - \rho_0 = \frac{1}{4\pi c_s^4} \frac{x_i x_j}{x^3} \int \frac{\partial^2}{\partial t^2} T_{ij}\left(\boldsymbol{y}, t - \frac{\xi}{c_s}\right) d\boldsymbol{y}. \tag{5.35}$$

Here, we assume that the turbulence is adiabatic to neglect the second term of Eq. (5.32) to lead to

$$\begin{aligned}\rho(\boldsymbol{x},t)-\rho_0 &= \frac{\rho_0}{4\pi c_s^4}\frac{x_i x_j}{x^3}\int\left[\frac{\partial^2}{\partial t^2}(v_i v_j)\right]_{t-\xi/c_s} d\boldsymbol{y} \\ &= \frac{\rho_0}{4\pi c_s^4 x}\int\left[\frac{\partial^2 v_x^2}{\partial t^2}\right]_{t-\xi/c_s} d\boldsymbol{y},\end{aligned} \tag{5.36}$$

where v_x is the x-component of turbulent velocity.

The strength of acoustic waves in the region far away from the turbulent region, I, is

$$I(\boldsymbol{x},t)=\frac{c_s^3}{\rho_0}\langle(\rho-\rho_0)^2\rangle, \tag{5.37}$$

where $\langle\ \rangle$ represents ensemble average. From Eqs. (5.36) and (5.37) we have thus

$$I(\boldsymbol{x},t)=\frac{\rho_0}{16\pi^2 c_s^5 x^2}\int\int\left\langle\left[\frac{\partial^2 v_x^2}{\partial t^2}\right]_{t-\xi/c_s}\left[\frac{\partial^2 v_x'^2}{\partial t^2}\right]_{t-\xi'/c_s}\right\rangle d\boldsymbol{y}d\boldsymbol{y}'. \tag{5.38}$$

To proceed further, it is necessary to introduce some assumptions concerning the characteristics of turbulence. For simplicity, the turbulence is assumed to be homogeneous. Furthermore, we assume that the correlation of turbulent quantities at two points rapidly decreases as the distance between them increases beyond the typical size of turbulence, ℓ. In this case, in calculating the correlation quantities in Eq. (5.38) the time difference between $t-\xi/c_s$ and $t-\xi'/c_s$ can be neglected. This comes from the following situations. First, the correlation between two points whose time difference is larger than ℓ/c_s is small and does not contribute to the integration. Next, let us consider the correlation between two points whose time difference is smaller than ℓ/c_s. The correlation time in turbulence is roughly ℓ/c_s. Furthermore, $u<c_s$. Hence, during the time difference of ℓ/c_s, the correlation changes little. Considering these situations and homogeneity of turbulence we can replace Eq. (5.38) to

$$I(\boldsymbol{x},t)=\frac{\rho_0}{16\pi^2 c_s^5 x^2}\int\left[\int U(\boldsymbol{r})d\boldsymbol{r}\right]_{t-\xi/c_s} d\boldsymbol{y}, \tag{5.39}$$

where

$$U(\boldsymbol{r})=\left\langle\frac{\partial^2 v_x^2(\boldsymbol{y},t)}{\partial t^2}\frac{\partial^2 v_x^2(\boldsymbol{y}+\boldsymbol{r},t)}{\partial t^2}\right\rangle \tag{5.40}$$

represents the correlation between two points at the same time.

The quantity $I(\boldsymbol{x}, t)$ is multiplied by $4\pi x^2$ and divided by mass to obtain the power, P, of generation of waves per unit time. The result is

$$P = \frac{1}{4\pi c_s^5} \int U(\boldsymbol{r}) d\boldsymbol{r}. \tag{5.41}$$

The dissipation rate of turbulence per unit mass is denoted by ε. Then, the decaying time of turbulence, τ, is $\tau \sim v^2/\varepsilon \sim l/v$. Since $\int U(\boldsymbol{r}) d\boldsymbol{r} \sim (v^4/l^2)^2 l^3$, we have

$$\int U(\boldsymbol{r}) d\boldsymbol{r} \sim \varepsilon v^5 \tag{5.42}$$

and

$$P = \alpha\varepsilon \left(\frac{v}{c_s} \right)^5, \tag{5.43}$$

where α is a dimensionless constant, depending on characteristics of turbulence. Detailed calculations to isotropic turbulence (Proudman 1952) show $\alpha = 10 \sim 40$.

In summary, turbulence radiates away a small fraction of the energy dissipated into thermal energy into acoustic waves. In the case of subsonic turbulence, the fraction is about $(v/c_s)^5$.

5.3.1.1 The Case of Stratified Medium

In the above studies the medium was assumed to have a uniform density. Stellar atmospheres, however, are stratified under the subject of gravitational force. In such cases, dipole and monopole radiations are present (Unno 1966).

For simplicity, an isothermal atmosphere is considered under a constant gravitational acceleration. In this case the equation corresponding to Eq. (5.31) is

$$\left(\frac{\partial^2}{\partial t^2} - c_s^2 \nabla^2 - g \frac{\partial}{\partial z} \right) \rho_1 = \frac{\partial^2}{\partial x_i \partial x_j} (\rho_0 v_i v_j). \tag{5.44}$$

Here, as the term of T_{ij}, the inertial term alone has been adopted, and ρ_1 is the density variation. The direction of the z-axis is taken in the direction opposite to the gravitational acceleration $\boldsymbol{g}$. By introducing ϕ defined by

$$\phi = \rho_1 / \rho_0^{1/2} \tag{5.45}$$

we can reform Eq. (5.44) to

$$\left(\frac{\partial^2}{\partial t^2} - c_s^2\nabla^2 + \frac{c_s^2}{4H^2}\right)\phi = \rho_0^{-1/2}\frac{\partial^2}{\partial x_i \partial x_j}(\rho_0 v_i v_j), \tag{5.46}$$

where H is the scale height of the atmosphere defined by

$$H = \frac{c_s^2}{g}. \tag{5.47}$$

The right-hand side of Eq. (5.46) can be written as

$$\frac{\partial^2}{\partial x_i \partial x_j}\left(\rho_0^{1/2} v_i v_j\right) - \frac{1}{H}\frac{\partial}{\partial x_i}\left(\rho_0^{1/2} v_i v_z\right) + \frac{1}{4H^2}\left(\rho_0^{1/2} v_z^2\right), \tag{5.48}$$

which shows the presence of monopole and dipole terms in addition to the quadrupole one. However, we must remember that in stratified atmospheres the low frequency fluctuations cannot propagate upwards as waves (see Sect. 4.1.2). This means that we need to make complicated analyses where the frequency spectrum of turbulent generation terms are considered.

For details of recent development of wave generation in stellar convection zone, see, for example, Samadi (2011) and Samadi and Goupil (2001).

5.3.1.2 Generation of MHD Waves

In medium with global magnetic fields, turbulent motions disturb the line of forces, which generates MHD waves. Hence, the efficiency of wave generation from magnetized medium will be high. This means, however, that there may be no clear separation between wave and turbulent motions, and evaluation of generation efficiency will be complicated. Another difficulty is the absence of a simple Green function describing solution of inhomogeneous wave equation.[1] Hence, Kulsrud (1955) studied the generation of MHD waves from turbulence by the following method: Turbulent fluctuations as well as wave motions are decomposed into Fourier components, and the time evolution of each Fourier component of wave equation with source term is examined. The source terms are assumed to be due to isotropic turbulence. Then, summing up these Fourier components, he obtained the generation rate of three MHD waves. This is very complicated procedures.

If we restrict only to Alfvén waves in incompressible media, however, we have a simple Green function, and the efficiency of generation of Alfvén waves can be roughly estimated (see Kato 1968), if structure of turbulence is specified. If isotropic

[1] In the case of acoustic waves we had the retarded potential as the solution of inhomogeneous wave equation, see Eq. (5.33).

turbulence is assumed, we have $P = \alpha\varepsilon(v/c_A)$, where α is a dimensionless quantity of the order of unity and c_A is the Alfvén speed.

5.3.2 *Stochastic Excitation of Oscillations by Goldreich and Keeley*

In Lighthill's model the observing point of generated waves is outside the turbulent region. In studying dynamical excitation of stellar oscillations by turbulence, the turbulent region and oscillating region should be taken to be the same. Furthermore, in order to examine whether oscillations are really excited against turbulent viscous damping and to evaluate the amplitude of oscillations, the effects of damping due to turbulent viscosity should be simultaneously taken into account. This issue was explored by Goldreich and Keeley (1977). They demonstrated that the stochastic processes of turbulent convection can excite the solar p-mode oscillations against the damping due to turbulent viscosity, and proposed that the stochastic processes of turbulence will be the origin of stellar non-radial oscillations. Here, we briefly present the essential part of their mathematical formulation.

As the zeroth-order approximations, stars are taken to have no rotation. The wave equation describing perturbations is then written as (see Sect. 2.5 and Appendix A)

$$\rho_0 \frac{\partial^2 \boldsymbol{\xi}}{\partial t^2} + \mathcal{L}(\boldsymbol{\xi}) = \boldsymbol{N}, \tag{5.49}$$

where $\boldsymbol{\xi}$ is the Lagrangian displacement vector describing fluid motions, and $\mathcal{L}(\boldsymbol{\xi})$ is a linear differential operator acting on $\boldsymbol{\xi}$. This equation is a special form of wave equation (2.46) in Sect. 2.5.1 in the sense that no motions are assumed in the unperturbed state, but is a generalization in the sense that nonlinear term $\boldsymbol{N}$ has been added. A quasi-nonlinear expression for $\boldsymbol{N}$ has been given in Appendix A. (In Appendix A the quasi-nonlinear term is written as $\boldsymbol{C}$, not $\boldsymbol{N}$.)

In the present problem, excitation of oscillations by turbulence is considered. Hence, $\boldsymbol{\xi}$ on the right-hand side of Eq. (5.49) is the displacement vector of wave motions and $\boldsymbol{N}$ is the source term resulting from turbulence motions, $\boldsymbol{\xi}_T$. Since the turbulence is assumed to be subsonic, a quasi-nonlinear expression for $\boldsymbol{N}$ is sufficient, and its k-component is given by (Appendix A)

$$\begin{aligned} N_k = & \frac{\partial}{\partial r_j}\left(p_0 \frac{\partial \xi_{T,i}}{\partial r_k}\frac{\partial \xi_{T,j}}{\partial r_i}\right) - \frac{\partial}{\partial r_j}\left[(\Gamma_1 - 1)p_0 \frac{\partial \xi_{T,j}}{\partial r_k}\frac{\partial \xi_{T,i}}{\partial r_i}\right] \\ & -\frac{1}{2}\frac{\partial}{\partial r_k}\left[(\Gamma_1 - 1)p_0 \frac{\partial \xi_{T,i}}{\partial r_j}\frac{\partial \xi_{T,j}}{\partial r_i}\right] - \frac{1}{2}\frac{\partial}{\partial r_k}\left[(\Gamma_1 - 1)^2 p_0 \frac{\partial \xi_{T,i}}{\partial r_i}\frac{\partial \xi_{T,j}}{\partial r_j}\right], \end{aligned} \tag{5.50}$$

where the terms with the same subscript twice in them should be added over three components of the Cartesian coordinates. In the simplest case where the adiabatic exponent, Γ_1, is unity, we have

$$N_k = \frac{\partial}{\partial r_j}\left(p_0 \frac{\partial \xi_{Ti}}{\partial r_k}\frac{\partial \xi_{Tj}}{\partial r_i}\right). \tag{5.51}$$

This special case corresponds to Lighthill's case, if c_s is taken to be constant [see the set of Eqs. (5.31) and (5.32)].

If relevant boundary conditions are imposed, normal mode oscillations of the linear part of Eq. (5.49) are a summation of eigenfunctions, $\boldsymbol{\xi}_q(\boldsymbol{r})$ $(q = 1, 2, 3...)$, satisfying

$$\rho_0 \omega_q^2 \boldsymbol{\xi}_q = \mathcal{L}(\boldsymbol{\xi}_q), \tag{5.52}$$

where ω_q is the eigenfrequency corresponding to eigenfunction $\boldsymbol{\xi}_q$,[2] i.e.,

$$\boldsymbol{\xi}_q(\boldsymbol{r}, t) = \Re \exp(i\omega_q t)\tilde{\boldsymbol{\xi}}_q(\boldsymbol{r}), \tag{5.53}$$

where $\Re$ represents the real part.

Since the operator $\mathcal{L}$ is Hermitian (self-adjoint) (see Sect. 2.5), we have the orthogonal relation [see Eq. (2.62) in Sect. 2.5]:

$$\int \rho_0 \tilde{\boldsymbol{\xi}}_q^* \tilde{\boldsymbol{\xi}}_{q'} d^3 r = 0 \quad \text{for} \quad q \neq q', \tag{5.54}$$

where $*$ represents the complex conjugate and the integration is performed over the whole volume.

Since the eigenfunctions, $\boldsymbol{\xi}_q$, form an orthogonal set (assuming further that they form a complete set), we expand the displacement vector, $\boldsymbol{\xi}_{\mathrm{T}}(\boldsymbol{r}, t)$, resulting from turbulence as

$$\boldsymbol{\xi}_{\mathrm{T}}(\boldsymbol{r}, t) = \Re \sum_q A_q(t)\tilde{\boldsymbol{\xi}}_q(\boldsymbol{r})\exp(i\omega_q t). \tag{5.55}$$

The equation that governs the time dependence of $A_q(t)$ is derived by substituting Eq. (5.55) into Eq. (5.49) and using the orthogonal relation (5.54). The result is then

$$\frac{d^2 A_q(t)}{dt^2} + 2i\omega_q \frac{dA_q(t)}{dt} = \exp(-i\omega_q t)\int^M \tilde{\boldsymbol{\xi}}_q^*(\boldsymbol{r}') \cdot \boldsymbol{N}(\boldsymbol{r}', t) dm', \tag{5.56}$$

[2]The subscript q should not be confused with a coordinate component of a vector.

where $dm \equiv \rho_0 d^3 r$ and the integration is performed over the whole volume. The first term on the left-hand side of Eq. (5.56) is usually small because of weak growth (or damping), and can be neglected compared with the second term.

Goldreich and Keeley (1977) model the damping due to turbulent viscosity and non-adiabatic processes by adding $2i\omega_q\gamma_q A_q$ to the left-hand side of Eq. (5.56), where γ_q is the damping rate of oscillations by turbulent viscosity and non-adiabatic processes. Then, from Eq. (5.56) we have

$$A_q(t) = -\frac{i}{2\omega_q}\int_{-\infty}^{t}\int_{0}^{M} dt'dm' \exp[\gamma_q(t'-t)-i\omega_q t']\tilde{\boldsymbol{\xi}}_q^*(\boldsymbol{r}')\cdot\boldsymbol{N}(\boldsymbol{r}',t'). \tag{5.57}$$

The expression for $A_q(t)$ cannot be directly evaluated, because $\boldsymbol{N}(\boldsymbol{r},t)$ is a fluctuating vector field and only its statistical properties are known. The expectation value of $|A_q(t)|^2$, denoted by $\langle|A_q(t)|^2\rangle$, is evaluated from

$$\langle|A_q(t)|^2\rangle = \frac{1}{4\omega_q^2}\int_{-\infty}^{t}\int_{-\infty}^{t}\int_{0}^{M}\int_{0}^{M} dt'dt''dm'dm''$$
$$\times\left[\exp[\gamma_q(t'+t''-2t)-i\omega_q(t'-t'')]\tilde{\boldsymbol{\xi}}_q^*(\boldsymbol{r}')\Big\langle\boldsymbol{N}(\boldsymbol{r}',t')\boldsymbol{N}(\boldsymbol{r}'',t'')\Big\rangle\tilde{\boldsymbol{\xi}}_q(\boldsymbol{r}'')\right]. \tag{5.58}$$

To proceed from this point, we require information about $\langle\boldsymbol{N}(\boldsymbol{r}',t')\boldsymbol{N}(\boldsymbol{r}'',t'')\rangle$. This is a difficult problem, since time-dependent structure of turbulence is required. Although it is a difficult problem, its specification is important, because it determines whether oscillations are excited or not. As the first step, Goldreich and Keeley (1977) considered a case based on the Kolmogorov picture of isotropic turbulence. For details, we recommend to see their original papers (Goldreich and Keeley 1977 and Goldreich and Kumar 1990).

References

Ando, H., Osaki, Y.: Publ. Astron. Soc. Jpn. **29**, 221 (1976)
Biermann, L.: Zeitschrift fur Astrophysik **28**, 304 (1951)
Böhm-Vitense, E.: Zeitschrift fur Astrophysik **46**, 108 (1958)
Canuto, V.M., Mazzitelli, I.: Astrophys. J. **370**, 295 (1991)
Goldreich, P., Keeley, D.A.: Astrophys. J. **212**, 243 (1977)
Goldreich, P., Kumar, P.: Astrophys. J. **363**, 694 (1990)
Kato, S.: Publ. Astron. Soc. Jpn. **18**, 201 (1966)
Kato, S.: Publ. Astron. Soc. Jpn. **20**, 59 (1968)
Kulsrud, R.M.: Astrophys. J. **121**, 461 (1955)
Leighton, R.B., Noyes, R.W., Simon G.W.: Astrophys. J. **135**, 474 (1962)
Lighthill, M.J.: Proc. Roy. Soc. (London) **A 211**, 564 (1952)
Pasetto, S., Chiosi, C., Cropper, M., Grebel, E.K.: Mon. Not. R. Astron. Soc. **445**, 3592 (2014)

Pellew, A., Southwell, R.V.: Proc. Roy. Soc. A. **176**, 312 (1940)
Proudman, L.: Proc. Roy. Soc. (London) **A214**, 119 (1952)
Samadi, R. (ed.): The Pulsation of the Sun and the Stars. Lecture Notes in Physics, vol. 832 (2011)
Samadi, R., Goupil, M.-J.: Astron. Astrophys. **370**, 136 (2001)
Spiegel, E.A., Veronis, G.: Astrophys. J. **131**, 442 (1960)
Unno, W.: Trans. IAU **12B**, 555 (1966)
Unsöld, A.: Physik der Sternatmosphären. Springer, Berlin, VIII Kapitel (1955)
Vitense, E.: Zeit. Astrophys. **32**, 135 (1953)

Chapter 6
Dynamical Instability and Dynamical Excitation of Oscillations

Astrophysical systems have many kinds of hydrodynamic and hydromagnetic instabilities, which are important in understanding their equilibrium configurations and activities. In this and subsequent two chapters we focus our attention to hydrodynamic instabilities. Hydromagnetic instabilities are discussed in Part II. Hydrodynamical instabilities are further classified into two classes. One is the instabilities which occur in the framework of non-dissipative processes, and is called *dynamical instabilities*. Instabilities of the other class are those which are realized by presence of dissipative processes. In this chapter we treat the former dynamical instabilities in the framework of linear analyses. Instabilities due to dissipative processes are considered in Chaps. 7 and 8.

6.1 Introductory Remarks

Astrophysical objects are not simple homogeneous systems. They have (1) density stratification, (2) rotation, (3) self-gravity, (4) magnetic fields, and (5) radiation fields. Related to each of them or combination of them, we have many instabilities.

For example, related to (1) we have the *Rayleigh–Taylor instability* (in addition to convective instability). Related to (2), the *Rayleigh criterion* for rotation law is known. These instabilities are described in Sect. 6.2. The *Solberg-Høiland criterion* for instability which is related to rotation will be argued in Sect. 6.3. The *gravitational instability* related to (3) will be described in Sect. 6.4. Arguments on instabilities related to (4) and (5) are made in Parts II and III, respectively.

In addition, there are some kinds of dynamical instabilities which can be interpreted in terms of concepts of resonant interactions among perturbations (waves). As these types of dynamical instabilities, we have the *Kelvin–Helmholtz instability*, *corotation instability*, and *wave-wave resonant instability* in deformed systems. They are described in Sects. 6.5, 6.6, and 6.7, in turn.

© Springer Nature Singapore Pte Ltd. 2020, corrected publication 2023
S. Kato, J. Fukue, *Fundamentals of Astrophysical Fluid Dynamics*,
Astronomy and Astrophysics Library,
https://doi.org/10.1007/978-981-15-4174-2_6

Finally we mention the presence of powerful and general methods for studying dynamical stability of gaseous systems with no internal motion. They are the variational method and the energy principle. Even when the systems have rotation, however, these methods are still applicable, if perturbations are restricted to axisymmetric ones. These general methods are described in Chap. 15, because these methods of stability analyses are applicable even in magnetized medium.

6.2 Rayleigh–Taylor Instability and Rayleigh Criterion

In this section we argue two instability criteria. The first one is the *Rayleigh–Taylor instability* criterion in stratified media, and the second is the *Rayleigh criterion* concerning rotation law.

6.2.1 Rayleigh–Taylor Instability

This instability is closely related to convective instability. To remember the convective instability described in Sect. 5.1, let us consider again a plane-parallel atmosphere stratified in the z-direction under the subject of a constant gravitational acceleration, $\boldsymbol{g}$, acting in the downward direction. That is, the density and pressure decrease upwards. In the plane perpendicular to the z-axis the medium (atmosphere) is uniform.

In Sect. 5.1 we found that even in such a stratified atmosphere as density and temperature decrease upwards, the atmosphere is unstable against convective motion if the rate of upward density decrease in the medium is weaker than a critical value. This is because in such a case an adiabatically rising element has a lower density compared with that of the surrounding gas since the element adiabatically expands during the rise. Then, the element rises further upwards by buoyancy. This is the convective instability.

The above arguments show that in a medium where the density increases upwards, the medium is convectively unstable even if the medium is incompressible (no density decrease in a rising element), because the rising element has a lower density compared with the surrounding one. The presence of this instability will be intuitively anticipated, since a heavier fluid is present on a lighter fluid. For example, water is on oil. This instability is called the *Rayleigh–Taylor instability*, and is shown schematically in Fig. 6.1.

As a limiting case, we consider the case where two incompressible gases with densities ρ_2 and ρ_1 contact on the $z = 0$ plane. The upper medium ($z > 0$) has ρ_2, and the lower medium ($z < 0$) has ρ_1. Under the boundary condition that at the infinity of z ($= \pm\infty$) there is no velocity in the z-direction, we have the growth rate

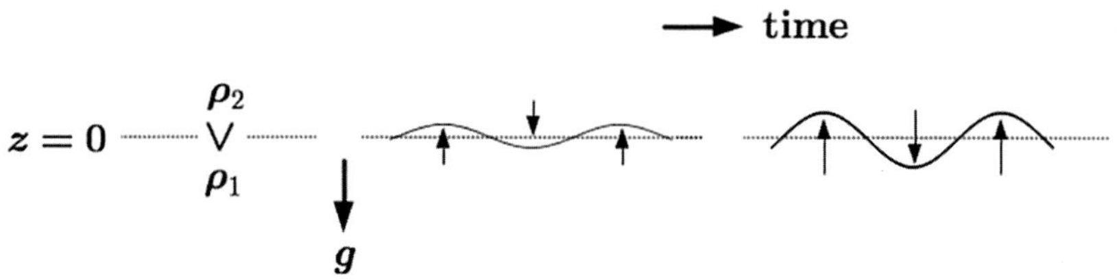

Fig. 6.1 Schematic picture showing the Rayleigh–Taylor instability and its time evolution. Two incompressible fluids contact on the surface at the height of $z = 0$. The fluid of the upper layer is assumed to have a higher density than that of the lower one, i.e., $\rho_2 > \rho_1$

n in the form:

$$n^2 = \frac{\rho_2 - \rho_1}{\rho_2 + \rho_1} gk, \tag{6.1}$$

where k is the wavenumber of perturbations in the horizontal direction (e.g., Chandrasekhar 1961). This equation shows that the medium is unstable if $\rho_2 > \rho_1$.

The growth rate, n, becomes larger as the wavelength becomes shorter. The fact that the growth rate infinitely increases as $n \propto k^{1/2}$ with increase of k is due to the fact that density is discontinuous. In the case where the density increases upwards smoothly with the scale height β, the growth rate tends to a finite value $(\beta g)^{1/2}$. This situation is the same as in the case of convection (see Sect. 5.1).

The Rayleigh–Taylor instability can occur in various situations and grows to highly nonlinear stages. For example, in the case of birth of early type stars the gases around the stars are ionized and swept away by strong radiation (Strömgren sphere). Then, high density regions are formed outside the sphere. Observations sometimes show a structure showing a drop of high density gases from the boundary of the sphere toward inside the spherical ionized region. This is called *globules*. One of the possible origins of such structure will be the Rayleigh-Taylor instability. One of another examples occurs in the case of supernova explosions. By explosions, shock fronts propagate outwards inside the stars. Numerical simulations show turbulent structure behind the shock fronts. The outward propagation of shock front is accelerated by passing through lower density region. The effective gravitational acceleration can be considered to be directed outwards. Hence, a turbulent structure is developed behind the shock by the Rayleigh–Taylor instability. Developments of the Rayleigh–Taylor instability to highly nonlinear stages have been examined by many authors in numerical simulations for various situations.

The Rayleigh–Taylor instability (including convective instability) is one of the elementary processes involved in magnetic buoyancy and Parker instability (see Chap. 16).

6.2.2 Rayleigh Criterion

The *Rayleigh criterion* is one of the basic criteria to judge whether a rotating system is dynamically stable or not. Let us consider a cylindrically rotating gaseous system. Cylindrical coordinates (r, φ, z) are introduced, where the z-axis is the axis of rotation of the system, and r is the distance from the rotating axis. The angular velocity of rotation of the system is taken to be $\Omega(r)$.

We consider a fluid element (cylindrical shell) which is at r, and displace it outward in the radial direction to $r + \Delta r$ under conservation of specific angular momentum $\ell\,(= r^2\Omega)$. Then, the centrifugal force acting on this fluid element (per unit mass) after the displacement at $r + \Delta r$ is $\ell^2/(r + \Delta r)^3$, which is $(\ell^2/r^3)(1-3\Delta r/r)$ for $\Delta r/r \ll 1$. Compared with this, the centrifugal force acting on a fluid element at $r + \Delta r$ in the unperturbed state is $[\ell(r + \Delta r)]^2/(r + \Delta r)^3$, which is $(\ell^2/r^3)[1 + (2rd\ell/\ell dr - 3)(\Delta r/r)]$.

Comparison of magnitude of the above two centrifugal forces shows that if $d\ell/dr < 0$, the displaced element moves further outward and the system is unstable. That is, $d\ell/dr > 0$ is a necessary condition for stability of the systems. This is called the *Rayleigh criterion*. If $d\ell/dr > 0$, the displaced element is returned to the original position and makes an oscillation around the original position. This oscillation is found to be a harmonic oscillation, and this oscillation called *epicyclic oscillation*. Its frequency is *(radial) epicyclic frequency*. Radial distribution of this epicyclic oscillation has important rolls in disks, especially in relativistic disks (see Sect. 9.4).

6.3 Solberg-Høiland Criterion

In the previous section the Rayleigh–Taylor instability in media with a density discontinuity and the Rayleigh criterion in differential rotating ones are qualitatively and independently argued. Here, we consider more general situations with more rigorous ways. That is, we consider axisymmetric systems where both differential rotation and density stratification are present, although perturbations are still restricted to axisymmetric ones.

We introduce again cylindrical coordinates, (r, φ, z), whose z-axis is the axisymmetric axis of the system. The rotation is not always a function of r alone. It can depend on z, i.e., $\Omega = \Omega(r, z)$. The unperturbed dynamically equilibrium state is then described by

$$-\frac{1}{\rho_0}\frac{\partial p_0}{\partial r} - \frac{\partial \psi_0}{\partial r} + r\Omega^2(r, z) = 0, \tag{6.2}$$

$$-\frac{1}{\rho_0}\frac{\partial p_0}{\partial z} - \frac{\partial \psi_0}{\partial z} = 0, \tag{6.3}$$

where ψ_0 is the gravitational potential, and the quantities of the unperturbed state are shown by attaching subscript 0.

Elimination of $\psi_0(r)$ from the above two equations gives

$$r\frac{\partial \Omega^2}{\partial z} = \frac{1}{\rho_0^2}\left(\frac{\partial \rho_0}{\partial r}\frac{\partial p_0}{\partial z} - \frac{\partial \rho_0}{\partial z}\frac{\partial p_0}{\partial r}\right). \tag{6.4}$$

This equation shows that in a barotropic gas where pressure is a function of ρ alone as $p = p(\rho)$, the rotation in steady state must be cylindrical, i.e., $\partial\Omega/\partial z = 0$. Our main interest here is the cases where the gases are not *barotropic*, i.e., iso-pressure and iso-density surfaces do not coincide. In general, the state where the above two surfaces do not coincide is called *baroclinic*.

6.3.1 Derivation of Dispersion Relation

Small amplitude perturbations are superposed on the unperturbed baroclinic systems. In order to avoid unnecessary complications, we filter out the acoustic motions as in the case of studies of convective instabilities in Chap. 5. The filtering can be made by introducing the Boussinesq approximation (see Chap. 5, and Spiegel and Veronis 1960).

We consider axially symmetric local perturbations. In order to filter out the acoustic waves, the equation of continuity is approximated as (Boussinesq approximatrion)

$$\frac{\partial v_r}{\partial r} + \frac{\partial v_z}{\partial z} = 0. \tag{6.5}$$

The r-, φ-, z-components of equation of motions are, respectively,

$$\frac{\partial v_r}{\partial t} - 2\Omega v_\varphi = -\frac{1}{\rho_0}\frac{\partial p_1}{\partial r} + \frac{\rho_1}{\rho_0^2}\frac{\partial p_0}{\partial r}, \tag{6.6}$$

$$\frac{\partial v_\varphi}{\partial t} + \frac{\kappa^2}{2\Omega}v_r + r\frac{\partial \Omega}{\partial z}v_z = 0, \tag{6.7}$$

$$\frac{\partial v_z}{\partial t} = -\frac{1}{\rho_0}\frac{\partial p_1}{\partial z} + \frac{\rho_1}{\rho_0^2}\frac{\partial p_0}{\partial z}, \tag{6.8}$$

where κ is the epicyclic frequency defined by

$$\kappa^2 = 2\Omega\left(2\Omega + r\frac{\partial \Omega}{\partial r}\right). \tag{6.9}$$

It should be noted that although we are considering local perturbations whose wavelengths are shorter than the characteristic lengths in the unperturbed state, terms of $\rho_1/\rho_0^2\nabla p_0$ are retained in the above equations. This is because we take into account the effects of the buoyancy force.[1] Furthermore, corresponding to filtering of acoustic oscillations [see Eq. (6.5)], we approximate the energy equation for adiabatic perturbations as

$$\frac{1}{\rho_0}\frac{\partial \rho_1}{\partial t} = v_r\frac{\partial}{\partial r}\ln\frac{p_0^{1/\gamma}}{\rho_0} + v_z\frac{\partial}{\partial z}\ln\frac{p_0^{1/\gamma}}{\rho_0}, \tag{6.10}$$

where γ is the ratio of specific heats. In the above energy equation, the term of time variation of p_1 has been neglected, because slow motions such as convection which occurs under the pressure balance with the surrounding media are considered here.[2]

Here we consider local perturbations whose radial and vertical wavelengths are short. Then, all perturbed quantities can be expressed in the form of $\exp[i(\omega t - k_r r - k_z z)]$ and the dispersion relation is derived. After a lengthy but straightforward calculations we have

$$\omega^2\left(k_r^2 + k_z^2\right) + ak_z^2 + bk_rk_z + ck_r^2 = 0, \tag{6.11}$$

where

$$a = -\frac{1}{r^3}\frac{\partial \ell^2}{\partial r} + \frac{1}{\gamma\rho_0}\frac{\partial p_0}{\partial r}\frac{\partial}{\partial r}\ln\frac{p_0}{\rho_0^\gamma}, \tag{6.12}$$

$$b = \frac{1}{r^3}\frac{\partial \ell^2}{\partial z} - \frac{1}{\gamma\rho_0}\frac{\partial p_0}{\partial r}\frac{\partial}{\partial z}\ln\frac{p_0}{\rho_0^\gamma} - \frac{1}{\gamma\rho_0}\frac{\partial p_0}{\partial z}\frac{\partial}{\partial r}\ln\frac{p_0}{\rho_0^\gamma}, \tag{6.13}$$

$$c = \frac{1}{\gamma\rho_0}\frac{\partial p_0}{\partial z}\frac{\partial}{\partial z}\ln\frac{p_0}{\rho_0^\gamma}, \tag{6.14}$$

where ℓ is specific angular momentum given by $\ell = r^2\Omega$.

Dispersion relation (6.11) shows that the condition for the system to be stable ($\omega^2 > 0$) is that $ax^2 + bx + c$ is negative for all real x. This requires that $a < 0$ and the discriminant, $b^2 - 4ac$, is negative. The latter condition of $b^2 - 4ac < 0$ can be reduced to a simple form if relation (6.4) is considered, which is

$$-\frac{1}{\rho_0}\frac{\partial p_0}{\partial z}\left[\frac{\partial \ell^2}{\partial r}\frac{\partial}{\partial z}\ln\frac{p_0}{\rho_0^\gamma} - \frac{\partial \ell^2}{\partial z}\frac{\partial}{\partial r}\ln\frac{p_0}{\rho_0^\gamma}\right] > 0. \tag{6.15}$$

---

[1] This procedure is a part of the Boussinesq approximation.

[2] This is also a part of Boussinesq approximation.

The Brunt–Väisälä frequency[3] in the radial direction, N_r, and the epicyclic frequency in the radial direction, κ, are defined, respectively, by

$$N_r^2 = -\frac{1}{\gamma\rho_0}\frac{\partial p_0}{\partial r}\frac{\partial}{\partial r}\ln\frac{p_0}{\rho_0^\gamma}, \quad \text{and} \quad \kappa^2 = \frac{1}{r^3}\frac{\partial \ell^2}{\partial r}. \tag{6.16}$$

Then, the condition of $a < 0$ is written as

$$N_r^2 + \kappa^2 > 0. \tag{6.17}$$

In summary, the set of two inequalities (6.15) and (6.17) is the condition of stability of the system. This is the *Solberg-Høiland criterion* for dynamical stability. If one of them is violated, the media is unstable.

The condition (6.15) has a geometrical meaning. Let us consider the case of $\partial p_0/\rho_0\partial z < 0$. Then, the condition implies that the inner product of vector $(\partial\ell/\partial r$, 0, $\partial\ell/\partial z)$ and vector $(\partial s/\partial z$, 0, $-\partial s/\partial r)$ is positive, where $s \equiv \ln(p_0/\rho_0^\gamma)$ is entropy. It is noted that in the barotropic gas, the scalar product is zero, since the isobaric, isodensity, and isentropic surfaces coincide. This situation is shown in Fig. 6.2.

If magnetic fields are present, the stability criterion is changed from the Solberg-Høiland criterion, even if the magnetic fields are weak (Balbus 1995). This will be described in Sect. 16.4.

---

[3]In Chap. 4 we have considered the Brunt–Väisälä frequency. In that chapter we have restricted our attention only to the case of a plane-parallel configuration. It is necessary, however, to introduce a more general expression for the Brunt–Väisälä frequency in order to consider cases of more general configurations. Let us consider an axially symmetric rotating system, and impose axially symmetric perturbations with velocity $\boldsymbol{v}$. Then, the energy equation for adiabatic perturbations is written as

$$i\omega\left(p_1 - c_s^2\rho_1\right) = \gamma p_0(\boldsymbol{v}\cdot\boldsymbol{A}),$$

where c_s is the sound speed defined by $c_s^2 = \gamma p_0/\rho_0$ and $\boldsymbol{A}$ is the Schwarzschild discriminant vector defined by

$$\boldsymbol{A} \equiv \nabla\ln\rho_0 - \frac{1}{\gamma}\nabla\ln p_0 = \nabla\left(\ln\frac{\rho_0}{p_0^{1/\gamma}}\right).$$

Then, Brunt–Väisälä frequency in the r- and z-direction is given by

$$N_r^2 = -(g_{\rm eff})_r A_r, \quad \text{and} \quad N_z^2 = -(g_{\rm eff})_z A_z,$$

where $(g_{\rm eff})_r$ and $(g_{\rm eff})_z$ are absolute values of effective gravitational accelerations in r and z directions:

$$(g_{\rm eff})_r = g_r - r\Omega^2 = -\frac{1}{\rho_0}\frac{\partial p_0}{\partial r}, \quad \text{and} \quad (g_{\rm eff})_z = g_z = -\frac{1}{\rho_0}\frac{\partial p_0}{\partial z}.$$

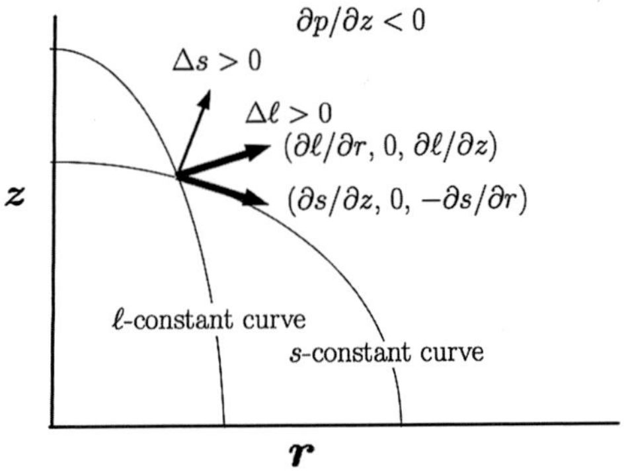

Fig. 6.2 An example of stable configuration against the Solberg-Høiland instability. The case where $-(1/\rho_0)\partial p_0/\partial z > 0$ is shown. The direction of increase of angular momentum, i.e., $(\partial \ell/\partial r, 0, \partial \ell/\partial z)$ and the direction perpendicular to entropy [$s \equiv \ln(p_0/\rho_o^\gamma)$] increase, i.e., $(\partial s/\partial z, 0, -\partial s/\partial r)$, are shown by thick arrows. The inner product of these two vectors is positive in the case shown here, and the stability condition (6.15) is satisfied

6.4 Gravitational Instability

One of the important characteristics of astrophysical objects is of importance of self-gravity, as often mentioned before. Different from cases of electrostatic forces, the gravitational force cannot be screened due to absence of forces acting in the opposite direction. This suggests that the effects of self-gravity become prominent as we consider large scale phenomena, and the presence of *gravitational instability* is easily speculated. Here we consider the presence of gravitational instability in the simplest case.

Let us consider an infinitely extended isothermal and homogeneous systems. It is said that consideration of such infinitely extended homogeneous systems as the unperturbed state is irrelevant, because such systems cannot be equilibrium. For simplicity, however, we consider such unperturbed systems, following the original consideration by Jeans (1902). Small amplitude perturbations are superposed. The velocity associated with the perturbations is $\boldsymbol{v}$, and temperature, pressure, and density perturbations are denoted, respectively, by T_1, p_1, ρ_1. The equation of continuity, the equation of motion, the Poisson equation, and the adiabatic relation are, respectively,

$$\frac{\partial \rho_1}{\partial t} + \rho_0 \mathrm{div}\boldsymbol{v} = 0, \tag{6.18}$$

$$\frac{\partial \boldsymbol{v}}{\partial t} = -\frac{1}{\rho}\mathrm{grad}\,p_1 - \mathrm{grad}\,\psi_1, \tag{6.19}$$

$$\nabla^2 \psi_1 = 4\pi G \rho_1, \tag{6.20}$$

$$p_1 = c_s^2 \rho_1, \tag{6.21}$$

where ψ_1 is the perturbation of gravitational potential and related to density perturbation ρ_1 through the Poisson equation (6.20), since we are now considering self-gravitating systems. By eliminating perturbed quantities other than ρ_1 from the above equations, we derive an equation with respect to ρ_1:

$$\left(\frac{\partial^2}{\partial t^2} - c_s\nabla^2 - 4\pi G\rho\right)\rho_1 = 0. \tag{6.22}$$

If $G = 0$ is adopted, this equation is obviously reduced to the equation describing acoustic waves in a homogeneous medium. In other words, the above equation shows how the acoustic waves in the homogeneous medium are modified by the effect of self-gravity. Now, we consider a plane wave as $\rho_1 \propto \exp[i(\omega t - kx)]$. Then, substituting this into Eq. (6.22), we have a dispersion relation:

$$\omega^2 = c_s^2 k^2 - 4\pi G\rho. \tag{6.23}$$

This equation shows that as the wavelength of perturbations increases, the frequency of acoustic oscillations decreases. This result is easily understood. In acoustic waves density variations are associated. In the phase where the gases are compressed, the pressure increases and this is the restoring force for oscillations. In this phase of compression, however, the density increase has another effect. That is, the density increase brings about increase of the effects of self-gravity. This acts in the direction opposite to the pressure restoring force, and the relative importance of the effect increases as wavelength of perturbations increases. Beyond a certain critical wavelength, the net restoring force becomes negative and perturbations grow.

Instead of a plane wave, we consider here a spherical perturbation as a more realistic one. The perturbation is taken to be proportional to $\rho_1 \propto \exp[i(\omega t - kr)]/r$ (r being the radial coordinate of spherical coordinates) and this is introduced to Eq. (6.22). We have then the same dispersion relation as Eq. (6.23). The critical wavenumber for which perturbations become gravitationally unstable is given by

$$k_c = \frac{(4\pi G\rho)^{1/2}}{c_s}. \tag{6.24}$$

By taking π/k_c as the critical radius, λ_c, for which the gravitational instability sets in, we have

$$\lambda_c = \frac{\pi c_s}{(4\pi G\rho)^{1/2}}. \tag{6.25}$$

The gaseous mass in this critical radius, say M_c, is

$$M_c = \frac{4}{3}\pi\lambda_c^3\rho = \frac{\pi^{5/2}}{6}\rho\left(\frac{c_s^2}{G\rho}\right)^{3/2}. \tag{6.26}$$

Table 6.1 Jeans radius and Jeans mass for some objects

Objects	Number density $n[\mathrm{cm}^{-3}]$	Temperature $T[K]$	Jeans radius $\lambda_c[\mathrm{pc}]$	Jeans mass $M_c[M_\odot]$
Molecular cloud	10^4	10^2	1	3×10^3
HI region	10^2	10^2	10	3×10^4
Gas-radiation Decoupling time in the universe	10^4	3×10^3	10	3×10^5

This gravitational instability is also called *Jeans instability*. The critical radius is called *Jeans radius*, and the critical mass *Jeans mass*.

The gravitational instability is one of the most important mechanisms for formation of stars. Usually, stars are formed in low-temperature, high density interstellar gases (i.e., molecular clouds) by gravitational instability. Galaxies and cluster of galaxies are also considered to be formed by the gravitational instability from density fluctuation in the Universe. In Table 6.1 the Jeans radius and the Jeans mass are listed for typical physical state.

Finally, fragmentation of gas clouds by gravitational instability is mentioned, as one of the applications of the instability. We examine how the Jeans mass given by Eq. (6.26) is changed by adiabatic contraction of gases. Since $c_s^2 \propto T$, M_c changes as

$$M_c \propto \frac{T^{3/2}}{\rho^{1/2}}. \tag{6.27}$$

Now, if the contraction of gases is adiabatic with the ratio of specific heats, γ, we have from Eq. (6.27)

$$M_c \propto \rho^{(3\gamma-4)/2}, \tag{6.28}$$

because $T \propto \rho^{\gamma-1}$.

Equation (6.28) shows that in the case of $\gamma < 4/3$, the Jeans mass decreases with contraction of the cloud. This means that in a contracting gas fragmentation of gas cloud can occur if $\gamma < 4/3$. The fact that the critical value of γ whether fragmentation occurs or not is $\gamma = 4/3$ is related to the fact that $\gamma = 4/3$ is the stability criterion of homologous contraction of gases (Sect. 15.1). As an example, let us consider contraction of low-density gas. At the initial stage gas density is low and optically thin. Then, radiative cooling is efficient and the effective value of γ is close to unity. As contraction proceeds, fragmentation of the cloud occurs, but at a certain stage of contraction, the cloud becomes optically thick. As a result γ becomes larger than 4/3, and fragmentation is stopped.

6.5 Kelvin–Helmholtz Instability: Resonant Instability I

In Sects. 6.1–6.4 we have presented instabilities whose origins are simple in the sense that their causes are related to the fact that net restoring forces of oscillations become negative. In this and the subsequent two sections (Sects. 6.5–6.7) we present another type of instabilities. They are interpreted by resonant interactions of oscillations.

The *Kelvin–Helmholtz instability* is presented here (see Cairns 1979). Cartesian coordinates (x, y, z) are introduced, and the z-axis is taken in the direction opposite to a constant gravitational acceleration, $\boldsymbol{g} = (0, 0, -g)$. On the plane of $z = 0$ two incompressible fluids with different densities, ρ_+ and ρ_-, are touched with a constant relative motion, $\boldsymbol{U}$, in the direction perpendicular to the z-axis. We take the direction of the flow is the y-direction. The densities in $z > 0$ and $z < 0$ are, respectively, ρ_+ and ρ_- (see Fig. 6.3). The system is unstable if $\rho_+ > \rho_-$ by the Rayleigh–Taylor instability (Sect. 6.2). Our purpose here is to examine whether the system becomes unstable even when $\rho_+ < \rho_-$, if there is a relative motion between the two fluids.

To consider the essential part, the perturbations are taken to have infinite wavelength in the x-direction. That is, the displacement vector, $\boldsymbol{\xi}$, associated with perturbations is $(0, \xi_y, \xi_z)$ and the perturbation of velocity, $\boldsymbol{v}$, over $\boldsymbol{U}$ is denoted by $(0, v_y, v_z)$. Pressure perturbation is denoted by p_1. All perturbed quantities are taken to be proportional to $\exp[i(\omega t - ky)]$.

Since an incompressible fluid is treated, the equation of continuity describing small amplitude perturbations is

$$-ik\xi_y + \frac{\partial \xi_z}{\partial z} = 0, \tag{6.29}$$

and y- and z-components of equation of motion are, respectively,

$$-\rho(\omega - kU)^2\xi_y = ikp_1, \tag{6.30}$$

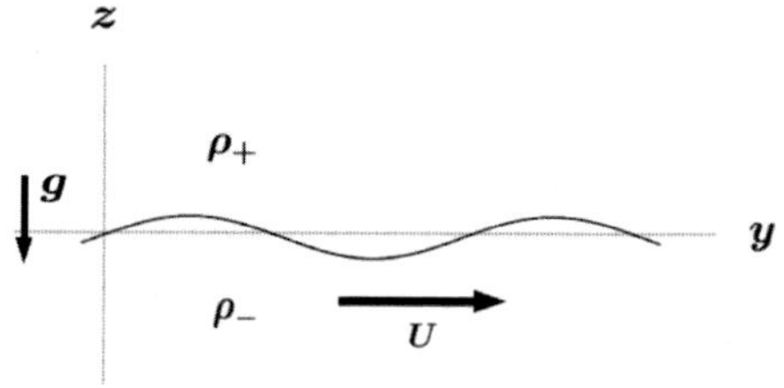

Fig. 6.3 Schematic diagram showing the Kelvin–Helmholtz instability. Two incompressible fluids with different densities (ρ_+ and ρ_-) are touched on the surface of $z = 0$. There is a relative motion between the two fluids, which is directed in the y-axis. On such configurations, a wavy perturbation propagating in the y-direction is considered

and

$$-\rho(\omega - kU)^2\xi_z = -\frac{\partial p_1}{\partial z}. \tag{6.31}$$

It is noted that $\boldsymbol{\xi}$ and $\boldsymbol{v}$ are related by

$$i(\omega - kU)\xi_y = v_y, \quad i(\omega - kU)\xi_z = v_z. \tag{6.32}$$

In the above Eqs. (6.30)–(6.32), we take $\rho = \rho_+$ and $U = 0$, when we treat the upper zone, while $\rho = \rho_-$ in the lower zone. This rule will be used hereafter without any mention.

Elimination of ξ_y and p_1 from Eqs. (6.29)–(6.31) gives an equation with respect to ξ_z, which is

$$\frac{\partial^2 \xi_z}{\partial z^2} - k^2\xi_z = 0. \tag{6.33}$$

Hence, by adopting the boundary conditions that $\xi_z = 0$ at $z = \pm\infty$, we have

$$\xi_z = A_\pm \exp[i(\omega t - ky)]\exp(\mp kz), \tag{6.34}$$

where $A_\pm$ are constants representing the amplitude of perturbations, and the upper and lower signs of $A_\pm$ and $\mp kz$ are for the upper and lower zones, respectively. Without loss of generality we take $k > 0$. By using ξ_z we can express other perturbed quantities:

$$v_y = \mp(\omega - kU)\xi_z, \quad v_z = i(\omega - kU)\xi_z, \tag{6.35}$$

$$p_1 = \mp\frac{\rho}{k}(\omega - kU)^2\xi_z, \quad \xi_y = \pm i\xi_z, \tag{6.36}$$

where the upper signs are for the upper zone and the lower ones for the lower zone, and for the upper zone $\rho = \rho_+, U = 0$ and for the lower zone $\rho = \rho_-$, as mentioned before.

Next, boundary conditions at $z = 0$ are considered. On the boundary of two zones, ξ_z and the Lagrangian change of pressure, δp, needs to be continuous. Since $\delta p = p_1 + \xi_z\partial p_0/\partial z = p_1 - \rho\xi_z g$, from Eq. (6.36) we have

$$\delta p = \frac{\rho}{k}\left[\mp(\omega - kU)^2 - kg\right]\xi_z, \tag{6.37}$$

where for the upper zone $\rho = \rho_+$, $U = 0$, and for the lower zone $\rho = \rho_-$. From the conditions that the above two boundary conditions lead to non-trivial A_+ and A_-, we have the dispersion relation:

$$(\rho_+ + \rho_-)\omega^2 - 2\rho_- kU\omega + \rho_- k^2U^2 - (\rho_- - \rho_+)gk = 0. \tag{6.38}$$

By solving this relation with respect to ω we have

$$\omega = \frac{\rho_-}{\rho_+ + \rho_-} kU \pm \left[\frac{C_0^2}{U^2} - \frac{\rho_+ \rho_-}{(\rho_+ + \rho_-)^2} \right]^{1/2} kU, \tag{6.39}$$

where

$$C_0^2 \equiv \frac{\rho_- - \rho_+}{\rho_+ + \rho_-} \frac{g}{k}. \tag{6.40}$$

It should be noted that the sign $\pm$ in Eq. (6.39) represents the presence of two ω's, and is not confused with the symbol for the upper and lower zones. Dispersion relation (6.39) shows if

$$U^2 > \frac{(\rho_+ + \rho_-)^2}{\rho_+ \rho_-} C_0^2 = \frac{\rho_-^2 - \rho_+^2}{\rho_+ \rho_-} \frac{g}{k}, \tag{6.41}$$

the flow is unstable, even if $\rho_+ < \rho_-$. This is the *Kelvin–Helmholtz instability*.

6.5.1 Wave Energy and Resonant Interaction

The Kelvin–Helmholtz instability can be interpreted as a result of interaction of two waves which have the same phase velocity but opposite signs of wave energy. To demonstrate this, wave energies are calculated here. In the upper zone the kinetic energy of perturbations per unit mass is $\rho_+(v_y^2 + v_z^2)/2$, while it is $\rho_-[(U + v_y)^2 + v_z^2 - U^2]/2$ in the lower zone. Thus, the kinetic energy of perturbations in unit column on the $x - y$ plane is

$$\frac{A^2}{2}\rho_+ \int_{A\cos(\omega t - ky)}^{\infty} \omega^2 \exp(-2kz)dz + \frac{A^2}{2}\rho_- \int_{-\infty}^{A\cos(\omega t - ky)} \Big[(\omega - kU)^2 \exp(2kz)$$

$$+ 2U(\omega - kU)\cos(\omega t - ky)\exp(kz) \Big] dz, \tag{6.42}$$

where $A_+ = A_- \equiv A$ has been used. By performing the integration, we have the kinetic energy in the form:

$$\frac{\rho_+}{4k}\omega^2 A^2 + \frac{\rho_-}{4k}\Big[(\omega - kU)^2 + 2kU(\omega - kU) \Big] A^2. \tag{6.43}$$

Next, the potential energy is calculated. In the lower layer the displacement of fluid element is

$$A\cos(\omega t - ky)\exp(kz). \tag{6.44}$$

The coordinate y changes with time as

$$y = y_0 + Ut + A\exp(kz)\sin(\omega t - ky). \tag{6.45}$$

Hence, by substituting this into Eq. (6.44) and taking the terms till the second order with respect to A, we have $A^2k\exp(2kz)/2$ after the time average. Multiplying ρ_-g to $A^2k\exp(2kz)/2$ and integrating it over z, we have the potential energy of perturbations in the lower layer, which is $A^2\rho_-g/4$. For the upper layer we can calculate in a similar way the potential energy. The sum is found to be

$$\frac{1}{4}(\rho_- - \rho_+)A^2g. \tag{6.46}$$

The sum of Eqs. (6.43) and (6.46) is the whole energy of perturbations per unit surface, which is denoted by W. By using dispersion relation (6.38) we can write W as

$$W = A^2\omega\left[\frac{\rho_+\omega}{2k} + \frac{\rho_-}{2k}(\omega - kU)\right]. \tag{6.47}$$

Furthermore, by using Eq. (6.39), we have

$$W = \pm A^2\frac{\rho_+ + \rho_-}{2}\omega U\left[\frac{C_0^2}{U^2} - \frac{\rho_+\rho_-}{(\rho_+ + \rho_-)^2}\right]^{1/2}. \tag{6.48}$$

This result shows the followings. In the case where the square of the flow velocity, U^2, is smaller than $[(\rho_+ + \rho_-)^2/\rho_+\rho_-]C_0^2$, we have two waves with different frequencies and different phase velocities [see Eq. (6.39)], for a given k. These waves have the different signs of wave energy, i.e., one has a positive energy and the other has a negative one [see Eq. (6.48)]. As U^2 increases, their frequencies approach, and finally become the same for $U^2 = [(\rho_+ + \rho_-)^2/\rho_+\rho_-]C_0^2$. The phase velocities are also the same. That is, two waves resonantly interact and one of them grows and the other damps. This situation is shown schematically in Fig. 6.4.

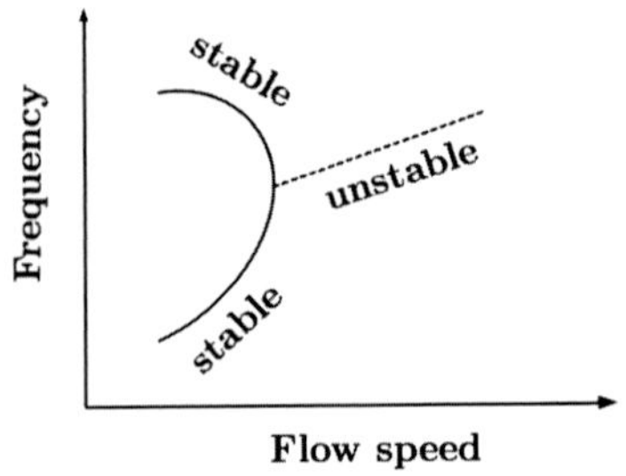

Fig. 6.4 Schematic diagram showing occurrence of instability by both frequency and phase velocity of two different waves coinciding

6.5.1.1 Wave Energy and Dispersion Relation

In the above arguments, the wave energy was derived directly from characteristics of wave motions. In some cases, wave energy and dispersion relation are closely related, and this is important in understanding instability (e.g., see the final part of Sect. 6.6.2). Following Cairns (1979), we derive here the relation between wave energy and dispersion relation in the present particular problem. To do so, we express the displacement of the boundary surface as

$$\xi_z = A(t)\exp[i(\omega t - ky)]. \tag{6.49}$$

Although ω and k are related so that the dispersion relation (6.38) is satisfied, the difference between Eqs. (6.49) and (6.34) is that the amplitude A is not a constant in Eq. (6.49), but depends weakly on time. That is, $A = 0$ at $t = -\infty$, but gradually increases with time and tends to $A = A_0$ at $t = \infty$ at $z = 0$. Due to this difference, δp is discontinuous at the boundary surface of $z = 0$ in the present case.

Let us consider the work done on the oscillations by this pressure discontinuity. The pressure difference at the upper side of the discontinuity is written as δp_+, while that at the lower side of the discontinuity is δp_-. The displacement velocity at the boundary is denoted by $\partial\xi_z/\partial t$. Then, on the oscillations in the region of $z > 0$, the rate of work done by this pressure discontinuity is δp_+ times the velocity at the boundary, $\partial\xi_z/\partial t$, on the unit surface. On the other hand, the oscillations in the region of $z < 0$ do work on the discontinuity on the rate of δp_- times $\partial\xi_z/\partial t$. Consequently, the net rate of work done on the oscillations by the pressure discontinuity on unit surface of the $x - y$ plane is $(\delta p_+ - \delta p_-)\partial\xi_z/\partial t$. The time integration of this till the time when the amplitude of the boundary surface becomes A_0 is the energy of the waves whose amplitude is A_0.

As a preparation for calculating the wave energy by the above procedures, we introduce D_+ and D_- defined by

$$D_+(\omega, k) \equiv \frac{\rho_+}{k}\left(-\omega^2 - gk\right), \quad D_-(\omega, k) \equiv \frac{\rho_-}{k}\left[(\omega - kU)^2 - gk\right]. \tag{6.50}$$

Then, the continuity of δp given by Eq. (6.37) is

$$D(\omega, k) \equiv D_+ - D_- = 0. \tag{6.51}$$

This is the dispersion relation. Since we are now considering the case where $A(t)$ increases gradually, we have at the upper side of the transition layer [see Eq. (6.37)]

$$\delta p_+ = D_+\left(\omega - i\frac{\partial}{\partial t}, k\right)A(t)\exp[i(\omega t - ky)], \tag{6.52}$$

where $\partial/\partial t$ is operated only on D_+. Similarly we have at the lower side

$$\delta p_- = D_-\left(\omega - i\frac{\partial}{\partial t}, k\right)A(t)\exp[i(\omega t - ky)]. \tag{6.53}$$

Since we can adopt the expansion:

$$D_\pm\left(\omega - i\frac{\partial}{\partial t}, k\right) = D_\pm(\omega, k) - i\frac{\partial D_\pm}{\partial \omega}\frac{\partial}{\partial t}, \tag{6.54}$$

we have

$$\delta p_+ - \delta p_- = -i\frac{\partial D}{\partial \omega}\frac{dA}{dt}\exp[i(\omega t - ky)]. \tag{6.55}$$

If Eq. (6.49) is considered, the work done by external force, dW/dt, is found to be

$$\frac{dW}{dt} = \left\langle \frac{\partial \xi_z}{\partial t}(\delta p_+ - \delta p_-)\right\rangle = -\frac{1}{4}\omega\frac{\partial D}{\partial \omega}\frac{|A|^2}{dt}. \tag{6.56}$$

Here, $\langle\ \rangle$ represents the average in phases. Therefore, the total energy necessary till A_0, i.e., W, is found to be

$$W = -\frac{1}{4}\omega\frac{\partial D}{\partial \omega}|A_0|^2. \tag{6.57}$$

Equations (6.57) and (6.47) are found to be the same, if the expressions for D_+ and D_- [Eqs. (6.50)] are used.

6.6 Corotation Resonance: Resonant Instability II

In Sect. 6.5, we have shown that the Kevin–Helmholtz instability can be interpreted as a result of interaction between two waves with positive and negative wave energies. There are many linear and nonlinear instability phenomena which can be classified in this category. One of them is the *corotation instability*.

Let us consider, for simplicity, an axisymmetric disk rotating differentially with angular velocity of rotation, $\Omega(r)$, where r is the distance from the rotation axis in cylindrical coordinates (r, φ, z) whose center is the disk center. On this disk non-axisymmetric perturbations which rotate with frequency ω are considered. The perturbations are taken to be proportional to $\exp[i(\omega t - m\varphi)]$. Here, m is the number of arms of perturbations in the azimuthal direction. The pattern of the perturbations rotates in the azimuthal direction with angular velocity of ω/m. Since the disk rotates differentially, the pattern angular velocity of the waves becomes equal to the angular velocity of disk rotation at a certain radius, i.e., at the radius where $\omega/m = \Omega$. This radius is called *corotation radius*.[4] At the corotation radius perturbations and unperturbed flows resonantly interact. This interaction is known to lead to overreflection or instability of perturbations.

In the fields of astrophysical fluid dynamics, the importance of corotation resonance was first recognized in studies of density wave theory developed by Lin and his group in 1960s to 1970s (e.g., Mark 1974, 1976a,b) as a mechanism for maintenance of density waves against their extinction as running waves (Toomre 1969). The original studies by Mark are made for collisionless stellar systems. Hydrodynamical version of the corotation instability in self-gravitating gaseous systems has been summarized by Lin and Lau (1979) by using the WKBJ method. Here, we present the essential parts of their arguments.

It is noted that in the case of nonself-gravitating disks (e.g., accretion disks around compact objects) the relation between the corotation radius and the wave propagation region is different from that of the self-gravitating disks. Hence, the case of nonself-gravitating disks is briefly mentioned in a separate subsection (Sect. 6.6.2).

6.6.1 Spiral Density Waves in Self-gravitating Disks

Self-gravitating gaseous disks are considered. For simplicity, the disks are assumed to be geometrically thin, and the vertically integrated or averaged quantities are treated. In Sect 4.5 we have considered local perturbations which have the form of

$$\exp[i(\omega t - kr - m\varphi)], \tag{6.58}$$

and derived the dispersion relation [see Eq. (4.95)]:

$$(\omega - m\Omega)^2 = \kappa^2 + k^2 c_s^2 - 2\pi G \Sigma_0 |k|, \tag{6.59}$$

where κ is the epicyclic frequency, and Σ_0 is the surface density of the gas.

In the present section, the radial dependence of perturbations is retained unspecified and the differential equation of $h_1 (\equiv \Pi_1/\Sigma_0)$ with respect to r is obtained, where $\Pi_1(r)$ is the vertical integration of p_1 and $\Sigma_0(r)$ is the surface density of the unperturbed disks. This differential equation for h_1 is reduced to the normal form using the standard transformation of h_1 to a variable, say u (a similar procedure is shown in Sect. 6.6.3). In the lowest order of approximations, the equation reduced

[4] In the fields of fluid mechanics and meteorology, the place where the phase velocity of perturbations becomes equal to the flow velocity of the unperturbed media is called *critical point*.

is (e.g., Lin and Lau 1979)

$$\frac{d^2u}{dr^2} + k^2 u = 0, \tag{6.60}$$

where

$$k^2 = \left(\frac{\kappa}{c_s}\right)^2 \left(Q^{-2} - 1 + \nu^2\right), \tag{6.61}$$

$\nu \equiv (\omega - m\Omega)/\kappa$, and Q is Toomre's Q parameter defined by Eq. (4.97). It is noticed that the set of Eqs. (6.60) and (6.61) leads to the same dispersion relationship as Eq. (6.59) if local perturbations are considered.[5]

Equation (6.60) is now solved in the region involving the corotation radius of $\nu = 0$. The propagation regions of oscillations are specified by $k^2(r) > 0$, which are, for a given frequency ω,

$$\omega > m\Omega + \kappa\left(1 - Q^{-2}\right)^{1/2} \quad \text{or} \quad \omega < m\Omega - \kappa\left(1 - Q^{-2}\right)^{1/2}. \tag{6.62}$$

In the followings we focus our attention to the waves in the propagation region specified by $\omega > m\Omega + \kappa(1 - Q^{-2})^{1/2}$ (see Fig. 6.5 and the next paragraph).

We consider a typical disk galaxy, and assume that in the propagation region of oscillations the rotation curve, $\Omega(r)$, increases outwards, and thus the epicyclic frequency, $\kappa(r)$, also increases outwards. The value of Q in the inner region of disks is larger than unity due to high velocity dispersion in nuclear bulge. It will decrease outwards, but its decrease less than unity is suppressed by generation of turbulence due to gravitational instability. Hence, the value of Q will be roughly unity in the outer region [see panel (a) of Fig. 6.5]. Then, the radial distributions of $m\Omega$ and $m\Omega \pm \kappa(1 - Q^{-2})^{1/2}$ will be those schematically shown in panel (b) of Fig. 6.5. (The curve of $m\Omega$ is shown only in the region of $Q = 1$.) For a given ω the radial distribution of k^2 becomes like that shown in panel (c) of Fig. 6.5. The curve of k^2 has a simple zero at radius r_i and a double zero at the corotation radius r_c where $\nu = 0$. The region inside r_i is an evanescent region of waves, while outside the corotation radius r_c a propagation region again extends.

In the region inside r_i, i.e., $r < r_i$, $k^2 < 0$ and thus we adopt the boundary condition that the oscillations are spatially damped in $r < r_i$. For $r > r_c$, the oscillations are still in the nature of propagating waves [see the second condition

[5] Equation (6.59) can be written as

$$\left(|k| - \frac{\pi G \Sigma_0}{c_s^2}\right)^2 = \left(\frac{\kappa}{c_s}\right)^2 \left(Q^{-2} - 1 + \nu^2\right).$$

This equation is reduced to Eq. (6.61) for local perturbations with $|k| > k_{crit}/Q^2$.

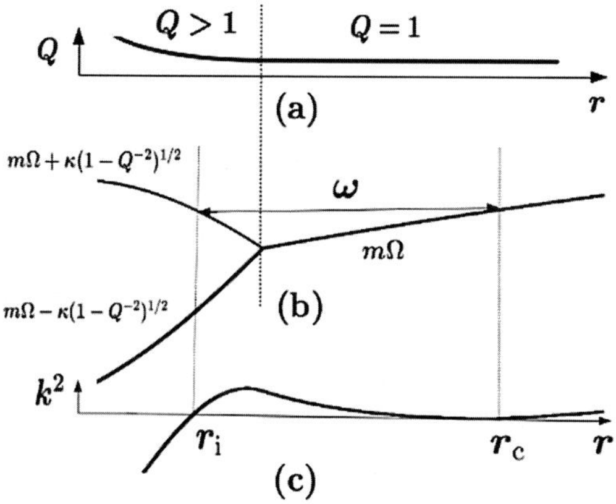

Fig. 6.5 The upper panel (**a**): Schematic picture showing radial distribution of $Q(r)$. In the inner region we have $Q > 1$ because of high velocity dispersion in nuclear bulge, while in the outer region $Q \sim 1$ since decrease of Q is suppressed by growth of turbulence due to gravitational instability. The middle panel (**b**): Schematic picture showing radial distribution of $m\Omega \pm \kappa(1 - Q^{-2})^{1/2}$ on frequency-radial diagram. Frequency of the density wave in consideration is shown by ω. The inner boundary of the propagation region of the wave is r_i and r_c is the corotation radius. The lower panel (**c**): Schematic picture showing radial distribution of $k^2(r)$. The region of $r < r_\mathrm{i}$ is an evanescent region. The radius r_c is the corotation radius where $\omega = m\Omega$. We consider density waves trapped between r_i and r_c, partially propagating out in the region of $r > r_\mathrm{c}$

in inequalities (6.62)]. We adopt a radiation condition at infinity, since the inward signal from infinity accounts to an external disturbance. With the above boundary conditions we solve the wave equation (6.60) by the WKBJ method and demonstrate that the waves are amplified by the corotation resonance, following Lin and Lau (1979).

6.6.1.1 Solution Around $r \sim r_\mathrm{i}$

First, we consider the solution around $r \sim r_\mathrm{i}$. Equation (6.60) can be approximated as $(a > 0)$

$$\frac{d^2u}{dx^2} + a(r - r_\mathrm{i})u = 0, \tag{6.63}$$

since k^2 is

$$k^2 \sim a(r - r_\mathrm{i}), \tag{6.64}$$

where $x = r - r_\mathrm{i}$. Two independent solutions of Eq. (6.63) in the region of $r \gtrsim r_\mathrm{i}$ are written as

$$u = (r - r_\mathrm{i})^{1/2} J_{\pm 1/3}\left[\frac{2}{3}a^{1/2}(r - r_\mathrm{i})^{3/2}\right], \tag{6.65}$$

where $J_{1/3}$ and $J_{-1/3}$ are the Bessel functions of index of $1/3$ and $-1/3$, respectively (see, e.g., Morse and Feshbach 1953).

As solutions which tend to Eq. (6.65) for $r \gtrsim r_\mathrm{i}$ and to the WKBJ solution:

$$k^{-1/4}\exp\left[\pm i \int^r kdr\right], \tag{6.66}$$

for a large value of k^2, we can adopt (e.g., Morse and Feshbach 1953)

$$u = \left(\frac{v}{k}\right)^{1/2} [AJ_{1/3}(v) + BJ_{-1/3}(v)], \tag{6.67}$$

where

$$v = \int_{r_\mathrm{i}}^r kdr, \tag{6.68}$$

and A and B are arbitrary constants.

For $r \lesssim r_\mathrm{i}$ both k and v become imaginary. Hence, to continue our analysis, we must carefully specify the branch of k and v for $r \lesssim r_\mathrm{i}$. Since function u is actually single-valued, either branch of k is to be used. We choose $k = \exp[i\pi/2]|k|$ for $r < r_\mathrm{i}$. Then, we have $v = \exp(3i\pi/2)|v|$ for $r < r_\mathrm{i}$. When arguments have a phase between $\pi/2$ and $3\pi/2$, the asymptotic dependences of the Bessel functions, $J_{1/3}$ and $J_{-1/3}$, are [e.g., equation (5.3.68) of Morse and Feshbach (1953)]

$$J_{1/3}(v) \sim \left(\frac{2}{\pi v}\right)^{1/2} \exp\left(i\frac{5\pi}{6}\right)\cos\left(v + \frac{5}{12}\pi\right), \tag{6.69}$$

$$J_{-1/3}(v) \sim \left(\frac{2}{\pi v}\right)^{1/2} \exp\left(i\frac{\pi}{6}\right)\cos\left(v + \frac{1}{12}\pi\right). \tag{6.70}$$

Substituting them into Eq. (6.67) under consideration of $v = \exp(i3\pi/2)|v|$, we have, for $r \ll r_\mathrm{i}$,

$$u = \frac{1}{2}\left(\frac{2}{\pi k}\right)^{1/2} \mathrm{e}^{i\pi/4}\left[(B - A)\mathrm{e}^{|v|} + \left(A\mathrm{e}^{i\pi/6} + B\mathrm{e}^{-i\pi/6}\right)\mathrm{e}^{-|v|}\right]. \tag{6.71}$$

For $r \ll r_\mathrm{i}$ the perturbations should spatially dampen, which means $B = A$. Hence, from Eq. (6.67) we can express the waves in the region of $r > r_\mathrm{i}$ in the form:

$$u \sim A\left(\frac{v}{k}\right)^{1/2}\left(J_{1/3}(v) + J_{-1/3}(v)\right), \tag{6.72}$$

which are rearranged as

$$u = A\left(u_\mathrm{i}^- + u_\mathrm{i}^+\right), \tag{6.73}$$

where u_i^- and u_i^+ are the inward and outward propagating waves, respectively, and are

$$u_\mathrm{i}^+ = \frac{\sqrt{3}}{2}\left(\frac{v}{k}\right)^{1/2} \mathrm{e}^{-i\pi/6}\mathcal{H}_{1/3}^{(2)}(v), \tag{6.74}$$

$$u_\mathrm{i}^- = \frac{\sqrt{3}}{2}\left(\frac{v}{k}\right)^{1/2} \mathrm{e}^{i\pi/6}\mathcal{H}_{1/3}^{(1)}(v). \tag{6.75}$$

Here, $\mathcal{H}$'s are Hankel functions, and in deriving equations (6.74) and (6.75) from Eq. (6.72) the relations between J's and $\mathcal{H}$'s for $\arg(v) \gtrsim 0$ have been adopted.[6]

6.6.1.2 Solution Around $r \sim r_\mathrm{c}$

Next, we consider solutions around $r \sim r_\mathrm{c}$. For $r \lesssim r_\mathrm{c}$, k^2 can be written as

$$k^2 = ax^2 \quad \text{with} \quad x \equiv r_\mathrm{c} - r, \tag{6.76}$$

where the coefficient a is different from a in Eq. (6.64), although the same symbol is used. Then, solutions of Eq. (6.60) are written as

$$u = x^{1/2}\mathcal{H}_{1/4}\left(\frac{1}{2}a^{1/2}x^2\right), \tag{6.77}$$

6

$$J_\nu(v) = \frac{1}{2}\left[\mathcal{H}_\nu^{(1)}(v) + \mathcal{H}_\nu^{(2)}(v)\right],$$

and

$$J_{-\nu}(v) = \frac{1}{2}\left[\mathrm{e}^{i\nu\pi}\mathcal{H}_\nu^{(1)}(v) + \mathrm{e}^{-i\nu\pi}\mathcal{H}_\nu^{(2)}(v)\right].$$

where $\mathcal{H}_{1/4}(z)$ represents, for simplicity, two Hankel functions of index 1/4 with argument z, i.e., $\mathcal{H}^{(1)}_{1/4}(z)$ and $\mathcal{H}^{(2)}_{1/4}(z)$. As in the case of solution of $r \sim r_\mathrm{i}$, asymptotic solutions which tend to Eq. (6.77) for x close to zero and to the WKBJ solution for large x can be written as

$$u = \left(\frac{z}{k}\right)^{1/2}\left[C\mathcal{H}^{(1)}_{1/4}(z) + D\mathcal{H}^{(2)}_{1/4}(z)\right], \tag{6.78}$$

where

$$z = \int_0^x k dx = -\int_{r_\mathrm{c}}^r k dr \tag{6.79}$$

and C and D are arbitrary constants at the present stage.

Solution (6.78) is now extended to the region of $r > r_\mathrm{c}$, i.e., to the region of $x < 0$. For this region we have two possibilities of $k = a^{1/2}x$ and $k = -a^{1/2}x$. Both selections are possible, if solutions are taken continuously through the point of $x = 0$. Here we adopt $\arg(x) \sim +0$ for $x > 0$. Then, for $x < 0$ we have $\arg(k) \sim -\pi$ and $\arg(z) \sim -2\pi$, i.e.,

$$z = \int_0^x k dx = \frac{1}{2}a^{1/2}|x|^2\exp(-i2\pi) \equiv |z|\exp(-i2\pi). \tag{6.80}$$

Then, corresponding to Eq. (6.78) we see that the solution in the region of $r > r_\mathrm{c}$ (i.e., $x < 0$) is

$$u = \left(\frac{z}{k}\right)^{1/2}\left[C\mathcal{H}^{(1)}_{1/4}\Big(|z|\exp(-i2\pi)\Big) + D\mathcal{H}^{(2)}_{1/4}\Big(|z|\exp(-i2\pi)\Big)\right]. \tag{6.81}$$

The use of relations among Hankel functions with different phases of arguments,[7] we can reduce equation (6.81) to

$$u = \left(\frac{z}{k}\right)^{1/2}\left(C - \sqrt{2}\mathrm{e}^{i\pi/4}D\right)\mathcal{H}^{(1)}_{1/4}(|z|) + \left(-C + \sqrt{2}\mathrm{e}^{-i\pi/4}D\right)\mathcal{H}^{(2)}_{1/4}(|z|). \tag{6.82}$$

7

$$\mathcal{H}^{(1)}_\nu(\mathrm{e}^{im\pi}z) = -\frac{\sin(m-1)\nu\pi}{\sin\nu\pi}\mathcal{H}^{(1)}_\nu(z) - \mathrm{e}^{-i\nu\pi}\frac{\sin\nu m\pi}{\sin\nu\pi}\mathcal{H}^{(2)}_\nu(z)$$

$$\mathcal{H}^{(2)}_\nu(\mathrm{e}^{im\pi}z) = \frac{\sin(m+1)\nu\pi}{\sin\nu\pi}\mathcal{H}^{(2)}_\nu(z) + \mathrm{e}^{i\nu\pi}\frac{\sin\nu m\pi}{\sin\nu\pi}\mathcal{H}^{(1)}_\nu(z).$$

Here we impose the boundary condition that far outside of the corotation radius there is no incoming waves (radiation condition). Then, from the asymptotic formula of the Hankel functions for large $|z|$ we see that the coefficient of the Hankel function $\mathcal{H}^{(1)}_{1/4}(|z|)$ needs to vanish, i.e.,

$$C = \sqrt{2}\exp(i\pi/4)D = (1+i)D. \tag{6.83}$$

That is, substituting this relation between C and D into Eq. (6.78), we have for $r < r_{\rm c}$

$$u = D\left(u_{\rm c}^{-} + u_{\rm c}^{+}\right), \tag{6.84}$$

where $u_{\rm c}^{-}$ and $u_{\rm c}^{+}$ are, respectively, the inward and outward propagating waves and are

$$u_{\rm c}^{+} = \left(\frac{z}{k}\right)^{1/2} \sqrt{2}{\rm e}^{i3\pi/4}\,\mathcal{H}^{(1)}_{1/4}(z), \tag{6.85}$$

$$u_{\rm c}^{-} = \left(\frac{z}{k}\right)^{1/2} {\rm e}^{i\pi/2}\mathcal{H}^{(2)}_{1/4}(z). \tag{6.86}$$

It is noted that for $r > r_{\rm c}$, there is the outgoing wave alone, which is

$$u = -2iD\left(\frac{z}{k}\right)^{1/2} \mathcal{H}^{(2)}_{1/4}(z). \tag{6.87}$$

6.6.1.3 Dispersion Relation and Growth Rate

The next subject is to connect solution (6.73) for $r > r_{\rm i}$ and solution (6.84) for $r < r_{\rm c}$ in an intermediate region between $r_{\rm i} < r < r_{\rm c}$. For large v, we have

$$\mathcal{H}^{(1)}_{1/3}(v) \propto \exp\left[i\left(v - \frac{5}{12}\pi\right)\right], \tag{6.88}$$

$$\mathcal{H}^{(2)}_{1/3}(v) \propto \exp\left[-i\left(v - \frac{5}{12}\pi\right)\right]. \tag{6.89}$$

On the other hand, for large z we have

$$\mathcal{H}^{(1)}_{1/4}(z) \propto \exp\left[i\left(z - \frac{3}{8}\pi\right)\right] = {\rm e}^{i\kappa^*}\exp\left[-i\left(v + \frac{3}{8}\pi\right)\right] \tag{6.90}$$

$$\mathcal{H}^{(2)}_{1/4}(z) \propto \left[-i\left(z - \frac{3}{8}\pi\right)\right] = {\rm e}^{-i\kappa^*}\exp\left[i\left(v + \frac{3}{8}\pi\right)\right], \tag{6.91}$$

where

$$v + z = \int_{r_i}^{r_c} k dr \equiv \kappa^* \tag{6.92}$$

have been used. Hence, expression (6.73) for u is smoothly connected to expression (6.84) for u by equating the terms proportional to e^{iv} in both equations. Similarly, the terms proportional to e^{-iv} in both equations are connected. From the terms proportional to e^{iv}, we have

$$\frac{\sqrt{3}}{2} A e^{i\pi/6} e^{-i5\pi/12} = D e^{-i\kappa^*} e^{i\pi/2} e^{i3\pi/8}. \tag{6.93}$$

similarly, from the terms proportional to e^{-iv}, we have

$$\frac{\sqrt{3}}{2} A e^{-i\pi/6} e^{i5\pi/12} = D\sqrt{2} e^{i\kappa^*} e^{i3\pi/4} e^{-i3\pi/8}. \tag{6.94}$$

Taking the ratio of these equations we obtain

$$\sqrt{2}\exp(2i\kappa^*) = -1. \tag{6.95}$$

This is the *quantum condition* for determination of the eigenfrequency ω, and is written in the form (Lau et al. 1976; Lin and Lau 1979)

$$\int_{r_i}^{r_c} k dr = \left(n + \frac{1}{2}\right)\pi + i\frac{1}{4}\ln 2, \quad n = 0, 1, 2, \ldots . \tag{6.96}$$

This equation shows that the frequency ω in k is not real. For example, when the growth rate $\gamma = -\omega_i$ is small, Eq. (6.96) leads to the approximate relations

$$\int_{r_i}^{r_c} k(\omega) dr = \left(n + \frac{1}{2}\right)\pi, \tag{6.97}$$

and

$$\gamma\tau = \frac{1}{4}\ln 2, \tag{6.98}$$

where τ is defined by

$$\tau = \int_{r_i}^{r_c} \frac{dr}{|c_g|}, \tag{6.99}$$

with

$$|c_{\mathrm{g}}| = \left| \frac{\partial k}{\partial \omega} \right|^{-1}. \tag{6.100}$$

Physically, $|c_{\mathrm{g}}|$ is the magnitude of the group velocity of the waves at a given location, and τ is the time it takes to propagate between r_{i} and r_{c}. Equation (6.98) shows that the growth of the amplitude of these waves over a round trip between r_{i} and r_{c} is

$$\exp(2\gamma\tau) = \sqrt{2}. \tag{6.101}$$

The above results are physically summarized as follows (Lin and Lau 1979). The outgoing wave u_{c}^{+} is turned back at r_{c} with a reinforcement (because there is another wave leaving r_{c} to go to infinity), which corresponds to a doubling of energy [see the factor $\sqrt{2}$ in Eq. (6.85)]. This incoming wave [equation (6.85)] propagates to r_{i} and is turned back at the same energy [cf. Eqs. (6.74) and (6.75)]. After a period of 2τ, the process is complete with a doubling of energy. This is precisely the assertion of Eq. (6.101).

6.6.2 P-Mode Oscillations in Nonself-Gravitating Disks

By analytical considerations, Drury (1985) showed the presence of corotation instability in nonself-gravitating disks. Almost at the same time, Papaloizou and Pringle (1984, 1985, 1987) found that the tori consisting of constant angular momentum gases ($\ell = r^2\Omega =$ const.) are dynamically unstable to non-axisymmetric perturbations. The latter studies took much attention of astronomers in the field of accretion disks, because the torus model of compact sources was challenged by this instability.

In tori of constant angular momentum, the epicyclic frequency, κ, vanishes throughout the system, and thus there is no evanescent region of waves around corotation radius. This brings about a strong instability of tori. In the case of Keplerian disks, however, the radius of corotation resonance for p-mode oscillations is outside their propagation region (see the next paragraphs). That is, the corotational interaction between the mean flow and the waves occurs only through penetrated component of the waves. Hence, the growth rate of the waves is much decreased by a penetration factor, compared with those in the cases of galactic density waves and waves in tori.

Hereafter, we restrict our attention to corotation instability of fundamental p-mode oscillations in geometrically thin disks, because in nonself-gravitating disks the counterpart of galactic density waves is the fundamental ($n = 0$) p-mode oscillations[8] (see Sect. 4.4 for classification of oscillations in geometrically thin nonself-gravitating disks). The second reason is that other oscillation modes are damped by corotation resonance; i.e., both g-modes (Kato 2003; Li et al. 2003) and c-modes (Lai and Tsang 2009) are all damped by corotation resonance, as well as the $n \geq 1$ p-modes.

In the study of corotation instability of the fundamental p-mode oscillations, many works have been done by using the vertically integrated quantities, as is done in the study of galactic density waves (see Sect. 6.6.1). In this section, however, we use the wave equation derived without vertical integration, because wave equations without vertical integration have been derived in Sect. 4.4. It is noticed that there is no essential difference in final results whether vertical integrated quantities are adopted or not, because fundamental ($n = 0$) p-mode oscillations have no node in the vertical direction.[9]

First, let us again emphasize that the corotation radius is outside the propagation region of the fundamental ($n = 0$) p-mode oscillations. As shown in Sect. 4.4, the dispersion relation of the fundamental p-mode oscillations in the local approximation is

$$(\omega - m\Omega)^2 = \kappa^2 + c_s^2 k^2, \tag{6.102}$$

where ω is the frequency of waves, k the radial wavenumber, m the azimuthal wavenumber, and c_s the acoustic speed in disks. This dispersion relation shows that the propagation regions of the oscillations are separated into two regions in the radial direction, i.e., $(\omega - m\Omega)^2 - \kappa^2 > 0$ are

$$\omega > m\Omega + \kappa \quad \text{and} \quad \omega < m\Omega - \kappa. \tag{6.103}$$

The corotation radius given by $\omega = m\Omega$ is between the above two propagation regions, as is shown in Fig. 6.6.

The above arguments based on the dispersion relation (6.102), which was derived from the local approximation, are insufficient to examine the behaviors of perturbations around the corotation radius. Without using the local approximations we have derived in Sect. 4.4 a more general form of wave equation, although some other simplifications are involved. The wave equation is given by the set of Eqs. (4.81) and (4.82). Since we are interested here in behavior of waves around the corotation radius, r_c, where $(\omega - m\Omega)^2 = 0$, the radial coordinate r is changed to dimensionless coordinate x which is zero at the corotation radius r_c, i.e., $x = (r - r_c)/r_c$. Then, including the term of corotation resonance [the term proportional

[8]The fundamental p-mode means the p-mode oscillations whose pressure and density variations have no node in the vertical direction.

[9]It is noted here that in the next subsection (Sect. 6.6.3) we consider the Rossby wave instability. In that study, effects of entropy gradient in the radial direction are considered (i.e., non-barotropic gas is considered). In that study, we will treat vertically integrated quantities.

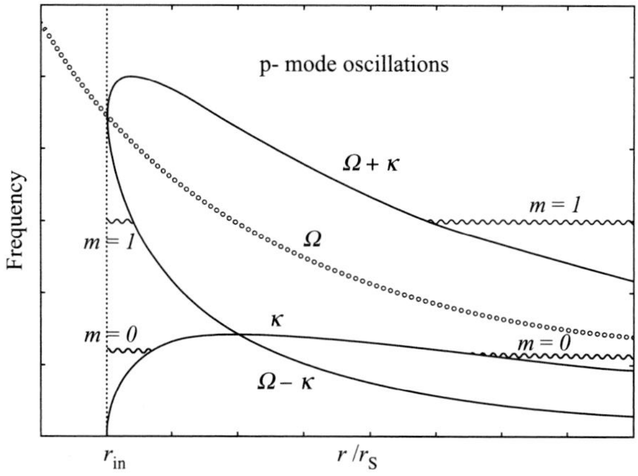

Fig. 6.6 A schematic figure showing that in the case of nonself-gravitating relativistic disks, the corotation radius is in the evanescent region of p-mode oscillations. This figure is shown, as a demonstration, for one-armed ($m = 1$) p-mode oscillations in relativistic disks with the Schwarzschild metric. $r_{\rm in}$ is the inner edge of disks, which is $3r_{\rm S}$ in the case of Schwarzschild metric, where $r_{\rm S}$ is the Schwarzschild radius defined by $r_{\rm S} = 2GM/c^2$. The epicyclic frequency κ vanishes at $r_{\rm in}$ (After Kato 2016)

to $(\omega - m\Omega)^{-1}$ in K^2 in Eq. (4.82), but neglecting terms related to apparent singularities at Lindblad resonances[10] [the terms proportional to $[(\omega - m\Omega)^2 - \kappa^2]^{-1}$ in K^2 in Eq. (4.82)], we can approximate the wave equation emphasizing the wave behavior around the corotation radius in the form:

$$\frac{d^2\bar{f}}{dx^2} + k^2\bar{f} = 0, \qquad (6.104)$$

where

$$k^2(r) = \frac{(\omega - m\Omega)^2 - \kappa^2}{c_{\rm s}^2} r_{\rm c}^2 + \frac{1}{(\omega - m\Omega)} \left[\frac{mr\rho_{00}}{(\kappa^2/2\Omega)^2} \frac{d}{dr} \left(\frac{\kappa^2/2\Omega}{\rho_{00}} \right) \right]_{\rm c}, \qquad (6.105)$$

where $\bar{f}$ is a variable associated with wave perturbations (its detailed expression is unnecessary here), and the subscript c denotes the values at the corotation radius. The density $\rho_{00}(r)$ represents the density on the equator.

The radial distribution of $k(x)$ is schematically shown in Fig. 6.7 for three cases of the sign of g_1, which is the radial gradient of *specific vorticity* $\kappa^2/2\Omega\rho_{00}$ at the

[10]The fact that the Lindblad resonances are apparent singularity in the present problem is demonstrated, for example, by Kato (2016).

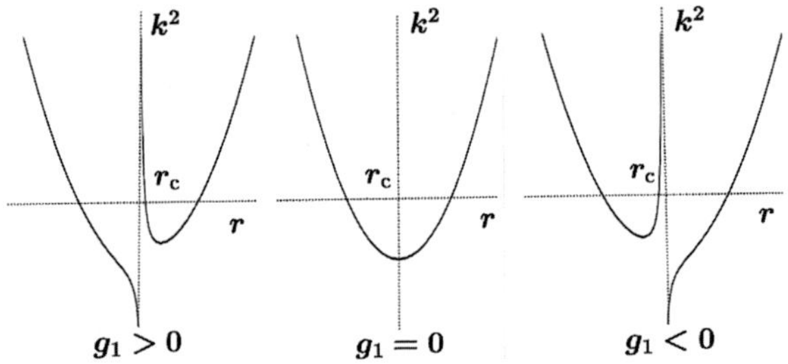

Fig. 6.7 A schematic figure showing the radial distribution of $k^2(r)$ in three typical cases. In this figure the radial gradient of specific vorticity $(\kappa^2/2\Omega)/\rho_{00}$at the corotation radius r_c is denoted by g_1, i.e., $g_1 \equiv d(\kappa^2/2\Omega\rho_{00})/dr$

corotation radius:

$$g_1 = \frac{d}{dr}\left[\frac{\kappa^2/2\Omega}{\rho_{00}}\right]_c. \tag{6.106}$$

If the specific vorticity vanishes at the corotation point, i.e., $g_1 = 0$, the corotation point is regular, not singular. The case of the constant angular momentum tori and the case of shearing sheet approximation (Goldreich and Lynden-Bell 1965) correspond to this case.

By using the wave equation described by the set of Eqs. (6.104) and (6.105) or similar wave equations, corotation resonance in nonself-gravitating disks have been extensively studied (e.g., Blaes 1985; Bleas 1987; Goldreich and Narayan 1985; Goldreich et al. 1986; Narayan et al. 1987, and so on). Recently, analyses by the WKBJ method were also made by Lai's group (Lai and Tsang 2009; Tsang and Lai 2008, 2009a,b). Here, we are satisfied only by presenting two important issues related to corotation resonance. They are (1) the sign of wave energy and (2) the condition of overreflection of waves at the corotation radius.

Waves in nonself-gravitating disks are separated into two classes. One is the waves which exist mainly in the region inside the corotation radius. The other is the one which exists mainly outside the corotation radius. Waves whose propagation region is inside the corotation radius have negative energy, while those outside the radius have positive energy (see, for example, Sect. 6.7). The instability at the corotation resonance is related to overreflection of waves at the corotation radius. The side where overreflection occurs (i.e., $r < r_c$ or $r > r_c$) depends on the sign of specific angular momentum at the corotation resonance. Figure 6.8 shows this situation. In the case where the radial gradient of specific vorticity is positive at the corotation radius, i.e., $d(\kappa^2/2\rho_{00}\Omega)_{r_c}/dr > 0$, the incident wave coming to the corotation point from the inside is overreflected by emitting a wave with positive energy to the outer region (see the upper left corner of Fig. 6.8). (Waves coming to the corotation radius from outside are underreflected.) In the case where $d(\kappa^2/2\rho_{00}\Omega)_{r_c}/dr < 0$, the incident wave coming to the corotation point from

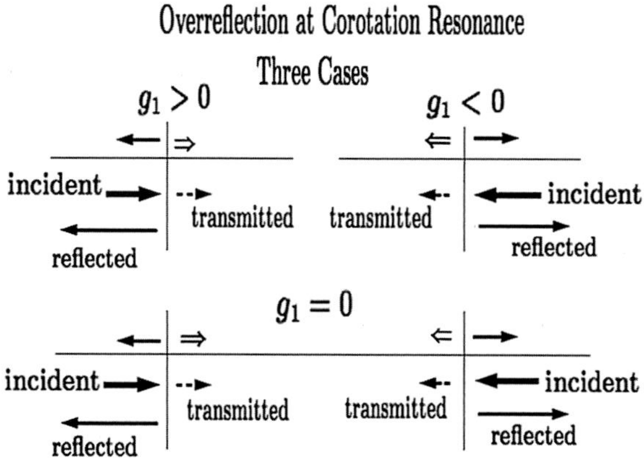

Fig. 6.8 A schematic figure showing overreflection of waves in three cases of $g_1 > 0$, $g_1 < 0$, and $g_1 = 0$, where g_1 represents the specific vorticity at the corotation resonance, i.e., $g_1 = [d(\kappa^2/2\rho_{00}\Omega)_{r_c}/dr]_{r_c}$ (After Kato 2016)

outside is overreflected (see the upper right corner of Fig. 6.8). (Waves coming to the corotation resonance from inside are underreflected.) In the special case of $d(\kappa^2/2\rho_{00}\Omega)_{r_c}/dr = 0$, incident waves coming to the corotation point from both sides are overreflected (see the lower part of Fig. 6.8).

The above results on overreflection (or overstability) at the corotation resonance can be interpreted in an instructive way. In remaining of this section, we present this picture, following Glatzel (1987). As shown before, the region around the corotation radius is the evanescent region of p-mode oscillations. Inside and outside of the region there are wave propagation regions. To make situations clear, we assume that the inner propagation region has a rigid inner boundary, and also the outer propagation region has a rigid outer boundary. Then, if penetration of oscillations into the evanescent region around the corotation radius is neglected, we have two dispersion relations representing two neutral trapped oscillations. One is the dispersion relation for oscillations in the inner propagation region, which is represented here by $D_1(\omega) = 0$. The other is the dispersion relation, $D_2(\omega) = 0$, for the oscillations in the outer region. These trapped oscillations interact through the evanescent region. If the interaction is weak, the dispersion relation for the whole oscillations will be written as

$$D_1(\omega)D_2(\omega) = \epsilon(\omega), \tag{6.107}$$

where the coupling term ϵ is a small quantity.

Let us denote ω_1 as one of solutions of $D_1 = 0$, i.e., $D_1(\omega_1) = 0$. Similarly, ω_2 is a solution of $D_2 = 0$, i.e., $D_2(\omega_2) = 0$. If ω_1 and ω_2 are close, two oscillations will interact through the coupling term ϵ. As the results of the interaction, an oscillation extended over two regions through the evanescent region will be realized. The frequency of the resulting oscillations is now written as ω, and we introduce small quantities δ and Δ by

$$\delta = \omega_2 - \omega_1, \quad \omega = \omega_1 + \Delta. \tag{6.108}$$

Here, we consider small quantities till the first order of δ and Δ. Then, we have

$$D_1(\omega) = \Delta \frac{\partial D_1}{\partial \omega}\bigg|_{\omega_1}, \tag{6.109}$$

$$D_2(\omega) = (\Delta - \delta)\frac{\partial D_2}{\partial \omega}\bigg|_{\omega_2}. \tag{6.110}$$

Substituting these relations to Eq. (6.107) and solving the resulting equation with respect to Δ, we obtain

$$\Delta = \frac{\omega_2 - \omega_1}{2} \pm \left[\frac{(\omega_2 - \omega_1)^2}{4} + \frac{\epsilon(\omega)}{(\partial D_1/\partial\omega)_{\omega_1} \cdot (\partial D_2/\partial\omega)_{\omega_2}}\right]^{1/2}. \tag{6.111}$$

Now we consider the case of $\epsilon > 0$. If $\partial D_1/\partial\omega|_{\omega_1}$ and $\partial D_2/\partial\omega|_{\omega_2}$ have opposite signs, Δ becomes complex in the case where ω_1 and ω_2 are sufficiently close. This means that the system is unstable. It is noticed that in some issues $\partial D/\partial\omega$ and wave energy W are related in the sense that $-\partial D/\partial\omega$ and W have the same sign. For example, in the case of Kelvin–Helmholtz instability, this is the case [see Eq. (6.57)]. This is consistent with the picture that oscillations become unstable through the corotation point when wave energy of the two waves is opposite. In the case where $(\partial D_1/\partial\omega)_1$ and $(\partial D_2/\partial\omega)_2$ have the same signs in wave energy, instability does not occur (when $\epsilon > 0$). This is a phenomenon called avoid crossing.

Finally, it is noted that there is an another instability which is related to the corotation instability, which is the Rossby wave instability (Lovelace and Hohlfeld 1978). This instability is argued in the next subsection (Sect. 6.6.3).

6.6.3 Rossby Wave Instability

The results in Sect. 6.6.2 show that the radial gradient of vortensity at the corotation radius, i.e., $[d\ln(\kappa^2/2\Omega\rho_{00})/dr]_c$, is an important quantity in determining characteristics of corotation instability of oscillation modes. In deriving this result, we treated the corotation term in wave equation [e.g., the second term on the right-hand side of Eq. (6.105)] as a subsidiary one in the sense that we examine how the characteristics of wave modes already known are modified by the effects of the term. This means that we did not examine a new wave mode which appears by considering the corotation term. The presence of a new mode can be understood, because Eq. (6.105) is a cubic equation with respect to $\omega - m\Omega$. The new mode has essentially complex frequency, because at the corotation radius $\omega - m\Omega$ cannot be zero except for special cases, and will be localized around the corotation radius. The importance of the new mode was pointed out first by Lovelace et al. (1999). The instability of the mode was called the *Rossby wave instability*. This naming seems to be due to analogy with the usual Rossby waves (Sect. 4.3) in the sense that the propagation direction of the new mode is one-way.[11]

In studying the Rossby wave instability, behaviors of gases around the corotation radius are important. Hence, it is desirable to extend the two-dimensional wave equation derived in Sect. 4.4 to cases of non-barotropic gases, although this is not essential for Rossby wave instability. Since this extension is complicated, we are satisfied here with deriving a wave equation expressed in terms of vertically integrated quantities (one-dimensional wave equation), although effects of entropy gradient in the radial direction are taking into account.

Because we are interested in oscillations with no node in the vertical direction, we treat vertically integrated density and pressure perturbations, Σ_1 and Π_1, assuming $v_z = 0$ and no vertical dependences of v_r and v_φ. Equations describing small amplitude perturbations are integrated in the vertical direction till the height where density and pressure vanish. In this case the integration and derivatives with respect to r, φ, and t are commutative. Then, the integrations of equation of continuity and the equation of motion are

$$i\tilde{\omega}\Sigma_1 + \frac{\partial}{r\partial r}(rU_r) - i\frac{m}{r}U_\varphi = 0, \tag{6.112}$$

$$i\tilde{\omega}U_r - 2\Omega U_\varphi = -\frac{\partial \Pi_1}{\partial r} + \alpha\frac{c_s^2}{r}\Sigma_1 \tag{6.113}$$

$$i\tilde{\omega}U_\varphi + \frac{\kappa^2}{2\Omega}U_r = i\frac{m}{r}\Pi_1, \tag{6.114}$$

$$i\tilde{\omega}\left(\Pi_1 - c_s^2\Sigma_1\right) = c_s^2\frac{\partial}{\partial r}\left(\ln\frac{p_0}{\rho_0^{1/\gamma}}\right)U_r, \tag{6.115}$$

where perturbations are assumed to be proportional to $\exp(i\omega t - m\varphi)$ and

$$\Sigma_1 = \int \rho_1 dz, \quad \Pi_1 = \int p_1 dz, \tag{6.116}$$

$$U_r = \int \rho_0 v_r dz, \quad U_\varphi = \int \rho_0 v_z dz, \tag{6.117}$$

and $\tilde{\omega} = \omega - m\Omega$. In deriving adiabatic relation (6.115) the vertical variations of c_s^2 and $\ln(p_0/\rho_0^{1/\gamma})$ have been neglected.

The second term on the right-hand side of Eq. (6.113) comes from the vertical integration of $(\rho_1/\rho_0)\partial p_0/\partial r$, which has been written as $\alpha c_s^2\Sigma_1/r$ by introducing a dimensionless quantity α. If $\partial p_0/\rho_0\partial r$ is approximated as $c_s^2\partial\rho_0/\rho_0\partial r$ with no

[11]It is notices that Eq. (6.105) is a cubic equation with respect to $\omega - m\Omega$.

r-dependence of disk thickness, we have

$$\alpha = r\frac{d\ln \Sigma_0}{dr}. \tag{6.118}$$

A more careful evaluation of $\partial p_0/\rho_0\partial r$ is $(1/\rho_0)\partial p_0/\partial r = r\Omega^2 - (GM)r/(r^2 + z^2)^{3/2} \sim r\Omega^2 - r\Omega_{\rm K}^2(1 - 3z^2/2r^2)$. If this is adopted, the vertical integration of $(\rho_1/\rho_0)\partial p_0/\partial r$ is approximately $(\Omega^2 - \Omega_{\rm K}^2 + 3\Omega_{\rm K}^2H^2/2r^2)r\Sigma_1$, where $H(r)$ is the half-thickness of the disk. In deriving this, the disk is approximated to be isothermal in the vertical direction as $\rho_0 \propto \exp[-z^2/2H^2(r)]$ and ρ_1 is also taken to be proportional to $\exp[-z^2/2H^2(r)]$ (see Sect. 4.4) since we are interested in wave modes with no node in the vertical direction. If the latter approximation is adopted, we have

$$\alpha = \frac{r^2\left(\Omega^2 - \Omega_{\rm K}^2\right)}{H^2\Omega_{\rm K}^2} + \frac{3}{2}. \tag{6.119}$$

By eliminating U_φ from Eqs. (6.112) and (6.114), we have an equation with respect to Σ_1, U_r, and Π_1. Similarly, from Eqs. (6.113) and (6.114) U_φ is eliminated to derive an equation with respect to Σ_1, U_r, and Π_1. From the resulting two equations Σ_1 is eliminated by using Eq. (6.115). Then, we have

$$\frac{d\Pi_1}{dr} = \frac{1}{r}\left(\alpha + \frac{2m\Omega}{\tilde{\omega}}\right)\Pi_1 + \frac{1}{i\tilde{\omega}}\left[\left(\tilde{\omega}^2 - \kappa^2\right) - \alpha c_{\rm s}^2\frac{d}{rdr}\left(\ln\frac{p_0}{\rho_0^{1/\gamma}}\right)\right]U_r, \tag{6.120}$$

$$\frac{dU_r}{dr} = -\frac{1}{r}\left[1 + \frac{m\kappa^2/2\Omega}{\tilde{\omega}} - r\frac{d}{dr}\left(\ln\frac{p_0}{\rho_0^{1/\gamma}}\right)\right]U_r + \frac{1}{i\tilde{\omega}}\left(\frac{\tilde{\omega}^2}{c_{\rm s}^2} - \frac{m^2}{r^2}\right)\Pi_1. \tag{6.121}$$

By eliminating U_r from the above two equations, we have a second-order differential equation of Π_1 with respect to r. By changing the dependent variable from Π_1 to $\tilde{\Pi}_1$ defined by

$$\tilde{\Pi}_1 = \left(\frac{r}{LSJ}\right)^{1/2}\Pi_1, \tag{6.122}$$

we have

$$\frac{d^2\tilde{\Pi}_1}{dr^2} + K^2\tilde{\Pi}_1 = 0, \tag{6.123}$$

where

$$K^2 = -\frac{1}{4}\left[\frac{d}{dr}\ln\left(\frac{rJ}{LS}\right)\right]^2 - \frac{1}{2}\left[\frac{d^2}{dr^2}\ln\left(\frac{rJ}{LS}\right)\right] + \left(\frac{L}{c_s^2} - \frac{m^2}{r^2}\right)$$
$$+\frac{2m\Omega}{\tilde{\omega}}\frac{d}{rdr}\ln\left(\frac{SL}{\Omega J}\right) + \alpha c_s^2 \frac{m^2}{r^2\tilde{\omega}^2}\frac{d}{rdr}\ln S. \tag{6.124}$$

In the above equations the following abbreviations have been used:

$$L = \tilde{\omega}^2 - \kappa^2 - \alpha c_s^2 \frac{d}{rdr}\ln\left(\frac{p_0}{\rho_0^{1/\gamma}}\right), \quad S = \frac{p_0}{\rho_0^{1/\gamma}}, \quad J = \exp\left(\int\frac{\alpha(r)}{r}dr\right). \tag{6.125}$$

The expression for K^2 given by Eq. (6.124) shows that near the corotation radius we have

$$K^2 \sim -\frac{\kappa^2}{c_s^2} + 2m\Omega\frac{d}{rdr}\ln\left(\frac{\kappa^2}{2\Omega\Sigma_0}\frac{p_0}{\rho_0^{1/\gamma}}\right)\frac{1}{\tilde{\omega}} + \frac{m^2c_s^2}{r^2}\frac{d}{dr}\ln\left(\frac{p_0}{\rho_0^{1/\gamma}}\right)\frac{d\ln\Sigma_0}{dr}\frac{1}{\tilde{\omega}^2}, \tag{6.126}$$

if $\alpha = rd\ln\Sigma_0/dr$ [i.e., Eq. (6.118)] is adopted.

As mentioned before, there are wavy perturbations around the corotation resonance, whose frequency is essentially complex. Our problem is whether there are growing (unstable) oscillations which are trapped near the corotation radius. This issue has been examined by solving equations similar to the set of Eqs. (6.123) and (6.126) with relevant boundary conditions. This is a complicated problem, but Lovelace et al. (1999) and Li et al. (2000) showed the presence of unstable trapped oscillations. They called it Rossby wave instability, because, as mentioned before, if we derive a local dispersion relation it is analogous to that of Rossby waves in the sense that the wave propagates only in one direction.[12] Subsequently, the instability was examined for barotropic disks by Tsang and Lai (2008) by the WKBJ method and by Ono et al. (2014, 2016, 2018) by analytical and numerical methods for various distributions of vortensity. Nonlinear numerical calculations also have been made (Li et al. 2001; Ono et al. 2018).

Here, we shall be satisfied with presenting a necessary condition of the instability, following Lovelace et al. (1999). Multiplying Eq. (6.123) by $\tilde{\Pi}_1^*$ (complex conjugate of $\tilde{\Pi}_1$) and integrating over the disk, we have

$$-\int\left|\frac{d\tilde{\Pi}_1}{dr}\right|^2 d^2r + \int K^2|\tilde{\Pi}_1|^2 d^2r = 0, \tag{6.127}$$

[12]Notice that K^2 given by Eq. (6.124) has an odd term with respect to $\tilde{\omega}$.

where we have assumed $\tilde{\Pi}_1^* d\tilde{\Pi}_1/dr$ tends to zero for $r \to 0$ and ∞. It is noted that the second integration of Eq. (6.127) is a complex quantity, since $\tilde{\omega}$ is complex. In the case where $\tilde{\omega}$ is a finite fraction of Ω and $c_s^2/r^2\Omega^2 \ll 1$ (geometrically thin disks), the contribution of the term with $1/\tilde{\omega}^2$ in K^2 to the second integral of Eq. (6.127) can be neglected compared with that with $1/\tilde{\omega}$ in evaluation of the second term in Eq. (6.127). Hence, for the imaginary part of relation (6.127) to vanish, we need

$$\Im \int 2m\Omega \frac{d}{rdr} \ln\left(\frac{\kappa^2}{2\Omega\,\Sigma_0} \frac{p_0}{\rho_0^{1/\gamma}}\right) \frac{1}{\tilde{\omega}} |\tilde{\Pi}_1|^2 d^2 r = 0. \tag{6.128}$$

For this to be realized, we see that $\mathcal{L}(r)$ defined by

$$\mathcal{L}(r) \equiv \frac{d}{rdr} \ln\left(\frac{\kappa^2}{2\Omega\,\Sigma_0} \frac{p_0}{\rho_0^{1/\gamma}}\right) \tag{6.129}$$

vanishes at some r. In other words, $\mathcal{L}(r)$ has a maximum or a minimum at a radius r. This is a necessary condition for presence of the Rossby wave instability (Lovelace et al. 1999, see also Ono et al. 2016 for a sufficient condition).

The Rossby wave instability is important as a possible origin of non-axisymmetric structure and angular momentum transport in planetary systems. For detailed studies on the Rossby wave instability we recommend to consult the above-mentioned references.

6.7 Wave-Wave Resonance Instability in Deformed Disks: Resonant Instability III

In Sect. 6.5 we have shown that the Kelvin–Helmholtz instability can be interpreted as a result of resonant interaction between two waves with opposite signs of wave energy. The corotation resonant instability described in Sect. 6.6 is also interpreted by the same concept, as mentioned in the last part of the section. This concept that an instability occurs when two resonant coupling waves have opposite signs of wave energy can be extended to more general cases.

For example, in dwarf novae disks the so-called 3:1 resonant instability (Whitehurst 1988) is known, which has been explained by a mode-mode coupling due to resonance (Hirose and Osaki 1990; Lubow 1991, 1992; Osaki 1996) (see Sect. 6.7.4). Furthermore, Goodman (1993) and Ryu and Goodman (1994) found a local instability (elliptical instability) in deformed disks in studying possible sources of turbulence in accretion disks. Lubow's idea of mode-mode coupling was extended to a global wave-wave coupling in deformed disks by Kato (2004, 2008), Ferreira and Ogilvie (2008), and Oktariani et al. (2010) in examining possible origins of quasi-periodic oscillations (QPOs) in X-ray binaries. Furthermore, we notice that the concept of wave-wave coupling instability with opposite signs of wave energies seems to be popular in fluid dynamics. For example, Khalzov et al. (2008) emphasized that such coupling can be interpreted as the causes of many MHD instabilities of flowing media.

The concept of resonant instability of two waves with opposite wave energies has been generalized to the case where the unperturbed flow is time-periodic (Kato 2013a,b, 2014a,b). The generalized instability condition is quite simple,[13] and in the case where the unperturbed flow is steady the instability condition is reduced to that mentioned above, i.e., the two resonantly interacting waves are unstable if their wave energies have opposite signs.

To demonstrate concrete situations we consider a disk around the primary star of a binary system. The disk is assumed, for simplicity, to be on the orbital plane of the secondary and deformed from an axisymmetric state by the tidal force resulting from the secondary star. We introduce here cylindrical coordinates (r, φ, z) whose origin is at the center of the star (also the disk center) and the z-axis is the rotating axis of the disk. If the orbit of the secondary star is eccentric, a few types of tidal waves are induced on the disk. We now consider a tidal wave whose frequency is $\omega_{\rm D}$ and azimuthal wavenumber $m_{\rm D}$. The displacement vector, $\boldsymbol{\xi}_{\rm D}(\boldsymbol{r}, t)$, associated with the deformation is denoted by

$$\boldsymbol{\xi}_{\rm D}(\boldsymbol{r}, t) = \Re\left[\hat{\boldsymbol{\xi}}_{\rm D}(\boldsymbol{r})\exp(i\omega_{\rm D}t)\right] = \Re\left[\breve{\boldsymbol{\xi}}_{\rm D}(r, z)\exp[i(\omega_{\rm D}t - m_{\rm D}\varphi)]\right], \tag{6.130}$$

where $\Re$ denotes the real part, and the subscript D is attached in order to emphasize the tidal deformation. Here, the tidal displacement of frequency $\omega_{\rm D}$ and azimuthal wavenumber $m_{\rm D}$ are considered. The vectors $\hat{\boldsymbol{\xi}}_{\rm D}(\boldsymbol{r})$ and $\breve{\boldsymbol{\xi}}_{\rm D}(r, z)$ are defined by Eq. (6.130). In addition to this tidal deformation, a set of two normal modes of oscillations are considered on the axisymmetric steady disks. The set of eigenfrequency and azimuthal wavenumber of these oscillations are denoted by (ω_1, m_1) and (ω_2, m_2). The displacement vectors, $\boldsymbol{\xi}_{\rm i}(\boldsymbol{r}, t)$ $({\rm i} = 1, 2)$, associated with these oscillations are expressed as

$$\boldsymbol{\xi}_{\rm i}(\boldsymbol{r}, t) = \Re\left[\hat{\boldsymbol{\xi}}_{\rm i}(\boldsymbol{r})\exp(i\omega_{\rm i}t)\right] = \Re\left[\breve{\boldsymbol{\xi}}_{\rm i}(r, z)\exp[i(\omega_{\rm i}t - m_{\rm i}\varphi)]\right] \quad ({\rm i} = 1, 2), \tag{6.131}$$

where the subscript i denotes two different oscillation modes of $\rm i = 1, 2$.

We now assume that the following three-wave resonant conditions among the above two oscillations and the tidal wave are present:

$$\omega_1 + \omega_2 + \omega_{\rm D} = 0 \quad \text{and} \quad m_1 + m_2 + m_{\rm D} = 0, \tag{6.132}$$

[13]The generalized instability condition is given by inequality (6.157).

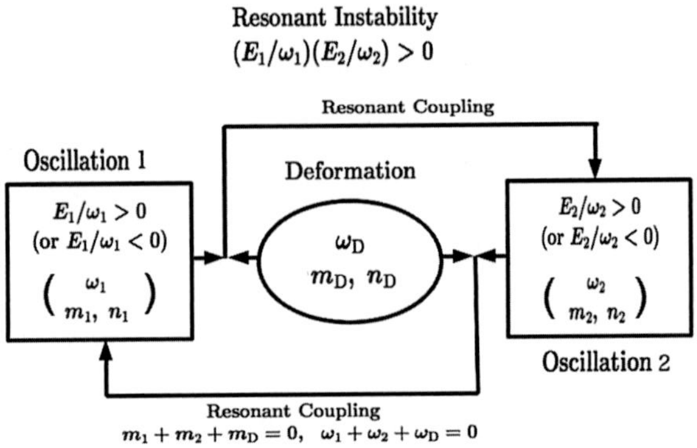

Fig. 6.9 Schematic diagram showing simultaneous amplification of two oscillations with (ω_1, m_1) and (ω_2, m_2) through resonant couplings with tidal wave with (ω_D, m_D). The conditions (6.132) are necessary conditions for resonant excitation. As shown in the text, $(E_1/\omega_1)(E_2/\omega_2) > 0$ is the necessary condition for the resonance to lead to instability [see inequality (6.157)] (After Kato 2016)

where m_i's (i $=$ 1, 2, and D) are integers. In order to represent the resonant conditions by simple forms without separately considering such cases as $\omega_2 = \omega_1 + \omega_D$ and $\omega_2 = -\omega_1 + \omega_D$, we have adopted equations (6.132). Hence, ω's and m's can be taken to be positive or negative as long as the conditions (6.132) are satisfied. The above resonant couplings are schematically shown in Fig. 6.9. The above conditions (6.132) are only necessary ones for resonant instability. To know whether the resonance really leads to instability and what is the condition of occurrence of instability, we must examine characteristics of couplings. As is shown below, this examination can be made rather generally, but the resulting conditions of instability have simple forms.

6.7.1 *Quasi-Nonlinear Coupling between Two Oscillations through Tidal Wave*

First, let us consider a small amplitude perturbation, $\boldsymbol{\xi}$, over an axially symmetric, purely rotating disk. The perturbation is governed by the linearized wave equation, which is, as mentioned in Chap. 2,

$$\rho_0 \frac{\partial^2 \boldsymbol{\xi}}{\partial t^2} + 2\rho_0 (\boldsymbol{v}_0 \cdot \nabla) \frac{\partial \boldsymbol{\xi}}{\partial t} + \mathcal{L}(\boldsymbol{\xi}) = 0, \tag{6.133}$$

where $\boldsymbol{v}_0$ is the unperturbed flow and $(0,\ r\Omega(r),\ 0)$ in the present case. Now we assume that the unperturbed disks are weakly deformed. In order to make situations clear, the deformation is a forced oscillation due to tidal force resulting from the tidal potential $\psi_{\rm D}(\boldsymbol{r})$ of the secondary. The displacement vector, $\boldsymbol{\xi}_{\rm D}$, associated with the tidal force is then governed by (no perturbation except for the tidal one)

$$\rho_0 \frac{\partial^2 \boldsymbol{\xi}_{\rm D}}{\partial t^2} + 2\rho_0(\boldsymbol{v}_0 \cdot \nabla)\frac{\partial \boldsymbol{\xi}_{\rm D}}{\partial t} + \mathcal{L}(\boldsymbol{\xi}_{\rm D}) = -\rho_0 \nabla \psi_{\rm D}. \tag{6.134}$$

Now, we consider how the perturbation $\boldsymbol{\xi}$ is affected by weak quasi-nonlinear coupling with the tidal perturbation. Hereafter, we represent the perturbations affected by the tidal perturbation by $\boldsymbol{\xi}$. It is noted that this is the same notation as $\boldsymbol{\xi}$ in Eq. (6.133), but different from that. This new displacement vector $\boldsymbol{\xi}$ is governed by the wave equation given by

$$\rho_0 \frac{\partial^2 \boldsymbol{\xi}}{\partial t^2} + 2\rho_0(\boldsymbol{v}_0 \cdot \nabla)\frac{\partial \boldsymbol{\xi}}{\partial t} + \mathcal{L}(\boldsymbol{\xi}) = \rho_0 \boldsymbol{C}(\boldsymbol{\xi}, \boldsymbol{\xi}_{\rm D}), \tag{6.135}$$

where the right-hand side is the coupling term with the tidal deformation, and represents how the perturbation is modified by tidal force. We assume that both the disk deformation due to tidal force, $\boldsymbol{\xi}_{\rm D}$, and the perturbation superposed on disk are weak, and consider only the quasi-nonlinear coupling among them.[14]

Even in the case of quasi-nonlinear coupling, derivation of the detailed expression for the coupling term, $\boldsymbol{C}$, is complicated. An expression for $\boldsymbol{C}$ is given in Eq. (A.25) in Appendix A, which is

$$\begin{aligned} C_i(\boldsymbol{\xi}, \boldsymbol{\xi}_{\rm D}) = & -\rho_0 \xi_j \xi_{{\rm D}k} \frac{\partial^3 \psi_0}{\partial r_i \partial r_j \partial r_k} - \frac{\partial}{\partial r_k}\left(p_0 \frac{\partial \xi_j}{\partial r_i}\frac{\partial \xi_{{\rm D}k}}{\partial r_j} + p_0 \frac{\partial \xi_{{\rm D}j}}{\partial r_i}\frac{\partial \xi_k}{\partial r_j}\right) \\ & + \frac{\partial}{\partial r_j}\left[(\Gamma_1 - 1)p_0\left(\frac{\partial \xi_j}{\partial r_i}{\rm div}\boldsymbol{\xi}_{\rm D} + \frac{\partial \xi_{{\rm D}j}}{\partial r_i}{\rm div}\boldsymbol{\xi}\right)\right] \\ & + \frac{\partial}{\partial r_i}\left[(\Gamma_1 - 1)p_0 \frac{\partial \xi_k}{\partial r_j}\frac{\partial \xi_{{\rm D}j}}{\partial r_k}\right] + \frac{\partial}{\partial r_i}\left[(\Gamma_1 - 1)^2 p_0 {\rm div}\boldsymbol{\xi} \cdot {\rm div}\boldsymbol{\xi}_{\rm D}\right] \\ & - \rho_0 \xi_j \frac{\partial^2 \psi_{\rm D}}{\partial r_j \partial r_i}, \end{aligned} \tag{6.136}$$

where Γ_1 is the barotropic index specifying the linear part of the relation between the Lagrangian variations, δp and $\delta \rho$, i.e., $(\delta p/p_0)_{\rm linear} = \Gamma_1(\delta\rho/\rho_0)_{\rm linear}$. This detailed expression for $\boldsymbol{C}(\boldsymbol{\xi}, \boldsymbol{\xi}_{\rm D})$ is unnecessary for a moment, but it has an important characteristic [see Eq. (6.155)], which will be used later to derive the instability condition.

[14] It is noticed that $\boldsymbol{\xi}_{\rm D}$ is assumed to be unchanged by the presence of wave motion $\boldsymbol{\xi}$, and thus Eq. (6.135) is a linear equation with respect to $\boldsymbol{\xi}$.

Now, we start from the state where two normal modes of oscillations, $\boldsymbol{\xi}_1$ and $\boldsymbol{\xi}_2$, with (ω_1, m_1) and (ω_2, m_2), are on axisymmetric purely rotating steady disks.

$$\begin{aligned}\boldsymbol{\xi}(\boldsymbol{r},t) &= A_1\boldsymbol{\xi}_1(\boldsymbol{r},t) + A_2\boldsymbol{\xi}_2(\boldsymbol{r},t)\\ &= \Re\Big[A_1\hat{\boldsymbol{\xi}}_1\exp[i(\omega_1 t)] + A_2\hat{\boldsymbol{\xi}}_2\exp[i(\omega_2 t)]\Big]\\ &= \Re\Big[A_1\breve{\boldsymbol{\xi}}_1\exp[i(\omega_1 t - m_1\varphi)] + A_2\breve{\boldsymbol{\xi}}_2\exp[i(\omega_2 t - m_2\varphi)]\Big], \end{aligned} \quad (6.137)$$

where A_1 and A_2 are amplitudes and are arbitrary constants at the linear stage. The second and third equalities in Eq. (6.137) present the definitions of $\hat{\boldsymbol{\xi}}$ and $\breve{\boldsymbol{\xi}}$, respectively. If a tidal deformation of the disk is considered, the perturbation $\boldsymbol{\xi}(\boldsymbol{r},t)$ given by Eq. (6.137) is modified so that they satisfy a quasi-nonlinear wave equation (6.135).

The disk oscillations, $\boldsymbol{\xi}(\boldsymbol{r},t)$, resulting from the quasi-nonlinear resonant coupling through disk deformation, $\boldsymbol{\xi}_{\rm D}(\boldsymbol{r},t)$, will be written generally in the form:

$$\begin{aligned}\boldsymbol{\xi}(\boldsymbol{r},t) = \Re\sum_{\rm i=1}^{2} A_{\rm i}(t)\hat{\boldsymbol{\xi}}_{\rm i}(\boldsymbol{r})\exp(i\omega_{\rm i}t) + \Re\sum_{i}^{2}\sum_{\alpha\neq 1,2} A_{i,\alpha}\hat{\boldsymbol{\xi}}_{\alpha}(\boldsymbol{r})\exp(i\omega_{\rm i}t)\\ + \text{oscillating terms with other frequencies.}\end{aligned} \quad (6.138)$$

The original two oscillations, $\boldsymbol{\xi}_1(\boldsymbol{r},t)$ and $\boldsymbol{\xi}_2(\boldsymbol{r},t)$, resonantly interact through the disk deformation. Hence, their amplitudes secularly change with time, which is taken into account in Eq. (6.138) by $A_{\rm i}$'s being taken to be slowly varying functions of time. The terms whose time-dependence is $\exp(i\omega_{\rm i}t)$ but spatial dependences are different from $\hat{\boldsymbol{\xi}}_{\rm i}$ $(i = 1.2)$ are expressed by a sum of a series of eigenfunctions, $\hat{\boldsymbol{\xi}}_\alpha$ ($\alpha \neq 1$ and 2), assuming that they make a complete set. The terms whose time-dependences are different from $\exp(i\omega_{\rm i}t)$ $(i = 1, 2)$ are not written down explicitly in Eq. (6.138), because these terms disappear by taking long-term time average when we are interested in phenomena with frequencies of $\omega_{\rm i}$ (i = 1 and 2).

In order to derive equations describing the time evolution of $A_{\rm i}(t)$, we substitute Eq. (6.138) into the left-hand side of Eq. (6.135). Then, considering that $\hat{\boldsymbol{\xi}}_{\rm i}$'s $(i = 1.2)$ and $\hat{\boldsymbol{\xi}}_\alpha$'s $(\alpha \neq 1, 2)$ are the displacement vectors associated with eigenfunctions of the linear wave equation (6.133), we find that the left-hand side of Eq. (6.135) is the real part of

$$\begin{aligned}&2\rho_0\sum_{i=1}^{2}\frac{dA_{\rm i}}{dt}\Big[i\omega_{\rm i} + (\boldsymbol{v}_0\cdot\nabla)\Big]\hat{\boldsymbol{\xi}}_{\rm i}\exp(i\omega_{\rm i}t)\\ &+\rho_0\sum_i\sum_\alpha A_{i,\alpha}\Big[\left(\omega_\alpha^2 - \omega_{\rm i}^2\right) - 2i(\omega_\alpha - \omega_{\rm i})(\boldsymbol{v}_0\cdot\nabla)\Big]\\ &\times\hat{\boldsymbol{\xi}}_\alpha\exp(i\omega_{\rm i}t),\end{aligned} \quad (6.139)$$

where d^2A_i/dt^2 has been neglected, since $A_i(t)(i = 1$ and $2)$ are slowly varying functions of time. Now, the real part of Eq. (6.139) is integrated over the whole volume of disks after being multiplied by $\boldsymbol{\xi}_1^*[= \Re\ \hat{\boldsymbol{\xi}}_1^* \exp(i\omega_1 t)]$.[15] Then, the term resulting from the second term of Eq. (6.139) vanishes. This comes from a generalized orthogonal relation given by Eq. (2.60) in Chap. 2. Using this result, we find that the above-mentioned integration of the left-hand side of Eq. (6.135) is reduced to

$$\Re\, i \frac{dA_1}{dt} \left\langle \rho_0 \hat{\boldsymbol{\xi}}_1^* [\omega_1 - i(\boldsymbol{v}_0 \cdot \nabla)] \hat{\boldsymbol{\xi}}_1 \right\rangle, \tag{6.140}$$

where $\langle X \rangle$ denotes the volume integration of $X(\boldsymbol{r})$ over the whole volume, and the asterisk * denotes the complex conjugate.

On the other hand, the integration of the right-hand side of Eq. (6.135) over the whole volume after being multiplied by $\hat{\boldsymbol{\xi}}_1^*$ gives

$$\frac{1}{2} \Re \left[A_2(t) A_{\rm D} \left\langle \hat{\boldsymbol{\xi}}_1^* \cdot \boldsymbol{C} \left(\hat{\boldsymbol{\xi}}_2, \hat{\boldsymbol{\xi}}_{\rm D} \right) \right\rangle \right]. \tag{6.141}$$

Hence, equating the above two equations we have

$$\Re\, i \frac{dA_1}{dt} \left\langle \rho_0 \hat{\boldsymbol{\xi}}_1^* [\omega_1 - i(\boldsymbol{v}_0 \cdot \nabla)] \hat{\boldsymbol{\xi}}_1 \right\rangle = \frac{1}{2} \Re \left[A_2(t) A_{\rm D} \left\langle \hat{\boldsymbol{\xi}}_1^* \cdot \boldsymbol{C} \left(\hat{\boldsymbol{\xi}}_2, \hat{\boldsymbol{\xi}}_{\rm D} \right) \right\rangle \right], \tag{6.142}$$

similarly, we multiply $\boldsymbol{\xi}_2^*(\boldsymbol{r})$ to the real part of Eq. (6.135) and integrate over the whole volume to lead to

$$\Re\, i \frac{dA_2}{dt} \left\langle \rho_0 \hat{\boldsymbol{\xi}}_2^* [\omega_1 - i(\boldsymbol{v}_0 \cdot \nabla)] \hat{\boldsymbol{\xi}}_2 \right\rangle = \frac{1}{2} \Re \left[A_1(t) A_{\rm D} \left\langle \hat{\boldsymbol{\xi}}_2^* \cdot \boldsymbol{C} (\hat{\boldsymbol{\xi}}_1, \hat{\boldsymbol{\xi}}_{\rm D}) \right\rangle \right]. \tag{6.143}$$

6.7.2 Wave Energy and Its Sign

The above two Eqs. (6.142) and (6.143) can be further reduced to simpler forms. To do so we shall remember the concept of wave energy. In Chap. 2 we have introduced

[15] The formula

$$\Re(A)\Re(B) = \frac{1}{2}\Re[AB + AB^*] = \frac{1}{2}\Re[AB + A^*B]$$

is used, where A and B are complex variables and B^* is the complex conjugate of B.

the wave energy, E, which is defined by

$$E = \frac{1}{2}\int \left[\rho_0\left(\frac{\partial \boldsymbol{\xi}}{\partial t}\right)^2 + \boldsymbol{\xi}\cdot\mathcal{L}(\boldsymbol{\xi})\right]dV. \tag{6.144}$$

An important point is that in linear perturbations in unperturbed disks, the wave energy E is time-independent [see Eq. (2.54) in Chap. 2]. In normal mode of oscillations, in particular, E can be expressed as

$$E = \frac{1}{2}\int \rho_0\left[\omega^2\rho_0\hat{\boldsymbol{\xi}}^*\hat{\boldsymbol{\xi}} + \hat{\boldsymbol{\xi}}^*\cdot\mathcal{L}(\hat{\boldsymbol{\xi}})\right]dV. \tag{6.145}$$

If $\mathcal{L}(\boldsymbol{\xi})$ is eliminated from Eq. (6.145) by using wave equation (6.133), we have

$$\begin{aligned} E &= \frac{1}{2}\int \rho_0\left[\omega^2\rho_0\hat{\boldsymbol{\xi}}^*\hat{\boldsymbol{\xi}} - i\omega\rho_0\hat{\boldsymbol{\xi}}^*(\boldsymbol{v}_0\cdot\nabla)\hat{\boldsymbol{\xi}}\right]dV \\ &= \frac{1}{2}\int \rho_0\left[\omega^2\rho_0\hat{\boldsymbol{\xi}}\hat{\boldsymbol{\xi}}^* + i\omega\rho_0\hat{\boldsymbol{\xi}}(\boldsymbol{v}_0\cdot\nabla)\hat{\boldsymbol{\xi}}^*\right]dV. \end{aligned} \tag{6.146}$$

We are now considering disks whose unperturbed state has no motion except for rotation, i.e., $\boldsymbol{v}_0 = (0, r\Omega, 0)$. Then, we have

$$\begin{aligned} \breve{\boldsymbol{\xi}}^*(\boldsymbol{v}_0\cdot\nabla)\breve{\boldsymbol{\xi}} &= -im\Omega\breve{\boldsymbol{\xi}}^*\breve{\boldsymbol{\xi}} - \Omega\breve{\xi}_r^*\breve{\xi}_\varphi + \Omega\breve{\xi}_\varphi^*\breve{\xi}_r \\ &= -im\Omega\breve{\boldsymbol{\xi}}^*\breve{\boldsymbol{\xi}} - i(\omega - m\Omega)\breve{\xi}_\varphi^*\breve{\xi}_\varphi. \end{aligned} \tag{6.147}$$

In deriving the last relation, we have used

$$i(\omega - m\Omega)\breve{\xi}_\varphi + 2\Omega\breve{\xi}_r = 0. \tag{6.148}$$

This comes from the φ-component of equation of motion (6.133) under the approximation that the perturbations are tightly wound and thus such terms as $\partial\delta p/r\partial\varphi$ can be neglected. If relation (6.147) is used, the wave energy given by Eq. (6.146) is written in the form:

$$E = \frac{1}{2}\int \rho_0\omega(\omega - m\Omega)(\breve{\xi}_r^*\breve{\xi}_r + \breve{\xi}_z^*\breve{\xi}_z)dV. \tag{6.149}$$

This form of wave energy is important and useful, because this expression shows that the sign of wave energy is directly related to the sign of $\omega - m\Omega$ in the region where the wave predominantly exists, if ω is taken to be positive. The radius where $\omega = m\Omega$ is the corotation radius, and the region around the radius is usually evanescent region of oscillations. That is, waves which exist inside the corotation radius ($\omega - m\Omega < 0$) penetrate little into outside the corotation one. Similarly, those outside the corotation radius ($\omega - m\Omega > 0$) penetrate little inside the corotation one.

Hence, if we consider waves with $\omega > 0$, the waves inside the corotation radius have negative energy, while those outside the corotation radius have positive energy.

By using expression (6.146), we can reduce equations (6.142) and (6.143) to

$$\Re\, i\frac{2E_1}{\omega_1}\frac{dA_1}{dt} = \frac{1}{2}\Re\left[A_2(t)A_{\rm D}\left\langle\hat{\boldsymbol{\xi}}_1^*\cdot \boldsymbol{C}\left(\hat{\boldsymbol{\xi}}_2,\hat{\boldsymbol{\xi}}_{\rm D}\right)\right\rangle\right], \tag{6.150}$$

$$\Re\, i\frac{2E_2}{\omega_1}\frac{dA_2}{dt} = \frac{1}{2}\Re\left[A_1(t)A_{\rm D}\left\langle\hat{\boldsymbol{\xi}}_2^*\cdot \boldsymbol{C}\left(\hat{\boldsymbol{\xi}}_1,\hat{\boldsymbol{\xi}}_{\rm D}\right)\right\rangle\right], \tag{6.151}$$

where E_1 and E_2 are wave energies of normal modes of oscillations, $\boldsymbol{\xi}_1$ and $\boldsymbol{\xi}_2$, respectively.

6.7.3 Conditions of Resonant Growth

Since the amplitude $A_{\rm D}$ is taken to be constant, the time evolutions of A_1 and A_2 are determined by solving simultaneously Eqs. (6.150) and (6.151). What are governed by these equations are the imaginary part of A_1 and A_2, i.e., $A_{1\rm i}$ and $A_{2\rm i}$. Their real parts are not related to the resonance, and we can neglect them in considering the exponential growth (or damping) of $A_{1\rm i}$ and $A_{2\rm i}$. Then, we have from Eqs. (6.150) and (6.151)

$$-\frac{2E_1}{\omega_1}\frac{A_{1,\rm i}}{dt} = -\frac{1}{2}A_{2,\rm i}\Im(A_{\rm D}W_1), \tag{6.152}$$

$$-\frac{2E_2}{\omega_2}\frac{A_{2,\rm i}}{dt} = -\frac{1}{2}A_{1,\rm i}\Im(A_{\rm D}W_2), \tag{6.153}$$

where

$$W_1 = \left\langle\hat{\boldsymbol{\xi}}_1^*\cdot \boldsymbol{C}\left(\hat{\boldsymbol{\xi}}_2,\hat{\boldsymbol{\xi}}_{\rm D}\right)\right\rangle, \qquad W_2 = \left\langle\hat{\boldsymbol{\xi}}_2^*\cdot \boldsymbol{C}\left(\hat{\boldsymbol{\xi}}_1,\hat{\boldsymbol{\xi}}_{\rm D}\right)\right\rangle. \tag{6.154}$$

An important issue to be noticed here is that $\hat{\boldsymbol{\xi}}_1$ and $\hat{\boldsymbol{\xi}}_2$ in W_1 and W_2 are commutative. We can show this by using the detailed expression for $\boldsymbol{C}$ given by Eq. (6.136), performing the volume integration by part, and neglecting surface integrals under the assumption that the unperturbed density and pressure vanish on the surface of the system (see Kato 2004). That is, we have

$$W_1 = W_2 \equiv W. \tag{6.155}$$

Equations (6.152) and (6.153) are thus a set of linear simultaneous differential equations with constant coefficients. Eliminating $A_{2,\rm i}$ from Eqs. (6.152) and (6.153), we have

$$\frac{d^2 A_{1,\mathrm{i}}}{dt^2} = \frac{1}{16}\left(\frac{E_1 E_2}{\omega_1 \omega_2}\right)^{-1} \left[\Im(A_\mathrm{D} W)\right]^2 A_{1,\mathrm{i}}. \tag{6.156}$$

A similar equation is obtained for $A_{2,\mathrm{i}}$ by eliminating $A_{1,\mathrm{i}}$ instead of $A_{2,\mathrm{i}}$.

Equation (6.156) shows that the condition of instability is

$$\frac{E_1 E_2}{\omega_1 \omega_2} > 0, \tag{6.157}$$

and the growth rate n is given by

$$n = \left(\frac{\omega_1 \omega_2}{16 E_1 E_2}\right)^{1/2} |\Im(A_\mathrm{D} W)|. \tag{6.158}$$

In the particular case of $\omega_\mathrm{D} = 0$, the resonant condition, $\omega_1 + \omega_2 + \omega_\mathrm{D} = 0$, is reduced to $\omega_1 = -\omega_2$ and thus the instability condition (6.157) is

$$E_1 E_2 < 0. \tag{6.159}$$

That is, in this particular case two resonantly interacting waves are amplified, if they have opposite signs of wave energy. The Kelvin–Helmholtz instability belongs to this case.

6.7.4 An Application: Superhumps of Dwarf Novae

Dwarf novae are binary systems: the primary star is a neutron star and the secondary is a red dwarf. In dwarf novae, in addition to normal outbursts which occur by thermal instability of shell burning (see Sect. 7.2), superoutbursts occur less frequently compared with normal outburst. The origin of superoutburst is the thermal instability of disks (see Sect. 7.3). During the superoutburst, periodic humps in luminosity, called *superhumps*, occur.

As is well-known, if the circular orbit of the secondary is sufficiently far from the primary and the disk is in the orbital plane of the secondary, the main tidal wave on the disk is that with $m_\mathrm{D} = 2$ and $\omega_\mathrm{D} = 2\Omega_\mathrm{orb}$, where Ω_orb is the orbital frequency of the secondary around the primary, observed from the primary. If the orbital radius is not sufficiently large, the tidal wave with $m_\mathrm{D} = 3$ and $\omega_\mathrm{D} = 3\Omega_\mathrm{orb}$ is also prominent. The superoutbursts are known to occur when the disk outer edge extends to the radius where the Keplerian angular velocity of disk rotation, $\Omega(r)$, decreases to $3\Omega_\mathrm{orb}$. Hence, the cause of superhumps is considered to be due to a disk instability at $\Omega = 3\Omega_\mathrm{orb}$, and this instability is called the *3:1 resonant instability*. This instability was found numerically (Whitehurst 1988) and explained theoretically (Hirose and Osaki 1990; Lubow 1991). This tidal instability can be explained as an example of the wave-wave resonant instability whose instability condition is given by inequality (6.157), as will be shown below.

Let us consider one-armed ($m = 1$) low frequency p-mode oscillations in tidally deformed disks as one of the pair of two oscillations. First, we remember that the local dispersion relation of p-mode oscillations is $(\omega - m\Omega)^2 - \kappa^2 = c_s^2 k^2$, where ω and k are the frequency and wavenumber of the oscillations, respectively, and c_s is the acoustic speed in the disk. Hence, the propagation regions of one-armed p-mode oscillations are specified by $(\omega - \Omega)^2 > \kappa^2$, and especially the propagation region of the low frequency one is $\omega < \Omega - \kappa$. In tidally deformed disks the epicyclic frequency, $\kappa(r)$, is slightly smaller than $\Omega(r)$, and thus $\Omega - \kappa$ is positive and increases with radius r (eg., see Kato 2016) as is demonstrated in Fig. 6.10. Then, if we consider an oscillation whose frequency is ω_1 (see Fig. 6.10), its propagation region is between r_c and r_t, and the oscillation is trapped there, as is shown in Fig. 6.10. Here, r_t is the outer edge of the disk and r_c is the radius where $\omega_1 = \Omega - \kappa$.

As the ω_2-oscillation, we take two-armed ($m_2 = 2$) p-mode oscillation. Then, one of its propagation region is specified by $\omega_2 < 2\Omega - \kappa$ (see Fig. 6.10). If we take its propagation region extends till the outer edge, r_t, of the disk so that ω_1- and ω_2-oscillations can strongly interact spatially (the propagation regions of the two oscillations overlap in the radial direction and W is large), we have $\omega_2 = (2\Omega - \kappa)_t$.

The above-mentioned two oscillations satisfy one of the resonant conditions, i.e., $m_1 + m_2 + m_D = 0$, if $m_D = -3$. The resonant condition for frequency, $\omega_1 + \omega_2 + \omega_D = 0$, is written as

$$(\Omega - \kappa)_c + (2\Omega - \kappa)_t + \omega_D = 0. \tag{6.160}$$

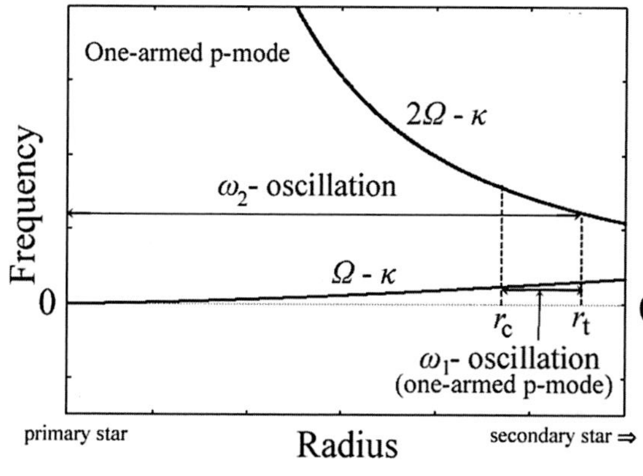

Fig. 6.10 Schematic diagram showing frequencies and propagation regions of the ω_1- and ω_2-oscillations in the case where the ω_1-oscillation is one-armed low frequency p-mode oscillation, and the ω_2-oscillation is two-armed p-mode oscillation. The scales of coordinates are arbitrary, and are not linear. The one-armed p-mode oscillation is trapped between r_c and r_t. The inside is the evanescent region. The ω_2-oscillation can propagate inside r_t (After Kato 2016)

Since $\kappa \sim \Omega$, this condition gives $\omega_D \sim -\Omega(r_t)$. The set of $m_D = -3$ and $\omega_D = -\Omega_t$ is equivalent to the set of $m_D = 3$ and $\omega_D = \Omega_t$, since the simultaneous changes of the signs of m_D and ω_D represent the same tidal wave. This result shows that if the tidal wave has a component of $m_D = 3$ and $\omega_D = \Omega(r_t)$, a resonant interaction between the ω_1- and ω_2-oscillations occurs. Hence, our problem is whether such situations can really occur. As mentioned before, the tidal wave has a component of $m_D = 3$ and $\omega_D = 3\Omega_{orb}$, if binary separation is not sufficiently far. If the disk size, r_t, is small, $\Omega(r_t)$ is much larger than Ω_{orb}, but as the disk size increases (by thermal instability), $\Omega(r_t)$ decreases and at a certain radius it can become equal to $3\Omega_{orb}$. That is, the required resonant condition $\omega_D = 3\Omega_{orb}$ is satisfied at the stage where $\Omega_t = 3\Omega_{orb}$. This is the so-called 3 : 1 resonance known as the cause of superoutburst in dwarf novae.

Next, we should examine whether the above resonantly interacting two waves (ω_1- and ω_2-oscillations) satisfy the resonant excitation condition of $(E_1/\omega_1)(E_2/\omega_2) > 0$. As shown in Eq. (6.149), the sign of wave energy of oscillations in thin disks is the same as the sign of $(\omega - m\Omega)$. This means that in the present case both of E_1 and E_2 are negative and thus $E_1/\omega_1 < 0$ and $E_2/\omega_2 < 0$, leading to $(E_1/\omega_1)(E_2/\omega_2) > 0$. That is, the resonant interaction works so as to excite the oscillations.

In summary, if the disk radius of dwarf novae extends to the radius where the so-called 3 : 1 resonance is realized, the disk becomes unstable (superoutburst) and one-armed p-mode oscillation with slow precession is excited by the wave-wave resonant process. It is noted that this picture of the superhumps suggests that a two-armed high-frequency p-mode oscillation is excited simultaneously when superhumps appear.

References

Balbus, S.A.: Astrophysical J. **453**, 380 (1995)
Blaes, O.M.: Mon. Not. R. Astron. Soc. **216**, 553 (1985)
Bleas, O.M.: Mon. Not. R. Astron. Soc. **227**, 975 (1987)
Cairns, R.A.: J. Fluid Mech. **92**, 1 (1979)
Chandrasekhar, S.: Hydrodynamics and Hydromagnetic Stability. Clarendon Press, Oxford (1961)
Drury, L.O.: Mon. Not. R. Astron. Soc. **217**, 821 (1985)
Ferreira, B., Ogilvie, G.: Mon. Not. R. Astron. Soc. **386**, 2297 (2008)
Glatzel, W.: Mon. Not. R. Astron. Soc. **228**, 77 (1987)
Goldreich, P., Lynden-Bell, D.: Mon. Not. R. Astron. Soc. **130**, 125 (1965)
Goldreich, P., Narayan, R.: Mon. Not. R. Astron. Soc. **213**, 7 (1985)
Goldreich, P., Goodman, J., Narayan, R.: Mon. Not. R. Astron. Soc. **221**, 339 (1986)
Goodman, J.: Astrophys. J. **406**, 596 (1993)
Hirose, M., Osaki, Y.: Publ. Astron. Soc. Jpn. **42**, 135 (1990)
Jeans, J.H.: Phil. Trans. R. Soc. London. Series A. **199**, 1 (1902)
Kato, S.: Publ. Astron. Soc. Jpn. **55**, 257 (2003)
Kato, S.: Publ. Astron. Soc. Jpn. **56**, 905 (2004)
Kato, S.: Publ. Astron. Soc. Jpn. **60**, 111 (2008)
Kato, S.: Publ. Astron. Soc. Jpn. **65**, 56 (2013a)

Kato, S.: Publ. Astron. Soc. Jpn. **65**, 75 (2013b)
Kato, S.: Publ. Astron. Soc. Jpn. **66**, 21 (2014a)
Kato, S.: Publ. Astron. Soc. Jpn. **66**, 24 (2014b)
Kato, S.: Oscillations of Disks. Astrophysics and Space Science Library, vol. 437. Springer, Tokyo (2016)
Khalzov, I.V., Smolyakov, A.I., Hgisonis, V.I.: Phys. Plasma **15**, 4501 (2008)
Lai, D., Tsang, D.: Mon. Not. R. Astron, Soc. **393**, 979 (2009)
Lau, Y.Y., Lin, C.C., Mark, J.W.-K.: Proc. Nat. Acad. Sci. U.S.A. **73**, 1379 (1976)
Li, H., Finn, J.M., Lovelace, R.V.E., Colgate, S.A.: Astrophys. J. **533**, 1023 (2000)
Li, H., Colgate S.A., Wendroff, B., Liska, R.: Astrophys. J. **551**, 874 (2001)
Li, L.X., Goodman, J., Narayan, R.: Astrophys. J. **593**, 980 (2003)
Lin, C.C., Lau, Y.Y.: Studies in Applied Mat. **60**, 97 (1979)
Lovelace, R.V.E., Hohlfeld, R.G.: Astrophys. J. **221**, 51 (1978)
Lovelace, R.V.E., Li, H., Colgate, A.A., Nelson, A.F.: Astrophys. J. **513**, 805 (1999)
Lubow, S.H.: Astrophys. J. **381**, 259 (1991)
Lubow, S.H. Astrophys. J. **398**, 525 (1992)
Mark, J.W.-K.: Astrophys. J. **193**, 539 (1974)
Mark, J.W.-K.: Astrophys. J. **203**, 81 (1976a)
Mark, J.W.-K. Astrophys. J. **205**, 363 (1976b)
Morse, P.M., Feshbach, H.: Method of Theoretical Physics. McGraw-Hill, New York (1953). Chap. 9
Narayan, R., Goldreich, P., Goodman, J.: Mon. Not. R. Astron. Soc. **228**, 1 (1987)
Oktariani, F., Okazaki, A.T., Kato, S.: Publ. Astron. Soc. Jpn. **62**, 709 (2010)
Ono, T., Nomura, H., Takeuchi, T.: Astrophys. J. **787**, 37 (2014)
Ono, T., Muto, T. Takeuchi, T. Nomura, H.: Astrophys. J. **823**, 84 (2016)
Ono, T., Muto, T., Tomida, K., Zhu, Z.: Astrophys. J. **864**, 70 (2018)
Osaki, Y.: Publ. Astron. Soc. Pac. **108**, 390 (1996)
Papaloizou, J.C.B., Pringle, J.E.: Mon. Not. R. Astron, Soc. **208**, 721 (1984)
Papaloizou, J.C.B., Pringle, J.E.: Mon. Not. R. Astron, Soc. **213**, 799 (1985)
Papaloizou, J.C.B., Pringle, J.E.: Mon. Not. R. Astron, Soc. **225**, 267 (1987)
Ryu, D., Goodman, J.: Astrophy. J. **422**, 269 (1994)
Spiegel, E.A., Veronis, G.: Astrophys. J. **131**, 442 (1960)
Toomre, A.: Astrophys. J. **158**, 899 (1969)
Tsang, D., Lai, D.: Mon. Not. R. Astron, Soc. **387**, 446 (2008)
Tsang, D., Lai, D.: Mon. Not. R. Astron, Soc. **393**, 992 (2009a)
Tsang, D., Lai, D.: Mon. Not. R. Astron. Soc. **400**, 470 (2009b)
Whitehurst, R.: Mon. Not. R. Astron. Soc. **232**, 35 (1988)

Chapter 7
Instabilities Due to Dissipative Processes I (Secular Instability)

In Chap. 6 we have considered various instabilities which occur in the framework of adiabatic processes. Distinct from such instabilities, there are other kinds of many important instabilities which occur by dissipative processes. There are two ways for classifying these instabilities. One is phenomenological, and the other is physical. In the former phenomenological classification, instabilities are classified by whether their growth is monotonous in idealized situation (*secular instability*) or oscillatory (*overstability*). In the latter classification, instabilities are divided by physical processes involved (for example, thermal or viscous). Here we classify, for convenience, instabilities in the former way. That is, we classify them into secular instability and overstability. Instabilities which are former in the idealized situations (secular instability) are described in this chapter (Chap. 7) and the latter (overstability) in the next chapter (Chap. 8). It is noted that by this classification some instabilities whose cause is thermal are described in this chapter, and some others in the next chapter (Chap. 8).

7.1 Thermal Instability in Optically Thin Medium

The *thermal instability* is one of the important instability processes in astrophysics. Here, we discuss thermal instability by separating it into three cases by the difference of circumstances where the instability sets in. They are instabilities in optically thin media (Sect. 7.1), those in optically thick media (Sects. 7.2–7.4), and those in self-gravitating media (Sect. 7.5).

First, we consider the thermal instability in optically thin, infinitely extended uniform media (e.g., Field 1965). The thermal instability is one of the processes forming low temperature, high density condensations in high temperature, low-density media (e.g., clouds and star formation in interstellar gases, and formation of prominence in solar corona). In optically thin gases, the thermal energy exchange

© Springer Nature Singapore Pte Ltd. 2020, corrected publication 2023

S. Kato, J. Fukue, *Fundamentals of Astrophysical Fluid Dynamics*,
Astronomy and Astrophysics Library,
https://doi.org/10.1007/978-981-15-4174-2_7

with surrounding media is determined by the physical states at the position in consideration, and not by diffusion processes. Hence, the net thermal energy loss per unit mass, $\mathcal{L}$ (cooling minus heating), is a function of local quantities as

$$\mathcal{L}(\rho, T) = 0, \tag{7.1}$$

where ρ and T are the gas density and temperature, respectively.

7.1.1 Qualitative Argument on Stability Criterion

First, the condition of the thermal instability of optically thin media is considered qualitatively. In the process of formation of condensations, the pressure in the condensing gas is assumed to be kept unchanged from that of the surrounding medium. This is based on the consideration that the condensations occur so slowly that pressure balance with the surrounding media is kept during the process of the condensation. This means that the stability criterion can be derived under the condition of no pressure change, because the surrounding media is considered to be a heat server. This is similar to the case of derivation of convective instability criterion (see Chap. 5).

Let us assume that the temperature of a part of the medium decreases by a certain reason. Then, the part is compressed and the density increases, keeping the pressure unchanged. If the cooling rate, $\mathcal{L}$, increases in such a state of higher density and lower temperature, the cooling from the condensed part increases and the condensation proceeds further, and the instability sets in. This condition is written as

$$\left(\frac{\partial \mathcal{L}}{\partial T}\right)_p < 0, \tag{7.2}$$

where the subscript p means the derivative under constant pressure. By changing the independent variables from T and p to T and ρ, we can rewrite the condition (7.2) in the form

$$\left(\frac{\partial \mathcal{L}}{\partial T}\right)_\rho - \frac{\rho}{T}\left(\frac{\partial \mathcal{L}}{\partial \rho}\right)_T < 0, \tag{7.3}$$

because

$$\left(\frac{\partial \mathcal{L}}{\partial T}\right)_p = \left(\frac{\partial \mathcal{L}}{\partial T}\right)_\rho + \left(\frac{\partial \mathcal{L}}{\partial \rho}\right)_T \left(\frac{\partial \rho}{\partial T}\right)_p, \quad \text{and} \quad p = \frac{\mathcal{R}}{\bar{\mu}}\rho T, \tag{7.4}$$

where $\mathcal{R}$ and $\bar{\mu}$ are, respectively, the gas constant and mean molecular weight.

Here, it will be necessary to mention why optically thin medium is considered. In optically thick media, heat flows from a higher temperature region to a lower temperature one by diffusion processes. In such cases a temperature decrease in the condensed region is suppressed by heat flow from a hot region. This acts in the direction of suppressing the growth of the instability in the present problem. This is the reason why we consider optically thin media. An issue to be examined here is whether there are gaseous media where relation (7.2) or (7.3) is realized in realistic situations. Let us consider a gaseous media whose temperature is higher than 10^4 K. Then, the main cooling processes are free-free and bound-free processes of hydrogen gases. In these processes, the cooling per unit mass is proportional to $\rho T^{-1/2}$. For simplicity, we assume that the heating rate is constant. Then, the left-hand side of Eq. (7.3) is negative. That is, the thermal instability is expected.

It is instructive to draw the equilibrium curve, $\mathcal{L}(p, T) = 0$, on the p–ρ plane. In many situations, the curve is like that shown schematically in Fig. 7.1. On this curve the middle part of thick one represents unstable equilibrium state. This is because if temperature decreases (density increases) under pressure constant, the cooling rate increases, i.e., $\mathcal{L} > 0$. Hence, the gases leave more from the equilibrium position. In the case where temperature increases, the situation is similar and the gases leave further the equilibrium position. As understood from the figure, there are two equilibrium states when pressure is between p_u and p_l. One is a low-density (high temperature) state and the other is a high density (low temperature) one. These two states correspond, for example, to HI and HII regions in interstellar gases, and to corona and prominence in the case of the solar outer atmosphere.

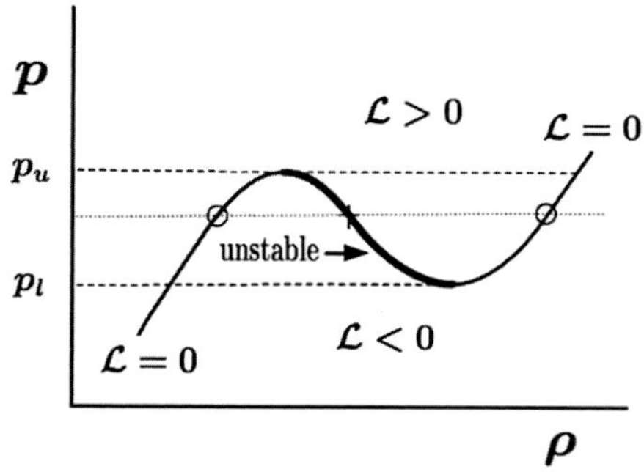

Fig. 7.1 Relation between thermally equilibrium series and instability. This figure schematically shows the relation between optically thin thermally equilibrium series on pressure-density diagram and thermal instability. The thermally equilibrium curve divides the p–ρ plane into two domains of $\mathcal{L} > 0$ and $\mathcal{L} < 0$. The part of the equilibrium state shown by thicker curve (the middle part of the equilibrium curve of $\mathcal{L}$) is thermally unstable. That is, in the case where $p_l < p < p_u$, there are three equilibrium states for a given p (shown by ∘ and ×), but the state shown by × is unstable, and there are two stable states for a given p

7.1.2 Analytical Derivation of Instability Criterion

In the following the thermal instability criterion in optically thin media, i.e., inequality (7.2), is analytically derived. As we are interested in thermal instability in an idealized situation, the medium is infinitely homogeneous in the unperturbed state and the effects of gravitational force are neglected. The equations describing small amplitude perturbations on such medium are

$$\frac{\partial \rho_1}{\partial t} + \rho_0 \, \mathrm{div} \boldsymbol{v} = 0, \tag{7.5}$$

$$\rho_0 \frac{\partial \boldsymbol{v}}{\partial t} = -\mathrm{grad}\, p_1, \tag{7.6}$$

$$\frac{\partial p_1}{\partial t} - c_\mathrm{s}^2 \frac{\partial \rho_1}{\partial t} = -(\gamma - 1)\rho(\mathcal{L}_\rho \rho_1 + \mathcal{L}_T T_1), \tag{7.7}$$

$$\frac{p_1}{p_0} = \frac{\rho_1}{\rho_0} + \frac{T_1}{T_0}, \tag{7.8}$$

where subscripts 0 and 1 denote, respectively, the unperturbed and perturbed quantities. The third equation is the equation of energy, and

$$\mathcal{L}_\rho \equiv \left(\frac{\partial \mathcal{L}}{\partial \rho}\right)_T, \quad \mathcal{L}_T \equiv \left(\frac{\partial \mathcal{L}}{\partial T}\right)_\rho. \tag{7.9}$$

Perturbations imposed on the equilibrium state are assumed to be plane waves proportional to $\exp(nt - ikx)$, where n is the growth rate, x is the direction of propagation of the plane wave, and k is the wave number of the wave. After some calculations and manipulations, we can derive a dispersion relation, which is

$$n^3 + c_\mathrm{s} k_T n^2 + c_\mathrm{s}^2 k^2 n + \frac{c_\mathrm{s}^3 k^2}{\gamma}(k_T - k_\rho) = 0, \tag{7.10}$$

where

$$k_T \equiv \frac{\bar{\mu}(\gamma - 1)\mathcal{L}_T}{\mathcal{R} c_\mathrm{s}}, \quad k_\rho \equiv \frac{\bar{\mu}(\gamma - 1)\rho \mathcal{L}_\rho}{\mathcal{R} c_\mathrm{s} T_0} \tag{7.11}$$

and $\mathcal{R}$ and $\bar{\mu}$ are the gas constant and the mean molecular weight, respectively. The condition of instability is that at least one of the solutions of Eq. (7.10) with respect to n has a real positive part, i.e., $\Re n > 0$.

This instability condition can be derived in a simple form by using the *Hurwitz theorem*. This theorem is powerful in studying stability criterion from a dispersion which is an algebraic equation with respect to growth rate n. Hence, the outline of the theorem is presented here.

- Hurwitz Theorem
  Let us consider a m-th order algebraic equation with respect to n with real coefficients. Hurwitz theorem tells us the necessary and sufficient conditions that all solutions of this algebraic equation have negative real parts, i.e., for all n, $\Re n < 0$. We describe here the condition in the case of the 4th order equation (i.e., $m = 4$):

$$a_0 n^4 + a_1 n^3 + a_2 n^2 + a_3 n + a_4 = 0 \quad (a_0 > 0), \tag{7.12}$$

since the procedure can be generalized straightly to algebraic equations of any order. In Eq. (7.12) the coefficients a_0 to a_4 are all real. Here we consider matrix consisting of the coefficients:

$$\begin{matrix} a_1 & a_0 & 0 & 0 \\ a_3 & a_2 & a_1 & a_0 \\ 0 & a_4 & a_3 & a_2 \\ 0 & 0 & 0 & a_4 \end{matrix}$$

Starting from the left-upper corner, we consider four matrixes: (i) 1 row and 1 column matrix, (ii) 2-rows and 2-columns matrix, (iii) 3-rows and 3-columns matrix, and (iv) 4-rows and 4-columns matrix. The necessary and sufficient conditions that all n's have negative real part are that the determinants of the above four matrixes are all positive.

In the present problem of thermal instability, the equation to be solved is Eq. (7.10), which is a third-order algebraic equation with respect to n, and

$$a_0 = 1, \quad a_1 = c_s k_T, \quad a_2 = c_s^2 k^2, \quad a_3 = \frac{c_s^3 k^2}{\gamma}(k_T - k_\rho). \tag{7.13}$$

What we need here for instability is that at least one of n's has a real positive value, $\Re n > 0$. This requires that at least one of the following three inequalities is realized:

$$a_1 < 0, \tag{7.14}$$

$$a_1 a_2 - a_3 < 0, \tag{7.15}$$

$$a_3(a_1 a_2 - a_3) < 0. \tag{7.16}$$

The condition $a_1 < 0$ is considered later. Then, from inequalities (7.15) and (7.16), we can rearrange the condition of instability is

$$a_3 < 0 \tag{7.17}$$

or

$$a_1 a_2 - a_3 < 0. \tag{7.18}$$

We see that in the present problem the condition $a_1 < 0$ is included in the condition of $a_3 < 0$, because in the present problem we have always $\mathcal{L}_\rho > 0$, and thus $k_\rho > 0$. That is, $a_1 < 0$ does not add any additional condition.

The condition $a_3 < 0$ is found to be

$$\left(\frac{\partial \mathcal{L}}{\partial T}\right)_p < 0, \tag{7.19}$$

since k_T and k_ρ are given by Eqs. (7.11). This is the criterion derived before from physical considerations.

Let us mention briefly what is the other instability condition of $a_1 a_2 - a_3 < 0$. By inserting detailed expressions for a_1, a_2, and a_3 into $a_1 a_2 - a_3 < 0$, we see that this condition is written as

$$(\gamma - 1)T\mathcal{L}_T + \rho\mathcal{L}_\rho < 0, \tag{7.20}$$

or in a compact form as

$$\left(\frac{\partial \mathcal{L}}{\partial T}\right)_\mathrm{s} < 0, \tag{7.21}$$

where s is entropy. $(\partial\mathcal{L}/\partial T)_\mathrm{s}$ is the derivative of cooling rate $\mathcal{L}$ with respect to temperature under the condition of constant entropy. This instability criterion is that known as the pulsational instability, which will be discussed more generally in Sect. 8.1 in relation to stellar pulsation.

7.2 Flash, Flicker, and Shell Burning in Stars

In Sect. 7.1 we have described the thermal instability in optically thin uniform media. The presence of thermal instability is not limited to optically thin media nor uniform ones. As described in Sect. 7.1, an essential part for presence of thermal instability is how pressure responds to a thermal perturbation. A necessary condition for the presence of thermal instability is that the pressure response to a temperature perturbation does not work so as to suppress the thermal perturbation. In this and next subsections we argue thermal instability in stars and disks, respectively.

First, we shall discuss why normal stars shine constantly without luminosity variation. In other words, why can the nuclear energy in stars burns constantly in time? If temperature of the central part of a star increases by chance, the rate of nuclear burning increases. If this increase of nuclear energy burning leads to a temperature increase, the burning increases further, leading to a runaway increase of luminosity.

Two situations in stars act so as to intercept this runaway process. The first is the difference between thermal and dynamical timescales in stars. The second is three-dimensional structure of stars. Let us consider the first issue. Inside stars energy transport is usually realized by radiative transport. The timescale of radiative energy transport is much longer than the dynamical timescale.[1] Hence, the thermal energy generation in normal stars occurs under the condition of dynamical equilibrium being kept during the thermal evolution.

The second issue is a nearly homologous expansion and contraction of the central part of stars (see Sect. 15.1 for homologous transformation). Homologous expansion (or contraction) means that an gas element at distance r from the center of a star moves to the position of ar after expansion (or contraction) with a constant expansion rate a which is independent of r. In such a homologous expansion, the gas density ρ of a gas element which was at r decreases after the expansion to ρa^{-3}. Since the dynamical balance is given by $\rho G M_r / r^2 = -dp/dr$, the pressure p of the gas element decreases to pa^{-4} after the expansion, where M_r is the mass in the sphere of radius r and unchanged during expansion. This shows that the temperature decreases after the expansion to Ta^{-1}. In summary, after a homologous expansion the temperature of the gas decreases by the factor a. This means that temperature increase due to increase of nuclear energy generation leads to expansion of the star and to decrease of temperature. In other words, the nuclear energy generation in the central part of normal stars is regulated by temperature adjustment. If we use the concept of "effective heat capacity" which will be introduced in Sect. 7.5, the central part of a star can be said to have a negative heat capacity.

The above arguments suggest in what cases thermal instability in stars is expected, or in what cases a temperature increase does not lead to increase of pressure. Two cases are conceivable. One is the case where the gases are degenerated (Sect. 7.2.1), and the other is the case where the energy generation zone is so thin in the radial direction that the expansion due to temperature increase is not three dimensional (Sect. 7.2.2).

7.2.1 *Helium and Carbon Flash*

In degenerate gas pressure is not directly related to temperature: In the extreme case the pressure is a function only of density. In such systems a temperature increase does not bring about an increase of pressure. There is no expansion of the system,

[1] If the thermometric conductivity due to radiative diffusion is written as $c\ell$, where c is the speed of light and ℓ is the mean free path of photons. The timescale of radiative heat diffusion is $r^2/c\ell$, where r is a typical linear size in stars. On the other hand, the dynamical timescale is r/c_s. Hence, the thermal timescale is longer than the dynamical one by $(r/\ell)(c_s/c)$, which is usually much larger than unity, because ℓ is much smaller than r due to large optical thickness inside stars.

and thus temperature increases further, leading to thermal instability. This kind of thermal instability in stars is known as *flash*.

The theory of stellar evolution tells us that the central part of hydrogen gases burns to helium, carbon, and so on. In stars of $M < 2M_\odot$ the temperature of the helium core is not high enough to burn the helium. Thus, the core continues to contract, accumulating helium ash resulting from the shell burning of hydrogen. The helium core finally degenerates (electron degeneracy), since the thermal temperature is not sufficient to counter the gravitational collapse. Then, the helium burning occurs explosively by the above-mentioned thermal instability. This is called *He flash*. The degenerate energy of electrons is, however, not so large in these stars. Hence, at a certain stage of the burning, the degeneracy dissolves, and the helium burning core expands. The burning is changed to stable one.

In more massive stars ($M < 8M_\odot$), electron gas is degenerated at the stage of Carbon burning. At this stage, *Carbon flash* will not occur as strong as in the case of He burning. This is because the helium temperature does not increase much by mass loss and neutrino loss. Strong Carbon flash, however, is expected in binary systems consisting of a C + O white dwarf and a secondary star. By mass accretion from the secondary star the temperature of the central part of the white dwarf increases to burn carbon again. Since electron gases are strongly degenerated, the burn occurs explosively. This is considered to be an origin of supernovae of type Ia.

7.2.2 Flicker and Shell Burning in Binary Stars

Another example of thermal instability in stars is explosive energy generation in thin shells. As mentioned at the beginning of Sect. 7.2, thermal runaway of energy generation in non-degenerated stars is usually restrained by three-dimensional expansion of stars. If nuclear burning occurs in a geometrically thin shell in stellar interior, however, the radial expansion of the burning region does not lead to temperature decrease, since the expansion is one dimensional. This comes from the following situation. The temperature increase by burning certainly expands the shell thickness. However, the pressure in the shell decreases little. This is because the radius of the shell from the center of the star changes little as the shell is thin (one-dimensional expansion). The pressure in the shell is determined by the mass above the shell, but it changes little by the expansion of the thin shell. Hence, the energy generation in the shell proceeds without suppression, which leads to thermal instability. This continues for a short time, but stops at a stage when the shell expansion changes the stellar structure. Then, the star returns to the original state of steady shell burning. Then, the next shell burning begins and this cycle is repeated.

This kind of thermal instability by shell burning inside stars is called *flicker* and was found by Schwarzschild and Härm (1965) as a sub-product in numerical studies of time evolution of giant stars. Usually, studies of time evolution of stars are made in units of the Kelvin–Helmholtz timescale. In studies of such a long-timescale interval, flicker cannot be found. This is because during a flicker stellar luminosity increases, but it continues only in a short-timescale of dynamical one.

The thermal instability due to shell burning is not limited to flicker. In binary stars, gas falls to the primary star from the secondary, forming an accretion disk around the primary. In the case where the primary star is compact and hot, the accretion gas can burn on the surface of the star. For example, the systems consisting of a white dwarf (primary star) and a red companion (from G type to M type star) show cataclysmic light variations. They are called *cataclysmic variables*, which include novae, dwarf novae, recurrent novae, nova-like variables, and polars as subclasses. If mass-accretion rate to the primary is low, hydrogen burning on the surface gives rise to shell burning and burst-like luminosity changes occur (nova). If the mass-accretion rate is sufficiently high, however, the hydrogen burning on the surface of the primary occurs steadily, and no burst-like luminosity variation occurs (super-soft X-ray sources). If the mass-accretion rate is moderate, small outburst (called normal outburst) frequently occurs (dwarf novae).

It is noted that in the normal outburst phase of dwarf novae, larger but less frequent outbursts (called *superoutburst*) appear semi-regularly. They are not the outbursts due to shell burning on the surface of primary stars, but those due to thermal instability of accretion disks surrounding the primary star. The thermal instability in accretion disks will be described in Sect. 7.3.

7.3 Thermal and Secular Instabilities in Disks

In accretion disks, different from stars, viscosity is one of the major contributors in energy balance and accretion flows. That is, the heat balance in disks is mainly realized between viscous heating and radiative cooling. Although the heating and cooling processes are different from those in stars, we can also expect thermal instability in disks. In dwarf-novae disks and black-hole (and neutron-star) ones, the cooling processes are different due to temperature differences. In spite of the differences, the mechanism of thermal instability is essentially the same in dwarf-novae disks and black-hole (and neutron-star) ones. Hence, in the following subsection (Sect. 7.3.1) we describe the thermal instability in black-hole (and neutron-star) accretion disks.

In addition to the thermal instability, another type of instability may occur in accretion disks, which is called *secular instability (viscous instability)*. This instability is related to the fact that angular momentum is transported by effects of viscosity, which will be described in Sect. 7.3.2. For details of thermal and secular instabilities in disks, see Chaps. 4, 5, and 10.4 of Kato et al. (2008).

7.3.1 Thermal Instability in Standard Disks

Different from stars, one of the important thermal processes in disks is viscous heating. Heat balance in disks is mainly realized between viscous heating and radiative cooling. Main thermal instabilities in disks thus come from the difference of parameter dependence of radiative cooling process and viscous heating process. There are two cases in radiative cooling, i.e., optically thin and optically thick coolings.

In this section we briefly summarize thermal instability processes in black-hole (and neutron-star) accretion disks. In studying the thermal instability we should remember the difference of thermal timescale by which thermal perturbations grow and the dynamical timescale by which dynamical phenomena occur. The dynamical timescale is roughly L/c_{s}, where L is the typical radial size of perturbations and c_{s} is acoustic speed in disks. Compared with this, the timescale by which thermal imbalance is adjusted is longer. Hence, considering thermal instability in disks, we can study the instability under the assumption that the perturbations occur under dynamical equilibrium.

In geometrically thin standard disks (Shakura and Sunyaev 1973), the force balance in the radial direction in the equatorial plane is realized between gravitational force of the central object and the centrifugal force due to disk rotation (Keplerian rotation). The pressure force resulting from disk temperature has negligible effects on this balance. This situation is unchanged during thermal perturbations are growing. That is, the angular velocity of disk rotation, Ω, is kept to be Keplerian one, i.e., $\Omega = \Omega_{\mathrm{K}} = (GM/r^3)^{1/2}$ during thermal perturbations are present on disks, where r is the radial distance from the central source. Furthermore, since thermal processes cannot change the above force balance in the radial direction, the change of disk structure by thermal perturbations is only in the vertical direction perpendicular to the equatorial plane. That is, if thermal perturbations are imposed on disks, the disks expand or contract in the vertical direction but the surface density, Σ, is unchanged, i.e.,

$$\Sigma = \text{const.} \tag{7.22}$$

Although a vertical expansion or contraction occurs, the vertical hydrostatic balance is maintained during the development of thermal perturbations, because as mentioned above the dynamical timescale is shorter than the thermal timescale by which thermal perturbations occur. The constant Σ means that the pressure on equator is proportional to disk half thickness, H, (see Sect. 3.2)

$$p \propto H. \tag{7.23}$$

In the standard disks the heat balance at a radius in the equilibrium state is realized between the heat generated by viscous shear motions at the radius and the radiative cooling at the same radius (toward the outside of the disk) (see Sect. 3.2).

Since we are considering geometrically thin disks, we are interested only in disk quantities which are vertically integrated. If the radiative cooling per unit surface of disks is denoted by $Q^-_{\rm rad}$, and viscous heating by $Q^+_{\rm vis}$, then the heat balance per unit surface in the equilibrium state is

$$Q^+_{\rm vis} = Q^-_{\rm rad}. \tag{7.24}$$

If the temperature of disks, $T(r)$, increases slightly over the equilibrium value, the disk responds to the perturbation under the condition of dynamical equilibrium and Σ = const. If the resulting adjustment of disks brings about higher heating rate than cooling rate, the temperature will increase further and the disks are thermally unstable. That is, the condition of thermal instability is (e.g., Pringle 1976)

$$\left[\frac{\partial}{\partial T}(Q^+_{\rm vis} - Q^-_{\rm rad})\right]_\Sigma > 0. \tag{7.25}$$

Let us consider three regions of standard accretion disks (see Sect. 3.2), separately. These three regions are (1) the innermost region, (2) the middle region, and (3) the outer region. First, in the innermost region pressure comes from radiation pressure, i.e., $p \simeq p_{\rm rad} \propto T^4$, and opacity is mainly due to electron scattering, i.e., $\bar{\kappa} \sim \kappa_{\rm es}$. Thus, the cooling rate per unit surface, $Q^-_{\rm rad}$, is given by $16cp_{\rm rad}/\bar{\kappa}\Sigma$ (see Sect. 3.2) and proportional to H since Σ can be taken to be constant and $p_{\rm rad} \propto H$ [see Eq. (7.23)], i.e.,

$$Q^-_{\rm rad} \propto H \propto T^4. \tag{7.26}$$

On the other hand, the viscous heating rate per unit surface, $Q^+_{\rm vis}$, is $-3\Omega T_{r\varphi}/2$ (see Sect. 3.2) and proportional to pH, where $T_{r\varphi}$ is the vertically integrated $r\varphi$-component of stress tensor. This means

$$Q^+_{\rm vis} \propto H^2 \propto T^8. \tag{7.27}$$

The temperature dependence of $Q^+_{\rm vis}$ is thus higher than that of $Q^-_{\rm rad}$. Hence, we see that the condition of thermal instability (7.25) is satisfied and the disk is thermally unstable.

Second, the middle region is considered. In this region temperature is low and density is high, compared with those in the innermost region, respectively. Thus, pressure comes from gas pressure although the opacity still comes from electron scattering (see Sect. 3.2). Since the expression $Q^-_{\rm rad} = 16cp_{\rm rad}/\bar{\kappa}\Sigma$ is unchanged, we have

$$Q^-_{\rm rad} \propto T^4. \tag{7.28}$$

On the other hand, although the expression of $Q^+_{\rm vis} = -3\Omega T_{r\varphi}/2$ is unchanged, p comes from gas pressure, and we have

$$Q^+_{\rm vis} \propto pH \propto T. \tag{7.29}$$

The above temperature dependences of $Q^-_{\rm rad}$ and $Q^+_{\rm vis}$ show that the middle region is thermally stable.

Third, let us consider the outer region of disks where pressure comes from gas pressure and the radiative cooling is due to free-free processes. In this region the temperature dependence of $Q^+_{\rm vis}$ is unchanged from that in the intermediate region, i.e., $Q^+_{\rm vis} \propto T$. However, $Q^-_{\rm rad}$ has a higher temperature dependence. That is, opacity is $\bar{\kappa} \propto \rho T^{-3.5} \propto T^{-4}$, and thus $Q^-_{\rm rad} \propto T^8$. These temperature dependences of $Q^+_{\rm vis}$ and $Q^-_{\rm rad}$ show that the disk is strongly stable against thermal perturbations.

The optically thick standard disk models by Shakura and Sunyaev (1973) have low temperature and cannot explain hard X-ray and gamma-rays from active galactic nuclei and compact objects. As an attempt to obtain disk models which can explain such high energy radiation, Shapiro et al. (1976) proposed an optically thin model (see Sect. 3.2). The model is known to be thermally unstable (for details, see Kato et al. 2008).

Other important models of accretion disks are optically thin advection-dominated accretion flow (ADAFs) and optically thick slim disks (see Sect. 3.2). These models are geometrically thick, but the aspect ratio is barely below unity. Stability analyses described above cannot be applied to these models with good accuracy. Furthermore, accretion flow are present, which contributes to energy balance. Roughly speaking, even if thermal perturbations glow, they cannot change the flow structure, because growing perturbations are transported inwards and swallowed into the central hole.

7.3.2 Secular (Viscous) Instability in Standard Disks

The secular instability of accretion disks brings about spatial modulation of the accretion rate, and leads to a coaxial density variation pattern. Since this density variation is the result of spatial change in the accretion rate, it is obvious that the timescale associated with this instability is the viscous timescale, which is longer than the thermal and dynamical ones in geometrically thin accretion disks. Hence, the essence of this instability can be derived by introducing approximations that the dynamical and thermal equilibria are maintained during the development of the perturbations. The essence of secular instability is briefly described here (for details see Kato et al. 2008).

In the case of axisymmetric flow, the angular momentum, $rv_\varphi = r^2\Omega$, is transported by[2]

$$\rho\frac{\partial}{\partial t}(r^2\Omega) + \rho\left(v_r\frac{\partial}{\partial r} + v_z\frac{\partial}{\partial z}\right)(r^2\Omega) = \frac{1}{r}\frac{\partial}{\partial r}(r^2 t_{r\varphi}), \tag{7.30}$$

where $t_{r\varphi}$ is the $r\varphi$-component of stress tensor. In the present problem the angular velocity of disk rotation, Ω, is time-independent, being kept to the Keplerian one. Hence, by integrating this equation in the vertical direction, we have

$$\dot{M}\frac{d}{dr}(r^2\Omega) = -\frac{d}{dr}(2\pi r^2 T_{r\varphi}), \tag{7.31}$$

where $T_{r\varphi}$ is the vertical integration of $t_{r\varphi}$, and $\dot{M}$ is the mass accretion rate at radius r, i.e.,

$$\dot{M}(r, t) = -2\pi r\Sigma v_r, \tag{7.32}$$

which is now not constant, but depends on time and radius.

The spatial change of mass-accretion rate is related to time change of accretion flows. The relation can be expressed as

$$2\pi r\frac{\partial\Sigma}{\partial t} = \frac{\partial\dot{M}}{\partial r}. \tag{7.33}$$

Elimination of $\dot{M}$ from Eqs. (7.32) and (7.33) gives

$$\frac{\partial\Sigma}{\partial t} = \frac{\partial}{r\partial r}\left\{\left[\frac{d}{dr}(r^2\Omega)\right]^{-1}\frac{\partial}{\partial r}\left(-r^2 T_{r\varphi}\right)\right\}. \tag{7.34}$$

---

[2]By using cylindrical coordinates (r, φ, z) and by expressing the flow in the corresponding directions by (v_r, v_φ, v_z), we have the momentum equation in the φ-direction in the form

$$\rho\left(\frac{\partial}{\partial t} + v_r\frac{\partial v_\varphi}{\partial r} + v_z\frac{\partial v_\varphi}{\partial z} + \frac{v_r v_\varphi}{r}\right) = \frac{1}{r^2}\frac{\partial}{\partial r}(r^2 t_{r\varphi}),$$

where $t_{r\varphi}$ is the $r\varphi$-component of viscous stress tensor. This equation can be arranged to the form:

$$\rho\frac{\partial}{\partial t}(rv_\varphi) + \rho\boldsymbol{v}\cdot\nabla(rv_\varphi) = \frac{1}{r}\frac{\partial}{\partial r}(r^2 t_{r\varphi}).$$

This is an equation representing angular momentum transport. It is noted that if this equation is combined with the continuity equation, we have an equation representing angular momentum conservation:

$$\frac{\partial(\rho r v_\varphi)}{\partial t} + \mathrm{div}\,(\boldsymbol{v}\rho r v_\varphi) = \frac{1}{r}\frac{\partial}{\partial r}(r^2 t_{r\varphi}).$$

This is a relation between Σ and $T_{r\varphi}$. If $T_{r\varphi}$ is expressed in terms of Σ and r, Eq. (7.34) is a kind of diffusion equation governing the viscous time evolution of Σ.

The next subject is thus to derive the relation between Σ and $T_{r\varphi}$. Since, as mentioned before, the viscous instability to be considered here occurs in a longer timescale than the dynamical and thermal ones, we can adopt the vertical hydrostatic balance:

$$p = \frac{1}{2}\Omega^2 \Sigma H, \tag{7.35}$$

and the heat balance between the viscous heating and the radiative cooling:

$$\frac{16cp_{\rm rad}}{\bar{\kappa}\Sigma} = -\frac{3}{2}\Omega T_{r\varphi}. \tag{7.36}$$

Furthermore, we adopt the conventional Shakura–Sunyaev type α-model:

$$T_{r\varphi} = -2\alpha p H. \tag{7.37}$$

As in the case of thermal instability, the secular stability of three regions of standard disks is considered separately. First, the innermost region is considered, where pressure comes from radiation pressure and the opacity is due to electron scattering. Since $\bar{\kappa} \sim \kappa_{\rm es} \sim$ const., combining Eqs. (7.36) and (7.35), we have $-T_{r\varphi} \propto \Omega H$, which, if combined with Eq. (7.37), leads to $p \propto \Omega \propto r^{-3/2}$. That is, p is a spatially given function, independent of time and the distributions of other physical quantities. The relations $p \propto \Omega$ and $p \propto \Omega^2 \Sigma H$ give $H \propto (\Omega\Sigma)^{-1}$. Equation (7.37) then leads to

$$-T_{r\varphi} \propto \frac{1}{\Sigma}. \tag{7.38}$$

This relation shows that when perturbations are imposed, the perturbed parts of $T_{r\varphi}$, i.e., $T_{r\varphi,1}$ and Σ_1, are related by

$$-T_{r\varphi,1} = \frac{T_{r\varphi,0}}{\Sigma_0}\Sigma_1, \tag{7.39}$$

where subscripts 0 and 1 denote the unperturbed and perturbed parts, respectively. Then, Eq. (7.34) gives for small amplitude perturbations

$$\frac{\partial \Sigma_1}{\partial t} = \frac{\partial}{r\partial r}\left\{\left[\frac{d}{dr}(r^2\Omega)\right]^{-1}\frac{\partial}{\partial r}\left(-r^2 T_{r\varphi,1}\right)\right\}. \tag{7.40}$$

Substitution of Eq. (7.39) into this equation gives

$$\frac{\partial \Sigma_1}{\partial t} = r\frac{T_{r\varphi,0}}{\Sigma_0}\left[\frac{d}{dr}(r^2\Omega)\right]^{-1}\frac{\partial^2 \Sigma_1}{\partial r^2} \tag{7.41}$$

for local perturbations, where the derivatives of the unperturbed quantities are neglected, since local perturbations are considered.

Equation (7.41) is a kind of diffusion equation describing the evolution of Σ_1. The diffusion coefficient, the coefficient of $\partial^2\Sigma_1/\partial r^2$ on the right-hand side, is negative since $T_{r\varphi,0} < 0$ and $d(r^2\Omega)/dr > 0$. This means an instability. An annulus region with $\Sigma_1 > 0$ increases Σ further and the region with $\Sigma_1 < 0$ behaved oppositely.

The above argument shows that the instability occurs if the changes of $T_{r\varphi}$ and Σ occur in phase under the condition of heat balance [see Eq. (7.38) or (7.39)], namely

$$\left(\frac{\partial T_{r\varphi}}{\partial \Sigma}\right)_{Q^+_{\rm vis}=Q^-_{\rm rad}} > 0. \tag{7.42}$$

In the middle region of standard disks, where the opacity still comes from electron scattering, but the gas pressure dominates over the radiation pressure. In this region the secular instability does not occur as shown below. In the case of $p \sim p_{\rm gas}$ the heat balance, Eq. (7.36), and pressure balance in the vertical direction, Eq. (7.37), give, respectively,

$$T_{r\varphi} \propto \frac{T^4}{\Omega\Sigma} \quad \text{and} \quad -T_{r\varphi} \propto \Sigma T. \tag{7.43}$$

Combining the two Eqs. (7.43), we have $-T_{r\varphi} \propto \Omega^{1/3}\Sigma^{5/3}$. Unlike the case of $p_{\rm rad} \gg p_{\rm gas}$, an increase of $-T_{r\varphi}$ leads to an increase of Σ. This means that the instability condition (7.42) is not satisfied, and the system is secularly stable.

Similar arguments easily show that the outer region of disks, where the gas pressure dominates over the radiation pressure and the opacity comes from the free-free processes, is also secularly stable.

7.3.3 Stability Analyses on Equilibrium Sequence

The equilibrium states of the standard disks are one-parameter family. For example, the equilibrium disks are along a curve on the T-Σ plane. Stability of disks can be examined on the diagram by studying the turning point of the curve.

First, we consider the thermal instability. We have already shown that the instability condition is given by inequality (7.25). The processes of derivation of this condition suggests that the condition is free from details of the heating and

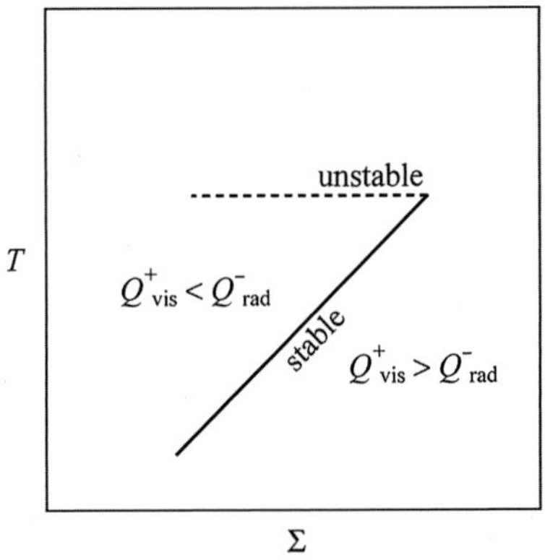

Fig. 7.2 Schematic picture showing thermal instability on the T-Σ plane. The disk models along the solid line are thermally stable, while the models on the part of dotted line are thermally unstable. (After Kato et al. 2008), *Black-Hole Accretion Disks—Toward a New Paradigm*)

cooling processes. That is, the condition of thermal instability can be generalized as

$$\left(\frac{\partial Q^+}{\partial T}\right)_\Sigma > \left(\frac{\partial Q^-}{\partial T}\right)_\Sigma, \tag{7.44}$$

independent of the detailed processes of heating and cooling. This instability condition shows that if the equilibrium sequence on the T-Σ diagram has a turning point as shown schematically in Fig. 7.2, the disk model which is at the turning point is marginally stable against perturbations, and the disk models on the part shown by dotted line are thermally unstable.

In the case of secular instability, the instability condition has been written by inequality (7.42). The $r\varphi$-component of stress tensor is an important quantity, but it is more convenient to rewrite inequality (7.42) by using an other physical quantity. This should be made under the assumptions of momentum and thermal equilibria, i.e., on the equilibrium sequence. Along the equilibrium sequence, the angular momentum balance, $\dot{M}(\ell - \ell_{\rm in}) = -2\pi r^2 T_{r\varphi}$, holds, where $\ell = r^2\Omega$ and $\ell_{\rm in}$ is the angular momentum which is finally absorbed to the central source (see Sect. 3.2). Hence, an increase of $-T_{{\rm r}\varphi}$ leads to an increase in $\dot{M}$ as long as the boundary condition is unchanged during the time evolution of the perturbations, since the angular momentum is fixed to be the Keplerian one. That is, along the equilibrium sequence changes of Σ and $\dot{M}$ occur with opposite signs. Hence, the instability condition (7.42) can be also expressed in the form:

$$\left(\frac{\partial \dot{M}}{\partial \Sigma}\right)_{Q^+_{\rm vis}=Q^-_{\rm rad}} < 0. \tag{7.45}$$

Let us consider the equilibrium sequence on the $\dot{M}$-Σ diagram (see Fig. 7.3). On the lower branch of the equilibrium curve, the change of $\dot{M}$ and Σ occurs with the same trend, but after passing the turning point they have an opposite trend.

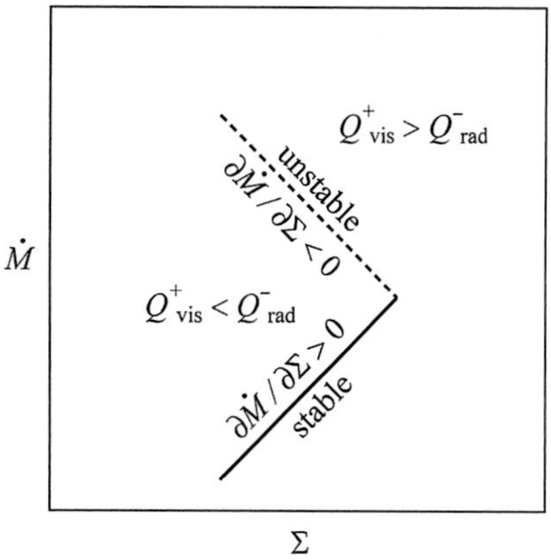

Fig. 7.3 Schematic picture showing secular instability on the $\dot{M}$-Σ plane. It is noted that the turning point in Fig. 7.2 is the same as that in Fig. 7.3 (see the text). (After Kato et al. 2008, *Black-Hole Accretion Disks—Toward a New Paradigm*)

This implies that the turning point represents just the marginally stable disk for the secular stability.

The above arguments are independent of details of heating and cooling processes, as long as thermal balances hold during secular instability proceeds. That is, the instability condition can be generally written as

$$\left(\frac{\partial \dot{M}}{\partial \Sigma}\right)_{Q^+=Q^-} < 0 \tag{7.46}$$

independent of the details of heating and cooling processes.

Finally it is noted that the turning points of the equilibrium disk sequence in Figs. 7.2 and 7.3 are the same. That is, the condition of marginal stability against thermal instability and that against secular instability are the same in the case of the standard disks. This comes from the following situations. An examination shows that the equilibrium sequence in Fig. 7.2 is the sequence of $\dot{M}$. That is, on the lower-left corner of the curve $\dot{M}$ is low, and $\dot{M}$ increases as the disk model moves upwards along the curve, and finally reaches the upper-left corner of the curve. This means that the sign of $(\partial\dot{M}/\partial\Sigma)_{Q^+=Q^-}$ is changed at the turning point in Fig. 7.2 as well as at the turning point in Fig. 7.3.

7.4 Goldreich–Schubert–Fricke Criterion

In Sect. 6.3 we have described the dynamical instability of differentially rotating non-baroclinic fluid systems, and presented the essence of the Solberg–Høiland criterion for stability. A next issue to be investigated is how the stability criterion is modified if the effects of dissipative processes are taken into account. Really, this issue was studied by Goldreich and Schubert (1967) and Fricke (1968). They examined how the Solberg–Høiland criterion derived in Sect. 6.3 is modified by taking into account both thermal conductivity and viscosity. Here, for simplicity, we examine the issue only by introducing the effects of thermal conductivity.

We consider the same situations as in Sect. 6.3, except that thermal conduction is taken into account in energy equation. That is, we consider axisymmetric system (stars). On this system, axially symmetric small amplitude perturbations are imposed, using the Boussinesq approximation.

Cylindrical coordinates (r, φ, z) are introduced, where z-axis is the axisymmetric axis of the system. The angular velocity of rotation Ω is a function of r and z, i.e., $\Omega = \Omega(r, z)$. The equation of continuity is then

$$\frac{\partial v_r}{\partial r} + \frac{\partial v_z}{\partial z} = 0, \tag{7.47}$$

and the r-, φ-, z-components of equation of motions are, respectively,

$$\frac{\partial v_r}{\partial t} - 2\Omega v_z = -\frac{1}{\rho_0}\frac{\partial p_1}{\partial r} + \frac{\rho_1}{\rho_0^2}\frac{\partial p_0}{\partial r}, \tag{7.48}$$

$$\frac{\partial v_\varphi}{\partial t} + \frac{\kappa^2}{2\Omega} v_r + r\frac{\partial \Omega}{\partial z} v_z = 0, \tag{7.49}$$

$$\frac{\partial v_z}{\partial t} = -\frac{1}{\rho_0}\frac{\partial p_1}{\partial z} + \frac{\rho_1}{\rho_0^2}\frac{\partial p_0}{\partial z}, \tag{7.50}$$

where the subscripts 0 and 1 denote unperturbed and perturbed quantities, respectively, and κ is the epicyclic frequency defined by $\kappa^2 = 2\Omega(2\Omega + r d\Omega/dr)$. It should be noted that although we are considering local perturbations whose wavelengths are shorter than the characteristic lengths in the unperturbed state, terms of $\rho_1/\rho_0^2 \nabla p_0$ are taken into account in the above equations. This is because we consider the effects of the buoyancy force (a part of the Boussinesq approximation). The above equations are the same as those adopted in Sect. 6.3.

The difference from Sect. 6.3 is energy equation. If thermal conduction is taken into account, a general form of energy equation is

$$\frac{dp}{dt} - c_s^2 \frac{d\rho}{dt} = (\gamma - 1) K \nabla^2 T, \tag{7.51}$$

where K is thermal conductivity and γ is the ratio of specific heats. Since we are considering local perturbations with the Boussinesq approximation, Eulerian pressure variation, p_1, is neglected in energy equation as mentioned in Sect. 5.1, and T_1/T_0 is approximated as $-\rho_1/\rho_0$. Then, as energy equation describing small amplitude perturbations, we have

$$\frac{1}{\rho_0}\left(\frac{\partial \rho_1}{\partial t} - \kappa_{\rm th}\nabla^2 \rho_1\right) = v_r \frac{\partial}{\partial r}\left(\ln \frac{p_0^{1/\gamma}}{\rho_0}\right) + v_z \frac{\partial}{\partial z}\left(\ln \frac{p_0^{1/\gamma}}{\rho_0}\right), \tag{7.52}$$

where $\kappa_{\rm th}$ is thermometric conductivity defined by $\kappa_{\rm th} = K/\rho_0 C_{\rm p}$, $C_{\rm p}$ being specific heat for constant pressure. In deriving the above equation, we have used the relation of $C_{\rm p} - C_{\rm v} = (\gamma - 1)C_{\rm v} = c_{\rm s}^2/\gamma T$, where $C_{\rm v}$ is specific heat for constant volume.

All perturbed quantities are represented in the form of $\exp[nt - i(k_r r + k_z z)]$ and the dispersion relation is derived by the same way as in Sect. 6.3. The result is then

$$n^3 + a_1 n^2 + a_2 n + a_3 = 0, \tag{7.53}$$

where

$$a_1 = \kappa_{\rm th} k^2, \tag{7.54}$$

$$\begin{aligned} a_2 = &\frac{1}{k^2}\left(\kappa^2 k_z^2 - \frac{1}{r^3}\frac{\partial \ell^2}{\partial z} k_r k_z\right) \\ &- \frac{1}{\gamma\rho_0}\frac{\partial p_0}{\partial r}\frac{\partial}{\partial r}\left(\ln\frac{p_0}{\rho_0^\gamma}\right)\frac{k_z^2}{k^2} - \frac{1}{\gamma\rho_0}\frac{\partial p_0}{\partial z}\frac{\partial}{\partial z}\left(\ln\frac{p_0}{\rho_0^\gamma}\right)\frac{k_r^2}{k^2} \\ &+ \frac{1}{\gamma\rho_0}\left[\frac{\partial p_0}{\partial r}\frac{\partial}{\partial z}\left(\ln\frac{p_0}{\rho_0^\gamma}\right) + \frac{\partial p_0}{\partial z}\frac{\partial}{\partial r}\left(\ln\frac{p_0}{\rho_0^\gamma}\right)\right]\frac{k_r k_z}{k^2}, \end{aligned} \tag{7.55}$$

$$a_3 = \kappa_{\rm th}\left(\kappa^2\frac{k_z^2}{k^2} - \frac{1}{r^3}\frac{\partial \ell^2}{\partial z}\frac{k_r k_z}{k^2}\right), \tag{7.56}$$

where ℓ is specific angular momentum defined by $\ell = r^2\Omega$, κ is epicyclic frequency, and $k^2 = k_r^2 + k_z^2$. In the limit of $\kappa_{\rm th} = 0$ Eq. (7.53) has a trivial solution $n = 0$. Except for this, Eq. (7.53) is reduced to the same dispersion relation (6.11) as that derived in the study of the Solberg–Høiland criterion.

The Hurwitz theorem (see Sect. 7.1) tells us that the condition of the real parts of all n's in dispersion relation (7.53) being negative (i.e., the condition of stability) is that both of $a_1 a_2 - a_3 > 0$ and $a_3 > 0$ are satisfied simultaneously (see Sect. 7.1.2). It is noted that $a_1 > 0$ is always satisfied. The condition of $a_1 a_2 - a_3 > 0$ gives

$$\begin{aligned} &-\frac{1}{\rho_0}\frac{\partial p_0}{\partial r}\frac{\partial}{\partial r}\left(\ln\frac{p_0}{\rho_0^\gamma}\right)x^2 - \frac{1}{\rho_0}\frac{\partial p_0}{\partial z}\frac{\partial}{\partial z}\left(\ln\frac{p_0}{\rho_0^\gamma}\right) \\ &+ \frac{1}{\rho_0}\left[\frac{\partial p_0}{\partial r}\frac{\partial}{\partial z}\left(\ln\frac{p_0}{\rho_0^\gamma}\right) + \frac{1}{\rho_0}\frac{\partial p_0}{\partial z}\frac{\partial}{\partial r}\left(\ln\frac{p_0}{\rho_0^\gamma}\right)\right]x > 0, \end{aligned} \tag{7.57}$$

where $x = k_z/k_r$. The condition of $a_3 > 0$ is written as

$$\kappa^2\frac{k_z^2}{k^2} - \frac{1}{r^3}\frac{\partial \ell^2}{\partial z}\frac{k_r k_z}{k^2} > 0. \tag{7.58}$$

Here we consider the latter condition (7.58). The former condition (7.57) is related to the double diffusive instability, and thus the meaning of the condition is discussed later in Chap. 8.

The latter condition (7.58) means that $\kappa^2 - (1/r^3)(\partial \ell^2/\partial z)(k_r/k_z)$ must be positive for any value of k_z/k_r. This requirement gives as necessary conditions of stability

$$\kappa^2 > 0 \quad \text{and} \quad \frac{\partial \ell^2}{\partial z} = 0. \tag{7.59}$$

This condition (7.59) was derived by Goldreich and Schubert (1967) and Fricke (1968), called the *Goldreich–Schubert–Fricke criterion.*

The necessary conditions (7.59) for stability are understandable if we consider the following situations. Let us consider first the condition $\kappa^2 > 0$. This means that if $\kappa^2 < 0$, the system is unstable even when the square of the Brunt–Väisälä frequency is positive (i.e., $N_r^2 > 0$). It is noted that in the case of adiabatic perturbations the condition corresponding to $\kappa^2 > 0$ is $\kappa^2 + N_r^2 > 0$ [see the Solberg–Høiland criteria given by inequality (6.17)]. This is because in the case of adiabatic perturbations, a fluid element displaced outward in the radial direction in the medium with $N_r^2 > 0$ has a lower temperature compared with that of the surrounding media and this acts so as to return the element to the original position (i.e., it acts in the direction to make the system stable). If thermal processes are present, however, the temperature difference between the displaced fluid element and the surrounding media is smoothed. If the temperature difference can be smoothed out completely, the stability criterion tends to $\kappa^2 > 0$. Such smoothing of temperature difference is really possible if there are slow oscillation modes. Such oscillation modes are always present when $N_r^2 > 0$. This means that the stability condition tends to $\kappa^2 > 0$, not $\kappa^2 + N_r^2 > 0$. For more careful arguments based on energy considerations for interchange of two mass rings, see Goldreich and Schubert (1967).

Next, let us consider the condition of $\partial \ell^2/\partial z = 0$. This means that if $\partial \Omega/\partial z \neq 0$, the system is unstable. In the case where $\partial \Omega/\partial z \neq 0$, the isobaric surface and the iso-density surface are different. They are inclined each other, i.e., the system is baroclinic [see Sect. 6.3 and Eq. (6.4)]. Temperature of the medium is not constant on isobaric surfaces. Then, gas motions occur along isobaric surface so that the temperature difference along isobaric surface is smoothed out. This motion readjusts angular momentum distribution on the medium, and finally leads to a barotropic gas, if there is no processes against this.

In summary, the Goldreich–Schubert–Fricke instability is a kind of rotational instability and contributes to the transfer of angular momentum in differentially rotating stars. Recently Caleo et al. (2016), however, emphasize that the instability is not so efficient in angular momentum transport in stars, because kinematic viscosity which acts against instability is not always negligibly small compared with thermometric conductivity and other reasons.

Finally, it is noted, as mentioned at the end of Sect. 6.3, that the Solberg–Høiland criterion is discontinuously changed if magnetic fields are present, even if they are weak (see Sect. 16.4). This revised stability criterion in Sect. 16.4 will be further modified if dissipative processes are taken into account, which will be different from the Goldreich–Schubert–Fricke stability criterion.

7.5 Gravothermal Instability

In nonself-gravitating systems the thermodynamically equilibrium state is isothermal, even if they are subject to external gravitational fields. These isothermal states are thermodynamically stable. They are in states where the entropy of the system is maximum. In self-gravitating systems, however, this well-known situation is violated. The thermodynamically equilibrium state is not always thermally stable, and the entropy is not always maximum. This is found by Antonov (1962) and developed by Lynden-Bell and Wood (1968). This instability is called the *gravothermal instability*. Importance of this instability is first recognized in stellar dynamics in relation to evolution of globular clusters. However, this instability is not only limited to stellar systems (collisionless systems), but also to gaseous systems, and is of importance in stars as a cause of their secular evolution.

The criterion of the instability depends on boundary conditions.[3] Here we assume, for simplicity, that the system is a non-rotating gaseous system with mass M, and is surrounded by an adiabatic spherical wall of a fixed radius R. Let the entropy of the system be S and the energy be E. The equilibrium states are those where the entropy has a stationary value for virtual changes of state under condition of constant E and M. These states are found to be isothermal, but the temperature is not unique. That is, the equilibrium states are a series of state where temperature is a parameter. Instead of temperature, it is instructive to take the central density as the parameter. Then, in an equilibrium system where the central concentration is weak we see that $\delta^2 S < 0$. That is, the system is a state where the entropy is local maximum, which is a stable state. However, if the central concentration increases beyond a certain limit, the equilibrium state is not a state of entropy maximum ($\delta^2 S < 0$), but becomes a state of entropy minimum ($\delta^2 S > 0$) and the system is unstable. This marginal value of density concentration (the ratio of the central density, ρ_c, to the surface density, ρ_s,) is $\sim$709, i.e., $\rho_c/\rho_s \sim 709$ (Antonov 1962). See Fig. 7.4 and Sect. 7.5.3 for physical interpretation of the instability.

[3]The instability criterion has been studied for various boundary conditions. Furthermore, the gravothermal instability has been extended to rotating systems (Inagaki and Hachisu 1978).

7.5.1 Derivation of Instability Criterion

First, we show that the state where entropy is stationary is an equilibrium state. Next, the relation between the second-order variation of entropy and stability is considered.

The entropy and energy of the whole system are defined, respectively, by

$$S = \int s dm, \tag{7.60}$$

and

$$E = \int \left(u + \frac{1}{2}\phi \right) dm, \tag{7.61}$$

where dm is a mass element, s entropy per unit mass, u internal energy per unit mass, and ϕ is the gravitational potential per unit mass:

$$\phi(\boldsymbol{r}) = -G \int \frac{dm'}{|\boldsymbol{r} - \boldsymbol{r}'|}. \tag{7.62}$$

The Lagrangian variation of s, δs, is given by the thermodynamical relation:

$$\delta s = \frac{1}{T}\delta u + \frac{p}{T}\delta\left(\frac{1}{\rho}\right). \tag{7.63}$$

1. Condition of $\delta S = 0$ Under $E =$ const.

First, let us consider stationary states of S under the condition of $E =$ const. This can be made by the method of Lagrange multipliers. We introduce W defined by $W = T_{\rm e}S - E$ and examine the state where W is stationary under virtual variations, where $T_{\rm e}$ is a parameter. That is, we examine the state where

$$\delta W = T_{\rm e}\delta S - \delta E = 0 \tag{7.64}$$

for virtual variations of S and E.[4] Since mass element is invariant to virtual displacements, the condition (7.64) can be written as

$$\delta W = \int \left(T_{\rm e}\delta s - \delta u - \frac{1}{2}\delta\phi \right) dm. \tag{7.65}$$

[4] In the conventional method of Lagrange multipliers, Lagrange function L defined by $L = S - \lambda E$, where λ is a Lagrange multiplier, is introduced and examine the state where L becomes stationary. This method of the Lagrange multipliers has been modified in the text in a slightly different form for convenience in the present study.

By using Eq. (7.62) we have[5]

$$\int \delta\phi dm = 2\int \frac{d\phi}{dr}\delta r dm. \tag{7.66}$$

Here, the variation has been taken for spherically symmetric systems, because the stationary state is obviously spherically symmetric. Hence, substituting (7.63) and (7.66) into Eq. (7.65) we have

$$\delta W = \int\int\left[\left(\frac{T_\mathrm{e}}{T}-1\right)\delta u - \frac{T_\mathrm{e}}{\rho}\frac{d}{dr}\left(\frac{p}{T}\right)\delta r - \frac{d\phi}{dr}\delta r\right]dm. \tag{7.67}$$

In deriving the above relation $\delta\rho/\rho = -(1/r^2)d(r^2\delta r)/dr$ (continuity), $dm = 4\pi r^2\rho dr$, and the integration by part with respect to r have been adopted. Since $\delta W = 0$ should be realized for arbitrary changes of δu and δr at stationary state, we have

$$T = T_\mathrm{e} \tag{7.68}$$

$$-\frac{1}{\rho}\frac{dp}{dr} - \frac{d\phi}{dr} = 0. \tag{7.69}$$

This is a state of isothermal and dynamical equilibrium, and obviously represents a state of thermodynamical equilibrium.

[5]From Eq. (7.62) we have

$$\delta\phi = -G\int\frac{dm'}{|(\boldsymbol{r}+\delta\boldsymbol{r})-(\boldsymbol{r}'+\delta\boldsymbol{r}')|} + G\int\frac{dm'}{|\boldsymbol{r}-\boldsymbol{r}'|} = G\int\frac{\delta r_i(r_i-r_i')}{|\boldsymbol{r}-\boldsymbol{r}'|^{3/2}}dm' - G\int\frac{\delta r_i'(r_i-r_i')}{|\boldsymbol{r}-\boldsymbol{r}'|^{3/2}}dm'$$

and

$$\int\delta\phi dm = G\int\int\frac{(\delta r_i-\delta r_i')(r_i-r_i')}{|\boldsymbol{r}-\boldsymbol{r}'|^{3/2}}dmdm'.$$

On the other hand,

$$\frac{\partial\phi}{\partial r_i} = G\int\frac{r_i-r_i'}{|\boldsymbol{r}-\boldsymbol{r}'|^{3/2}}dm',$$

and

$$\begin{aligned}\int\frac{\partial\phi}{\partial r_i}\delta r_i dm &= G\int\int\frac{\delta r_i(r_i-r_i')}{|\boldsymbol{r}-\boldsymbol{r}'|^{3/2}}dmdm' = G\int\int\frac{\delta r_i'(r_i'-r_i)}{|\boldsymbol{r}-\boldsymbol{r}'|^{3/2}}dmdm' \\ &= \frac{1}{2}G\int\int\frac{(\delta r_i-\delta r_i')(r_i-r_i')}{|\boldsymbol{r}-\boldsymbol{r}'|^{3/2}}dmdm'.\end{aligned}$$

2. Condition of Entropy Maximum Under E = const.
Next, we should examine whether the thermodynamically equilibrium state given by Eqs. (7.68) and (7.69) is stable. This can be studied by examining whether the thermodynamically equilibrium state derived above is really a state of entropy maximum. The necessary and sufficient conditions for this is that the second-order variation of W is negative. That is, for an arbitrary virtual displacement around the equilibrium state ($\delta W = 0$) the second-order variation of W is negative at $\delta S = 0$. In other words, the necessary and sufficient condition for stability is

$$\delta^2 W \equiv T_e \delta^2 S - \delta^2 E < 0 \tag{7.70}$$

for arbitrary virtual displacements. Antonov (1962) calculated $\delta^2 W$ by using Eulerian quantities, but here following Inagaki and Hachisu (1978) we calculate $\delta^2 W$ by using Lagrangian quantities. As an expression for s (entropy per unit mass) we use $s = \int (C_v/T)dT - (k_B/\bar{\mu}m_H)\ln \rho + \text{const}$, where k_B is the Boltzmann constant. Then, considering till the second-order variations, we have

$$\delta s = C_v\left[\frac{\delta T}{T} - \frac{1}{2}\left(\frac{\delta T}{T}\right)^2\right] - \frac{k_B}{\bar{\mu}m_H}\left[\frac{\delta\rho}{\rho} - \frac{1}{2}\left(\frac{\delta\rho}{\rho}\right)^2\right]. \tag{7.71}$$

In the above equation $\delta\rho$ is the virtual variation of ρ. In order to derive an explicit expression for the second-order variation of $\delta^2\rho$, in this paragraph, the density in the equilibrium state in consideration is expressed by attaching the subscript 0 as $\rho_0(r)$ and that in a virtual displaced equilibrium state by $\rho(r + \delta\rho)$. Then, since the Lagrangian variation of density, i.e., $\delta\rho$, and the Eulerian one, i.e., ρ_1, are given, respectively, by $\delta\rho = \rho(r + \delta r) - \rho_0(r)$ and $\rho_1 = \rho(r) - \rho_0(r)$, we have

$$\delta\rho = \rho_1 + \frac{d}{dr}(\rho_0 + \rho_1)\delta r + \frac{1}{2}\frac{d^2}{dr^2}(\rho_0 + \rho_1)(\delta r)^2 + \ldots \tag{7.72}$$

This gives that $\delta^2\rho$ is

$$\delta^2\rho = \frac{d\rho_1}{dr}\delta r + \frac{1}{2}\frac{d^2\rho_0}{dr^2}(\delta r)^2. \tag{7.73}$$

Substituting Eq. (7.73) into Eq. (7.71) and performing the integration of dm by part under consideration of the continuity, $\delta^1\rho/\rho + (1/r^2)[d(r^2\delta r)/dr] = 0$, we have after some manipulation

$$\delta^2 W = -\frac{1}{2}\int\left[\frac{C_v}{T}(\delta T)^2 + \frac{p}{\rho}\left(\frac{1}{r^2}\frac{d}{dr}(r^2\delta r)\right)^2 + \delta^2\phi - \frac{2}{r}\frac{d\phi}{dr}(\delta r)^2\right]dm. \tag{7.74}$$

Since we restrict our attention only to axisymmetric perturbations as mentioned before, $\int \delta^2\phi dm$ can be rearranged in the following way [cf. the derivation of

Eq. (7.66)]. Since the gravitational potential ϕ is written as

$$\phi[r(m)] = -\frac{G}{r}\int_0^{r(m)} dm' - G\int_{r(m)}^{\infty} \frac{1}{r'}dm', \tag{7.75}$$

we have

$$\delta^2\phi = -\frac{G}{r^3}(\delta r)^2 \int_0^{r(m)} dm' - G\int_{r(m)}^{\infty} \frac{(\delta r')^2}{r'^3}dm'. \tag{7.76}$$

In integrating this expression for $\delta^2\phi$ over the whole mass, the order of the double integration with respect to m and m' is changed. Then, we have easily

$$\int \delta^2\phi dm = -2\int \frac{d\phi}{rdr}(\delta r)^2 dm. \tag{7.77}$$

Hence, from Eqs. (7.74) and (7.77), we have finally

$$\delta^2 W = \frac{1}{2}\int\left[-\frac{C_{\rm v}}{T}(\delta T)^2 - \frac{k_{\rm B}T}{\bar{\mu}m_{\rm H}}\left(\frac{1}{r^2}\frac{\partial}{\partial r}(r^2\delta r)\right)^2 + \frac{4}{r}\frac{d\phi}{dr}(\delta r)^2\right]dm. \tag{7.78}$$

It should be noticed that in non-gravitating systems $\delta^2 W$ is always negative, and thus the thermally equilibrium system is stable.

An equilibrium state, $\delta W = 0$, is stable, if we can show that the value of $\delta^2 W$ is negative for any virtual displacement around the state. An intuitive consideration suggests that in the marginally stable state virtual displacements which make $\delta^2 W$ maximum is slow displacements (not an oscillatory mode). In such slow displacements a spatial difference in temperature is smoothed out by thermal diffusion. In other words, the virtual displacement which makes $\delta T = \text{const.}$ in space is the virtual displacement which makes $\delta^2 W$ maximum. Using this and considering the fact that we are now imposing the boundary condition $\delta E = 0$, we see from Eq. (7.61) that in the study of marginally stable case we can adopt

$$\delta T = -\frac{\int (d\phi/dr)\delta r dm}{\int C_{\rm v} dm} = \frac{k_{\rm B}T}{\bar{\mu}m_{\rm H}C_{\rm v}M}\int \frac{d\ln\rho}{dr}\delta r dm. \tag{7.79}$$

Substituting Eq. (7.79) into (7.78) we have

$$\begin{aligned}\delta^2 W &= -\int\left[\frac{k_{\rm B}T}{2\bar{\mu}m_{\rm H}}\left(\frac{1}{r^2}\frac{d}{dr}(r^2\delta r)\right)^2 - \frac{2}{r}\frac{d\phi}{dr}(\delta r)^2\right]dm \\ &\quad -\frac{1}{2}\left(\frac{k_{\rm B}}{\bar{\mu}m_{\rm H}}\right)^2\frac{T}{C_{\rm v}M}\left(\int \frac{d\ln\rho}{dr}\delta r dm\right)^2. \end{aligned}\tag{7.80}$$

In summary, the system is stable, if $\delta^2 W$ given by Eq. (7.80) is negative for any arbitrary displacement δr, while the system is unstable if there is a displacement δr which makes $\delta^2 W$ given by Eq. (7.80) positive.

Now we consider the condition that $\delta^2 W$ given by Eq. (7.80) has a stationary value under a normalization condition:

$$K \equiv 2\int \frac{1}{r}\frac{d\phi}{dr}(\delta r)^2 dm (= \text{const.}) > 0. \tag{7.81}$$

The method of Lagrangian multipliers tells us that a stationary state of $\delta^2 W$ is realized when there is a real number λ (Lagrange multiplier) for which $\delta^2 W - \lambda K$ becomes stationary.

By performing integration with respect to r in the expression for $\delta^2 W - \lambda K$ by part by adopting boundary condition

$$\delta r = 0 \tag{7.82}$$

at $r = 0$ and $r = R$, we can reduce $\delta^2 W - \lambda K$ in the form of $\int 2\pi r^2 G(r)\delta r dr$, where $G(r)$ is a differential equation with respect to r. Hence, the condition of $\delta^2 W - \lambda K$ becoming stationary, i.e., $G = 0$, gives (the so-called Euler equation)

$$\frac{k_{\rm B}T}{\bar{\mu} m_{\rm H}}\frac{d}{dr}\left[\frac{\rho}{r^2}\frac{d}{dr}(r^2\delta r)\right] + 4(1-\lambda)\rho\frac{1}{r}\frac{d\phi}{dr}\delta r - \frac{k_{\rm B}}{\bar{\mu} m_{\rm H}}\delta T\frac{d\rho}{dr} = 0. \tag{7.83}$$

In deriving Eq. (7.83), we have adopted Eqs. (7.79) and (7.80). This result shows that if Eq. (7.83) is solved as an eigenvalue problem with the boundary condition (7.82), we obtain the cases where $\delta^2 W$ has stationary values. By substituting Eq. (7.83) into (7.80) we see that the stationary value of $\delta^2 W$ is related to the eigenvalue λ by

$$\delta^2 W = 2\lambda \int \frac{1}{r}\frac{d\phi}{dr}(\delta r)^2 dm = \lambda K. \tag{7.84}$$

In summary, the stability of the system is related to the eigenvalues of Eq. (7.83). That is, if the largest eigenvalue is negative, the system is stable, while if it is positive, the system is unstable. In other words, the critical state of stability is realized in the system for which the largest eigenvalue of Eq. (7.83) is zero. It is noted that the eigenvalue λ is guaranteed to be real, since Eq. (7.83) is self-adjoint.

7.5.2 Relation Between Stability Criterion and Equilibrium Sequence

As mentioned above, the stability problem was reduced to an eigenvalue problem. In their original work by Lynden-Bell and Wood (1968), however, they examined the stability by examining the equilibrium series. That is, they used the fact that in the equilibrium sequence of systems the point where energy $E(<0)$ becomes the minimum is the critical state of stability (see Fig. 7.4). The fact that the both methods (the method of eigenvalue problem and the method of equilibrium sequence) give the same instability criterion is presented here, following Inagaki and Hachisu (1978).

The equilibrium state is described by Eq. (7.69). Now, we consider Lagrangian changes of r and T along the equilibrium sequence. They are now expressed by Δr and ΔT in order to avoid confusion with δr and δT.

From Eq. (7.75) we see that $d\Delta\phi/dr$ can be written as[6]

$$\frac{d\Delta\phi}{dr} = -\left(\frac{\Delta\rho}{\rho} + 4\frac{\Delta r}{r}\right)\frac{d\phi}{dr}. \tag{7.85}$$

Since we are considering the equilibrium sequence, from (7.69) we have

$$-\frac{d\Delta p}{dr} + \left(\frac{\Delta\rho}{\rho}\right)\frac{dp}{dr} + \Delta\rho\frac{d\phi}{dr} + 4\rho\frac{\Delta r}{r}\frac{d\phi}{dr} = 0, \tag{7.86}$$

where Eq. (7.85) has been used to rewrite $d\Delta\phi/dr$ by $\Delta\rho$ and Δr. Considering that $dT/dr = 0$ in the equilibrium state, we can reduce the above equation to

$$\frac{k_B T}{\bar{\mu} m_H}\frac{d}{dr}\left[\frac{\rho}{r^2}\frac{d}{dr}(r^2\Delta r)\right] + 4\rho\frac{\Delta r}{r}\frac{d\phi}{dr} - \frac{k_B}{\bar{\mu} m_H}\Delta T\frac{d\rho}{dr} = 0, \tag{7.87}$$

To derive the above equation, $p = (k_B/\bar{\mu} m_H)\rho T$ and $d(\Delta r)/dr = -2\Delta r/r - \Delta\rho/\rho$ (continuity) have been used.

This equation is apparently the same as Eq. (7.83) with $\lambda = 0$. For them to be exactly the same, the relation between Δr and ΔT must be the same as that between δr and δT, i.e., Eq. (7.79). This can be easily shown if we consider

$$\Delta E = 0. \tag{7.88}$$

Furthermore, in the case where we are considering the equilibrium sequence bounded by a rigid wall of radius R, the boundary conditions are $r = 0$ and $r = R$,

[6] From Eq. (7.75) we have

$$\Delta\phi = -\frac{G}{r+\Delta r}\int_0^{r+\Delta r} dm' - G\int_{r+\Delta r}^{\infty}\frac{1}{r'}dm' - \phi,$$

which gives for small Lagrangian variations

$$\frac{d}{dr}\Delta\phi = \left(-2\frac{\Delta r}{r} + \frac{d\Delta r}{dr}\right)\frac{d\phi}{dr}.$$

The above equation is written as Eq. (7.85) by use of the mass conservation given by $2\Delta r/r + \Delta\rho/\rho + d\Delta r/dr = 0$, which comes from $4\pi(r+\Delta r)^2\rho(r+\Delta r)d(r+\Delta r) = 4\pi r^2\rho(r)dr$.

and thus

$$\Delta r = 0. \tag{7.89}$$

This is the same as the boundary condition (7.82) when Eq. (7.83) is solved as an eigenvalue problem.

The above consideration shows that the critical configuration for thermodynamical equilibrium (i.e., marginal state of stability) is the one of equilibrium series where energy E becomes stationary ($\Delta E = 0$).

Here, we consider the equilibrium series in the case where gases with mass M are bounded in a sphere of radius R. As a parameter representing the equilibrium series, we consider the concentration to the center and adopt here $v_1 \equiv \ln(\rho_c/\rho_e)$, where ρ_c and ρ_e are the density of gases at the center and on the surface, respectively. Furthermore, as a dimensionless energy of the system $-E/(GM^2/R)$ is adopted. The relation between $-E/(GM^2/R)$ and v_1 on the equilibrium series is schematically shown in Fig. 7.4 for ideal gases. Equilibrium solutions for weak central concentration are those that have high temperature ($E > 0$) and their expansion is suppressed by the wall of radius R. As central concentration increases,

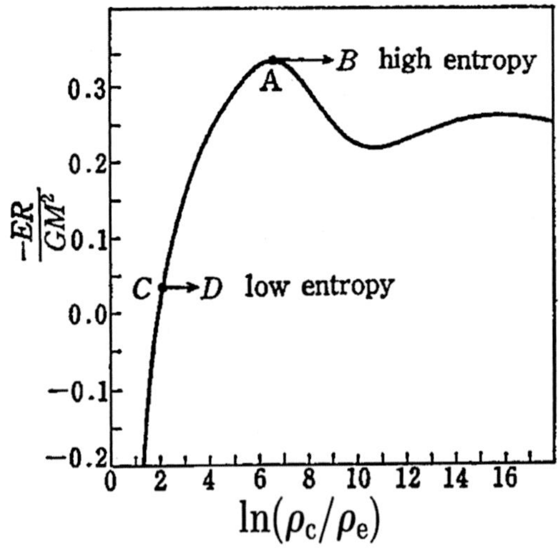

Fig. 7.4 Schematic diagram showing the thermodynamical equilibrium sequence on the plane of total energy of the system (E) versus a degree of central concentration (ρ_c/ρ_e), ρ_c and ρ_e being the density at center and surface, respectively. Let us first consider a low concentration system (say point C). As the central density increases, the energy of the equilibrium system, E, decreases due to increase of gravitational energy, i.e., $E < 0$ and $|E|$ increases. If the density concentration increases further, there is a minimum of E (a maximum of $|E|$) at the configuration shown by A, where $\Delta E = 0$ along the equilibrium sequence. This is the marginary state of stability, $\delta^2 W = 0$ and $\delta^2 S = 0$. If the density concentration increases beyond this point, the system is unstable. (Based on Fig. 2 of Lynden-Bell and Wood 1968, Monthly Not. Roy. Astron. Soc. 138, 495 with some changes)

the energy of the system decreases and becomes negative ($E < 0$). Figure 7.4 shows that there is the minimum of E (the maximum of $-E$) at point A, and E increases again as the central concentration increases further. That is, we have $\Delta E = 0$ at point A, and the system corresponding to this minimum of E is the state of critical stability. The value of v_1 for this solution is 6.55. The value of ρ_c/ρ_e for this critical state is $\rho_c/\rho_e = 709$. It is noticed that the critical point for stability is a turning point of the equilibrium series, but the inverse of this does not always hold.

In the case where there is no outer boundary, an isothermal self-gravitating system has infinite radius with infinite mass. Therefore, when we consider thermodynamical equilibrium of finite mass a wall with finite radius is necessary. If boundary conditions at the wall are changed, we have different instability condition (for details see Lynden-Bell and Wood 1968). For gravothermal instability in rotating cylinder, see Inagaki and Hachisu (1978).

7.5.3 Physical Processes of Gravothermal Instability

The physical processes of the gravothermal instability can be understood by considering the following processes.

Let us consider an isothermal spherical system with self-gravity, and assume that the temperature of the central part is increased by a perturbation, compared with that of the outer part of the system, keeping E. Then, heat flows from the central part to the outer one. In the case of nonself-gravitating system, this heat flow from the hotter part to a cooler one acts so as to decrease the temperature at the hotter place, and the perturbation is smoothed out. In the case of self-gravitating systems, however, situations are different. The central part contracts by heat being transported outwards, since the dynamical timescale is much shorter than the thermal timescale. By this contraction, the temperature in the central part increases rather than decreases. This characteristic is often said: The effective heat capacity in the central part of stars is negative.[7]

Distinct from this, in the outer part of stars with weak self-gravity, temperature increases with heat input. Hence, temperatures of both the central and outer parts of stars increase by heat transport from the central part to the outer part. If the temperature increase in the inner part is higher than that in the outer part, heat further flows from the inner part to the outer part, and the initial temperature perturbation is further amplified. That is, the system is thermally unstable. This really occurs, if central concentration of gases is stronger beyond a certain limit.

[7] As will be noticed in the Virial theorem (Sect. 15.1) this characteristic of negative heat capacity is demonstrated as follows. The Virial theorem tells us that in dynamically equilibrium state of self-gravitating system $2T + \Omega = 0$ holds, where T and Ω are, respectively, thermal and gravitational energies of the system. Since the total energy of the system, E, is given by $E = T + \Omega$, we have

$$T = -E.$$

In the above arguments the case where the temperature of the central part increases is considered. The situation is similar even when the temperature of the central part decreases.

Let us explain the instability by using Fig. 7.4. First, we consider the case where central concentration is weak. The system is assumed to be at state C in the figure. If the central concentration increases by a perturbation, the state of the system moves from C to D, keeping E constant. As the figure shows, the state D has a higher energy, compared with the equilibrium state required by the central concentration of D. That is, the system at point D has energy larger than that required for thermodynamical equilibrium system. Hence, the system expands and returns to the original state C. That is, the system is stable.

Next, let us consider the case where perturbations are imposed at the system specified by point A so that the system moves towards a state specified by B, keeping E unchanged. Then, the perturbed state has a lower energy compared with that necessary to keep the system at an equilibrium state. This means that kinetic energy (thermal energy) of the system is smaller than that required to keep the system at an equilibrium state against gravitational force. Hence, the system cannot expand to the initial state and thus the central concentration of density further increases. This is the reason why the turning point A of the equilibrium series is the critical point of stability.

The reason why the energy for equilibrium state becomes higher (i.e., $|E|$ decreases), when the central concentration is higher beyond the critical value is that the gravitational force is so strong that any equilibrium state cannot be kept without changing energy of the system. To keep the system to be equilibrium, energy input from outside is necessary.

References

Antonov, V.A.: Vestn. Leningr. Gos. Univ. **7**, 135 (1962)
Caleo, A., Balbus, S.A., Tognelli, E.: Mon. Not. R. Astron. Soc. **460**, 338 (2016)
Field, G.: Astrophys. J. **142**, 531 (1965)
Fricke, K.: Z. Astrophys. **68**, 317 (1968)
Goldreich, P., Schubert, G.: Astrophys. J. **150**, 571 (1967)
Inagaki, S., Hachisu, I.: Publ. Astron. Soc. Jpn. **30**, 39 (1978)
Kato, S., Fukue, J., Mineshige, S.: Black-Hole Accretion Disks – Towards a New Paradigm. Kyoto University Press, Kyoto (2008)
Lynden-Bell, D., Wood, R.: Mon. Not. R. Astron. Soc. **138**, 495 (1968)
Pringle, J.E.: Mon. Not. R. Astron. Soc. **177**, 65 (1976)
Schwarzschild, M., Härm, R.: Astrophys. J. **142**, 855 (1965)
Shakura, N.I., Sunyaev R.A.: Astron. Astrophys. **24**, 337 (1973)
Shapiro, S.L., Lightman, A.P., Eardley, D.H.: Astrophys. J. **204**, 187 (1976)

Chapter 8
Overstability Due to Dissipative Processes II: Excitation of Oscillations

In Chap. 7 we have presented monotonically growing instabilities which are caused by dissipative processes. Instabilities caused by dissipative processes are not limited to those which grow monotonically. Dissipative processes can amplify oscillations with time. The growth of oscillations is called *overstability*. In this chapter we present dissipative processes which make oscillations overstable.

8.1 Excitation of Stellar Oscillations by Thermal Processes

Before discussing in detail the pulsational instability of stars, we shall mention here the meaning of the instability condition (7.21) obtained in Sect. 7.1, in addition to the thermal instability condition (7.19).

The set of Eqs. (7.5)–(7.8) describe two modes of perturbations. One is, of course, thermal mode. The other is acoustic mode. This is shown in the dispersion relation (7.10). In the limit of adiabatic motions [i.e., the second and forth terms on the left-hand side of Eq. (7.10) vanish] Eq. (7.10) is separated into modes: $n = 0$ and $n^2 + c_s^2k^2 = 0$. The first is a trivial mode in this limit of adiabatic motions, but by non-adiabatic effects it becomes $n \neq 0$. This is the thermal mode, which was discussed in detail in Sect. 7.1. The other mode which tends to $n^2 + c_s^2k^2 = 0$ is obviously the acoustic mode. If non-adiabatic effects are taken into account, the amplitude of the acoustic mode damps or grows, depending on the nature of non-adiabatic processes. Inequality (7.21) should be the condition of amplification of the acoustic mode.

The reason why the condition of overstability of acoustic mode is given by inequality (7.21) can be understood as follows. In non-adiabatic oscillations, there are phase dependent heat input and output to oscillations. If in the phase where the gas is compressed (i.e., the phase where temperature and pressure increase) there is heat input by non-adiabatic processes, the oscillations expand in the next

© Springer Nature Singapore Pte Ltd. 2020, corrected publication 2023

S. Kato, J. Fukue, *Fundamentals of Astrophysical Fluid Dynamics*,
Astronomy and Astrophysics Library,
https://doi.org/10.1007/978-981-15-4174-2_8

expansion phase stronger than in the previous expansion phase. Similarly, if in the expanded phase (i.e., the phase where temperature and pressure are decreased) heat escapes from the region by non-adiabatic processes, the oscillation is compressed in the next compressed phase stronger than in the previous compressed phase. Then, oscillations are amplified. This condition of overstability is expressed in the form of instability (7.21).

The condition (7.21) is a result obtained for optically thin systems. The above arguments, however, are not limited to optically thin systems. The essence is that in the phase of temperature increase energy saving occurs. This is the condition of pulsational instability of stars.

Following Eddington (1959), a general criterion of stellar pulsational instability is presented here. Let us assume that thermal energy input to an oscillation at a phase of the oscillation is dQ. Then, the net energy input to the oscillation during one cycle of oscillation, W, is

$$W = \oint dQ. \tag{8.1}$$

In the zeroth-order approximation we have $\oint dQ = 0$, because the oscillation returns to the same state as the one cycle before. To proceed to the next order of approximation, we use the fact that the entropy of the system returns to the same state of one cycle before, i.e.,

$$\oint \frac{dQ}{T} = 0. \tag{8.2}$$

Then, expressing the Lagrangian variation of temperature as δT, i.e., $T = T_0 + \delta T$ (T_0 represents temperature of the unperturbed state), we write $1/T$ as $1/T = 1/T_0 - \delta T/T_0^2$. Then, from Eq. (8.2) we have

$$\oint dQ = \oint \frac{\delta T}{T_0} dQ. \tag{8.3}$$

This shows that the condition of instability ($W > 0$) is

$$\oint \frac{\delta T}{T_0} dQ > 0. \tag{8.4}$$

This expression for pulsational instability is general, as long as the oscillation energy of pulsation is positive (this is generally true in stellar oscillations).

In the case of stars, heat generation per unit mass is $\epsilon - (1/\rho)\mathrm{div}\boldsymbol{F}$, where ϵ is the rate of heat generation by nuclear reaction and $\boldsymbol{F}$ is the radiative heat flux through the star. The use of them leads condition (8.4) to the form:

$$\oint \int \frac{\delta T}{T_0} \delta\left(\epsilon - \frac{1}{\rho}\mathrm{div}\boldsymbol{F}\right) dm > 0, \tag{8.5}$$

where dm is a mass element and the integration is performed over one cycle as well as over the whole mass. This is the most popular expression for pulsational instability.

In practical problems the processes of excitation of oscillations can be classified into a few processes. They are ϵ-, κ-, δ- mechanisms and so on. The ϵ*-mechanism* is that due to nuclear energy generation, and comes from the term of $\delta\epsilon$ in expression (8.5). The κ*-mechanism* is that due to modulation of radiative flux $\boldsymbol{F}$ by phase dependent change of opacity, and is involved in the second term of inequality (8.5).

There are excellent text books on stellar pulsation and non-radial oscillations (Cox 1980; Unno et al. 1989). We recommend to see them for further studies on excitation mechanisms and on their applications to various types of stellar oscillations. In the following, we only present an alternative derivation of stability condition (8.5), directly based on dynamical point of view. In this formulation the unperturbed stars are not unnecessary to be static, but are allowed to have steady motions such as rotation. Oscillations are also not limited to radial pulsation, but non-radial oscillations can be involved (see also Aizenman and Cox 1975; Cox and Everson 1983).

In Sect. 2.5 we have presented equations describing small amplitude perturbations on equilibrium states in the Lagrangian formulation. If viscous processes are neglected, the equation of motion describing the perturbations is

$$\frac{d_0^2\boldsymbol{\xi}}{dt^2}+\delta\left(\frac{1}{\rho}\nabla p+\nabla\psi\right)=0, \tag{8.6}$$

where d_0/dt is the Lagrangian time derivative along the unperturbed flow, δ is the Lagrangian variation, $\boldsymbol{\xi}$ is the displacement vector, and ψ is the gravitational potential. For other notations, see Sect. 2.5.

In the case of non-adiabatic perturbations, the energy equation describing perturbations should be written as

$$\frac{d_0}{dt}\delta p-c_{\rm s}^2\frac{d_0}{dt}\delta\rho=\delta[(\Gamma_3-1)(-{\rm div}\boldsymbol{F}+\epsilon)], \tag{8.7}$$

where $\boldsymbol{F}$ and ϵ are thermal energy flux and thermal energy generation rate (per unit volume), respectively. Here, as the ratio of specific heats, Γ_3 is adopted so that the cases where radiation pressure is involved can be considered. In Eq. (8.7) $c_{\rm s}$ is the speed of sound, defined by $c_{\rm s}^2=\Gamma_1 p_0/\rho_0$, where Γ_1 is another exponent specifying the ratio of the specific heats [see, e.g., Chandrasekhar (1938) for adiabatic exponents and Sect. 19.6]. If the Lagrangian pressure variation is decomposed into the adiabatic part and non-adiabatic part, $(\delta p)_{\rm na}$, defined by

$$(\delta p)_{\rm na}\equiv\delta p-\Gamma_1\frac{p_0}{\rho_0}\delta\rho, \tag{8.8}$$

energy equation (8.7) is written as

$$\frac{d_0}{dt}(\delta p)_{\rm na} = \delta[(\Gamma_3 - 1)(-{\rm div}\boldsymbol{F} + \varepsilon)]. \tag{8.9}$$

Finally, the equation of continuity is

$$\delta\rho + \rho_0 {\rm div}\boldsymbol{\xi} = 0, \tag{8.10}$$

which is unchanged from the expression in the cases of adiabatic and inviscid perturbations.

By performing the same procedure as in Sect. 2.5, we can reduce Eq. (8.6) to[1]

$$\rho_0 \frac{\partial^2 \boldsymbol{\xi}}{\partial t^2} + 2\rho_0(\boldsymbol{v}_0\cdot\nabla)\frac{\partial\boldsymbol{\xi}}{\partial t} + \mathcal{L}(\boldsymbol{\xi}) = -\nabla(\delta p)_{\rm na}, \tag{8.11}$$

where $\boldsymbol{v}_0$ is the unperturbed flow such as rotation, and $\mathcal{L}$ is a Hermitian operator. For a detailed expression for $\mathcal{L}$ see Sect. 2.5. This equation is an extension of Eq. (2.46) in Sect. 2.5 to cases of non-adiabatic perturbations.

Hereafter, all perturbed quantities are assumed to be proportional to $\exp[i(\omega t - m\varphi)]$, and the displacement vector, $\boldsymbol{\xi}$, is written as

$$\boldsymbol{\xi} = \Re[\hat{\boldsymbol{\xi}}\exp(i\omega t)] = \Re\Big[\breve{\boldsymbol{\xi}}\exp[i(\omega t - m\varphi)]\Big], \tag{8.12}$$

where $\Re$ represents the real part. Then, Eq. (8.11) is written as

$$-\omega^2\rho_0\hat{\boldsymbol{\xi}} + 2i\omega\rho_0(\boldsymbol{v}_0\cdot\nabla)\hat{\boldsymbol{\xi}} + \mathcal{L}(\hat{\boldsymbol{\xi}}) = -\nabla(\hat{\delta p})_{\rm na}. \tag{8.13}$$

Multiplying equation (8.13) by the complex conjugate of $\hat{\boldsymbol{\xi}}$, i.e., $\hat{\boldsymbol{\xi}}^*$, and integrating the resulting equation over the whole volume, we have

$$\Im\Big[-\omega^2\int\rho_0\hat{\boldsymbol{\xi}}^*\hat{\boldsymbol{\xi}}d^3r + 2i\omega\int\rho_0\hat{\boldsymbol{\xi}}^*(\boldsymbol{v}_0\cdot\nabla)\hat{\boldsymbol{\xi}}d^3r\Big] = -\Im\int\hat{\boldsymbol{\xi}}^*\cdot\nabla(\hat{\delta p})_{\rm na}d^3r, \tag{8.14}$$

[1] It is noted that

$$\delta\Big(\frac{1}{\rho}\nabla p\Big) = -\frac{\delta\rho}{\rho_0^2}\nabla p_0 + \frac{1}{\rho_0}[\nabla p_1 + (\boldsymbol{\xi}\cdot\nabla)\nabla p_0]$$
$$= -\frac{\delta\rho}{\rho_0^2}\nabla p_0 + \frac{1}{\rho_0}[\nabla(\delta p) - \nabla(\boldsymbol{\xi}\cdot\nabla)p_0 + (\boldsymbol{\xi}\cdot\nabla)\nabla p_0].$$

If δp in the above equation is written as $\delta p = \Gamma_1(p_0/\rho_0)\delta\rho + (\delta p)_{\rm na}$, we have $\rho_0\delta[(1/\rho_0)\nabla p] = \nabla(\delta p)_{\rm na} +$ parts of $\mathcal{L}(\boldsymbol{\xi})$.

where $\Im$ represents the imaginary part and we have used the fact that $\mathcal{L}$ is a Hermite operator under the condition that ρ_0 vanishes on the surfaces of the systems in consideration.

Due to non-adiabatic processes, the frequency of oscillation, ω, becomes complex with an imaginary part $\omega_{\rm i}$. If we write ω as[2]

$$\omega = \omega_0 + i\omega_{\rm i}, \tag{8.15}$$

Equation (8.14) is written as

$$-2\omega_{\rm i}\biggl[\omega_0 \int \rho_0 \hat{\boldsymbol{\xi}}^* \cdot \hat{\boldsymbol{\xi}} d^3r - i \int \rho_0 \hat{\boldsymbol{\xi}}^* (\boldsymbol{v}_0 \cdot \nabla)\hat{\boldsymbol{\xi}} d^3r\biggr] = -\Im \int \hat{\boldsymbol{\xi}}^* \cdot \nabla(\hat{\delta p})_{\rm na} d^3r, \tag{8.16}$$

where the fact that $i\rho_0(\boldsymbol{v}_0 \cdot \nabla)$ is a Hermite operator has been used.

In Sect. 2.5.3 we have defined wave energy E:

$$E = \frac{1}{2}\int \rho_0 \biggl[\biggl(\frac{\partial \boldsymbol{\xi}}{\partial t}\biggr)^2 + \boldsymbol{\xi} \cdot \mathcal{L}(\boldsymbol{\xi})\biggr] d^3r, \tag{8.17}$$

which is time-independent in adiabatic oscillations (see Sect. 2.5.3). By using $\hat{\boldsymbol{\xi}}$ and eliminating $\mathcal{L}$ by use of wave equation for adiabatic oscillations, we can express E in the form [see Eq. (6.146) in Sect. 6.7.2 and also Sect. 3.1.2 of Kato (2016)]:

$$E = \frac{1}{2}\int \biggl[\omega_0^2 \rho_0 \hat{\boldsymbol{\xi}}^* \cdot \hat{\boldsymbol{\xi}} - i\omega_0 \rho_0 \hat{\boldsymbol{\xi}}^* (\boldsymbol{v}_0 \cdot \nabla)\hat{\boldsymbol{\xi}}\biggr] d^3r. \tag{8.18}$$

By using this expression for wave energy, we find that the growth rate of oscillations, $-\omega_{\rm i}$, is written as

$$-\omega_{\rm i} = -\frac{\omega_0}{4E}\Im \int \hat{\boldsymbol{\xi}}^* \cdot \nabla(\hat{\delta p})_{\rm na} d^3r. \tag{8.19}$$

In order to compare this expression with the stability criterion (8.5), the integration in Eq. (8.19) is performed by part. Then, using Eq. (8.9) and $(\Gamma_3 - 1)\delta\rho/\rho_0 = \delta T/T_0$, we have

$$-\Im \int \hat{\boldsymbol{\xi}}^* \cdot \biggl[\nabla(\hat{\delta p})_{\rm na}\biggr] d^3r = \Re \int (\omega_0 - m\Omega)^{-1} \biggl(\frac{\breve{\delta T}}{T_0}\biggr)^* \delta(-{\rm div}\breve{\boldsymbol{F}} + \breve{\epsilon}) d^3r, \tag{8.20}$$

where the unperturbed flow $\boldsymbol{v}_0$ is taken to be a pure rotation, i.e., $\boldsymbol{v}_0 = (0, r\Omega, 0)$ in the cylindrical coordinates.

[2] Non-adiabatic processes introduce not only an imaginary part of frequencies but also a slight change of the real part. By neglecting the latter change, however, we regard ω_0 as the frequency of non-adiabatic and inviscid oscillations.

Equation (8.20) shows that the growth rate by non-adiabatic processes, $-\omega_{\rm i}$, is given by

$$- \omega_{\rm i} = \frac{1}{4E}\Re\int \frac{\omega_0}{\omega_0 - m\Omega}\left(\frac{\delta\breve{T}}{T_0}\right)^* \delta(-{\rm div}\breve{\boldsymbol{F}} + \breve{\epsilon})d^3r. \tag{8.21}$$

In the case of non-rotating stars, $\omega_0/(\omega_0 - m\Omega)$ is unity, and the real part of the integral in Eq. (8.21) is two times the mechanical work done on stars by thermal processes. Hence, the quantity dividing it by $4E$ is the growth rate of oscillations, which is consistent with Eq. (8.5).

8.2 Excitation of Disk Oscillations by Viscous Processes

In Sect. 8.1 we have mentioned that one of main processes of excitation of stellar oscillations (pulsation) is κ-mechanism. In the κ-mechanism energy input to oscillations comes from outward energy flow (mainly radiative energy flow) through stars. This steady radiation flow in stars is modulated by time variation of opacity caused by oscillations. If the modulation occurs in a phase favorable for the phase of oscillations, a part of the energy flowing in stars is transferred to the oscillations and the oscillations are amplified.

In addition to the κ-mechanism, a similar but different process is expected in accretion disks. One of important characteristics of accretion disks is steady angular momentum flow through disks from inner to outer parts of the disks by the effects of viscosity. This steady angular momentum flow is modulated, if oscillations are present on the disks. If this angular momentum modulation occurs in the phase favorable for oscillations, a part of the angular momentum is transferred to that of oscillations and the oscillations are amplified (Kato 1978).

It is noticed that the above viscous excitation process of disk oscillations, which will be discussed here, is one of examples that in non-equilibrium open systems dissipative processes can excite oscillations, as well as the κ-mechanism.

We derive the criterion of viscous excitation of disk oscillations by a similar method as that in Sect. 8.1. If viscous processes are taken into account, the equation of motion describing perturbations is

$$\frac{d_0\boldsymbol{\xi}}{dt^2} + \delta\left(\frac{1}{\rho}\nabla p + \nabla\psi\right) = \delta\boldsymbol{N}, \tag{8.22}$$

where $\boldsymbol{N}$ is the viscous stress force per unit mass. Then, the equation of motion corresponding to Eq. (8.11) is now

$$\rho_0\frac{\partial^2\boldsymbol{\xi}}{\partial t^2} + 2\rho_0(\boldsymbol{v}_0\cdot\nabla)\frac{\partial\boldsymbol{\xi}}{\partial t} + \mathcal{L}(\boldsymbol{\xi}) = \rho_0\delta\boldsymbol{N}, \tag{8.23}$$

where the effects of the non-adiabatic processes have been neglected in order to concentrate our attention here to viscous processes. As in Sect. 8.1, all perturbed quantities are taken to be proportional to $\exp[i(\omega t - m\varphi)]$ as shown in Eq. (8.12), where cylindrical coordinates whose center is at the disk center have been adopted and m is the azimuthal wavenumber of perturbations. Then, we have, as the equation corresponding to Eq. (8.13),

$$-\omega^2\rho_0\hat{\boldsymbol{\xi}} + 2i\omega\rho_0(\boldsymbol{v}_0\cdot\nabla)\hat{\boldsymbol{\xi}} + \mathcal{L}(\hat{\boldsymbol{\xi}}) = \rho_0\delta\hat{\boldsymbol{N}}, \tag{8.24}$$

where the hats attached to $\boldsymbol{\xi}$ and $\boldsymbol{N}$ have the same meanings as in Eq. (8.12).

Multiplying equation (8.24) by the complex conjugate of $\hat{\boldsymbol{\xi}}$, i.e., $\hat{\boldsymbol{\xi}}^*$, and integrating the resulting equation over the whole volume, we have

$$\Im\left[-\omega^2\int\rho_0\hat{\boldsymbol{\xi}}^*\hat{\boldsymbol{\xi}}d^3r + 2i\omega\int\rho_0\hat{\boldsymbol{\xi}}^*(\boldsymbol{v}_0\cdot\nabla)\hat{\boldsymbol{\xi}}d^3r\right] = \Im\int\hat{\boldsymbol{\xi}}^*\cdot\rho_0\delta\hat{\boldsymbol{N}}d^3r. \tag{8.25}$$

Using again that the wave energy of oscillations is given by Eq. (8.18), we have

$$-\omega_{\rm i} = \frac{\omega_0}{4E}\Im\int\hat{\boldsymbol{\xi}}^*\cdot\rho_0\delta\hat{\boldsymbol{N}}d^3r. \tag{8.26}$$

This equation and Eq. (8.19) are a set. The growth rate due to thermal processes is Eq. (8.19), while that due to viscous processes is by Eq. (8.26).

Equation (8.26) shows that if the time variation of viscous force associated with an oscillation, $\delta\hat{\boldsymbol{N}}$, has a component in phase with the oscillation, $i\omega\hat{\boldsymbol{\xi}}$, a work is done on the oscillation and it makes the oscillation grow with time. This can be realized in accretion disks by the following situations. In accretion disks, there is angular momentum flow in the unperturbed disk due to viscous force. This angular momentum flow is modulated by superposition of oscillations. If this modulation of angular momentum flow has a favorable phase with oscillations, a part of the modulated part of the angular momentum flow is transferred to the oscillations to amplify them. This is similar to the κ-mechanism in stellar pulsation. In the κ-mechanism, the radiation flowing inside stars is modulated by the presence of oscillations, and a part of modulated energy is transferred to oscillations. In the present issue a part of the angular momentum flows in disks is transferred to disks oscillations by the effects of modulation of viscous force by the oscillations themselves.

Lagrangian variation of viscous force per unit mass, $\delta\hat{\boldsymbol{N}}$, consists of two parts as

$$\delta\hat{\boldsymbol{N}} = \hat{\boldsymbol{N}}_1 + (\hat{\boldsymbol{\xi}}\cdot\nabla)\boldsymbol{N}_0, \tag{8.27}$$

where $\hat{\boldsymbol{N}}_1$ is the Eulerian variation of $\hat{\boldsymbol{N}}$, and $\boldsymbol{N}_0$ is the imbalance of viscous force in the unperturbed state. It is noted that $\boldsymbol{N}_0 \neq 0$ is the cause of accretion flow (advection flow) in the unperturbed disks.

The components of $\delta\hat{\boldsymbol{N}}$ in the cylindrical coordinates are

$$(\delta\hat{\boldsymbol{N}})_r = \hat{N}_{r1}+\left(\hat{\xi}_r\frac{\partial}{\partial r}+\hat{\xi}_z\frac{\partial}{\partial z}\right)N_{0r}-\frac{\hat{\xi}_\varphi}{r}N_{0\varphi}, \tag{8.28}$$

$$(\delta\hat{\boldsymbol{N}})_\varphi = \hat{N}_{\varphi 1}+\left(\hat{\xi}_r\frac{\partial}{\partial r}+\hat{\xi}_z\frac{\partial}{\partial z}\right)N_{0\varphi}+\frac{\hat{\xi}_\varphi}{r}N_{0r}, \tag{8.29}$$

$$(\delta\hat{\boldsymbol{N}})_z = \hat{N}_{z1}+\left(\hat{\xi}_r\frac{\partial}{\partial r}+\hat{\xi}_z\frac{\partial}{\partial z}\right)N_{0z}. \tag{8.30}$$

Among the various terms including components of $\boldsymbol{N}_0$ in Eqs. (8.28)–(8.30), the main ones are $-(\hat{\xi}_\varphi/r)N_{0\varphi}$ in the r-component and $\hat{\xi}_r\partial N_{0\varphi}/\partial r$ in the φ-component. Further, $\hat{\xi}_r$ and $\hat{\xi}_\varphi$ are related by[3]

$$(\omega_0 - m\Omega)\breve{\xi}_\varphi = i2\Omega\breve{\xi}_r. \tag{8.31}$$

Based on the above considerations, we decompose $-\omega_{\rm i}$ given by Eq. (8.26) into two terms:

$$-\omega_{\rm i} = (-\omega_{\rm i})_{\rm viscosity} + (-\omega_{\rm i})_{\rm imbalance}, \tag{8.32}$$

where

$$(-\omega_{\rm i})_{\rm viscosity} = \frac{\omega_0}{4E}\Im\int\rho_0\left(\breve{\xi}_r^*\breve{N}_{1r}+\breve{\xi}_\varphi^*\breve{N}_{1\varphi}+\breve{\xi}_z^*\breve{N}_{1z}\right)d^3r, \tag{8.33}$$

$$(-\omega_{\rm i})_{\rm imbalance} = -\frac{\omega_0}{4E}\Re\int\rho_0\frac{2\Omega}{\omega_0-m\Omega}\breve{\xi}_r^*\breve{\xi}_r\frac{\partial}{r\partial r}(rN_{o\varphi})d^3r. \tag{8.34}$$

Equation (8.33) is the main term of $-\omega_{\rm i}$, and Eq. (8.34) is supplementary.

Hereafter, we consider $(-\omega_{\rm i})_{\rm viscosity}$ alone, since it is the main term. It is convenient to change variables from displacement vector $\boldsymbol{\xi}$ to velocity perturbation $\boldsymbol{v}$. A relation between $\boldsymbol{\xi}$ and $\boldsymbol{v}_1$ is given by Eq. (2.40) in Sect. 2.5. In the case where

[3] It is noted that in the case where the force acting on perturbations in the azimuthal direction is neglected, $\breve{v}_\varphi$ and $\breve{v}_r$ are related by

$$i(\omega-m\Omega)\breve{v}_\varphi+\frac{\kappa^2}{2\Omega}\breve{v}_r=0,$$

but $\breve{\xi}_\varphi$ and $\breve{\xi}_r$ are related by

$$i(\omega-m\Omega)\breve{\xi}_\varphi+2\Omega\breve{\xi}_r=0.$$

cylindrical coordinates (r, φ, z) are adopted, the relation gives

$$
\begin{aligned}
i(\omega - m\Omega)\breve{\xi}_r &= \breve{v}_r, \\
i(\omega - m\Omega)\breve{\xi}_\varphi &= \breve{v}_\varphi + \breve{\xi}_r r\frac{d\Omega}{dr} = \frac{4\Omega^2}{\kappa^2}\breve{v}_\varphi, \\
i(\omega - m\Omega)\breve{\xi}_z &= \breve{v}_z.
\end{aligned} \tag{8.35}
$$

Using these relations, we have

$$
(-\omega_{\rm i})_{\rm viscosity} = \frac{\omega_0}{4E}\Re\int\rho_0\frac{1}{\omega_0 - m\Omega}\left(\breve{v}_r^*\breve{N}_{1r} + \frac{4\Omega^2}{\kappa^2}\breve{v}_\varphi^*\breve{N}_{1\varphi} + \breve{v}_z^*\breve{N}_{1z}\right)d^3r. \tag{8.36}
$$

To examine what kinds of disk oscillations are excited by this viscous processes, we need to know detailed expressions for viscous stress tensor $\boldsymbol{N}_1$ and for eigenfunctions of disk oscillations. The former depends on the form of (turbulent) viscosity and its time variation. In the case of conventional α viscosity or in the case where the viscous stress tensor is expressed with use of viscosity and the viscosity is proportional to density (i.e., kinematic viscosity is constant), the p-mode oscillations are excited (e.g., Kato 1978). If the effects of accretion flows in the unperturbed disks are taken into account, there appears a difference on growth rate in outgoing and ingoing waves. The outgoing p-mode oscillations have a higher growth rate than those of ingoing ones (Wu and Li 1996).

It is noted that when we consider non-axisymmetric oscillations the factor, $(\omega - m\Omega)^{-1}$, appears in the denominator. This means that the above perturbation method cannot be applied to oscillations with corotation resonance in their propagation region. This is related to the fact that the corotation radius is a resonant point. See Sect. 6.6 for corotation instability.

8.3 Double Diffusive Instability

The *double diffusive instability* is an instability (overstability) resulting from coexistence of two dissipative processes. This instability is known in various fields of physics (e.g., plasma physics), oceanography, and astrophysical fluid dynamics. In the field of astrophysical fluid dynamics, Cowling (1958) will be the first who pointed out the mechanism of this kind of overstability. Cowling showed the presence of this kind of overstability by considering convection in rotating media and in magnetized ones.

Roughly speaking, magnetic fields and rotation act so as to suppress the onset of convective motions. Hence, one may think that for convection to occur in such media, the temperature gradient, β, in the media needs to be larger than the adiabatic one, $\beta_{\rm ad}$, i.e., $\beta > \beta_{\rm ad}$, by a certain finite amount. This is, however, not the case, if there are dissipative processes in the media. In rotating media, for example, convection (overstable convection) can set in when β is just over $\beta_{\rm ad}$, if the media are thermally conductive (more exactly, if thermometric conductivity in the media, $\kappa_{\rm th}$, is (roughly) larger than kinematic viscosity, ν, in the media). The situation is similar even in the case of magnetized media. That is, even in magnetized media, convection (overstable convection) can occur in cases where $\beta > \beta_{\rm ad}$ and $\kappa_{\rm th}$ is (roughly) larger than magnetic diffusivity, η, in the media. These are special examples of *double diffusive instability*.

Mathematical derivation of *overstable convection* in magnetized and rotating media are presented by Chandrasekhar (1961). In this section overstable convection in rotating medium and that in chemically inhomogeneous one will be presented. The cause of overstable convection will be explained somewhat in detail in the case of rotating media.

8.3.1 Overstable Convection in Uniformly Rotating Fluid

Let us consider a plane-parallel gas system heated from below, which is uniform in the x–y direction but subject to a constant gravitational field, $\boldsymbol{g}$, directed in the negative z-direction. In the z-direction, temperature decreases with a constant rate $dT_0/dz = -\beta$. The system is further assumed to be rotating around the z-direction with a constant angular velocity, Ω. It is noted that if there is no rotation and no dissipative processes, the system is convectively unstable when the temperature decreasing rate, β, is larger than the adiabatic temperature decreasing rate, $\beta_{\rm ad}$, i.e., $\beta > \beta_{\rm ad}$. The value of $\beta_{\rm ad}$ is given by $\beta_{\rm ad} = g/C_{\rm p}$, $C_{\rm p}$ being the specific heat for constant pressure (see Chap. 5).[4]

On such a plane-parallel system small amplitude perturbations are superposed, taking into account the effects of two dissipative processes due to viscosity and conduction. In studying the onset of convection, we can describe the perturbations by using the Boussinesq approximation (see Sect. 5.1 for details):

$$\mathrm{div}\boldsymbol{v} = 0 \tag{8.37}$$

$$\frac{\partial \boldsymbol{v}}{\partial t} + 2\boldsymbol{\Omega} \times \boldsymbol{v} = -\frac{1}{\rho_0}\mathrm{grad}p_1 - \alpha T_1 \boldsymbol{g} - \nu\nabla^2\boldsymbol{v}, \tag{8.38}$$

$$\frac{\partial T_1}{\partial t} - (\beta - \beta_{\rm ad})v_z = \kappa_{\rm th}\nabla^2 T_1, \tag{8.39}$$

[4] In Chap. 5 we have used ∇ and $\nabla_{\rm ad}$ instead of β and $\beta_{\rm ad}$, respectively. They are related to β and $\beta_{\rm ad}$, respectively, by $\beta = -T(d\ln p/dz)\nabla$ and $\beta_{\rm ad} = -T(d\ln p/dz)\nabla_{\rm ad}$, because $\beta = -dT/dz = -(T/p)(d\ln T/d\ln p)(dp/dz)$. In particular, $\beta_{\rm ad} = (\rho T/p)g\nabla_{\rm ad} = [\gamma/(\gamma-1)](g/C_{\rm p})\nabla_{\rm ad} = g/C_{\rm p}$.

where $\boldsymbol{v} = (v_x, v_y, v_z)$ is velocity perturbation, $\mathbf{g} = (0, 0, -g)$ is the gravitational acceleration, and p_1 and T_1 are, respectively, pressure and temperature perturbations over the unperturbed ones, p_0, and T_0. The quantity α is defined by $\alpha = 1/T_0$, and $\kappa_{\rm th}$ and ν are, respectively, thermometric conductivity and kinematic viscosity. Hereafter, the angular velocity of rotation, $\boldsymbol{\Omega}$, is taken to be $\boldsymbol{\Omega} = (0, 0, \Omega)$ with $\Omega = \text{const}$.

Let us take the derivative of x-component of Eq. (8.38) with respect to x and the derivative of y-component of Eq. (8.38) with respect to y. Summing the resulting two equations, we have

$$\left(\frac{\partial}{\partial t} - \nu\nabla^2\right)\frac{\partial v_z}{\partial z} - 2\Omega\left(\frac{\partial v_x}{\partial y} - \frac{\partial v_y}{\partial x}\right) = \frac{1}{\rho_0}\nabla_1^2 p_1, \tag{8.40}$$

where $\nabla_1^2 = \partial^2/\partial x^2 + \partial^2/\partial y^2$. Eliminating p_1 from this equation and the z-component of Eq. (8.38) we have

$$\left(\frac{\partial}{\partial t} - \nu\nabla^2\right)\nabla^2 v_z - 2\Omega\frac{\partial}{\partial z}\left(\frac{\partial v_x}{\partial y} - \frac{\partial v_y}{\partial x}\right) = \alpha g \nabla_1^2 T_1. \tag{8.41}$$

Another relation where p_1 has been eliminated is obtained from x- and y-components of Eq. (8.38):

$$\left(\frac{\partial}{\partial t} - \nu\nabla^2\right)\left(\frac{\partial v_x}{\partial y} - \frac{\partial v_y}{\partial x}\right) + 2\Omega\frac{\partial v_z}{\partial z} = 0. \tag{8.42}$$

Elimination of $\partial v_x/\partial y - \partial v_y/\partial x$ from Eqs. (8.41) and (8.42) leads to

$$\left(\frac{\partial}{\partial t} - \nu\nabla^2\right)^2 \nabla^2 v_z + 4\Omega^2\frac{\partial^2 v_z}{\partial z^2} = \alpha g \nabla_1^2\left(\frac{\partial}{\partial t} - \nu\nabla^2\right)T_1. \tag{8.43}$$

Finally, eliminating T_1 from Eqs. (8.39) and (8.43), we have

$$\left(\frac{\partial}{\partial t} - \kappa_{\rm th}\nabla^2\right)\left[\left(\frac{\partial}{\partial t} - \nu\nabla^2\right)^2\nabla^2 + 4\Omega^2\frac{\partial^2}{\partial z^2}\right]v_z = \alpha(\beta - \beta_{\rm ad})g\nabla_1^2\left(\frac{\partial}{\partial t} - \nu\nabla^2\right)v_z. \tag{8.44}$$

By assuming that the perturbations are proportional to $\exp(nt)\exp(i\boldsymbol{k}\cdot\boldsymbol{r})$ we have the dispersion relation:

$$\left(n + \kappa_{\rm th}k^2\right)\left[\left(n + \nu k^2\right)^2 k^2 + 4\Omega^2 k_z^2\right] - \alpha(\beta - \beta_{\rm ad})g k_1^2\left(n + \nu k^2\right) = 0. \tag{8.45}$$

This dispersion relation is written in the form:

$$n^3 + a_1 n^2 + a_2 n + a_3 = 0, \tag{8.46}$$

where

$$a_1 = (\kappa_{\rm th} + 2\nu)k^2, \quad a_2 = \nu(\nu + 2\kappa_{\rm th})k^4 + 4\Omega^2\frac{k_z^2}{k^2} - \alpha(\beta - \beta_{\rm ad})g\frac{k_1^2}{k^2},$$

$$a_3 = \kappa_{\rm th}k^2\left(\nu^2k^4 + 4\Omega^2\frac{k_z^2}{k^2}\right) - \alpha(\beta - \beta_{\rm ad})g\nu k_1^2. \tag{8.47}$$

In the case of no dissipation, i.e., $\nu = \kappa_{\rm th} = 0$, the dispersion relation (8.45) reduces to

$$n\left[n^2 + 4\Omega^2\frac{k_z^2}{k^2} - \alpha(\beta - \beta_{\rm ad})g\frac{k_1^2}{k^2}\right] = 0. \tag{8.48}$$

This shows that except the mode which tends to trivial one of $n = 0$, in the limit of $\nu = \kappa_{\rm th} = 0$ we have the convection mode which grows when

$$\beta - \beta_{\rm ad} > 4\Omega^2\frac{T_0}{g}\frac{k_z^2}{k_1^2} \equiv \beta_{\rm crit}. \tag{8.49}$$

This equation shows that the rotation suppresses the onset of convection until β becomes larger than $\beta_{\rm ad}$ by a certain amount, $\beta_{\rm crit}$, depending upon the strength of rotation. If $\beta - \beta_{\rm ad}$ is smaller than $\beta_{\rm crit}$, the convective mode is an oscillatory mode.

Hereafter, we concentrate our attention to the oscillatory mode, assuming that $\beta_{\rm ad} < \beta < \beta_{\rm ad} + \beta_{\rm crit}$, and examine what happens when viscosity and thermal conductivity are present. As mentioned in Sect. 7.1.1, the Hurwitz theorem tells us that the oscillatory mode becomes overstable if $a_1a_2 - a_3 < 0$. In the present problem a_1, a_2, and a_3 are given by Eq. (8.47) and leads to

$$a_1a_2 - a_3 = 2\nu k^2\left[(\kappa_{\rm th} + \nu)^2k^4 + 4\Omega^2\frac{k_z^2}{k^2}\right] - (\kappa_{\rm th} + \nu)\alpha(\beta - \beta_{\rm ad})gk_1^2. \tag{8.50}$$

It is convenient to rewrite the above equation to

$$a_1a_2 - a_3 = \nu k^2\left[-\alpha(\beta - \beta_{\rm ad} - \beta_{\rm crit})g\frac{k_1^2}{k^2} + 4\Omega^2\frac{k_z^2}{k^2} + 2(\kappa_{\rm th} + \nu)^2k^4\right]$$
$$-\kappa_{\rm th}\alpha(\beta - \beta_{\rm ad})gk_1^2. \tag{8.51}$$

Equation (8.51) shows that when $\kappa_{\rm th} = 0$, $a_1a_2 - a_3$ is positive as long as $\beta_{\rm ad} < \beta < \beta_{\rm ad} + \beta_{\rm crit}$. That is, the oscillatory mode is damped by the effects of viscosity. In the case of $\nu = 0$, however, $a_1a_2 - a_3$ is negative unless $\kappa_{\rm th} = 0$, i.e., the oscillatory mode is overstable. In summary, in the case of $\beta_{\rm ad} < \beta < \beta_{\rm ad} + \beta_{\rm crit}$, the convection mode is a growing oscillatory motion, if $\kappa_{\rm th}$ is larger than ν by a certain amount. It is noted that in many astrophysical objects, $\kappa_{\rm th}$ is much larger than ν, because thermal diffusion is due to radiation processes.

The reason why oscillatory convection becomes overstable in the case of $\kappa_{th} \gg \nu$ can be understood in the following ways. In the limit of no dissipative processes, the restoring force due to rotation is stronger than the buoyancy force when $\beta < \beta_{ad} + \beta_{crit}$, and thus the perturbation is a harmonic oscillation (oscillatory convection). If κ_{th} is larger than ν by a certain amount, the decrease of buoyancy force during the oscillation is larger than the decrease of the restoring force of rotation due to viscosity. This means that in the next cycle of the oscillation the net restoring force increases, and the oscillation is amplified. This is the reason why we have overstable convection in the case of $\beta > \beta_{ad}$, even if inequality (8.49) is not satisfied. This situation is schematically shown in Fig. 8.1.

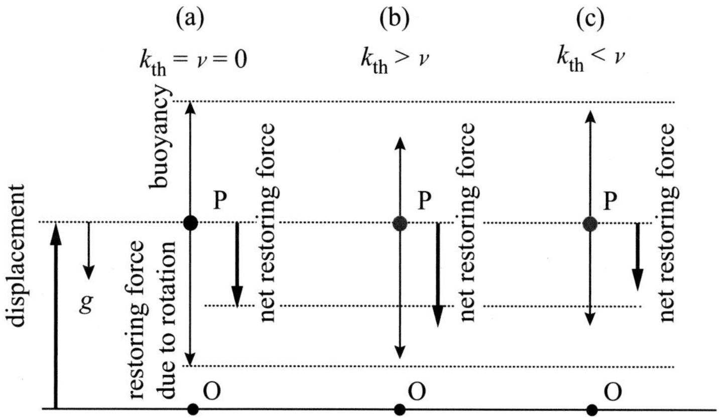

Fig. 8.1 Schematic diagram showing why oscillatory convection ($\beta_{ad} < \beta < \beta_{ad} + \beta_{crit}$) grows with time (overstable) when $\kappa_{th} \gg \nu$ in rotating media. Panel (**a**) shows the case of no dissipation ($\nu = 0$ and $\kappa_{th} = 0$). The point O represents the equilibrium position of a fluid element, and the point P the position of the maximum displacement of the fluid element upwards. At point P, the forces acting on the fluid element are a buoyancy force by superadiabatic temperature gradient and a restoring force by rotation. The former is in the direction upwards and the latter is downwards. The net restoring force is the difference of these two forces and is directed downwards, since the amount of restoring force by rotation is larger than that of the buoyancy force. Due to this net restoring force the fluid element makes a harmonic oscillation around the equilibrium position O (the net restoring force is shown by thick downward arrow). The panel (**c**) represents the case of $\nu \gg \kappa_{th}$. At the maximum displacement position P (this position is not always equal to the position in the case of $\nu = \kappa_{th} = 0$, but shown by the same symbol P), the fluid element still have a net restoring force acting downwards, but the amount is less than that in the case of one cycle before, because the restoring force by rotation much decreases by the effects of viscosity compared with the decrease of buoyancy force by the effects of conductivity. This means that the fluid element returns to the original position with a slower speed compared with one cycle before. That is, the oscillation is damped with time. Panel (**b**) represents the case of $\nu \ll \kappa_{th}$. In this case the fluid element receives much stronger restoring force at point P, because the buoyancy force much decreases compared with the decrease of the restoring force by rotation. This means that the fluid element returns to the original position O with a stronger restoring force compared with the one cycle before. That is, the oscillation is amplified (overstable). It should be noticed that in the case of panel (**b**) the net restoring force becomes larger than that in the case of panel (**a**)

8.3.2 *Overstable Convection in Chemically Inhomogeneous Media*

In an evolutional stage of a star, chemical inhomogeneity is realized inside the star by shrink of nuclear burning region by mass coordinate. In such a region, the chemical inhomogeneity acts against the onset of convection, because the lower medium is heavier in weight than the upper one is. That is, the onset of convection is suppressed until the temperature gradient becomes larger than the adiabatic temperature gradient by a certain amount.

Let us first consider the condition of onset of adiabatic convection n her media with chemical inhomogeneity. The condition of the onset of convection is, as derived in Sect. 5.1,

$$\frac{1}{\Gamma_1}\frac{d\ln p}{dz} < \frac{d\ln\rho}{dz}, \tag{8.52}$$

where the ratio of specific heats is written as Γ_1, not γ, since radiation pressure is not negligible in the cases where chemical inhomogeneity becomes important.

The equation of state gives

$$\frac{d\ln p}{dz} = (4-3\beta)\frac{d\ln T}{dz} + \beta\frac{d\ln\rho}{dz} - \beta\frac{d\ln\bar{\mu}}{dz}, \tag{8.53}$$

where β is the ratio of gas pressure, $p_{\text{gas}}[=(k_{\text{B}}/\bar{\mu}m_{\text{H}})\rho T]$, to the total pressure, $p[=p_{\text{rad}}+p_{\text{gas}}=(1/3)aT^4+(k_{\text{B}}/\bar{\mu}m_{\text{H}})\rho T]$:[5]

$$\beta = \frac{p_{\text{gas}}}{p}. \tag{8.54}$$

Here and hereafter in this subsection, β is used to represent the fraction of gas pressure, which should not be confused with the temperature gradient used in the previous subsection (Sect. 8.3.1). Here, $\bar{\mu}$ is the mean molecular weight and m_{H} is the hydrogen mass.

Eliminating ρ from Eq. (8.52) by use of Eq. (8.53), we can rewrite Eq. (8.52) in the form:

$$\nabla \equiv \frac{d\ln T}{d\ln p} > \frac{\Gamma_2-1}{\Gamma_2} + \frac{\beta}{4-3\beta}\frac{d\ln\bar{\mu}}{d\ln p}. \tag{8.55}$$

[5] In Sect. 8.3.1, β was used to represent temperature gradient, i.e., $\beta=-dT_0/dz$. In this subsection (Sect. 8.3.2), however, we use β to represent p_{gas}/p. The temperature gradient, $-dT_0/dz$, is represented by $-(T/p)(dp/dz)\nabla$ (see Eq. (4.62) and (8.57)).

This is the criterion of onset of convection in the case where the effects of gradient of mean molecular weight is taken into account (*Ledoux criterion*). It is noted that $d\ln\bar{\mu}/d\ln p$ is generally positive. It is noted that this criterion of convective instability has been derived under the assumption that no dissipative process works in the onset of instability.

Equation (8.55) shows that the criterion of onset of convection is severer in chemically inhomogeneous medium than $\nabla \geq \nabla_{\rm ad} (\equiv (\Gamma_2 - 1)/\Gamma_2)$. However, it is important to notice here that even if Eq. (8.55) is not satisfied, a fluid element rising upwards has a higher temperature than the surrounding medium, if $d\ln T/d\ln p$ is larger than $(\Gamma_2 - 1)/\Gamma_2$. Hence, if we consider thermal diffusion, overstable convection is expected by the same process considered in the previous Sect. 8.3.1. This is really the case, and this kind of overstable convection is called *semi-convection*.

In the following we present the overstable convection in the case of $\nabla_{\rm ad} < \nabla < \nabla_{\rm ad} + [\beta/(4-3\beta)](d\ln\bar{\mu}/d\ln p)$ by deriving the dispersion relation (Kato 1966). If the Boussinesq approximation is adopted, the hydrodynamical equations describing perturbations are

$$\mathrm{div}\boldsymbol{v} = 0, \tag{8.56}$$

$$\rho_0\frac{\partial \boldsymbol{v}}{\partial t} = -\mathrm{grad}p_1 + \rho_1\boldsymbol{g}, \tag{8.57}$$

$$\frac{\partial T_1}{\partial t} + T_0\frac{d\ln p_0}{dz}v_z(\nabla - \nabla_{\rm ad}) = \kappa_{\rm th}\nabla^2 T_1, \tag{8.58}$$

$$\beta\frac{\rho_1}{\rho_0} + (4-3\beta)\frac{T_1}{T_0} - \beta\frac{\bar{\mu}_1}{\bar{\mu}_0} = 0, \tag{8.59}$$

where the subscripts 0 and 1 represent, respectively, the unperturbed and perturbed quantities. The last equation is derived from the equation of state under the assumption that there is no pressure variation (a part of the Boussinesq approximation). Since the timescale by which chemical inhomogeneity is smoothed is much longer than the the thermal timescale, we adopt[6]

$$\frac{\partial\bar{\mu}}{\partial t} + \bar{\mu}_0\frac{d\ln\bar{\mu}_0}{dz}v_z = 0. \tag{8.60}$$

[6] Since the timescale of nuclear reaction is long, the Lagrangian time variation of the mean molecular weight is neglected, i.e., $d\bar{\mu}/dt = 0$.

Perturbations are assumed to be proportional to $\exp(nt - i\boldsymbol{k}\boldsymbol{r})$. Then, from Eqs. (8.56)–(8.60) we can derive a dispersion relation, which is

$$(n + \kappa_{\rm th}k^2)\left(n^2k^2 - g\frac{d\ln\bar{\mu}_0}{dz}k_1^2\right) + ng\frac{d\ln p_0}{dz}(\nabla - \nabla_{\rm ad})k_1^2 = 0, \tag{8.61}$$

where $k_1^2 = k_x^2 + k_y^2$ and $k^2 = k_1^2 + k_z^2$. In the adiabatic case ($\kappa_{\rm th} = 0$), the condition of the onset of instability is

$$\nabla > \nabla_{\rm ad} + \frac{\beta}{4 - 3\beta}\frac{d\ln\bar{\mu}_0}{d\ln p_0} \quad \textit{(Ledoux criterion)} \tag{8.62}$$

and coincides with Eq. (8.55). If thermal conduction is taken into account, the Hurwitz theorem (see Sect. 7.1.2) tells that the condition of overstability is

$$\nabla > \nabla_{\rm ad} \quad \textit{(Schwarzschild criterion)}. \tag{8.63}$$

In oceanography the mixing of ocean water is an important issue. In this issue salinity has important effects. A problem similar to semi-convection due to mean molecular weight is known in oceanography.

8.3.3 Goldreich–Schubert–Fricke Criterion-II

As described in Sect. 7.4, the Goldreich–Schubert–Fricke criterion (Goldreich and Schubert 1967; Fricke 1968) for stability of rotating fluid systems consists of two conditions. The first one is inequality (7.57), and the second one is inequality (7.58). At least one of these criteria is violated, the systems are unstable. The meaning of the second criterion was examined in Sect. 7.4. Here, we examine the meaning of the first criterion (7.57). The criterion can be reduced to simpler forms.

Let us consider three cases separately, i.e.,

- case (i) $-(1/\rho_0)(\partial p_0/\partial r)\partial\ln(p_0/\rho_0^\gamma)/\partial r < 0$,
- case (ii) $-(1/\rho_0)(\partial p_0/\partial r)\partial\ln(p_0/\rho_0^\gamma)/\partial r > 0$,
- case (iii) $-(1/\rho_0)(\partial p_0/\partial r)\partial\ln(p_0/\rho_0^\gamma)/\partial r = 0$.

In case (i), inequality (7.57) shows that the right-hand side of inequality (7.57) becomes negative for large values of x^2. That is, inequality (7.57) is not always satisfied, and the system is unstable. In case (ii), for inequality (7.57) to be satisfied for all x, the discriminant of Eq. (7.57), say D, needs to be zero or negative. Since the discriminant, D, is expressed in the form:

$$D = \left[\frac{1}{\rho_0}\frac{\partial p_0}{\partial r}\frac{\partial}{\partial z}\left(\ln\frac{p_0}{\rho_0^\gamma}\right) - \frac{1}{\rho_0}\frac{\partial p_0}{\partial z}\frac{\partial}{\partial r}\left(\ln\frac{p_0}{\rho_0^\gamma}\right)\right]^2, \tag{8.64}$$

the system is stable and only marginally stable when

$$\frac{1}{\rho_0}\frac{\partial p_0}{\partial r}\frac{\partial}{\partial z}\left(\ln\frac{p_0}{\rho_0^\gamma}\right)-\frac{1}{\rho_0}\frac{\partial p_0}{\partial z}\frac{\partial}{\partial r}\left(\ln\frac{p_0}{\rho_0^\gamma}\right)=0. \tag{8.65}$$

That is, for the system to be (marginally) stable the normal direction of the isobaric surface (constant pressure surface) [i.e., $(-\partial p_0/\rho_0\partial r,\ 0,\ -\partial p_0/\rho_0\partial z)$], and the isentropic surface (constant entropy surface) [i.e., $(-\partial \ln s/\partial z,\ 0,\ \partial \ln s/\partial r)$, where $s\equiv p_0/\rho_0^\gamma$] must cross at right angles. In other words, the isobaric and isentropic surfaces must coincide. The condition of case (ii) implies that the direction of pressure increase and that of entropy one are in the opposite direction (see Fig. 8.2).

Next, let us consider case (iii). In this case condition (7.57) is satisfied only when

$$-\frac{1}{\rho_0}\frac{\partial p_0}{\partial r}\frac{\partial \ln(p_0/\rho_0^\gamma)}{\partial z}-\frac{1}{\rho_0}\frac{\partial p_0}{\partial z}\frac{\partial \ln(p_0/\rho_0^\gamma)}{\partial r}=0 \tag{8.66}$$

and

$$-\frac{1}{\rho_0}\frac{\partial p_0}{\partial z}\frac{\partial \ln(p_0/\rho_0^\gamma)}{\partial z}\geq 0. \tag{8.67}$$

Equality in Eq. (8.67) is for marginally stability. They are involved in case (ii) as a special case.

Summing the above results, we can say that for the system to be stable, the directions of pressure decrease and entropy increase need to be parallel and in the same direction. In other words, the system is stable only when the classical Schwarzschild criterion of convective stability is satisfied. The presence of restoring force of any rotation cannot prevent the overstable convection due to thermal conduction. This is an extension of the results in Sect. 8.3.1.

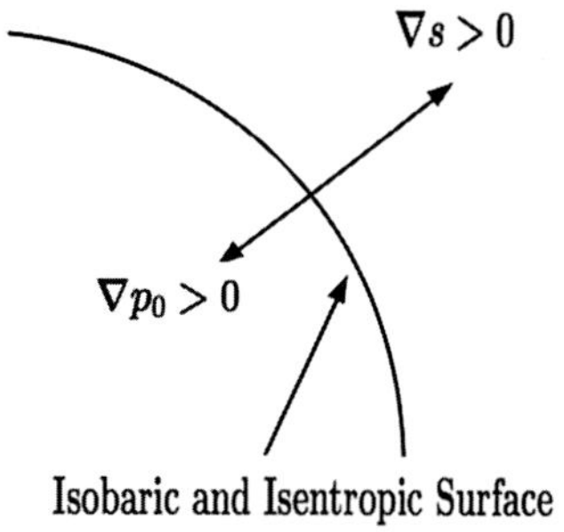

Fig. 8.2 Schematic picture showing one of the stability condition of Goldreich–Schubert–Fricke (7.57). For systems to be stable, the isobaric surface and the isentropic one need to be the same, and the directions of pressure gradient and of entropy one need to be in the opposite direction. In other words, this is a generalization of the Schwarzschild stability criterion to arbitrary rotating systems

The results in Sect. 7.4 and those in the present Sect. 8.3.3 are summarized as follows: Rotating astrophysical systems with $\kappa_{\rm th} \gg \nu$ are stable only when conditions (7.59) and the Schwarzshild criterion of convective stability are simultaneously satisfied. If one of them is violated the systems are unstable.

References

Aizenman, M.L., Cox, J.P.: Astrophys. J. **202**, 137 (1975)
Chandrasekhar, S.: Stellar Structure. University Chicago Press, Chicago (1938)
Chandrasekhar, S.: Hydrodynamic and Hydromagnetic Stability. Clarendon Press, Oxford (1961)
Cowling,T.G.: Magnetohydrodynamics. Inderscience Publishers, New York (1958)
Cox, J.P.: Theory of Stellar Pulsation. Princeton University Press, Princeton (1980)
Cox, J.P., Everson, B.L.: Astrophys. J. Suppl. **52**, 451 (1983)
Eddington, A.S.: The Internal Constitution of the Stars. Dover, Mineola (1959)
Fricke, K.: Z. Astrophys. **68**, 317 (1968)
Goldreich, P., Schubert, G.: Astrophys. J. **150**, 571 (1967)
Kato, S.: Publ. Astron. Soc. Jpn. **18**, 374 (1966)
Kato, S.: Mon. Not. R. Astron. Soc. **185**, 629 (1978)
Kato, S.: Oscillations of Disks. Astrophysics and Space Science Library, vol. 437. Springer, Tokyo (2016)
Unno, Y., Osaki, Y., Ando, H., Saio, H., Shibahashi, H.: Nonradial Oscillations of Stars, 2nd edn. University of Tokyo Press, Tokyo (1989)
Wu, X.-B.. Li, Q.-B.: Astrophys. J. **469**, 776 (1996)

Chapter 9
General Relativistic Hydrodynamics

Up to Chap. 8 we concentrated our attention only to the Newtonian hydrodynamics. In this chapter we shall briefly summarize the basic equations of general relativistic fluid dynamics, and present a few applications to astrophysical problems.

9.1 Relativistic Hydrodynamic Equations

Let us first derive the form of energy-momentum tensor $T_{\mu\nu}$ for fluid in motion.[1] Now we consider a reference system where a given element of volume of body is at rest (proper frame). In this reference system, the energy-momentum tensor has the form (see, e.g., Landau and Lifshitz 1975 for derivation):

$$T_0^{\mu\nu} = \begin{pmatrix} \varepsilon & 0 & 0 & 0 \\ 0 & p & 0 & 0 \\ 0 & 0 & p & 0 \\ 0 & 0 & 0 & p \end{pmatrix},$$

where p is the gas pressure and ε is the energy density of the body and related to density ρ by

$$\varepsilon = \rho c^2 + \frac{1}{\Gamma - 1} p, \tag{9.1}$$

[1]In this chapter three-dimensional tensor indices are denoted by Latin letters $i, j, k, \ldots$ while four-dimensional tensor indices are denoted by Greek letters $\alpha, \beta, \gamma, \ldots$ and take on the values 0, 1, 2, 3. The summation abbreviation is used when the same letter appears in a term twice. The metric tensor is given by $ds^2 = g_{\mu\nu} dx^\mu dx^\nu$ with $g_{00} > 0$.

© Springer Nature Singapore Pte Ltd. 2020, corrected publication 2023

S. Kato, J. Fukue, *Fundamentals of Astrophysical Fluid Dynamics*,
Astronomy and Astrophysics Library,
https://doi.org/10.1007/978-981-15-4174-2_9

Γ being the ratio of specific heats. It is noted that the subscript 0 has been attached to the energy-momentum tensor, $T^{\mu\nu}$, in order to emphasize the proper frame.

Now this expression is extended to that in an arbitrary reference system. To do so we introduce the four-velocity u^μ for the macroscopic motion of an element of volume of the body. A general form of $T^{\mu\nu}$ is

$$T^{\mu\nu} = \varepsilon\left(au^\mu u^\nu + bg^{\mu\nu}\right) + p\left(cu^\mu u^\nu + dg^{\mu\nu}\right), \tag{9.2}$$

where a, b, c, and d are coefficients to be determined. In the rest frame under consideration, we have $u^0 = 1$, $u^i = 0$ ($i = 1, 2, 3$), $g^{00} = 1$, $g^{0i} = 0$, $g^{ij} = -1$ for $i = j$ and 0 for $i \neq j$. From them and from the expression for $T^{\mu\nu}$ in the rest frame, we can determine the value of a, b, c, and d, and a general expression for $T^{\mu\nu}$ is found to be

$$T^{\mu\nu} = (p + \varepsilon)u^\mu u^\nu - pg^{\mu\nu}. \tag{9.3}$$

In Galilean systems, the energy-momentum conservation can be written in the form of $\partial T^{\mu\nu}/\partial x^\nu = 0$ by adopting Cartesian coordinates. In the systems with gravitational fields, the ordinary derivatives are replaced by the covariant ones. That is, the energy-momentum conservation of a fluid system is

$$T^{\mu\nu}_{\ \ ;\nu} = 0, \tag{9.4}$$

where a semicolon (;) represents the covariant derivative.

It is usually convenient to use the following equations, instead of Eq. (9.4). The component of $T^{\mu\nu}_{\ \ ;\nu}$ which is perpendicular to u_μ is $T^{\mu\nu}_{\ \ ;\nu} - u^\mu u_\nu T^{\nu\rho}_{\ \ ;\rho}$. Hence, by substituting Eq. (9.3) into $T^{\mu\nu}_{\ \ ;\nu} - u^\mu u_\nu T^{\nu\rho}_{\ \ ;\rho} = 0$ with use of $u_\mu u^\mu = 1$ and $u_\nu u^\nu_{\ ;\rho} = 0$, we obtain

$$(\varepsilon + p)u^\nu u^\mu_{\ ;\nu} = \left(g^{\mu\nu} - u^\mu u^\nu\right)\frac{\partial p}{\partial x^\nu}. \tag{9.5}$$

The three components of Eq. (9.5), i.e., $\mu = 1, 2, 3$, are three components of equation of motion.

Next, we should consider the component of $\mu = 0$ of Eq. (9.5). Usually, instead of it, $u_\mu T^{\mu\nu}_{\ \ ;\nu} = 0$ is adopted. By using the energy-momentum tensor (9.3), this equation is written as

$$[(\varepsilon + p)u^\mu]_{;\mu} - u^\mu \frac{\partial p}{\partial x^\mu} = 0. \tag{9.6}$$

This is the energy conservation in fluid motions.

To fully describe fluid motions additional relations are necessary, one of which is the conservation of baryon number. If baryon number in a unit volume is written as n, its conservation is written as $(nu^{\mu})_{;\mu} = 0$. If $n \propto \rho$ is adopted, we have

$$(\rho u^{\mu})_{;\mu} = 0, \tag{9.7}$$

which is the equation of continuity. By using this continuity equation and Eq. (9.1), we can change Eq. (9.6) to

$$\left(\frac{p}{\Gamma - 1} u^{\mu}\right)_{;\mu} + p u^{\mu}_{\ ;\mu} = 0. \tag{9.8}$$

This is the equation of adiabatic motions.

In summary, basic equations describing general relativistic adiabatic fluid motions are $\mu = 1, 2, 3$ components of Eq. (9.5) (equation of motion), Eq. (9.7) (equation of continuity), Eq. (9.8) (adiabatic relation), and a subsidiary relation (9.1) between ρ and p (equation of state).

9.1.1 Relativistic Equation of State

As the equation of state we have adopted Eq. (9.1), where Γ is assumed to be constant (5/3 or 4/3). This expression is for the nonrelativistic gas. Our main interest in this textbook is low temperature gases in the sense that the effects of special relativity are not so strong, i.e., $k_{\rm B}T_{\rm e} \le m_{\rm e}c^2$ (see Chap. 1), where $T_{\rm e}$ and $m_{\rm e}$ are electron temperature and electron mass, respectively. We mention here, however, (special) relativistic expression for the equation of state, because in many fluid dynamical phenomena in astrophysics the temperature range is very wide from the nonrelativistic regime to the relativistic one; i.e., Γ varies from 5/3 to 4/3.

The special relativity tells us that the total energy, $\mathcal{E}$, and the kinetic energy, E, of a particle with mass m, are, respectively,

$$\mathcal{E} = c(P^2 + m^2c^2)^{1/2}, \quad E = \mathcal{E} - mc^2 = mc^2\left[\left(1 + \frac{P^2}{m^2c^2}\right)^{1/2} - 1\right], \tag{9.9}$$

where P is the momentum of the particle. In both nonrelativistic and ultra-relativistic limits we have

$$E = \begin{cases} P^2/2m & \text{for } P \ll mc, \quad \text{(non-relativistic limit)}, \\ Pc & \text{for } P \gg mc, \quad \text{(untra-relativistic limit)}. \end{cases}$$

Velocity of a particle is given in general by

$$v = \frac{d\mathcal{E}}{dP} = \frac{P}{m}\left[1 + \left(\frac{P}{mc}\right)^2\right]^{-1/2}. \tag{9.10}$$

In both nonrelativistic and ultra-relativistic limits we have

$$v = \begin{cases} P/m & \text{for } P \ll mc, \quad \text{(non-relativistic limit),} \\ c & \text{for } P \gg mc, \quad \text{(untra-relativistic limit).} \end{cases}$$

The energy density of the gas, ε, is thus given by

$$\varepsilon = \rho c^2 + \int_0^\infty E(P)N(P)dP, \tag{9.11}$$

where $N(P)$ is the number density of particles of momentum P. Similarly, the pressure p of the gas is given by

$$p = \frac{1}{3}\int_0^\infty Pv(P)N(P)dP, \tag{9.12}$$

where the factor $1/3$ comes from the assumption that particle angular distribution is isotropic.

In the limit of nonrelativistic gas, we have

$$p = \frac{1}{3}\int \frac{P^2}{m}N(P)dP, \tag{9.13}$$

and

$$\varepsilon = \rho c^2 + \int_0^\infty \frac{P^2}{2m}N(P)dP = \rho c^2 + \frac{3}{2}p. \tag{9.14}$$

Γ in Eq. (9.1) is thus $\Gamma = 5/3$. In the another limit of ultra-relativistic gas, we have

$$p = \frac{1}{3}\int_0^\infty cPN(P)dP, \tag{9.15}$$

and

$$\varepsilon = \rho c^2 + \int_0^\infty cPN(P)dP = \rho c^2 + 3p. \tag{9.16}$$

Γ in Eq. (9.1) is thus $\Gamma = 4/3$, which is the same as that of photon gas.

To know a general relation between ε and p in the intermediate region, an detailed expression for the distribution function $N(P)$ (relativistic Maxwell function) is necessary. This was made by Cox and Giuli (1968). Let us write p and ε in the form

$$p = nk_{\rm B}T, \tag{9.17}$$

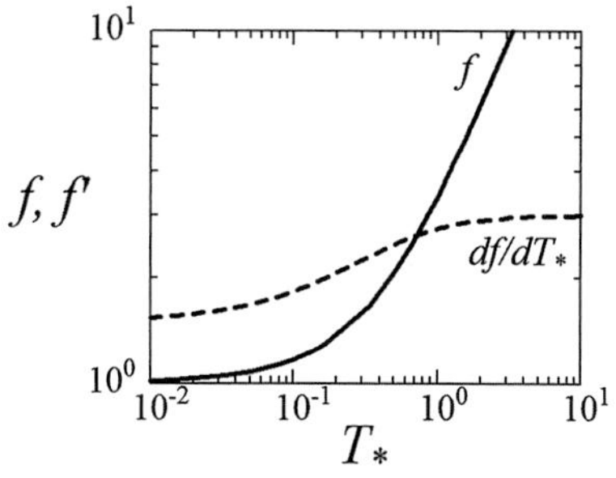

Fig. 9.1 Temperature dependences of f and df/dT. The abscissa is the dimensionless temperature $k_\mathrm{B}T/(mc^2)$, whereas the ordinates are $f/(mc^2)$ (thick curve) and $df/(k_\mathrm{B}dT)$ (thick-dashed curve)

and

$$\varepsilon = nf(T). \tag{9.18}$$

Then, we have

$$f(T) = mc^2\left[\frac{3k_\mathrm{B}T}{mc^2} + \frac{K_1(mc^2/k_\mathrm{B}T)}{K_2(mc^2/k_\mathrm{B}T)}\right], \tag{9.19}$$

and

$$\frac{df(T)}{k_\mathrm{B}dT} = 3 + 3\frac{f}{k_\mathrm{B}T} - \left(\frac{f}{k_\mathrm{B}T}\right)^2 + \left(\frac{mc^2}{k_\mathrm{B}T}\right)^2, \tag{9.20}$$

where K_n's are the modified Bessel functions of the second kind of order n.[2] The functional forms of $f(T)$ and $df/k_\mathrm{B}dT$ are shown in Fig. 9.1.

Finally, it is noted that ionized gases consist of at least two species (e.g., ions and electrons). The pressure p and energy density ε of the system are thus the sum of pressure and energy density coming from each species, respectively. The magnitude of $k_\mathrm{B}T/mc^2$ is different for different species. Hence, Γ of the whole gas is a weighted mean of Γ coming from each species.

[2] In terms of the dimensionless temperature Θ defined by $\Theta = k_\mathrm{B}T/mc^2$, the function $\Phi \equiv f/(mc^2)$ and its derivative are expressed as

$$\Phi(\Theta) = 3\Theta + \frac{K_1(1/\Theta)}{K_2(1/\Theta)} \quad \text{and} \quad \frac{d\Phi}{d\Theta} = 3 + \frac{3\Phi}{\Theta} - \frac{\Phi^2 - 1}{\Theta^2}.$$

9.2 General Relativistic Spherical Stars

We consider static, spherically symmetric self-gravitating stars by using hydrodynamical equations derived in Sect. 9.1. One of the important characteristics of these stars is that they collapse to black holes, if they are compressed beyond a certain critical value.

9.2.1 *Static Line Element with Spherical Symmetry*

In the case of spherically symmetry, we can write the line element in the standard form:

$$ds^2 = e^{\nu}(cdt)^2 - e^{\lambda}dr^2 - r^2d\theta^2 - r^2\sin^2\theta\, d\varphi^2, \tag{9.21}$$

where $\nu = \nu(r)$ and $\lambda = \lambda(r)$. The nonzero components of the metric tensor are

$$g_{00} = e^{\nu}, \quad g_{11} = -e^{\lambda}, \quad g_{22} = -r^2, \quad g_{33} = -r^2\sin^2\theta. \tag{9.22}$$

Corresponding to them, we have

$$g^{00} = e^{-\nu}, \quad g^{11} = -e^{-\lambda}, \quad g^{22} = -r^{-2}, \quad g^{33} = -r^{-2}\sin^{-2}\theta. \tag{9.23}$$

Using these values, we calculate Christoffel three-index symbols $\Gamma^{\sigma}_{\mu\nu}$ defined by

$$\Gamma^{\sigma}_{\mu\nu} = \frac{1}{2}g^{\sigma\rho}\left(\frac{\partial g_{\rho\mu}}{\partial x^{\nu}} + \frac{\partial g_{\rho\nu}}{\partial x^{\mu}} - \frac{\partial g_{\mu\nu}}{\partial x^{\rho}}\right). \tag{9.24}$$

Nonzero components of $\Gamma^{\sigma}_{\mu\nu}$ are found to be (the prime means differentiation with respect to r) (e.g., Landau and Lifshitz 1975)

$$\Gamma^1_{00} = (\nu'/2)e^{\nu-\lambda},\ \Gamma^1_{11} = \lambda'/2,\ \Gamma^1_{22} = -re^{-\lambda}, \tag{9.25}$$

$$\Gamma^1_{33} = -r\sin^2\theta\, e^{-\lambda},\ \Gamma^0_{10} = \nu'/2,\ \Gamma^2_{12} = \Gamma^3_{13} = 1/r, \tag{9.26}$$

$$\Gamma^2_{33} = -\sin\theta\cos\theta,\ \Gamma^3_{23} = \cot\theta. \tag{9.27}$$

All other components (except for those which differ from the ones we have written by transposition of the indices μ and ν) are zero.

The Einstein equation,

$$R_{\mu\nu} - \frac{1}{2}g_{\mu\nu} = \frac{8\pi G}{c^4}T_{\mu\nu}, \tag{9.28}$$

gives then

$$e^{-\lambda}\left(\frac{1}{r}\frac{d\nu}{dr}+\frac{1}{r^2}\right)-\frac{1}{r^2}=\frac{8\pi G}{c^4}p, \tag{9.29}$$

$$\frac{1}{2}e^{-\lambda}\left[\frac{d^2\nu}{dr^2}+\frac{1}{2}\left(\frac{d\nu}{dr}\right)^2+\frac{1}{r}\left(\frac{d\nu}{dr}-\frac{d\lambda}{dr}\right)-\frac{1}{2}\frac{d\nu}{dr}\frac{d\lambda}{dr}\right]=\frac{8\pi G}{c^4}p, \tag{9.30}$$

$$e^{-\lambda}\left(\frac{1}{r}\frac{d\lambda}{dr}-\frac{1}{r^2}\right)+\frac{1}{r^2}=\frac{8\pi G}{c^4}\varepsilon. \tag{9.31}$$

Although the equation of motion of fluid, Eq. (9.5), is involved in the above equations, we write it explicitly in the form:

$$\frac{\varepsilon+p}{2}\frac{d\nu}{dr}+\frac{dp}{dr}=0. \tag{9.32}$$

9.2.2 *Tolman–Oppenheimer–Volkoff Equation and Collapse of Stars*

Equation (9.31) can be integrated to give

$$e^{-\lambda}=1-\frac{2GM(r)}{c^2r}, \tag{9.33}$$

where

$$M(r)=\frac{4\pi}{c^2}\int_0^r \varepsilon r^2dr. \tag{9.34}$$

If ν is eliminated from Eqs. (9.29) and (9.32), we have

$$\frac{dp}{dr}+\frac{\varepsilon+p}{2}\left(\frac{8\pi G}{c^4}rpe^{\lambda}+\frac{e^{\lambda}-1}{r}\right)=0. \tag{9.35}$$

By eliminating e^{λ} from the above equation by using Eq. (9.33), we have finally

$$-\frac{dp}{dr}-\frac{\varepsilon+p}{c^2}\frac{G}{r^2}\left[M(r)+\frac{4\pi r^3p}{c^2}\right]\frac{1}{1-2GM(r)/c^2r}=0. \tag{9.36}$$

This is *Tolman–Oppenheimer–Volkoff equation* (Oppenheimer and Volkov 1939), and describes the hydrostatic balance of general relativistic gaseous sphere under self-gravity. In the limit of $\varepsilon = \rho c^2 \gg p$ and $GM \ll c^2 r$ Eq. (9.36) tends to the well-known hydrostatic equation describing spherical Newtonian stars.

If a relation between ε and p is specified, we obtain in principle $p(r)$ and $\rho(r)$ by solving Eq. (9.36). Then, we have $\nu(r)$ from Eq. (9.32)

$$\nu = -\int_0^p \frac{2}{\varepsilon + p} dp + \nu(R), \tag{9.37}$$

where R is the radius of the star, where $p = 0$. Equations (9.29) and (9.31) show that in the empty space we have $d(\lambda + \nu)/dr = 0$, which gives $\lambda + \nu = \text{const.}$ Without any loss of generality, we can take $\lambda + \nu = 0$. Then, considering Eq. (9.33), we have

$$\mathrm{e}^{\nu(R)} = \mathrm{e}^{-\lambda(R)} = 1 - \frac{2GM(R)}{c^2 R}. \tag{9.38}$$

Hence, Eq. (9.37) is written in the form:

$$e^{\nu} = \left(1 - \frac{2GM(R)}{c^2 R}\right) \exp\left(-2\int_0^p \frac{dp}{\varepsilon + p}\right). \tag{9.39}$$

Hereafter, we shall consider a special case of $\varepsilon = \text{const.}$, because in this case we can obtain an analytical expression for $p(r)$. Even in this special case the results are important. If $\varepsilon = \text{const.}$, $M(r)$ is given by

$$M(r) = \frac{4\pi \varepsilon r^3}{3c^2}, \tag{9.40}$$

and the integration of Eq. (9.36) gives

$$p(r) = \varepsilon \frac{\left(1 - r^2/a^2\right)^{1/2} - \left(1 - R^2/a^2\right)^{1/2}}{3(1 - R^2/a^2)^{1/2} - (1 - r^2/a^2)^{1/2}}, \tag{9.41}$$

where the boundary condition $p = 0$ at $r = R$ has been adopted, and a^2 is defined by

$$a^2 = \frac{3c^4}{8\pi G\varepsilon}. \tag{9.42}$$

In this special case of $\varepsilon = \text{const.}$, Eq. (9.33) gives

$$e^{-\lambda} = 1 - \frac{r^2}{a^2}. \tag{9.43}$$

To obtain an explicit expression for e^ν, we must integrate Eq. (9.39). The integration with respect to p can be changed to an integration with respect to r by using $dp = (dp/dr)dr$ and dp/dr given by Eq. (9.36). The results give

$$e^\nu = \frac{1}{4}\left[3\left(1 - \frac{R^2}{a^2}\right)^{1/2} - \left(1 - \frac{r^2}{a^2}\right)^{1/2}\right]^2. \tag{9.44}$$

In summary, the radial structure of general relativistic spherical stars with $\varepsilon = \text{const.}$ are given by Eqs. (9.41), (9.43), and (9.44).

Equation (9.41) shows that $p(r)$ is zero at $r = R$ as imposed by the boundary condition, and increases inwards. In the case where R is smaller than a critical value, however, $p(r)$ becomes infinite before $r = 0$, since the denominator of Eq. (9.41) becomes zero. The value of e^ν then also becomes zero. In other words, $3(1 - R^2/a^2)^{1/2} - 1 > 0$ is needed for presence of static solution. This condition gives

$$R < \left(\frac{8}{9}\right)^{1/2} a. \tag{9.45}$$

If a given by Eq. (9.42) and $M(R)$ defined by $M(R) = (4\pi/3)(\varepsilon/c^2)R^3$ are adopted, inequality (9.45) is written as

$$R > \frac{9GM(R)}{4c^2}. \tag{9.46}$$

By defining the Schwarzschild radius for mass M by $r_S \equiv 2GM/c^2$ (see the next Sect. 9.3), inequality (9.46) is reduced to

$$R > \frac{9}{8} r_S. \tag{9.47}$$

That is, if a star with mass M is compressed with uniform ε to the radius $(9/8)r_S$, the star has no equilibrium state and collapses to a black hole.

9.3 Schwarzschild Metric and Kerr Metric

In the previous Sect. 9.2, we have considered self-gravitating spherically symmetric static objects (relativistic stars), and saw how the self-gravity affects the space-time structure and derived a condition of gravitational collapse of the stars in a simplified case. Hereafter, we consider nonself-gravitating objects which are in static, spherically symmetric or axisymmetric given metrics.

(a) Schwarzshild Metric

First, the metric around a spherically symmetric non-rotating point mass M is described. In Sect. 9.2 we mentioned that in empty space $\lambda + \nu = 0$. Considering this and Eq. (9.33), we obtain the Schwarzschild (exterior) metric by using spherical space coordinates r, θ, and φ (see, e.g., Landau and Lifshitz 1975 for rigorous derivation):

$$ds^2 = \left(1 - \frac{r_S}{r}\right)c^2dt^2 - \left(1 - \frac{r_S}{r}\right)^{-1} dr^2 - r^2\left(d\theta^2 + \sin^2\theta d\varphi^2\right), \tag{9.48}$$

where

$$r_S \equiv \frac{2GM}{c^2} \tag{9.49}$$

is the *Schwarzschild radius* and M is the mass of the central object. The symbol r_g is used for GM/c^2, which is called the *gravitational radius*.

(b) Kerr Metric

As well as the Schwarzschild metric, the metric around a rotating point mass is of importance, which is known as the Kerr metric (Kerr 1963) and written as

$$\begin{aligned} ds^2 = &\left(\frac{\Delta}{\rho^2}\right)\left[cdt - a\sin^2\theta d\varphi\right]^2 - \left(\frac{\sin^2\theta}{\rho^2}\right)\left[\left(r^2 + a^2\right)d\varphi - acdt\right]^2 \\ &- \left(\frac{\rho^2}{\Delta}\right)dr^2 - \rho^2 d\theta^2, \end{aligned} \tag{9.50}$$

where

$$\Delta = r^2 - rr_S + a^2, \quad \rho^2 = r^2 + a^2\cos^2\theta, \tag{9.51}$$

and a is a parameter representing the spin of the central object. (This "a" is different from "a" used in Sect. 9.2, although the same notation is used.) If dimensionless parameter a_* defined by

$$a_* = \frac{2a}{r_S} = \frac{a}{r_g} \tag{9.52}$$

is introduced, the allowed range of a_* is

$$-1 \le a_* \le 1. \tag{9.53}$$

Here, $a_* = 0$ is the case of no-rotation, and in this case the metric (9.50) coincides with metric (9.48)

Next, let us consider singular surfaces of the Kerr metric. The coefficient of $(dr)^2$, i.e., g_{11}, becomes infinite on the surface of

$$r^2 - rr_S + a^2 = 0. \tag{9.54}$$

This equation represents two spherical surfaces:

$$r_\pm = \frac{1}{2}\left[r_S \pm \left(r_S^2 - 4a^2\right)^{1/2}\right] = \frac{r_S}{2}\left[1 \pm \left(1 - a_*^2\right)^{1/2}\right]. \tag{9.55}$$

The surface corresponding to the "+" sign is denoted by S_0^+, and the one corresponding to the "−" sign by S_0^- (see Fig. 9.2). The surface S_0^+ is called the *(outer) horizon* r_H, and

$$r_H(a_*) = \frac{r_S}{2}\left[1 + \left(1 - a_*^2\right)^{1/2}\right]. \tag{9.56}$$

It is noted that the a_*-dependence of r_H is shown later in Fig. 9.5 with the a_*-dependences of some other characteristic quantities in accretion disks.

The coefficient of $(dt)^2$, i.e., g_{00}, vanishes at

$$r^2 - rr_S + a^2\cos^2\theta = 0. \tag{9.57}$$

The two closed surfaces specified by the above equation are

$$r = \frac{1}{2}\left[r_S \pm \left(r_S^2 - 4a^2\cos^2\theta\right)^{1/2}\right] = \frac{r_S}{2}\left[1 \pm \left(1 - a_*^2\cos^2\theta\right)^{1/2}\right]. \tag{9.58}$$

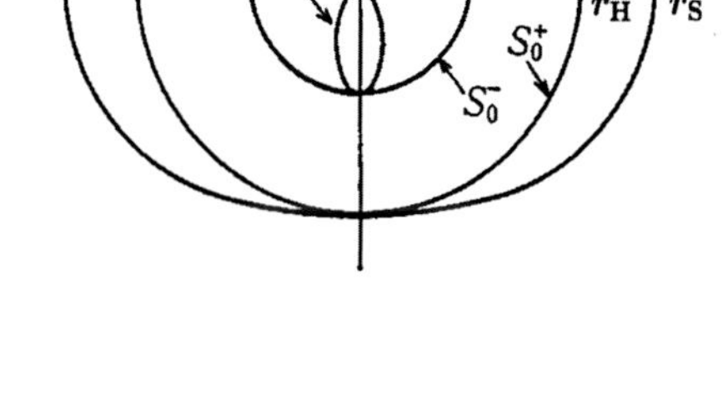

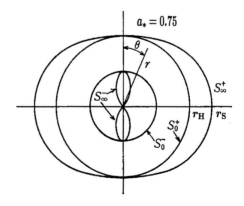

Fig. 9.2 Kerr metric around rotating black hole. S_0^+ is the event horizon, and S_∞^+ is the surface of infinite redshift. The region between S_0^+ and S_∞^+ is the ergosphere

The surface corresponding to the "+" sign is denoted by S_∞^+, and the one with the "−" sign by S_∞^- (see Fig. 9.2). The surface S_∞^+ is called the *static limit* $r_{\rm E}$, and

$$r_{\rm E}(a_*, \theta) = \frac{r_{\rm S}}{2}\left[1 + \left(1 - a_*^2\cos^2\theta\right)^{1/2}\right]. \tag{9.59}$$

Distinct from the case of Schwarzschild metric, the event horizon S_0^+ and the surface of infinite redshift, S_∞^+, are separated. The region between the above two surfaces is called the *ergosphere*. The ergosphere is of importance, because the rotational energy of a black hole is considered to be able to be extracted from the region.

9.4 Particle Motions in Schwarzschild and Kerr Metrics

When we study active phenomena of compact stellar objects and galactic nuclei, accretion disks are important ingredients and studies on their structure and stabilities are indispensable. Accretion disks in such objects are usually nonself-gravitating. Hence, their studies can be made in the framework of Schwarzschild or Kerr metrics. In this section, as one of the most important quantities in such disks, we describe the radial dependence of angular velocity of circular rotation of a particle around a compact source and the (radial) epicyclic frequency of a particle around the circular motions.

9.4.1 Angular Velocity of Circular Motions

We assume that a particle makes circular motion on the equator, i.e., $\theta = \pi/2$ in the Schwarzschild metric. By this assumption we have $u^r = u^\theta = 0$. If the angular velocity of rotation, $\tilde{\Omega}$, is defined by $u^\varphi = \tilde{\Omega}/c$, we have $u_\varphi = -r^2\tilde{\Omega}/c$. Furthermore, we have $u_t = c^2(1 - r_{\rm S}/r)u^t$. Hence, the relation $u_i u^i = 1$ gives $u^t = [(1 + \tilde{\Omega}^2 r^2/c^2)/(1 - r_{\rm S}/r)]^{1/2}/c$. Then, Eq. (9.5) gives, after some manipulations,

$$\frac{\varepsilon + p}{c^2}\left(1 - \frac{r_{\rm S}}{r}\right)^{-1}\left[\left(1 + \frac{\tilde{\Omega}^2 r^2}{c^2}\right)\frac{GM}{r^2} - \left(1 - \frac{r_{\rm S}}{r}\right)\tilde{\Omega}^2 r\right] = -\frac{\partial p}{\partial r}. \tag{9.60}$$

Since we are interested in a particle motion, using $p = 0$ we have

$$\left(1 + \frac{\tilde{\Omega}^2 r^2}{c^2}\right)\frac{GM}{r^2} = \left(1 - \frac{r_{\rm S}}{r}\right)\tilde{\Omega}^2 r, \tag{9.61}$$

or

$$\tilde{\Omega} = \left[\frac{GM}{(1 - 3r_S/2r)r^3} \right]^{1/2}. \tag{9.62}$$

In the above argument the angular velocity is defined by $\tilde{\Omega} = cu^\varphi$. However, it is more natural to define angular velocity of rotation by the quantity observed at infinity. This angular velocity Ω is given by $\Omega = d\varphi/dt = u^\varphi/u^t = \tilde{\Omega}/(cu^t)$. Since u^t is given by $u^t = [(1 + \tilde{\Omega}^2 r^2/c^2)/(1 - r_S/r)]^{1/2}/c$ as mentioned above, we find that

$$\Omega = \left(\frac{GM}{r^3} \right)^{1/2}. \tag{9.63}$$

This expression for angular velocity is formally the same as that in the Newtonian one.

In the case of the Kerr metric the angular velocity of rotation, Ω, is generalized to

$$\Omega = \frac{(GM/r^3)^{1/2}}{1 + (r_S/r)^{3/2} a_*}. \tag{9.64}$$

Equation (9.64) tends to Eq. (9.63) in the limit of $a_* = 0$. The radial dependence of $\Omega(r)$ is shown in Fig. 9.3 for some values of spin parameter a_*.

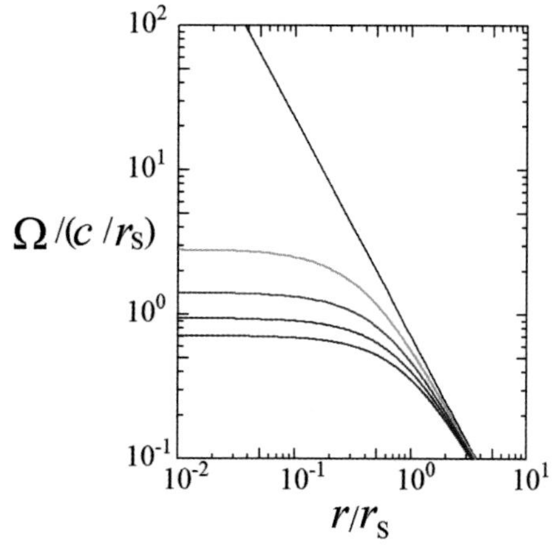

Fig. 9.3 The radial dependence of angular velocity of rotation, $\Omega(r; a_*)$, for several values of spin parameter a_*. From the upper to lower curves, $a_* = 0$, 025, 0.50, 0.75, 0.99 in turn

9.4.2 Stability of Circular Motions and Epicyclic Frequency

In Sect. 9.4.1, we have considered circular motions. Here, we examine the stability of the circular motion against perturbations within the plane of the circular motions.

Let us consider free motion of a particle of mass m in the Schwarzschild metric. The four momentum vector of the particle is written as[3]

$$p^\mu = mcu^\mu. \tag{9.65}$$

The particle moves in a plane. Without loss of generality, the plane is taken to be $\theta = \pi/2$. Then, obviously $p^\theta = 0$.

To proceed further, it is simple to start from the equation which can be derived by integrating equation of motion. The integration constants are energy and angular momentum. That is, for displacement of $t \to t + \Delta t$, $\varphi \to \varphi + \Delta\varphi$, the Schwarzschild metric is invariant, and t and φ are cyclic coordinates. Hence, p_t and p_φ are conserved during motions. They are written, respectively, as $\mathcal{E}$ and $-\mathcal{L}$. i.e., $p_t = \mathcal{E}$ and $p_\varphi = -\mathcal{L} \equiv -mL$, where L is the angular momentum per unit mass.[4]

Since the requirement of $u_\mu u^\mu = 1$ is written as $p_\mu p^\mu = g^{\mu\nu} p_\mu p_\nu = m^2c^2$, the above considerations on p_t and p_φ lead to

$$\frac{(\mathcal{E}/mc)^2}{1 - r_S/r} - \frac{1}{1 - r_S/r}\left(\frac{\partial r}{\partial \tau}\right)^2 - \frac{L^2}{r^2} - c^2 = 0, \tag{9.66}$$

where τ is the proper time.

By introducing $\psi(r)$ defined by

$$\psi(r) = -\frac{GM}{r}\left(1 + \frac{L^2}{c^2r^2}\right) + \frac{L^2}{2r^2}, \tag{9.67}$$

Equation (9.66) is written in the form

$$\frac{1}{2}\left(\frac{\partial r}{\partial \tau}\right)^2 = \frac{(\mathcal{E}/m)^2 - c^4}{2c^2} - \psi. \tag{9.68}$$

The set of Eqs. (9.67) and (9.68) has forms which can be compared with the Newtonian case. That is, the sum of the left-hand side of Eq. (9.68) (kinetic energy)

---

3 In this subsection p^μ is used to represent a component of four momentum vector of particle with mass m. In other subsections in this chapter, however, p is used to represent pressure of gas.

4 In the next Sect. 9.5 we use ℓ defined by $\ell \equiv -u_\varphi/u_t$ as definition of specific angular momentum. In empty space L is conserved, but not conserved in fluid systems. In fluid systems what is conserved is ℓ defined by $\ell \equiv -u_\varphi/u_t$ (see Kozłowski et al. 1978).

and ψ (potential energy) is constant. The constant is the first term on the right-hand side of equation. It is noted that the constant is $\mathcal{E}/m - c^2$ in the Newtonian limit. In the case of the Newtonian dynamics, the effective potential ψ is the sum of gravitational potential $-GM/r$ and the positive potential of centrifugal force, $L/2r^2$. In the relativistic case the gravitational potential, $-GM/r$, is multiplied by a factor $(1 + L^2/c^2r^2)$, which can be interpreted as a result of increase of mass of the particle by rotational motion.

The radius of steady circular orbit is the radius where potential $\psi(r)$ is minimum. In the case of Newtonian dynamics there is always such a radius, since the effective potential ψ is the sum of gravitational potential ($\propto -1/r$) and the centrifugal potential ($\propto 1/r^2$). That is, for any angular momentum we have stable circular orbit where the effective potential is minimum. In the case of general relativity, however, situation is changed by the presence of the term of proportional to $-1/r^3$ in the effective potential. The gravitational potential overcomes the centrifugal one at small r [see Eq. (9.67)]. Because of this, when angular momentum of the particle is smaller than a certain value, there is no minimum point of $\psi(r)$, as is shown in Fig. 9.4.

Let us consider quantitatively. Fixing the value of $\mathcal{E}$, we change the value of L. The radius where the gravitational potential proportional to $1/r$ and the centrifugal potential force proportional to $1/r^2$ are balanced is $r = L^2/2GM = (L^2/c^2r_S^2)r_S$. On the other hand, the radius where the $-1/r^3$ component of the gravitational force overcomes the centrifugal potential of $L^2/2r^2$ is r_S. Hence, for a particle of $L \leq cr_S$, the effective potential has no extremum, and decreases monotonically with decrease of r. That is, there is no stable circular motion (see Fig. 9.4). The critical value of L is found from $d\psi/dr = d^2\psi/dr^2 = 0$. By imposing this condition to ψ

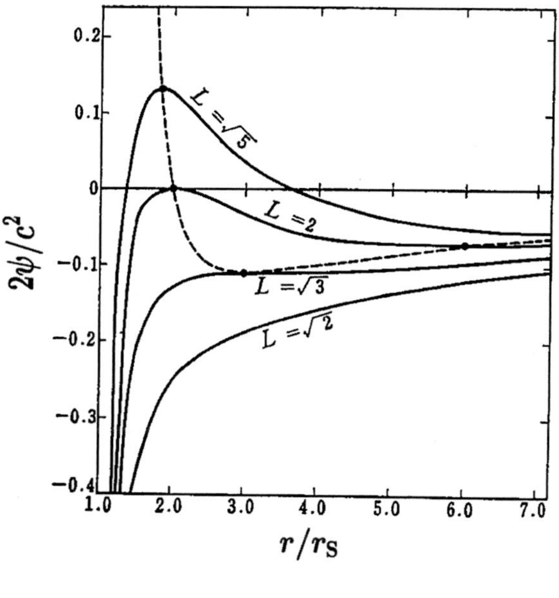

Fig. 9.4 Effective potential for circular motions in the Schwarzschild metric. The radial dependence of the effective potential $\psi(r)$ is shown for some values of L. In the case where L is large, the effective potential $\psi(r)$ has a maximum and a minimum. As L decreases these two radii approach and finally merge for $L < \sqrt{3}r_Sc$, and there is no minimum and maximum. The dotted curve shows the loci of the maximum and minimum points

given by Eq. (9.67), we have $r = 3r_S$ and $L = 3^{1/2}cr_S$. The particles with angular momentum larger than $\sqrt{3}cr_S$ have a stable circular orbit at a radius larger than $3r_S$, but a particle with smaller angular momentum has no (stable) circular orbit. That is, $r = 3r_S$ is the radius of marginally stable circular orbit, say r_{ms}, and the orbit at r_{ms} is called a *innermost stable circular orbit (ISCO)*. The specific angular momentum of circular motion at r_{ms} is written hereafter L_{ms}. That is, in the case of Schwarzschild metric, we have

$$r_{ms} = 3r_S, \quad L_{ms} = 3^{1/2}cr_S. \tag{9.69}$$

Next, let us consider the binding energy of particle at $r = 3r_S$. Substitution of Eq. (9.69) into Eq. (9.67) gives $\psi = -c^2/18$. Hence, by taking the left-hand side of Eq. (9.68) to be zero, we see that the energy of a unit mass, $\mathcal{E}/m$, is $(8/9)^{1/2}c^2$. Thus, the binding energy of an unit mass, E_b, is

$$E_b = c^2 - \mathcal{E}/m = \left[1 - \left(\frac{8}{9}\right)^{1/2}\right]c^2 = 0.0572c^2. \tag{9.70}$$

If the radius of orbital motion of a particle becomes smaller than $3r_S$, the particle falls towards the central object. If the central object is a black hole, the particle is swallowed to the central object. Hence, no energy can be extracted from the particle at such a stage. In this sense, E_b given by Eq. (9.70) is the maximum binding energy which can be extracted from the particle falling towards a black hole. It is noted that this energy is much larger than the binding energy $0.007c^2$ released by hydrogen burning.

Next, the case of the Kerr metric is briefly mentioned. If a particle rotates around a rotating central object with the same direction as the rotation of the central object, the radius of the marginally stable circular orbit becomes smaller than $3r_S$. This is naturally understood, since the particle rotates around the central object by being dragged by the rotation of the central object. Hence, by a force analogous to the Coriolis force, the particle receives a force directed outwards. Due to this, steady circular orbit exist till a radius smaller than $3r_S$, i.e., r_{ms} is smaller than $3r_S$. If the particle rotates in the direction opposite to the spin of the central object, the situation is opposite, and the radius of the marginally stable circular orbit r_{ms} becomes larger than $3r_S$. These results are summarized in Table 9.1 and Fig. 9.5.

Table 9.1 Radius of the innermost stable circular orbit (ISCO) and related quantities at ISCO

		Extreme Kerr	
	Schwarzschild	$a_* = 1$	$a_* = -1$
Innermost stable circular orbit (ISCO) (r_{ms}/r_S)	3	1/2	9/2
Angular momentum at ISCO (L_{ms}/cr_S)	$\sqrt{3}$	$1/\sqrt{3}$	$11/(3\sqrt{3})$
Binding energy at ISCO (E_b/c^2)	0.0572	0.4226	0.0377

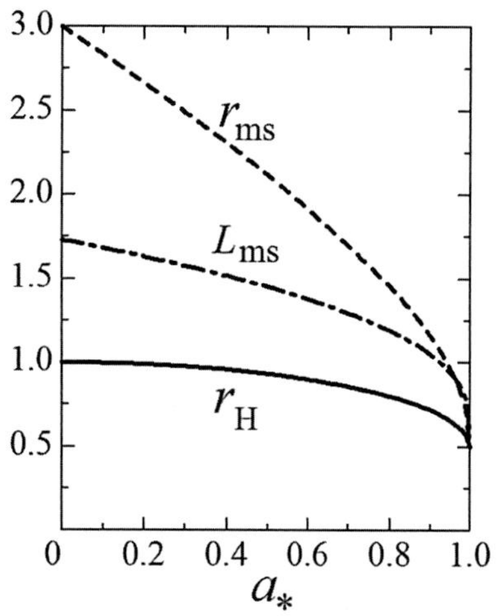

Fig. 9.5 The radius of event horizon, r_H, the radius of marginally stable circular orbit, r_{ms}, and the specific angular momentum at r_{ms}, L_{ms}, as functions of the spin parameter a_*. The radii r_H and r_{ms} are shown in units of r_S, and L_{ms} is in units of cr_S

Finally, we refer to marginally bound radius r_{mb}. Let us consider a particle whose binding energy is just zero at infinity. The purpose here is to examine how much angular momentum is required for the particle not to be swallowed onto the central source. In the case of the Schwarzschild metric the right-hand side of Eq. (9.68) is $-\psi$ for $\mathcal{E}/m = c^2$. Hence, what we require here is the value of L for which the maximum value of ψ defined by Eq. (9.67) becomes just zero. The radius where this is realized is called a *marginally bound radius* r_{mb}. This radius is obtained from the condition of $\psi = \partial\psi/\partial r = 0$. These are realized for $L = 2cr_S$ at $r = 2r_S$. That is, in the case of Schwarzschild metric, we have

$$r_{mb} = 2r_S, \quad L_{mb} = 2cr_S. \tag{9.71}$$

9.4.3 Epicyclic Frequency

When we consider oscillations and stability of rotating fluid systems, (radial) epicyclic frequency has important roles in many situations. For example, epicyclic frequency is one of the important ingredients in determining the oscillation frequencies in disks (see Chap. 4) and also in determining stability criterions against perturbations (e.g., see Chap. 7). Hence, it is important to examine how the epicyclic frequency is affected by general relativity. Especially, the disappearance of steady circular orbit at ISCO (i.e., the presence of r_{ms}) is related to vanishing of epicyclic frequency due to general relativity.

Let us consider, for simplicity, a particle rotating around a central source of mass M in the Schwarzschild metric. The particle rotates in a fixed plane, and the radial variation of the particle in the plane is described by Eq. (9.68). Hence, by taking the time derivative of the equation, we have

$$\frac{\partial^2 r}{\partial \tau^2} = -\frac{\partial \psi}{\partial r}, \tag{9.72}$$

where τ is the proper time, and $\psi(r)$ is given by Eq. (9.67). This equation shows that the radius of the circular orbit can be derived from $\partial\psi/\partial r = 0$ (see Sect. 9.4.2). Using expression (9.67) for $\psi(r)$, we find that the angular momentum $L(r)$ of the particle which makes a circular motion at radius r is

$$L^2 = \frac{GMr}{1 - 3r_S/2r}. \tag{9.73}$$

It is noted that for $r = 3r_S$ we have $L = \sqrt{3}cr_S$, which is, of course, the same as L given in Eq. (9.69).

Now, we assume that a particle is perturbed from the above circular orbit within the plane of the circular orbit, the angular momentum being kept unchanged. Then, the time variation of small amplitude radial displacement Δr from the circular orbit is governed by

$$\frac{\partial^2 \Delta r}{\partial \tau^2} = -\frac{\partial^2 \psi}{\partial r^2}\Delta r. \tag{9.74}$$

This equation shows that Δr makes a harmonic oscillation around the circular radius with frequency $\tilde{\kappa}$ defined by

$$\tilde{\kappa}^2 \equiv \frac{\partial^2 \psi}{\partial r^2}. \tag{9.75}$$

After calculating $\partial\psi^2/\partial r^2$ under the condition of $L =$ const. and substituting L given by Eq. (9.73) we have

$$\tilde{\kappa}^2 = \left(1 - 3\frac{r_S}{r}\right)\frac{GM/r^2}{1 - 3r_S/2r}. \tag{9.76}$$

The above expression for epicyclic frequency has been measured by the proper time τ. Usually, the coordinate time t is used to measure the frequency, instead of the proper time. Since $\partial/\partial\tau = cu^t\partial/\partial t$, the epicyclic frequency defined by the coordinate time, κ, is related to $\tilde{\kappa}$ by $\kappa(r) = \tilde{\kappa}(r)/cu^t(r)$. Considering $(u^t)^2 = [1 + (u_\varphi)^2/r^2]/[c^2(1 - r/r_S)]$ and $u_\varphi = -L/c$ (see a part of the beginning of Sect. 9.4.1) we have

$$cu^t = \frac{1}{(1 - 3r_S/2r)^{1/2}} \tag{9.77}$$

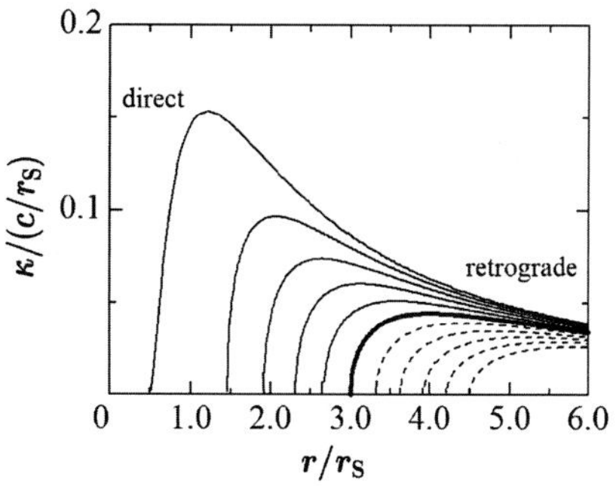

Fig. 9.6 Radial dependence of epicyclic frequency, $\kappa(r)$. $a_* = 0$ shows the case of the Schwarzschild metric (thick curve). The dotted curves denote the cases of $a_* < 0$ (retrograde rotation)

and thus (Kato and Fukue 1980, see also Aliev and Galtsov 1981)

$$\kappa^2 = \left(1 - \frac{3r_S}{r}\right)\frac{GM}{r^3}. \tag{9.78}$$

This shows the radial dependence of epicyclic frequency. It should be noticed that κ becomes zero at $r = 3r_S$ (i.e., at ISCO).

Finally, epicyclic frequency κ in the Kerr metric is presented here without derivation (see Okazaki et al. 1987; see Aliev and Galtsov 1981 for more general cases).

$$\kappa^2(r) = \frac{c^2 r_S}{2r^3}\frac{1 - 6(r_S/2r) + 8a_*(r_S/2r)^{3/2} - 3a_*^2(r_S/2r)^2}{[1 + a_*(r_S/2r)^{3/2}]^2}. \tag{9.79}$$

Here, the range of variation of parameter a_* is $-1 \le a_* \le 1$, and $a_* = 0$ represents the case of the Schwarzschild metric. The radial distribution of $\kappa(r)$ is shown in Fig. 9.6.

9.5 Relativistic Tori

In Newtonian systems, if hot gases fall to a central point source, keeping their angular momenta, the centrifugal force finally overcomes the gravitational force of the central source at a certain distance from the central source. Thus, a donut-type torus will be formed around the central gravitational source with a funnel in the direction of the rotation axis of the torus. In relativistic systems, however, gases can fall to a central object, if their angular momentum is less than a certain value, as mentioned in Sect. 9.4. Hence, in such systems the torus will have a cusp structure in the equatorial region, and gases fall to the central source through the cusp. To demonstrate this, Abramowicz's group (Abramowicz et al. 1978; Kozłowski et al. 1978) studied, in the general relativity, steady configurations of an axially symmetric torus with no internal motion except for rotation. Here we present the essence of their torus, although their idealized models are known to be strongly unstable against axially asymmetric perturbations (*Papaloizou–Pringle instability*, i.e., *corotation instability*).

Let us assume that the external gravitational force due to a central source is stationary and axisymmetiric, and the metric is written as

$$ds^2 = g_{tt}dt^2 + 2g_{t\varphi}dtd\varphi + g_{\varphi\varphi}d\varphi^2 + g_{rr}dr^2 + g_{\theta\theta}d\theta^2, \tag{9.80}$$

where $\partial g_{\mu\nu}/\partial t = 0$ and $\partial g_{\mu\nu}/\partial\varphi = 0$. The fluid system is assumed to consist of perfect gas and rotates in the φ-direction, i.e., $u_r = u_\theta = 0$. The metric signature is taken so that $g_{tt} > 0$, and thus the normalization condition is $u_t u^t = 1$.

The angular velocity of rotation of the fluid, $\Omega(r, \theta)$, and the angular momentum per unit mass, $l(r, \theta)$, are defined by[5]

$$\Omega = \frac{d\varphi}{dt} = \frac{u^\varphi}{u^t} \quad \text{and} \quad l = -\frac{u_\varphi}{u_t}. \tag{9.81}$$

From the above definitions we have

$$(u_t)^{-2} = \frac{g_{\varphi\varphi} + 2lg_{t\varphi} + l^2 g_{tt}}{g_{tt}g_{\varphi\varphi} - g_{t\varphi}^2}, \tag{9.82}$$

$$\Omega = -\frac{g_{tt}l + g_{t\varphi}}{g_{t\varphi}l + g_{\varphi\varphi}}, \qquad l = -\frac{g_{t\varphi} + \Omega g_{\varphi\varphi}}{g_{tt} + \Omega g_{t\varphi}}. \tag{9.83}$$

[5] In Sect. 9.4 we have adopted L defined by $L = -cu_\varphi$ as angular momentum (see a footnote in Sect. 9.4), but here adopt l defined by $l = -u_\varphi/u_t$ as angular momentum. Here, the relation between L and l on the equatorial plane (i.e., $\sin\theta = 1$) in the case of Schwarzschild metric is summarized. Since u_t is given by Eq. (9.82), we have

$$l^2 = \frac{L^2}{c^2}\frac{r^2 - l^2(1 - r_S/r)c^2}{(1 - r_S/r)c^2r^2}.$$

Rearranging of this equation leads to

$$l = \frac{rL}{c(1 - r_S/r)^{1/2}(c^2r^2 + L^2)^{1/2}}.$$

This is the relation between L and ℓ. At the marginally stable circular orbit ($r = r_{\rm ms} = 3r_S$), $L_{\rm ms} = \sqrt{3}cr_S$. Hence, substituting these values to the above equation of l, we have

$$l_{\rm ms} = \left(\frac{3}{2}\right)^{3/2}\frac{r_S}{c} = 1.837\ldots\frac{r_S}{c}.$$

Similarly, at the marginally bound circular orbit ($r_{\rm mb} = 2r_S$ and $L_{\rm mb} = 2cr_S$),

$$l_{\rm mb} = 2\frac{r_S}{c}.$$

The energy-momentum tensor of the fluid is, as mentioned in Sect. 9.1,

$$T_\mu^\nu = (\varepsilon + p)u_\mu u^\nu - \delta_\mu^\nu p. \tag{9.84}$$

Hence, the equation of motion of the fluid, $T_{\mu\,;\nu}^{\nu} - u_\mu u_\nu T^{\nu\rho}{}_{;\rho} = 0$, leads to (Abramowicz et al. 1978)

$$\frac{1}{p+\varepsilon}\frac{\partial}{\partial x^\mu}p = -\frac{\partial}{\partial x^\mu}\ln u_t + \frac{\Omega}{1-\Omega l}\frac{\partial l}{\partial x^\mu}, \quad (\mu = r \text{ and } \theta). \tag{9.85}$$

This expression for fluid motions is simple, but its derivation is rather troublesome. Hence, a derivation will be presented in the last part of this section.

In barotropic gases, i.e., $p = p(\varepsilon)$, the left-hand side of Eq. (9.85) is a gradient of a term. Hence, the right-hand side of Eq. (9.85) needs to be also a gradient of a term, which means that Ω is a function of l, i.e., $\Omega = \Omega(l)$.

Let us first consider the Newtonian limit of Eq. (9.85). In this limit, $l = -\Omega(g_{\varphi\varphi}/g_{tt}) \sim (r\sin\theta)^2\Omega/c^2$. Hence, r-component of the second term on the right-hand side of Eq. (9.85) is approximated as

$$\frac{\Omega}{1-\Omega l}\frac{\partial l}{\partial r} \sim \frac{\Omega\sin^2\theta}{c^2}\frac{\partial}{\partial r}(r^2\Omega). \tag{9.86}$$

On the other hand, $(u_t)^2$ is given Eq. (9.82), which can be approximated as $(u_t)^2 = c^2(1-r_S/r)/(1-(r^2\Omega^2/c^2)\sin^2\theta)$. Hence we have

$$-\frac{\partial \ln u_t}{\partial r} \sim -\frac{r_S}{2r^2} - \frac{\sin^2\theta}{2c^2}\frac{\partial}{\partial r}(r^2\Omega^2). \tag{9.87}$$

The sum of the above two equations represents the sum of the gravitational force, $-GM/c^2r^2$ and the r-component of the centrifugal force, $r\Omega^2\sin^2\theta/c^2$, as expected.

To demonstrate the structure of relativistic tori, let us consider their equi-potential surfaces, i.e., the contour of the surfaces where p and ρ are constant. First, we integrate equation (9.85) to lead

$$\int_0^p \frac{dp}{\varepsilon+p} = -\ln\frac{(u_t)}{(u_t)_{\rm out}} + F(l) \equiv -W, \tag{9.88}$$

where

$$F(l) = \int_{l_{\rm out}}^{l} \frac{\Omega}{1-\Omega l}dl. \tag{9.89}$$

More, we have assumed that far outside of the central object the pressure in the tori tend to zero. The subscript "out" attached to some quantities denote their values in the outer edge of the tori where p tends to zero.

Equation (9.88) can be solved in the following ways. First, the functional form of $\Omega(l)$ is specified. Then, we have the functional form of $F(l)$ from Eq. (9.89). Next, the functional forms of $\Omega(r, \theta)$ and $l(r, \theta)$ are obtained from Eq. (9.83), and that of $u_t(r, \theta)$ from Eq. (9.82). Then, we have the functional form of $W(r, \theta)$. Integrating equation (9.88) with use of the above functional forms, we have the functional form of $p(r, \theta)$ [and $\rho(r, \theta)$].

Here, we present the simplest but important case where l is constant in the tori. In this case $F(r, \theta) = 0$. Furthermore, we assume that in the outer part of tori with large r, $(u_t)_{\rm out}$ tends to c. Then, we have

$$\int_0^p \frac{dp}{\varepsilon + p} = -\ln(u_t/c). \tag{9.90}$$

If $\varepsilon + p$ is approximated as $c^2\rho$ and $p = K\rho^\gamma$ is adopted with the ratio of specific heat γ, we have

$$\frac{\gamma}{\gamma - 1} K^{1/\gamma} p^{(\gamma-1)/\gamma} = -c^2 \ln(u_t/c). \tag{9.91}$$

The topological structure of equi-potential surfaces of tori is obtained by deriving r–θ relation for a given p. Let us demonstrate the equi-potential surface in the case of the Schwarzschild metric. If the left-hand side of Eq. (9.91) is written as $-W(p)$, an explicit form of the equi-potential surface is given by

$$\frac{(1 - r_{\rm S}/r) r^2 \sin^2\theta}{r^2 \sin^2\theta - l^2 (1 - r_{\rm S}/r) c^2} = \exp(2W), \tag{9.92}$$

where an explicit form of u_t has been used. For some typical values of l the contours of equi-potential surfaces on the r–θ plane are shown in Fig. 9.7a–d.

If $l > l_{\rm mb}$, the gas in torus cannot fall from the equator toward the central source [see Fig. 9.7a], because there is a strong barrier of centrifugal force. This means that a surface of torus (i.e., the equi-potential surface of $p = 0$ or $W = 0$) crosses the equator as shown in Fig. 9.7a. In other words, the torus has a clear inner boundary (funnel structure), as in the case of Newtonian torus.

If l decreases below $l_{\rm mb}$, however, situations are changed by relativistic effects: There is no inner surface of $p = 0$ on the equator. The assumption of no radial flows is violated, and the gas falls towards the central object by the combined effects of general relativity and pressure force. A critical value of l of this transition is at $l = l_{\rm mb}$. As Eq. (9.90) shows, the equi-potential surface of $p = 0$ (i.e., $W = 0$) is the surface where the effective potential u_t/c becomes just unity. On the equator, this is the radius where the net potential energy (the sum of gravitational energy and rotational energy) is zero. That is, ψ defined by Eq. (9.67) becomes just zero there.

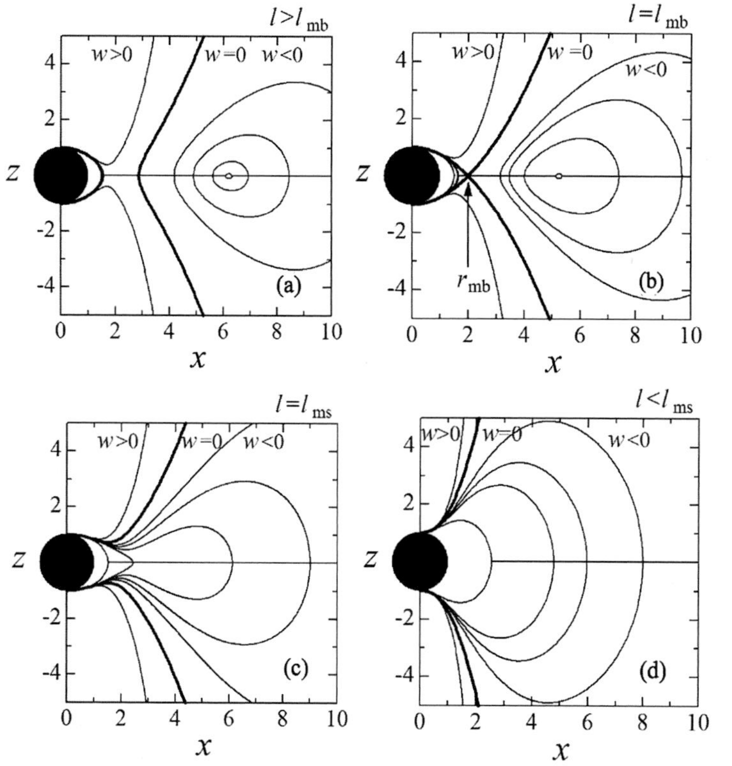

Fig. 9.7 Schematic picture showing topological structure of equi-potential surfaces in tori with a constant angular momentum l for four cases of $l > l_{mb}$, $l = l_{mb}$, $l = l_{ms}$, and $l < l_{ms}$. Panels (**a**)–(**d**) are for $l = 2.1 r_S/c$, $l = l_{mb} = 2.0 r_S/c$, $l = l_{ms} = 1.837 r_S/c$, and $l = 1.0 r_S/c$ in turn. It is noticed that in the case of $l > l_{mb}$ the surface of torus (equi-potential surface of $W = 0$) crosses the equator near the central source. This means that the torus has a funnel structure. In the case of $l = l_{mb}$, the funnel has a cusp structure on the equator (panel **b**). If $l < l_{mb}$, the cusp disappears, and the torus is opened toward the central source

This can be realized for $L = L_{mb}$, i.e., $l = l_{mb}$: The surface of the torus crosses the equator when $l = l_{mb}$, which has a cusp structure [see Fig. 9.7b].

If $l < l_{mb}$ (i.e., $L < L_{mb}$), the surface of $p = 0$ can no longer cross the equator. For example, in the case of $l = l_{ms}$, the effective potential ψ at r_{ms} is negative. Hence, to keep $u_t = c$ on the equator, a positive value of p at $r = r_{ms}$ is required in order to keep a steady rotating state (see Fig. 9.3). That is, the cusp on the equator has been opened toward the central object [see Fig. 9.7c]. On the equator at $r = r_{ms}$ there is pressure. Hence, the torus has a finite thickness at r_{ms}, although it is thin [see Fig. 9.7c]. It is noted that to keep a purely rotating state, pressure needs to increase inwards. In other words, this is unrealistic and in real situations fluids fall towards the central source.

9.5.1 Derivation of Eq. (9.85)

Since $u^r = u^\theta = 0$ and $\partial/\partial t = \partial/\partial\varphi = 0$ in the present problem, $T^{\nu}_{\mu\,;\nu} - u_\mu u_\nu T^{\nu\mu}_{\ \ ;\rho} = 0$ gives

$$\frac{1}{\varepsilon + p}\frac{\partial p}{\partial x^\mu} = u_{\mu;\nu}u^\nu, \tag{9.93}$$

where $\mu = r$ or $\mu = \theta$, while $\nu = t, r, \theta$, and φ.

Using the following relations:

$$g^{tt}\left(g_{tt}g_{\varphi\varphi} - g_{\varphi t}^2\right) = g_{\varphi\varphi}, \tag{9.94}$$

$$g^{t\varphi}\left(g_{tt}g_{\varphi\varphi} - g_{\varphi t}^2\right) = -g_{t\varphi}, \tag{9.95}$$

$$g^{\varphi\varphi}\left(g_{tt}g_{\varphi\varphi} - g_{\varphi t}^2\right) = g_{tt}, \tag{9.96}$$

we have

$$\begin{aligned} u_{\mu;\nu}u^\nu = &-\frac{1/2}{g_{tt}g_{\varphi\varphi} - g_{t\varphi}^2} \\ &\times\left[\left(g_{\varphi\varphi}\frac{\partial g_{tt}}{\partial x^\mu} - g_{t\varphi}\frac{\partial g_{\varphi t}}{\partial x^\mu}\right)u_t u^t + \left(-g_{t\varphi}\frac{\partial g_{tt}}{\partial x^\mu} + g_{tt}\frac{\partial g_{\varphi t}}{\partial x^\mu}\right)u_\varphi u^t\right. \\ &\left.+\left(g_{\varphi\varphi}\frac{\partial g_{t\varphi}}{\partial x^\mu} - g_{t\varphi}\frac{\partial g_{\varphi\varphi}}{\partial x^\mu}\right)u_t u^\varphi + \left(-g_{t\varphi}\frac{\partial g_{t\varphi}}{\partial x^\mu} + g_{tt}\frac{\partial g_{\varphi\varphi}}{\partial x^\mu}\right)u_\varphi u^\varphi\right]. \end{aligned} \tag{9.97}$$

Next, let us consider to represent the right-hand side of Eq. (9.85) in terms of $g_{\mu\nu}$ and l. Taking the derivative of Eq. (9.82) with respect to x^μ, we have

$$\begin{aligned} \frac{\partial \ln u_t}{\partial x^\mu} = &\frac{1/2}{g_{tt}g_{\varphi\varphi} - g_{t\varphi}^2}\left[\frac{\partial g_{tt}}{\partial x^\mu}g_{\varphi\varphi} + g_{tt}\frac{\partial g_{\varphi\varphi}}{\partial x^\mu} - 2g_{t\varphi}\frac{\partial g_{t\varphi}}{\partial x^\mu}\right. \\ &\left. - (u_t)^2\left(\frac{\partial g_{\varphi\varphi}}{\partial x^\mu} + 2l\frac{\partial g_{t\varphi}}{\partial x^\mu} + l^2\frac{\partial g_{tt}}{\partial x^\mu}\right) + 2\Omega(g_{t\varphi}l + g_{\varphi\varphi})\frac{\partial l}{\partial x^\mu}\right]. \end{aligned} \tag{9.98}$$

Furthermore, we have

$$\begin{aligned} \frac{\Omega}{1 - \Omega l}\frac{\partial l}{\partial x^\mu} &= (u_t)^2\frac{1}{u_t}(g^{tt}u_t + g^{t\varphi}u_\varphi)\Omega\frac{\partial l}{\partial x^\mu} \\ &= (u_t)^2\frac{1}{g_{tt}g_{\varphi\varphi} - g_{t\varphi}^2}(g_{\varphi\varphi} + l g_{t\varphi})\Omega\frac{\partial l}{\partial x^\mu}. \end{aligned} \tag{9.99}$$

Combining equations (9.98) and (9.99), we have

$$-\frac{\partial \ln u_t}{\partial x^\mu}+\frac{\Omega}{1-\Omega l}\frac{\partial l}{\partial x^\mu}=-\frac{1/2}{g_{tt}g_{\varphi\varphi}-g_{t\varphi}^2}$$
$$\times\left[\frac{\partial g_{tt}}{\partial x^\mu}g_{\varphi\varphi}+g_{tt}\frac{\partial g_{\varphi\varphi}}{\partial x^\mu}-2g_{t\varphi}\frac{\partial g_{t\varphi}}{\partial x^\mu}-(u_t)^2\left(\frac{\partial g_{\varphi\varphi}}{\partial x^\mu}+2l\frac{\partial g_{t\varphi}}{\partial x^\mu}+l^2\frac{\partial g_{tt}}{\partial x^\mu}\right)\right].\tag{9.100}$$

Considering Eq. (9.82) we see that the right-hand sides of Eqs. (9.100) and (9.97) are equal, and thus we have proofed equation (9.85).

References

Abramowicz, M.A., Jaroszyński, M., Sikora, M.: Astron. Astrophys. **63**, 221 (1978)
Aliev, A.N., Galtsov, D.V.: Gen. Rela. Gravitation **13**, 899 (1981)
Cox, J.P., Giuli, R.T.: Principles of Stellar Structure, chap. 24, vol. 2. Gordon and Breach, New York (1968)
Kato, S., Fukue, J.: Publ. Astron. Soc. Jpn. **32**, 377 (1980)
Kerr, R.P.: Phys. Rev. Lett. **11**, 237 (1963)
Kozłowski, M., Jaroszyński, M., Abramowicz, M.A.: Astron. Astrophys. **63**, 200 (1978)
Landau, L.D., Lifshitz, E.M.: The Classical Theory of Fields. Pergamon Press, Oxford (1975)
Okazaki, A.T., Kato, S., Fukue, J.: Publ. Astron. Soc. Jpn. **39**, 473 (1987)
Oppenheimer, J.R., Volkov, G.M.: Phys. Rev. **55**, 3740 (1939)

Part II
Magnetohydrodynamical Phenomena in Astrophysical Objects

Chapter 10
Derivation of Magnetohydrodynamical Equations from Boltzmann Equation

In this chapter magnetohydrodynamical (MHD) equations are derived from the Boltzmann equation. First, we discuss under what conditions particle systems can be treated as magnetohydrodynamical systems. Next, the set of magnetohydrodynamical equations are derived from the Boltzmann equation by taking the moments of the equation. In this derivation of MHD equations, we examine ion and electron systems separately, because it is instructive to see the origin of electromagnetic force on fluid. Generalized Ohm's is also derived. In this chapter gravitational force is neglected for simplicity.

10.1 Conditions for Ionized Gases Being Fluid

In Chap. 1 we have derived $n_{\rm e}\lambda_{\rm D}^3 \gg 1$ as a condition for ionized gases being an ensemble of free particles. Furthermore, we showed that in such systems the *mean free path* ℓ of electrons is much longer than the *Debye length* $\lambda_{\rm D}$, i.e., $\ell \gg \lambda_{\rm D}$. Our interest in this chapter is in such systems. The next issue is in what cases such systems can be regarded as fluids. It will be intuitively obvious that an additional condition is $L \gg \ell$, where L is the size of the phenomena which we are interested in. The curve where the mean free path ℓ is constant on the temperature–density diagram is shown in Fig. 1.1 in Chap. 1. This figure shows that the condition $L \gg \ell$ is sufficiently satisfied in many global phenomena in astronomical gaseous objects.

Let us summarize, from the viewpoint of timescales, the conditions we adopt here. Because of $\ell > \lambda_{\rm D}$ and $L > \ell$, we can define temperature locally. Hence, by using the acoustic speed $c_{\rm s}$, we define the following three timescales: $\tau_{\rm corr} = \lambda_{\rm D}/c_{\rm s}$, $\tau_{\rm rel} = \ell/c_{\rm s}$, and $\tau_{\rm dyn} = L/c_{\rm s}$. Then, the conditions by which gaseous systems can

© Springer Nature Singapore Pte Ltd. 2020, corrected publication 2023

S. Kato, J. Fukue, *Fundamentals of Astrophysical Fluid Dynamics*,
Astronomy and Astrophysics Library,
https://doi.org/10.1007/978-981-15-4174-2_10

be regarded as fluid systems are

$$\tau_{\rm corr} \ll \tau_{\rm rel} \ll \tau_{\rm dyn}. \tag{10.1}$$

Here, $\tau_{\rm corr}$ is the timescale of ion particles passing through the Debye radius and represents the timescale during which the correlation between particles is important. The time $\tau_{\rm rel}$ is the timescale of collisions of ion particles, representing the timescale of the ion velocity distribution approaching to the Maxwell distribution (*relaxation time*). In other words, this is the timescale by which temperature is defined.

If we treat magnetized gaseous systems, the *gyration radius*, $r_{\rm gyr}$, needs to be also longer than the mean free path ℓ for the systems being regarded as fluid systems. That is, we need $\ell \ll r_{\rm gyr}$, or if the gyration timescale $\tau_{\rm gyr}$ (inverse of gyration frequency) is used,

$$\tau_{\rm rel} \ll \tau_{\rm gyr} \tag{10.2}$$

is needed.

In addition to the above timescales, there is one more important timescale, which is *diffusion timescale*, say $\tau_{\rm dif}$. In many gaseous systems, the thermal energy is transported by diffusion processes to different parts of the systems. If this occurs by thermal processes, the timescale is on the order of L^2/κ, where κ is thermometric conductivity. In the case where the thermal conduction is due to gas particles, we have $\kappa \sim c_{\rm s}\ell$. If it is due to diffusion of radiation, we have $\kappa \sim c\ell_{\rm p}$, where $\ell_{\rm p}$ is the mean free path of photon and c is the speed of light. If we adopt the former κ, we have

$$\tau_{\rm dif} \sim \frac{L^2}{\kappa} \sim \tau_{\rm dyn}\left(\frac{\tau_{\rm dyn}}{\tau_{\rm rel}}\right) \gg \tau_{\rm dyn}. \tag{10.3}$$

In the inner region of stars, the diffusion of thermal energy by radiation is much faster than that by thermal conduction, i.e., $c\ell_{\rm p} \gg c_{\rm s}\ell$. In this case, the diffusion timescale $\tau_{\rm dif}$ is shorter than that estimated by Eq. (10.3) by a factor of $c_{\rm s}\ell/c\ell_{\rm p}$. Even in this case, if we consider that the optical depth, τ $(= L/\ell_{\rm p})$, is much larger than unity, we still have

$$\tau_{\rm dif} \sim \frac{L^2}{c\ell_{\rm p}} \sim \tau_{\rm dyn}\tau\frac{c_{\rm s}}{c} \gg \tau_{\rm dyn}. \tag{10.4}$$

In summary, in many gaseous systems, we have

$$\tau_{\rm corr} \ll \tau_{\rm rel} \ll \tau_{\rm dyn} \ll \tau_{\rm dif}. \tag{10.5}$$

These inequalities imply that in these gaseous systems, the concept of temperature is locally defined (*local thermodynamical equilibrium*), since $\tau_{\rm corr} \ll \tau_{\rm rel} \ll \tau_{\rm dyn}$. Such systems with local thermodynamical equilibrium gradually evolve since $\tau_{\rm dyn} \ll \tau_{\rm dif}$. These characteristics of gaseous systems allow us to derive hydrodynamical and magnetohydrodynamical equations by taking moments of the Boltzmann equation. Such derivation of magnetohydrodynamical equations is presented in the next section (Sect. 10.2).

Finally it should be noted that the relations among timescales given by inequality (10.5) seem to be satisfied in many gaseous systems, but it is not always the case in other astronomical objects. A typical example is stellar systems which consist of stars, e.g., globular clusters and elliptical galaxies. In such systems, local thermalization does not occur in a timescale shorter than the timescale of evolution of the systems, because $\tau_{\rm dyn} \ll \tau_{\rm rel}$, and relaxation processes among particles (stars) are due to evolution of the systems. Because of this, the orbit-averaged Fokker–Planck equation is used to study the dynamical evolution of such stellar systems.

10.2 Derivation of MHD Equations from the Boltzmann Equation

We start from the Boltzmann equation describing time evolution of distribution functions in six-dimensional phase space. By taking moments of the equation with respect to velocity, we derived basic fluid dynamical equations. We assume, for simplicity, that the gases consist of fully ionized ions and electrons and there are no ionization and recombination processes in collisions. Collisions are also assumed to be elastic.

10.2.1 Boltzmann Equations for Ion and Electron Gases

Let us introduce distribution functions in six-dimensional phase space, $(\boldsymbol{x}, \boldsymbol{V})$, for ions and electrons, respectively, by $f_+(\boldsymbol{x}, \boldsymbol{V}, t)$ and $f_-(\boldsymbol{x}, \boldsymbol{V}, t)$, where the subscripts $+$ and $-$ represent ions and electrons, respectively. Then, their time evolutions are described by the Boltzmann equation and are expressed, respectively, as

$$\frac{\partial f_+}{\partial t} + \boldsymbol{V} \cdot \frac{\partial f_+}{\partial \boldsymbol{x}} + \boldsymbol{a}_+ \cdot \frac{\partial f_+}{\partial \boldsymbol{V}} = \frac{\delta f_+}{\delta t}, \tag{10.6}$$

and

$$\frac{\partial f_-}{\partial t} + \boldsymbol{V} \cdot \frac{\partial f_-}{\partial \boldsymbol{x}} + \boldsymbol{a}_- \cdot \frac{\partial f_-}{\partial \boldsymbol{V}} = \frac{\delta f_-}{\delta t}, \tag{10.7}$$

where f_+ and f_- are normalized so that their integrations over the whole velocity space are densities of ions and electrons, respectively [see Eqs. (10.14)], and $\boldsymbol{a}_+$ and $\boldsymbol{a}_-$ are accelerations (per unit mass) acting on ions and electrons, respectively. The right-hand sides of the above two equations are collision terms. The collision term for ions consists of ion–ion collisions and ion–electron collisions:

$$\frac{\delta f_+}{\delta t} = \frac{\delta f_{++}}{\delta t} + \frac{\delta f_{+-}}{\delta t}. \tag{10.8}$$

Similarly, the term for electrons consists of electron–ion collisions and electron–electron collisions:

$$\frac{\delta f_-}{\delta t} = \frac{\delta f_{-+}}{\delta t} + \frac{\delta f_{--}}{\delta t}. \tag{10.9}$$

Because ionization and recombination in collisions are not considered for simplicity, numbers of ions and electrons are conserved during collisions. Thus, integrations of $\delta f_+/\delta t$ and $\delta f_-/\delta t$ over the whole velocity space vanish, respectively:

$$\int \frac{\delta f_+}{\delta t} d\boldsymbol{V} = \int \frac{\delta f_-}{\delta t} d\boldsymbol{V} = 0. \tag{10.10}$$

Linear momenta of ion and electron gases do not change during collisions when the collisions are those among the same kinds of particles. Furthermore, if collisions are elastic, kinetic energies of ion and electron gases do not change, respectively, during the collisions between the same kind of particles. That is, we have

$$\int \boldsymbol{V} \frac{\delta f_{++}}{\delta t} d\boldsymbol{V} = 0, \quad \int \boldsymbol{V} \frac{\delta f_{--}}{\delta t} d\boldsymbol{V} = 0, \tag{10.11}$$

$$\int \frac{1}{2} V^2 \frac{\delta f_{++}}{\delta t} d\boldsymbol{V} = 0, \quad \int \frac{1}{2} V^2 \frac{\delta f_{--}}{\delta t} d\boldsymbol{V} = 0. \tag{10.12}$$

10.2.2 Equation of Continuity

From the zeroth-order moment of Eqs. (10.6) and (10.7) we have, since Eq. (10.10) holds,

$$\frac{\partial \rho_\pm}{\partial t} + \frac{\partial}{\partial x_i}(\rho_\pm v_{\pm i}) = 0. \tag{10.13}$$

Two equations for ions and electrons are written in one Eq. (10.13). This abbreviation is used hereafter. The upper signs of the subscript $\pm$ in Eq. (10.13) are for ion gas, and the lower ones are for electron gas. In this equation, $\rho_\pm$ and $\rho_\pm v_{\pm i}$ are defined, respectively, by

$$\rho_\pm \equiv \int f_\pm d\boldsymbol{V}, \quad \rho_\pm v_{\pm i} \equiv \int V_i f_\pm d\boldsymbol{V}, \tag{10.14}$$

and $\rho_\pm$ is mass density of ion gas and of electron one, and $\boldsymbol{v}_\pm$ is macroscopic velocities of ion and electron gases. Equation (10.13) is equations of continuity for ion gas and electron one.

10.2.2.1 Equation of Continuity for the Whole System

In order to derive the equation of continuity for the whole gas consisting of ions and electrons, two Eqs. (10.13) are summed. The result gives the equation of continuity for the whole system:

$$\frac{\partial \rho}{\partial t} + \frac{\partial}{\partial x_i}(\rho v_i) = 0, \tag{10.15}$$

where the total mass density, ρ, and macroscopic velocity of the whole system, $\boldsymbol{v}$, are defined, respectively, by

$$\rho \equiv \rho_+ + \rho_-, \tag{10.16}$$

$$\rho \boldsymbol{v} \equiv \rho_+ \boldsymbol{v}_+ + \rho_- \boldsymbol{v}_-. \tag{10.17}$$

10.2.3 *Equation of Motion*

After taking the first-order moment of Eqs. (10.6) and (10.7) with respect to velocity, and performing the integrations by part we have, after some manipulations, the equation of motion for ion gas and electron one:

$$\frac{\partial}{\partial t}(\rho_\pm v_{\pm i}) + \frac{\partial}{\partial x_j}(\rho_\pm v_{\pm i} v_{\pm j}) = -\frac{\partial p_{\pm ij}}{\partial x_j} + \rho_\pm F_{\pm i} + (\Delta p)_{\pm i}, \tag{10.18}$$

where

$$p_{\pm ij} \equiv \int c_{\pm i} c_{\pm j} f_\pm dV \equiv \rho_\pm \langle c_{\pm i} c_{\pm j} \rangle, \quad c_{\pm i} \equiv V_{\pm i} - v_{\pm i}, \tag{10.19}$$

$$\rho_\pm F_{\pm i} \equiv \int a_{\pm i} f_\pm dV. \tag{10.20}$$

If the deviation of particle velocity $\boldsymbol{V}$ from $\boldsymbol{v}_\pm$ is isotropic, $p_{\pm ij}$ is zero except for the diagonal components of $i = j$ $(i = 1, 2, 3)$ and represents pressure. The off-diagonal components are related to viscous force, which is not discussed here further.

Deviation of macroscopic velocities of ion gas and electron gas (i.e., $\boldsymbol{v}_+$ and $\boldsymbol{v}_-$) from the mean velocity, $\boldsymbol{v}$, is denoted here as $\Delta\boldsymbol{v}_\pm$:

$$\Delta\boldsymbol{v}_\pm \equiv \boldsymbol{v}_\pm - \boldsymbol{v}. \tag{10.21}$$

By assuming electrical neutrality at every points (i.e., $n_+e_+ + n_-e_- = 0$) we can define electric current $\boldsymbol{j}$ by

$$\boldsymbol{j} \equiv (ne)_+\boldsymbol{v}_+ + (ne)_-\boldsymbol{v}_- = (ne)_+\Delta\boldsymbol{v}_+ + (ne)_-\Delta\boldsymbol{v}_- \tag{10.22}$$

or by

$$\boldsymbol{j} = ne(\boldsymbol{v}_+ - \boldsymbol{v}_-), \tag{10.23}$$

with $n = n_+ = n_-$ and $e = e_+ = -e_-$. In the case where forces acting on ions and electrons are electromagnetic force alone, we have $\boldsymbol{a}_\pm = e_\pm(\boldsymbol{E} + \boldsymbol{V} \times \boldsymbol{B}/c)/m_\pm$, where $m_\pm$ are masses of ion and electron, $e_\pm$ are charges of ion and electron. Then, the forces acting on ion gas and electron gas in unit volume, $\rho_\pm\boldsymbol{F}_\pm$, can be expressed as

$$\rho_\pm\boldsymbol{F}_\pm = n_\pm e_\pm\left(\boldsymbol{E} + \boldsymbol{v}_\pm \times \frac{\boldsymbol{B}}{c}\right). \tag{10.24}$$

The last term on the right-hand side of Eq. (10.18), $(\Delta p)_{\pm i}$, is defined by

$$\begin{aligned}(\Delta p)_{+i} &\equiv \int V_i \frac{\delta f_{+-}}{\delta t} d\boldsymbol{V},\\ (\Delta p)_{-i} &\equiv \int V_i \frac{\delta f_{-+}}{\delta t} d\boldsymbol{V}.\end{aligned} \tag{10.25}$$

This shows that $(\Delta p)_{+i}$ is the rate of gain of the i-component of momentum of ion gas by collisions with electrons. Similarly, $(\Delta p)_{-i}$ is the rate of gain of i-component of momentum of electron gas by collisions with ions. It is noted that the conservation of momentum in collisions gives

$$(\Delta p)_{+i} + (\Delta p)_{-i} = 0. \tag{10.26}$$

10.2.3.1 Equation of Motion of the Whole System

Next, we take the summation of two equations of Eq. (10.18). Considering Eqs. (10.24) and (10.26), we have

$$\rho\left(\frac{\partial}{\partial t} + v_j\frac{\partial}{\partial x_j}\right)v_i = -\frac{\partial}{\partial x_j}p_{ij} + \frac{1}{c}(\boldsymbol{j} \times \boldsymbol{B})_i, \tag{10.27}$$

where the pressure tensor, p_{ij}, is given by

$$p_{ij} \equiv p_{+ij} + p_{-ij} + \rho_+ \Delta v_{+i} \Delta v_{+j} + \rho_- \Delta v_{-i} \Delta v_{-j}. \tag{10.28}$$

Equation (10.27) is the equation of motion to the whole gaseous system consisting of ions and electrons. As a characteristic of ionized gases, the Lorenz force, $\boldsymbol{j} \times \boldsymbol{B}/c$, exists on the right-hand side of Eq. (10.27). This is due to the fact that we have $\rho_+ \boldsymbol{F}_+ + \rho_- \boldsymbol{F}_- = \boldsymbol{j} \times \boldsymbol{B}/c$ under the assumption of electrical neutrality and definition of $\boldsymbol{j}$.

10.2.4 Energy Equation

Equations (10.6) and (10.7) are multiplied, respectively, by $V_+^2/2$ and $V_-^2/2$, and integrated over the whole velocity space to obtain

$$\frac{\partial}{\partial t}\left(\frac{1}{2}\rho_\pm v_\pm^2 + \rho_\pm C_{V\pm} T_\pm\right) + \frac{\partial}{\partial x_j}\left[\left(\frac{1}{2}\rho_\pm v_\pm^2 + \rho_\pm C_{V\pm} T_\pm\right) v_{\pm j} + p_{\pm ij} v_{\pm i}\right]$$
$$= -\frac{\partial q_{\pm i}}{\partial x_i} + \boldsymbol{j}_\pm \cdot \boldsymbol{E} + (\Delta E)_\pm, \tag{10.29}$$

where $\rho_\pm \boldsymbol{v}_\pm \cdot \boldsymbol{F}_\pm = \boldsymbol{j}_\pm \cdot \boldsymbol{E}$ has been adopted. Equations (10.29) are energy equations for ion gas and electron gases. In the above equations,

$$\rho_\pm C_{V\pm} T_\pm \equiv \int \frac{1}{2} c_\pm^2 f_\pm d\boldsymbol{V} \equiv \frac{1}{2}\rho_\pm \langle c_\pm^2 \rangle, \quad \boldsymbol{q}_\pm \equiv \frac{1}{2}\rho_\pm \langle \boldsymbol{c}_\pm c_\pm^2 \rangle, \tag{10.30}$$

$$(\Delta E)_+ \equiv \int \frac{1}{2} V_+^2 \frac{\delta f_{+-}}{\delta t} d\boldsymbol{V}, \tag{10.31}$$

$$(\Delta E)_- \equiv \int \frac{1}{2} V_-^2 \frac{\delta f_{-+}}{\delta t} d\boldsymbol{V}. \tag{10.32}$$

Here, $C_{V\pm} T_\pm$ is the internal energy per unit mass, $\boldsymbol{q}_\pm$ is thermal energy flux, and $(\Delta E)_\pm$ is the rate of gain of kinetic energy of ion and electron gases by collisions with other kind of particles per unit volume. As the collisions are taken to be elastic, we have

$$(\Delta E)_+ + (\Delta E)_- = 0. \tag{10.33}$$

By multiplying Eq. (10.18) by $v_{\pm i}$, we have, after some manipulation, the equation of mechanical energy conservation:

$$\rho_{\pm}\left(\frac{\partial}{\partial t}+v_{\pm j}\frac{\partial}{\partial x_{\pm j}}\right)\left(\frac{1}{2}v_{\pm}^{2}\right)=v_{\pm j}\left(-\frac{\partial p_{\pm ij}}{\partial x_i}+\rho_{\pm}F_{\pm j}+(\Delta p)_{\pm j}\right). \quad (10.34)$$

It is noted that in the case of the Lorentz force the term $v_{\pm ij}\rho_{\pm}F_{\pm j}$ on the right-hand side of Eq. (10.34) can be written as $j_{\pm j}E_j$. If this equation is substituted from Eq. (10.29), we have an equation representing thermal energy conservation for ion and electron gases:

$$\rho_{\pm}\left(\frac{\partial}{\partial t}+v_{\pm i}\frac{\partial}{\partial x_i}\right)(C_{\mathrm{V}\pm}T_{\pm})+p_{\pm ij}\frac{\partial v_{\pm i}}{\partial x_j}$$
$$=-\frac{\partial q_{\pm i}}{\partial x_i}+(\Delta E)_{\pm}-v_{\pm i}(\Delta p)_{\pm i}. \quad (10.35)$$

10.2.4.1 Equation of Energy for the Whole System

Finally, we consider energy equation for the whole system. Vector $(\boldsymbol{\Delta}p)_+$ is the momentum which the ion gas obtains from collision with the electron gas in unit volume per unit time. Hence, $(\boldsymbol{\Delta}p)_+$ will be proportional to the relative velocity between the ion gas and the electron gas, i.e., $(\boldsymbol{\Delta}p)_+ \propto \boldsymbol{v}_- - \boldsymbol{v}_+$ with a positive coefficient. Since $\boldsymbol{j} \propto \boldsymbol{v}_+ - \boldsymbol{v}_-$ with a positive proportional coefficient, we can write

$$(\Delta p)_{+i}=-(\Delta p)_{-i}=-n_+e_+\eta j_i, \quad (10.36)$$

with a proportional coefficient η (*electric resistivity*). Using this relation, we see that the sum of the last term on the right-hand side of Eq. (10.35) for ion gas and electron gas gives ηj^2. Furthermore, for simplicity, under the assumption of $p_{+ij}=p_{-ij}=p_{ij}$, the sum of two equations of (10.35) then leads to

$$\rho\left(\frac{\partial}{\partial t}+v_i\frac{\partial}{\partial x_i}\right)(C_{\mathrm{V}}T)+p_{ij}\frac{\partial v_i}{\partial x_j}=-\frac{\partial q_i}{\partial x_i}+\eta j^2. \quad (10.37)$$

This is the equation of thermal energy conservation. The second term on the right-hand side is the term of thermalization due to Joule loss. Derivation of detailed expressions for internal energy per unit volume, $C_{\mathrm{V}}T$, and thermal flux, $\boldsymbol{q}$, is somewhat complicated, and thus omitted here. (For details, see, e.g., Thompson 1962.)

Equations (10.15), (10.27), and (10.37) are basic equations describing fluid motions. In cases where the fluid systems are in subject to other forces such as gravitational one, additional terms are added on the right-hand side Eq. (10.27). If collisions among particles are inelastic, heating (and cooling) terms appear on the right-hand side of Eq. (10.37).

The set of Eqs. (10.15), (10.27), and (10.37) are coupled with electromagnetic fields. To describe the whole systems including electromagnetic fields we need equations describing time variations of the fields. They are the Maxwell equations (Chap. 11). Fluid equations and Maxwell equations are coupled through electric current $\boldsymbol{j}$. In the next section (section 10.3), we derive the equation connecting fluid motions and the electromagnetic fields, which is generalized Ohm's law.

10.3 Generalized Ohm's Law

Equations of motions for ion gas and electron one are given by Eq. (10.18). Now, we compare these two equations. The terms on the right-hand side of the equations for ions and for electrons are comparable in their magnitudes. Compared with this, the left-hand side for the electron gas is much smaller than that for the ion gas, because the mass of electron is much smaller than that of ion. This means that when we consider fluid motions whose timescale is determined by the ion gas, the terms on the right-hand side of Eq. (10.18) for electron gas can be taken to be practically zero. In other words, in the case of electron gas the terms on the right-hand side of the equation are balanced without the terms on the left-hand side, due to short timescale of electron system. That is, we can adopt

$$0 = -\frac{\partial p_{-ij}}{\partial x_j} + \rho_- F_{-i} + (\Delta p)_{-i}. \tag{10.38}$$

In Eq. (10.38), the term $(\Delta p)_{-i}$ represents the rate of increase of i-component of momentum of the electron gas per unit volume by collisions with the ion gas. In the lowest order of approximation this rate will be proportional to the velocity difference between the electron gas and the ion gas, i.e., $(\Delta p)_{-i} \propto v_{+i} - v_{-i}$, as mentioned in Sect. 10.2.4. Discussions in Sect. 10.2.4 further show $(\Delta p)_{-i} = ne\eta j_i$. Then, using $\rho_- \boldsymbol{F}_- = -ne(\boldsymbol{E} + \boldsymbol{v}_- \times \boldsymbol{B}/c)$ [see Eq. (10.24)], and the approximation of $p_{-ij} = p_- \delta_{ij}$, we have

$$\boldsymbol{j} = \sigma\left(\boldsymbol{E} + \boldsymbol{v}_- \times \frac{\boldsymbol{B}}{c} + \frac{1}{ne}\nabla p_-\right), \tag{10.39}$$

where $\sigma = 1/\eta$ is *electric conductivity*.

When we treat fluid motions consisting of ions and electrons, the use of the total velocity $\boldsymbol{v}$ is convenient, compared with the use of $\boldsymbol{v}_-$. Since both of them are related by Eq. (10.23) and $\boldsymbol{v}_+ \sim \boldsymbol{v}$, we have

$$\boldsymbol{v}_- \sim \boldsymbol{v} - \frac{\boldsymbol{j}}{en}. \tag{10.40}$$

Hence, substitution of Eq. (10.40) into (10.39) leads finally to (see also Chap. 17)

$$\boldsymbol{j} = \sigma\left(\boldsymbol{E} + \boldsymbol{v} \times \frac{\boldsymbol{B}}{c} - \frac{1}{ne}\boldsymbol{j} \times \frac{\boldsymbol{B}}{c} + \frac{1}{ne}\nabla p_-\right). \tag{10.41}$$

This equation is called the *generalized Ohm's law*. The four terms on the right-hand side of Eq. (10.41) can be interpreted as currents due to, respectively, electromotive force, electromotive force induced by motions through a magnetic field, the Hall electromotive force, and the thermo-electric force.

An expression for generalized Ohm's law in gases consisting of three species (neutral gas is included) is presented in Sect. 17.1.

Reference

Thompson, W.B.: An Intoduction to Plasma Physics, Chap. 8. Pergamon Press, Oxford (1962)

Chapter 11
MHD Equations and Basic Characteristics of Magnetic Fields

In Chap. 10 the equations describing gaseous motions in subject to magnetic fields were derived from the Boltzmann equation. To fully describe such gaseous motions, however, the feedback of the gaseous motions to electromagnetic fields should be considered. In other words, interactions between gaseous motions and electromagnetic fields need to be considered. In this chapter we describe such set of equations under the MHD approximations, and summarized some fundamental characteristics of magnetized fluids under the MHD approximations.

11.1 MHD Approximations and Induction Equation

In Sect. 10.2 we have derived fluid dynamical equations of ionized gases in subject to magnetic fields. The results show that the Lorentz force $\boldsymbol{j} \times \boldsymbol{B}/c$ acts on ionized gases. In return, magnetic fields are affected by fluid motions. To describe these interactions the dynamical equations of fluid motions, derived in Sect. 10.2, should be combined with the Maxwell equations.

Since we are interested in dynamical phenomena of fluids, the timescale concerned here is L/c_s or L/c_A, where L is the characteristic length of phenomena, c_s is the acoustic speed, and c_A is the Alfvén speed which will be described later. The timescales mentioned above are generally much longer than the timescale of electromagnetic waves. As is well known, the presence of electromagnetic waves comes from the presence of displacement current in the Maxwell equations. Hence, in the present treatment of gaseous systems, we filter out the electromagnetic waves by neglecting the displacement current in the Maxwell equations. The Maxwell

© Springer Nature Singapore Pte Ltd. 2020, corrected publication 2023

S. Kato, J. Fukue, *Fundamentals of Astrophysical Fluid Dynamics*,
Astronomy and Astrophysics Library,
https://doi.org/10.1007/978-981-15-4174-2_11

equations are thus approximated as[1]

$$\mathrm{curl}\boldsymbol{B} = \frac{4\pi}{c}\boldsymbol{j}, \quad \mathrm{curl}\boldsymbol{E} = -\frac{\partial \boldsymbol{B}}{c\partial t}. \tag{11.1}$$

The treatment of ionized fluid motions described by Eqs. (10.15), (10.27), and (10.37), with the above approximated Maxwell equations and the simplified Ohm's law:

$$\boldsymbol{j} = \sigma\left(\boldsymbol{E} + \frac{\boldsymbol{v}}{c} \times \boldsymbol{B}\right) \tag{11.2}$$

is called magnetohydrodynamic approximations or *MHD approximations*, where σ is the electric conductivity. In some times, in a narrower sense, the case where Ohm's law is further approximated by $\boldsymbol{E} + \boldsymbol{v} \times \boldsymbol{B}/c = 0$ is called *MHD approximations.*

If $\boldsymbol{E}$ is eliminated from Eqs. (11.1) and (11.2), we have an equation describing time variation of magnetic fields:

$$\frac{\partial \boldsymbol{B}}{\partial t} = \mathrm{curl}\left(\boldsymbol{v} \times \boldsymbol{B} - \frac{c}{\sigma}\boldsymbol{j}\right). \tag{11.3}$$

For simplicity, electric conductivity σ is taken to be constant. Then, by using the first relation of Eq. (11.1), we can reduce Eq. (11.3) to

$$\frac{\partial \boldsymbol{B}}{\partial t} = \mathrm{curl}(\boldsymbol{v} \times \boldsymbol{B}) + \eta\nabla^2\boldsymbol{B}, \tag{11.4}$$

where η is given by

$$\eta = \frac{c^2}{4\pi\sigma} \tag{11.5}$$

and called *magnetic diffusivity*.[2] Equation (11.4), or more generally Eq. (11.3), is called *induction equation*, and one of the important equations describing characteristics of MHD fluids.

In the following sections, some important characteristics of MHD fluids are argued. Before that, we should mention on validity of Ohm's law. In the usual MHD approximations, Eq. (11.2) is adopted. As Eq. (10.39) shows, however, the following equation:

$$\boldsymbol{j} = \sigma\left(\boldsymbol{E} + \frac{\boldsymbol{v}_-}{c} \times \boldsymbol{B}\right) \tag{11.6}$$

[1] In Part II "curl" (not "rot") is used, although the latter may be popular.

[2] η given here is different from η used in Sects. 10.2 and 10.3, although the same notation is used.

is more accurate than Eq. (11.2), where $\boldsymbol{v}_-$ is the velocity of electron gas. In the conventional MHD approximations, this difference is not taken into account. To understand some astrophysical phenomena, however, this difference is important, and should be considered (e.g., see Sects. 14.1 and 17.1).

11.2 Frozen-in of Magnetic Fields to Fluids

In Sect. 11.1 we have derived induction Eq. (11.4), using the MHD approximations. In this subsection an important characteristic of induction Eq. (11.4) is argued.

11.2.1 Case of No Fluid Motions ($\boldsymbol{v} = 0$)

First, let us consider the case of $\boldsymbol{v} = 0$. Equation (11.4) is then reduced to

$$\frac{\partial \boldsymbol{B}}{\partial t} = \eta \nabla^2 \boldsymbol{B}. \tag{11.7}$$

This equation is a diffusion equation, showing that magnetic fields weaken with time by diffusion. This is due to the fact that the electric current producing magnetic fields is weakened by electric resistivity. The timescale, $\tau_{\rm dif}$, by which magnetic fields are dissipated is given by, if typical scale length of magnetic fields is represented by L,

$$\tau_{\rm dif} \sim L^2/\eta = 4\pi\sigma L^2/c^2. \tag{11.8}$$

Table 11.1 shows typical values of $\tau_{\rm dif}$ for various objects. As is shown in this table, the timescale of diffusion of magnetic fields is long, and much longer than the age of the Universe in typical astrophysical objects. This result of long timescale comes from large scale of astrophysical phenomena, not from smallness of η.

Let us estimate the condition of unimportance of the diffusion term in Eq. (11.4) in a different way. The timescale associated with the first term on the right-hand side of Eq. (11.4) is L/v, where L and v are, respectively, the typical size and typical velocity associated with fluid phenomena. Then, the condition of the diffusion term

Table 11.1 Timescale of diffusion, $\tau_{\rm dif}$, of magnetic fields

Objects	Timescale of diffusion
Sphere of copper with radius of 1 m	10 s
Magnetic fields in solar convection zone	$\sim 10^8$ years
General magnetic fields in the Sun	$\sim 10^{10}$ years
Fields in interstellar gases	$> 10^{10}$ years
Fields in intergalactic space	$\gg 10^{10}$ year

being negligible is $L/v < L^2/\eta$, which is

$$L > \frac{\eta}{v}. \tag{11.9}$$

This condition is usually satisfied by large scales of dynamical phenomena of astrophysical fluids.

11.2.2 Case of No Diffusion of Magnetic Field

The arguments in Sect. 11.2.1 show that in many astrophysical objects dissipation of magnetic fields can be neglected, because the timescale of their dissipation is long. If the dissipation is neglected, Eq. (11.4) is reduced to

$$\frac{\partial \boldsymbol{B}}{\partial t} = \mathrm{curl}(\boldsymbol{v} \times \boldsymbol{B}). \tag{11.10}$$

In the followings, we argue the characteristics of this equation.

Equation (11.10) is analogous to the equation for conservation of vorticity in inviscid barotropic gas [$p = p(\rho)$]:

$$\frac{\partial \boldsymbol{\zeta}}{\partial t} = \mathrm{curl}\ (\boldsymbol{v} \times \boldsymbol{\zeta}), \tag{11.11}$$

where $\boldsymbol{\zeta}$ is the vorticity defined by $\boldsymbol{\zeta} = (1/2)\mathrm{curl}\, \boldsymbol{v}$. This similarity shows that magnetic fields are frozen to gases. This can be derived as follows. As is shown in Fig. 11.1 we consider a closed curve l in fluid, and examine the flux of magnetic fields passing through a surface S bounded by the curve l. The close curve l is assumed to move with fluid motions.

Since the curve l moves with fluid motions, we have

$$\frac{d}{dt}\int\!\!\int_S \boldsymbol{B} \cdot d\boldsymbol{S} = \int\!\!\int_S \frac{\partial \boldsymbol{B}}{\partial t} \cdot d\boldsymbol{S} + \int_l \boldsymbol{B} \cdot (\boldsymbol{v} \times d\boldsymbol{l}), \tag{11.12}$$

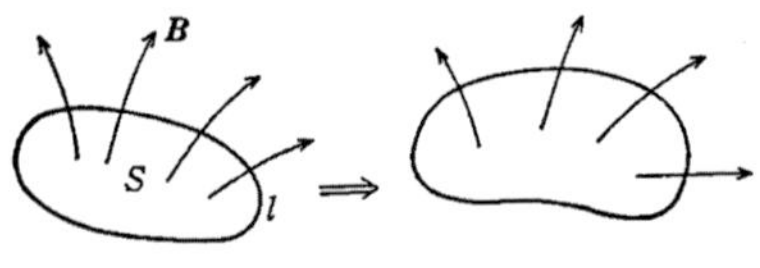

Fig. 11.1 Showing that magnetic fields are frozen to gases. The magnetic field lines passing through an arbitrary surface S are unchanged even if the boundary of the surface, l, is deformed with time

where $d\boldsymbol{S}$ is a surface element of the surface S, and $d\boldsymbol{l}$ is a line element along l. Using the relation $\boldsymbol{B}\cdot(\boldsymbol{v}\times d\boldsymbol{l}) = -(\boldsymbol{v}\times\boldsymbol{B})\cdot d\boldsymbol{l}$ and changing the line integration to surface integration by the Stokes theorem, we can rewrite the right-hand side of Eq. (11.12) as

$$\int\!\!\int_{\mathrm{S}}\left[\frac{\partial \boldsymbol{B}}{\partial t} - \mathrm{curl}(\boldsymbol{v}\times\boldsymbol{B})\right]\cdot d\boldsymbol{S}. \tag{11.13}$$

If Eq. (11.10) is used, the above integration vanishes. That is, we have

$$\frac{d}{dt}\int\!\!\int_{\mathrm{S}}\boldsymbol{B}\cdot d\boldsymbol{S} = 0. \tag{11.14}$$

Equation (11.14) shows that the number of the magnetic field lines through an arbitrary surface, S, are unchanged with time in the case where the boundary of the surface, l, moves with fluid motions. In this sense we can say that magnetic fields flow with fluid. This is called that magnetic fields are *frozen* to fluids.

We have examined two limiting cases of induction Eq. (11.4) in Sects. 11.2.1 and 11.2.2 (present subsection). An important dimensionless quantity to measure which process is prominent in real systems is *magnetic Reynolds number* $\mathcal{R}_m$. Let us denote a typical size and a typical speed of phenomena in consideration by l and u, respectively. Then, a rough ratio of the frozen term in Eq. (11.4) to the diffusion one is $uB/l : \eta B/l^2$, which is ul/η:

$$\mathcal{R}_m \equiv \frac{ul}{\eta}. \tag{11.15}$$

That is, in the systems of $\mathcal{R}_m \gg 1$, the magnetic fields are frozen in fluids and their dissipation can be neglected.

Finally, it is important to note here that as mentioned in Sect. 11.1, $\boldsymbol{E} + \boldsymbol{v}_- \times \boldsymbol{B}/c = 0$ is more general than $\boldsymbol{E} + \boldsymbol{v}\times\boldsymbol{B}/c = 0$, where $\boldsymbol{v}_-$ is the macroscopic velocity of electron gases. If $\boldsymbol{E} + \boldsymbol{v}_- \times \boldsymbol{B}/c = 0$ is adopted, we obtain

$$\frac{\partial \boldsymbol{B}}{\partial t} = \mathrm{curl}(\boldsymbol{v}_- \times \boldsymbol{B}) \tag{11.16}$$

instead of Eq. (11.10). This means that it is general to say that magnetic fields are frozen to electron gases, not to fluid systems. In the usual MHD approximations, however, $\boldsymbol{v}$ and $\boldsymbol{v}_-$ are not distinguished.

11.3 Maxwell Stress Tensor and Magnetic Torque

In Sect. 11.2 we have shown that magnetic fields in many astrophysical objects are frozen in gaseous fluids. This implies that not only magnetic fields are deformed by fluid motions, but magnetic fields also react to fluid motions, because they are coupled. Here, we examine how magnetic fields affect fluid motions by the freezing. The gaseous fluids receive the Lorentz force, $\boldsymbol{j} \times \boldsymbol{B}/c$, per unit volume from magnetic fields [see Eq. (10.27)]. Hence, the purpose of this section is to examine the characteristics of this force.

11.3.1 Maxwell Stress Tensor

Using $\mathrm{curl}\boldsymbol{B} = 4\pi \boldsymbol{j}/c$ and formulae of vector analyses, we can write the Lorenz force in the form:

$$\frac{\boldsymbol{j} \times \boldsymbol{B}}{c} = \frac{1}{4\pi}\mathrm{curl}\boldsymbol{B} \times \boldsymbol{B} = -\mathrm{grad}\left(\frac{B^2}{8\pi}\right) + \frac{1}{4\pi}(\boldsymbol{B} \cdot \nabla)\boldsymbol{B}, \tag{11.17}$$

which is further written in a form of divergence of a tensor as

$$\frac{1}{c}(\boldsymbol{j} \times \boldsymbol{B})_i = \frac{\partial M_{ij}}{\partial x_j}, \tag{11.18}$$

where M_{ij} is the tensor given by

$$M_{ij} = -\frac{1}{8\pi}B^2\delta_{ij} + \frac{1}{4\pi}B_i B_j, \tag{11.19}$$

and called the *Maxwell stress tensor.*[3]

Equation (11.18) implies that the Lorentz force acting on fluid of volume V (the part shown by a rectangle in the central part of Fig. 11.2) is replaced to a surface force acting on the volume:

$$\iiint \frac{(\boldsymbol{j} \times \boldsymbol{B})_i}{c} dV = \iint M_{ij} dS_j, \tag{11.20}$$

where the Gauss theorem has been used to replace the volume integration to the surface integration. Here, dS_j is the j-component of a surface element $d\boldsymbol{S}$.

[3]The Maxwell stress tensor M_{ij} should not be confused with the energy momentum tensor which will be introduced in Chap. 18, although the same notation is used.

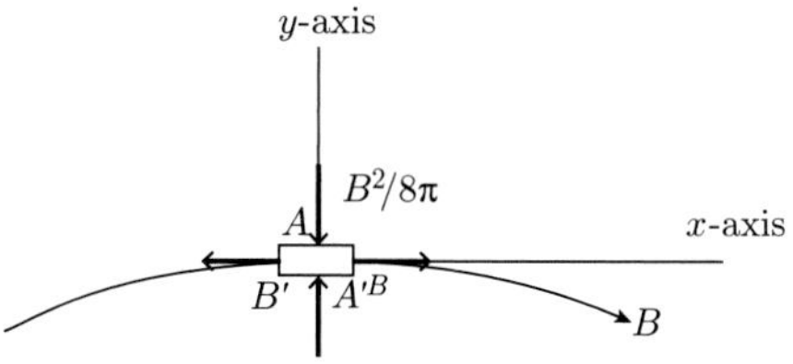

Fig. 11.2 A figure showing characteristics of the Maxwell stress tensor. A fluid element receives a pressure force, $B^2/8\pi$, from the direction perpendicular to the line of force, and does a tension, $B^2/8\pi$, in the direction along the line of force

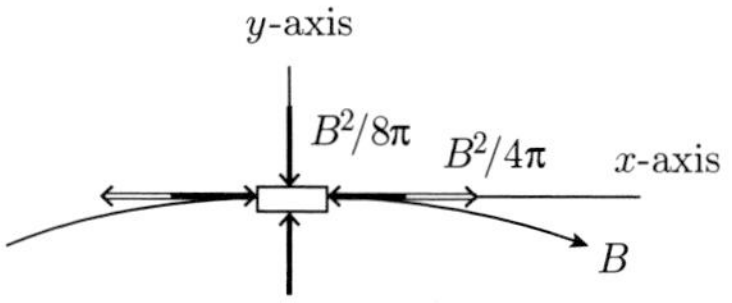

Fig. 11.3 A figure showing the characteristics of Maxwell stress tensor. The Lorentz force consists of isotropic pressure, $B^2/8\pi$ (shown by thick arrows), and tension, $B^2/4\pi$ (shown by long hollow arrows), along the line of force

As is shown in Fig. 11.2, we consider a small volume element, dV, along a magnetic field line, and examine the magnetic force acting on the volume element. Let us take Cartesian coordinates, x-axis being along the field line, y-axis in the vertical direction to x-axis on the paper plane, and z-axis in the direction perpendicular to the paper plane. In this case the Maxwell stress tensor, M_{ij}, is written as

$$M_{ij} = \begin{pmatrix} B^2/8\pi & 0 & 0 \\ 0 & -B^2/8\pi & 0 \\ 0 & 0 & -B^2/8\pi \end{pmatrix}. \tag{11.21}$$

Let us now consider the surface of the volume element dV. As is shown in Fig. 11.2, the surface A whose normal is in the y-direction is $dS_j = (0, dS, 0)$. Hence, $M_{ij}dS_j = (0, -B^2/8\pi, 0)dS$ and we see that through the surface A a force, $B^2/8\pi$, acts per unit surface from outside. Similarly, to the surface A' a force of $B^2/8\pi$ acts from outside. Next, we consider the surface B (see Fig. 11.2). The surface element is $(dS, 0, 0)$ and $M_{ij}dS_j = (B^2/8\pi, 0, 0)dS$. Hence, to the surface B a force $B^2/8\pi$ directed outward works per unit surface. Similarly to the surface B' an force $B^2/8\pi$ directed outwards works. These results show that on the surface of volume element dV, forces act as shown by thick arrows in Fig. 11.2.

The above results can be interpreted as a sum of two kinds of forces, as is shown in Fig. 11.3. That is, one is the isotropic pressure force of $B^2/8\pi$, and the other is the tension $B^2/4\pi$ along the line of force. This characteristic of the Lorentz force, of course, can be directly observed from the form of Eq. (11.17).

11.3.2 Magnetic Torque and Angular Momentum Transport

As shown in Sect. 11.3.1, the Lorentz force is an anisotropic force. Hence, this brings about various important effects on fluid motions which are different from the gas pressure force. One of them is that magnetic fields can transport angular momentum along the line of forces.

To demonstrate angular momentum transport by the Lorentz force, we introduce cylindrical coordinates (r, φ, z). Then, the φ-component of equation of motion is written as

$$\rho\frac{\partial v_\varphi}{\partial t} + \rho[(\boldsymbol{v}\cdot\nabla)\boldsymbol{v}]_\varphi = -\frac{\partial p}{r\partial\varphi} + \frac{1}{c}(\boldsymbol{j}\times\boldsymbol{B})_\varphi. \tag{11.22}$$

If this equation is multiplied by r and the equation of continuity is used, we have after some manipulations

$$\frac{\partial A}{\partial t} + \mathrm{div}\left(A\boldsymbol{v} - r\frac{B_\varphi}{4\pi}\boldsymbol{B}\right) = -\frac{\partial}{\partial\varphi}\left(p + \frac{1}{8\pi}B^2\right), \tag{11.23}$$

where A is defined by $A \equiv \rho r v_\varphi$ and represents angular momentum per unit volume in the z-direction. That is, this equation is an equation describing time evolution of A.

We now consider an axisymmetric system rotating around the z-axis, and examine how the angular momentum around the z-axis, A, is transported in the r-direction. Since axially symmetric systems $(\partial/\partial\varphi = 0)$ are considered, the pressure and the diagonal components of the Maxwell stress tensor do not contribute to the angular momentum transport [see Eq. (11.23)]. What contribute to the transport are the off-diagonal components of $v_i v_j$ and $B_i B_j$. Equation (11.23) shows that the angular momentum flux in the r- and z-directions, F_r and F_z, are given, respectively, by

$$F_r = \rho r v_r v_\varphi - \frac{r}{4\pi}B_r B_\varphi, \quad F_z = \rho r v_\varphi v_z - \frac{r}{4\pi}B_\varphi B_z. \tag{11.24}$$

The first terms of F_r and F_z represent the angular momentum transports by inertial forces and the second ones are those by magnetic ones.

The fact that the angular momentum flux transported by magnetic fields in the r-direction is $-(r/4\pi)B_r B_\varphi$ is demonstrated in Fig. 11.4. Let us consider a cylinder of radius r, and examine the angular momentum transport outward through the surface of r (unit thickness in the z-direction). As shown by the set of Eqs. (11.18)–(11.20), the Lorentz force can be expressed as a force acting on the surface. In the case of the configuration shown in Fig. 11.4, the force acting on the cylindrical surface in the φ-direction is $(B_r B_\varphi/4\pi)dS$, where dS is a surface element on the cylinder of radius

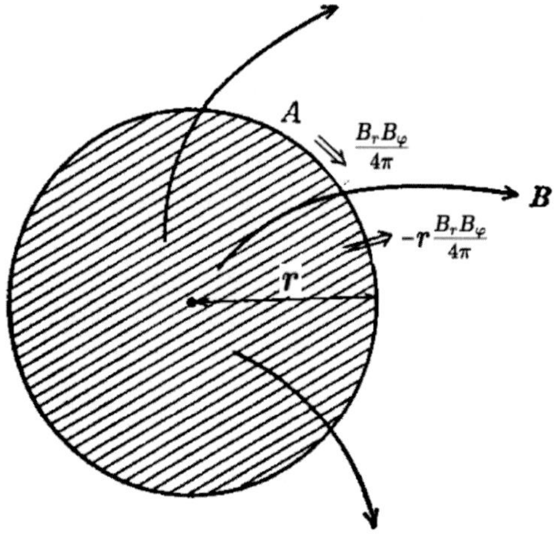

Fig. 11.4 Magnetic torque and angular momentum transport. A cylinder of radius r is considered. The z-axis is perpendicular to the paper plane and the φ-direction is taken in the anti-clockwise direction. The magnetic force acting on an unit surface of the cylinder is $B_r B_\varphi/4\pi$, which is negative in the case shown here, because the line of force is trailing. By the torque due to this force, angular momentum inside the cylinder is transported outwards by the rate of $-r B_r B_\varphi/4\pi$

r.[4] It is noted that $B_r B_\varphi/4\pi$ is negative in the case of Fig. 11.4, because the line of force is trailing. By the effect of torque of $r B_r B_\varphi/4\pi$, the angular momentum inside the cylinder of radius r increases by the rate of $(r B_r B_\varphi/4\pi) \times 2\pi r$. In other words, the inward angular momentum flow through the surface of radius r is $(r B_r B_\varphi/4\pi) \times 2\pi r$. In other words, the outward angular momentum flux through a unit surface is $-r B_r B_\varphi/4\pi$. In the case of the configuration of Fig. 11.4, angular momentum is transported outward, because $B_r B_\varphi < 0$. This is understandable, since the pattern of the magnetic fields is trailing in the case of Fig. 11.4 and thus the fields outside r drag the gas inside r in the clockwise direction. This means that the rotation of gas inside radius r is braked by the magnetic fields pulling out angular momentum outward.

11.3.2.1 Magnetic Braking of Rotation and Ambipolar Diffusion

Angular momentum transport by magnetic fields is very important in various levels of astrophysical objects such as planetary systems to galaxies. For example, in the process of formation of stars from interstellar clouds, the contraction of the clouds is suppressed if the initial angular momentum of the clouds is conserved. For example, the gravitational force due to self-gravity increases with decrease of

[4]If we introduce local Cartesian coordinates at a point on the surface, Eq. (11.20) shows that the force in the φ-direction at the surface is $(1/4\pi)B_\varphi B_i dS_i = (1/4\pi)B_\varphi B_r dS$.

the cloud size, r, as $1/r^2$, while the centrifugal force due to rotation increases as $1/r^3$, if angular momentum is conserved. Hence, eventually the latter overcomes the former and suppresses the contraction. Observational evidence shows, however, that the observed rotational velocities of normal stars are much slower than those of centrifugal equilibrium. This suggests that much angular momentum is extracted from star-forming clouds in the star formation stage by magnetic fields (magnetic breaking).

In accretion disks the gaseous accretion to central objects is realized by outward angular momentum transport in disks. This angular momentum transport is now understood to be due to magnetic turbulence resulting from magneto-rotational instability (see Sect. 16.3). The angular momentum transport by turbulence in the radial direction is $-r\langle b_r b_\varphi \rangle/4\pi$, where b_r and b_φ are, respectively, the r- and φ-components of turbulent magnetic field $\boldsymbol{b}$, and $\langle\ \rangle$ is ensemble average.

It should be noticed, however, that in some astrophysical objects the angular momentum extraction by magnetic fields is too efficient to explain some observational evidence, if the gas is assumed to be fully ionized. For example, galaxies formed from intergalactic matter become too compact compared with observational ones, if angular momentum extraction by magnetic braking is efficient. This suggests that the frozen-in of magnetic fields to gases is partial in intergalactic space because of low temperature. In other words, separation of magnetic fields to gas motions is really present in formation of galaxies by ambipolar diffusion. Ambipolar diffusion will be briefly described in Sect. 17.1.

11.4 Force-Free Fields

In magnetized gas the Lorentz force, $\boldsymbol{j} \times \boldsymbol{B}/c$, acts on the gas. The forces acting on the gas are not only the Lorentz force, but also the pressure one and the gravitational one and so on. In the case where the magnetic fields are strong enough, however, forces other than the Lorentz one are negligible. Hence, in such systems a static state will be realized only when the Lorentz force is adjusted itself so that

$$\boldsymbol{j} \times \boldsymbol{B} = 0. \tag{11.25}$$

The magnetic fields which satisfy Eq. (11.25) is called *force-free magnetic fields*. The most characteristic example of force-free fields is a dipole field in the empty space. The magnetic fields in the solar corona will be also close to force-free fields.

The magnetic field configurations of force-free fields are studied in various cases. As is clear from Eq. (11.25), the simplest case of force-free fields is that with $\boldsymbol{j} = 0$. That is, $\mathrm{curl}\boldsymbol{B} = 0$. This is the case where $\boldsymbol{B}$ can be expressed in terms of a potential ψ as $\boldsymbol{B} = \mathrm{grad}\psi$. By considering that div $\boldsymbol{B} = 0$, we see that ψ should be a solution of $\nabla^2\psi = 0$.

One of another examples of force-free fields is curl$\boldsymbol{B} = \alpha\boldsymbol{B}$, where α is an arbitrary function with respect to time and space. Since the divergence of curl$\boldsymbol{B}(= \alpha\boldsymbol{B})$ gives $\boldsymbol{B}\cdot \text{grad}\,\alpha = 0$, α must be constant along each line of forces, although it can be different for a different magnetic line. In particular, in the case where α is constant in the whole space, we have $\Delta^2\boldsymbol{B} + \alpha^2\boldsymbol{B} = 0$ by taking rotation of curl$\boldsymbol{B} = \alpha\boldsymbol{B}$. That is, in the case of $\alpha = \text{const.}$, the solutions of force-free fields are involved in the solutions of $\Delta^2\boldsymbol{B} + \alpha^2\boldsymbol{B} = 0$.

For details on force-free fields and their applications to solar physics, see, for example, Wiegelmann and Sakurai (2012).

11.5 Magnetic Helicity and Cross Helicity

Magnetic fields are frozen to fluid motions in ideal MHD fluids. Hence, in complicated flows the magnetic fields are tangled. However, there is a measure of the topological complexity of a magnetic field. This is the *magnetic helicity*, which is a scalar quantity to measure the extent to which the field lines wrap and coil around each other. In other words, magnetic helicity is important as a measure of linkage of magnetic fields.

In barotropic non-magnetized inviscid flows, vorticity is frozen to fluid flows (Helmholtz theorem) (see Sect. 2.4.2). Due to this, we had the concept of conservation of fluid helicity in an isolated system (see Sect. 2.4.2). The concept of magnetic helicity conservation mentioned above is an extension of the concept of fluid helicity conservation. As an intermediate concept between fluid helicity and magnetic helicity, we have that of *cross helicity*. The cross helicity is also conserved, which is briefly mentioned below.

11.5.1 Magnetic Helicity and Its Conservation

The magnetic helicity, H, is defined by

$$H \equiv \iiint (\boldsymbol{A}\cdot\boldsymbol{B})dV, \tag{11.26}$$

where $\boldsymbol{B}$ is the magnetic field strength, and $\boldsymbol{A}$ is the vector potential of $\boldsymbol{B}$ defined by

$$\boldsymbol{B} = \text{curl}\,\boldsymbol{A}. \tag{11.27}$$

The volume V is taken to move with fluid motions.

Important characteristics of the magnetic helicity are that in isolated systems it is i) gauge invariance and ii) a conserved quantity. In the absence of diffusion processes, magnetic field lines cannot reconnect. Hence, all their knots and linkages must be preserved (Moffatt 1978).

Let us change the vector potential $\boldsymbol{A}$ to $\boldsymbol{A} + \nabla\phi$. Then, the helicity is written as

$$H = \iiint [(\boldsymbol{A} \cdot \boldsymbol{B}) + (\nabla\phi) \cdot \boldsymbol{B}]\, dV = \iiint [(\boldsymbol{A} \cdot \boldsymbol{B}) + \mathrm{div}(\phi\boldsymbol{B})]\, dV, \quad (11.28)$$

where $\mathrm{div}\boldsymbol{B} = 0$ has been used. The second term on the right-hand side of Eq. (11.28) vanishes when the system is closed in the sense that there is no magnetic line of force penetrating the surface of the system. That is, the magnetic helicity has gauge invariance.

The second important characteristics is $dH/dt = 0$ in isolated systems, where the time variation is taken so that the volume in consideration changes with fluid motions. First, we introduce *magnetic helicity density*:

$$h_0 = \boldsymbol{A} \cdot \boldsymbol{B}. \quad (11.29)$$

Since the Maxwell equation, $\mathrm{curl}\boldsymbol{E} = -\partial\boldsymbol{B}/c\partial t$, is written as $\partial\boldsymbol{A}/c\partial t = -\boldsymbol{E} - \nabla\tilde{\phi}$, where $\tilde{\phi}$ is a scalar potential, we have[5]

$$\frac{\partial h_0}{\partial t} + \mathrm{div}\boldsymbol{h} = -2c\boldsymbol{E} \cdot \boldsymbol{B}, \quad (11.30)$$

where $\boldsymbol{h}/c = \boldsymbol{E} \times \boldsymbol{A} + \tilde{\phi}\boldsymbol{B}$.

In the case of ideal MHD of $\boldsymbol{E} = -\boldsymbol{v} \times \boldsymbol{B}/c$, we have $\boldsymbol{E} \cdot \boldsymbol{B} = 0$ and $\boldsymbol{h} = -(\boldsymbol{v} \times \boldsymbol{B}) \times \boldsymbol{A} + c\tilde{\phi}\boldsymbol{B}$. Hence, Eq. (11.30) is reduced to

$$\frac{\partial h_0}{\partial t} + \mathrm{div}[\boldsymbol{A} \times (\boldsymbol{v} \times \boldsymbol{B}) + c\tilde{\phi}\boldsymbol{B}] = 0. \quad (11.31)$$

By using a formula of vector analyses[6] we have

$$\frac{\partial h_0}{\partial t} + \mathrm{div}(h_0\boldsymbol{v}) + \mathrm{div}[-\boldsymbol{B}(\boldsymbol{A} \cdot \boldsymbol{v}) + c\tilde{\phi}\boldsymbol{B}] = 0. \quad (11.32)$$

[5] A formula of vector analyses:

$$\mathrm{div}\,(\boldsymbol{A}_1 \times \boldsymbol{A}_2) = \boldsymbol{A}_2\mathrm{curl}\boldsymbol{A}_1 - \boldsymbol{A}_1\mathrm{curl}\boldsymbol{A}_2,$$

has been used, where $\boldsymbol{A}_1$ and $\boldsymbol{A}_2$ are arbitrary vectors.

[6] A formular of vector analyses:

$$\boldsymbol{A}_1 \times (\boldsymbol{A}_2 \times \boldsymbol{A}_3) = \boldsymbol{A}_2(\boldsymbol{A}_1 \cdot \boldsymbol{A}_3) - \boldsymbol{A}_3(\boldsymbol{A}_1 \cdot \boldsymbol{A}_2),$$

has been used, where $\boldsymbol{A}_1$, $\boldsymbol{A}_2$, and $\boldsymbol{A}_3$ are arbitrary vectors.

Let us now consider the time derivative of H, say dH/dt, assuming that the volume in consideration moves with fluid motions. Then, we have

$$\frac{dH}{dt} = \iiint \frac{\partial h_0}{\partial t} dV + \iint h_0 \boldsymbol{v} \cdot d\boldsymbol{S} = \iiint \left[\frac{\partial h_0}{\partial t} + \mathrm{div}(h_0 \boldsymbol{v}) \right] dV, \tag{11.33}$$

where $d\boldsymbol{S}$ is a vector specifying surface element, directed to the normal direction to the surface. Substituting Eqs. (11.32) into the above equation, we have

$$\frac{dH}{dt} = \iint \left[(\boldsymbol{A} \cdot \boldsymbol{v}) - c\tilde{\phi} \right] \boldsymbol{B} \cdot d\boldsymbol{S}. \tag{11.34}$$

This equation shows that magnetic helicity, H, defined by Eq. (11.26) is conserved in the isolated systems in the sense that magnetic fields are confined in the system as $\boldsymbol{B} \cdot \boldsymbol{n} = 0$, where $\boldsymbol{n}$ is the unit vector normal to the surface of the system.

In the above arguments, magnetic helicity was found to be gauge invariant and a conserved quantity in isolated systems. In practical applications in solar physics and in dynamo problems, however, it becomes necessary to treat open systems. Hence, a concept of *relative helicity* has been introduced (Berger and Field 1984). By knowing velocity and magnetic fields on the boundary surface of systems we can know the time change of relative helicity, i.e., injection of magnetic helicity to the system in consideration.

It is noted here that corresponding to magnetic helicity, there are concepts of fluid helicity (Sect. 2.2.1) and of cross helicity. These concepts are often used in studies of dynamo processes.

11.5.2 Cross Helicity and Its Conservation

The cross helicity H_c is defined by

$$H_\mathrm{c} \equiv \iiint (\boldsymbol{v} \cdot \boldsymbol{B}) dV. \tag{11.35}$$

We consider time variation of H_c under the condition that the surface of the volume flows with fluid motion. Then, we have

$$\begin{aligned} \frac{dH_\mathrm{c}}{dt} &= \iiint \frac{\partial}{\partial t}(\boldsymbol{v} \cdot \boldsymbol{B}) dV + \iint (\boldsymbol{v} \cdot \boldsymbol{B}) \boldsymbol{v} \cdot d\boldsymbol{S} \\ &= \iiint \left\{ \frac{\partial}{\partial t}(\boldsymbol{v} \cdot \boldsymbol{B}) + \mathrm{div}[(\boldsymbol{v} \cdot \boldsymbol{B})\boldsymbol{v}] \right\} dV, \end{aligned} \tag{11.36}$$

where $d\boldsymbol{S}$ is the vector specifying surface element, and in deriving the second equality the Gauss theorem has been used.

In barotropic inviscid flows the equation of motion can be written as (e.g., Sect. 2.2.4)

$$\frac{\partial \boldsymbol{v}}{\partial t} + \frac{1}{2}\nabla v^2 - \boldsymbol{v} \times \mathrm{curl}\boldsymbol{v} = -\frac{\gamma}{\gamma - 1}\nabla\left(\frac{p}{\rho}\right) - \nabla\psi + \frac{1}{\rho c}\boldsymbol{j} \times \boldsymbol{B}, \tag{11.37}$$

where ψ is the gravitational potential. Substituting this equation and the induction equation, $\partial \boldsymbol{B}/\partial t = \mathrm{curl}(\boldsymbol{v} \times \boldsymbol{B})$, into the above equation, we have[7]

$$\begin{aligned}\frac{dH_{\mathrm{c}}}{dt} &= -\iiint \mathrm{div}\left[\left(\frac{1}{2}v^2 + \frac{\gamma}{\gamma - 1}\frac{p}{\rho} + \psi\right)\boldsymbol{B} + \boldsymbol{v} \times (\boldsymbol{v} \times \boldsymbol{B}) - (\boldsymbol{v} \cdot \boldsymbol{B})\boldsymbol{v}\right] dV \\ &= -\iint \left(-\frac{1}{2}v^2 + \frac{\gamma}{\gamma - 1}\frac{p}{\rho} + \psi\right)\boldsymbol{B} \cdot d\boldsymbol{S}, \end{aligned} \tag{11.38}$$

where $\boldsymbol{v} \times (\boldsymbol{v} \times \boldsymbol{B}) = \boldsymbol{v}(\boldsymbol{v} \cdot \boldsymbol{B}) - \boldsymbol{B}(\boldsymbol{v} \cdot \boldsymbol{v})$ has been used. Equation (11.38) shows that the cross helicity is conserved, if the system is isolated in the sense that $\boldsymbol{B} \cdot d\boldsymbol{S} = 0$.

11.6 Magnetic Energy

The energy density of magnetic fields can be regarded to be $B^2/8\pi$. This can be shown by the following considerations. Let us consider a fixed volume V in magnetized gases and introduce W_{B} defined by

$$W_{\mathrm{B}} \equiv \iiint \frac{1}{8\pi} B^2 dV. \tag{11.39}$$

By using Eq. (11.3), we can write the time variation of W_{B} in the form

$$\frac{\partial W_{\mathrm{B}}}{\partial t} = \frac{1}{4\pi}\iiint \boldsymbol{B} \cdot \left[\mathrm{curl}(\boldsymbol{v} \times \boldsymbol{B}) - \mathrm{curl}\left(\frac{c}{\sigma}\boldsymbol{j}\right)\right] dV. \tag{11.40}$$

[7] There are formula of vector analyses for arbitrary vectors $\boldsymbol{A}$, $\boldsymbol{B}$, and $\boldsymbol{C}$:

$$\mathrm{div}(\boldsymbol{A} \times \boldsymbol{B}) = \boldsymbol{B} \cdot \mathrm{curl}\, \boldsymbol{A} - \boldsymbol{A} \cdot \mathrm{curl}\, \boldsymbol{B},$$

and

$$\boldsymbol{A} \cdot (\boldsymbol{B} \times \boldsymbol{C}) = -\boldsymbol{C} \cdot (\boldsymbol{B} \times \boldsymbol{A}).$$

If formulae of vector analyses are used, we have

$$\boldsymbol{B}\cdot\mathrm{curl}\left(\frac{\boldsymbol{j}}{\sigma}\right) = \mathrm{div}\left(\frac{\boldsymbol{j}}{\sigma}\times\boldsymbol{B}\right)+\frac{1}{\sigma}\boldsymbol{j}\cdot\mathrm{curl}\boldsymbol{B}, \tag{11.41}$$

$$\boldsymbol{B}\cdot\mathrm{curl}(\boldsymbol{v}\times\boldsymbol{B}) = \mathrm{div}[(\boldsymbol{v}\times\boldsymbol{B})\times\boldsymbol{B}]-\boldsymbol{v}\cdot(\mathrm{curl}\boldsymbol{B}\times\boldsymbol{B}). \tag{11.42}$$

Hence, by using the above equations, the Gauss theorem, and Eq. (11.1), we can rearrange Eq. (11.40) in the form:

$$\frac{\partial W_{\mathrm{B}}}{\partial t} = -\iiint \boldsymbol{v}\cdot\left(\frac{1}{c}\boldsymbol{j}\times\boldsymbol{B}\right)dV-\iiint\frac{j^2}{\sigma}dV-\frac{c}{4\pi}\iint(\boldsymbol{E}\times\boldsymbol{B})\cdot d\boldsymbol{S}, \tag{11.43}$$

where $\boldsymbol{j}=\sigma(\boldsymbol{E}+\boldsymbol{v}\times\boldsymbol{B}/c)$ has been used, and $d\boldsymbol{S}$ represents a surface element of the volume V.

The meaning of Eq. (11.43) is examined. The first term on the right-hand side of Eq. (11.43) is the rate of work done to gases by the Lorentz force. This is obvious, since from the equation of motion of gases we see that the force acting on gases by magnetic fields is $\boldsymbol{j}\times\boldsymbol{B}/c$. The second term on the right-hand side of Eq. (11.43) is the rate of Joule loss. From the above consideration we can regard W_{B} as the energy of magnetic fields in the volume in consideration and $(c/4\pi)(\boldsymbol{E}\times\boldsymbol{B})_n$ as the energy flux transported outside from the surface of the volume by magnetic fields, where the subscript n represents the direction normal to the surface. The vector, $(c/4\pi)(\boldsymbol{E}\times\boldsymbol{B})$, is called *Poynting vector*. Equation (11.43) holds for arbitrary volume, and thus we can write it in a differential form as

$$\frac{\partial}{\partial t}\left(\frac{B^2}{8\pi}\right) = -\boldsymbol{v}\cdot(\boldsymbol{j}\times\boldsymbol{B}/c)-\frac{j^2}{\sigma}-\mathrm{div}\left(\frac{c}{4\pi}\boldsymbol{E}\times\boldsymbol{B}\right). \tag{11.44}$$

Let us here consider the whole system consisting of gas and magnetic field. If the force acting on the gas is the Lorentz force alone, the first term on the right-hand side of Eq. (2.13) is $\boldsymbol{v}\cdot(\boldsymbol{j}\times\boldsymbol{B}/c)$. The second term is j^2/σ, since the gas is heated by this rate. Considering them, we summarize Eqs. (2.13) and (11.44) to give

$$\frac{\partial}{\partial t}\left[\rho\left(U+\frac{1}{2}v^2\right)+\frac{B^2}{8\pi}\right]+\mathrm{div}\left[\rho\left(H+\frac{1}{2}v^2\right)\boldsymbol{v}+\frac{c}{4\pi}\boldsymbol{E}\times\boldsymbol{B}\right]=0. \tag{11.45}$$

This is the expected form of conservation of whole energy.

Finally, a comment is added here. In the above treatment, the displacement current has been neglected because the MHD approximation has been adopted. Hence, in terms of energy density, $\rho(U+v^2/2)+B^2/8\pi$, in Eq. (11.45), energy density of electric fields does not appear. If the displacement current is taken into account in the first equation of Eqs. (11.1), $\partial(E^2/8\pi)/\partial t$ is added on energy density in Eq. (11.45).

References

Berger, M.A., Field, G.B.: J. Fluid Mech. **147**, 133 (1984)

Moffatt, H.K.: Magnetic Field Generation in Electrically Conducting Fluids. Cambridge University Press, Cambridge (1978), p. 353

Wiegelmann, T., Sakurai, T.: Living Rev. Solar Phys. **9**(1), 5 (2012)

Chapter 12
Astrophysical MHD Flows

In Chap. 11 angular momentum transport by magnetic fields was examined. The Sun is really believed to be decelerated by magnetic torque due to the magnetized solar wind to its present angular momentum. Similar processes may govern the contraction of protostellar gas clouds. In accretion disks, angular momentum transport from their inner part to outer part is usually done by magnetic turbulent viscosity. Furthermore, in the case of binary systems the angular momentum of stellar rotations will be extracted to orbital motions. In the case of a single black-hole system, the angular momentum extraction from disks by magnetic fields will be of important. In such systems angular momentum may be eventually carried off in a jet moving perpendicular to the disk. Considering such situations, we present the essence of the solar MHD winds, and algebraic relations describing physical and dynamical quantities in steady axisymmetric MHD flows, as important bases for studying MHD flows in various astrophysical objects.

12.1 Equatorial Flows in Axisymmetric Magnetic Stellar Winds

Along the main sequence on the HR diagram, the rotational velocity of stars decreases abruptly below the F-type stars. This is considered to be due to angular momentum drawing by magnetic fields (Schatzman 1962), because in late type stars below F-type ones there are surface convection zones where magnetic fields are generated by dynamo processes. As an example of steady hydromagnetic flows which extract angular momentum from central stars by magnetic torque, we consider here the hydromagnetic stellar wind examined first by Weber and Davis (1967). This is an extension of Parker's solar wind considered in Chap. 3 to cases with magnetic fields.

© Springer Nature Singapore Pte Ltd. 2020, corrected publication 2023

S. Kato, J. Fukue, *Fundamentals of Astrophysical Fluid Dynamics*,
Astronomy and Astrophysics Library,
https://doi.org/10.1007/978-981-15-4174-2_12

Stars which we consider here are steady and axisymmetric, and plane-symmetric with respect to the equatorial plane. Our attention is concentrated only on equatorial plane. Thus, in spherical coordinates (r, φ, θ), all physical quantities of winds are taken to be functions only of r. Due to this, many flow equations are ordinary differential equations with respect to r alone, and can be integrated. First, we summarize integrated forms of such equations in the following subsections.

12.1.1 Algebraic Relations Among Physical Quantities

Six r-dependent physical or dynamical variables are necessary to describe the structure of winds. They are v_r, v_φ, B_r, B_φ, ρ, and p, where $v_r(r)$ and $v_\varphi(r)$ are, respectively, the r- and φ-components of the flow, $B_r(r)$ and $B_\varphi(r)$, respectively, r- and φ- components of magnetic fields, and $\rho(r)$ and $p(r)$, respectively, density and gas pressure.

Many differential equations describing the relations among the above-mentioned quantities can be easily integrated. First, the equation of continuity gives that $\rho v_r r^2$ is constant. By writing the constant as f, we have

$$\rho v_r r^2 = f. \tag{12.1}$$

The equation of div $\boldsymbol{B} = 0$ can be integrated to give

$$r^2 B_r = \Phi, \tag{12.2}$$

where Φ is a constant. The φ-component of the induction equation is $[\mathrm{curl}(\boldsymbol{v} \times \boldsymbol{B})]_\varphi = d[r(v_r B_\varphi - v_\varphi B_r)]/rdr = 0$, which leads to $r(v_r B_\varphi - v_\varphi B_r) = \mathrm{const.}$ This constant is written here as $-\Omega r^2 B_r$ by introducing a constant, Ω. It is noted that $r^2 B_r$ is a constant, as mentioned above. Then, we have

$$(v_\varphi - r\Omega)B_r = v_r B_\varphi. \tag{12.3}$$

This equation shows that the constant Ω can be interpreted as the angular velocity of rotation of the central star.

The φ-component of equation of motion can be written as

$$\rho \frac{v_r}{r}\frac{d}{dr}(r v_\varphi) = \frac{1}{4\pi}[\mathrm{curl}\, \boldsymbol{B} \times \boldsymbol{B}]_\varphi = \frac{B_r}{4\pi r}\frac{d}{dr}(r B_\varphi). \tag{12.4}$$

Since

$$\frac{B_r}{4\pi \rho v_r} = \frac{B_r r^2}{4\pi \rho v_r r^2} = \mathrm{const.}, \tag{12.5}$$

equation (12.4) can be integrated to give

$$r\left(v_\varphi - \frac{B_r B_\varphi}{4\pi\rho v_r}\right) = L, \tag{12.6}$$

where L is a constant, which is specified later. In addition we introduce the polytropic relation:

$$p = K\rho^\gamma, \tag{12.7}$$

where K is a constant and γ is the polytropic index.

We have obtained five algebraic relations among six quantities with constants f, Φ, Ω, L, and K. The remaining differential equation is the r-component of the equation of motion, which is

$$\rho v_r \frac{dv_r}{dr} = -\frac{dp}{dr} - \rho\frac{GM}{r^2} + \frac{1}{4\pi}[\mathrm{curl}\,\boldsymbol{B} \times \boldsymbol{B}]_r + \rho\frac{v_\varphi^2}{r}, \tag{12.8}$$

where M is the mass of the star. This equation cannot be integrated directly, but as an alternative equation, we have the equation of total energy conservation [see Eq. (11.45)], which is integrated easily to lead to

$$\rho v_r r^2\left[\frac{1}{2}(v_r^2 + v_\varphi^2) + H - \frac{GM}{r}\right] + \frac{c}{4\pi}r^2(\boldsymbol{E} \times \boldsymbol{B})_r = \mathrm{const}, \tag{12.9}$$

where H is the enthalpy defined by $H = [\gamma/(\gamma-1)]p/\rho$, and the last term on the left-hand side is the r-component of the Poynting vector. The gravitational potential energy of the central star, $-GM/r$, has been added in Eq. (12.9), although it is neglected in Eq. (11.45).

Since $\boldsymbol{E} = -(1/c)\boldsymbol{v} \times \boldsymbol{B}$, we have

$$\begin{aligned}\frac{c}{4\pi}r^2(\boldsymbol{E} \times \boldsymbol{B})_r &= -\frac{1}{4\pi}r^2[(\boldsymbol{v} \times \boldsymbol{B}) \times \boldsymbol{B}]_r = -\frac{1}{4\pi}r^2[\boldsymbol{B}(\boldsymbol{v} \cdot \boldsymbol{B}) - \boldsymbol{v}(\boldsymbol{B} \cdot \boldsymbol{B})]_r \\ &= -\frac{1}{4\pi}r^2 B_\varphi(B_r v_\varphi - v_r B_\varphi) = -\frac{1}{4\pi}r B_r B_\varphi \Omega r^2. \end{aligned} \tag{12.10}$$

In deriving the final equality, Eq. (12.3) has been used. This final result is related to the fact that the angular momentum flux through a unit surface is $-rB_rB_\varphi/4\pi$ [see Eq. (11.24) in Sect. 11.3]. The energy flux due to magnetic fields is thus the angular momentum flux times Ω. The final expression in Eq. (12.10) is further rewritten by using Eq. (12.6) as

$$-\rho v_r(rv_\varphi - L)\Omega r^2. \tag{12.11}$$

Then, using $\rho v_r r^2 = \text{const.}$, from Eq. (12.9) we can derive[1]

$$\frac{1}{2}(v_r^2 + v_\varphi^2) + H - \frac{GM}{r} - (r v_\varphi - L)\Omega = \text{const.}, \tag{12.12}$$

which is further rewritten in the form:

$$\frac{1}{2}v_r^2 + \frac{1}{2}(v_\varphi - \Omega r)^2 + H - \frac{GM}{r} - \frac{1}{2}\Omega^2 r^2 = E, \tag{12.13}$$

where E is a constant. Equation (12.13) has the form of the Bernoulli constant in the rotating frame with potential $-GM/r - r^2\Omega^2/2$. Equation (12.13) was derived by Sakurai (1985), and has been used to examine the flow structure on the ρ–r diagram.

We have obtained six algebraic relations among the six dependent variables describing the flows. These relations are Eqs. (12.1)–(12.3), (12.6), (12.7), and (12.13). By solving these equations as functions of r, we can determine the flow structure. At this stage, however, we should mention that there is an important restriction on the flows. Let us eliminate B_φ from Eqs. (12.3) and (12.6) to express v_φ in terms of r, v_r, ρ, and B_r. Then, we have

$$(v_r^2 - c_{\mathrm{Ar}}^2)v_\varphi = \Omega r\left(\frac{L}{\Omega r^2}v_r^2 - c_{\mathrm{Ar}}^2\right), \tag{12.14}$$

where c_{Ar} is the radial Alfvénic speed defined by

$$c_{\mathrm{Ar}}^2(r) \equiv \frac{B_r^2}{4\pi\rho}. \tag{12.15}$$

Equation (12.14) shows that at the radius, r_{A}, where $v_r^2 = c_{\mathrm{Ar}}^2$ (defined as the *Alfvén point*), we must have

$$L = r_{\mathrm{A}}^2\Omega. \tag{12.16}$$

This means that all the steady flows which go far outside of the star must arrange themselves so that the radius of r_{A} is at the radius of $(L/\Omega)^{1/2}$ (see Figures 1 and 2 of Weber and Davis (1967).

[1] If the last term on the right-hand side of Eq. (12.12) is neglected (the term comes from the Lorentz force), Eq. (12.12) is the same as the Bernoulli equation (see Sect. 2.2).

12.1.2 Flow Behaviors on v_r-r Diagram

Since algebraic relations among physical and dynamical quantities describing wind flows have been obtained, we can study the wind structure by using these relations. As in the case of hydrodynamical winds studied in Sect. 3.1, however, there are critical points where the flow must pass there. To see this and the topology of the flow around the critical points, it is instructive to return to the differential equation describing the flows on the v_r–r diagram, although all differential equations are already integrated. This situation is the same as in the case of hydrodynamical winds (accretions) studied in Sect. 3.1. To do so, it is convenient to start from Eq. (12.8).

As a preparation, the Lorentz force in the radial direction is expressed as

$$\frac{1}{c}(\boldsymbol{j}\times\boldsymbol{B})_r = \frac{1}{4\pi}(\mathrm{curl}\,\boldsymbol{B}\times\boldsymbol{B})_r = -\frac{1}{4\pi}B_\varphi^2\frac{d}{dr}\ln(rB_\varphi). \tag{12.17}$$

Eliminating v_φ from Eqs. (12.3) and (12.6) with $L = \Omega r_{\rm A}^2$, we have

$$rB_\varphi = \Omega(r_{\rm A}^2 - r^2)\frac{4\pi\rho v_r B_r}{4\pi\rho v_r^2 - B_r^2} = \Omega(r_{\rm A}^2 - r^2)\frac{4\pi f\rho\Phi}{4\pi f^2 - \rho\Phi^2}. \tag{12.18}$$

This relation gives

$$\begin{aligned}\frac{d}{dr}\ln(rB_\varphi) &= -\frac{2r}{r_{\rm A}^2 - r^2} + \frac{4\pi f^2}{4\pi f^2 - \rho\Phi^2}\frac{d}{dr}\ln\rho \\ &= -\frac{2r}{r_{\rm A}^2 - r^2} - \frac{v_r^2}{v_r^2 - c_{\rm Ar}^2}\left(\frac{2}{r} + \frac{1}{v_r}\frac{dv_r}{dr}\right).\end{aligned} \tag{12.19}$$

If Eq. (12.18) is further used to eliminate $r_{\rm A}$ from Eq. (12.19), Eq. (12.19) is reduced to

$$\frac{d}{dr}\ln(rB_\varphi) = -\frac{v_r}{v_r^2 - c_{\rm Ar}^2}\left(\frac{2v_\varphi}{r}\frac{B_r}{B_\varphi} + \frac{dv_r}{dr}\right). \tag{12.20}$$

Furthermore, for the polytropic gases we have

$$\frac{1}{\rho}\frac{dp}{dr} = -c_{\rm s}^2\left(\frac{2}{r} + \frac{1}{v_r}\frac{dv_r}{dr}\right). \tag{12.21}$$

Substituting the above expressions for the Lorentz force and the pressure force into Eq. (12.8), we have, after some manipulations,

$$\frac{r}{v_r}\frac{dv_r}{dr} = \frac{N}{D}, \tag{12.22}$$

where

$$D = v_r^4 - (c_{\mathrm{Ar}}^2 + c_{\mathrm{A}\varphi}^2 + c_{\mathrm{s}}^2)v_r^2 + c_{\mathrm{Ar}}^2 c_{\mathrm{s}}^2 \equiv (v_r^2 - c_{\mathrm{Af}}^2)(v_r^2 - c_{\mathrm{As}}^2), \tag{12.23}$$

and

$$N = 2v_r v_\varphi c_{\mathrm{Ar}} c_{\mathrm{A}\varphi} + (v_r^2 - c_{\mathrm{Ar}}^2)\left(v_\varphi^2 + 2c_{\mathrm{s}}^2 - \frac{GM}{r}\right), \tag{12.24}$$

c_{Af} and c_{As} being, respectively, the speeds of the fast and slow modes.[2] Furthermore, $c_{\mathrm{A}\varphi}$ is the tangential Alfvénic speed defined by

$$c_{\mathrm{A}\varphi}^2 \equiv \frac{B_\varphi^2}{4\pi\rho}. \tag{12.25}$$

The expression for D, i.e., Eq. (12.23), shows that at the radii where $v_r^2 = c_{\mathrm{Af}}^2$ and $v_r^2 = c_{\mathrm{As}}^2$ the denominator of the differential equation of v_r [Eq. (12.22)] becomes zero. This means that for the wind to go far outside through the radii where $v_r^2 = c_{\mathrm{As}}^2$ and $v_r^2 = c_{\mathrm{Af}}^2$, the numerator N also needs to be zero at these radii. That is, these radii are critical points (see Chap. 3 for critical points). It is noted that in the case of hydrodynamical stellar winds (Parker 1958), the critical point was the sonic point only (see Sect. 3.1), where $v_r^2 = c_{\mathrm{s}}^2$. In the present hydromagnetic case we have critical points at $v_r^2 = c_{\mathrm{As}}^2$ and $v_r^2 = c_{\mathrm{Af}}^2$. Detailed examinations show that as in the case of the hydrodynamical wind, the topologies around these points are X-type critical points. It should be noticed, however, that we have another type critical point at the Alfvén radius. This critical point does not appear in Eq. (12.22), since the condition of passing the Alfvén radius [i.e., Eq. (12.16)] is already taken into account in deriving Eq. (12.22).

In summary, for a wind from the stellar surface to go far outside, it must adjust itself so that it can pass three critical points at $v_r^2 = c_{\mathrm{As}}^2$, $v_r^2 = c_{\mathrm{Ar}}^2$, and $v_r^2 = c_{\mathrm{Af}}^2$. Examples of the topological structures of flows in the v_r–r diagram are shown in Figures 1 and 2 of Weber and Davis (1967). Their Figure 1 shows the topological structure of flows around the radii, where $v_r^2 = c_{\mathrm{As}}^2$ and $v_r^2 = c_{\mathrm{Ar}}^2$. Their Figure 2 shows that around the radii where $v_r^2 = c_{\mathrm{Ar}}^2$ and $v_r^2 = c_{\mathrm{Af}}^2$. It is noted that the critical point at $v_r^2 = c_{\mathrm{Ar}}^2$ (i.e., the Alfvén point) is not a X-type critical point but a higher order critical point.

By this model of solar winds, Weber and Davis (1967) evaluated the timescale of the spin down of the Sun to be $\sim 7 \cdot 10^9$ year provided the angular momentum is mixed uniformly within the Sun.

[2] See the next chapter (Chap. 13) for the fast and slow modes of MHD waves.

12.1.3 Flow Behaviors on ρ-r Diagram

In the above subsection the wind flows are studied on the v_r–r diagram. The studies on the ρ–r diagram are, however, more convenient (Sakurai 1985).

The energy E is a constant in a steady flow, but now we regard it as a function of ρ and r, although other integration constants are really fixed. Then, from the expression for E given by Eq. (12.13), we have

$$\begin{aligned}\rho\frac{\partial E(r,\rho)}{\partial\rho} &= v_r^2\frac{\partial \ln v_r}{\partial \ln\rho}+(v_\varphi-r\Omega)^2\frac{\partial\ln(v_\varphi-r\Omega)}{\partial\ln\rho}+c_s^2\\ &= -v_r^2+\left(\frac{B_\varphi}{B_r}v_r\right)^2\frac{\partial}{\partial\ln\rho}\ln\left(\frac{B_\varphi}{B_r}v_r\right)+c_s^2,\\ &= -v_r^2+\left(\frac{B_\varphi}{B_r}v_r\right)^2\left(\frac{\partial\ln B_\varphi}{\partial\ln\rho}-1\right)+c_s^2.\end{aligned}\tag{12.26}$$

where Eqs. (12.1), (12.3), and $(B_\varphi/B_r)v_r = (B_\varphi/B_r r^2)(r^2 v_r) \propto B_\varphi/\rho$ have been used. Since B_φ is expressed as Eq. (12.18), we have after some manipulations

$$\frac{\partial\ln B_\varphi}{\partial\ln\rho}-1=\frac{\rho\Phi^2}{4\pi f^2-\rho\Phi^2}=-\frac{1}{1-v_r^2/c_{Ar}^2}.\tag{12.27}$$

Substituting this relation into Eq. (12.26), we have finally (Sakurai 1985)

$$\rho\frac{\partial E(r,\rho)}{\partial\rho}=-\frac{(v_r^2-c_{\rm Af}^2)(v_r^2-c_{\rm As}^2)}{v_r^2-c_{\rm Ar}^2},\tag{12.28}$$

where $c_{\rm Af}^2$ and $c_{\rm As}^2$ are defined by Eq. (12.23).

Let us draw the curve of $E(r,\rho)$ = const. on the r-ρ plane. Then, Eq. (12.28) shows that the critical points (the fast and slow Alfvénic points) arise at the places where $\partial E(\rho,r)/\partial\rho=0$.

Along the steady flows, $E(\rho,r)$ = const, since E is an integration constant, i.e., $dE(\rho(r),r)/dr=(\partial E/\partial\rho)(d\rho(r)/dr)+\partial E/\partial r=0$. This means that the critical points on the ρ-r plane are also at positions where $\partial E(r,\rho)/\partial r = 0$. That is, the critical points are positions on the r-ρ plane where the curve of $\partial E/\partial\rho = 0$ and that of $\partial E/\partial r = 0$ cross. In other words, let us consider a three-dimensional figure showing the plane of $E(\rho,r)$ above the r-ρ plane. The critical points are then the places where the plane of $E = E(\rho,r)$ is locally flat, i.e.,

$$\frac{\partial E}{\partial\rho}=\frac{\partial E}{\partial r}=0.\tag{12.29}$$

Finally, it is noted that Weber and Davis's model of stellar wind is essentially one dimensional. The poloidal magnetic field is assumed to be purely radial. The force balance across the poloidal magnetic fields is not considered. Relaxing this assumption has been made by Sakurai (1985). He showed that the magnetic fields of solar wind are deflected polewards and are collimated in the direction of the rotation axis at large distances.

12.2 Two-Dimensional MHD Flows from Axisymmetric Objects

When configuration of magnetic fields in the two-dimensional space is not known and should be determined by calculations, it is convenient to adopt an alternative method. That is, we examine what quantities are constant along a magnetic field line in general in order to reduce the order of differential equations to be solved. The final process is then to determine the configuration of magnetic field lines by solving a differential equation involving the constants. Here, following Mestel (1961, 1968), we summarize such algebraic relations along a magnetic field. Then, we briefly mention the work describing the field lines in winds from accretion disks by Blandford and Payne (1982).

12.2.1 Integration along a Magnetic Field Line

First, following Mestel (1961, 1968) we derive algebraic relations among physical and dynamical quantities along a magnetic field line. The flows are assumed to be steady and axisymmetric with no dissipative processes. The cylindrical coordinates (r, φ, z) are adopted when coordinates are necessary. The magnetic and velocity fields are written as

$$\boldsymbol{B} = \boldsymbol{B}_\mathrm{p} + \boldsymbol{B}_\mathrm{t}, \tag{12.30}$$

$$\boldsymbol{v} = \boldsymbol{v}_\mathrm{p} + \boldsymbol{v}_\mathrm{t} = \boldsymbol{v}_\mathrm{p} + r\Omega\boldsymbol{t}, \tag{12.31}$$

where the subscripts p and t refer, respectively, to poloidal and toroidal components, and $\boldsymbol{t}$ is the unit vector in the toroidal direction. The angular velocity of rotation is denoted by $\Omega(r, z)$, i.e., $v_\varphi = r\Omega$. It should be noted here that in Sect. 12.1, Ω is used to represent a constant (angular velocity at the solar surface), but here Ω is a variable representing angular velocity of rotation at a point, i.e., $\Omega = \Omega(r, z)$.

The steady induction equation is decomposed into poloidal and toroidal components, respectively, as

$$\mathrm{curl}\,(\boldsymbol{v}_\mathrm{p} \times \boldsymbol{B}_\mathrm{p}) = 0, \tag{12.32}$$

$$\mathrm{curl}\,(\boldsymbol{v}_\mathrm{t} \times \boldsymbol{B}_\mathrm{p} + \boldsymbol{v}_\mathrm{p} \times \boldsymbol{B}_\mathrm{t}) = 0. \tag{12.33}$$

Equation (12.32) implies that $\boldsymbol{v}_\mathrm{p} \times \boldsymbol{B}_\mathrm{p}$ must be the gradient of a potential, say ϕ, as $\boldsymbol{v}_\mathrm{p} \times \boldsymbol{B}_\mathrm{p} = \mathrm{grad}\,\phi$. However, $\boldsymbol{v}_\mathrm{p} \times \boldsymbol{B}_\mathrm{p}$ is toroidal, leading to $\mathrm{grad}\,\phi = 0$. That is, we have

$$\boldsymbol{v}_\mathrm{p} = \kappa^* \boldsymbol{B}_\mathrm{p}, \tag{12.34}$$

where κ^* is an arbitrary scalar quantity (not a constant). The left-hand side of Eq. (12.33) has only a toroidal component, which is

$$\frac{\partial}{\partial z}(v_\varphi B_z - v_z B_\varphi) - \frac{\partial}{\partial r}(-v_\varphi B_r + v_r B_\varphi). \tag{12.35}$$

By using div $\boldsymbol{B} = 0$ and Eq. (12.34), we can reduce Eq. (12.33) to

$$(\boldsymbol{B} \cdot \nabla)\left(\Omega - \frac{\kappa^* B_\varphi}{r}\right) = 0. \tag{12.36}$$

This leads to

$$\Omega - \frac{\kappa^* B_\varphi}{r} = \omega, \tag{12.37}$$

where ω is constant along each magnetic line. In the special case of $\kappa^* = 0$, Ω is constant along a magnetic field line, which is *Ferraro's isorotation law* (Ferraro 1937). Combination of Eqs. (12.34) and (12.37) leads to

$$\boldsymbol{v} = \kappa^* \boldsymbol{B} + r\omega \boldsymbol{t}. \tag{12.38}$$

The equation of continuity, $\mathrm{div}(\rho \boldsymbol{v}) = 0$, gives $\mathrm{div}(\rho\kappa^* \boldsymbol{B}) = (\boldsymbol{B} \cdot \nabla)(\rho\kappa^*) = 0$, which leads to

$$\rho\kappa^* = \frac{k}{4\pi}, \tag{12.39}$$

where $k/4\pi$ is constant on a field line. Equation (12.38) is written in the form, by using Eq. (12.39),

$$\boldsymbol{v} = \frac{k}{4\pi\rho}\boldsymbol{B} + r\omega \boldsymbol{t}. \tag{12.40}$$

The toroidal component of the equation of motion is now written in the form of angular momentum conservation, which is

$$(\rho \boldsymbol{v} \cdot \nabla)(r^2 \Omega) = \frac{r}{4\pi}[(\boldsymbol{B} \cdot \nabla)\boldsymbol{B}]_\varphi = \frac{1}{4\pi}(\boldsymbol{B} \cdot \nabla)(r B_\varphi). \tag{12.41}$$

Using Eq. (12.40), we reduce Eq. (12.41) to

$$(\boldsymbol{B} \cdot \nabla)\left(k r^2 \Omega - r B_\varphi\right) = 0. \tag{12.42}$$

This equation gives

$$r\left(v_\varphi - \frac{1}{k} B_\varphi\right) = L, \tag{12.43}$$

where L is constant on a field line.

Next, we consider the equation of total energy conservation, which is written in the form [see Eq. (2.16) and also Eq. (12.9)]

$$\mathrm{div}\left[\rho \boldsymbol{v}\left(\frac{1}{2}v_{\mathrm{p}}^2 + \frac{1}{2}\Omega^2 r^2 + H + \psi\right) + \frac{c}{4\pi}\boldsymbol{E} \times \boldsymbol{B}\right] = 0, \tag{12.44}$$

where H is the enthalpy and ψ is the gravitational potential. If $\mathrm{div}[(c/4\pi)\boldsymbol{E} \times \boldsymbol{B}]$ is written in the form of $(\rho \boldsymbol{v} \cdot \nabla)F$ with a scalar function F, we have

$$\frac{1}{2}v_{\mathrm{p}}^2 + \frac{1}{2}\Omega^2 r^2 + H + \psi + F = \mathrm{const.} \tag{12.45}$$

along a field line.

As derived in Sect. 12.1, the r-component of the Poynting vector is

$$\frac{c}{4\pi}(\boldsymbol{E} \times \boldsymbol{B})_r = -\frac{1}{4\pi}B_\varphi(B_r v_\varphi - v_r B_\varphi) = -\frac{1}{4\pi} r\omega B_r B_\varphi, \tag{12.46}$$

where the last expression is slightly different from that in Eq. (12.10), since the situation in consideration is different. Similarly, we have

$$\frac{c}{4\pi}(\boldsymbol{E} \times \boldsymbol{B})_z = -\frac{1}{4\pi} r\omega B_z B_\varphi, \tag{12.47}$$

Hence, we have

$$\mathrm{div}\left[\frac{c}{4\pi}(\boldsymbol{E} \times \boldsymbol{B})\right] = -\frac{1}{4\pi}\mathrm{div}(r\omega B_\varphi \boldsymbol{B}). \tag{12.48}$$

Using $\boldsymbol{B}_{\rm p} = \boldsymbol{v}_{\rm p}/\kappa^* = (4\pi/k)\rho\boldsymbol{v}_{\rm p}$, we can rewrite Eq. (12.48) as

$$\mathrm{div}\left[\frac{c}{4\pi}(\boldsymbol{E}\times\boldsymbol{B})\right] = -\frac{1}{k}\mathrm{div}(\rho\boldsymbol{v}r\omega B_\varphi) = -(\rho\boldsymbol{v}\cdot\nabla)\frac{r\omega B_\varphi}{k}. \tag{12.49}$$

Consequently, from Eq. (12.44) we have

$$\frac{1}{2}v_{\rm p}^2 + \frac{1}{2}\Omega^2 r^2 + H + \psi - \frac{r\omega B_\varphi}{k} = E, \tag{12.50}$$

where E is constant along a field line. It should be noted that this equation is an extension of the Bernoulli equation (2.23) to axisymmetric magnetized systems.

The algebraic relations obtained in this subsection are Eqs. (12.40), (12.37), (12.43), and (12.50), where k, ω, L, and E are constants along a magnetic field line.

12.2.2 MHD Winds from Accretion Disks

The remaining equations to be used to determine flow structures are a relation between density and pressure, the equation of continuity, and the equation of motion in the z-direction. As the relation between density and pressure, we adopt the polytropic relation. The spatial change of density is related to the field geometry. Hence, the important equation remained to be studied is the equation describing the z-component of equation of motion, which is

$$\rho(\boldsymbol{v}\cdot\nabla)v_z = -\frac{\partial p}{\partial z} - \rho\frac{\partial\psi}{\partial z} - \frac{1}{8\pi}\frac{\partial B^2}{\partial z} + \frac{1}{4\pi}(\boldsymbol{B}\cdot\nabla)B_z, \tag{12.51}$$

where p is the gas pressure and specified by using the polytropic relation.

Magnetic field configurations in disk winds have been examined by Blandford and Payne (1982) by solving Eqs. (12.51) with algebraic relations obtained in Sect. 12.2.1. Here, we shall be satisfied with noticing their results. Let us consider a field line which passes the disk at position, say $\boldsymbol{r} = (r_0, \phi, 0)$, and examine how the field line departs from the position. To study this, the field line is described by a function $\xi(\chi)$ with variable χ:

$$\boldsymbol{r} = [r_0\xi(\chi), \phi, r_0\chi]. \tag{12.52}$$

Our problem is to derive a differential equation describing $\xi(\chi)$ as a function of χ and solve the equation with suitable boundary conditions. Derivation of the differential equation is very complicated. Solving the equation is also troublesome, because there appear critical points as in the case of solar winds. This issue, however, was studied by Blandford and Payne (1982). Their results show that disk MHD winds are one of the important processes extracting angular momentum and energy from disks. At large distances from the disks, toroidal components of the magnetic fields become important and collimate the outflow into a pair of anti-parallel jets driven by magnetic fields in a hot magnetically dominated corona.

Numerical simulations of formation of collimated magnetic jet from accretion disks and tori have been made by many researchers. See, for example, Uchida and Shibata (1985), Shibata and Uchida (1986), Koide et al. (1999), Kato et al. (2004), Ohsuga et al. (2009).

References

Blandford, R.D., Payne, D.G.: Mon. Not. R. Astron. Soc. **199**, 883 (1982)
Ferraro, V.C.A.: Mon. Not. R. Astron. Soc. **97**, 458 (1937)
Kato, Y., Mineshige, S., Shibata, K.: Astrophys. J. **605**, 307 (2004)
Koide, S., Shibata, K., Kudoh, T.: Astrophys. J. **522**, 727 (1999)
Mestel, L.: Mon. Not. R. Astron. Soc. **122**, 473 (1961)
Mestel, L.: Mon. Not. R. Astron. Soc. **138**, 359 (1968)
Ohsuga, K., Mineshige, S., Mori, M., Kato, Y.: Publ. Astron. Soc. Jpn. **61**, 7 (2009)
Parker, E.N.: Astrophys. J. **128**, 664 (1958)
Sakurai, T.: Astron. Astrophys. **152**, 121 (1985)
Schatzman, E.: AnAp **25**, 18 (1962)
Shibata, K., Uchida, Y.: Publ. Astron. Soc. Jpn. **38**, 631 (1986)
Uchida, Y., Shibata, K.: Publ. Astron. Soc. Jpn. **37**, 515 (1985)
Weber, E.J., Davis, L.: Astrophys. J. **148**, 217 (1967)

Chapter 13
Waves and Shocks in Magnetohydrodynamical Fluids

As mentioned in Sect. 11.3, magnetic fields have tension. Hence, it will be easily supposed that when a line of force is shaked, a perturbation is transported along the line. This is really shown by Alfvén, and the wave is called the Alfvén wave. The pure Alfvén wave is a transverse wave. However, fluids are compressible gases and pressure perturbations can coexist. Hence, in general situations wavy perturbations can have both characteristics of transverse and longitudinal waves. In this chapter we describe basic characteristics of MHD waves in homogeneous MHD fluids. MHD shock waves are also described.

13.1 Alfvén Wave

First, the presence of Alfvén waves in an idealized situation is shown. The unperturbed state is assumed to be uniform with constant density, ρ_0, and constant magnetic field $\boldsymbol{B}_0$. On this medium a small amplitude adiabatic perturbation whose motion is perpendicular to the magnetic fields is superposed. We examine how the perturbation propagates along the field lines.

Let the small amplitude gaseous motion be $\boldsymbol{v}$, and the variation of magnetic fields be $\boldsymbol{b}$. The equation describing the motion is (see Chap. 10)

$$\rho_0 \frac{\partial \boldsymbol{v}}{\partial t} = -\text{grad}\left(p_1 + \frac{1}{4\pi} \boldsymbol{B}_0 \cdot \boldsymbol{b} \right) + \frac{1}{4\pi} (\boldsymbol{B}_0 \cdot \nabla) \boldsymbol{b}, \tag{13.1}$$

where p_1 is the change of the gas pressure. Since the magnetic fields are frozen to the gaseous motions, we have, from the induction equation,

$$\frac{\partial \boldsymbol{b}}{\partial t} = \text{curl}(\boldsymbol{v} \times \boldsymbol{B}_0). \tag{13.2}$$

© Springer Nature Singapore Pte Ltd. 2020, corrected publication 2023

S. Kato, J. Fukue, *Fundamentals of Astrophysical Fluid Dynamics*,
Astronomy and Astrophysics Library,
https://doi.org/10.1007/978-981-15-4174-2_13

Furthermore, from the Maxwell equations we have

$$\mathrm{div}\boldsymbol{b} = 0. \tag{13.3}$$

Assuming incompressible perturbations, we adopt

$$\mathrm{div}\boldsymbol{v} = 0. \tag{13.4}$$

The validity of assumption of incompressibility, i.e., Eq. (13.4), is confirmed in the final results.

If Eq. (13.4) is adopted, Eq. (13.2) is written as

$$\frac{\partial \boldsymbol{b}}{\partial t} = (\boldsymbol{B}_0 \cdot \nabla)\boldsymbol{v}. \tag{13.5}$$

Moreover, the divergence of Eq. (13.1) gives

$$\nabla^2\left(p_1 + \frac{1}{4\pi}\boldsymbol{B}_0 \cdot \boldsymbol{b}\right) = 0. \tag{13.6}$$

If $p_1 + \boldsymbol{B}_0 \cdot \boldsymbol{b}/4\pi = 0$ is adopted as the solution of Eqs. (13.6), Eq. (13.1) becomes

$$\frac{\partial \boldsymbol{v}}{\partial t} = \frac{1}{4\pi\rho_0}(\boldsymbol{B}_0 \cdot \nabla)\boldsymbol{b}. \tag{13.7}$$

Equations (13.5) and (13.7) then show the presence of the solution of $\boldsymbol{b}/(4\pi\rho_0)^{1/2} = \boldsymbol{v}$. This shows that in a particular perturbation of $\boldsymbol{v} \perp \boldsymbol{B}_0$, we have $\boldsymbol{b} \perp \boldsymbol{B}_0$. In this case $p_1 + \boldsymbol{B}_0 \cdot \boldsymbol{b}/4\pi = 0$ leads to $p_1 = 0$. In incompressible motions there is no density variation, and thus $p_1 = 0$. From the above considerations we find that incompressible perturbations are really possible and these motions are described by the set of Eqs. (13.5) and (13.7).

Elimination of $\boldsymbol{b}$ from Eqs. (13.5) and (13.7) gives

$$\frac{\partial^2 \boldsymbol{v}}{\partial t^2} - \frac{B_0^2}{4\pi\rho_0}\frac{\partial^2 \boldsymbol{v}}{\partial z^2} = 0, \tag{13.8}$$

where the direction of the unperturbed magnetic field is taken to be the direction of z-axis. It is noted that $\boldsymbol{v}$ in Eq. (13.8) can be changed to $\boldsymbol{b}$. Equation (13.8) is an one-dimensional wave equation with respect to $\boldsymbol{v}$, and show that perturbations propagates along $\boldsymbol{B}_0$ with velocity $(B_0^2/4\pi\rho_0)^{1/2}$. This wave is called *Alfvén waves*, and the propagation speed c_{A} is

$$c_{\mathrm{A}} \equiv \left(\frac{B_0^2}{4\pi\rho_0}\right)^{1/2}, \tag{13.9}$$

and called the *Alfvén speed*.

Table 13.1 Comparison of Alfvén waves with acoustic waves

	Restoring force for waves	Characteristics of waves
Acoustic waves	Gas pressure	Compressible waves (longitudinal waves)
Alfvén waves	Tension of magnetic fields	Incompressible waves (transverse waves)

In the above argument, we have shown that motions perpendicular to magnetic fields propagate in the direction parallel to magnetic fields with the Alfvén speed. Here we show a more direct example of propagation of twisted magnetic fields along unperturbed magnetic fields. In general, a measure of distortion of fluid motions, $\boldsymbol{v}$, is vorticity, which is $(1/2)\mathrm{curl}\,\boldsymbol{v}$. Now we write its z-component as ζ_v. Similarly, $(1/2)\mathrm{curl}\,\boldsymbol{b}$ is a measure of distortion of magnetic fields. We write its z-component as ζ_b. By taking the rotation of Eq. (13.1), we see that its z-component gives

$$\rho_0 \frac{\partial \zeta_v}{\partial t} = \frac{1}{4\pi}(\boldsymbol{B}_0 \cdot \nabla)\zeta_b. \tag{13.10}$$

Next, taking curl of Eq. (13.2), we have from its z-component by using a formula of vector analyses,

$$\frac{\partial \zeta_b}{\partial t} = (\boldsymbol{B}_0 \cdot \nabla)\zeta_v. \tag{13.11}$$

Combining the above two equations, we have

$$\frac{\partial^2 \zeta_b}{\partial t^2} - \frac{B_0^2}{4\pi\rho_0}\frac{\partial^2 \zeta_b}{\partial z^2} = 0. \tag{13.12}$$

(The same equation is obtained for ζ_v.) The Alfvén waves describing the distortion are called *torsional Alfvén waves*. As will be shown in Sect. 16.3, magneto-rotational instability (MRI) in the case of vertical magnetic fields can be regarded as an instability where this torsional wave becomes unstable by the effects of differential rotation of systems.

Some fundamental differences of the Alfvén waves from the acoustic waves are summarized in Table 13.1.

13.2 Three MHD Waves

In Sect. 13.1 we have shown the presence of Alfvén waves which are purely transversal. In compressible media, however, acoustic waves which are longitudinal exist. Hence, except for particular types of perturbations, perturbations have generally mixed characteristics of transversal and longitudinal waves. The purpose here is to examine such wave motions. As in Sect. 13.1, we consider a medium with uniform density ρ_0 and uniform magnetic fields $\boldsymbol{B}_0$. Small amplitude, adiabatic perturbations are superposed on the media. By taking into account compressibility of the media, we have equations describing small amplitude adiabatic perturbations in the forms:

$$\frac{\partial \rho_1}{\partial t} + \rho_0 \mathrm{div}\, \boldsymbol{v} = 0, \tag{13.13}$$

$$\rho_0 \frac{\partial \boldsymbol{v}}{\partial t} = -\mathrm{grad}\, p_1 + \frac{1}{4\pi} \mathrm{curl}\, \boldsymbol{b} \times \boldsymbol{B}_0, \tag{13.14}$$

$$\frac{\partial \boldsymbol{b}}{\partial t} = \mathrm{curl}(\boldsymbol{v} \times \boldsymbol{B}_0), \tag{13.15}$$

$$\mathrm{div}\, \boldsymbol{b} = 0, \tag{13.16}$$

$$p_1 = c_\mathrm{s}^2 \rho_1. \tag{13.17}$$

The subscript 1 attached to pressure p and density ρ represent perturbed quantities, while $\boldsymbol{v}$ and $\boldsymbol{b}$ are perturbed velocity and magnetic fields, respectively. The final equation is that of adiabatic perturbation and c_s is the acoustic speed.

Here, we eliminate ρ_1, p_1 and $\boldsymbol{b}$ from the above equations. First, we take the time derivative of Eq. (13.14) to eliminate p_1 and ρ_1 by using Eqs. (13.17) and (13.13), and to eliminate $\boldsymbol{b}$ by using Eq. (13.15). Then, we have

$$\frac{\partial^2 \boldsymbol{v}}{\partial t^2} - c_\mathrm{s}^2 \mathrm{grad}(\mathrm{div} \boldsymbol{v}) - c_\mathrm{A}^2 [\mathrm{curl}\, \mathrm{curl}(\boldsymbol{v} \times \boldsymbol{i}_z)] \times \boldsymbol{i}_z = 0. \tag{13.18}$$

In this equation, the direction of $\boldsymbol{B}_0$ is taken to be the z-direction, and $\boldsymbol{i}_z$ is the unit vector in the z-direction.

Let us consider a small amplitude plane wave specified by

$$\boldsymbol{v} = \boldsymbol{A} \exp[i(\omega t - \boldsymbol{k} \cdot \boldsymbol{r})], \tag{13.19}$$

where $\boldsymbol{A}$ is the amplitude of waves, and ω and $\boldsymbol{k}$ are, respectively, the frequency and wavenumber vector of the waves. Without loss of generality, as is shown in Fig. 13.1, we take the x-axis in the direction perpendicular to the plane specified by the wavenumber vector $\boldsymbol{k}$ and z-axis, and y-axis in the plane specified by the wavenumber vector $\boldsymbol{k}$ and z-axis. That is, $\boldsymbol{k} = (0, k_y, k_z)$. After taking the axes in this way, we write down Eqs. (13.18)–(13.19). Then, x-, y-, and z-components of

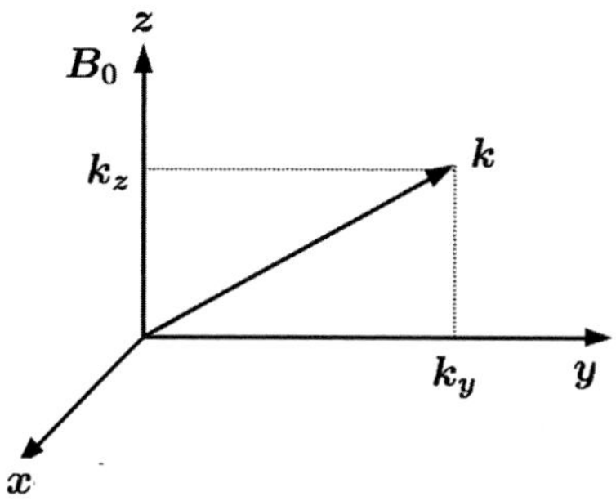

Fig. 13.1 Directions of Cartesian coordinates, magnetic fields $\boldsymbol{B}_0$, and wavenumber vector $\boldsymbol{k}$

the resulting equation are, respectively,

$$-\omega^2 A_x + c_A^2 k_z^2 A_x = 0, \tag{13.20}$$

$$-\omega^2 A_y + \left(c_s^2 k_y^2 + c_A^2 k^2\right) A_y + c_s^2 k_y k_z A_z = 0, \tag{13.21}$$

$$-\omega^2 A_z + c_s^2 k_z^2 A_z + c_s^2 k_y k_z A_y = 0, \tag{13.22}$$

where $k^2 = k_y^2 + k_z^2$.

Here, we examine how A_x, A_y, and A_z are coupled in the above equations. A_x appears only in Eq. (13.20), and this equation consists of A_x alone. This means that the oscillations in the x-direction are independent of the oscillations in the other directions. Furthermore, the wave propagates in the z-direction with speed $\pm c_A$ with motions perpendicular to the $\boldsymbol{k}$-direction ($\boldsymbol{v} \cdot \boldsymbol{k} = 0$). That is, this is incompressible Alfvén waves described in Sect. 13.1.

Equations (13.21) and (13.22) show that A_y and A_z are coupled. The condition of existence of solutions other than $A_y = A_z = 0$ is that the determinant made from the coefficients of A_y and A_z vanishes, which gives the dispersion relation:

$$\omega^4 - \omega^2 \left(c_s^2 + c_A^2\right) k^2 + c_s^2 c_A^2 k^2 k_z^2 = 0. \tag{13.23}$$

This equation is a second-order equation with respect to ω^2, which represents the presence of two wave modes. By introducing dimensionless quantity:

$$\zeta = c_A / c_s, \quad \alpha = k_z / k, \tag{13.24}$$

we can derive from dispersion relation (13.23) the square of phase velocity, ω^2/k^2, in the form:

$$\frac{\omega^2}{k^2} = c_s^2 \frac{1 + \zeta^2 \pm \left[\left(1 + \zeta^2\right)^2 - 4\alpha^2 \zeta^2\right]^{1/2}}{2}. \tag{13.25}$$

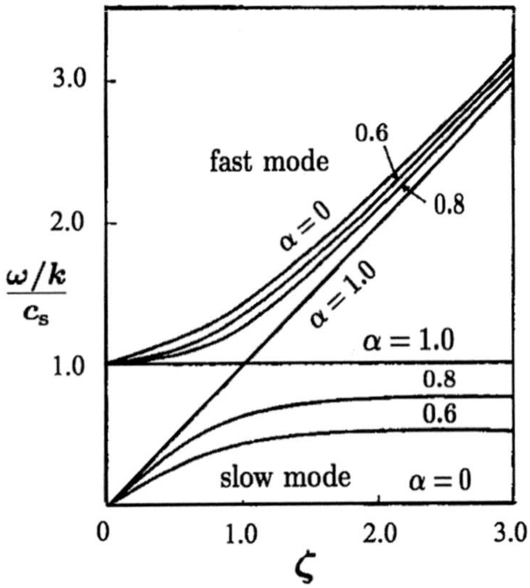

Fig. 13.2 Dispersion relation for MHD waves. There are two wave modes for a given parameter $\zeta (= c_A/c_s)$. The higher frequency mode is called the fast mode and its phase velocity is always higher than c_s, while the lower one the slow mode and its phase speed is always lower than c_s. The parameter α is k_z/k

In this equation the parameter specifying characteristics of the medium is ζ. When ζ is fixed, we have two wave modes for a given α.

Dependence of ω/k on ζ is shown in Fig. 13.2 for some values of α. In the case of $\alpha = 0$ (the case where the wave plane propagates in the y-direction, which is perpendicular to the magnetic fields), $\omega^2/k^2 = c_s^2(1+\zeta^2) = c_s^2+c_A^2$ and $\omega^2/k^2 = 0$. In the case of $\alpha = 1$, $\omega^2/k^2 = c_s^2$ and $\omega^2/k^2 = c_s^2\zeta^2 = c_A^2$. One of the two modes always has a propagation speed higher than the acoustic one and is called the *fast mode*, while the other mode always has a propagation speed lower than the acoustic one and is called the *slow mode*.

Next, we examine how the phase velocity depends on the direction of wavenumber vector for a given ζ. The results are shown in Fig. 13.3 for two cases of $\zeta < 1$ and $\zeta > 1$, separately. In this figure, the direction to a point on curves from the origin represents the direction of the wavenumber vector, $\boldsymbol{k}$, and the length to the point from the origin does the magnitude of the phase velocity. The outer curve represents the fast mode, while the inner one with a character of "8" does the slow mode.

Finally, we examine more in detail the characteristics of wave modes. From Eqs. (13.21) and (13.22) we derive the ratio of A_y to A_z as

$$\frac{A_y}{A_z} = \frac{\left(\omega^2 - c_s^2 k_z^2\right)/k^2}{c_s^2 k_y k_z/k^2}. \tag{13.26}$$

This gives for the fast mode

$$\frac{A_y}{A_z} = \begin{cases} k_y/k_z; & \zeta \ll 1 \\ \infty \quad ; & \zeta \gg 1. \end{cases}$$

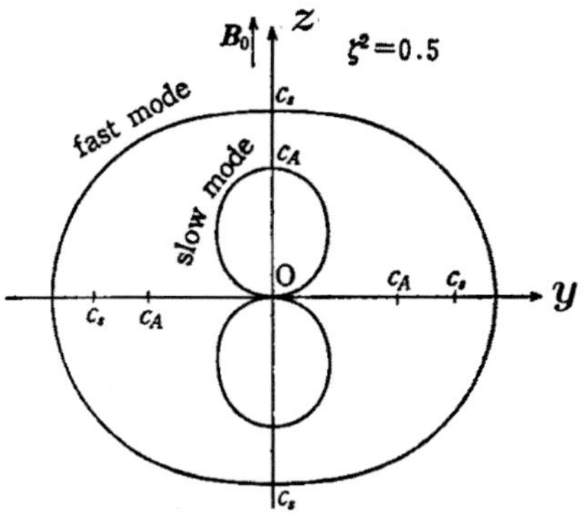

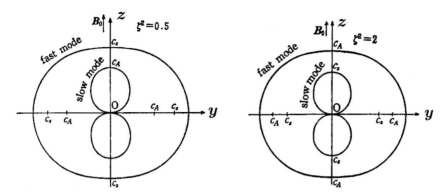

Fig. 13.3 Left panel: The relation between the direction of phase velocity and its magnitude in the case of $c_s > c_A$. The direction to a point on curves from the origin represents the direction of the wavenumber vector, $\boldsymbol{k}$, and the length to the point from the origin does the amount of phase velocity, i.e., ω/k. Right panel: The same as the left panel, except for $c_A > c_s$

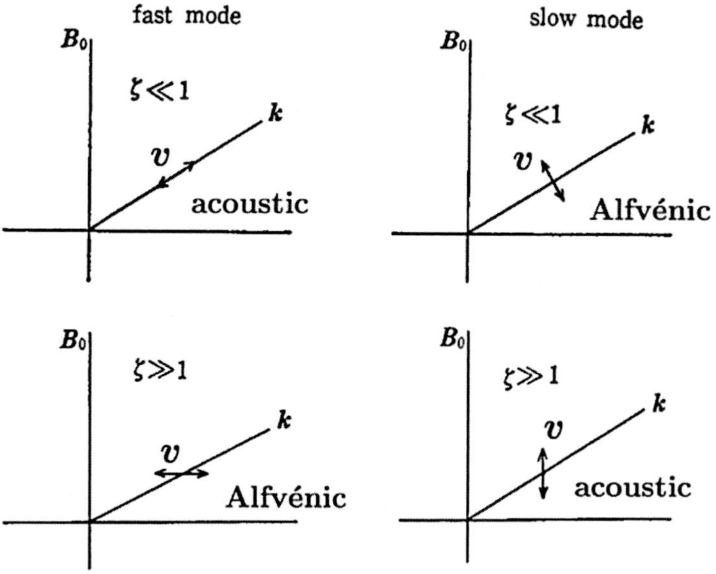

Fig. 13.4 Relation between the direction of wavenumber vector k and that of oscillatory motions $\boldsymbol{v}$ for the fast mode (left panel) and the slow mode (right panel). Two cases of weak and strong magnetic fields are shown for each mode. It should be noticed that the characteristics of oscillations are changed by the magnitude of ζ. That is, the fast mode changes its characteristic from acoustic to Alfvénic one as ζ increases, while the slow mode changes from Alfvénic to acoustic one as ζ increases

The above results for the fast mode can be derived, since $\omega^2/k^2 \sim c_s^2$ for $\zeta \ll 1$, and $\omega^2/k^2 \sim c_A^2 \gg c_s^2$ for $\zeta \gg 1$. This result is shown in the left panel of Fig. 13.4 as the relation between the wavenumber vector $\boldsymbol{k}$ and the direction of oscillations. That is, in the case of $c_s \gg c_A$ (i.e., $\zeta \ll 1$), $\boldsymbol{v} \| \boldsymbol{k}$ and the oscillations are longitudinal (like acoustic waves). Distinct from this, in the case of $c_s \ll c_A$ (i.e., $\zeta \gg 1$), $\boldsymbol{v}$ is perpendicular to $\boldsymbol{B}_0$. Since $\boldsymbol{B}_0$ is strong and the wave motions are perpendicular to this strong fields, we can say that the oscillations are like Alfvén waves. In this sense, the fast mode is acoustic-like when magnetic fields are weak, while it becomes like Alfvén waves in the case where the fields is strong.

For the slow mode we have, from Eq. (13.26),

$$\frac{A_y}{A_z} = \begin{cases} -k_z/k_y; & \zeta \ll 1 \\ 0 \quad ; & \zeta \gg 1. \end{cases}$$

This result can be understood, since in the case of $\zeta \ll 1$, we have $\omega^2/k^2 \sim 0$, while in the case of $\zeta \gg 1$ we have $\omega^2/k^2 \sim c_s^2 k_z^2/k^2$. The relation between wavenumber vector and direction of oscillations is shown in the right panel of Fig. 13.4. It is noticed that in the case of $c_s \gg c_A$ (i.e., $\zeta \ll 1$), $\boldsymbol{v} \perp \boldsymbol{k}$ and the oscillations are transverse, while in the case of $c_s \ll c_A$ (i.e., $\zeta \gg 1$), we have $\boldsymbol{v} \| \boldsymbol{B}_0$ and the gas moves so that they can avoid to cross the magnetic field lines. In this sense, the oscillations are not like Alfvénic (rather like acoustic).

As discussed above, in the case of weak fields, the fast mode is like an acoustic wave and the slow mode is Alfvénic. In the opposite case of $c_s \ll c_A$, however, the fast mode is Alfvénic and the slow mode is like acoustic. That is, the characteristics of oscillations are exchanged by change of magnitude of c_s and c_A.

13.3 MHD Shocks

As in the case of acoustic waves, MHD waves can grow to shock waves as they propagate to low-density regions. Formation of MHD shock waves and energy dissipation in shock regions are important in various active phenomena in astrophysical objects. Here, we describe the structure of steady MHD shocks.

13.3.1 Relations Among Discontinuous Quantities at Shock Front

Before focusing our attention on shocks, we consider conditions to be satisfied on discontinuous surfaces in general. For simplicity, the position of discontinuous surface is assumed to be static, i.e., there is no deceleration nor acceleration. The relative motion of fluid to the surface is denoted by $\boldsymbol{u}$. We consider connection conditions at the surface. The conditions can be derived by integrating MHD equations in the direction perpendicular to the surface in a narrow width sandwiching the surface. The limiting case of infinitesimally thin width gives the conditions.

As in the case of acoustic waves, the integration of equation of continuity gives that the normal component of $\rho \boldsymbol{u}$ to a discontinuous surface must be continuous at the surface, i.e.,

$$[\rho u_{\rm n}] = 0, \tag{13.27}$$

where ρ is the density, and $u_{\rm n}$ is the normal component of fluid velocity, $\boldsymbol{u}$, to discontinuous surface. The brackets [ ] show the difference of the quantity inside the brackets at the surface. For convenience, the mass flow through discontinuity is denoted by j:

$$j = \rho_1 u_{1{\rm n}} = \rho_2 u_{2{\rm n}}. \tag{13.28}$$

Here and hereafter, the subscripts 1 and 2 represent, respectively, the quantities in the region ahead of the discontinuity and those behind the one.

The i-component of equation of motion for MHD flows is written in the Cartesian coordinates as

$$\frac{\partial \rho u_i}{\partial t} + \frac{\partial}{\partial x_j}\left(\rho u_i u_j + p\delta_{ij} + \frac{B^2}{8\pi}\delta_{ij} - \frac{1}{4\pi}B_i B_j\right) = 0. \tag{13.29}$$

Let us now consider the normal component of this equation to a discontinuous surface, and integrate it in the direction normal to the surface. We have then in the limit of infinitely thin width

$$\left[\rho u_{\rm n}^2 + p + \frac{1}{8\pi}\boldsymbol{B}_{\rm t}^2\right] = 0, \tag{13.30}$$

where $\boldsymbol{B}_{\rm t}$ is the tangential component of $\boldsymbol{B}$ to the surface. In deriving equation (13.30)

$$[B_{\rm n}] = 0 \tag{13.31}$$

has been used, which comes from div $\boldsymbol{B} = 0$ and $B_{\rm n}$ is the normal component of $\boldsymbol{B}$. The integration of the tangential component of Eq. (13.29) to the surface gives

$$\left[\rho u_{\rm n}\boldsymbol{u}_{\rm t} - \frac{1}{4\pi}B_{\rm n}\boldsymbol{B}_{\rm t}\right] = 0, \tag{13.32}$$

where $\boldsymbol{u}_{\rm t}$ is the tangential component (vector) of $\boldsymbol{u}$.

Next, we consider the conservation of total energy, which is given in the form of Eq. (11.45) in Sect. 11.6. Furthermore, $\boldsymbol{E} = -\boldsymbol{u} \times \boldsymbol{B}/c$ is adopted, since we restrict our attention to MHD flows. Then, from the resulting equation we have

$$\left[\rho u_{\rm n}\left(\frac{1}{2}u_{\rm n}^2+\frac{1}{2}u_{\rm t}^2+\frac{\gamma}{\gamma-1}\frac{p}{\rho}\right)+\frac{1}{4\pi}\left(u_{\rm n}\boldsymbol{B}^2-(\boldsymbol{u}\cdot\boldsymbol{B})B_{\rm n}\right)\right]=0, \tag{13.33}$$

where the enthalpy H is written as $[\gamma/(\gamma-1)]p/\rho$ with the ratio of specific heats, γ.

In addition to the above relations we need the continuity of curl$\boldsymbol{E}=-\partial\boldsymbol{B}/c\partial t$ at the discontinuous surface. This gives $[\boldsymbol{E}_{\rm t}]=0$. By using $\boldsymbol{E}=-\boldsymbol{u}\times\boldsymbol{B}/c$, we can write $[\boldsymbol{E}_{\rm t}]=0$ as

$$[u_{\rm n}\boldsymbol{B}_{\rm t}-B_{\rm n}\boldsymbol{u}_{\rm t}]=0. \tag{13.34}$$

In summary, relations necessary for describing discontinuities are Eqs. (13.27), (13.30)–(13.34), and the equation of state, which is written as

$$p=\frac{1}{\gamma}\rho c_{\rm s}^2, \tag{13.35}$$

where $c_{\rm s}$ is the acoustic speed.

13.3.2 *Types of Discontinuity*

In the above subsection we have derived conditions to be satisfied at a discontinuity. Flows which satisfy these conditions are not only shocks. Discontinuities are classified into four types (e.g., see Landau and Lifshitz 1960). They are (1) contact discontinuity, (2) tangential discontinuity, (3) rotational discontinuity, and (4) shock.

13.3.2.1 Contact Discontinuity ($j=0$, $B_{1\rm n}\neq 0$)

In this case of $j=0$, we have $u_{1\rm n}=u_{2\rm n}=0$, and the fluid moves parallel to the surface of discontinuity. If $B_n\neq 0$, from Eqs. (13.30), (13.32), and (13.33) we find that velocity $\boldsymbol{u}$, pressure p, and magnetic field $\boldsymbol{B}$ are continuous. Density and temperature, however, can have any discontinuity. This is simply the boundary between two media, and called *contact discontinuity*.

13.3.2.2 Tangential Discontinuity ($j=0$, $B_{1\rm n}=0$)

In this case equations (13.32), (13.33), and (13.34) are satisfied identically. Velocity and magnetic fields are tangential, and can have any discontinuity in both magnitude and direction. Pressure discontinuity is, however, related to that of $\boldsymbol{B}_{\rm t}$ by $[p+\boldsymbol{B}_{\rm t}^2/8\pi]=0$ [see Eq. (13.30)]. This discontinuity is called *tangential discontinuity*.

13.3.2.3 Rotational Discontinuity ($j \neq 0$, $[\rho] = 0$, $[u_n] = 0$)

In this case there is mass flow through discontinuity, but no discontinuity in u_{n} and density. From Eq. (13.32) we have

$$j[\boldsymbol{u}_{\mathrm{t}}] = \frac{1}{4\pi} B_{\mathrm{n}}[\boldsymbol{B}_{\mathrm{t}}]. \tag{13.36}$$

Furthermore, Eq. (13.34) leads to

$$B_{\mathrm{n}}[\boldsymbol{u}_{\mathrm{t}}] = u_{\mathrm{n}}[\boldsymbol{B}_{\mathrm{t}}]. \tag{13.37}$$

From the above two equations, by eliminating $[\boldsymbol{B}_{\mathrm{t}}]$ we have

$$u_{\mathrm{n}} = \frac{1}{(4\pi\rho)^{1/2}} B_{\mathrm{n}}. \tag{13.38}$$

If this relation between u_{n} and B_{n} is used, Eq. (13.37) gives

$$[\boldsymbol{u}_{\mathrm{t}}] = \frac{1}{(4\pi\rho)^{1/2}} [\boldsymbol{B}_{\mathrm{t}}]. \tag{13.39}$$

In the present case Eq. (13.33) is reduced to

$$\frac{1}{2} j \left[\boldsymbol{u}_{\mathrm{t}}^2\right] + u_n \left[p + \frac{B_{\mathrm{t}}^2}{8\pi}\right] + \frac{1}{\gamma - 1} u_{\mathrm{n}}[p] + \frac{1}{8\pi}\left(u_{\mathrm{n}}\left[\boldsymbol{B}_{\mathrm{t}}^2\right] - 2B_{\mathrm{n}}[\boldsymbol{u}_{\mathrm{t}}\boldsymbol{B}_{\mathrm{t}}]\right) = 0. \tag{13.40}$$

By using identity:

$$\left[\left(\boldsymbol{u}_{\mathrm{t}} - \frac{1}{(4\pi\rho)^{1/2}}\boldsymbol{B}_{\mathrm{t}}\right)^2\right] = \left[\boldsymbol{u}_{\mathrm{t}}^2\right] - \frac{2}{(4\pi\rho)^{1/2}}[\boldsymbol{u}_{\mathrm{t}}\boldsymbol{B}_{\mathrm{t}}] + \frac{1}{4\pi\rho}\left[\boldsymbol{B}_{\mathrm{t}}^2\right] = 0 \tag{13.41}$$

we have

$$\begin{aligned} B_n[\boldsymbol{u}_{\mathrm{t}}\boldsymbol{B}_{\mathrm{t}}] &= \frac{1}{2}(4\pi\rho)^{1/2}\left(\mathrm{B}_{\mathrm{n}}\left[\boldsymbol{u}_{\mathrm{t}}^2\right] + \frac{B_{\mathrm{n}}}{4\pi\rho}\left[\boldsymbol{B}_{\mathrm{t}}^2\right]\right) \\ &= 2\pi\rho u_{\mathrm{n}}\left[\boldsymbol{u}_{\mathrm{t}}^2\right] + \frac{1}{2}u_{\mathrm{n}}\left[\boldsymbol{B}_{\mathrm{t}}^2\right], \end{aligned} \tag{13.42}$$

where in deriving the second equality in Eqs. (13.42), Eq. (13.38) has been used. Substituting Eq. (13.42) and $[p + \boldsymbol{B}_{\mathrm{t}}^2/8\pi] = 0$ [see Eq. (13.30)], into Eq. (13.40), we have

$$[p] = 0. \tag{13.43}$$

and thus

$$\left[\boldsymbol{B}_{\mathrm{t}}^{2} \right] = 0. \tag{13.44}$$

All other thermodynamical quantities are continuous.

The fact that $\boldsymbol{B}_{\mathrm{t}}^2$ and B_{n} are both continuous means that the magnitude of $\boldsymbol{B}$ and its angle to the surface are continuous. However, the magnetic fields are turned through an angle about the normal, its magnitude remaining unchanged. The vector $\boldsymbol{B}_{\mathrm{t}}$ and thus $\boldsymbol{u}_{\mathrm{t}}$ are discontinuous [see Eq. (13.39)], but the normal components are continuous. In this sense this is called *rotational discontinuity*.

13.3.2.4 Shock Waves ($j \neq 0$, $[\rho] \neq 0$)

Another type of discontinuity is the one where both $j \neq 0$ and $[\rho] \neq 0$ and thus $[u_{\mathrm{n}}] \neq 0$. This is an MHD shock. Our main interest here is this type of discontinuity, which will be considered in the followings.

13.3.3 MHD Shocks

We consider here the case (shocks) where both $\boldsymbol{u}$ and ρ are discontinuous at the discontinuous surface. First, let us show that vectors $\boldsymbol{B}_{t1}$, $\boldsymbol{B}_{t2}$ and the normal to the surface are coplanar, unlike what happens in tangential and rotational discontinuities. Here, as mentioned before, the subscripts 1 and 2 represent, respectively, the quantities in the region ahead of shock propagation and those behind the shock (see Fig. 13.5). From Eqs. (13.32) and (13.34) we obtain

$$[\boldsymbol{B}_{\mathrm{t}}] = \frac{4\pi j}{B_{\mathrm{n}}^2}[u_{\mathrm{n}}\boldsymbol{B}_{\mathrm{t}}]. \tag{13.45}$$

This relation shows that $\boldsymbol{B}_{\mathrm{t}1}$, $\boldsymbol{B}_{\mathrm{t}2}$ and the normal to the surface are coplanar.

Next, we determine configuration of shocks. By adopting coordinates suitably, the flow in the front of shock can be taken to be perpendicular to the shock front, i.e., $u_{1t} = 0$. Magnetic field lines in the front of shock are generally not perpendicular to the shock front. The inclination angle of magnetic field lines to the normal direction of the shock surface is taken to be θ_1, as is shown in Fig. 13.5. That is,

$$B_{1\mathrm{n}} = B_1 \cos\theta_1, \quad B_{1\mathrm{t}} = B_1 \sin\theta_1. \tag{13.46}$$

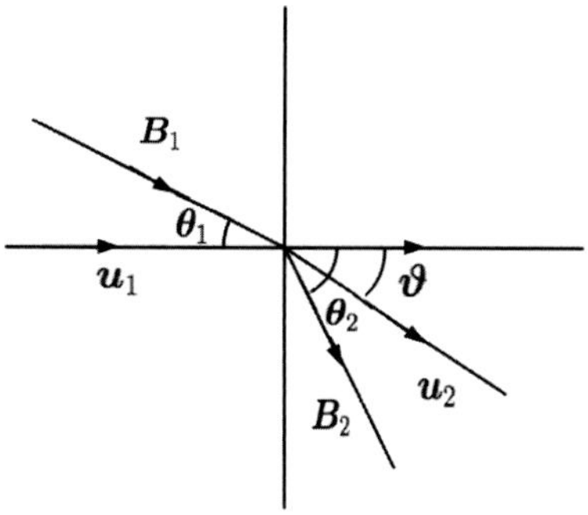

Fig. 13.5 A figure showing side 1 (ahead shock) and side 2 (behind shock). The vector $\boldsymbol{B}_1$ is assumed to enter to the shock front with angle θ_1, and leaves the front with angle θ_2. The fluid flow vector $\boldsymbol{u}_1$ is perpendicular to the surface, and leaves the surface with angle ϑ

As parameters specifying the flows ingoing to shock front we adopt[1]

$$\beta = \frac{c_{1s}^2}{c_{1A}^2}, \quad \mathcal{M}_A = \frac{u_1}{c_{1A}}, \tag{13.47}$$

where u_{1n} is simply written as u_1, since $\boldsymbol{u}$ has been assumed to have no tangential component. Here, c_{1s} and c_{1A} are acoustic and Alfvén speeds in side 1, i.e., $c_{1s}^2 = \gamma p_1/\rho_1$ and $c_{1A}^2 = B_1^2/4\pi\rho_1$. As in the case of fluid shocks in Sect. 4.6, the set of shock conditions given in the previous Sect. 13.3.1 and parameters specifying the ingoing flows [Eqs. (13.46) and (13.47)] determine the whole structure of outgoing fluids. In some cases, however, instead of $\mathcal{M}_A$, an another parameter is useful to specify shock strength. Here, we introduce the density ratio s defined by

$$s = \frac{\rho_2}{\rho_1}, \tag{13.48}$$

which is larger than unity.[2] The relation between $\mathcal{M}_A$ and s is given later, which is Eq. (13.55).

Now, we express ratios of various physical or dynamical quantities before and after shock in terms of quantities before shock and shock strength s. First, from Eqs. (13.32) and (13.34) we have

$$\frac{B_{2t}}{B_{1t}} = A, \tag{13.49}$$

[1] In this section the flow speed u_1 is normalized by Alfvén speed, not by acoustic speed, different from the case of fluid shocks in Sect. 4.6.

[2] When the gas passes through the shock, its entropy can only increase. In the case of weak MHD shocks, it is shown that the entropy increase implies $s > 1$ (e.g., Landau and Lifshitz 1960). They say that this result seems to hold for shock waves of any intensity.

where

$$A = \frac{\cos^2\theta_1 - \mathcal{M}_\mathrm{A}^2}{\cos^2\theta_1 - \mathcal{M}_\mathrm{A}^2/s} \tag{13.50}$$

and

$$\frac{u_{2\mathrm{t}}}{u_1} = \left(\frac{A}{s} - 1\right)\tan\theta_1. \tag{13.51}$$

It is noticed that A is larger than unity when $s > 1$.

Next, from Eq. (13.30) we have

$$\frac{p_2}{p_1} = 1 + \left[\left(1 - \frac{1}{s}\right) + \frac{1}{2\mathcal{M}_\mathrm{A}^2}\left(1 - A^2\right)\sin^2\theta_1\right]\frac{\rho_1 u_1^2}{p_1}. \tag{13.52}$$

The angle θ_2 is now compared with θ_1. Using Eq. (13.49), we have

$$\tan\theta_2 \equiv \frac{B_{2\mathrm{t}}}{B_{2\mathrm{n}}} = \frac{AB_{1\mathrm{t}}}{B_{1\mathrm{n}}} = A\tan\theta_1. \tag{13.53}$$

This result show that $\theta_2 > \theta_1$ for $s > 1$ (see Fig. 13.5). Behind shock $\boldsymbol{u}$ has generally a tangential component, i.e., $u_{2t} \neq 0$. The angle ϑ between $\boldsymbol{u}_2$ and the normal direction of the shock surface is given by

$$\tan\vartheta \equiv \frac{u_{2\mathrm{t}}}{u_{2\mathrm{n}}} = s\left(\frac{A}{s} - 1\right)\tan\theta_1 = \left(1 - \frac{s}{A}\right)\tan\theta_2. \tag{13.54}$$

This equation shows that ϑ is smaller than θ_2, when $1 - s/A < 1$. In other words, $\vartheta < \theta_2$ is realized when $\cos^2\theta_1 > M_A^2$ or $\cos^2\theta_1 < M_A^2/s$, since $s > 1$.

The remaining equation is Eq. (13.33). By substituting Eqs. (13.49), (13.51), and (13.52) into Eq. (13.33), we obtain, after some manipulations,

$$\left(a\frac{\mathcal{M}_\mathrm{A}^2}{s} - \beta\right)\left(\frac{\mathcal{M}_\mathrm{A}^2}{s} - \cos^2\theta_1\right)^2 + \frac{\mathcal{M}_\mathrm{A}^2}{s}\sin^2\theta_1\left(\frac{\mathcal{M}_\mathrm{A}^2}{s}\left[\frac{\gamma}{2}(s-1) - s\right] + a\cos^2\theta_1\right) = 0, \tag{13.55}$$

where a is defined by

$$a = \frac{\gamma + 1 - (\gamma - 1)s}{2}. \tag{13.56}$$

Equation (13.55) can be regarded as an equation giving the relation between the Alfvén Mach number $\mathcal{M}_{\rm A}$ and strength of shock, s. For a given $\mathcal{M}_{\rm A}^2$ this equation is a third-order algebraic equation with respect to s (or for a give s this equation is a third order with respect to $\mathcal{M}_{\rm A}^2$). This corresponds to the presence of three MHD modes. To check this, let us consider a special case of $s = 1$. In this case, Eq. (13.55) gives

$$\mathcal{M}_{\rm A}^2 = \cos^2\theta_1, \tag{13.57}$$

and

$$\mathcal{M}_{\rm A}^4 - (1+\beta)\mathcal{M}_{\rm A}^2 + \beta\cos^2\theta_1 = 0. \tag{13.58}$$

The above two equations really have the same forms as equations of the Alfvén wave and the fast and slow modes [see Eq. (13.23)].[3]

13.3.4 Special Cases

We consider here two special cases. First, the case is considered where the magnetic fields in the ingoing side to shock front are perpendicular to the shock front, i.e., $\cos\theta_1 = 1$, or $\boldsymbol{B}_1 \parallel \boldsymbol{u}_1$ (parallel shock). In this case solutions of Eq. (13.55) are

$$\mathcal{M}_{\rm A}^2 = \frac{\beta s}{a}, \quad \mathcal{M}_{\rm A}^2 = s. \tag{13.59}$$

The former is further reduced to $u_1^2/c_{1s}^2 = s/a$. This can be rearranged as

$$s = \frac{\rho_2}{\rho_1} = \frac{(\gamma+1)\left(u_1^2/c_{1s}^2\right)}{2+(\gamma-1)\left(u_1^2/c_{1s}^2\right)}, \tag{13.60}$$

which is the result known in fluid shocks with no magnetic fields [see Eq. (4.142)]. This is understandable since the gaseous flows are parallel to magnetic fields and are not affected by magnetic fields. In the latter case of $\mathcal{M}_{\rm A}^2 = s$, A is infinity [see Eq. (13.50)], i.e., $A = \infty$. This means $B_{2\rm t} \neq 0$, although $B_{1\rm t} = 0$ [see Eq. (13.49)]. In this sense this is called a switch-on shock.

Second, let us consider the case where magnetic fields are parallel to the shock surface, i.e., $\cos\theta_1 = 0$, or $\boldsymbol{B}_1 \perp \boldsymbol{u}_1$ (normal shock). In this case Eq. (13.50) shows $A = s$ and we have $B_{2t}/B_{1t} = s$. That is, the compression rate of magnetic fields at

[3] $\mathcal{M}_{\rm A}^2$ corresponds to $\omega^2/k^2c_{\rm A}^2$ and $\cos^2\theta_1$ to k_z^2/k^2.

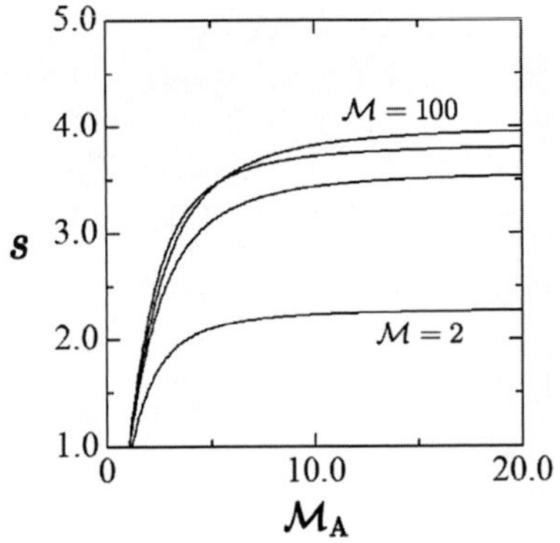

Fig. 13.6 The s-$\mathcal{M}_A$ relation calculated by using Eq. (13.61) is shown for some values of $\mathcal{M}^2 \equiv u_1^2/c_{1s}^2$ when $\gamma = 5/3$. From the uppermost curve to the lowermost one the value of $\mathcal{M}$ is 100, 10, 5, and 2 in turn

the shock is the same as that of the fluids. This is obvious since magnetic fields are frozen to fluids. Furthermore, Eq. (13.55) gives

$$\mathcal{M}_A^2 = 0, \quad \left(\frac{a}{s} - \frac{c_{1s}^2}{u_1^2}\right)\mathcal{M}_A^2 + \frac{\gamma}{2}(s-1) - s = 0. \tag{13.61}$$

The former is trivial. In the limit of no magnetic fields the latter gives $a/s - (c_{1s}^2/u_1^2) = 0$, which leads again to Eq. (13.60), as is expected. The $s - \mathcal{M}_A$ relation given by the latter relation of Eq. (13.61) is shown in Fig. 13.6 for some values of $\mathcal{M}(\equiv u_1/c_{1s})$ when $\gamma = 5/3$. As magnetic fields become strong ($\mathcal{M}_A$ becomes small), the compression ratio s decreases because an effective shock strength, $u_1^2/(c_{1s}^2 + c_{1A}^2)$, decreases with decrease of $\mathcal{M}_A^2$, and finally $s = 1$ is realized for $\mathcal{M}_A^2 = \mathcal{M}^2/(\mathcal{M}^2 - 1)$.

Reference

Landau, L.D., Lifshitz, E.M.: Electrodynamics of Continuous Media. Pergamon Press, Oxford (1960)

Chapter 14
Astrophysical Dynamo

Magnetic fields of planets and stars are believed to be generated and maintained by dynamo action, because the timescale of their dissipation is shorter than their lifetime. For dynamo processes to act, however, a seed field is required, which will be brought in from outside when the objects are formed or generated inside them. In this chapter, as preliminary studies of dynamo processes, Biermann's battery mechanism and Cowling's anti-dynamo theorem are introduced. Then, the essences of the mean field dynamo by Steenbeck–Krause–Rädler and Parker's cyclonic dynamo are described.

14.1 Battery Mechanism

Possible origins of seed fields for dynamo action are usually not considered seriously, since in many astrophysical objects, the seed fields will be brought in the systems from the outside when they are formed. However, how magnetic fields can be generated in astrophysical objects from the state of no fields is one of theoretically interesting problems. Here, we describe the battery mechanism presented by Biermann (1950).

In Chap. 10 we have derived the equation of motion of electron gases, which is Eq. (10.38). From this equation we have further derived the generalized Ohm's law (10.41). In this procedure thermo-electric force $-\nabla p_-$ is minor on the force balance of electron gas, where p_- is the pressure of electron gases. In steady systems with no electromagnetic fields, however, the thermo-electric force can have non-negligible secular effects, depending on the configuration of the systems.

Let us consider spherically symmetric stars. The particles which directly receive the gravitational force from the central region of the stars are ions. Hence, the ions are slightly shifted toward the central region, relative to the electron gases. The electron gases are attracted to the stellar center by the electric field, $\boldsymbol{E}$, generated

© Springer Nature Singapore Pte Ltd. 2020, corrected publication 2023

S. Kato, J. Fukue, *Fundamentals of Astrophysical Fluid Dynamics*,
Astronomy and Astrophysics Library,
https://doi.org/10.1007/978-981-15-4174-2_14

by charge separation. The force balance on electron gas is thus realized between $-n_e e\boldsymbol{E}$ and $-\nabla p_-$ per unit volume. That is, $\boldsymbol{E} = -(1/n_e e)\nabla p_-$, where n_e is the number density of electron.

Since curl $\boldsymbol{E} = (n_e^2 e)^{-1}\nabla n_e \times \nabla p_-$, we have $\boldsymbol{E} = 0$ in spherically symmetric star. In rotating stars, however, the direction of ∇n_e is in general not parallel to that of ∇p_-. This means that in rotating stars, curl$\boldsymbol{E} \neq 0$ and the Maxwell equation, curl$\boldsymbol{E} = -\partial \boldsymbol{B}/c\partial t$, show that magnetic fields in the azimuthal direction are generated. Although curl$\boldsymbol{E}$ is small, magnetic fields increases secularly in a longtime. This is the *battery mechanism* by Biermann (1950). In his original studies a considerable strength of toroidal magnetic fields is supposed to be generated in rotating stars.

Mestel and Roxburgh (1962), however, showed that the presence of a weak poloidal magnetic field suppresses greatly the generation of the toroidal magnetic field. Subsequently, Dolginov (1977) pointed out that if the mean molecular weight, μ, in stellar radiative zone has a non-spherical distribution, the battery process will efficiently act. Dolginov's study is, however, a kinematical study. Mestel and Moss (1983) examined this problem in a self-consistent stellar structure. Their conclusion is that the generated fields are weaker than those anticipated by Biermann in his original study, but in rapid rotators the fields can grow to strength comparable with observed fields in Ap stars.

14.2 Cowling's Anti-Dynamo Theorem

Cowling (1933) showed that no motion can maintain a perfectly axisymmetric magnetic field. In other words, if magnetic fields can be maintained by dynamo processes, they are non-axisymmetric and fluid motions required are also non-axisymmetric. This is called *Cowling's anti-dynamo theorem.* This anti-dynamo theorem is now shown under rather general conditions, compared with the original ones considered by Cowling. Here, however, we demonstrate the theorem under a simple situation of incompressible fluids, following Jones (2007).[1]

We introduce magnetic vector potential, $\boldsymbol{A}$, as

$$\boldsymbol{B} = \mathrm{curl}\,\boldsymbol{A}, \quad \mathrm{div}\,\boldsymbol{A} = 0. \tag{14.1}$$

Then, assuming a constant magnetic diffusivity, η, we can write the induction equation:

$$\frac{\partial \boldsymbol{B}}{\partial t} = \mathrm{curl}\,(\boldsymbol{u} \times \boldsymbol{B}) + \eta\nabla^2\boldsymbol{B}, \tag{14.2}$$

[1] Chris A. Jones's Les Houches Summer School 2007 Notes on Dynamo Theory. Available electrically at http://www.maths.leeds.ac.uk/cajones/LesHouches.html.

as

$$\frac{\partial \boldsymbol{A}}{\partial t} = \boldsymbol{u} \times \text{curl}\, \boldsymbol{A} + \eta \nabla^2 \boldsymbol{A} + \nabla \Psi, \tag{14.3}$$

where the scalar potential Ψ is defined so that

$$\nabla^2 \Psi = \text{div}(\boldsymbol{u} \times \boldsymbol{B}). \tag{14.4}$$

In addition, as mentioned above, for simplicity, fluid motions are taken to be incompressible, i.e., $\text{div}\, \boldsymbol{u} = 0$.

We consider axisymmetric configurations. That is, both fluid flows and magnetic fields are taken to be axisymmetric, and we take $\partial/\partial\varphi = 0$ under cylindrical coordinates (r, φ, z). Then, magnetic fields $\boldsymbol{B}$ and fluid velocity $\boldsymbol{u}$ can be written in the form:

$$\boldsymbol{B} = B_\varphi \boldsymbol{i}_\varphi + \boldsymbol{B}_p = B_\varphi \boldsymbol{i}_\varphi + \text{curl}(A_\varphi \boldsymbol{i}_\varphi), \quad \boldsymbol{u} = r\Omega \boldsymbol{i}_\varphi + \text{curl}(\psi \boldsymbol{i}_\varphi), \tag{14.5}$$

where A_φ and B_φ are scalar quantities representing φ-components of $\boldsymbol{A}$ and $\boldsymbol{B}$, respectively, and $\boldsymbol{B}_\text{p}$ is the poloidal component of $\boldsymbol{B}$. The symbol $\boldsymbol{i}_\varphi$ is the unit vector in the φ-direction, and $\psi(r, z)$ is a scalar potential for velocity. The rotation, $r\Omega$, has been separated from ψ.

Now, the expressions for $\boldsymbol{B}$ and $\boldsymbol{u}$ given by Eq. (14.5) are substituted into Eqs. (14.2) and (14.3) lead to[2]

$$\frac{\partial B_\varphi}{\partial t} + r(\boldsymbol{u}_\text{p} \cdot \nabla)\left(\frac{B_\varphi}{r}\right) = \eta\left(\nabla^2 - \frac{1}{r^2}\right)B_\varphi + r(\boldsymbol{B}_\text{p} \cdot \nabla)\Omega, \tag{14.6}$$

$$\frac{\partial A_\varphi}{\partial t} + \frac{1}{r}(\boldsymbol{u}_\text{p} \cdot \nabla)(rA_\varphi) = \eta\left(\nabla^2 - \frac{1}{r^2}\right)A_\varphi. \tag{14.7}$$

---

[2]A formula of vector analyses gives, for arbitrary $\boldsymbol{u}$ and $\boldsymbol{B}$,

$$\text{curl}(\boldsymbol{u} \times \boldsymbol{B}) = (\boldsymbol{B} \cdot \nabla)\boldsymbol{u} - (\boldsymbol{u} \cdot \nabla)\boldsymbol{B} + \boldsymbol{u}\, \text{div}\, \boldsymbol{B} - \boldsymbol{B}\, \text{div}\, \boldsymbol{u}.$$

Furthermore, for arbitrary $\boldsymbol{u}$ and $\boldsymbol{B}$ we have

$$[(\boldsymbol{B} \cdot \nabla)\boldsymbol{u}]_\varphi = \left(B_r \frac{\partial}{\partial r} + B_\varphi \frac{\partial}{r\partial\varphi} + B_z \frac{\partial}{\partial z}\right)u_\varphi + \frac{B_\varphi}{r}u_r.$$

For arbitrary vector $\boldsymbol{A}$, $\nabla^2 \boldsymbol{A}$ is defined by

$$\nabla^2 \boldsymbol{A} \equiv \text{grad}(\text{div}\, \boldsymbol{A}) - \text{curl}(\text{curl}\, \boldsymbol{A}).$$

Also, for example, we have

$$(\text{curl}\, \boldsymbol{A})_z = \frac{1}{r}\frac{\partial}{\partial r}(rA_\varphi) - \frac{\partial A_r}{r\partial\varphi},$$

where A_r and A_φ are, respectively, r- and φ-components of $\boldsymbol{A}$.

Equation (14.6) shows the time evolution of toroidal component of magnetic fields, B_φ. We now consider the meaning of each term of this equation. The second term on the left-hand side is the term of advection of B_φ, representing the transport of B_φ. The first term on the right-hand side is the term of dissipation of B_φ. The generation of B_φ comes only from the second term on the right-hand side. That is, the generation of the toroidal component of magnetic field, B_φ, comes from the coupling of the poloidal component of magnetic field, $\boldsymbol{B}$, with differential rotation. Next, Eq. (14.7) is considered. This equation shows how the poloidal component of $\boldsymbol{B}$ evolves with time.[3] The second term on the left-hand side of this equation represents the advection of A_φ, while the right-hand side of this equation shows dissipation of A_φ. There is no generation term of A_φ. Hence, with time A_φ decreases and finally disappears. This implies that $\boldsymbol{B}_\mathrm{p}$ also finally disappears, and there is no maintenance of B_φ.

This is Cowling's theorem, and shows that how the poloidal component of magnetic fields (in other words, the φ-component of $\boldsymbol{A}$) is generated from the toroidal component of magnetic fields is the main subject of dynamo theory.

14.3 Dynamo Theories

Almost all astrophysical objects have magnetic fields. Concerning their origin, there are two possibilities. One is that they are brought in from outside with the materials of the objects. The second one is that they are made in the objects after the formation. The diffusion timescale of magnetic fields is very long especially in large scale objects. Hence, the first possibility is not excluded at all. However, if we consider the possibilities of their escape from the systems or dissipations due to turbulence, we must consider the second possibility.

The dynamo theories have been extensively developed in various situations, but here we introduce only the classical standard theories by Steenbeck et al. (1966) (*turbulent dynamo*) and by Parker (1970) (*cyclonic dynamo*). It is noticed that these dynamo theories are ideas that small scale motions maintain large scale magnetic fields. Distinct from these, there are ideas that relatively large scale motions can sustain the magnetic fields.

To show that magnetic fields are maintained by dynamo actions, it is necessary to demonstrate that the whole systems of magnetic fields and fluid motions are self-sustained. It is not enough to show that given fluid motions can maintain the magnetic fields (kinematic theory). It is necessary to show that the feedback of the resulting magnetic fields to fluid motions is really consistent with the fluid motions (dynamical theory).

[3] It is noticed that $\boldsymbol{B}_\mathrm{p}$ is related to the r- and z-components of A_φ.

14.3.1 Outline of Turbulent Dynamo by Steenbeck et al.

As a preparation, the equations describing the mean magnetic fields are derived. The induction equation is

$$\frac{\partial \boldsymbol{B}}{\partial t} = \mathrm{curl}(\boldsymbol{v} \times \boldsymbol{B}) + \eta \nabla^2 \boldsymbol{B}. \tag{14.8}$$

Here, the fluid velocity $\boldsymbol{v}$ is separated into rotational motion, $\boldsymbol{U}$, and small scale motion, $\boldsymbol{u}$, as

$$\boldsymbol{v} = \boldsymbol{U} + \boldsymbol{u}, \quad \langle \boldsymbol{u} \rangle = 0, \tag{14.9}$$

where $\langle\ \rangle$ means the ensemble average of small scale motions (it can be considered to be an average in a scale longer than that of turbulence, but shorter than that of the global structure of the system). The magnetic fields $\boldsymbol{B}$ are also separated into the global component, $\boldsymbol{B}_0$, and turbulent one, $\boldsymbol{b}$, as

$$\boldsymbol{B} = \boldsymbol{B}_0 + \boldsymbol{b}, \quad \langle \boldsymbol{b} \rangle = 0. \tag{14.10}$$

By taking the ensemble average of Eq. (14.8), we have

$$\frac{\partial \boldsymbol{B}_0}{\partial t} = \mathrm{curl}(\boldsymbol{U} \times \boldsymbol{B}_0) + \eta \nabla^2 \boldsymbol{B}_0 + \mathrm{curl}(\langle \boldsymbol{u} \times \boldsymbol{b} \rangle). \tag{14.11}$$

What we are interested in is the behavior of large scale fields $\boldsymbol{B}_0$. Equation (14.11) shows that the turbulent fields have influences on large scale magnetic fields, $\boldsymbol{B}_0$, through the last term on the right-hand side of Eq. (14.11). Our purpose is to study how this term affects the behavior of $\boldsymbol{B}_0$.

If the ensemble average of Ohm's equation is taken, using Eqs. (14.9) and (14.10), we have

$$\langle \boldsymbol{i} \rangle = \sigma \left(\langle \boldsymbol{E} \rangle + \frac{1}{c} \boldsymbol{U} \times \boldsymbol{B}_0 + \frac{1}{c} \langle \boldsymbol{u} \times \boldsymbol{b} \rangle \right), \tag{14.12}$$

where $\langle \boldsymbol{i} \rangle$ and $\langle \boldsymbol{E} \rangle$ are, respectively, the ensemble average of electric current and the one of electric field. This equation shows that turbulent motions can produce a global current through a nonlinear term $\sigma \langle \boldsymbol{u} \times \boldsymbol{b} \rangle / c$. In the following we examine whether this global current induced by turbulence has a component parallel to the global magnetic field, $\boldsymbol{B}_0$. The reasons for examining this come from the following (see Fig. 14.1). Let us consider a loop of magnetic field $\boldsymbol{B}_{0(0)}$ as shown in Fig. 14.1. If current $\boldsymbol{i}_{(1)}$ flows along the magnetic field $\boldsymbol{B}_{0(0)}$, magnetic field $\boldsymbol{B}_{0(1)}$ is generated. Then, by the assumption current $\boldsymbol{i}_{(2)}$ flows along the field line $\boldsymbol{B}_{0(1)}$. Then, the original magnetic field $\boldsymbol{B}_{0(0)}$ is amplified by this current.

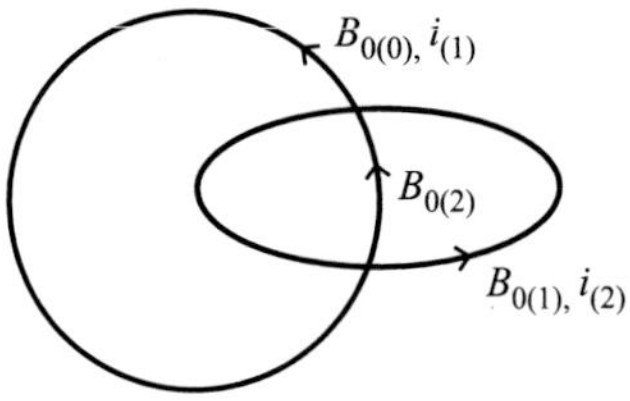

Fig. 14.1 Current in the direction of magnetic fields and amplification of the magnetic fields

By subtracting Eq. (14.11) from the induction equation (14.8), we obtain an equation describing the time evolution of $\boldsymbol{b}$, which is

$$\frac{\partial \boldsymbol{b}}{\partial t} - \eta \nabla^2 \boldsymbol{b} = \mathrm{curl}[\boldsymbol{U} \times \boldsymbol{b} + \boldsymbol{u} \times \boldsymbol{B}_0 + (\boldsymbol{u} \times \boldsymbol{b} - \langle \boldsymbol{u} \times \boldsymbol{b} \rangle)]. \tag{14.13}$$

To proceed further, characteristics of turbulence is specified. The characteristic size and timescale of turbulence are denoted, respectively, by l and δt. The characteristic timescale δt is assumed to be shorter than the turnover time, l/u, of turbulent element, but longer than the timescale of propagation of acoustic waves:

$$l/c_s \ll \delta t \ll l/u. \tag{14.14}$$

Since $u < c_s$ is involved in the above inequalities, the turbulent flows can be approximated to be incompressible in the zeroth-order approximations (cf. Boussinesq approximation).

Related to $\delta t \ll l/u$, we neglect $\boldsymbol{u} \times \boldsymbol{b} - \langle \boldsymbol{u} \times \boldsymbol{b} \rangle$ in Eq. (14.13) (the *first-order smoothing approximation*). Furthermore, since the effects of rotation can be understood easily, the term in Eq. (14.13) is neglected. Then, we have

$$\frac{\partial \boldsymbol{b}}{\partial t} - \eta \nabla^2 \boldsymbol{b} = \mathrm{curl}(\boldsymbol{u} \times \boldsymbol{B}_0). \tag{14.15}$$

This equation shows that if the turbulent velocity $\boldsymbol{u}$ is given, we can calculate $\boldsymbol{b}$ and thus $\boldsymbol{u} \times \boldsymbol{b}$ and $\langle \boldsymbol{u} \times \boldsymbol{b} \rangle$.

To further simplify, we assume that the size of turbulence, l, is so small that in describing turbulent magnetic flows their diffusion is more important than their transport:

$$l^2 \ll \eta \delta t. \tag{14.16}$$

The use of inequality (14.16) leads Eq. (14.15) to

$$\eta \nabla^2 \boldsymbol{b} = -\mathrm{curl}(\boldsymbol{u} \times \boldsymbol{B}_0). \tag{14.17}$$

It is noted that the adoption of inequalities $\delta t \ll l/u$ and $l^2 \ll \eta\delta t$ implies that we are considering the cases where the *magnetic Reynolds number*, $\mathcal{R}_m (\equiv ul/\eta)$, is smaller than unity, i.e., $ul/\eta \ll 1$.[4]

Next, the effects of rotation on turbulence are considered. We assume that the rotation is weak and its effects on turbulent motion can be treated as perturbations. That is, $\boldsymbol{u}$ is written as

$$\boldsymbol{u} = \boldsymbol{u}^{(0)} + \boldsymbol{u}^{(1)}, \tag{14.18}$$

where $\boldsymbol{u}^{(1)}$ is the perturbed part of $\boldsymbol{u}$, resulting from rotation. The equation of motion is now expanded with respect to rotation. In this equation the inertial term is neglected compared with the term of time variation by the assumption of Eq. (14.14). The viscous term is also neglected for simplicity. This can be really neglected if the *Reynolds number* $\mathcal{R}(\equiv ul/\nu)$ is not too small.[5] Then, we have

$$\frac{\partial \rho \boldsymbol{u}^{(1)}}{\partial t} = 2\rho \boldsymbol{u}^{(0)} \times \boldsymbol{\Omega} - \text{grad}\, p^{(1)}. \tag{14.19}$$

This equation shows that if $\rho\boldsymbol{u}^{(0)}$ is given, we can calculate $\rho\boldsymbol{u}^{(1)}$ by integrating Eq. (14.19).[6] Then, by using this $\rho\boldsymbol{u}^{(1)}$, we have $\boldsymbol{u}\ (= \boldsymbol{u}^{(0)} + \boldsymbol{u}^{(1)})$, and thus substituting this $\boldsymbol{u}$ into Eq. (14.17) and performing the integration, we can derive an expression for $\boldsymbol{b}$.

In this way, for given ρ and $\rho\boldsymbol{u}^{(0)}$, quantities $\boldsymbol{u}$ and $\boldsymbol{b}$ are derived. Then, we can know $\langle \boldsymbol{u} \times \boldsymbol{b} \rangle$. If this procedure is done for relevant forms of ρ and $\rho\boldsymbol{u}^{(0)}$, we have an expression for $\langle \boldsymbol{u} \times \boldsymbol{b} \rangle$ after detailed calculations (Steenbeck et al. 1966):

$$\begin{aligned} \langle \boldsymbol{u} \times \boldsymbol{b} \rangle = \gamma \frac{l^2}{\eta} \langle u^2 \rangle^{1/2} \Bigg[& -\frac{1}{3} \langle u^2 \rangle^{1/2} \text{grad}\left(\ln \left\langle u^2 \right\rangle^{1/2} \right) \times \boldsymbol{B}_0 \\ & - \frac{4}{15} \delta t \Bigg\{ 4 \left(\boldsymbol{\Omega} \cdot \text{grad} \left(\ln \rho \left\langle u^2 \right\rangle^{1/2} \right) \right) \boldsymbol{B}_0 - (\boldsymbol{B}_0 \cdot \boldsymbol{\Omega}) \text{grad} \left(\ln \rho \left\langle u^2 \right\rangle^{1/2} \right) \\ & - \left(\boldsymbol{B}_0 \cdot \text{grad} \left(\ln \rho \left\langle u^2 \right\rangle^{1/2} \right) \right) \boldsymbol{\Omega} \Bigg\} \Bigg], \end{aligned} \tag{14.20}$$

where γ is a dimensionless quantity of the order of unity.

Equation (14.20) shows that $\langle \boldsymbol{u} \times \boldsymbol{b} \rangle$ has a term parallel to the direction of $\boldsymbol{B}_0$. The important point to be emphasized here is that for the term to be present, both

[4] It is noted that the cyclonic dynamo which will be considered in Sect. 14.3.3 corresponds to the opposite case, i.e., $\mathcal{R}_m \gg 1$.

[5] The condition of neglecting the viscous term is $\nu/l^2 < 1/\delta t$, which is $\delta t < l^2/\nu = (l/u)\mathcal{R}$. The inequality (14.14) is satisfied if $\mathcal{R}$ is not too small.

[6] The term of $\rho\boldsymbol{u}^{(1)}$ can be derived from the zeroth-order quantities, since $\nabla^2 p^{(1)}$ can be expressed by the zeroth-order quantities by use of the assumption of incompressibility.

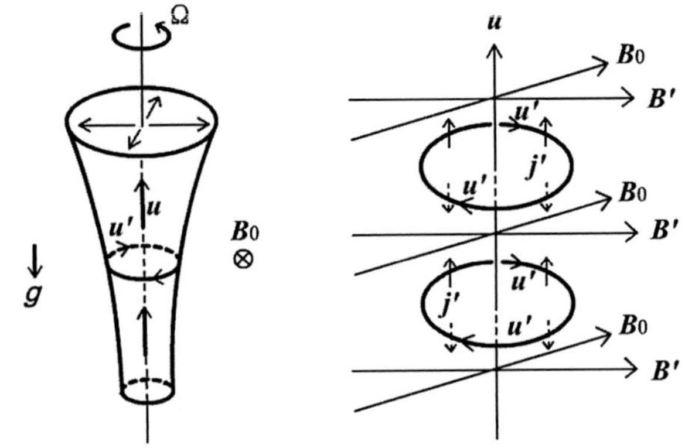

Fig. 14.2 Schematic picture showing dynamo action by convective elements in a medium where density is stratified in the direction of rotation axis (see a figure by Steenbeck et al. 1966)

the presence of rotation and inhomogeneity of $\rho\langle u^2\rangle^{1/2}$ in the direction of rotation vector are necessary [see the coefficient of the term of $\boldsymbol{B}_0$ in Eq. (14.20)]. The reason of the appearance of such a term can be understood from Fig. 14.2, as will be discussed below.

Let us consider a fluid element rising in an inhomogeneous medium whose density decreases upwards. The direction of the gravitational acceleration is downwards, the rotation axis is upwards, and the magnetic line of force is directed in a direction perpendicular to the rotation axis (see the left panel of Fig. 14.2). The fluid element expands as it rises upwards. By this expansion, a fluid motion around the rotation axis is generated. This rotation flow is denoted by $\boldsymbol{u}'$. Since this flow crosses the magnetic field line $\boldsymbol{B}_0$, a current $\boldsymbol{j}'$ flows in the direction of $\boldsymbol{u}' \times \boldsymbol{B}_0$ (see the right panel of Fig. 14.2). This current produces magnetic field $\boldsymbol{B}'$ in the direction perpendicular to the original field $\boldsymbol{B}_0$. Since the rising motion $\boldsymbol{u}$ of the fluid element crosses $\boldsymbol{B}'$, a current flows. The direction of this current is in the same direction as the initial magnetic fields $\boldsymbol{B}_0$. As mentioned before, if a current flows in the same direction as the initial magnetic field, the initial magnetic fields are amplified (see Fig. 14.1). In the case where a fluid element makes downward motion, situations are the same and the initial magnetic fields are amplified.

14.3.2 $\alpha\omega$ and α^2 Dynamos

Since the above arguments showed that $\langle \boldsymbol{u} \times \boldsymbol{b}\rangle$ has a component parallel to $\boldsymbol{B}_0$, we now focus our attention only to that term and write the term as $\alpha\boldsymbol{B}_0$, i.e., we write $\langle \boldsymbol{u} \times \boldsymbol{b}\rangle = \alpha\boldsymbol{B}_0$. Then, Eq. (14.11) is written as

$$\frac{\partial \boldsymbol{B}_0}{\partial t} = \mathrm{curl}(\boldsymbol{U} \times \boldsymbol{B}_0) + \eta\nabla^2\boldsymbol{B}_0 + \mathrm{curl}(\alpha\boldsymbol{B}_0). \tag{14.21}$$

Here, we consider the case where the global structure is axisymmetric and the mean flow is rotational motion alone, i.e., $\boldsymbol{U} = r\Omega \boldsymbol{i}_\varphi$. The global magnetic fields and global velocity fields are decomposed into toroidal and poloidal components as

$$\boldsymbol{B}_0 = \boldsymbol{B}_\mathrm{t} + \boldsymbol{B}_\mathrm{p}, \quad \boldsymbol{U} = \boldsymbol{U}_\mathrm{t}, \tag{14.22}$$

where the subscripts t and p represent the toroidal and poloidal components, respectively. Then, Eq. (14.21) can be decomposed into toroidal and poloidal components as

$$\frac{\partial \boldsymbol{B}_\mathrm{t}}{\partial t} - \eta \nabla^2 \boldsymbol{B}_\mathrm{t} = \mathrm{curl}(\boldsymbol{U}_\mathrm{t} \times \boldsymbol{B}_\mathrm{p}) + \mathrm{curl}(\alpha \boldsymbol{B}_\mathrm{p}), \tag{14.23}$$

$$\frac{\partial \boldsymbol{B}_\mathrm{p}}{\partial t} - \eta \nabla^2 \boldsymbol{B}_\mathrm{p} = \mathrm{curl}(\alpha \boldsymbol{B}_\mathrm{t}). \tag{14.24}$$

Equations (14.23) and (14.24) are the fundamental equations for studying dynamo processes.

Equations (14.23) and (14.24) show that $\boldsymbol{B}_\mathrm{t}$ and $\boldsymbol{B}_\mathrm{p}$ are interacting with each other. This means that as demonstrated in Fig. 14.3, there are possibility that magnetic fields are maintained by this coupling. That is, if $\boldsymbol{B}_\mathrm{p}$ is given, $\boldsymbol{B}_\mathrm{t}$ is generated through the first and second terms on the right-hand side of Eq. (14.23). The generated $\boldsymbol{B}_\mathrm{t}$ then generates $\boldsymbol{B}_\mathrm{p}$ through the term on the right-hand side of Eq. (14.24).

It is noted here that in the case where the global state is axisymmetric, the first term on the right-hand side of Eq. (14.23) is written as (see the formula given in footnotes in Sect. 14.2)

$$\mathrm{curl}(\boldsymbol{U}_\mathrm{t} \times \boldsymbol{B}_\mathrm{p}) = (\boldsymbol{B}_\mathrm{p} \cdot \nabla)\boldsymbol{U}_\mathrm{t} - (\boldsymbol{U}_\mathrm{t} \cdot \nabla)\boldsymbol{B}_\mathrm{p} = \boldsymbol{i}_\varphi (\boldsymbol{B}_\mathrm{p} \cdot \nabla)\Omega. \tag{14.25}$$

Then, Eq. (14.23) is written as

$$\frac{\partial \boldsymbol{B}_\mathrm{t}}{\partial t} - \eta \nabla^2 \boldsymbol{B}_\mathrm{t} = \boldsymbol{i}_\varphi (\boldsymbol{B}_\mathrm{p} \cdot \nabla)\Omega + \mathrm{curl}(\alpha \boldsymbol{B}_\mathrm{p}), \tag{14.26}$$

and Eqs. (14.26) and (14.24) are the set of equations describing dynamo processes. That is, $\boldsymbol{B}_\mathrm{p}$ is generated from the α-term in Eq. (14.24), and $\boldsymbol{B}_\mathrm{t}$ is generated from

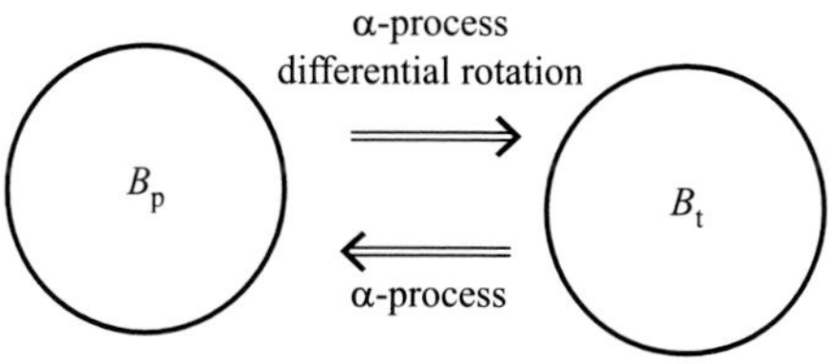

Fig. 14.3 Interaction between poloidal and toroidal fields

$\boldsymbol{B}_\mathrm{p}$ by two processes. One comes from differential rotation, while the other does from α-term.

The idea that toroidal magnetic fields, $\boldsymbol{B}_\mathrm{t}$, are generated from poloidal fields $\boldsymbol{B}_\mathrm{p}$ by differential rotation, and the poloidal fields $\boldsymbol{B}_\mathrm{p}$ are generated from the toroidal fields by the effects of α process is called *$\alpha\omega$ dynamo*. The case where generation of toroidal magnetic fields is also due to α process is called *α^2 dynamo*. Since many astrophysical objects have differential rotation, the α^2-process is minor in many objects compared with the $\alpha\omega$ one.

Here, we summarize the basic equations describing $\alpha\omega$ dynamo in terms of cylindrical coordinates (r, φ, z). Let us introduce the magnetic vector potential $\boldsymbol{A}$ by $\boldsymbol{B} = \mathrm{curl}\,\boldsymbol{A}$, and write the tangential components of $\boldsymbol{A}$ and $\boldsymbol{B}$ by A_φ and B_φ, respectively. Then, from Eqs. (14.24) and (14.23), we have

$$\left(\frac{\partial}{\partial t} - \eta\nabla^2\right)A_\varphi = \alpha(r, z, t)B_\varphi, \tag{14.27}$$

$$\left(\frac{\partial}{\partial t} - \eta\nabla^2\right)B_\varphi = -r\frac{\partial\Omega}{\partial r}\frac{\partial A_\varphi}{\partial z} + \frac{\partial\Omega}{\partial z}\frac{\partial}{\partial r}(rA_\varphi), \tag{14.28}$$

where A_φ, B_φ, and α are in general functions of r, z, and t. The above two equations are the set of equations describing $\alpha\omega$ dynamo in terms of A_φ and B_φ. The r- and z-components of $\boldsymbol{B}$ can be derived from A_φ by using the following relations:

$$B_r = -\frac{\partial A_\varphi}{\partial z}, \quad B_z = \frac{1}{r}\frac{\partial}{\partial r}(rA_\varphi). \tag{14.29}$$

If polar coordinates (r, θ, φ) are used, the set of equations describing $\alpha\omega$ dynamo in terms of A_φ and B_φ are

$$\left(\frac{\partial}{\partial t} - \eta\nabla^2\right)A_\varphi = \alpha(r, z, t)B_\varphi, \tag{14.30}$$

$$\left(\frac{\partial}{\partial t} - \eta\nabla^2\right)B_\varphi = \frac{\partial\Omega}{\partial r}\frac{\partial}{\partial\theta}(A_\varphi\sin\theta) - \frac{\partial\Omega}{r\partial\theta}\frac{\partial}{\partial r}(rA_\varphi\sin\theta). \tag{14.31}$$

Equations (14.30) and (14.31) are the set of equations describing $\alpha\omega$ dynamo in terms of A_φ and B_φ. The r- and z-components of $\boldsymbol{B}$ can be derived from A_φ by using the following relations:

$$B_r = \frac{1}{r\sin\theta}\frac{\partial}{\partial\theta}(A_\varphi\sin\theta), \quad B_\theta = -\frac{\partial}{r\partial r}(rA_\varphi). \tag{14.32}$$

14.3.3 Outline of Parker's Cyclonic Dynamo

From a viewpoint different from Steenbeck et al. (1966), Parker (1970) presented an intuitive model of dynamo actions. In his model, distinct from the turbulent dynamo by Steenbeck et al., convective motions with large magnetic Reynolds number ($\mathcal{R}_m \gg 1$) are considered. This model is called *cyclonic dynamo*.

The important issue to be considered in dynamo theory is, as argued in the previous Sect. 14.3.1, how the poloidal component of global magnetic fields is generated from the toroidal component. Parker (1970) presented this process by considering turbulent convective motions. Its essence is demonstrated by Fig. 14.4. Let us consider a rising fluid element in a rotating stratified medium. The element expands as it rises because of density stratification. Then, by the presence of rotation the element rotates as shown in Fig. 14.4. By this rotation, the magnetic fields are twisted and poloidal fields are generated.

Let us express the above argument in terms of vector potential of magnetic fields. We now introduce vector potential $\boldsymbol{A}_0$ for global magnetic fields by $\boldsymbol{B}_0 = \text{curl}\,\boldsymbol{A}_0$. Then, our present issue is to study how the toroidal component of $\boldsymbol{A}_0$, say A_φ, can be generated from fluid motions. In the case where magnetic fields are frozen to convective motions, small scale toroidal component of vector potential $\boldsymbol{A}$ is generated. The problem is whether such small scale fluctuating vector potential can produce a global component of A_φ. The issue needs to be examined quantitatively. This is done by Parker. Its outline is presented below.

Let us introduce assumptions concerning the turbulent convection:

$$l \ll L, \quad \delta t \ll l^2/\eta. \tag{14.33}$$

The first assumption is that the size l of turbulence is much shorter than the global scale of the system, L, and the second one is that the turbulent magnetic fields are frozen-in to the fluid motions, i.e., the diffusion of magnetic fields is neglected. It is

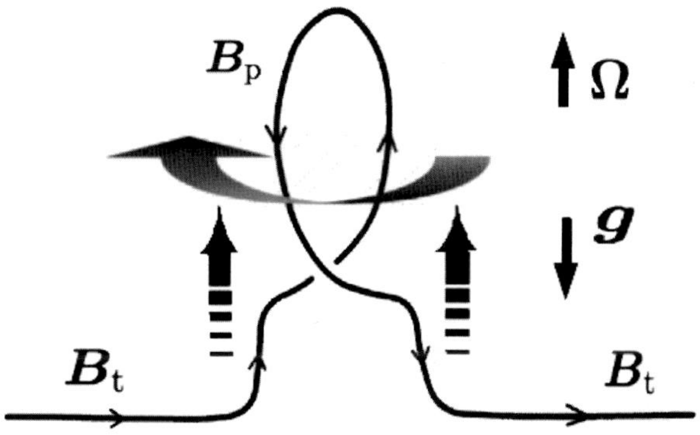

Fig. 14.4 Schematic diagram showing Parker's cyclonic dynamo

noted that the second one corresponds to the opposite case of Steenbeck et al. [see Eq. (14.16)]. Due to the second assumption (frozen-in), the magnetic flux through an arbitrary surface is unchanged with time if the boundary curve of the surface moves with the fluid. Hence, by changing surface integral to line integral by use of the Stokes theorem, we have

$$\int \boldsymbol{A}\cdot d\boldsymbol{l} = \int \boldsymbol{A}'\cdot d\boldsymbol{l}', \tag{14.34}$$

where $\boldsymbol{A}$ is a vector potential for $\boldsymbol{B}$ (not for $\boldsymbol{B}_0$), $d\boldsymbol{l}$ is a line element of boundary curve of the surface, and $d\boldsymbol{l}'$ is the corresponding line element when the boundary curve of the surface moves with fluids. Equation (14.34) implies that if a fluid element at point $\boldsymbol{X}$ moves to a point $\boldsymbol{x}(\boldsymbol{X})$, $\boldsymbol{A}(\boldsymbol{X})$ before the movement and $\boldsymbol{A}'(\boldsymbol{x})$ after the movement are related by

$$A_i'(\boldsymbol{x}) = A_j(\boldsymbol{X})\frac{\partial X_j}{\partial x_i}. \tag{14.35}$$

We now assume that the origin of the coordinates $\boldsymbol{X}$ is at the center of a turbulent element and that the vector potential $\boldsymbol{A}(\boldsymbol{X})$ can be expanded as

$$A_i(\boldsymbol{X}) = a_i + a_{im}X_m + \cdots, \tag{14.36}$$

where the coefficients a_i, a_{im}, ... are constants. Further, we assume that the fluid element which was at $\boldsymbol{X}$ before the movement is at $\boldsymbol{x}$ after the movement. The displacement is denoted by Δx_i as

$$x_i = X_i + \Delta x_i. \tag{14.37}$$

Let us denote the change of $\boldsymbol{A}$ at $\boldsymbol{x}$ due to fluid motions by $\Delta\boldsymbol{A}(\boldsymbol{x})$. Then, we have

$$\begin{aligned}\Delta A_i(\boldsymbol{x}) &= A_i'(\boldsymbol{x}) - A_i(\boldsymbol{x}) = A_j(\boldsymbol{X})\frac{\partial X_j}{\partial x_i} - A_i(\boldsymbol{x})\\ &= [a_{jk}(\Delta x_k - x_k) - a_j]\frac{\partial \Delta x_j}{\partial x_i} - a_{ik}\Delta x_k.\end{aligned} \tag{14.38}$$

Next, the volume integration of $\Delta A_i(\boldsymbol{x})$ is considered. We assume that outside the turbulent element Δx_i vanishes, and the center of the turbulent element is unchanged, i.e., $\int \Delta x_i dV = 0$. Then, from Eq. (14.38) we have

$$\int \Delta A_i(\boldsymbol{x})dV = a_{jk}\int \Delta x_k \frac{\partial \Delta x_j}{\partial x_i}dV. \tag{14.39}$$

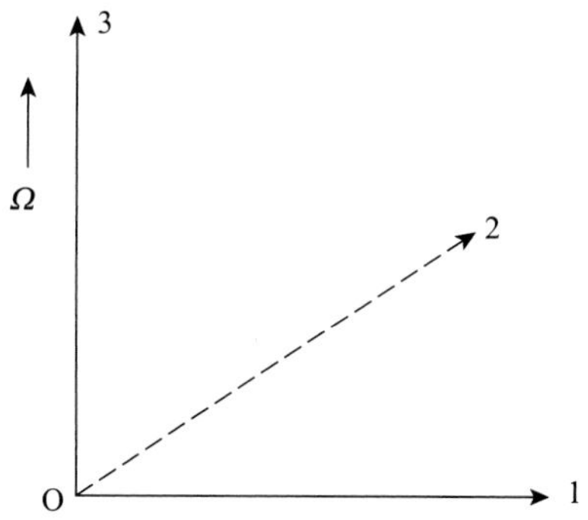

Fig. 14.5 Local coordinate systems. The 1-axis is in the toroidal direction, the 3-axis is in the direction of rotation axis, and the 2-axis is in the direction orthogonal to the 1-axis and the 3-axis

Now, Q_{ijk} defined by

$$Q_{ijk} \equiv \left\langle \int \Delta x_k \frac{\partial \Delta x_j}{\partial x_i} dV \right\rangle \tag{14.40}$$

is introduced. We have then

$$\left\langle \int \Delta A_i(\boldsymbol{x}) dV \right\rangle = a_{jk} Q_{ijk} = -\frac{1}{2} \varepsilon_{rst} B_r Q_{ist}, \tag{14.41}$$

where the brackets are the ensemble average, and in deriving the second equality

$$Q_{ijk} = -Q_{ikj} \tag{14.42}$$

has been used. The symbol ε_{ijk} is the unit alternative tensor.[7]

The frequency of occurrence of the cyclone, $\nu(\boldsymbol{x}, t)$, is taken to be independent of the strength of magnetic fields. Then, as the rate of generation of the i-component of $\boldsymbol{A}$ on Cartesian coordinates, say s_i, we have from Eq. (14.41)

$$s_i = -\frac{1}{2} \nu(\boldsymbol{x}, t) \epsilon_{kmn} Q_{ikm}(\boldsymbol{x}, t) B_{0n}. \tag{14.43}$$

This is a formal result concerning an expression for s_i (Parker 1970).

To have a concrete value of $\boldsymbol{s}$, we need ν and Q_{ikm}, for which global structures of the medium and a model of turbulence are necessary. Here, we follow again Parker (1970). As shown in Fig. 14.5, local Cartesian coordinates are introduced, where the 1-axis is in the toroidal direction, the 3-axis is in the direction of rotation axis, and the 2-axis is in the direction perpendicular to the 1- and 2-axes. In the basic

[7] ε_{ijk} is the unit alternative tensor. When there is the same subscripts among i, j, k, we have $\varepsilon_{ijk} = 0$. When i, j, and k are different, and they are the same cycle as 1, 2, 3, it is unity, while if others, it is -1.

equations of $\alpha\omega$ dynamo, i.e., the set of Eqs. (14.27) and (14.28), the dynamo term is present only in Eq. (14.27). Hence, what we need here is s_1 alone. Furthermore, what is needed for s_1 is a term proportional to B_1. Considering them we have, from (14.42) and (14.43),

$$s_1 = -\nu Q_{123} B_1. \tag{14.44}$$

If cylindrical coordinates (or polar coordinates) are adopted, s_1 and B_1 are s_φ and B_φ, respectively. That is, in the case where cylindrical coordinates are adopted, the basic set of dynamo equations, corresponding to the set of Eqs. (14.27) and (14.28), are

$$\left(\frac{\partial}{\partial t} - \eta\nabla^2\right)A_\varphi = -\nu Q_{123} B_\varphi, \tag{14.45}$$

$$\left(\frac{\partial}{\partial t} - \eta\nabla^2\right)B_\varphi = -r\frac{\partial\Omega}{\partial r}\frac{\partial A_\varphi}{\partial z} + \frac{\partial\Omega}{\partial z}\frac{\partial}{\partial r}(rA_\varphi). \tag{14.46}$$

The term $-\nu Q_{123}$ corresponds to α in the $\alpha\omega$ dynamo.

Let us estimate the magnitude of $-\nu Q_{123}$. As mentioned before and shown in Fig. 14.2, the important process for dynamo action is expansion (or contraction) and subsequent rotation of a rising (or descending) fluid element to generate poloidal component of magnetic fields. If the coordinate axes are taken as shown in Fig. 14.5, we can take the velocity of turbulent motion to u_2, and the velocity resulting from Coriolis force to u_1. Then, from the definition of Q_{ijk} given by Eq. (14.40), we have

$$Q_{123} \sim (u_1\delta t)(u_2\delta t)l^2. \tag{14.47}$$

If $u_2\delta t \sim l$ (in general cases $u_2\delta t \leq l$), we have

$$u_1 \sim \Omega_{\rm loc} u_2 \delta t \sim \Omega_{\rm loc} l, \tag{14.48}$$

where $\Omega_{\rm loc}$ is the local rotational velocity and if the Keplerian rotation is adopted, $\Omega_{\rm loc} = \Omega/4$. The frequency of occurrence, ν, is estimated as

$$\nu \sim \frac{1}{l^3\delta t}. \tag{14.49}$$

Then, from Eqs. (14.47)–(14.49), we see that

$$\alpha \sim -\nu Q_{123} \sim u_1 \sim l\Omega. \tag{14.50}$$

14.4 Dynamo Number and Dynamo Waves

In this section some important properties of dynamo action are mentioned. Magnetic diffusivity works so as to dissipate magnetic fields. Thus, for magnetic fields to be really amplified and sustained against diffusion, the dynamo processes need to be stronger than a certain degree. This issue will be argued in Sect. 14.4.1. In Sect. 14.4.2, we show that magnetic fields generated by dynamo processes can spatially propagate.

14.4.1 Dynamo Number

For magnetic fields to be maintained, the dynamo action needs to be stronger than the dissipation due to magnetic diffusivity η. One of the measures for this is the dynamo number, which represents the ratio of these two processes.

Cylindrical coordinates are introduced, and, for simplicity, the rotation is assumed to be cylindrical, i.e., $\Omega = \Omega(r)$. Then, the set of dynamo equations [see, for example, Eqs. (14.27) and (14.28)] are

$$\left(\frac{\partial}{\partial t} - \eta\nabla^2\right)A_\varphi = \alpha B_\varphi \tag{14.51}$$

$$\left(\frac{\partial}{\partial t} - \eta\nabla^2\right)B_\varphi = -r\frac{d\Omega}{dr}\frac{\partial A_\varphi}{\partial z}. \tag{14.52}$$

Elimination of A_φ from the above equations with η =const. leads to

$$\left(\frac{\partial}{\partial t} - \eta\nabla^2\right)^2 B_\varphi = \alpha G\frac{\partial B_\varphi}{\partial z}, \tag{14.53}$$

where $G \equiv -rd\Omega/dr$. For simplicity, we consider here a plane wave of $B_\varphi = \exp(nt - ikz)$ to derive a dispersion relation:

$$\left(n + \eta k^2\right)^2 + i\alpha Gk = 0. \tag{14.54}$$

By solving this equation with respect to n, we have, for $\alpha G > 0$,

$$n = -\eta k^2 \pm \frac{1}{\sqrt{2}}(1 - i)(\alpha G)^{1/2}k^{1/2} \tag{14.55}$$

and for $\alpha G < 0$

$$n = -\eta k^2 \mp \frac{1}{\sqrt{2}}(1 + i)|\alpha G|^{1/2}k^{1/2}, \tag{14.56}$$

where k has been taken to be positive. In spite of the sign of αG, magnetic fields grow if

$$N \equiv \left| \frac{\alpha G}{\eta^2 k^3} \right| > 2. \tag{14.57}$$

In the above arguments the wavelength is taken to be so short that boundary conditions are unnecessary. In the cases where long wave perturbations are considered, the critical value for N will be changed. Roughly speaking, the condition of maintenance of magnetic fields by the dynamo action will be $N > 1 \sim 10$. The quantity N defined by Eq. (14.57) is called the *dynamo number*.

14.4.2 Dynamo Waves

The arguments in Sect. 14.4.1 show that the magnetic fields generated by dynamo actions propagate in space, as well as they grow with time. Really, Eqs. (14.55) and (14.56) show that in the case of $\alpha G > 0$, the growing mode propagates in the negative z-direction, while in the case of $\alpha G < 0$, the growing mode does in the positive z-direction. An important point is that the diffusion determines whether waves grow or not, but does not contribute to the direction of propagation of the waves. The propagation of dynamo-generated magnetic fields is called *dynamo waves*. The dynamo waves are of importance in understanding observational evidence in solar magnetic fields and the solar cycle. The dynamo waves was first discussed by Parker (1970). Here, the propagation of dynamo waves along the iso-rotation surface is presented, following Yoshimura (1975).

Since η is not so important in propagation of dynamo waves, as mentioned just above. Hence, $\eta = 0$ is now adopted for simplicity. Considering applications to stellar dynamo, we introduce here polar coordinates (r, θ, φ), and the angular velocity of rotation, Ω, is taken to be a function of both r and θ, i.e., $\Omega = \Omega(r, \theta)$. In this case, from Eqs. (14.30)–(14.31) with $\eta = 0$, dynamo equations are written as

$$\frac{\partial A_\varphi}{\partial t} = \alpha B_\varphi, \tag{14.58}$$

$$\frac{\partial B_\varphi}{\partial t} = \frac{\partial \Omega}{\partial r}\frac{\partial}{\partial \theta}(A_\varphi \sin\theta) - \sin\theta \frac{\partial \Omega}{r\partial \theta}\frac{\partial}{\partial r}(r A_\varphi). \tag{14.59}$$

To rewrite the above equations in simple forms, variables A_φ and B_φ are changed to Ψ and Φ defined by

$$\Psi = r A_\varphi \sin\theta, \quad \Phi = r B_\varphi \sin\theta. \tag{14.60}$$

Then, Eqs. (14.58) and (14.59) are written in the form:

$$\frac{\partial \Psi}{\partial t} = \alpha \Phi, \tag{14.61}$$

$$\frac{\partial \Phi}{\partial t} = r\sin\theta\left(\frac{\partial \Omega}{\partial r}\frac{\partial \Psi}{r\partial \theta} - \frac{\partial \Omega}{r\partial \theta}\frac{\partial \Psi}{\partial r}\right) = r\sin\theta(\nabla\Omega \times \nabla\Psi)_\varphi. \tag{14.62}$$

If Φ is eliminated, we have an equation for Ψ:

$$\frac{\partial^2 \Psi}{\partial t^2} = \alpha r\sin\theta(\nabla\Omega \times \nabla\Psi)_\varphi. \tag{14.63}$$

This equation shows that Ψ propagates along equi-Ω surfaces. To clear this more, new curved coordinate systems (ξ, ζ) are introduced on the meridional plane, as is shown in Fig. 14.6. The ζ-axis is taken in the direction perpendicular to the curve of $\Omega(r, \theta)$ =const. and in the direction of increasing Ω. The direction of ξ is taken along the curve of $\Omega(r, \theta)$ =const. as is shown in Fig. 14.6. If this coordinate systems are adopted, Eq. (14.63) becomes

$$\frac{\partial^2 \Psi}{\partial t^2} = N_D(\xi, \zeta)\frac{\partial \Psi}{\partial \xi}, \tag{14.64}$$

where

$$N_D(\xi, \zeta) = \alpha r\sin\theta\frac{\partial \Omega}{\partial \zeta}. \tag{14.65}$$

By comparing equations (14.64) and (14.53), we see that in the case of $\alpha < 0$, the growing wave propagates in the positive direction of ξ, and in the case of $\alpha > 0$, the growing wave propagates in the negative direction of ξ.

It is noted that Eq. (14.64) has a form of a standard diffusion equation, but time and space are exchanged. That is, in the usual diffusion equation, the time

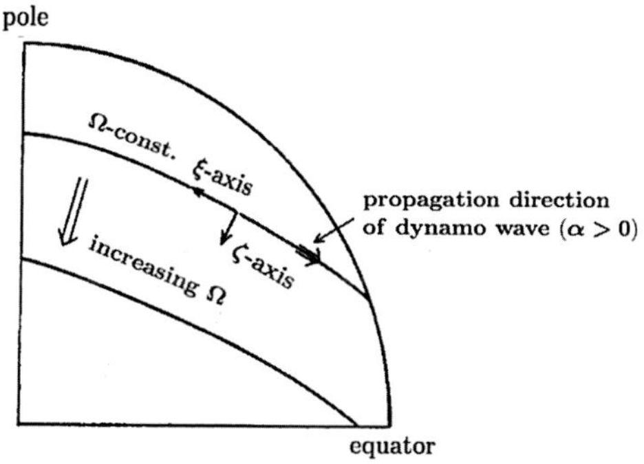

Fig. 14.6 Iso-Ω plane and propagation direction of dynamo waves

is irreversible and physical quantities have wavy forms in space. Different from this, in the case of dynamo equation, physical quantities have oscillatory structure in time and the spatial coordinate is a definite direction of propagation. From this consideration we see that what determines the solar cycle is not the magnitude of η, but the strength of dynamo action. The diffusivity η contributes to make the oscillations coherent.

References

Biermann, L.: Zeit. Naturforsch. **A5**, 65 (1950)
Cowling, T.G.: Mon. Not. Roy. Astron. Soc. **94**, 39 (1933)
Dolginov, A.Z.: Astr. Astrophys. **54**, 17 (1977)
Mestel, L., Moss, D.L.: Mon. Not. R. Astron. Soc. **1983**(204), 557 (1983)
Mestel, L., Roxburgh, I.W.: Astrophys. J. **136**, 615 (1962)
Parker, E.N.: Astrophys. J. **162**, 665 (1970)
Steenbeck, M., Krause, F., Rädler, K.-H.: Zeit. Naturforsch. **A21**, 369 (1966)
Yoshimura, H.: Astrophys. J. **201**, 740 (1975)

Chapter 15
General Stability Theorem for MHD Systems

Astrophysical objects are not always static nor steady. They are rather active even in hydrodynamical levels as already shown in Part I. In hydromagnetic systems there are many additional instability processes. Before proceeding to arguments concerning such processes, we summarize here basic stability theorems which are known in the fields of hydrodynamic and hydromagnetic systems. They are virial theorem, variational principle, and energy principle.

15.1 Virial Theorem

Isolated magnetized fluid systems are considered. In these systems various energies, such as kinetic, gravitational, thermal, and magnetic energies, are involved. The virial theorem tells us what relations are in these energies, when the systems are in a dynamically equilibrium state.

Let us consider a self-gravitating, magnetized isolated system. We use Cartesian coordinates and express the i-component of momentum equation in the form:

$$\rho\frac{dv_i}{dt} = -\frac{\partial}{\partial r_i}\left(p + \frac{B^2}{8\pi}\right) - \rho\frac{\partial\psi}{\partial r_i} + \frac{1}{4\pi}\frac{\partial}{\partial r_j}(B_i B_j), \tag{15.1}$$

where summation abbreviation has been adopted for the same indexes in a term, v_i is the i-component of velocity, and ψ is the gravitational potential. Since self-gravitating systems are considered, ψ is related to density distribution $\rho(\boldsymbol{r})$ in the system by

$$\psi(\boldsymbol{r}) = -G\int\frac{\rho(\boldsymbol{r}')}{|\boldsymbol{r}-\boldsymbol{r}'|}dV', \tag{15.2}$$

© Springer Nature Singapore Pte Ltd. 2020, corrected publication 2023
S. Kato, J. Fukue, *Fundamentals of Astrophysical Fluid Dynamics*,
Astronomy and Astrophysics Library,
https://doi.org/10.1007/978-981-15-4174-2_15

where the integration is performed over the whole volume, and other notations have their usual meanings.

Equation (15.1) is multiplied by r_i in both sides of the equation and is summed up. The resulting equation is then integrated over the whole space. First, we consider the left-hand side of the equation. Since

$$r_i \frac{dv_i}{dt} = r_i \frac{d^2 r_i}{dt^2} = \frac{1}{2} \frac{d^2}{dt^2} r^2 - v^2, \tag{15.3}$$

we have

$$\int \rho r_i \frac{dv_i}{dt} dV = \int r_i \frac{dv_i}{dt} dm = \frac{1}{2} \frac{d^2}{dt^2} \int r^2 dm - \int v^2 dm, \tag{15.4}$$

where $r^2 = \Sigma r_i^2$ and in deriving the last relation of Eq. (15.4) we have used the fact that a volume element dm ($\equiv \rho dV$) is time-independent.

Here, we define the moment of inertia around the origin by

$$I = \int r^2 dm, \tag{15.5}$$

and the total kinetic energy of the system, T, by

$$T = \frac{1}{2} \int v^2 dm. \tag{15.6}$$

Then, Eq. (15.4) is written as

$$\int \rho r_i \frac{dv_i}{dt} dV = \frac{1}{2} \frac{d^2 I}{dt^2} - 2T. \tag{15.7}$$

Next, let us consider the integration of the gas-pressure term in Eq. (15.1). By performing the integration by part, we have

$$-\int r_i \frac{\partial p}{\partial r_i} dV = \int 3p dV, \tag{15.8}$$

where the surface integral has been taken to vanish, since the whole volume of the system is considered. Since the total internal energy of the systems, U, is given by

$$U = \int \rho C_V dV, \tag{15.9}$$

where C_V is the specific heat per constant volume and is related to p by $p = (\gamma - 1)\rho C_V T$, γ being the ratio of specific heats. Hence, using Eq. (15.9), we can reduce Eq. (15.8) to

$$-\int r_i \frac{\partial p}{\partial r_i} dV = 3(\gamma - 1)U. \tag{15.10}$$

Thirdly, the term of the gravitational force in Eq. (15.1) is considered. By considering Eq. (15.2), we have

$$\begin{aligned} -\int \rho r_i \frac{\partial \psi}{\partial r_i} dV &= -G \int\int \frac{\rho(\boldsymbol{r})\rho(\boldsymbol{r}')r_i\left(r_i - r_i'\right)}{|\boldsymbol{r} - \boldsymbol{r}'|} dV dV' \\ &= -\frac{1}{2}G \int\int \frac{\rho(\boldsymbol{r})\rho(\boldsymbol{r}')}{|\boldsymbol{r} - \boldsymbol{r}'|} dV dV' \equiv \Omega. \end{aligned} \tag{15.11}$$

Here, the second equality is derived by replacing r_i and r_i'. The derived result is the gravitational potential energy Ω.

Finally, for the magnetized terms in Eq. (15.1), we have

$$\int r_i \left[-\frac{\partial}{\partial r_i}\left(\frac{B^2}{8\pi}\right) + \frac{1}{4\pi}\frac{\partial}{\partial r_i} B_i B_j \right] dV = \int \frac{B^2}{8\pi} dV = \mathcal{M}, \tag{15.12}$$

where $\mathcal{M}$ is the magnetic energy of the whole volume.

Summing the above results, from Eqs. (15.7), (15.10), (15.11) and (15.12), we have

$$\frac{1}{2}\frac{d^2 I}{dt^2} = 2T + 3(\gamma - 1)U + \Omega + \mathcal{M}. \tag{15.13}$$

In the case where the system is in a dynamically equilibrium, the left-hand side of the equation vanishes, and we have

$$2T + 3(\gamma - 1)U + \Omega + \mathcal{M} = 0. \tag{15.14}$$

This equation is called the *virial theorem*. It is noted here that this equation is realized only for isolated systems. If not isolated, terms of surface integrals are added.

The virial theorem is applied to various problems. Here we consider an application to stability of stars. If a star has no internal motion and is at a dynamically equilibrium state, we have from Eq. (15.14)

$$3(\gamma - 1)U + \mathcal{M} + \Omega = 0. \tag{15.15}$$

The total energy of the star, E, is given by

$$E = U + \mathcal{M} + \Omega. \tag{15.16}$$

By subtracting internal energy U from the above two equations, we have

$$E = -\frac{3\gamma - 4}{3(\gamma - 1)}(|\Omega| - \mathcal{M}). \tag{15.17}$$

Equation (15.17) shows that for stars with $\gamma = 4/3$ a steady state is realized when $E = 0$ (independent of the radius of the stars). This means that if we consider a sequence of equilibrium configurations in which γ varies continuously, a change from stability to instability sets in at $\gamma = 4/3$ (Jeans 1919). In other words, a necessary condition for stability of stars is

$$\gamma > 4/3 \tag{15.18}$$

in the case where $|\Omega| > \mathcal{M}$.

Let us consider somewhat in detail why a star is dynamically unstable (more accurately, absence of equilibrium state) if $\gamma < 4/3$. For simplicity, a non-magnetized star is considered. In the case of spherically symmetry, the condition of dynamical equilibrium is

$$-\frac{1}{\rho}\frac{\partial p}{\partial r} = \frac{GM_r}{r^2}, \tag{15.19}$$

where M_r is the mass of the star till the radius r. Now, homologous expansion or contraction of the star is considered. The homologous expansion (contraction) means that an expansion (contraction) occurs in the way that a fluid element at r moves to the radius of $r^*(= ar)$, where a is a constant independent of r and $a > 1$ for expansion and $a < 1$ for contraction. Therefore, the condition that by a homologous expansion and contraction the dynamical equilibrium is not violated is found to be $p(r^*) = p(r)a^{-4}$, which can be found by substituting $r^*(= ar)$ and $\rho \propto a^{-3}$ into Eq. (15.19). The gas which has such a ρ–p relation is a polytropic gas with $\gamma = 4/3$. This means that the polytropic gas with $\gamma = 4/3$ is just neutral to expansion and contraction. In the gases with $\gamma < 4/3$ the decrease of pressure force by expansion is weaker than the decrease of gravitational force, and thus the gas expands further. In the case of contraction the situation is opposite. That is, the increase of pressure force by contraction is weaker than the increase of gravitational force, and thus the contraction proceeds further.

In monatomic gases, $\gamma = 5/3$, but there are cases where in a wide interior region of a star γ becomes smaller than 4/3. As will be described in Part III, in the gases where radiation pressure is stronger than gaseous pressure the effective γ tends to 4/3. Furthermore, in the temperature–density domain where ionization and decombination occur predominately, γ can decrease as close as unity.

It is noted that if $\mathcal{M} > |\Omega|$ the system is unstable because the gravitational force cannot confine magnetic fields which have tendency to expand infinity by their own repulsive forces. Let us roughly estimate the magnetic fields necessary for $\mathcal{M} > |\Omega|$. If we adopt $|\Omega| \sim GM^2/R$ and $\mathcal{M} \sim \langle B^2 \rangle R^3$, the above condition is $\langle B^2 \rangle^{1/2} > G^{1/2}M/R^2$, where M and R are mass and radius of the star, and $\langle B^2 \rangle$ is the mean value of B^2 in the star. By introducing numerical values, we have

$$\langle B^2 \rangle^{1/2} > 1.0 \times 10^8 \frac{M/M_\odot}{(R/R_\odot)^2} \text{gauss}. \tag{15.20}$$

Such stars as magnetors have strong magnetic fields as high as 10^{12} gauss, but their radius is much smaller than $R_\odot$, and the condition given by (15.20) is not realized.

Finally, some comments related to the virial theorem are presented. Equation of the virial theorem (15.13) has been derived by neglecting terms of surface integrals. Such terms cannot be neglected in cases where non-isolated systems are studied. Second, the virial theorem has been extended to relations among tensor quantities. The main purpose of such extension is to apply it to stability analyses.

Third, it is noted that the fact that the marginal state of dynamical instability of spherical stars is $\gamma = 4/3$ is modified if general relativistic effects are considered. As mentioned before, if a spherical star contracts, the gravitational force increases, while the pressure force against it also increases. The balance of these two forces cannot be realized when $\gamma < 4/3$ and the star contracts further (gravitational instability). The marginal value of γ is thus 4/3. In the case of general relativistic stars, however, the increase of the gravitational force due to contraction of the star is stronger than the case of the Newtonian dynamics. Hence, it is easily supposed that the marginal state of the dynamical instability is not at $\gamma = 4/3$, but at a value of γ larger than $\gamma = 4/3$, depending on the structure of the star. In other words, there is a critical radius, R_c, of stars over which the stars are dynamically unstable when γ and the mass of the stars, M, are given. The results for non-rotating stars show (Chandrasekhar 1964)

$$R_c = \frac{K}{\gamma - 4/3} \frac{2GM}{c^2}, \tag{15.21}$$

where K is a dimensionless quantity of the order of unity, depending on the structure of stars. This result shows that in the case where γ is close to 4/3, the stars are unstable even when their radius is much larger than the Schwarzschild radius ($= 2GM/c^2$).

In supermassive stars ($M \sim 10^8 M_\odot$) the radiation pressure predominates over gas pressure and γ is close to 4/3. Hence, these stars are dynamically unstable and collapse with dynamical timescale even if the radius is much large compared with the Schwarzschild radius. This is one of the main reasons why supermassive stars were excluded as a possible candidate of quasars.

15.2 Lagrangian Description of MHD Perturbations

This subsection is a preparation for studying variational and energy principles which will be made in Sects. 15.3 and 15.4, respectively. The equation of motion describing MHD motions is

$$\frac{D\boldsymbol{v}}{Dt} = -\nabla\psi - \frac{1}{\rho}\nabla p + \frac{1}{4\pi\rho}\mathrm{curl}\,\boldsymbol{B}\times\boldsymbol{B}. \tag{15.22}$$

On such systems small amplitude perturbations are superposed. The equation describing the perturbations is written as (see Chap. 2)

$$\frac{D_0^2\boldsymbol{\xi}}{Dt^2} = \delta\Big(-\nabla\psi - \frac{1}{\rho}\nabla p + \frac{1}{4\pi\rho}\mathrm{curl}\boldsymbol{B}\times\boldsymbol{B}\Big), \tag{15.23}$$

where δ represents the Lagrangian variation and $\boldsymbol{\xi}(\boldsymbol{r}, t)$ is a displacement vector associated with the perturbations, $\boldsymbol{B}$ is the magnetic field flux density, and other notations are the same as those in Sect. 2.5. This is a direct extension of the formulation for hydromagnetic systems.

Hereafter, the unperturbed state is assumed to be steady with flow $\boldsymbol{v}_0(\boldsymbol{r})$. Perturbations are assumed to have small amplitudes and are non-dissipative. Then, Eq. (15.23) is written as (see Chap. 2)

$$\rho_0\frac{\partial^2\boldsymbol{\xi}}{\partial t^2} + 2\rho_0(\boldsymbol{v}_0\cdot\nabla)\frac{\partial\boldsymbol{\xi}}{\partial t} + \mathcal{L}(\boldsymbol{\xi}) = 0, \tag{15.24}$$

where $\mathcal{L}(\boldsymbol{\xi})$ consists of hydrodynamic and hydromagnetic parts as

$$\mathcal{L}(\boldsymbol{\xi}) = \mathcal{L}^{\mathrm{G}}(\boldsymbol{\xi}) + \mathcal{L}^{\mathrm{B}}(\boldsymbol{\xi}). \tag{15.25}$$

The hydrodynamical part of $\mathcal{L}$, say $\mathcal{L}^{\mathrm{G}}$, has been derived in Sect. 2.5 and is for nonself-gravitating systems [see Eq. (2.47)]

$$\begin{aligned}\mathcal{L}^{\mathrm{G}}(\boldsymbol{\xi}) = {} & \rho_0(\boldsymbol{v}_0\cdot\nabla)(\boldsymbol{v}_0\cdot\nabla)\boldsymbol{\xi} + \rho_0(\boldsymbol{\xi}\cdot\nabla)\nabla\psi_0 \\ & +\nabla\Big[(1-\Gamma_1)p_0\,\mathrm{div}\boldsymbol{\xi}\Big] - p_0\nabla\mathrm{div}\boldsymbol{\xi} - \nabla\Big[(\boldsymbol{\xi}\cdot\nabla)p_0\Big] \\ & +(\boldsymbol{\xi}\cdot\nabla)\nabla p_0. \end{aligned} \tag{15.26}$$

Here, we try to express $\mathcal{L}^{\rm B}$ in terms of the displacement vector $\boldsymbol{\xi}$. Since we are interested only in the first-order quantities with respect to perturbations, we have

$$\begin{aligned}\mathcal{L}^{\rm B} &= \rho_0\delta\biggl[\frac{1}{4\pi\rho}\biggl(-\frac{1}{2}\nabla\boldsymbol{B}^2+(\boldsymbol{B}\cdot\nabla)\boldsymbol{B}\biggr)\biggr]\\ &= -\frac{1}{4\pi}\biggl[\frac{\delta\rho}{\rho_0}\biggl(-\frac{1}{2}\nabla\boldsymbol{B}_0^2+(\boldsymbol{B}_0\cdot\nabla)\boldsymbol{B}_0\biggr)\biggr]\\ &\quad+\frac{1}{4\pi}\biggl[-\frac{1}{2}\delta\left(\nabla\boldsymbol{B}^2\right)+\delta[(\boldsymbol{B}\cdot\nabla)\boldsymbol{B}]\biggr].\end{aligned} \tag{15.27}$$

Because the second-order terms with respect to $\boldsymbol{\xi}$ are neglected, we have

$$\frac{\delta\rho}{\rho_0}=-\mathrm{div}\boldsymbol{\xi} \quad \text{and} \quad \delta B_i = B_\ell\frac{\partial\xi_i}{\partial r_\ell}-B_i\mathrm{div}\,\boldsymbol{\xi}, \tag{15.28}$$

where the summation abbreviation is adopted, and the second equation in Eq. (15.28) comes from the induction equation. The induction equation, $\partial\boldsymbol{B}/\partial t = \mathrm{curl}\,(\boldsymbol{v}\times\boldsymbol{B})$, is written when the higher order term is neglected as

$$\frac{d\boldsymbol{B}}{dt}=\mathrm{curl}\,(\boldsymbol{v}\times\boldsymbol{B})+(\boldsymbol{v}\cdot\nabla)\boldsymbol{B}=(\boldsymbol{B}\cdot\nabla)\boldsymbol{\xi}-\boldsymbol{B}\,\mathrm{div}\boldsymbol{\xi}. \tag{15.29}$$

By substituting $\delta\rho$ and δB_i given by Eqs. (15.28) into Eqs. (15.27), we have an expression for $\mathcal{L}^{\rm B}$ expressed explicitly in terms of $\boldsymbol{\xi}$. The explicit expression is omitted here, since it is somewhat lengthy. For a detailed expression, see, for example, Bernstein et al. (1958).

An important characteristic of $\mathcal{L}^{\rm B}$ is that it is Hermitian (self-adjoint) in the sense that

$$\int\boldsymbol{\eta}\cdot\mathcal{L}^{\rm B}(\boldsymbol{\xi})dV=\int\boldsymbol{\xi}\cdot\mathcal{L}^{\rm B}(\boldsymbol{\eta})dV, \tag{15.30}$$

where $\boldsymbol{\xi}$ and $\boldsymbol{\eta}$ are arbitrary functions of $\boldsymbol{r}$, but are assumed to vanish on the surface of the system. Since we have already shown that the operator $\mathcal{L}^{\rm G}$ is Hermitian (Sect. 2.5.2), $\mathcal{L}=\mathcal{L}^{\rm G}+\mathcal{L}^{\rm B}$ is Hermitian.

15.2.1 Conservation of Wave Energy and Orthogonality of Normal Modes of Oscillations

15.2.1.1 Conservation of Wave Energy

In Chap. 2, the concept of wave energy and its conservation in adiabatic perturbations were shown for non-magnetized fluid systems. Here we emphasize that they are extended to magnetized MHD systems, as a preparation for studying energy principle in Sect. 15.4.

It is obvious that the concept of wave energy can be generalized to MHD systems as

$$E = \frac{1}{2}\int \rho_0\left[\left(\frac{\partial \boldsymbol{\xi}}{\partial t}\right)^2 + \boldsymbol{\xi}\cdot\mathcal{L}(\boldsymbol{\xi})\right]dV, \tag{15.31}$$

where $\mathcal{L}(\boldsymbol{\xi})$ consists of hydrodynamic and hydromagnetic parts as [see Eq. (15.25)]

$$\mathcal{L}(\boldsymbol{\xi}) = \mathcal{L}^{\mathrm{G}}(\boldsymbol{\xi}) + \mathcal{L}^{\mathrm{B}}(\boldsymbol{\xi}). \tag{15.32}$$

In the case of non-magnetized fluid systems, the conservation of wave energy was derived by use of the Hermitian properties of $\mathcal{L}^{\mathrm{G}}(\boldsymbol{\xi})$ (see Chap. 2). As mentioned at the beginning of this section, the operator $\mathcal{L}^{\mathrm{B}}(\boldsymbol{\xi})$ is also Hermitian. Hence, by the same procedures as in Chap. 2, we can show that the wave energy defined by Eq. (15.31) is also time-independent. The first and second terms on the right-hand side of Eq. (15.31) can be interpreted as kinetic energy and potential energy of perturbation, respectively. As a preparation for energy principle in the next section (Sect. 15.4), the potential energy is denoted by W:

$$W = \frac{1}{2}\int \rho_0\boldsymbol{\xi}\cdot\mathcal{L}(\boldsymbol{\xi})dV. \tag{15.33}$$

15.2.1.2 Orthogonality of Normal Modes

Let us consider solutions of Eq. (15.24) with relevant boundary conditions. The solutions are taken to be a set of time-periodic functions. Each solution, $\boldsymbol{\xi}_\alpha(\boldsymbol{r}, t)$, is written as

$$\boldsymbol{\xi}_\alpha(\boldsymbol{r}, t) = \Re\left[\hat{\boldsymbol{\xi}}_\alpha(\boldsymbol{r})\exp(i\omega_\alpha t)\right], \quad (\alpha = 1, 2, 3, \ldots). \tag{15.34}$$

In the case of non-magnetized fluid gases we have shown the presence of orthogonal relations among these normal modes of oscillations, when the unperturbed state has no motion (Sect. 2.5.4). What was used in Chap. 2 to derive the orthogonality was the fact that the operator $\mathcal{L}^{\mathrm{G}}(\boldsymbol{\xi})$ is Hermitian. In the present case of MHD systems $\mathcal{L}^{\mathrm{G}}(\boldsymbol{\xi})$ is extended to $\mathcal{L}^{\mathrm{G}}(\boldsymbol{\xi}) + \mathcal{L}^{\mathrm{B}}(\boldsymbol{\xi})$. As was mentioned at the beginning of this section, $\mathcal{L}^{\mathrm{B}}(\boldsymbol{\xi})$ is also Hermitian. Hence, the arguments on orthogonality among eigenfunctions can be extended to the present case of MHD systems. That is, as in the case of hydrodynamical perturbations, we have [see Eq. (2.60)]

$$(\omega_\alpha + \omega_\beta)\langle\rho_0\hat{\boldsymbol{\xi}}_\alpha\cdot\hat{\boldsymbol{\xi}}^*_\beta\rangle + 2i\left\langle\rho_0\hat{\boldsymbol{\xi}}_\alpha\cdot[(\boldsymbol{v}_0\cdot\nabla)\hat{\boldsymbol{\xi}}^*_\beta]\right\rangle = 0, \tag{15.35}$$

where the brackets $\langle\ \rangle$ means the integration of the quantities in the brackets over the whole volume. In the particular case of no motion in the unperturbed state, i.e., $\boldsymbol{v}_0 = 0$, we have again

$$\left\langle \rho_0 \hat{\boldsymbol{\xi}}_\alpha \cdot \hat{\boldsymbol{\xi}}_\beta^* \right\rangle = 0 \qquad \text{for} \quad \alpha \neq \beta \tag{15.36}$$

It should be further noticed that if the flow in the unperturbed state is a purely rotational motion, i.e., $\boldsymbol{v}_0 = r\Omega \boldsymbol{i}_\varphi$, and the perturbations considered are restricted only to axisymmetric motions, the second term on the right-hand side of Eq. (15.35) vanishes and we have again the orthogonal relation (15.36).

15.3 Variational Principle

The most popular method to study stability of systems is "normal mode analyses," as demonstrated in some chapters in Part I. In this method, differential equations describing small amplitude perturbations are solved with proper boundary condition as eigenvalue problems. From the set of eigenvalues and eigenfunctions, we know the types of perturbations to which the system becomes unstable. The analyses in Chaps. 6–8 are based on this method. This method is powerful and useful to understand the physical causes of instability, because eigenfunctions as well as eigenvalues are known. In some complicated problems, however, their calculations are troublesome. Compared with the normal mode analyses, variational methods are sometimes simpler and have been developed since Rayleigh and Ritz. The variational method is certainly simpler and powerful in some cases compared with the normal mode analyses, but to obtain accurate eigenvalues and eigenfunctions many parameters are necessary in trial functions.

We should notice here that the variational method is applicable only when the basic equations describing perturbations are self-adjoint (Hermitian). There are many attempts to generalize the classical (conventional) variational principles (e.g., Clement 1964; Lynden-Bell and Ostriker 1967).

15.3.1 Outline of Variational Principle

Let us consider isolated MHD systems with no internal motion ($\boldsymbol{v}_0 = 0$) in the unperturbed state. In such systems, as mentioned before, the equation describing small amplitude perturbations can be written as [see Eq. (15.24)]

$$\rho_0 \frac{\partial^2 \boldsymbol{\xi}}{\partial t^2} + \mathcal{L}(\boldsymbol{\xi}) = 0, \tag{15.37}$$

where $\boldsymbol{\xi}$ is the perturbation and $\mathcal{L}$ is a self-adjoint (Hermitian) operator. If perturbations are written as $\boldsymbol{\xi}(\boldsymbol{r}, t) = \Re[\hat{\boldsymbol{\xi}}(\boldsymbol{r})\exp(i\omega t)]$, we have

$$-\omega^2\rho_0\hat{\boldsymbol{\xi}} + \mathcal{L}(\hat{\boldsymbol{\xi}}) = 0. \tag{15.38}$$

Including self-gravitating cases the operator $\mathcal{L}(\hat{\boldsymbol{\xi}})$ is generally Hermitian (self-adjoint operator), and we have

$$\int \hat{\boldsymbol{\eta}}^* \cdot \mathcal{L}(\hat{\boldsymbol{\xi}})dV = \int \hat{\boldsymbol{\xi}}^* \cdot \mathcal{L}(\hat{\boldsymbol{\eta}})dV, \tag{15.39}$$

for arbitrary displacement vectors $\hat{\boldsymbol{\xi}}$ and $\hat{\boldsymbol{\eta}}$ under the condition that the surface integrals vanish, where the asterisk represents the complex conjugate.

Now, we consider an arbitrary function $\hat{\boldsymbol{\xi}}$ which satisfies boundary conditions[1] and introduce λ^2 defined in terms of $\hat{\boldsymbol{\xi}}$ by

$$\lambda^2 = \int \hat{\boldsymbol{\xi}}^* \cdot \mathcal{L}(\hat{\boldsymbol{\xi}})dV \bigg/ \int \rho_0\hat{\boldsymbol{\xi}}^* \cdot \hat{\boldsymbol{\xi}}dV. \tag{15.40}$$

The variational principle tells us that functional forms $\hat{\boldsymbol{\xi}}$'s which make the value of λ^2 stationary are the eigenfunctions and the corresponding stationary values of λ's are eigenvalues.

The proof of this variational principle is as follows. The change of λ^2 by variation of $\hat{\boldsymbol{\xi}}$, say $\delta\hat{\boldsymbol{\xi}}$, is denoted $\delta\lambda^2$. Then, we have

$$\begin{aligned}\delta\lambda^2\left(\int \rho_0\hat{\boldsymbol{\xi}}^* \cdot \hat{\boldsymbol{\xi}}dV\right)^2 &= \int \left(\delta\hat{\boldsymbol{\xi}}^* \cdot \mathcal{L}(\hat{\boldsymbol{\xi}}) + \hat{\boldsymbol{\xi}}^* \cdot \mathcal{L}(\delta\hat{\boldsymbol{\xi}})\right)dV \cdot \int \rho_0\hat{\boldsymbol{\xi}}^* \cdot \hat{\boldsymbol{\xi}}dV \\ &\quad - \int \hat{\boldsymbol{\xi}}^* \cdot \mathcal{L}(\hat{\boldsymbol{\xi}})dV \int \rho_0\left(\delta\hat{\boldsymbol{\xi}}^* \cdot \hat{\boldsymbol{\xi}} + \hat{\boldsymbol{\xi}}^* \cdot \delta\hat{\boldsymbol{\xi}}\right)dV. \end{aligned} \tag{15.41}$$

Considering that $\mathcal{L}$ is Hermitian (self-adjoint) [see Eq. (15.39)], the above equation can be written as

$$\delta\lambda^2\left(\int \rho_0\hat{\boldsymbol{\xi}}^* \cdot \hat{\boldsymbol{\xi}}dV\right) = 2\int \delta\hat{\boldsymbol{\xi}}^* \cdot \left[\mathcal{L}(\hat{\boldsymbol{\xi}}) - \lambda^2\rho_0\hat{\boldsymbol{\xi}}\right]dV. \tag{15.42}$$

We require that for arbitrary variation $\delta\hat{\boldsymbol{\xi}}$, $\delta\lambda^2$ vanishes. This is realized only when $\mathcal{L}(\hat{\boldsymbol{\xi}}) - \lambda^2\rho_0\hat{\boldsymbol{\xi}} = 0$ holds. That is, comparing this condition with Eq. (15.38), we see that a stationary value of λ is one of eigenvalues ω's and the function which makes λ^2 stationary is one of eigenfunctions.

[1]It is noticed that $\hat{\boldsymbol{\xi}}$ given here is not an eigenfunction of Eq. (15.38), but a trial function although the same notation is used.

This result can be expressed as a variational problem:

$$\int \hat{\boldsymbol{\xi}}^* \cdot \mathcal{L}(\hat{\boldsymbol{\xi}}) dV = \text{stationary} \quad \text{for} \int \rho_0 \hat{\boldsymbol{\xi}}^* \cdot \hat{\boldsymbol{\xi}} dV = 1. \tag{15.43}$$

The functions $\hat{\boldsymbol{\xi}}$ which make the first integral of Eq. (15.43) stationary are eigenfunctions and the stationary values are eigenvalues. The second equation of Eq. (15.43) is a normalization condition.

This variational problem is usually solved by introducing trial functions, $\hat{\boldsymbol{\eta}}_m$ $(m = 1, 2, 3, \ldots)$, which satisfy boundary conditions. For simplicity, the set of these functions is taken to be orthogonal[2] :

$$\langle \rho_0 \hat{\boldsymbol{\eta}}_m \cdot \hat{\boldsymbol{\eta}}_n^* \rangle = \delta_{nm}. \tag{15.44}$$

By use of these functions a trial function (an approximate function), $\hat{\boldsymbol{\xi}}$, is approximated as

$$\hat{\boldsymbol{\xi}}(\boldsymbol{r}) = \sum_{m=1}^{n} a_m \hat{\boldsymbol{\eta}}_m(\boldsymbol{r}), \tag{15.45}$$

where a_m's are arbitrary parameters to be determined below. Substitution of Eq. (15.45) into Eq. (15.43) leads to a binomial equation of a_m's:

$$\sum_{i,k} A_{ik} a_i a_k = \text{stationary}, \quad \sum a_i^2 = 1, \tag{15.46}$$

where A_{ik} is symmetric with respect to i and k and defined by

$$A_{ik} = \int \hat{\boldsymbol{\eta}}_i^* \cdot \mathcal{L}(\hat{\boldsymbol{\eta}}_k) dV. \tag{15.47}$$

The problem of Eq. (15.46) can be solved by use of Lagrange's method of undetermined multiplier. That is, the problem of Eq. (15.46) is rewritten in the following form:

$$\sum_{i.k} A_{ik} a_i a_k - \lambda \sum_i a_i^2 = \sum_{i,k} (A_{ik} - \lambda \delta_{ik}) a_i a_k = \text{stationary}. \tag{15.48}$$

---

[2] It is possible to make these trial functions orthogonal, since the equation describing perturbations is self-adjoint.

This problem is reduced to solve the following first-order simultaneous algebraic equations

$$\sum_i (A_{ik} - \lambda\delta_{ik})a_k = 0 \quad (i, k = 1, 2, 3 \ldots). \tag{15.49}$$

This problem is to determine the eigenvalue, λ, of the following determinant:

$$\begin{vmatrix} A_{11} - \lambda & A_{12} & \ldots. & A_{1n} \\ A_{21} & A_{22} - \lambda & \ldots. & A_{2n} \\ \ldots. & \ldots. & \ldots. & \ldots. \\ A_{n1} & A_{2n} & \ldots. & A_{nn} - \lambda \end{vmatrix} = 0.$$

By solving this equation we obtain approximate values of λ's. Then, substituting one of them into Eq. (15.49) we obtain the ratio of a_1,..., and a_n.

Finally, it is noted that the above variational method can be applied even to rotating systems if perturbations are restricted to axisymmetric ones. This is because in this case the term of $\partial/\partial t$ in Eq. (15.24) vanishes and the equation describing perturbations is reduced to Eq. (15.37). An extension to arbitrary perturbations (not restricted to axisymmetric perturbations) in uniform rotating systems has been made by Clement (1964) (see also Tassoul 1978). A more general extension with arbitrary motions in unperturbed systems is made by Lynden-Bell and Ostriker (1967), although the meaning of stationary value is a little restricted.

15.4 Energy Principle

The energy principle has a close relation to the variational principle. Different from the variational principle, the energy principle cannot give detailed information on eigenfunctions which make the system unstable. However, the energy principle is sometimes simpler in judging whether a system is unstable or not.

15.4.1 Potential Energy of Perturbations and Stability

We consider the same situations as those in the cases where the variational principle is applied. That is, (1) unperturbed MHD systems are isolated and have no internal motions, or (2) the unperturbed systems can have rotation, but the perturbations are restricted to be axisymmetric ones. We examine stability of such systems against adiabatic perturbations.

As mentioned in Sect. 15.2.1, energy of perturbations is conserved with time, and it can be interpreted as consisting of kinetic energy and potential energy.

The potential energy W has been expressed in the form of Eq. (15.33). Now, we introduce an arbitrary complex variable $\hat{\boldsymbol{\xi}}$ which satisfies boundary conditions and define W in terms of $\hat{\boldsymbol{\xi}}$ as

$$W = \frac{1}{2}\int \rho_0 \hat{\boldsymbol{\xi}}^* \cdot \mathcal{L}(\hat{\boldsymbol{\xi}}) dV, \tag{15.50}$$

where the asterisk * denotes complex conjugate. It is noted that W is a real quantity, because the operator $\mathcal{L}$ is Hermitian.

The eigenfunctions of oscillations which satisfy the wave equation (15.37) are expressed as $\hat{\boldsymbol{\xi}}_n$'s, and taken to be a complete set. Then, $\hat{\boldsymbol{\xi}}$ is expressed as

$$\hat{\boldsymbol{\xi}} = \sum_m a_m \hat{\boldsymbol{\xi}}_m(\boldsymbol{r}), \tag{15.51}$$

where a_m's are expansion coefficients. Then, we can write W defined by Eq. (15.50) in the form:

$$W = \frac{1}{2}\sum_n\sum_m a_m^* a_n \int \rho_0 \hat{\boldsymbol{\xi}}_m^* \cdot \mathcal{L}(\hat{\boldsymbol{\xi}}_n) dV = \frac{1}{2}\sum_n\sum_m a_m^* a_n \omega_n^2 \int \rho_0 \hat{\boldsymbol{\xi}}_m^* \cdot \hat{\boldsymbol{\xi}}_n dV$$
$$= \frac{1}{2}\sum_n \omega_n^2 |a_n|^2 \int \rho_0 \hat{\boldsymbol{\xi}}_n^* \cdot \hat{\boldsymbol{\xi}}_n dV. \tag{15.52}$$

In deriving the second equality the wave equation (15.38) has been used. Furthermore, in deriving the third equality, the fact that eigenfunctions can be taken to be orthogonal [see Eq. (15.36)] is adopted.

The above Eq. (15.52) presents an important result called *energy principle*. That is, if there is an eigenmode with $\omega_n^2 < 0$ (i.e., unstable mode) we can make W negative by adopting a relevant form of $\hat{\boldsymbol{\xi}}$. Conversely, if we can make W negative for a trial function, $\hat{\boldsymbol{\xi}}$, under the condition that $\hat{\boldsymbol{\xi}}$ satisfies boundary conditions, there is an eigenmode with $\omega^2 < 0$. In other words, let us consider arbitrary perturbations which satisfy necessary boundary conditions. If there is a perturbation which makes the potential energy W negative, the system is unstable. Inversely, if the system is unstable, there is always such a perturbation.

This energy principle is a powerful method to judge whether the system in consideration is dynamically unstable or not, but we have no definite information concerning eigenfunctions which make the system unstable. This is a different point, compared with the variational methods.

It is noticed that for this method of stability analyses to be valid, the operator $\mathcal{L}$ should be Hermitian when boundary conditions required are satisfied. Trial functions should be taken so that they satisfy the boundary conditions. This latter requirement is sometimes troublesome in applying the energy principle. This troublesome has been removed in Bernstein et al. (1958). In their formulations some additional terms, $W_{\rm S}$ and $W_{\rm U}$, are added to W. Here, $W_{\rm S}$ is a term related to the surface condition and W_U is that related to the outside (empty space) of the fluid system. See Bernstein et al. (1958) for details.

References

Bernstein, I.B., Frieman, E.A., Kruskal, M.D., Kulsrud, R.M.: Proc. Roy. Soc. (London) **A244**, 17 (1958)
Chandrasekhar, S.: , Astrophys. J. **139**, 664 (1964)
Clement, M.J.: Astrophys. J. **140**, 1045 (1964)
Jeans, J.H.: Problems of Cosmology and Stellar Dynamics. Cambridge University Press (1919)
Lynden-Bell, D., Ostriker, J.P.: Monthly Not. Roy, Astron. Soc. **136**, 293 (1967)
Tassoul, J.-L.: Theory of Rotating Stars. Princeton University Press Princeton (1978)

Chapter 16
Instabilities Related to Magnetic Fields

In Chaps. 6, 7, and 8, various types of hydrodynamical instabilities related to astrophysical phenomena have been described. In many active phenomena, however, instabilities caused by characteristics of magnetized fields have important roles. The purpose of this chapter is to explore magnetic fields-oriented instabilities.

16.1 Magnetic Buoyancy and Interchange Instability

In this section, let us first discuss the concept of magnetic buoyancy. Next, the interchange instability is considered.

16.1.1 Magnetic Buoyancy

For simplicity, an isolated magnetic bubble is considered in a stratified medium which is in subject to gravitational fields (see the upper panel of Fig. 16.1). The pressure in the bubble, p', consists of gas pressure $p'_{\rm gas}$ and magnetic pressure $p'_{\rm mag}$, i.e., $p' = p'_{\rm gas} + p'_{\rm mag}$. (The quantities inside the bubble are denoted by attaching a prime.) On the other hand, the surrounding medium is assumed to consist of gas alone, whose pressure is thus $p_{\rm gas}$. If the heat exchange between the bubble and the surrounding gas is efficient, the temperature in the bubble, T', is equal to that of the surrounding medium, T, i.e., $T' = T$. Then, the gas density in the bubble becomes lower than that of the surroundings, i.e., $\rho' < \rho$, since a part of the pressure inside the bubble is carried by the magnetic pressure $p'_{\rm mag}$. A lower gas density in the bubble leads to a rising of the gas due to buoyancy. This is a typical demonstration of *magnetic buoyancy*.

© Springer Nature Singapore Pte Ltd. 2020, corrected publication 2023

S. Kato, J. Fukue, *Fundamentals of Astrophysical Fluid Dynamics*,
Astronomy and Astrophysics Library,
https://doi.org/10.1007/978-981-15-4174-2_16

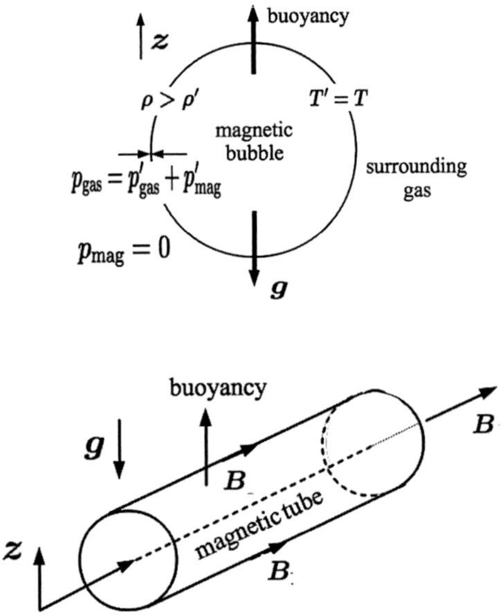

Fig. 16.1 Schematic picture showing the magnetic buoyancy. The upper panel: Inside the circle (sphere) is a magnetic bubble. The quantities inside the bubble are shown by attaching prime. Due to pressure balance the gas density is lower in the bubble than that in the surrounding gas, if the temperature in the bubble is the same as that in the surrounding medium. This brings about a rise of the bubble by magnetic buoyancy. The lower panel: A two-dimensional isolated magnetic tube is shown. The tube rises up by magnetic buoyancy

A more realistic situation than a magnetic bubble may be a two-dimensional isolated magnetic tube, which is shown on the lower panel of Fig. 16.1. This tube will rise by the same reason mentioned above, if the temperature in the tube is the same as that of the surrounding medium.

The above statements on magnetic buoyancy, however, are not arguments on stability criterion of dynamical equilibrium system against small amplitude perturbations. Our next issues are to examine how the presence of magnetic buoyancy affects on stability criterion of dynamically equilibrium systems. There are two types of instabilities related to magnetic buoyancy. The first one is that where line of forces is not deformed, and the second one is that deformation of line of forces has essential effects on stability. The former (*interchange instability*) is described in this section, because it is directly related to the magnetic buoyancy. The latter (the *Parker instability*) is described in the next section.

16.1.2 Interchange Instability

To simplify situations, a vertically stratified, horizontally infinite medium is considered, where a constant gravitational acceleration $\boldsymbol{g}$ acts in the negative z-direction. Although the magnetic field runs parallel in the horizontal direction (taken hereafter to be the y-direction), its strength is a function of the vertical direction (z-direction) alone; i.e.,

$$\boldsymbol{B} = (0,\ B(z),\ 0). \tag{16.1}$$

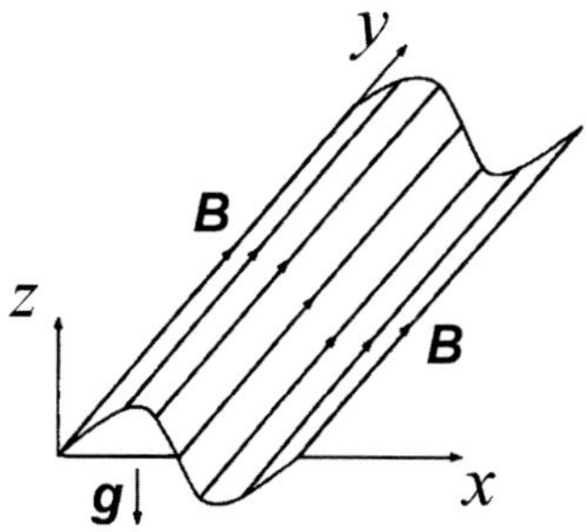

Fig. 16.2 Schematic picture showing the interchange instability with no deformation of the line of forces. (After Kato et al. (1998), *Black-Hole Accretion Disks*)

Then, the force balance in the vertical direction is given by

$$\frac{d}{dz}\left(p+\frac{1}{8\pi}B^2\right)=-\rho g, \tag{16.2}$$

where p and ρ are the gas pressure and density, respectively.

Here, we consider perturbations whose wavelength is infinite in the y-direction ($k_y = 0$), but not in other directions, $k_x \neq 0$ and $k_z \neq 0$ (see Fig. 16.2). This kind of perturbations does not deform the line of forces. A line of force has its position changed without any deformation. Hence, no force resulting from magnetic tension acts on the perturbations. Because of this, the stability criterion to this type of perturbations can be derived by simple physical considerations on *magnetic buoyancy*, without considering complications resulting from a deformation of the line of forces.

We consider a magnetic tube infinitely extending in the y-direction. In the unperturbed state the tube is static at height z under both pressure and thermal balances with the surrounding medium. The tube is perturbed adiabatically to rise to height $z + dz$ without being deformed. The rise is assumed to be so slow that the pressure balance with the surrounding gas is maintained during the rise. The Lagrangian changes of the physical quantities in the tube are denoted by attaching δ, while the changes with height of the physical quantities of the unperturbed surrounding gas by attaching d. Then, the pressure balance after the rise is

$$\delta p+\frac{B}{4\pi}\delta B=dp+\frac{B}{4\pi}dB. \tag{16.3}$$

On the other hand, the adiabatic change implies that δp and $\delta\rho$ are related by

$$\delta p=c_{\rm s}^2\delta\rho, \tag{16.4}$$

where $c_{\rm s}$ is the adiabatic sound speed, defined by $c_{\rm s}^2 \equiv (dp/d\rho)_{\rm s} = \gamma p/\rho$, where γ is the ratio of specific heats.

Due to the displacement with pressure balance, the cross section S of the tube changes to $S + \delta S$, but the mass within the tube remains unchanged; i.e., $\delta(\rho S) =$

0. The magnetic flux through the tube is also unchanged; i.e., $\delta(BS) = 0$. The combination of these relations leads to

$$\frac{\delta\rho}{\rho} = \frac{\delta B}{B}. \tag{16.5}$$

Substitution of relations (16.4) and (16.5) into Eq. (16.3) gives

$$\left(c_s^2 + c_A^2\right)\delta\rho = \left(\frac{dp}{d\rho} + \frac{B}{4\pi}\frac{dB}{d\rho}\right)d\rho, \tag{16.6}$$

where c_A^2 is the square of the Alfvén speed, defined by

$$c_A^2 \equiv \frac{B^2}{4\pi\rho}. \tag{16.7}$$

Equation (16.6) is a relation between $\delta\rho$ and $d\rho$. In the case where the tube rises upwards, both $\delta\rho$ and $d\rho$ are negative; however, if the decrease in the density in the tube is larger than that in the surrounding region (i.e., $|\delta\rho| > |d\rho|$), the tube rises further due to buoyancy, and departs from the initial position. In the case when the tube sinks, conversely, both $\delta\rho$ and $d\rho$ are positive; however, if $\delta\rho > d\rho$, the tube sinks further. In summary, the condition for an instability is $\delta\rho/d\rho > 1$. From Eq. (16.6), this instability condition is written as

$$\frac{d\ln p}{d\ln\rho} > \gamma - \frac{\gamma c_A^2}{c_s^2}\left(\frac{d\ln B}{d\ln\rho} - 1\right). \tag{16.8}$$

Equation (16.8) shows that in a medium with no magnetic field ($c_A^2/c_s^2 = 0$), the condition of instability is $d\ln p/d\ln\rho > \gamma$. This is the well-known condition of convective instability in non-magnetized medium (Schwarzschild criterion)(see Chap. 5). Equation (16.8) further shows that the medium is unstable even if $d\ln p/d\ln\rho < \gamma$, if the decrease of magnetic field with height is stronger than that of gas density (i.e., $d\ln B/d\ln\rho > 1$). This instability due to magnetic buoyancy is called the *interchange instability*.

It is noted that inequality (16.8) shows that magnetic fields have no effects on stability criterion in the case where $d\ln B/d\ln\rho = 1$. This comes from the following situations. Let us consider the case where a magnetic tube rises from the height of z to $z + dz$. The density change in the tube by this rise is denoted by $\delta\rho$, while that in the surrounding medium by $d\rho$. Then, the change of magnetic pressure in the tube is $\delta(B^2/8\pi)$, which is $c_A^2(\delta\rho/\rho)$, because the change of B and ρ in the tube is related by $B \propto \rho$. On the other hand, the change of magnetic fields in the surrounding medium is $d(B^2/8\pi)dz$, which is $c_A^2(d\ln B/d\ln\rho)(d\rho/\rho)$, because $d\rho = (d\rho/dz)dz$. This result shows that in the case where $d\ln B/d\ln\rho = 1$, we have $\delta\rho = d\rho$. That is, the magnetic fields do not contribute to generation of buoyancy force, and the criterion of convective instability is governed only by the behavior of gas as shown in inequality (16.8). For magnetic fields to contribute positively to instability, the condition $d\ln B/d\ln\rho > 1$ is necessary.

16.2 Parker Instability in Isothermal Atmosphere

Another important instability related to magnetic buoyancy is the *Parker instability* (Parker 1966). In the interchange instability described in Sect. 16.1, unstable perturbations are those which do not deform magnetic fields. In perturbations which deform magnetic field lines, the magnetic tension acts against the instability. Even in the cases where magnetic tension works against instability, however, the gaseous downward flow along the line of forces will increase buoyancy force (see Fig. 16.3). If this increase of buoyancy force overcomes the magnetic force against it, an instability can be expected.

In order to have a rough picture of the instability, let us consider an isothermal atmosphere which is subject to a constant gravitational field (directed in the negative z-direction) and to magnetic fields which are directed in a horizontal direction (y-direction) (see Fig. 16.3). Then, in the unperturbed state the hydrostatic balance in the vertical direction is described by

$$\frac{dp_0}{dz} + \frac{B_0}{4\pi}\frac{dB_0}{dz} + \rho_0 g = 0. \tag{16.9}$$

The unperturbed density $\rho_0(z)$ is thus stratified in the vertical direction as

$$\rho_0(z) \propto \exp\left(-\frac{z}{H}\right), \tag{16.10}$$

where H is the scale height of density stratification, and roughly given by

$$H \sim \frac{c_\mathrm{s}^2}{g}\left(1 + \alpha\frac{c_\mathrm{A}^2}{c_\mathrm{s}^2}\right), \tag{16.11}$$

with a constant α of the order of unity.

Now, a part of the magnetic fields is assumed to be raised up over a small distance Δz with horizontal scale L, as is shown schematically in Fig. 16.3. The magnetic tension resists this deformation. This resisting force is roughly $B_0^2/(4\pi R)$ per unit volume, where R is the radius of curvature of the magnetic fields.

The raised part of gases can fall along the deformed magnetic field lines. Then, along the magnetic fields, the upward density decrease is sharper than that of the surrounding one. That is, the density inside the deformed tube, $\rho^*(z + \Delta z)$ is $\sim \rho_0(z)\exp(-\Delta z/H^*)$, where $H^* \sim c_\mathrm{s}^2/g$ [the case of $\alpha = 0$ in Eq. (16.11)]. Compared with this, the density of the surrounding medium is $\sim \rho_0(z)\exp(-\Delta z/H)$. The raised element in the tube has a lower density compared with that of the

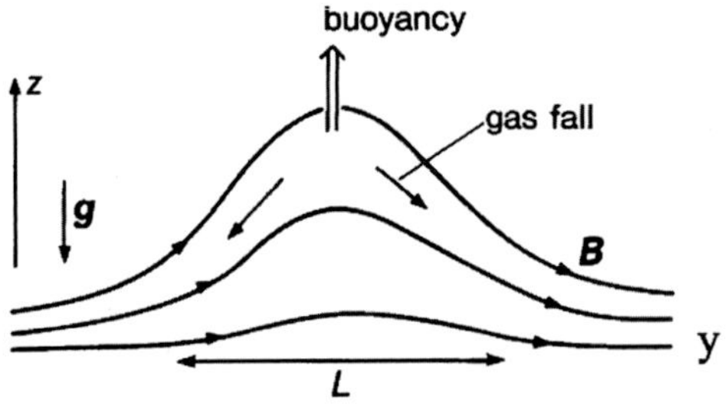

Fig. 16.3 Schematic picture showing the mechanism of the Parker instability. In the x-direction, which is in the direction perpendicular to the y-z plane, the system is assumed to be uniform. (After Kato et al. (1998), *Black-Hole Accretion Disks*)

surrounding gases by the amount $\Delta\rho$, which is

$$\frac{\Delta\rho}{\rho_0} \sim \exp\left(-\frac{\Delta z}{H}\right) - \exp\left(-\frac{\Delta z}{H^*}\right) \sim \alpha \frac{c_A^2}{g}\frac{1}{HH^*}\Delta z. \tag{16.12}$$

Hence, by adopting $R \sim L^2/\Delta z$, we can write the instability condition $g\Delta\rho > (B_0^2/4\pi)(\Delta z/L^2)$ (buoyancy is larger than tension) in the form

$$L \gtrsim \left(\frac{1}{\alpha} + \frac{c_A^2}{c_s^2}\right)^{1/2} H^*. \tag{16.13}$$

This result shows that when the wavelength of perturbations along the line of force is short, the restoring force due to magnetic tension is stronger than the buoyancy force and no instability is expected. If the wavelength of perturbations is long, however, we can expect the instability. Parker (1966) showed that this kind of instability really exists. This instability is called the *Parker instability*. It is also called the *magnetic Rayleigh–Taylor instability*, since the cause of the instability is due to the fall of dense gas along the line of force.

In the following, the presence of this instability is demonstrated by examining the dispersion relation for a particular case.

(a) **Equilibrium configuration**

As is mentioned above, we consider an isothermal atmosphere which is subject to a constant gravitational acceleration and to magnetic fields directed in a horizontal direction. It is further assumed that the strength of the magnetic field decreases upwards so that the Alfvén speed is constant with height.[1] Here, as before, Cartesian coordinates (x, y, z) are introduced so that the z-axis is in the direction anti-parallel to a constant external gravitational field $\boldsymbol{g}$, i.e., $\boldsymbol{g} = (0, 0, -g)$. The direction of the y-axis is taken in the direction of the unperturbed field $\boldsymbol{B}_0$; i.e., $\boldsymbol{B}_0 = (0, B_0(z), 0)$ [see Eq. (16.1)]. Since the Alfvén speed c_A is given by $c_A^2 = B_0^2(z)/[4\pi\rho_0(z)]$, the assumption that c_A =const. implies $B_0^2(z) \propto \rho_0(z)$. In this isothermal atmosphere,

[1] This is for mathematical simplicity.

$\Gamma \equiv d\ln p_0/d\ln\rho_0 = 1$ and $d\ln B_0/d\ln\rho_0 = 1/2$; thus, inequality (16.8) does not hold, implying that the atmosphere is stable against interchange-type perturbations. In spite of this, this atmosphere is unstable against perturbations with $k_y \neq 0$, as is shown below.

Since the atmosphere is isothermal and $c_{\rm A}^2 = $ const., we can integrate the equation of the vertical force balance (16.9) to give

$$\rho_0(z) = \rho_0(0)\exp\left(-\frac{z}{H}\right), \quad p_0(z) = p_0(0)\exp\left(-\frac{z}{H}\right) \tag{16.14}$$

$$B_0(z) = B_0(0)\exp\left(-\frac{z}{2H}\right), \tag{16.15}$$

where H is the scale height of density stratification and is given by

$$H = \left(\frac{1}{\gamma} + \frac{\alpha_{\rm B}}{2}\right)\frac{c_{\rm s}^2}{g}. \tag{16.16}$$

Here, γ is the ratio of specific heats and $\alpha_{\rm B}$ is a constant ratio of $c_{\rm A}^2$ to $c_{\rm s}^2(\equiv \gamma p_0/\rho_0)$:

$$\alpha_{\rm B} = \frac{c_{\rm A}^2}{c_{\rm s}^2}. \tag{16.17}$$

(b) **Perturbations and dispersion relation**

On the isothermal atmosphere described above, a small amplitude adiabatic perturbation is superposed. The perturbation is assumed to occur only in the y-z plane, i.e., the wavelength of the perturbation in the x-direction is infinite. Velocity and magnetic field perturbations are denoted by $(0, v_y, v_z)$ and $(0, b_y, b_z)$, respectively.

The equation of continuity describing the small amplitude perturbations is

$$\frac{\partial \rho_1}{\partial t} + \frac{\partial}{\partial y}(\rho_0 v_y) + \frac{\partial}{\partial z}(\rho_0 v_z) = 0, \tag{16.18}$$

where ρ_1 is the density perturbation over $\rho_0(z)$.

The y- and z-components of equation of motion are, respectively,

$$\frac{\partial v_y}{\partial t} = -\frac{1}{\rho_0}\frac{\partial p_1}{\partial y} + \frac{b_z}{4\pi\rho_0}\frac{dB_0}{dz}, \tag{16.19}$$

$$\frac{\partial v_z}{\partial t} = -\frac{1}{\rho_0}\frac{\partial}{\partial z}\left(p_1 + \frac{B_0 b_y}{4\pi}\right) + \frac{B_0}{4\pi\rho_0}\frac{\partial b_z}{\partial y} - \frac{\rho_1}{\rho_0}g, \tag{16.20}$$

where p_1 is the Eulerian pressure perturbation.

As the energy equation we adopt an adiabatic relation, which is written here in the form

$$\frac{\partial p_1}{\partial t} + v_z \frac{dp_0}{dz} + \rho_0 c_s^2 \left(\frac{\partial v_y}{\partial y} + \frac{\partial v_z}{\partial z} \right) = 0. \tag{16.21}$$

Finally, from the induction equation we have

$$\frac{\partial b_y}{\partial t} = -\frac{\partial}{\partial z}(B_0 v_z), \tag{16.22}$$

$$\frac{\partial b_z}{\partial t} = B_0 \frac{\partial v_z}{\partial y}. \tag{16.23}$$

The next procedure is to derive a dispersion relation. Elimination of p_1, b_y, and b_z from the above set of equations can be carried out easily. That is, by taking the time derivative of Eq. (16.19) to eliminate p_1 and b_z by using Eqs. (16.21) and (16.23), we obtain

$$\left(\frac{\partial^2}{\partial t^2} - c_s^2 \frac{\partial^2}{\partial y^2} \right) \left(\rho_0^{1/2} v_y \right) = \left(c_s^2 \frac{\partial}{\partial z} + \frac{c_s^2}{2H} - g \right) \frac{\partial}{\partial y} \left(\rho_0^{1/2} v_z \right). \tag{16.24}$$

Similarly, from Eq. (16.20), we have another relation between v_y and v_z:

$$\left[\frac{\partial^2}{\partial t^2} - c_A^2 \frac{\partial^2}{\partial y^2} - \left(c_s^2 + c_A^2 \right) \frac{\partial^2}{\partial z^2} + \frac{1}{4H^2} \left(c_s^2 + c_A^2 \right) \right] \left(\rho_0^{1/2} v_z \right)$$
$$= \left(c_s^2 \frac{\partial}{\partial z} - \frac{c_s^2}{2H} + g \right) \frac{\partial}{\partial y} \left(\rho_0^{1/2} v_y \right). \tag{16.25}$$

By eliminating $\rho_0^{1/2} v_y$ from the above two equations, we have

$$\left(\frac{\partial^2}{\partial t^2} - c_s^2 \frac{\partial^2}{\partial y^2} \right) \left[\frac{\partial^2}{\partial t^2} - c_A^2 \frac{\partial^2}{\partial y^2} - \left(c_s^2 + c_A^2 \right) \frac{\partial^2}{\partial z^2} + \frac{c_s^2 + c_A^2}{4H^2} \right] \left(\rho_0^{1/2} v_z \right)$$
$$= \left[c_s^4 \frac{\partial^2}{\partial z^2} - \left(\frac{c_s^2}{2H} - g \right)^2 \right] \frac{\partial^2}{\partial y^2} \left(\rho_0^{1/2} v_z \right). \tag{16.26}$$

Equation (16.26) shows that $\rho_0^{1/2} v_z$ can be taken to be proportional to $\exp[i(\omega t - k_y y - k_z z)]$, since the coefficients of the differential equation are constant with respect to both space and time. Then, as the dispersion relation, we finally have

$$\omega^4 - \left(k_y^2 + k_z^2 + \frac{1}{4H^2} \right) \left(c_s^2 + c_A^2 \right) \omega^2$$
$$+ c_s^2 k_y^2 \left[c_A^2 \left(k_y^2 + k_z^2 \right) + \frac{c_s^2 + c_A^2}{4H^2} - \frac{1}{c_s^2} \left(\frac{c_s^2}{2H} - g \right)^2 \right] = 0. \tag{16.27}$$

In the limit of $H = \infty$ and $g = 0$ (uniform medium), this dispersion relation tends to that representing the fast and slow MHD waves in homogeneous media. In the limit of $k_y = 0$, the dispersion relation (16.27) gives

$$\omega^2 = \left(c_{\rm s}^2 + c_{\rm A}^2\right)\left(k_z^2 + \frac{1}{4H^2}\right) \quad \text{and} \quad \omega^2 = 0. \tag{16.28}$$

The former is the fast MHD mode propagating vertically in the stratified medium and is not our concern here. The mode which we are interested in is the latter slow mode. This mode is trivial ($\omega^2 = 0$) in the limit of $k_y = 0$, but can become unstable ($\omega^2 < 0$) when $k_y^2 \neq 0$, as shown below.

(c) **Instability criterion**

The dispersion relation (16.27) indicates that it has always two real solutions with respect to ω^2. Hence, the condition of $\omega^2 < 0$ for one of its two solutions (with respect to ω^2) is $k_y^2 \neq 0$ and

$$c_{\rm A}^2\left(k_y^2 + k_z^2\right) < \frac{1}{c_{\rm s}^2}\left(\frac{c_{\rm s}^2}{2H} - g\right)^2 - \frac{c_{\rm s}^2 + c_{\rm A}^2}{4H^2}. \tag{16.29}$$

Inequality (16.29) can be rewritten as

$$k_y^2 + k_z^2 < \frac{1}{\gamma c_{\rm T}^2 c_{\rm A}^2 H^2}\left[-\gamma c_{\rm T}^2\left(c_{\rm T}^2 + \frac{3}{4}c_{\rm A}^2\right) + \left(c_{\rm T}^2 + \frac{1}{2}c_{\rm A}^2\right)^2\right], \tag{16.30}$$

where $c_{\rm T}$ is the isothermal sound speed:

$$c_{\rm T}^2 = \frac{c_{\rm s}^2}{\gamma} = \frac{p_0}{\rho_0}. \tag{16.31}$$

Inequality (16.30) is satisfied for long-wavelength perturbations, if the right-hand side is positive; namely, if

$$\gamma < \frac{\left(c_{\rm T}^2 + c_{\rm A}^2/2\right)^2}{c_{\rm T}^2\left(c_{\rm T}^2 + 3c_{\rm A}^2/4\right)}. \tag{16.32}$$

In the limit of $c_{\rm A}^2 = 0$, the right-hand side of inequality (16.32) is unity,[2] but increases monotonically along with an increase in $c_{\rm A}^2/c_{\rm T}^2$. For example, the right-

[2]The isothermal medium is considered here. In this medium $d\ln p/d\ln\rho$ specifying the structure of the medium is unity. Hence, $\gamma < 1$ is necessary for instability. In other words, the medium is highly convectively stable, when $c_{\rm A}^2 = 0$.

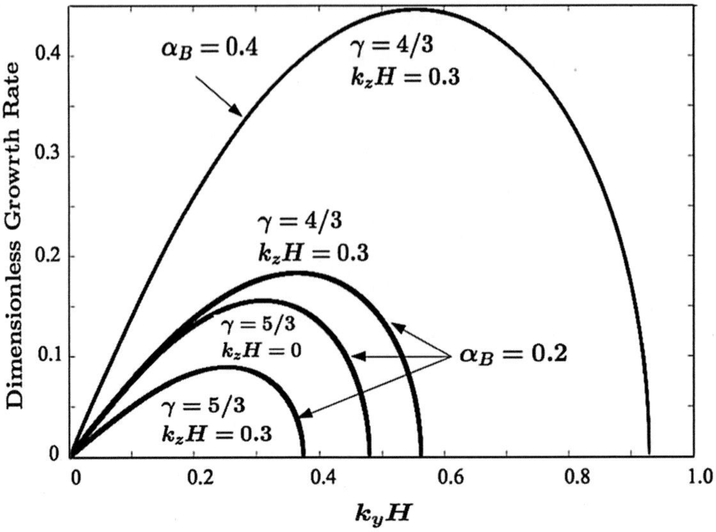

Fig. 16.4 Dimensionless growth rate, $-\Im[\omega/(c_s/H)]$, of the Parker instability as a function of $k_y H$ for some parameters sets of γ, $k_z H$ and α_B. Notice that the growth rate increases with decrease of γ and $k_z H$ and increase of c_A^2/c_s^2

hand side is $8/5$ for $c_A^2/c_T^2 = 2$, and $9/4$ for $c_A^2/c_T^2 = 4$. This means that the system becomes unstable if the magnetic field is stronger beyond a certain critical value.

The inequality (16.30) shows that the instability occurs for perturbations whose wavelength along the magnetic field line is longer than a certain critical value. For perturbations whose wavelength along the magnetic field is too long, however, the growth rate is low [see that the term of ω^0 in Eq. (16.27) is proportional to k_y^2]. This means that the growth rate has the maximum at a certain finite wavelength. Roughly speaking, this wavelength is comparable to the vertical scale height.

Figure 16.4 shows the dimensionless growth rate, $-\Im[\omega/(c_s/H)]$, as a function of k_y for some typical cases of γ and k_z. The growth rate is generally large when the vertical scale of perturbations is long and γ is small.

In the above arguments $k_x = 0$ has been assumed. There is no difficulty in including the effects of k_x, although the equations become complicated. In the case that $k_x \neq 0$, the buoyancy force per unit mass increases. Hence, this acts so as to increase the growth rate. In this sense, the essence of the Parker instability is already shown by considering the case $k_x = 0$.

In the above formulation an isothermally stratified atmosphere subject to a constant gravitational acceleration is assumed. In accretion disks, however, the gravitational acceleration acting on disks is not constant but varies with height from the equator. The Parker instability in such configurations has been studied analytically (Horiuchi et al. 1988) and by numerical simulations (Matsumoto et al. 1988). In these cases, the mode with the largest growth rate is found for what the magnetic fields cross the equator (antisymmetric mode with respect to equator) (Horiuchi et al. 1988).

The Parker instability is important in many active phenomena in astrophysical objects. For example, the emergency of magnetic fields on the solar photosphere and subsequent magnetic activities will be related to the Parker instability. In dynamo processes, rising of magnetic fields by the Parker instability and subsequent twisting of them by rotation will be one of the important elementary processes of the $\alpha\Omega$ dynamo.

16.3 Magneto-Rotational Instability

In the previous two Sects. 16.1 and 16.2 we have described two magnetic instabilities related to magnetic buoyancy: the interchange instability and the Parker instability. In the latter instability deformation of magnetic field lines is essential. Another important magnetic instability which results from deformation of magnetic field lines is the *magneto-rotational instability (MRI)* (or the *magnetic shearing instability*).

This instability was originally found by Velihov (1959) and Chandrasekhar (1960), and examined further by Fricke (1969). Recently, Balbus and Hawley (1991) and Hawley and Balbus (1991) rediscovered it and recognized its significance in relation to the origin of turbulence in accretion disks. Importantly, this is an instability in differentially rotating systems with specific angular momentum increasing outward (a stable system in the sense of the Rayleigh criterion). Furthermore, even an extremely weak magnetic field can lead to a rapidly growing dynamical instability. After Balbus and Hawley (1991) and Hawley and Balbus (1991), this instability has been examined extensively by three-dimensional numerical simulations. They show that the instability grows to a highly nonlinear regime and provides the most promising mechanism for turbulence in magnetized accretion disks (for review see Balbus 2003).

Here, after an brief overview of the instability mechanism, we consider the simplest case (i.e., axisymmetric perturbations in a disk with uniform vertical magnetic fields) in order to understand the essence of the instability. A non-axisymmetric instability in a disk with toroidal magnetic fields is mentioned only briefly in the final subsection.[3]

16.3.1 Overview of Instability Mechanism

As is well-known, a magnetic line of force has elasticity, like a rubber string; i.e., the Maxwell stress tensor can be decomposed into two parts: an isotropic pressure

[3]The main part of this section (Sect. 16.3) is a reproduction of Sect. 17.3 of Kato et al. (1998).

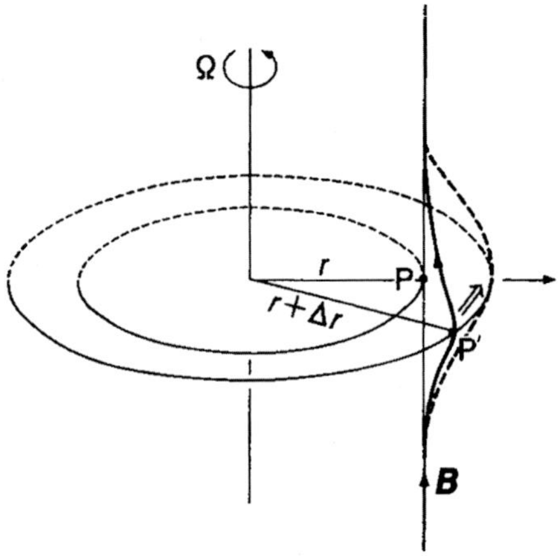

Fig. 16.5 Schematic picture showing magneto-rotational instability in the case of axisymmetric perturbations in disks with vertical magnetic field. (After Kato et al. (1998), *Black-Hole Accretion Disks*)

and a tension along the line of force (see Sect. 11.3). The latter force resists against a deformation of the line of force, and acts so as to return a perturbed fluid element to its original position. In a differentially rotating system, this tensional force can lead the system to instability.

Let us consider an axially symmetric, cylindrically rotating, steady system with uniform vertical magnetic fields (see Fig. 16.5). The system is rotating so that $r^2\Omega$ increases outwards (stable in the sense of the Rayleigh criterion). If a fluid element at a point P(r) is perturbed outwards to a point P$'(r + \Delta r)$ under conservation of specific angular momentum, the angular velocity of the perturbed element becomes smaller than that in the unperturbed one at $r + \Delta r$. That is, we have $\Omega^*(r + \Delta r) < \Omega(r + \Delta r)$, the star being attached to $\Omega(r + \Delta r)$ to emphasize that it is a quantity of the fluid element. The fluid element displaced outwards is, however, elastically tethered by the magnetic field lines, which tend to cause rigid rotation. Hence, angular momentum is transported along the distorted magnetic line of force so that the deformed part tries to co-rotate[4] with the unperturbed part of the line of force (this trend is shown in Fig. 16.4 by arrow). If this is realized, the angular velocity of rotation of the perturbed element becomes larger than that of the surrounding gas, since we are considering the system which is stable in the sense of Rayleigh criterion as mentioned above. Thus, the fluid element has a tendency to go further outward by excess centrifugal force. On the other hand, a deformation of magnetic fields in the meridional plane acts so as to resist to the deformation (magnetic tension). The latter resisting force is stronger than the excess centrifugal force when the vertical size of perturbations is short (steep curvature), and the perturbations cannot grow. However, if the size of perturbations in the z-direction is moderately large, conversely, the former excess centrifugal force can dominate over the latter force of

[4]Recall Ferraro's isorotation law (Ferraro 1937); in magnetized axially symmetric steady rotating systems, the gas along a magnetic field line must rotate with the same angular velocity [Sect. 12.2 and Eq. (12.37)].

magnetic tension, and instability sets in. This is the main process of the magneto-rotational instability.

The magneto-rotational instability occurs in more general circumstances. For example, the instability is present even when the magnetic fields in the unperturbed state are toroidal, considering non-axisymmetric perturbations. Examination in general cases is complicated. Hereafter, we derive the instability condition and the growth rate in detail only in the case of the above-considered simplest case.

16.3.2 Dispersion Relation for Axisymmetric Perturbations

As the simplest example of magneto-rotational instability, we consider an axisymmetric disk threaded by uniform, vertical magnetic fields (the case shown in Fig. 16.5). To describe perturbations, a cylindrical coordinate system (r, φ, z) is introduced, where r is the perpendicular distance from the rotation axis (z-axis) of the disk. The unperturbed magnetic field $\boldsymbol{B}_0$ is $\boldsymbol{B}_0 = (0,\ 0,\ B_0)$, and the unperturbed gas flow in the disk is $(0,\ r\Omega(r),\ 0)$. For simplicity, B_0 is taken to be constant.

Axially symmetric perturbations are superposed on the disks described above. The wavelength in the vertical direction is taken to be finite. Then, there are three modes of perturbations. The first one is the acoustic mode (see Chap. 4 for details), resulting from the restoring force of pressure. Since this mode is not our concern here, we filter out this mode by introducing the *Boussinesq approximation*.[5] The second is the gravity mode (the mode of inertial-gravity oscillations). This mode is still present under the Boussinesq approximation, since the compressibility of the gas is not the cause of this mode (see Chap. 4 for details). This mode becomes oscillatory, if there is rotation and convectively stable density stratification. The third is the Alfvén modes. If compressibility of the system is filtered out by Boussinesq approximation, the fast and slow modes of MHD waves disappear and the remaining mode related to magnetic fields is Alfvén waves or *torsional Alfvén waves*. (In the absence of both rotation and magnetic fields, this mode is only a trivial motion with zero frequency; i.e., $\omega = 0$.) We are interested here in this mode, since it becomes unstable if the wavelength in the vertical direction is not very short.

In the Boussinesq approximation, the gas motion (velocity $\boldsymbol{v}$) is taken to be incompressible; i.e., div $\boldsymbol{v} = 0$, and the Eulerian pressure variation is neglected in the energy equation. It is noticed that the terms of pressure variation (density variation) in the momentum equation are, however, retained in the Boussinesq approximation, although div $\boldsymbol{v} = 0$ is adopted.

Under the above approximations, we consider axially symmetric small amplitude (linearized) perturbations on axially symmetric rotating systems. The velocity and magnetic field perturbations are denoted by $\boldsymbol{v} = (v_r,\ v_\varphi,\ v_z)$ and $\boldsymbol{b} = (b_r,\ b_\varphi,\ b_z)$,

[5]For details of the Boussinesq approximation, see Sect. 5.1 and also Spiegel and Veronis (1960).

respectively. The density and pressure perturbations over the unperturbed ones (ρ_0 and p_0) are denoted by ρ_1 and p_1, respectively.

(a) **Equations describing perturbations**

Following the Boussinesq approximation, we adopt div $\boldsymbol{v} = 0$:

$$\frac{\partial}{r\partial r}(rv_r) + \frac{\partial v_z}{\partial z} = 0. \tag{16.33}$$

The radial, azimuthal, and vertical components of equation of motion are written, respectively, as

$$\frac{\partial v_r}{\partial t} - 2\Omega v_\varphi = -\frac{\partial}{\rho_0 \partial r}\left(p_1 + \frac{B_0 b_z}{4\pi}\right) + \frac{B_0}{4\pi\rho_0}\frac{\partial b_r}{\partial z} - \frac{\rho_1}{\rho_0}\left(g_r - r\Omega^2\right), \tag{16.34}$$

$$\frac{\partial v_\varphi}{\partial t} + \frac{\kappa^2}{2\Omega}v_r = \frac{B_0}{4\pi\rho_0}\frac{\partial b_\varphi}{\partial z}, \tag{16.35}$$

$$\frac{\partial v_z}{\partial t} = -\frac{\partial p_1}{\rho_0 \partial z} - \frac{\rho_1}{\rho_0}g_z, \tag{16.36}$$

where $\boldsymbol{g} = (-g_r,\ 0,\ -g_z)$ is the gravitational acceleration, and we hereafter write $g_r - r\Omega^2$ as $g_{r,\mathrm{eff}}$, the effective gravitational acceleration in the radial direction.

The three components of the induction equation are

$$\frac{\partial b_r}{\partial t} = B_0 \frac{\partial v_r}{\partial z}, \tag{16.37}$$

$$\frac{\partial b_\varphi}{\partial t} = B_0 \frac{\partial v_\varphi}{\partial z} + r\frac{d\Omega}{dr}b_r, \tag{16.38}$$

$$\frac{\partial b_z}{\partial t} = -\frac{B_0}{r}\frac{\partial}{\partial r}(rv_r). \tag{16.39}$$

The condition of div $\boldsymbol{b} = 0$ is satisfied automatically.

Since p_1 is neglected in the energy equation as is mentioned before, the energy equation becomes

$$\frac{1}{\rho_0}\frac{\partial \rho_1}{\partial t} = v_r S_r + v_z S_z, \tag{16.40}$$

where

$$S_r \equiv \frac{\partial}{\partial r}\ln\left(\frac{p_0^{1/\gamma}}{\rho_0}\right) \quad \text{and} \quad S_z \equiv \frac{\partial}{\partial z}\ln\left(\frac{p_0^{1/\gamma}}{\rho_0}\right). \tag{16.41}$$

(b) **Local perturbations**

Here, we consider local perturbations which have such space-time dependence as $\exp[i(\omega t - k_r r - k_z z)]$. Then, the above Eqs. (16.33)–(16.40) reduce to

$$k_r v_r + k_z v_z = 0, \tag{16.42}$$

$$i\omega v_r - 2\Omega v_\varphi = ik_r \frac{p_1}{\rho_0} + i\frac{B_0}{4\pi\rho_0}(k_r b_z - k_z b_r) - \frac{\rho_1}{\rho_0} g_{r,\mathrm{eff}}, \tag{16.43}$$

$$i\omega v_\varphi + \frac{\kappa}{2\Omega} v_r = -i\frac{B_0}{4\pi\rho_0} k_z b_\varphi, \tag{16.44}$$

$$i\omega v_z = ik_z \frac{p_1}{\rho_0} - \frac{\rho_1}{\rho_0} g_z, \tag{16.45}$$

$$i\omega b_r = -iB_0 k_z v_r, \tag{16.46}$$

$$i\omega b_\varphi = -iB_0 k_z v_\varphi + r\frac{d\Omega}{dr} b_r, \tag{16.47}$$

$$i\omega b_z = iB_0 k_r v_r, \tag{16.48}$$

$$i\omega\rho_1 = \rho_0(S_r v_r + S_z v_z). \tag{16.49}$$

It is convenient to express all perturbed quantities in terms of v_z, using Eqs. (16.42) and (16.44)–(16.49). The results are

$$v_r = -\frac{k_z}{k_r} v_z, \tag{16.50}$$

$$v_\varphi = -i\frac{k_z}{k_r} \frac{(\kappa^2/2\Omega)\,\omega^2 - r(d\Omega/dr)c_{\mathrm{A}}^2 k_z^2}{\omega\left(\omega^2 - c_{\mathrm{A}}^2 k_z^2\right)} v_z, \tag{16.51}$$

$$\frac{b_r}{B_0} = \frac{k_z^2}{\omega k_r} v_z, \tag{16.52}$$

$$\frac{b_\varphi}{B_0} = i\frac{k_z^2}{k_r} \frac{2\Omega}{\omega^2 - c_{\mathrm{A}}^2 k_z^2} v_z, \tag{16.53}$$

$$\frac{b_z}{B_0} = -\frac{k_z}{\omega} v_z, \tag{16.54}$$

$$\frac{\rho_1}{\rho_0} = i\frac{k_z}{\omega}\left(\frac{S_r}{k_r} - \frac{S_z}{k_z}\right) v_z, \tag{16.55}$$

$$\frac{p_1}{\rho_0} = \left[\frac{\omega}{k_z} + \frac{1}{\omega}\left(\frac{S_r}{k_r} - \frac{S_z}{k_z}\right) g_z\right] v_z. \tag{16.56}$$

(c) Dispersion relation

Substitution of the above relations into Eqs. (16.43) gives a dispersion relation. After rearrangement we have (Balbus and Hawley 1991)

$$\omega_*^4 + \frac{k_z^2}{k^2}\left[\frac{k_r}{k_z}\left(\frac{S_r}{k_r} - \frac{S_z}{k_z}\right)(k_r g_z - k_z g_{r,\mathrm{eff}}) - \kappa^2\right]\omega_*^2 - 4\Omega^2\frac{k_z^4}{k^2}c_\mathrm{A}^2 = 0, \tag{16.57}$$

where

$$\omega_*^2 = \omega^2 - c_\mathrm{A}^2 k_z^2, \tag{16.58}$$

$$k^2 = k_r^2 + k_z^2. \tag{16.59}$$

Let us introduce the Brunt–Väisälä frequency N defined by (see Sect. 4.1)

$$N^2 = N_r^2 + N_z^2, \tag{16.60}$$

where

$$N_r^2 \equiv -S_r g_{r,\mathrm{eff}} \quad \text{and} \quad N_z^2 \equiv -S_z g_z. \tag{16.61}$$

It is noted here that between S_r and S_z there is the relation:

$$\frac{S_r}{g_{r,\mathrm{eff}}} = \frac{S_z}{g_z}. \tag{16.62}$$

This relation can be found from the following consideration. In the case where Ω is a function of r alone, the unperturbed state is described by

$$\frac{1}{\rho_0}\nabla p_0 + \nabla\psi - \nabla\left(\int^r r\Omega^2 dr\right) = 0. \tag{16.63}$$

This equation shows that the pressure needs to be a function of the density, i.e., $p_0 = p_0(\rho_0)$. In other words, there is a barotropic relation between the unperturbed pressure and density. Because of this, the equi-pressure (i.e., equi-density), equi-entropy, and equi-potential surfaces coincide with each other. The effective gravitational force is perpendicular to these surfaces.

If relations (16.61) and (16.62) are used, the dispersion relation (16.57) is simplified as (Balbus and Hawley 1991)

$$\frac{k^2}{k_z^2}\omega_*^4 - \left[\left(\frac{k_r}{k_z}N_z - N_r\right)^2 + \kappa^2\right]\omega_*^2 - 4\Omega^2 c_\mathrm{A}^2 k_z^2 = 0. \tag{16.64}$$

This dispersion relation is a quadratic equation with respect to ω_*^2, implying that there are two modes of perturbations. If there is no magnetic field, one of the two modes tends to a trivial mode of $\omega_*^2 = \omega^2 = 0$. This is the mode of a torsional Alfvén wave when magnetic fields are present and becomes, as will be shown below, unstable for sufficiently long-wavelength perturbations in the vertical direction.

It is noted that in the case of $c_{\rm A}^2 = 0$, a non-trivial solution of the dispersion relation (16.64) is

$$k^2\omega^2 = (k_r N_z - k_z N_r)^2 + \kappa^2 k_z^2. \tag{16.65}$$

This is the inertial-gravity mode in the limit of infinite sound speed.

16.3.3 Instability Criterion and Growth Rate

From Eq. (16.64) we derive the condition of instability and the growth rate, following the procedures by Balbus and Hawley (1991).

(a) **Instability criterion**

The left-hand side of Eq. (16.64) is a quadratic form with respect to ω_*^2. The coefficient of square of ω_*^2 is positive, while the constant term (independent of ω_*) is negative. Hence, the dispersion relation (16.64) always has two real solutions with respect to ω_*^2: One is negative while the other is positive. If the negative solution with respect to ω_*^2 is smaller than $-c_{\rm A}^2 k_z^2$, ω^2 is negative and the perturbation is unstable. The condition of negative ω^2 is thus equivalent to that the left-hand side of Eq. (16.64) is negative when $\omega_*^2 = -c_{\rm A}^2 k_z^2$ is inserted. The instability criterion is thus

$$c_{\rm A}^2 k^2 + \left(\frac{k_r}{k_z}N_z - N_r\right)^2 + 2\Omega r\frac{d\Omega}{dr} < 0. \tag{16.66}$$

This inequality can be rearranged as being a quadratic inequality with respect to k_r:

$$\left(N_z^2 + c_{\rm A}^2 k_z^2\right)k_r^2 - 2k_r k_z N_r N_z + k_z^2\left[c_{\rm A}^2 k_z^2 + N_r^2 + \frac{d\Omega^2}{d\ln r}\right] < 0. \tag{16.67}$$

This inequality is realized for certain range of k_r, if

$$N_r^2 N_z^2 - \left(N_z^2 + c_{\rm A}^2 k_z^2\right)\left[c_{\rm A}^2 k_z^2 + N_r^2 + \frac{d\Omega^2}{d\ln r}\right] > 0. \tag{16.68}$$

That is, instability occurs for k_z which satisfies inequality (16.68). This inequality can be written as

$$\left(c_{\rm A}^2 k_z^2\right)^2 + \left(N^2 + \frac{d\Omega^2}{d\ln r}\right)\left(c_{\rm A}^2 k_z^2\right) + N_z \frac{d\Omega^2}{d\ln r} < 0. \tag{16.69}$$

Let us consider, for simplicity, the case where $N_z^2 d\Omega^2/d\ln r < 0$. Then, inequality (16.69) is satisfied for $k_z^2 < k_{\rm crit}^2$, where $k_{\rm crit}^2$ is given by

$$k_{z,{\rm crit}}^2 = \frac{1}{2c_{\rm A}^2}\left\{\left[\left(N^2 + \frac{d\Omega^2}{d\ln r}\right)^2 - 4N_z^2\frac{d\Omega^2}{d\ln r}\right]^{1/2} - \left(N^2 + \frac{d\Omega^2}{d\ln r}\right)\right\} \tag{16.70}$$

(Balbus and Hawley 1991).

In geometrically thin disks, we have $N_r^2 \ll N_z^2$, except for a narrow region around the mid-plane of $z = 0$, where $N_z = 0$. Except for such a special case, the critical wavenumber $k_{z,{\rm crit}}$ given by Eq. (16.70) is found to be

$$k_{z,{\rm crit}}^2 = \frac{1}{c_{\rm A}^2}\left|\frac{d\Omega^2}{d\ln r}\right|. \tag{16.71}$$

Since $|d\Omega^2/d\ln r| \sim \Omega^2 \sim c_{\rm s}^2/H^2$, where H is the vertical thickness of the disk, Eq. (16.71) leads to

$$k_{z,{\rm crit}} H \sim \frac{c_{\rm s}}{c_{\rm A}}. \tag{16.72}$$

It is noted that $k_{z,{\rm crit}}$ is much larger than $1/H$ when the magnetic field is weak.

(b) **Growth rate**

Figure 16.6 shows the growth rate as a function of k_z in two cases of $k_r/k_z = 0$ and $k_r/k_z = 1.0$, with $\kappa^2 = \Omega^2$, and $N_r^2 = 0.008\Omega^2$. The instability occurs for small k_z. The critical k_z below which the instability set in is consistent with Eq. (16.72).

Next, let us consider the maximum growth rate $n_{\rm max}$ and the corresponding wavenumber $k_{z,{\rm max}}$. Although the growth rate n, defined by $n^2 = -\omega^2 = -(\omega_*^2 + c_{\rm A}^2 k_z^2)$, is a function of k_r and k_z, it is convenient to regard it as a function of $\alpha_{\rm k}$ ($\equiv k_r/k_z$) and k_z. Let us change $\alpha_{\rm k}$ after fixing k_z. Then, Eq. (16.64) says that the maximum growth (where $\partial\omega_*^2/\partial\alpha_{\rm k} = 0$) occurs when

$$\alpha_{\rm k} = \frac{N_r N_z}{-\omega_*^2 + N_z^2}. \tag{16.73}$$

For this $\alpha_{\rm k}$, $(\alpha_{\rm k} N_z - N_r)^2 = [\omega_*^2/(\omega_*^2 - N_z^2)]^2 N_r^2$, and this is usually much smaller than κ^2 in accretion disks. This is because in accretion disks $N_r^2 \ll N_z^2$, and N_z^2 and κ^2 are of the order of Ω^2. Hence, the term $[(k_r/k_z)N_z - N_r]^2$ in Eq. (16.64) can

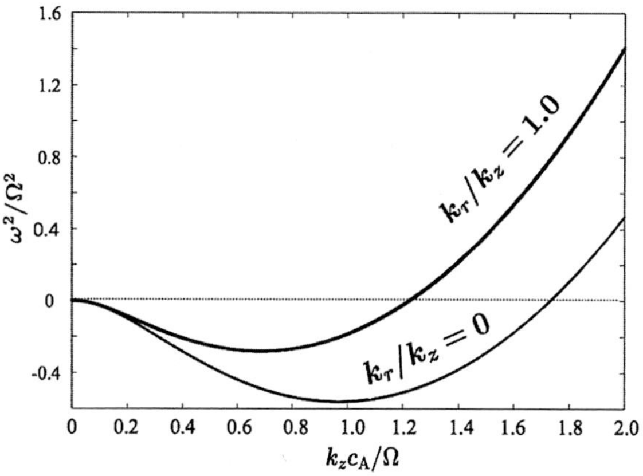

Fig. 16.6 Dimensionless growth rate of magneto-rotational instability in the case where the magnetic field is vertical. The growth rate is shown as a function of dimensionless vertical wavelength k_z for $k_r/k_z = 0$ and $k_r/k_z = 1.0$. Other parameters adopted are $\kappa/\Omega = 1.0$ and $N_r^2/\Omega^2 = 0.008$. Notice that $\omega^2 < 0$ corresponds to instability

be neglected in comparison with κ^2 when the maximum growth rate is examined. If this is done, Eq. (16.64) can be written explicitly by using n^2 ($\equiv \omega^2$) as

$$\left(1+\alpha_{\rm k}^2\right)n^4 + \left[2\left(1+\alpha_{\rm k}^2\right)c_{\rm A}^2 k_z^2 + \kappa^2\right]n^2 + \left(1+\alpha_{\rm k}^2\right)c_{\rm A}^4 k_z^4 + \frac{d\Omega^2}{d\ln r}c_{\rm A}^2 k_z^2 = 0. \tag{16.74}$$

This equation shows that when k_z is changed, while k_r/k_z is fixed, the maximum of n^2 (i.e., $\partial n^2/\partial k_z = 0$) occurs for

$$2\left(1+\alpha_{\rm k}^2\right)n^2 + 2\left(1+\alpha_{\rm k}^2\right)c_{\rm A}^2 k_z^2 + \frac{d\Omega^2}{d\ln r} = 0. \tag{16.75}$$

Substituting n^2 given by Eq. (16.75) into Eq. (16.74), we find that the maximum growth occurs at $k_z = k_{z,\max}$ given by

$$k_{z,\max}^2 = \frac{1}{\left(1+\alpha_{\rm k}^2\right)c_{\rm A}^2}\left[-\frac{1}{2}\frac{d\Omega^2}{d\ln r} - \frac{1}{4}\left(\frac{d\Omega}{d\ln r}\right)^2\right]. \tag{16.76}$$

Substituting this $k_{z,\max}^2$ into Eq. (16.75), we have the maximum growth rate, $n_{\max}$, in the form:

$$n_{\max}^2 = \frac{1}{4\left(1+\alpha_{\rm k}^2\right)}\left(\frac{d\Omega}{d\ln r}\right)^2. \tag{16.77}$$

Here, $1+\alpha_{\rm k}^2$ in the denominators of the above two equations can be taken to be ~ 1, since $\alpha_{\rm k}$ given by Eq. (16.73) is much smaller than unity.

Important and remarkable results obtained above are that the maximum growth rate is almost independent of the strength of magnetic field and always of the order of Ω, although $k_{z,\max}$ depends upon the strength of the magnetic field. The fact that $n_{\max}$ is of the order of Ω holds more generally (Balbus and Hawley 1992a). The reason behind this remarkably general behavior of $n_{\max}$ has been explored by Balbus and Hawley (1992a) (see also a review article by Balbus 1998).

(c) **Remarks on non-axisymmetric perturbations**

We have considered the case where the unperturbed magnetic field is uniform and vertical. Another limiting case is that the field lines are toroidal. Even in this case a magneto-rotational instability occurs for non-axisymmetric perturbations, as mentioned before. After Balbus and Hawley (1991) and Hawley and Balbus (1991), detailed examination in this case were also made extensively both analytically and numerically (e.g., Balbus and Hawley 1992b; Matsumoto and Tajima 1995; Ogilvie and Pringle 1996; Trequem and Papaloizou 1996; Papaloizou and Trequem 1997). This case is, however, very complicated compared with the case of axisymmetric perturbations. A reason for this is that there are limitations of normal-mode analyses, since the pattern of the perturbations is distorted due to differential rotation. Non-modal stability analyses will be of importance.

We shall be satisfied here with only pointing out that under the shearing sheet approximation (Goldreich and Lynden-Bell 1965), Balbus and Hawley (1992b) solved linearized differential equations as initial-value problems and found that non-axisymmetric perturbations really grow with time. The maximum growth rate in the case of a toroidal magnetic field seems to be smaller than that in the case of the vertical one considered above by nearly one order.

Finally, it is noted that there are many excellent numerical simulations of accretion disks with turbulent magnetic fields in order to obtain realistic accretion disks without introducing α viscosity (e.g., Machida et al. 2021, in preparation).

16.4 Modification of Solberg–Høiland Stability Criterion by Magnetic Fields

In Sect. 6.3 we have studied the Solberg–Høiland stability criterion in stratified rotating media. A next issue to be examined here is how the stability criterion is modified if magnetic fields are present. This issue has been examined by Papaloizou and Szuszukiewicz (1992) and Balbus (1995) with interesting results that the stability criterion is not a simple extension of that in non-magnetized cases. That is, the stability criterion does not tend to the Solberg–Høiland criterion even in the limit of weak magnetic fields. This situation is similar to the case of magneto-rotational instability described in Sect. 16.3.

16.4.1 Adiabatic Perturbations

Let us consider the case of adiabatic perturbations, following Balbus (1995): The unperturbed system to be considered here is the same as that in Sect. 6.3, except that there are weak magnetic fields. That is, the system is axisymmetric and differentially rotating with no dissipative processes. The magnetic fields are assumed to be weak in the sense that the Alfvén velocity is small compared to both the local acoustic speed as well as the local rotational velocity. This implies that the equilibrium state is determined by non-magnetic forces, as in Sect. 6.3. We introduce cylindrical coordinates (r, φ, z). The angular velocity of rotation Ω is a function of r and z as $\Omega = \Omega(r, z)$.

Small amplitude axisymmetric perturbations are superposed on the equilibrium system mentioned above. As in Sects 6.3 and 16.3, the Boussinesq approximation is introduced to describe the perturbations. The Boussinesq approximation filters out the acoustic oscillations and is valid as long as low frequency perturbations are considered.[6] The perturbations are assumed to be axisymmetric with perturbed velocity, $\boldsymbol{v} = (v_r, v_\varphi, v_z)$, perturbed magnetic fields, $\boldsymbol{b} = (b_r, b_\varphi, b_z)$, perturbed pressure and density, p_1 and ρ_1, respectively. Unperturbed magnetic field, pressure, and density are denoted, respectively, by $\boldsymbol{B}_0 = (B_{0r}, B_{0\varphi}, B_{0z})$, p_0, and ρ_0.

The equation of continuity describing perturbations is the incompressibility relation:

$$\frac{\partial v_r}{\partial r} + \frac{\partial v_z}{\partial z} = 0. \tag{16.78}$$

The r-, φ-, and z-components of equation of motions are, respectively,

$$\frac{\partial v_r}{\partial t} - 2\Omega v_\varphi = -\frac{\partial p_1}{\rho_0 \partial r} + \frac{\rho_1}{\rho_0^2}\frac{\partial p_0}{\partial r} - \frac{B_{0\varphi}}{4\pi\rho_0}\frac{\partial b_\varphi}{\partial r} + \frac{B_{0z}}{4\pi\rho_0}\left(-\frac{\partial b_z}{\partial r} + \frac{\partial b_r}{\partial z}\right), \tag{16.79}$$

$$\frac{\partial v_\varphi}{\partial t} + \frac{\kappa^2}{2\Omega} v_r + r\frac{\partial \Omega}{\partial z} v_z = \frac{1}{4\pi\rho_0}\mathcal{D} b_\varphi, \tag{16.80}$$

$$\frac{\partial v_z}{\partial t} = -\frac{\partial p_1}{\rho_0 \partial z} + \frac{\rho_1}{\rho_0^2}\frac{\partial p_0}{\partial z} - \frac{1}{4\pi\rho_0} B_{0\varphi}\frac{\partial b_\varphi}{\partial z} + \frac{B_{0r}}{4\pi\rho_0}\left(\frac{\partial b_z}{\partial r} - \frac{\partial b_r}{\partial z}\right), \tag{16.81}$$

where $\mathcal{D}$ is an operator defined by

$$\mathcal{D} \equiv B_{0r}\frac{\partial}{\partial r} + B_{0z}\frac{\partial}{\partial z}. \tag{16.82}$$

The r-, φ-, and z-components of the induction equation are, respectively,

$$\frac{\partial b_r}{\partial t} = \mathcal{D} v_r, \tag{16.83}$$

[6] The approximation is not always accurate when overstable oscillations with relatively high frequency are studied, e.g., Kato (1966).

$$\frac{\partial b_\varphi}{\partial t} = \mathcal{D}v_\varphi + r\frac{\partial \Omega}{\partial r}b_r + r\frac{\partial \Omega}{\partial z}b_z, \tag{16.84}$$

$$\frac{\partial b_z}{\partial t} = \mathcal{D}v_z. \tag{16.85}$$

The energy equation is written as, using the Boussinesq approximation,

$$\frac{\partial \rho_1}{\rho_0 \partial t} = \frac{\partial S}{\partial r}v_r + \frac{\partial S}{\partial z}v_z, \tag{16.86}$$

where S is defined by

$$S = \ln \frac{p_0^{1/\gamma}}{\rho_0}. \tag{16.87}$$

In addition, we have $\partial b_r/\partial r + \partial b_z/\partial z = 0$.

A conventional way to eliminate p_1 from the above set of equations is to take the derivative of Eq. (16.79) with respect to z and the derivative of Eq. (16.81) with respect to r, and to take the difference of the resulting two equations to lead to

$$\frac{\partial}{\partial z}\left(\frac{\partial v_r}{\partial t} - 2\Omega v_\varphi\right) - \frac{\partial^2 v_z}{\partial r \partial t} = \frac{1}{\rho_0^2}\left(\frac{\partial \rho_1}{\partial z}\frac{\partial p_0}{\partial r} - \frac{\partial \rho_1}{\partial r}\frac{\partial p_0}{\partial z}\right) + \frac{1}{4\pi\rho_0}(B_{0z}\nabla^2 b_r - B_{0r}\nabla^2 b_z), \tag{16.88}$$

where $\nabla^2 = \partial^2/\partial r^2 + \partial^2/\partial z^2$.

To eliminate v_φ from the above equation, we derive an expression for v_φ by eliminating b_φ from Eqs. (16.80) and (16.84):

$$\left(\frac{\partial^2}{\partial t^2} - \frac{1}{4\pi\rho_0}\mathcal{D}^2\right)v_\varphi = -\frac{1}{r}\frac{\partial}{\partial t}\left(\frac{\partial \ell}{\partial r}v_r + \frac{\partial \ell}{\partial z}v_z\right) + \frac{1}{4\pi\rho_0}\mathcal{D}\left(r\frac{\partial \Omega}{\partial r}b_r + r\frac{\partial \Omega}{\partial z}b_z\right), \tag{16.89}$$

where ℓ is angular momentum defined by $\ell = r^2\Omega$, and $\kappa^2 = (2\Omega/r)\partial\ell/\partial r$ has been used.

Eliminating v_φ from the above two equations, we have an equation among v_r, v_z, ρ_1, b_r, and b_z. From this equation we can easily obtain an equation between v_r and v_z, by expressing ρ_1, b_r, and b_z in terms of v_r and v_z by using Eqs. (16.86), (16.83), and (16.85), respectively. Then, the resulting equation between v_r and v_z can be expressed as an equation for v_r by using Eq. (16.78). After lengthy manipulations, we have

$$\left(\frac{\partial^2}{\partial t^2}-\frac{1}{4\pi\rho_0}\mathcal{D}^2\right)^2\nabla^2 v_r+\left(\frac{\partial^2}{\partial t^2}-\frac{1}{4\pi\rho_0}\mathcal{D}^2\right)$$
$$\times\left[-\left(\frac{\partial p_0}{\rho_0\partial r}\frac{\partial}{\partial z}-\frac{\partial p_0}{\rho_0\partial z}\frac{\partial}{\partial r}\right)\left(\frac{\partial S}{\partial r}\frac{\partial}{\partial z}-\frac{\partial S}{\partial z}\frac{\partial}{\partial r}\right)\right.$$
$$\left.+\frac{1}{r^3}\left(\frac{\partial \ell^2}{\partial r}\frac{\partial}{\partial z}-\frac{\partial \ell^2}{\partial z}\frac{\partial}{\partial r}\right)\frac{\partial}{\partial z}\right]v_r+4\Omega^2\frac{1}{4\pi\rho_0}\mathcal{D}^2\frac{\partial^2 v_r}{\partial z^2}=0. \quad (16.90)$$

Equation (16.90) is now solved by taking $v_r \propto \exp(nt - ik_r r - ik_z z)$. Then, as the dispersion relation we have

$$\tilde{n}^4 k^2 \; -\tilde{n}^2\left[\left(\frac{\partial p_0}{\rho_0\partial r}k_z-\frac{\partial p_0}{\rho_0\partial z}k_r\right)\left(\frac{\partial S}{\partial r}k_z-\frac{\partial S}{\partial z}k_r\right)-\frac{1}{r^3}\left(\frac{\partial \ell^2}{\partial r}k_z-\frac{\partial \ell^2}{\partial z}k_r\right)k_z\right]$$
$$-4\Omega^2\frac{k_z^2}{4\pi\rho_0}(\boldsymbol{k}\cdot\boldsymbol{B_0})^2=0, \quad (16.91)$$

where $k^2 = k_r^2 + k_z^2$ and $\tilde{n}^2$ is defined by

$$\tilde{n}^2 \equiv n^2 + \frac{1}{4\pi\rho_0}(\boldsymbol{k}\cdot\boldsymbol{B_0})^2. \quad (16.92)$$

Equation (16.91) is a quadratic equation with respect to $\tilde{n}^2$ and the zeroth-order term with respect to $\tilde{n}^2$ is negative. Hence, two solutions of Eq. (16.91) with respect to $\tilde{n}^2$ are real and one is negative and the other is positive. The stability condition is that the positive solution of $\tilde{n}^2$ is smaller than $(1/4\pi\rho_0)(\boldsymbol{k}\cdot\boldsymbol{B_0})^2$. This means that if $\tilde{n}^2 = (1/4\pi\rho_0)(\boldsymbol{k}\cdot\boldsymbol{B_0})^2$ is inserted into Eq. (16.91), the left-hand side of Eq. (16.91) is positive. Since $(1/4\pi\rho_0)(\boldsymbol{k}\cdot\boldsymbol{B_0})^2$ is a positive quantity and further the magnetic fields in the unperturbed state are assumed to be weak, the condition for stability is reduced to (Balbus 1995)

$$\left[\left(\frac{\partial p_0}{\rho_0\partial r}k_z-\frac{\partial p_0}{\rho_0\partial z}k_r\right)\left(\frac{\partial S}{\partial r}k_z-\frac{\partial S}{\partial z}k_r\right)-\frac{1}{r^3}\left(\frac{\partial \ell^2}{\partial r}k_z-\frac{\partial \ell^2}{\partial z}k_r\right)k_z\right]+4\Omega^2 k_z^2<0. \quad (16.93)$$

Let us now set $x = k_r/k_z$. Then, the above inequality is written as

$$x^2 N_z^2 + x\left(\frac{\partial p_0}{\rho_0\partial r}\frac{\partial S}{\partial z}+\frac{\partial p_0}{\rho_0\partial z}\frac{\partial S}{\partial r}-r\frac{\partial \Omega^2}{\partial z}\right)+N_r^2+r\frac{\partial \Omega^2}{\partial r}>0, \quad (16.94)$$

where N_r^2 and N_z^2 are the squares of the Brundt–Väisärä frequencies in r- and z-directions, respectively, and are defined by

$$N_r^2 = -\frac{\partial p_0}{\rho_0 \partial r}\frac{\partial S}{\partial r}, \quad N_z^2 = -\frac{\partial p_0}{\rho_0 \partial z}\frac{\partial S}{\partial z}. \tag{16.95}$$

Two conditions are necessary for inequality (16.94) to be satisfied. One is $N_z^2 > 0$ and $N_r^2 + r\partial\Omega^2/\partial r > 0$. That is, one of the necessary conditions for stability is

$$N_z^2 > 0 \quad \text{and} \quad N_r^2 + r\frac{\partial \Omega^2}{\partial r} > 0. \tag{16.96}$$

The other one is that the quadratic discriminant is negative, i.e.,

$$\left(\frac{\partial p_0}{\rho_0 \partial r}\frac{\partial S}{\partial z} + \frac{\partial p_0}{\rho_0 \partial z}\frac{\partial S}{\partial r} - r\frac{\partial \Omega^2}{\partial z}\right)^2 - 4N_z^2\left(N_r^2 + r\frac{\partial \Omega^2}{\partial r}\right) < 0. \tag{16.97}$$

In the case where the unperturbed system is in a differentially rotating steady state, the following relation holds [see Eq. (6.4) in Sect. 6.3]:

$$r\frac{\partial \Omega^2}{\partial z} = \frac{1}{\rho_0^2}\left(\frac{\partial \rho_0}{\partial r}\frac{\partial p_0}{\partial z} - \frac{\partial \rho_0}{\partial z}\frac{\partial p_0}{\partial r}\right). \tag{16.98}$$

If this relation and definitions of N_r^2 and N_z^2 are adopted, inequality (16.97) can be expressed, after some manipulations, as (Balbus 1995)

$$\left(-\frac{\partial p_0}{\partial z}\right)\left(\frac{\partial \Omega^2}{\partial r}\frac{\partial S}{\partial z} - \frac{\partial \Omega^2}{\partial z}\frac{\partial S}{\partial r}\right) > 0. \tag{16.99}$$

In summary, the stability conditions are the set of inequalities (16.96) and (16.99). This stability condition should be compared with the Solberg–Høiland condition [inequality (6.15) in in Sect. 6.3]. The angular momentum gradient in the Solberg–Høiland criterion is now changed to the angular velocity gradient (Papaloizou and Szuszukiewicz 1992; Balbus 1995). The present criterion will be more general than that of Solberg–Høiland. The geometrical structure of condition (16.99) is shown in Fig. 16.7.

16.4.2 *Effects of Thermal Conduction*

The above analyses in Sect. 16.4.1 are under the assumption that the perturbations are adiabatic. Balbus (2001) relaxed this assumption and examined the effects of electron heat conduction along magnetic fields. This extension is done by including

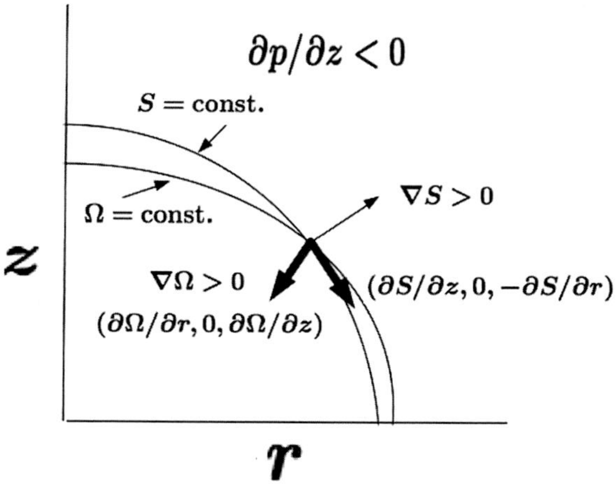

Fig. 16.7 Schematic picture showing stability condition (16.99). The case where $-(1/\rho)\partial p_0/\partial z > 0$ is shown. The equi-Ω surface has been taken to be oblate so that a case of equatorial acceleration is demonstrated. The direction of increase of angular velocity, i.e., $(\partial\Omega/\partial r,\ 0,\ \partial\Omega/\partial z)$ and the direction perpendicular to entropy ($s \equiv p_0/\rho_o^\gamma$) increase, i.e.,$(\partial s/\partial z,$ $0,\ -\partial s/\partial r)$, are shown by thick arrows. In this figure the inner product of these two vectors are positive, and the system is stable

a term of thermal heat conduction along the magnetic fields in energy Eq. (16.86). Other equations describing perturbations are unchanged.

The stability analyses can be made in parallel with those in Sect. 16.4.1. The results of his analyses show that the stability conditions consist of two: The first condition is

$$-\frac{1}{\rho}\nabla p_0 \cdot \nabla T_0 + \frac{\partial \Omega^2}{\partial \ln r} > 0, \tag{16.100}$$

where T_0 is the temperature in the unperturbed medium, and the second condition is

$$-\frac{\partial p_0}{\partial z}\left(\frac{\partial \Omega^2}{\partial r}\frac{\partial \ln T_0}{\partial z} - \frac{\partial \Omega^2}{\partial z}\frac{\partial \ln T_0}{\partial r}\right) > 0. \tag{16.101}$$

It should be noticed that the stability criterion in the case of non-magnetized medium (the Solberg–Høiland criterion) is changed in magnetized medium to the set of Eqs. (16.96) and (16.99). They are further changed to the set of inequalities (16.100) and (16.101).

Finally, the case where radiative conduction is considered is briefly mentioned (see also Balbus 2001). For simplicity, the electron heat conduction parallel to the magnetic fields is neglected. All equations describing perturbations are the same as those in Sect. 16.4.1, except that the energy equation describing perturbations is changed from Eq. (16.86) to

$$\frac{1}{\rho_0}\left(\frac{\partial \rho_1}{\partial t} - \kappa_{\rm th}\nabla^2\rho_1\right) = \frac{\partial S}{\partial r}v_r + \frac{\partial S}{\partial z}v_z, \tag{16.102}$$

where $\kappa_{\rm th}$ is thermometric conductivity due to radiation.

Using Eq. (16.102) instead of Eq. (16.86), we perform the same procedures to derive dispersion relation. Then, instead of Eq. (16.91) we obtain

$$\begin{aligned}\tilde{n}^4(n+\kappa_{\rm th}k^2)k^2 \; &-\tilde{n}^2 n\left(\frac{\partial p_0}{\rho_0\partial r}k_z - \frac{\partial p_0}{\rho_0\partial z}k_r\right)\left(\frac{\partial S}{\partial r}k_z - \frac{\partial S}{\partial z}k_r\right)\\ &+\tilde{n}^2(n+\kappa_{\rm th}k^2)\frac{1}{r^3}\left(\frac{\partial \ell^2}{\partial r}k_z - \frac{\partial \ell^2}{\partial z}k_r\right)k_z\\ &-4\Omega^2(n+\kappa_{\rm th}k^2)\frac{k_z^2}{4\pi\rho_0}(\boldsymbol{k}\cdot\boldsymbol{B}_0)^2 = 0.\end{aligned} \tag{16.103}$$

This equation, of course, tends to Eq. (16.91) in the limit of $\kappa_{\rm th} = 0$.

A full examination of stability condition from dispersion relation (16.103) is troublesome, although there is no difficulty in principle if the Hurwitz theorem (see Sect. 7.1) is used. Here, however, we are satisfied only by presenting a necessary stability condition. Equation (16.103) is a fifth-order algebraic equation with respect to n. The term of n^5 is positive. Hence, if the coefficient of the term of n^0, say A, is negative, we have at least one real positive n, i.e., we have a monotonically growing perturbation. In other word, a necessary condition of stability is $A > 0$.

Dispersion relation (16.103) shows that

$$A = \frac{(\boldsymbol{k}\cdot\boldsymbol{B}_0)^2}{4\pi\rho_0}\kappa_{\rm th}k^2\left[\frac{(\boldsymbol{k}\cdot\boldsymbol{B}_0)^2}{4\pi\rho_0}k^2 + 2r\Omega\left(\frac{\partial\Omega}{\partial r}k_z^2 - \frac{\partial\Omega}{\partial z}k_rk_z\right)\right]. \tag{16.104}$$

Since we are considering the system where magnetic fields are weak, the above result implies that in the limit of weak magnetic fields, a necessary condition for stability is

$$\frac{\partial\Omega}{\partial r}k_z^2 - \frac{\partial\Omega}{\partial z}k_rk_z > 0. \tag{16.105}$$

In other words, a necessary condition of stability is

$$\frac{\partial\Omega}{\partial z} = 0 \quad \text{and} \quad \frac{\partial\Omega}{\partial r} > 0. \tag{16.106}$$

The first condition shows that the rotation of the system needs to be cylindrical for stability. This requirement for stability is the same as that of the Goldreich–Schubert–Fricke criterion [see Eq. (7.59)]. It is also noticed that the unstable mode in the case where inequalities (16.106) are violated is a mode which appears by the presence of both magnetic fields and thermal conduction. If $\boldsymbol{B}_0 = 0$ or $\kappa_{\rm th} = 0$, the mode is a trivial mode of $n = 0$.

References

Balbus, S.A.: Astrophys. J. **453**, 380 (1995)
Balbus, S.A.: Rev. Mod. Phys. **70**, 1 (1998)
Balbus, S.A.: Astrophys. J. **562**, 909 (2001)
Balbus, S.A.: ARA&A **41**, 555 (2003)
Balbus, S.A., Hawley, J.F.: Astrophys. J. **376**, 214 (1991)
Balbus, S.A., Hawley, J.F.: Astrophys. J. **392**, 662 (1992a)
Balbus, S.A., Hawley, J.F.: Astrophys. J. **400**, 610 (1992b)
Chandrasekhar, S.: Proc. Nat. Acad. Sci. **46**, 253 (1960)
Ferraro, V.C.A.: Mon. Not. R. Astron. Soc. **97**, 458 (1937)
Fricke, K.: A&A **1**, 388 (1969)
Goldreich, P., Lynden-Bell, D.: Mon. Not. R. Astron. Soc. **130**, 125 (1965)
Hawley, J.F., Balbus, S.A.: Astrophys. J. **376**, 223 (1991)
Horiuchi, T., Matsumoto, R., Hanawa, T., Shibata, K.: Publ. Astron. Soc. Jpn. **40**, 147 (1988)
Kato, S.: Publ. Astron. Soc. Jpn. **18**, 261 (1966)
Kato, S., Fukue, J., Mineshige, S.: Black-Hole Accretion Disks. Kyoto University Press, Kyoto (1998)
Matsumoto, R., Tajima, T.: Astrophys. J. **445**, 767 (1995)
Matsumoto, R., Horiuchi, T., Hanawa, T., Shibata, K.: Publ. Astron. Soc. Jpn. **40**, 171 (1988)
Ogilvie, G.I., Pringle, J.E.: Mon. Not. R. Astron. Soc. **279**, 15 (1996)
Papaloizou, J.C.B., Szuszukiewicz, E.: Geophys. Astrophys. Fluid Dynamics **66**, 223 (1992)
Papaloizou, J.C.B., Trequem, C.: Mon. Not. R. Astron. Soc. **287**, 771 (1997)
Parker, E.N.: Astrophys. J. **145**, 811 (1966)
Spiegel, E.A., Veronis, G.: Astrophys. J. **131**, 442 (1960)
Trequem, C., Papaloizou, J.C.B.: Mon. Not. R. Astron. Soc. **279**, 767 (1996)
Velihov, E.: Sov. Phys. JETP **36**, 1398 (1959)

Chapter 17
Important Non-Ideal MHD Processes

Up to the previous chapters we have focused our attention mainly on dynamical processes which are realized in the framework of MHD approximations. In this chapter some important processes which are not described by idealized MHD equations are mentioned. We choose two issues. One is ambipolar diffusion and the other is reconnection processes of magnetic fields.

17.1 Ambipolar Diffusion

In studying star and galaxy formations, how angular momentum is transport inside the systems or toward outside the systems is an important issue. In ideal MHD systems, contraction of astrophysical objects is made under pile-up of magnetic fields. Hence, the formation of large scale Keplerian protostar and protogalactic disks, which will be required from observational evidence, will be a delicate issue. If the magnetic braking is too strong, for example, the central object formed will be too compact. If the magnetic braking is too weak, on the other hand, the angular momentum involved in the system resists against formation of central objects. Since Mestel and Spitzer (1956), necessity of ambipolar diffusion has been argued in the fields of star and galactic formation.

(a) **Equation of motion in gas consisting of three species**

Let us consider gas systems consisting of three species: neutral gases, ionized gases, and electrons. For simplicity, these gases are taken to be neutral hydrogen, ionized hydrogen, and electrons. The set of their mass and number density are denoted by (m, n), $(m_{\rm i}, n_{\rm i})$, and $(m_{\rm e}, n_{\rm e})$, with $n_{\rm i} = n_{\rm e}$. The mass densities of neutral hydrogen, ionized hydrogen, and electrons are $\rho(= nm)$, $\rho_{\rm i}(= n_{\rm i}m_{\rm i})$, and $\rho_{\rm e}(= n_{\rm e}m_{\rm e})$, respectively. The charges of an ion and an electron are expressed as e and $-e$,

© Springer Nature Singapore Pte Ltd. 2020, corrected publication 2023
S. Kato, J. Fukue, *Fundamentals of Astrophysical Fluid Dynamics*,
Astronomy and Astrophysics Library,
https://doi.org/10.1007/978-981-15-4174-2_17

respectively. The velocities of hydrogen, ionized hydrogen, and electron gases are written, respectively, as $\boldsymbol{u}$, $\boldsymbol{u}_i$, and $\boldsymbol{u}_e$.

We assume that the ionization degree of the gas is weak, i.e., $n \gg n_i, n_e$. Then, the force balances in each species can be approximated as

$$\rho \frac{d\boldsymbol{u}}{dt} = -\text{grad}\, p - \rho\, \text{grad}\, \psi + \boldsymbol{f}_{ni}, \tag{17.1}$$

$$0 = n_i e \left(\boldsymbol{E} + \frac{\boldsymbol{u}_i}{c} \times \boldsymbol{B} \right) + \boldsymbol{f}_{in} + \boldsymbol{f}_{ie}, \tag{17.2}$$

$$0 = -n_e e \left(\boldsymbol{E} + \frac{\boldsymbol{u}_e}{c} \times \boldsymbol{B} \right) + \boldsymbol{f}_{ei}, \tag{17.3}$$

where $\boldsymbol{f}_{ni}$ is the rate of momentum gain of hydrogen gas per unit volume from collisions with ions, and $\boldsymbol{f}_{in}$ is that of ion gas per unit volume from collision with hydrogen. Similarly, $\boldsymbol{f}_{ie}$ and $\boldsymbol{f}_{ei}$ are the rate of momentum gain of ion gas per unit volume from collisions with electron gas, and that of the inverse process, respectively. The terms of $\boldsymbol{f}_{ne}$ in Eq. (17.1) and $\boldsymbol{f}_{en}$ in Eq. (17.3) have been neglected, since they are small quantities. The pressure and gravitational forces in Eqs. (17.2) and (17.3) have been neglected, since they are small compared with the corresponding terms in Eq. (17.1) by the factor of n_i/n. The inertial terms in Eqs. (17.2) and (17.3) have been neglected, because $n_i, n_e \ll n$ and $m_e \ll m_i$.

The interaction terms $\boldsymbol{f}_{ni}$ and $\boldsymbol{f}_{in}$ can be written as, respectively,

$$\boldsymbol{f}_{ni} = n\nu m(\boldsymbol{u}_i - \boldsymbol{u}), \quad \boldsymbol{f}_{in} = n_i \nu_i m(\boldsymbol{u} - \boldsymbol{u}_i), \tag{17.4}$$

where ν is the rate of collision of a hydrogen with ions in unit volume, and ν_i is the rate of collision of an ion with hydrogen gas in unit volume. Since ν and ν_i are related by $n\nu = n_i \nu_i$, we have $\boldsymbol{f}_{ni} + \boldsymbol{f}_{in} = 0$. The force $\boldsymbol{f}_{ni}$ is sometimes written by use of the drag coefficient γ_{AD} as

$$\boldsymbol{f}_{ni} = -\gamma_{AD} \rho \rho_i (\boldsymbol{u} - \boldsymbol{u}_i). \tag{17.5}$$

Then, ν_i and γ_{AD} are related by

$$\nu_i = \rho \gamma_{AD}. \tag{17.6}$$

The interaction terms $\boldsymbol{f}_{ie}$ and $\boldsymbol{f}_{ei}$ are related by $\boldsymbol{f}_{ie} + \boldsymbol{f}_{ei} = 0$. As mentioned in Sect. 10.3 in deriving generalized Ohm's law, $\boldsymbol{f}_{ei}$, which is proportional to $\boldsymbol{u}_i - \boldsymbol{u}_e$, is written as

$$\boldsymbol{f}_{ei} = n_e \nu_e m_e (\boldsymbol{u}_i - \boldsymbol{u}_e) = \frac{m_e \nu_e}{e} \boldsymbol{j} = \frac{1}{\sigma} (n_e e) \boldsymbol{j}, \tag{17.7}$$

where $n_e \nu_e$ is the rate of collision of electron gas with ion gas in unit volume, and ν_e is related to electric conductivity σ by

$$\nu_e = \frac{n_e e^2}{m_e \sigma}. \tag{17.8}$$

Finally, it is noted that the sum of Eqs. (17.2) and (17.3) gives

$$\boldsymbol{f}_{\rm in} + \frac{1}{c}\boldsymbol{j} \times \boldsymbol{B} = 0. \tag{17.9}$$

(b) **Induction equation**

From Eqs. (17.3) and (17.7), we obtain

$$\boldsymbol{j} = \sigma\left(\boldsymbol{E} + \frac{1}{c}\boldsymbol{u}_e \times \boldsymbol{B}\right). \tag{17.10}$$

This is a fundamental form of generalized Ohm's law, as mentioned in Sect. 10.3. Since $\boldsymbol{u}_e$ is expressed as

$$\boldsymbol{u}_e = \boldsymbol{u} + (\boldsymbol{u}_e - \boldsymbol{u}_i) + (\boldsymbol{u}_i - \boldsymbol{u}), \tag{17.11}$$

using Eqs. (17.5), (17.7), and (17.9), we have

$$\boldsymbol{u}_e = \boldsymbol{u} - \frac{1}{n_e e}\boldsymbol{j} + \frac{1}{\rho\rho_i\gamma_{\rm AD}c}\boldsymbol{j} \times \boldsymbol{B}. \tag{17.12}$$

Substitution of this expression for $\boldsymbol{u}_e$ into Eq. (17.10) leads to

$$\boldsymbol{j} = \sigma\left[\boldsymbol{E} + \frac{1}{c}\boldsymbol{u} \times \boldsymbol{B} - \frac{1}{n_e e c}\boldsymbol{j} \times \boldsymbol{B} + \frac{1}{\rho\rho_i\gamma_{\rm AD}c^2}(\boldsymbol{j} \times \boldsymbol{B}) \times \boldsymbol{B}\right]. \tag{17.13}$$

This is an expression for generalized Ohm's law [cf., Eq. (10.41) in Sect. 10.3]. In Sect. 10.3 we have considered the gases consisting only of two species (electrons and ions) and $\boldsymbol{v}$ in Eq. (10.41) is the mean velocity of the whole gas. Here, we consider gases consisting of three species and $\boldsymbol{u}$ denotes the velocity only of the neutral gas.

Now, we derive the induction equation. Substituting $\boldsymbol{E}$ given by the above equation into the Maxwell equation, $-(1/c)\partial\boldsymbol{B}/\partial t = {\rm curl}\,\boldsymbol{E}$, under the use of $\boldsymbol{j} = (c/4\pi){\rm curl}\,\boldsymbol{B}$, we have

$$\begin{aligned}\frac{\partial \boldsymbol{B}}{\partial t} &= -{\rm curl}(\eta\,{\rm curl}\,\boldsymbol{B}) + {\rm curl}\,(\boldsymbol{u} \times \boldsymbol{B}) \\ &\quad -{\rm curl}\left[\eta_{\rm H}{\rm curl}\,\boldsymbol{B} \times \frac{\boldsymbol{B}}{B}\right] + {\rm curl}\left[\eta_{\rm AD}\left({\rm curl}\,\boldsymbol{B} \times \frac{\boldsymbol{B}}{B}\right) \times \frac{\boldsymbol{B}}{B}\right]. \end{aligned} \tag{17.14}$$

The first term on the right-hand side of Eq. (17.14) is the well-known diffusion term. Note that $-\mathrm{curl}\,\mathrm{curl}\,\boldsymbol{B} = \nabla^2 \boldsymbol{B}$ and that the magnetic diffusivity η is given by

$$\eta = c^2/4\pi\sigma. \tag{17.15}$$

The coefficients η_{H} and η_{AD} are given by

$$\eta_{\mathrm{H}} = \frac{cB}{4\pi n_{\mathrm{e}} e} = \frac{\omega_{\mathrm{e}}}{\nu_{\mathrm{e}}}\eta, \qquad \eta_{\mathrm{AD}} = \frac{B^2}{4\pi\rho\rho_{\mathrm{i}}\gamma_{\mathrm{AD}}} = \frac{\omega_{\mathrm{i}}}{\nu_{\mathrm{i}}}\frac{\omega_{\mathrm{e}}}{\nu_{\mathrm{e}}}\eta, \tag{17.16}$$

where ω_{e} and ω_{i} are cyclotron frequencies of electrons and ions, respectively, and given by

$$\omega_{\mathrm{e}} = \frac{eB}{m_{\mathrm{e}}c}, \qquad \omega_{\mathrm{i}} = \frac{eB}{m_{\mathrm{i}}c}. \tag{17.17}$$

The third term on the right-hand side of Eq. (17.14) is the one due to *Hall current*, and the fourth one is that due to *ambipolar diffusion*. The ratios among the terms of Ohmic dissipation, Hall current, and ambipolar diffusion are roughly

$$1 \; : \; \frac{\omega_{\mathrm{e}}}{\nu_{\mathrm{e}}} \; : \; \frac{\omega_{\mathrm{e}}}{\nu_{\mathrm{e}}}\frac{\omega_{\mathrm{i}}}{\nu_{\mathrm{i}}}. \tag{17.18}$$

The importance of the terms of Hall current and ambipolar diffusion compared with the Ohmic dissipation term can be measured by $\omega_{\mathrm{e}}/\nu_{\mathrm{e}}$ and $(\omega_{\mathrm{e}}/\nu_{\mathrm{e}})(\omega_{\mathrm{i}}/\nu_{\mathrm{i}})$. The inequality $\omega_{\mathrm{e}}/\nu_{\mathrm{e}} > 1$ is realized in the medium where the gas density is so low that an electron gyrates around magnetic fields many times before making collisions with ions.

In general, in low-density gases the terms of Hall current and ambipolar diffusion become important compared with that of Ohmic dissipation, since $\omega_{\mathrm{e}}/\nu_{\mathrm{e}}$ and $\omega_{\mathrm{i}}/\nu_{\mathrm{i}}$ are large in such gases. The density-, temperature-, and $c_{\mathrm{A}}/c_{\mathrm{s}}$-dependences of the ratios of $\eta \,:\, \eta_{\mathrm{H}} \,:\, \eta_{\mathrm{AD}}$ are given numerically by Balbus and Terquem (2001).

(c) **Generalized Ohm's law and electric conductivity**

Generalized Ohm's law is given by Eq. (17.13):

$$\boldsymbol{j} = \sigma\left(\boldsymbol{E} + \frac{1}{c}\boldsymbol{u} \times \boldsymbol{B}\right) - \frac{\eta_{\mathrm{H}}}{\eta}\boldsymbol{j} \times \frac{\boldsymbol{B}}{B} + \frac{\eta_{\mathrm{AD}}}{\eta}\left(\boldsymbol{j} \times \frac{\boldsymbol{B}}{B}\right) \times \frac{\boldsymbol{B}}{B}. \tag{17.19}$$

Starting from this equation, we consider how the relation between electric current $\boldsymbol{j}$ and electric fields $\boldsymbol{E}$ depends on the strength and direction of magnetic fields. In this study the induction term is neglected for simplicity, i.e., $\boldsymbol{u} = 0$ is adopted.

(i) **The case of $\boldsymbol{B} \parallel \boldsymbol{E}$**
The vector product of Eq. (17.19) with $\boldsymbol{B}$ leads to

$$\left(1+\frac{\eta_{\mathrm{AD}}}{\eta}\right)(\boldsymbol{j}\times\boldsymbol{B}) = -\frac{\eta_{\mathrm{H}}}{\eta}\left[(\boldsymbol{j}\cdot\boldsymbol{B})\left(\frac{\boldsymbol{B}}{B}\right) - B\boldsymbol{j}\right], \tag{17.20}$$

where $(\boldsymbol{j}\times\boldsymbol{B})\times\boldsymbol{B} = (\boldsymbol{j}\cdot\boldsymbol{B})\boldsymbol{B} - B^2\boldsymbol{j}$ has been used. Equation (17.20) shows $\boldsymbol{j} \parallel \boldsymbol{B}$, and thus $\boldsymbol{j}\times\boldsymbol{B} = 0$, which leads to

$$\boldsymbol{j} = \sigma\boldsymbol{E}. \tag{17.21}$$

(ii) **The case of $\boldsymbol{B} \perp \boldsymbol{E}$**
The inner product of Eq. (17.19) with $\boldsymbol{B}$ shows that in this case we have $\boldsymbol{j}\cdot\boldsymbol{B} = 0$. That is, no current flows in the direction of magnetic fields. If this relation used, Eq. (17.19) can be reduced to[1]

$$\left(1+\frac{\eta_{\mathrm{AD}}}{\eta}\right)\boldsymbol{j} = \sigma\boldsymbol{E} - \frac{\eta_{\mathrm{H}}}{\eta}\boldsymbol{j}\times\frac{\boldsymbol{B}}{B}. \tag{17.22}$$

The above equation is again substituted into Eq. (17.19). Then we have

$$\left[\left(1+\frac{\eta_{\mathrm{AD}}}{\eta}\right)^2 + \left(\frac{\eta_{\mathrm{H}}}{\eta}\right)^2\right]\boldsymbol{j} = \sigma\left(1+\frac{\eta_{\mathrm{AD}}}{\eta}\right)\boldsymbol{E} + \sigma\frac{\eta_{\mathrm{H}}}{\eta}\frac{\boldsymbol{B}}{B}\times\boldsymbol{E}. \tag{17.23}$$

In summary, the above equation is written in the form:

$$\boldsymbol{j} = \sigma_{\parallel}\boldsymbol{E} + \sigma_{\perp}\frac{\boldsymbol{B}}{B}\times\boldsymbol{E}, \tag{17.24}$$

where

$$\sigma_{\parallel} = \frac{1+\eta_{\mathrm{AD}}/\eta}{[1+(\eta_{\mathrm{AD}}/\eta)]^2 + (\eta_{\mathrm{H}}/\eta)^2}\sigma \tag{17.25}$$

and

$$\sigma_{\perp} = \frac{\eta_{\mathrm{H}}/\eta}{[1+(\eta_{\mathrm{AD}}/\eta)]^2 + (\eta_{\mathrm{H}}/\eta)^2}\sigma. \tag{17.26}$$

The relations among directions of $\boldsymbol{E}$, $\boldsymbol{B}$, and $\boldsymbol{j}$ are shown in Fig. 17.1.

---

[1] A formula of vector analyses:

$$(\boldsymbol{A}\times\boldsymbol{B})\times\boldsymbol{C} = \boldsymbol{B}(\boldsymbol{A}\cdot\boldsymbol{C}) - \boldsymbol{A}(\boldsymbol{B}\cdot\boldsymbol{C}).$$

has been adopted, where $\boldsymbol{A}$, $\boldsymbol{B}$, and $\boldsymbol{C}$ are arbitrary vectors.

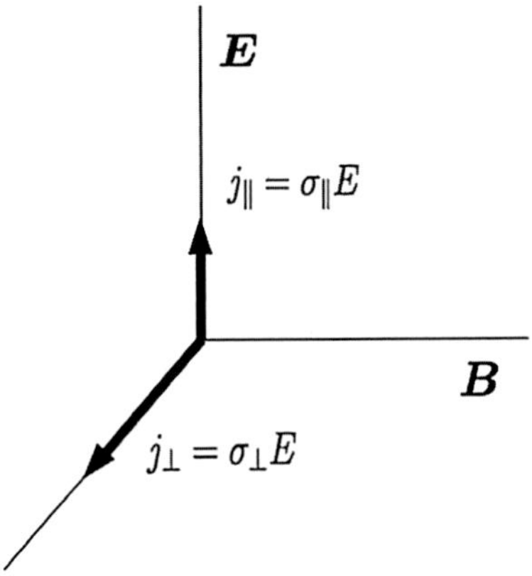

Fig. 17.1 Directions of electric current in the case where magnetic field ***B*** is perpendicular to electric field ***E***

The collision frequencies, ν_e and ν_i, are proportional to n_i and n, respectively. Hence, in an early collapse stage of gases with low-density and low ionization degree, we have $\eta < \eta_H < \eta_{AD}$, although ω_i is much smaller than ω_e since $m_i > m_e$. In this case, we can neglect η and η_H compared with η_{AD}, and have

$$\sigma_\parallel \sim \frac{\eta}{\eta_{AD}}\sigma, \qquad \sigma_\perp \sim \frac{\eta_H}{\eta_{AD}}\frac{\eta}{\eta_{AD}}\sigma. \tag{17.27}$$

This result shows that the current flows mainly in the direction perpendicular both to magnetic fields and electric fields, not in the direction of $\boldsymbol{j}$. This is due to drift motions of charged particles under magnetic fields.

17.2 Magnetic Reconnection Models

Let us consider diffusion time of magnetic fields. If the characteristic size of magnetic fields is L, the induction equation shows that the characteristic timescale of diffusion of magnetic fields, τ_D, is

$$\tau_D = \frac{L^2}{\eta}, \tag{17.28}$$

where η is the magnetic diffusivity. In astrophysical phenomena there are many violent phenomena which show that magnetic fields are dissipated with much faster timescale than τ_D. For example, the timescale of solar flares is 10–100 s, while τ_D is million years if the classical value of η is applied.

To overcome this basic difficulty, some reconnection models have been developed, taking into account the effects of gas flows. The well-known models are Sweet–Parker model and Petschek one. Here, we will briefly describe the Sweet–Parker model.

17.2.1 Sweet–Parker Model

A short timescale of energy release can be realized if phenomena in consideration have small scales, but an important issue is how we can release a large amount of energy in a short timescale. To solve this issue, Sweet (1958) and Parker (1957) proposed a model of steady driven reconnection, taking into account the effects of continuous gas supply. Here, we present the essence of the model. Details can be found, for example, in Zweibel and Yamada (2009).

We consider a situation shown in Fig. 17.2. That is, on the $x = 0$ plane a current sheet exists. The current j is in the z-direction (perpendicular to the plane of the paper). Due to this, around the $x = 0$ plane the magnetic fields are in the y-direction, but change their sign in a thin width 2δ, i.e., a *neutral current sheet*. Then, the Maxwell equation shows that the current j, B_0, and δ are related by

$$j \sim \frac{cB_0}{4\pi\delta}. \tag{17.29}$$

The heat generation rate by Ohmic heating, Q, is then

$$Q = \frac{1}{\sigma}j^2 = \frac{4\pi\eta j^2}{c^2} \sim \frac{\eta B_0^2}{4\pi\delta^2}, \tag{17.30}$$

where σ is the electric conductivity and related to magnetic diffusivity η by $\eta = c^2/4\pi\sigma$.

Next, let us consider the force balances in the perpendicular and parallel directions to the magnetic fields. In the perpendicular direction (z-direction), the force balance is realized between gas pressure p (inside of current), and magnetic

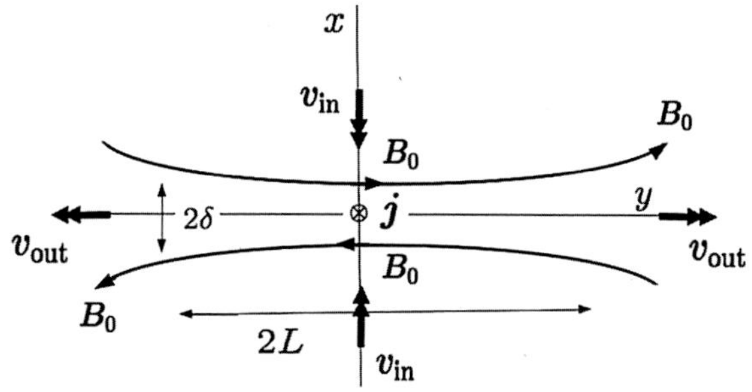

Fig. 17.2 Schematic picture showing the Sweet-Parker reconnection model. The sheet current on the x-plane is j. B_0 is the magnetic fields outside the current region. 2δ and $2L$ are the thickness and length of the region of the current sheet region. v_{in} and v_{out} are the fluid flows toward and away from the reconnection region, respectively

pressure $B_0^2/8\pi$ (outside the current), i.e.,

$$p = \frac{B_0^2}{8\pi}. \tag{17.31}$$

On the other hand, the force balance in the parallel direction (y-direction) is

$$\rho v \frac{dv}{dy} = -\frac{dp}{dy}, \tag{17.32}$$

where the y-direction has been taken in the parallel direction, and v is the outflow speed in the reconnection region. If the size of the reconnection region is taken to be $2L$ (see Fig. 17.2), the integration of the above equation in the y-direction gives

$$\frac{1}{2}\rho v_{\rm out}^2 \sim \frac{B_0^2}{8\pi}, \tag{17.33}$$

where $v_{\rm out}$ is the outgoing flow of gases from the reconnection region. This equation shows that

$$v_{\rm out} \sim c_{\rm A}, \tag{17.34}$$

which demonstrates that the outgoing flow speed of gases from the reconnection region is on the order of the Alfvén speed $c_{\rm A}$.

Next, let us consider heat balance. The heating of the reconnection region occurs during the timescale of the flow passing through the reconnection region, which is $L/c_{\rm A}$. On the other hand, the pressure increase at the reconnection region, δp, is

$$\delta p \sim Q\frac{L}{c_{\rm A}} = \frac{\eta B_0^2}{4\pi\delta^2}\frac{L}{c_{\rm A}}. \tag{17.35}$$

By taking $\delta p = B_0^2/8\pi$, we find that the thickness of the reconnection region, δ, to be

$$\delta \sim \left(\frac{\eta L}{c_{\rm A}}\right)^{1/2} = L\mathcal{R}_{\rm m}^{-1/2}, \tag{17.36}$$

where $\mathcal{R}_{\rm m}$ is the magnetic Reynolds number defined by[2]

$$\mathcal{R}_{\rm m} = \frac{c_{\rm A}L}{\eta}. \tag{17.37}$$

[2]The magnetic Reynolds number is usually defined by $\mathcal{R}_{\rm m} = vL/\eta$, where v is a typical flow velocity. Here, Alfvén speed is adopted as v.

Finally, assuming nearly incompressible flows, we find that the incoming flow speed to the reconnection region, $v_{\rm in}$, (see Fig. 17.2) is

$$v_{\rm in} \sim \frac{\delta}{L} c_{\rm A}. \tag{17.38}$$

The reconnection timescale, $\tau_{\rm SP}$, due to this model is roughly estimated as follows. The timescale $\tau_{\rm SP}$ is not δ^2/η, but longer since for magnetic energy $B_0^2/8\pi$ to be converted to gas energy, a gas element need to propagate to the edge of the reconnection region. Hence, we can estimate $\tau_{\rm SP}$ as

$$\tau_{\rm SP} \sim \frac{\delta^2}{\eta}\frac{L}{\delta} \sim \tau_{\rm A}\mathcal{R}_{\rm m}^{1/2} \sim \tau_{\rm D}\mathcal{R}_{\rm m}^{-1/2}, \tag{17.39}$$

where $\tau_{\rm A} = L/c_{\rm A}$. That is, the reconnection time, $\tau_{\rm SP}$, is longer than the Alfvén time $\tau_{\rm A}$ by a factor of $\mathcal{R}_{\rm m}^{1/2}$, but shorter than the diffusion time $\tau_{\rm D}$ in Eq. (17.28) by a factor of $\mathcal{R}_{\rm m}^{-1/2}$. The timescale $\tau_{\rm SP}$ is still longer to explain observational evidence of flares and of some other astrophysical plasmas.

Petschek (1964) proposed a model of increase of reconnection rate by considering the effect of slow MHD shock on the region outside the diffusion region. For more dertails on the Petschek model and references concerning the Petschek model and related numerical simulations, see Tajima and Shibata (1997).

17.3 Tearing Instability

In Sect. 17.2, we have considered the steady reconnection processes. The current sheet is, however, unstable against small amplitude perturbations (Furth et al. 1963), and can grow to island structure by reconnection of field lines, which will increase the reconnection rate. This instability is called the *tearing instability*.

For simplicity, we consider two-dimensional structure in the x-y plane, i.e., in the z-direction the system is assumed to be homogeneous. Furthermore, the gas is assumed to be homogeneous and incompressible. As is shown in Fig. 17.3, the magnetic fields are assumed to be in the y-direction and its sign changes on the plane of $x = 0$. i.e., the $x = 0$ plane is a neutral sheet of magnetic fields. That is, the unperturbed static state is described by

$$\boldsymbol{v}_0 = 0, \quad \text{and} \quad \boldsymbol{B}_0 = (0, B_0(x), 0), \tag{17.40}$$

and $B_0(x) > 0$ for $x > 0$, while $B_0(x) < 0$ for $x < 0$. Unperturbed gas pressure $p_0(x)$ and $B_0(x)$ are related by

$$\frac{d}{dx}\left(p_0 + \frac{B_0^2}{8\pi}\right) = 0. \tag{17.41}$$

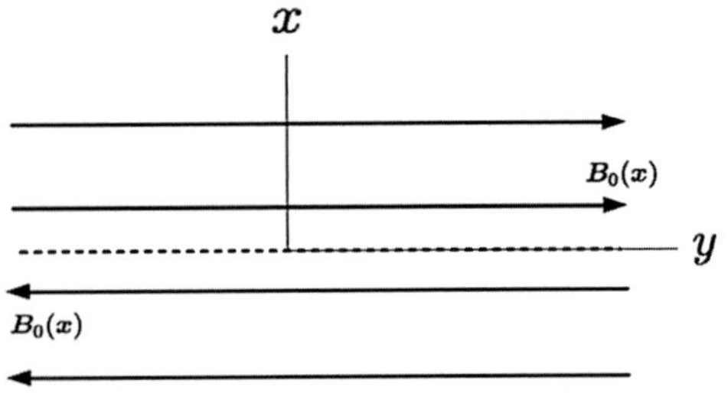

Fig. 17.3 Schematic picture showing magnetic field configuration where tearing instability can occur. In the region of current sheet on the y-axis (dashed line), perturbations with finite wavelength in the y-direction grows with time, and magnetic islands are formed by magnetic reconnection

(a) **Small amplitude perturbations**
Small amplitude perturbations of velocity and magnetic field superposed on the above unperturbed state are denoted by

$$\boldsymbol{v}_1 = (v_x, v_y, 0) \quad \text{and} \quad \boldsymbol{B}_1 = (B_{1x}, B_{1y}, 0). \tag{17.42}$$

Then, x- and y-components of equation of motion for perturbations are

$$\rho_0 \frac{\partial v_x}{\partial t} = -\frac{\partial}{\partial x}\left(p_1 + \frac{1}{4\pi} B_0 B_{1y}\right) + \frac{1}{4\pi} B_0 \frac{\partial B_{1x}}{\partial y}, \tag{17.43}$$

$$\rho_0 \frac{\partial v_y}{\partial t} = -\frac{\partial}{\partial y}\left(p_1 + \frac{1}{4\pi} B_0 B_{1y}\right) + \frac{1}{4\pi} B_0 \frac{\partial B_{1y}}{\partial y} + \frac{1}{4\pi} B_{1x} \frac{\partial B_0}{\partial x}. \tag{17.44}$$

Similarly, the induction equation gives

$$\frac{\partial B_{1x}}{\partial t} = B_0 \frac{\partial v_x}{\partial y} + \eta \nabla^2 B_{1x}, \tag{17.45}$$

$$\frac{\partial B_{1y}}{\partial t} = -\frac{\partial}{\partial x}(v_x B_0) + \eta \nabla^2 B_{1y}, \tag{17.46}$$

where the magnetic diffusivity η is taken to be constant.

To eliminate the pressure terms from Eqs. (17.43) and (17.44), the y-derivative of the former equation and the x-derivative of the latter equation are taken. Their difference then gives

$$\rho_0 \frac{\partial}{\partial t}\left(\frac{\partial v_x}{\partial y} - \frac{\partial v_y}{\partial x}\right) = \frac{1}{4\pi} B_0 \frac{\partial^2 B_{1x}}{\partial y^2} + \frac{1}{4\pi}\frac{\partial}{\partial x}\left(B_0 \frac{\partial B_{1x}}{\partial x} - B_{1x}\frac{d B_0}{dx}\right). \tag{17.47}$$

Here, the x-dependence of B_0 is specified by $f(x)$, introducing a reference value B_{00}, and $\boldsymbol{B}_1$ is also normalized by B_{00} as

$$B_0(x) = f(x)B_{00}, \quad \boldsymbol{B}_1(x, y; t) = \boldsymbol{b}(x, y; t)B_{00}. \tag{17.48}$$

Then, after using div $\boldsymbol{v} = 0$, we can reduce Eq. (17.47) to a relation between v_x and b_x. By considering perturbations proportional to

$$\exp(nt - iky), \tag{17.49}$$

we reduce the relation between v_x and b_x finally to

$$n\left(\frac{d^2}{dx^2} - k^2\right)v_x = -ikc_A^2 f\left(\frac{d^2}{dx^2} - k^2 - \frac{f''(x)}{f(x)}\right)b_x, \tag{17.50}$$

where n is the growth rate, k is the wavenumber in the y-direction, $c_A^2 = B_{00}^2/4\pi\rho_0$, and $f''(x) = d^2 f(x)/dx^2$.

Another relation between v_x and b_x is derived from Eq. (17.45), which is

$$nb_x - \eta\left(\frac{d^2}{dx^2} - k^2\right)b_x = -ikfv_x. \tag{17.51}$$

(b) **Growth rate**

The set of Eqs. (17.50) and (17.51) are the basic equations to be solved. In the region away from the reconnection region the magnetic diffusivity has no important effects and thus we look for solutions which tend to ideal MHD waves with relevant outer boundary conditions. In the reconnection region, on the other hand, the diffusion processes are important, and we look for solutions where diffusion processes have main effects. The solutions in the outer region and those in the reconnection region are matched in an intermediate region to obtain the solutions extending in the whole regions. This procedure is analogous to the boundary layer problem known in fluid dynamics.

After complicated analyses, Furth et al. (1963) derived the growth rate n of the tearing mode. The timescale of the growth, $\tau_T \sim 1/n$, is

$$\tau_T \sim \left(\frac{a^2}{\eta}\right)^{3/5}\left(\frac{a}{c_A}\right)^{2/5}, \tag{17.52}$$

for large $\mathcal{R}_m$, where a is a measure of the thickness of the current layer and $\mathcal{R}_m$ is the magnetic Reynolds number defined by $\mathcal{R}_m = c_A a/\eta$. By using $\mathcal{R}_m$ and τ_A defined by $\tau_A = a/c_A$, Eq. (17.52) is written as

$$\tau_T \sim \tau_A \mathcal{R}_m^{3/5}. \tag{17.53}$$

It will be instructive to compare this with τ_{SP} given by Eq. (17.39).

Linear stability of the tearing mode has been studied extensively (e.g., Steinolfson and Hoven 1984; Horton and Tajima 1988). For long wavelength perturbations, Steinolfson and Hoven (1984) (see also Horton and Tajima 1988) showed that

$$\tau_T \sim \tau_A \alpha^{-2/3} \mathcal{R}_m^{1/3} \tag{17.54}$$

$$\alpha_{max} \sim \mathcal{R}_m^{-1/4}, \tag{17.55}$$

where $\alpha = ka$. For most unstable wavelength we have

$$\tau_T \sim \tau_A \mathcal{R}_m^{1/2}. \tag{17.56}$$

The timescale τ_T has been estimated to be

$$\tau_T \sim 10\,\mathrm{s} \tag{17.57}$$

for relevant parameter values for solar flares (Tajima and Shibata 1997).

References

Balbus, S.A., Terquem, C.: Astrophys. J. **552**, 235 (2001)
Furth, H.P., Killeen, J., Rosenbluth, M.N.: Phys. Fluids **6**, 459 (1963)
Horton, W., Tajima, T.: J. Geophys. Res. **93**, 2741 (1988)
Mestel, L., Spitzer, L. Jr.: Mon. Not. R. Astron. Soc. **116**, 503 (1956)
Parker, E.N.: J. Geophys. Res. **62**, 509 (1957)
Petschek, H.E.: In: W.N. Hess (ed.) Physics of Solar Flares, p. 425. NASA, SP-50 (1964)
Steinolfson, R.S., Hoven, G.: Phys. Fluids **27**, 1207 (1984)
Sweet, P.A.: In: Lehnert, B. (ed.) Electromagnetic Phenomena in Cosmic Physics. IAU Symp. No. 6, p. 123. Cambridge University Press, Cambridge (1958)
Tajima, T., Shibata, K.: Plasma Astrophysics. Addison-Wesley, Massachusetts (1997)
Zweibel, E. G., Yamada, M.: Ann. Rev. A&A. **47**, 291 (2009)

Chapter 18
Relativistic Magnetohydrodynamics

As the final chapter of Part II, magnetohydrodynamical (MHD) equations are generalized to tensor forms so that they can be applied to general relativistic systems. After deriving basic general relativistic MHD equations (GRMHD equations), we rewrite these equations in forms which are analogous to forms of nonrelativistic MHD equations, so that they can be easily compared with nonrelativistic ones. The readers are supposed to have fundamental knowledge of tensor analyses.

18.1 Relativistic MHD Equations

The Maxwell equations are expressed in tensor forms. Energy-momentum tensor for magnetic fields is also derived.

18.1.1 Four-Dimensional Expression for Maxwell Equations

In this textbook we are interested in macroscopic fluid motions. Hence, Maxwell equations are written in the forms:

$$\mathrm{curl}\boldsymbol{E} = -\frac{1}{c}\frac{\partial \boldsymbol{B}}{\partial t}, \quad \mathrm{div}\boldsymbol{B} = 0, \tag{18.1}$$

and

$$\mathrm{curl}\boldsymbol{B} = \frac{1}{c}\frac{\partial \boldsymbol{E}}{\partial t} + \frac{4\pi}{c}\boldsymbol{j}, \quad \mathrm{div}\,\boldsymbol{E} = 4\pi\rho_e \tag{18.2}$$

by using $\boldsymbol{B}$ and $\boldsymbol{E}$ (not $\boldsymbol{H}$ and $\boldsymbol{D}$). Here, ρ_e is charge density.

© Springer Nature Singapore Pte Ltd. 2020, corrected publication 2023
S. Kato, J. Fukue, *Fundamentals of Astrophysical Fluid Dynamics*,
Astronomy and Astrophysics Library,
https://doi.org/10.1007/978-981-15-4174-2_18

By introducing a vector potential $\boldsymbol{A}$ defined by $\boldsymbol{B} \equiv \operatorname{curl} \boldsymbol{A}$, we rewrite the set of Eqs. (18.1) in the forms:

$$\boldsymbol{E} = -\frac{1}{c}\frac{\partial \boldsymbol{A}}{\partial t}, \quad \boldsymbol{B} = \mathrm{curlA}, \tag{18.3}$$

where the scalar potential in the expression for $\boldsymbol{E}$ has been taken to be zero. Our purpose here is to express Eqs. (18.2) and (18.3) in four-dimensional forms.[1]

(1) **The pair of Eqs. (18.3)**
Let us first consider flat Cartesian space-time, and introduce the *electromagnetic field tensor* $F_{\mu\nu}$ defined by

$$F_{\mu\nu} = \frac{\partial A_\nu}{\partial x^\mu} - \frac{\partial A_\mu}{\partial x^\nu}. \tag{18.4}$$

The matrix forms of $F_{\mu\nu}$ and $F^{\mu\nu}$ are

$$F_{\mu\nu} = \begin{pmatrix} 0 & -E_1 & -E_2 & -E_3 \\ E_1 & 0 & B_3 & -B_2 \\ E_2 & -B_3 & 0 & B_1 \\ E_3 & B_2 & -B_1 & 0 \end{pmatrix},$$

and

$$F^{\mu\nu} = \begin{pmatrix} 0 & E_1 & E_2 & E_3 \\ -E_1 & 0 & B_3 & -B_2 \\ -E_2 & -B_3 & 0 & B_1 \\ -E_3 & B_2 & -B_1 & 0 \end{pmatrix},$$

where the index μ (= 0, 1, 2, 3) labels the rows, and the index ν the columns.
By use of the electromagnetic field tensor defined above, we can summarize the set of Eqs. (18.3) in the form:

$$\frac{\partial F_{\mu\nu}}{\partial x^\sigma} + \frac{\partial F_{\nu\sigma}}{\partial x^\mu} + \frac{\partial F_{\sigma\mu}}{\partial x^\nu} = 0. \tag{18.5}$$

The expression on the left-hand side of Eq. (18.5) is a tensor of rank three, which is antisymmetric in all three indices. The only components which are

[1] In this chapter the same symbols as in Chap. 9 are used. That is, three-dimensional tensor indices are denoted by Latin letters $i, j, k, \ldots$ while four-dimensional tensor indices are denoted by Greek letters $\alpha, \beta, \gamma, \ldots$ and take on the values 0, 1, 2, 3. The summation abbreviation is used when the same letter appears in a term twice. The metric tensor is given by $ds^2 = g_{\mu\nu}dx^\mu dx^\nu$ with $g_{00} > 0$.

not identically zero are those with $\mu \neq \nu \neq \sigma$. Thus there are altogether four different equations, which can be easily shown to be three components of the first equation of Eq. (18.3) and div $\boldsymbol{B} = 0$.

We can construct the four-vector which is dual to antisymmetric four-tensor of rank three, i.e., Eq. (18.5), by multiplying the tensor by the antisymmetric unit (pseudo) tensor of rank four, $e^{\mu\nu\alpha\beta}$, and contracting on three pairs of indices. Then, Eq. (18.5) can be written in the form

$$e^{\mu\nu\alpha\beta}\frac{\partial F_{\alpha\beta}}{\partial x^{\nu}} = 0, \tag{18.6}$$

where $e^{\mu\nu\alpha\beta}$ is defined so that it changes sign under any transposition of a pair of indices. The value of $e^{0123} = 1$ and an even permutation of indices give 1, while an odd permutation leads to -1.[2] It is noticed that Eq. (18.6) shows explicitly that there are only four independent equations.

Equations (18.5) and (18.6) can be easily generalized in an arbitrary four-dimensional curvilinear systems of coordinate, i.e., in systems of general relativity. To extend to curvilinear systems, derivatives of vectors and tensors are changed to covariant derivatives, and we have

$$F_{\mu\nu;\sigma} + F_{\nu\sigma;\mu} + F_{\sigma\mu;\nu} = 0. \tag{18.7}$$

It is noted that Eq. (18.7) is the same as Eq. (18.5), since $\Gamma^{\mu}_{\nu\sigma} = \Gamma^{\mu}_{\sigma\nu}$.

Equation (18.6) can be generalized to any curvilinear coordinate systems by use of the dual pseudotensor, $^{*}F^{\mu\nu}$, defined by

$$^{*}F^{\mu\nu} = \frac{1}{2}E^{\mu\nu\alpha\beta}F_{\alpha\beta} \tag{18.8}$$

as

$$^{*}F^{\mu\nu}{}_{;\nu} = 0, \tag{18.9}$$

where $E^{\mu\nu\alpha\beta}$ is the antisymmetric unit tensor of rank four (the Levi-Civita tensor) defined by

$$E^{\mu\nu\alpha\beta} = (-g)^{-1/2}e^{\mu\nu\alpha\beta}. \tag{18.10}$$

It is noted that $E_{\mu\nu\alpha\beta;\xi} = \sqrt{-g}\, e_{\mu\nu\alpha\beta,\xi} = 0$.[3,4]

---

[2] It is noticed that $e_{0123} = -1$.

[3] It is further noticed that $E_{\mu\nu\alpha\beta} = \sqrt{-g}\, e_{\mu\nu\alpha\beta}$ and $e_{\mu\nu\alpha\beta} = -e^{\mu\nu\alpha\beta}$.

[4] Here, some important identities for contractions of the Levi-Civita tensor are summarized, since they will be used in subsequent sections.

(2) **The pair of Eqs. (18.2)**
Equation (18.2) is written in a tensor form. In Cartesian coordinates it is written as

$$\frac{\partial F^{\mu\nu}}{\partial x^{\nu}} = -\frac{4\pi}{c} j^{\mu}, \tag{18.15}$$

where j^{μ} is four current vector:

$$j^{\mu} = (c\rho_{\mathrm{e}}, \boldsymbol{j}). \tag{18.16}$$

For $\mu = 1, 2,$ and 3 this equation gives the three components of the first equation of Eqs. (18.2), and for $\mu = 0$ we have div $\boldsymbol{E} = 4\pi\rho_e$ by using four current vector:

This expression (18.15) is generalized to curvilinear coordinate system as

$$F^{\mu\nu}_{\ \ ;\nu} = -\frac{4\pi}{c} j^{\mu}. \tag{18.17}$$

The set of Eqs. (18.9) and (18.17) are the set of Maxwell equations.

(3) **Ohm's law**
In addition to the above Maxwell equations, we need to use Ohm's law

$$\boldsymbol{j} = \sigma\left(\boldsymbol{E} + \frac{\boldsymbol{u}}{c} \times \boldsymbol{B}\right), \tag{18.18}$$

to connect electric current and electromagnetic fields. Our main interest here is on the case where the MHD approximations are applicable. In this case the electric current is assumed to vanish in the fluid rest frame as a result of the high electric conductivity, i.e.,

$$\boldsymbol{E} + \frac{\boldsymbol{u}}{c} \times \boldsymbol{B} = 0. \tag{18.19}$$

$$E^{\mu\nu\alpha\beta} E_{\mu\xi\sigma\rho} = -\delta^{\nu\alpha\beta}_{\xi\sigma\rho}, \tag{18.11}$$

$$E^{\mu\nu\alpha\beta} E_{\mu\nu\sigma\rho} = -2\delta^{\alpha\beta}_{\sigma\rho}, \tag{18.12}$$

where the δ-symbols are given by

$$\delta^{\mu\nu}_{\sigma\rho} = \delta^{\mu}_{\sigma}\delta^{\nu}_{\rho} - \delta^{\mu}_{\rho}\delta^{\nu}_{\sigma}, \tag{18.13}$$

$$\delta^{\mu\nu\xi}_{\alpha\beta\gamma} = \delta^{\mu}_{\alpha}\delta^{\nu}_{\beta}\delta^{\xi}_{\gamma} - \delta^{\mu}_{\beta}\delta^{\nu}_{\alpha}\delta^{\xi}_{\gamma} + \delta^{\mu}_{\beta}\delta^{\nu}_{\gamma}\delta^{\xi}_{\alpha} - \delta^{\mu}_{\gamma}\delta^{\nu}_{\beta}\delta^{\xi}_{\alpha} + \delta^{\mu}_{\gamma}\delta^{\nu}_{\alpha}\delta^{\xi}_{\beta} - \delta^{\mu}_{\alpha}\delta^{\nu}_{\gamma}\delta^{\xi}_{\beta}. \tag{18.14}$$

It is noted that $E^{\mu\nu\alpha\beta} = (-g)^{-1/2}e^{\mu\nu\alpha\beta}$ and $E_{\mu\xi\sigma\rho} = (-g)^{1/2}e_{\mu\xi\sigma\rho}$, and thus the terms of $(-g)$ disappear in calculations of Eqs. (18.11) and (18.12).

This condition is written as

$$u_\mu F^{\mu\nu} = 0. \tag{18.20}$$

In the case of flat space, $\nu = 1, 2, 3$ components of this equation give Eq. (18.19). The $\nu = 0$ component is trivial when Eq. (18.19) is satisfied.

18.1.2 Energy-Momentum Tensor for Magnetic Field

Next, the energy-momentum tensor for magnetic fields is considered. Since a derivation of the energy-momentum tensor by Landau and Lifshitz (1973) is instructive, we shall present here the outline of the derivation in the case of Galilean coordinates. We consider any system whose action integral has the form

$$S = \int \Lambda\left(q, \frac{\partial q}{\partial x^\mu}\right) dV dt, \tag{18.21}$$

where Λ is some function of the quantities q, describing the state of the system, and of their first derivatives with respect to coordinates and time (in the case of the electromagnetic field, the components of the four-potential, A_ν, correspond to the quantities q). For brevity we write here only one of the q's. The integration in Eq. (18.21) is performed over four volume.

As is known in mechanics, "equation of motion" is obtained in accordance with the principle of least action by varying S. That is, we consider arbitrary variation of q, say δq, satisfying boundary condition $\delta q = 0$ on the boundary of the system. Then, after performing integration by part with respect to x^μ so that such quantities as $\partial \delta q / \partial x^\mu$ are replaced to terms with no derivative, we have

$$\begin{aligned} \delta S &= \int \left(\frac{\partial \Lambda}{\partial q} \delta q + \frac{\partial \Lambda}{\partial q_{,\mu}} \delta q_{,\mu} \right) dV dt \\ &= \int \left(\frac{\partial \Lambda}{\partial q} - \frac{\partial}{\partial x^\mu} \frac{\partial \Lambda}{\partial q_{,\mu}} \right) \delta q dV dt, \end{aligned} \tag{18.22}$$

where $q_{,\mu}$ represents $\partial q / \partial x_\mu$. The extremal of $\delta S = 0$ is realized for

$$\frac{\partial \Lambda}{\partial q} - \frac{\partial}{\partial x^\mu} \frac{\partial \Lambda}{\partial q_{,\mu}} = 0, \tag{18.23}$$

which is an equation called *Euler equation* in the field of calculus of variations.

This Euler equation is rearranged in the form

$$\frac{\partial M_\mu^\nu}{\partial x^\nu} = 0, \tag{18.24}$$

with M_μ^ν defined by

$$M_\mu^\nu = q_{,\mu}\frac{\partial \Lambda}{\partial q_{,\nu}} - \delta_\mu^\nu \Lambda. \tag{18.25}$$

Equation (18.25) can be obtained because

$$\begin{aligned}\frac{\partial}{\partial x^\nu}\left(q_{,\mu}\frac{\partial \Lambda}{\partial q_{,\nu}}\right) &= \frac{\partial q_{,\mu}}{\partial x^\nu}\frac{\partial \Lambda}{\partial q_{,\nu}} + q_{,\mu}\frac{\partial}{\partial x^\nu}\left(\frac{\partial \Lambda}{\partial q_{,\nu}}\right) = \frac{\partial q_{,\mu}}{\partial x^\nu}\frac{\partial \Lambda}{\partial q_{,\nu}} + q_{,\mu}\frac{\partial \Lambda}{\partial q} \\ &= \frac{\partial q_{,\nu}}{\partial x^\mu}\frac{\partial \Lambda}{\partial q_{,\nu}} + q_{,\mu}\frac{\partial \Lambda}{\partial q} \\ &= \frac{\partial \Lambda}{\partial x^\mu}.\end{aligned} \tag{18.26}$$

In deriving the second equality in Eq. (18.26), Eq. (18.23) has been adopted, and $q_{,\mu\nu} = q_{,\nu\mu}$ has been used in deriving the third equality.

The next issue is what Λ is. After some arguments we find that Λ can be taken as (see Landau and Lifshitz 1973)

$$\Lambda = -\frac{1}{16\pi}F_{\mu\nu}F^{\nu\mu}. \tag{18.27}$$

Then, substituting equation (18.27) into Eq. (18.25), after some manipulations, we have finally the following expression for the energy-momentum tensor $M^{\mu\nu}$ of the electromagnetic field (see, e.g., Landau and Lifshitz 1973)

$$M^{\mu\nu} = \frac{1}{4\pi}\left(- F^{\mu\sigma}F_\sigma^{\ \nu} + \frac{1}{4}g^{\mu\nu}F_{\sigma\rho}F^{\sigma\rho}\right). \tag{18.28}$$

This expression is valid even in curvilinear coordinate systems. It is noted that $M^{\mu\nu}$ is a symmetric tensor, since $F^{\mu\sigma}F^{\nu}_{\ \sigma} = F^{\mu\sigma}F^{\nu\rho}g_{\rho\sigma} = F^\mu_{\ \rho}F^{\nu\rho}$.

The energy-momentum tensor for fluid, $T^{\mu\nu}$, is given by (see Chap. 9)

$$T^{\mu\nu} = (\varepsilon + p)u^\mu u^\nu - pg^{\mu\nu}. \tag{18.29}$$

The energy and momentum conservation of the whole system is thus expressed as

$$(T^{\mu\nu} + M^{\mu\nu})_{;\nu} = 0. \tag{18.30}$$

18.2 GRMHD Equations with Close Appearance to Newtonian MHD Equations

In Sect. 18.1, we have presented general relativistic MHD (GRMHD) equations. In studies of active astrophysical objects such as galactic nuclei or compact stellar objects, numerical simulations of GRMHD are required, since phenomena in such objects are highly nonlinear and complicated. Various codes of numerical calculations of GRMHD have been developed, for example, by Koide and his collaborators (Koide et al. 1999, and subsequent papers). In numerical simulations, however, we are accustomed with hydrodynamical equations and thus a set of equations analogous to Newtonian MHD equations will be easy to handle. Considering this, Gammie et al. (2003) and De Villiers and Hawley (2003) have formulated GRMHD equations in forms analogous to MHD equations. We present here the basic parts of their formulations.

The above authors use the metric with signature $(-+++)$. In this textbook, however, we have used the metric with signature $(+---)$. Hence, in order to keep consistency we use here the metric with $(+---)$ in this subsection. Some expressions presented below will be thus different from those in the above authors.

Since GRMHD phenomena seem to be studied usually by using GRMHD equations of the metric with $(-+++)$, the GRMHD equations derived in this subsection are converted to those expressed in terms of the metric with $(-+++)$ in the next subsection (Sect. 18.3) for convenience.

(1) **Particle number conservation**
Let us introduce particle flux four-vector nu^μ, where n is the baryon number density. The conservation of baryon number is

$$\left(nu^\mu\right)_{;\mu} = 0. \tag{18.31}$$

This is written as

$$\frac{\partial}{\partial t}\left(\sqrt{-g}nu^t\right) + \frac{\partial}{\partial x^i}\left(\sqrt{-g}nu^i\right) = 0, \tag{18.32}$$

where g is the determinant of the metric tensor and $-g > 0$.

(2) **Energy-momentum tensor**
By using ${}^*F^{\mu\nu}$ defined by Eq. (18.8), we define the magnetic field b^μ in the rest frame of the fluid by

$$b^\mu = -{}^*F^{\mu\nu}u_\nu = -\frac{1}{2}E^{\mu\nu\alpha\beta}F_{\alpha\beta}u_\nu. \tag{18.33}$$

It is noted that there is the orthogonality:

$$b^\mu u_\mu = 0, \tag{18.34}$$

which results directly from Eq. (18.33). By using the definition of b^μ given above and the condition of infinite conductivity, $u_\mu F^{\mu\nu} = 0$, we obtain[5] (see also the appendix of De Villiers and Hawley (2003)

$$F_{\mu\nu} = E_{\alpha\beta\mu\nu} b^\alpha u^\beta. \tag{18.36}$$

The use of b^μ defined above leads the electromagnetic portion of the energy-momentum tensor, $M^{\mu\nu}$, given by Eq. (18.28), to the following form[6] (see also the appendix of De Villiers and Hawley (2003):

$$M^{\mu\nu} = \frac{1}{4\pi}\left[\left(\frac{1}{2}g^{\mu\nu}b_\sigma b^\sigma - b^\mu b^\nu\right) - b^\sigma b_\sigma u^\mu u^\nu\right]. \tag{18.38}$$

[5]Following De Villiers and Hawley (2003), we show Eq. (18.36) by working backward from the result, using identities given by Eqs. (18.11) and (18.14), infinite conductivity (18.20), and antisymmetry of $F_{\mu\nu}$. That is, expanding the right-hand side of Eq. (18.36), we have

$$\begin{aligned}
E_{\alpha\beta\mu\nu} b^\alpha u^\beta &= -\frac{1}{2} E_{\alpha\beta\mu\nu} E^{\alpha\sigma\rho\xi} u_\sigma u^\beta F_{\rho\xi} \\
&= \frac{1}{2}\delta^{\sigma\rho\xi}_{\beta\mu\nu} u_\sigma u^\beta F_{\rho\xi} \\
&= \frac{1}{2}\left(\delta^\sigma_\beta \delta^\rho_\mu \delta^\xi_\nu - \delta^\sigma_\mu \delta^\rho_\beta \delta^\xi_\nu + \delta^\sigma_\mu \delta^\rho_\nu \delta^\xi_\beta - \delta^\sigma_\nu \delta^\rho_\mu \delta^\xi_\beta + \delta^\sigma_\nu \delta^\rho_\beta \delta^\xi_\mu - \delta^\sigma_\beta \delta^\rho_\nu \delta^\xi_\mu\right) u_\sigma u^\beta F_{\rho\xi} \\
&= \frac{1}{2}(u_\beta u^\beta F_{\mu\nu} - u_\mu u^\rho F_{\rho\nu} + u_\mu u^\xi F_{\nu\xi} - u_\nu u^\xi F_{\mu\xi} + u_\nu u^\rho F_{\rho\mu} - u_\beta u^\beta F_{\nu\mu}) \\
&= u_\beta u^\beta F_{\mu\nu} = F_{\mu\nu}.
\end{aligned} \tag{18.35}$$

[6]To prove Eq. (18.38), extend the energy-momentum tensor (18.28) with use of Eq. (18.36), formulae (18.11)–(18.14), and Eq. (18.34):

$$\begin{aligned}
M^{\mu\nu} &= -\frac{1}{4\pi}\left(g^{\mu\xi} F_{\xi\alpha} F^{\nu\alpha} - \frac{1}{4} F_{\alpha\beta} F^{\alpha\beta} g^{\mu\nu}\right) \\
&= -\frac{1}{4\pi}\left[g^{\mu\xi}\left(E_{\sigma\rho\xi\alpha} b^\sigma u^\rho\right)\left(E^{\kappa\lambda\nu\alpha} b_\kappa u_\lambda\right) - \frac{1}{4}\left(E_{\sigma\rho\alpha\beta} b^\sigma u^\rho\right)\left(E^{\kappa\lambda\alpha\beta} b_\kappa u_\lambda\right) g^{\mu\nu}\right] \\
&= \frac{1}{4\pi}\left(g^{\mu\xi}\delta^{\kappa\lambda\nu}_{\sigma\rho\xi} b^\sigma u^\rho b_\kappa u_\lambda - \frac{1}{2} g^{\mu\nu}\delta^{\kappa\lambda}_{\sigma\rho} b^\sigma u^\rho b_\kappa u_\lambda\right) \\
&= \frac{1}{4\pi}\left[\left(g^{\mu\nu} b_\kappa b^\kappa - b^\mu b^\nu\right)\left(u_\lambda u^\lambda\right) - b_\kappa b^\kappa u^\mu u^\nu - \frac{1}{2} g^{\mu\nu}\left(b^\kappa b_\kappa\right)\left(u^\lambda u_\lambda\right)\right] \\
&= \frac{1}{4\pi}\left[\left(\frac{1}{2} g^{\mu\nu} b_\alpha b^\alpha - b^\mu b^\nu\right)\left(u_\lambda u^\lambda\right) - b^\kappa b_\kappa u^\mu u^\nu\right] \\
&= \frac{1}{4\pi}\left[\left(\frac{1}{2} g^{\mu\nu} b_\alpha b^\alpha - b^\nu b^\mu\right) - b^\alpha b_\alpha u^\mu u^\nu\right].
\end{aligned} \tag{18.37}$$

It will be instructive to rearrange this energy-momentum tensor in the form

$$M_{\mu\nu} = \frac{1}{4\pi}\left(- b^2 u_\mu u_\nu - b_\mu b_\nu + \frac{b^2}{2} g_{\mu\nu} \right), \tag{18.39}$$

where b^2 has been defined by

$$b^2 = b^\mu b_\mu = g_{\mu\sigma} b^\mu b^\sigma. \tag{18.40}$$

Expression (18.39) is analogous to $T_{\mu\nu}$ for fluids [see Eq. (18.29)]:

$$T_{\mu\nu} = (\varepsilon + p) u_\mu u_\nu - p g_{\mu\nu}. \tag{18.41}$$

The total of energy-momentum tensor, $T^{(\mathrm{tot})\mu}_{\ \ \nu}$, is the sum of M^μ_ν and T^μ_ν, i.e.,

$$T^{(\mathrm{tot})\mu}_{\ \ \nu} = T^\mu_\nu + M^\mu_\nu. \tag{18.42}$$

Its conservation, i.e., $T^{(\mathrm{tot})\nu}_{\ \ \mu;\nu} = 0$, can be written as

$$\frac{\partial}{\partial t}\left(\sqrt{-g}T^{(\mathrm{tot})t}_{\ \ \mu}\right) + \frac{\partial}{\partial x^i}\left(\sqrt{-g}(T^{(\mathrm{tot})\,i}_{\ \ \mu})\right) - \sqrt{-g}\Gamma^\beta_{\alpha\mu} T^{(\mathrm{tot})\alpha}_{\ \ \beta} = 0. \tag{18.43}$$

The $\mu = 1, 2, 3$ components of Eq. (18.43) lead to equation of motion, and the $\mu = 0$ component gives the energy conservation.

(3) **Induction equation**
The Maxwell equation (18.9) can be written under the use of infinite conductivity as[7]

$$\left(u^\mu b^\nu - b^\mu u^\nu\right)_{;\nu} = 0. \tag{18.45}$$

Using $\Gamma^\mu_{\nu\sigma} = \Gamma^\mu_{\sigma\nu}$, we can reduce Eq. (18.45) to

$$\frac{\partial}{\partial x^\nu}\left[\sqrt{-g}\left(u^\mu b^\nu - u^\nu b^\mu\right)\right] = 0. \tag{18.46}$$

[7]To prove Eq. (18.45) we use the definition of $^*F^{ik}$ and identities given by Eqs. (18.12) and (18.13) to lead to

$$\begin{aligned} {}^*F^{\mu\nu}_{\ \ ;\mu} &= \frac{1}{2}\left(E^{\mu\nu\alpha\beta}F_{\alpha\beta}\right)_{;\mu} = \frac{1}{2}\left(E^{\mu\nu\alpha\beta}E_{\rho\sigma\alpha\beta}b^\rho u^\sigma\right)_{;\mu} = -\left(\delta^{\mu\nu}_{\rho\sigma}b^\rho u^\sigma\right)_{;\mu} \\ &= \left(b^\nu u^\mu - b^\mu u^\nu\right)_{;\mu}. \end{aligned} \tag{18.44}$$

If the coordinate velocity V^μ defined by

$$V^\mu = \frac{u^\mu}{u^t} \tag{18.47}$$

is introduced, Eq. (18.46) is rewritten as

$$\frac{\partial}{\partial x^\nu}\left[\sqrt{-g}u^t\left(V^\mu b^\nu - V^\nu b^\mu\right)\right] = 0. \tag{18.48}$$

Now, we introduce $\mathcal{B}^i$ $(i = 1, 2, 3)$ defined by

$$\mathcal{B}^i \equiv u^t\left(b^i - V^i b^t\right). \tag{18.49}$$

Then, after some manipulations, we find that Eq. (18.48) can be expressed as

$$\frac{\partial}{\partial x^t}\left(\sqrt{-g}\mathcal{B}^i\right) - \frac{\partial}{\partial x^k}\left[\sqrt{-g}\left(V^i\mathcal{B}^k - V^k\mathcal{B}^i\right)\right] = 0 \quad (i,\ k = 1, 2, 3) \tag{18.50}$$

and

$$\frac{\partial}{\partial x^k}\left(\sqrt{-g}\mathcal{B}^k\right) = 0 \quad (k = 1, 2, 3). \tag{18.51}$$

Equations (18.50) and (18.51) correspond, respectively, to induction equation and no monopole of magnetic fields. This formulation is due to Evans and Hawley (1988). It is noted that Eqs. (18.50) and (18.51) have a Cartesian appearance even though they are valid for arbitrary curvilinear coordinates. The metric describing coordinates does not directly appear, although it is hidden in definition (18.49) of $\mathcal{B}^i$.

18.3 GRMHD Equations in the Metric with Signature $(-+++)$

In order to avoid confusion, the metric tensor in the metric with signature $(-+++)$ is denoted as $\tilde{g}_{\mu\nu}$ by attaching tilde. Furthermore, all quantities expressed in this metric are represented by attaching tilde, say $\tilde{u}_\mu$. Then, the GRMHD equations derived in Sect. 18.2 can be transformed in the GRMHD equations in the $(-+++)$ signature by the following rules. That is, $g_{\mu\nu}$ is changed to $-\tilde{g}_{\mu\nu}$. In addition, all dynamical quantities with superscript in Sect. 18.2 are converted to quantities with tilde with the same sign, but those with subscript is converted to those with tilde with the opposite sign. For example, u^μ is converted to $\tilde{u}^\mu$, but u_μ is converted to $-\tilde{u}_\mu$.

The quantity $b^2 \equiv b^\mu b_\mu = g_{\mu\nu} b^\mu b^\nu$ in Sect. 18.2 is converted to $-\tilde{g}_{\mu\nu}\tilde{b}^\mu\tilde{b}^\nu = -\tilde{b}_\mu\tilde{b}^\mu \equiv -\tilde{b}^2$.

(1) **Particle number conservation**
The continuity is the same as that in Sect. 18.2:

$$\frac{\partial}{\partial t}\left(\sqrt{-g}n\tilde{u}^t\right) + \frac{\partial}{\partial x^i}\left(\sqrt{-g}n\tilde{u}^i\right) = 0. \qquad (18.52)$$

(2) **Energy-momentum tensor**
The energy-momentum tensor of magnetic fields $\tilde{M}_{\mu\nu}$ is written as

$$\tilde{M}_{\mu\nu} = \frac{1}{4\pi}\left(\tilde{b}^2\tilde{u}_\mu\tilde{u}_\nu - \tilde{b}_\mu\tilde{b}_\nu + \frac{1}{2}\tilde{b}^2\tilde{g}_{\mu\nu}\right). \qquad (18.53)$$

For comparison the energy-momentum tensor of fluid, $\tilde{T}_{\mu\nu}$, is presented here:

$$\tilde{T}_{\mu\nu} = (\varepsilon + p)\tilde{u}_\mu\tilde{u}_\nu + p\tilde{g}_{\mu\nu}. \qquad (18.54)$$

It is noticed that the above two expressions for $\tilde{M}_{\mu\nu}$ and $\tilde{T}_{\mu\nu}$ are analogous. The energy-momentum tensor of the total system (fluid and magnetic fields) is the sum of the above two quantities, i.e., $\tilde{T}^{\rm tot}_{\mu\nu} = \tilde{T}_{\mu\nu} + \tilde{M}_{\mu\nu}$. The conservations of momentum and energy are expressed as

$$\frac{\partial}{\partial t}\left(\sqrt{-g}\tilde{T}^{({\rm tot})t}{}_{\mu}\right) + \frac{\partial}{\partial x^i}\left(\sqrt{-g}\left(\tilde{T}^{({\rm tot})\,i}{}_{\mu}\right)\right) - \sqrt{-g}\tilde{\Gamma}^\beta_{\alpha\mu}\tilde{T}^{({\rm tot})\alpha}{}_{\beta} = 0. \qquad (18.55)$$

(3) **Induction equation and div $\boldsymbol{B} = 0$**
Expressions for these equations are the same as those in Sect. 18.2. That is, we have

$$\frac{\partial}{\partial x^t}\left(\sqrt{-g}\tilde{\mathcal{B}}^i\right) - \frac{\partial}{\partial x^k}\left[\sqrt{-g}\left(\tilde{V}^i\tilde{\mathcal{B}}^k - \tilde{V}^k\tilde{\mathcal{B}}^i\right)\right] = 0 \quad (i,\ k = 1, 2, 3) \qquad (18.56)$$

and

$$\frac{\partial}{\partial x^k}\left(\sqrt{-g}\tilde{\mathcal{B}}^k\right) = 0 \quad (k = 1, 2, 3). \qquad (18.57)$$

(4) **Ideal MHD condition**
Finally, the ideal MHD equation is mentioned. Corresponding to Eq. (18.20), we have

$$\tilde{u}_\mu\tilde{F}^{\mu\nu} = 0. \qquad (18.58)$$

References

De Villiers, J.-P., Hawley, J.: Astrophys. J. **589**, 458 (2003)
Evans, C.R., Hawley, J.H.: Astrophys. J. **332**, 659 (1988)
Gammie, C., McKinney, J.C., Tóth, G.: Astrophys. J. **589**, 447 (2003)
Koide, S., Shibata, K., Kudoh, T.: Astrophys. J.: **522**, 727 (1999)
Landau, L. D., Lifshitz, E.M.: The Classical Theory of Fields, 4th edn. Elsevier, Oxford (1973)

Part III
Astrophysical Radiation Hydrodynamics

Radiation plays important roles in various astrophysical objects, which include first stars and present massive stars, radiative-driven or dust-driven stellar winds, nova outbursts, X-ray bursters, supernovae and hypernovae, accretion disks around compact objects, protoplanetary disks, interstellar matter, starbursts and active galaxies, astrophysical jets, gamma-ray bursts, and very early universe. In addition, radiation is quite essential in a sense that the radiative transfer determines the observational properties of gaseous astronomical objects, e.g., stellar atmosphere, interstellar matter, black-hole environments, and exoplanets.

In this Part III, we describe the fundamentals for astrophysical radiation hydrodynamics. We first summarize the basic concepts in radiation hydrodynamics. Then, we derive the basic equations for radiation hydrodynamics, apply them to radiation hydrodynamical flow, and consider waves and instabilities in radiative fluids. Finally, we derive the basic equations for relativistic radiation hydrodynamics, and briefly discuss several applications.

Chapter 19
Basic Concepts of Radiative Fluids

Similar to gaseous materials, which consist of numberless particles, radiation also consists of numberless photons. Hence, in the viewpoint of statistical dynamics, radiation is similar to fluid in some sense. Indeed, the fundamental equation and its moment equations are parallel for radiation and matter, except that photons are massless particles. One of the essential differences between radiation and matter is the speed of "particles" and, therefore, the interaction distance is remarkably different. Gaseous particles can travel at sub-light speed, and in usual gaseous objects, the gas pressure is isotropic and local thermodynamical equilibrium is quickly established, and the fluid approximation holds. On the other hand, radiation in the free space travels at the speed of light, and, therefore, can interact with gaseous particles at great distance. As a result, we sometimes consider an anisotropic radiation field and non-local interaction. Another essential difference is that the fluid particles are fermions, but photons are bosons; photons are produced and destroyed by gaseous matter, and, therefore, the numbers of photons change, while gaseous particles do not. As a result, we must consider emission, absorption, and scattering of radiation by gaseous matter in the radiation hydrodynamical situations. In this chapter, we first summarize the basic concepts in radiation hydrodynamics; mean free path and optical depth, photon creations (emission) and destructions (absorption) and scattering. We further review the concepts of the local thermodynamical equilibrium (LTE) and blackbody radiation. We then show the radiation force and the Eddington luminosity in relating to the momentum exchange between radiation and matter. We also briefly describe radiative viscosity and heat conduction. Finally, we introduce the adiabatic exponents of radiative fluids.

© Springer Nature Singapore Pte Ltd. 2020, corrected publication 2023

S. Kato, J. Fukue, *Fundamentals of Astrophysical Fluid Dynamics*,
Astronomy and Astrophysics Library,
https://doi.org/10.1007/978-981-15-4174-2_19

19.1 Mean Free Path and Optical Depth

We first describe the mean free path of photons and optical depth of the medium. Photons propagating through a gaseous medium would eventually collide with particles of the medium. The *mean free path of photons* is defined as the mean length during which photons collide with a gaseous particle.

Let us consider a light-ray propagating in a uniform medium, where the particle number density is n and its cross section σ (Fig. 19.1). If the condition $n\sigma\ell = 1$ is satisfied, where ℓ is the light path length, then the number of particles in the cylinder with volume $\sigma\ell$ becomes unity. This is also the condition that one photon collides with one particle, and the distance ℓ at this condition is just the mean free path of photons:

$$\ell = \frac{1}{\sigma n}. \tag{19.1}$$

Using the mass density ρ $(= nm)$, where m is the particle mass, the mean free path is also written as

$$\ell = \frac{1}{\sigma(\rho/m)} = \frac{1}{\kappa\rho}. \tag{19.2}$$

Here, κ defined by $\kappa \equiv \sigma/m$ is the cross section per unit mass and called *opacity* of the medium. The unit of opacity is [cm^2 g^{-1}], and it generally depends on frequency of radiation.

For example, inside the solar interior, the mean mass density is about $\rho \sim 1.4\,\mathrm{g\,cm^{-3}}$, and the opacity is $\kappa \sim 1\,\mathrm{cm^2\,g^{-1}}$. Hence, the mean free path of photons in the solar interior is roughly $\ell \sim 0.5\,\mathrm{cm}$.

When a photon collides with one particle on an average after propagating the mean free path, the real distance in space depends on the mass density or opacity.

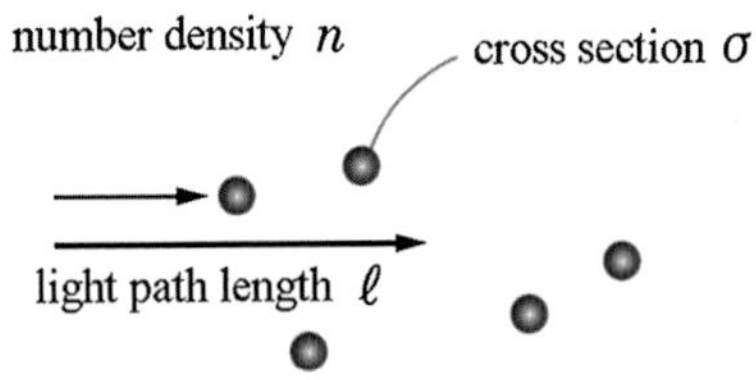

Fig. 19.1 Collisional cross section, light path length, and mean free path of photons

For photons, therefore, the dimensionless distance $\sigma n\ell = \kappa\rho\ell = 1$ (or generally $\sigma ns = \kappa\rho s$, s being the real distance) is physically important. We thus define the *optical depth*/optical thickness τ_ν so that

$$\tau_\nu(s) \equiv \int_{s_0}^{s} \kappa_\nu \rho(s')ds', \tag{19.3}$$

where s is the coordinate along the light path, and the frequency dependence of opacity is explicitly expressed.

When the optical depth is greater than unity ($\tau_\nu > 1$), the medium is called *optically thick* or *opaque*, while it is called *optically thin* or *translucent* when the optical depth is less than unity ($\tau_\nu < 1$).[1]

19.2 Emission, Absorption, and Scattering

A light-ray propagating through the gaseous medium suffers from various interaction processes and changes its intensity (Fig. 19.2). The precise definition of the specific intensity (bundle of light-ray) and the propagation equation will be introduced in the next section and Chap. 20. Before it, we roughly show these interaction processes.

The interaction between matter and radiation is mainly two; the momentum exchange due to radiative force, and energy one via radiative heating and cooling. From the viewpoint of microscopic physical processes, these interactions are mainly caused by three-type processes. *Emission*, which includes free-free, bound-free, bound-bound transitions, and non-thermal processes like synchrotron radiation, causes radiative cooling of photon gas, but does not exchange momentum, since emission usually takes place isotropically. *Absorption* is roughly the inverse process of emission and causes radiative heating of photon gas. In contrast to emission, however, it exchanges momentum, since the distribution of incident photons is not isotropic in general. *Scattering*, which includes electron (Thomson) one, causes momentum exchange, but the elastic electron scattering does not exchange energy.

19.2.1 Emission

Emission of photons (photon creation) from the gaseous medium is caused by atomic bound-bound transition, free-bound one (radiative recombination), free-free

[1] The word *transparent* is not adequate for an optically thin case.

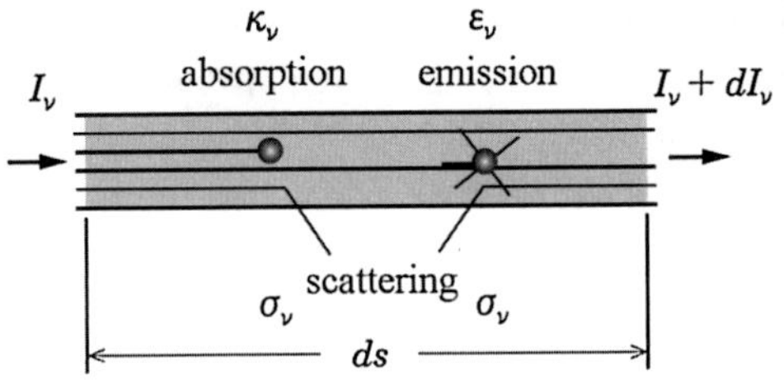

Fig. 19.2 Light-ray traveling through the gaseous medium under emission, absorption, and scattering processes

one (bremsstrahlung), synchrotron radiation (magnetic bremsstrahlung) under the existence of magnetic field, electron–positron pair annihilations in the very high energy regime, molecular emission in the low energy one, metal bound-bound (band) emission, dust thermal radiation, and others.

The *emission coefficient* or *emissivity* η_ν [erg cm^{-3} s^{-1} Hz^{-1}sr^{-1}] is defined by the radiation energy emitted per unit volume, unit time, unit frequency, and unit solid angle. The *mass emission coefficient* or *mass emissivity* ϵ_ν [erg g^{-1} s^{-1} Hz^{-1}sr^{-1}] per unit mass is also used ($\eta_\nu = \rho\epsilon_\nu$). Similarly, the mass emissivity in all direction j_ν [erg g^{-1} s^{-1} Hz^{-1}] is also used:

$$\eta_\nu = \epsilon_\nu \rho = \frac{j_\nu}{4\pi}\rho. \tag{19.4}$$

One of the emission processes is free-free emission, which is important in many astrophysical situations. The free-free mass emissivity $j_\nu^{\rm ff}$ and its frequency-averaged one $j^{\rm ff}$ are, respectively, expressed by

$$j_\nu^{\rm ff} = 2.4 \times 10^{10} \rho T^{-1/2} e^{-h\nu/kT}\ [\text{erg g}^{-1}\ \text{s}^{-1}\ \text{Hz}^{-1}], \tag{19.5}$$

$$j^{\rm ff} = 5.1 \times 10^{20} \rho T^{1/2}\ [\text{erg g}^{-1}\ \text{s}^{-1}]. \tag{19.6}$$

The synchrotron emissivity is found in, e.g., Rybicki and Lightman (1979).

19.2.2 Absorption

Absorption of photons (photon destruction) in the medium is taken place by, e.g., atomic bound-bound transition, bound-free one (ionization), free-free one, synchrotron processes, electron–positron pair creations in the very high energy regime, molecular processes in the low energy regime, photoelectric processes by metal, dust processes, and other various processes.

The *absorption coefficient* α_ν [cm^{-1}] is defined by the relative amount of the incident radiation absorbed per unit path length, whereas the *mass absorption coefficient* or *opacity* κ_ν [cm^2 g^{-1}] is defined by the relative amount of the incident radiation absorbed per unit mass and unit path length:

$$\alpha_\nu = \rho\kappa_\nu. \tag{19.7}$$

In particular, the free-free absorption is important in many astrophysical situations. The free-free opacity $\kappa_\nu^{\rm ff}$ and its frequency-averaged one $\kappa^{\rm ff}$ are, respectively, expressed by

$$\begin{aligned}\kappa_\nu^{\rm ff} &= 1.3\times10^{56}\rho T^{-1/2}\nu^{-3}(1-e^{-h\nu/kT})\ [\mathrm{cm^2\ g^{-1}}]\\ &= 1.5\times10^{25}\rho T^{-7/2}(h\nu/kT)^{-3}(1-e^{-h\nu/kT})\ [\mathrm{cm^2\ g^{-1}}], \qquad (19.8)\\ \kappa^{\rm ff} &= 6.2\times10^{22}\rho T^{-7/2}\ [\mathrm{cm^2\ g^{-1}}]. \qquad (19.9)\end{aligned}$$

19.2.3 Scattering

Scattering includes electron/Thomson scattering by free electrons (Compton scattering in the high energy regime) (e.g., solar corona, accretion disks), Rayleigh one by corpuscles smaller than the wavelength of light (e.g., blue sky, reflection nebulae, protoplanetary disks), Mie scattering by small particles (e.g., interstellar dusts).

In scattering process, a photon in the incident radiation is scattered off and removed; i.e., in this sense scattering works as same as the true absorption. In contrast to the true absorption, however, another photon is scattering-in into the incident radiation and added.

Similar to absorption, the *scattering coefficient* β_ν [cm^{-1}] is defined by the relative amount of the incident radiation scattered per unit path length, whereas the *mass scattering coefficient* or *scattering opacity* σ_ν [cm^2 g^{-1}] is defined by the relative amount of the incident radiation scattered per unit mass and unit path length:

$$\beta_\nu = \rho\sigma_\nu. \tag{19.10}$$

The electron scattering opacity is independent of frequency and given by

$$\kappa_{\rm es} = 0.20(1+X)\,\mathrm{cm^2\ g^{-1}} \sim 0.4\,\mathrm{cm^2\ g^{-1}}. \tag{19.11}$$

On the other hand, the Rayleigh scattering cross section depends on wavelength λ as $\propto 1/\lambda^4$, and this nature causes blue sky, as is well-known, whereas the Mie scattering does not depend on wavelength.

The sum of the absorption coefficient and scattering, χ_ν, is called the *extinction coefficient*:

$$\chi_\nu = \alpha_\nu + \beta_\nu = (\kappa_\nu + \sigma_\nu)\rho. \tag{19.12}$$

19.2.4 Scattering Probability Function

As was stated, the difference between scattering and true absorption is that in scattering there is a scattering-in process as well as a scattering-out one. When a part of radiation propagating toward the direction $\boldsymbol{l}'$ in unit solid angle $d\Omega'$ is scattered off into the direction $\boldsymbol{l}$ in unit solid angle $d\Omega$, the radiative intensity in the direction $\boldsymbol{l}$ increases via ds by the following amount:

$$dI_\nu(\boldsymbol{l}) = ds \cdot \sigma_\nu \rho \oint \phi_\nu(\boldsymbol{l}, \boldsymbol{l}') I_\nu(\boldsymbol{l}') d\Omega' = \eta_\nu^{\text{sca}} ds. \tag{19.13}$$

Here, $\phi_\nu(\boldsymbol{l}, \boldsymbol{l}')$ is a *scattering probability function* (*scattering redistribution function, phase function*), which expresses the probability that the light-ray is scattered from direction $\boldsymbol{l}$ to direction $\boldsymbol{l}'$. This function satisfies the following normalized condition:

$$\oint \phi_\nu(\boldsymbol{l}, \boldsymbol{l}') d\Omega' = \oint \phi_\nu(\boldsymbol{l}, \boldsymbol{l}') d\Omega = 1. \tag{19.14}$$

In the case of *isotropic scattering*, where scattering takes place isotropically,

$$\phi_\nu = \frac{1}{4\pi}, \tag{19.15}$$

and in this case

$$\oint \phi_\nu I_\nu(\boldsymbol{l}') d\Omega' = J_\nu, \tag{19.16}$$

where the mean intensity J_ν will be defined in Sect. 20.1.2.

Both the Thomson and Rayleigh scatterings have a weak anisotropy, and the phase function is expressed as

$$\phi_\nu(\boldsymbol{l}, \boldsymbol{l}') = \frac{3}{4}\frac{1}{4\pi}\left[1 + (\boldsymbol{l} \cdot \boldsymbol{l}')^2\right], \tag{19.17}$$

and in this case

$$\oint \phi_\nu I_\nu(\boldsymbol{l}')d\Omega' = \frac{3}{4}\frac{1}{4\pi}\oint I_\nu\left(1+\cos^2\theta'\right)d\Omega' = \frac{3}{4}(J_\nu + K_\nu), \tag{19.18}$$

where the K-integral K_ν will be defined in Sect. 20.1.2.

The Mie scattering is anisotropic, and its phase function is complicated. Hence, in the anisotropic scattering by interstellar dusts or aerosol particles, the Henyey–Greenstein phase function $\phi_{\rm HG}$ is often approximately used:

$$\phi_{\rm HG} = \frac{1}{4\pi}\frac{1-g^2}{(1+g^2-2g\cos\theta)^{3/2}} \tag{19.19}$$

(Henyey and Greenstein 1941; see also Wendish and Yang 2012). Here, g is a parameter and becomes the first moment of the phase function: $\langle\cos\theta\rangle = \int \phi_{\rm HG}\cos\theta d\Omega = g$. It should be noted that in the limit of $g \to 0$ this approximate phase function becomes not the correct Rayleigh scattering, but the isotropic one.

19.2.5 Scattering and Effective Optical Depth

During scattering, photon's direction is changed, but the photon number is not changed. After repeated scatterings, photons are eventually absorbed. According to Rybicki and Lightman (1979), we describe the effective mean path and effective optical depth in the scattering process.

Let us suppose a photon which repeats scattering and travels in a homogeneous medium (Fig. 19.3). The scattered direction is isotropic, and, therefore, the movement of a photon becomes a *random walk* process. When the mean free path of a single scattering is ℓ, we can estimate the *root mean square displacement* ℓ_* after N times scattering, as follows.

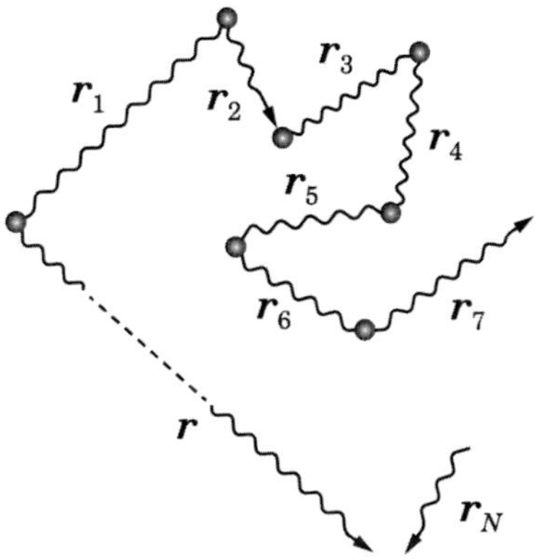

Fig. 19.3 Scattering and random walk. Reprinted from "Radiative transfer and radiation hydrodynamics" (in Japanese), with the permission of Nihon-Hyoronsha

The net displacement vector $\boldsymbol{r}$ of the photon from the photon-creation point (emission) to the destruction one (absorption) is expressed as

$$\boldsymbol{r} = \boldsymbol{r}_1 + \boldsymbol{r}_2 + \cdots + \boldsymbol{r}_N, \tag{19.20}$$

where $\boldsymbol{r}_i$ is a i-displacement vector. Since the mean square displacement ℓ_* is the mean square of the net displacement vector $\boldsymbol{r}$, we have

$$\ell_*^2 = \langle \boldsymbol{r}^2 \rangle = \langle \boldsymbol{r}_1^2 \rangle + \langle \boldsymbol{r}_2^2 \rangle + \cdots \langle \boldsymbol{r}_N^2 \rangle + 2\langle \boldsymbol{r}_1 \cdot \boldsymbol{r}_2 \rangle + 2\langle \boldsymbol{r}_1 \cdot \boldsymbol{r}_3 \rangle + \cdots = N\ell^2, \tag{19.21}$$

where the square of a single displacement, $\langle \boldsymbol{r}_i^2 \rangle$, is set to be the square of the mean free path, ℓ^2, and the cross terms vanish for isotropic scattering. Thus, the root mean square net displacement (*effective displacement*) ℓ_* of the photon is expressed as

$$\ell_* = \sqrt{N}\ell. \tag{19.22}$$

Using this result, we can relate the number N to the optical depth τ of the system with a finite size L. When the optical depth of the system is sufficiently small, the mean number N of scattering is on the order of τ;

$$N \sim \tau \quad (\tau \ll 1), \tag{19.23}$$

since the optical depth corresponding with the mean free path of a single scattering is unity. When the optical depth is large, on the other hand, a photon can escape the system for $\ell_* \sim L$, whereas L/ℓ is on the order of the optical depth τ;

$$N \sim \frac{\ell_*^2}{\ell^2} \sim \frac{L^2}{\ell^2} \sim \tau^2 \quad (\tau \gg 1). \tag{19.24}$$

Combining equation (19.23) and (19.24), we have a rough expression for N:

$$N \sim \tau^2 + \tau, \tag{19.25}$$

or $N \sim \mathrm{Max}(\tau, \tau^2)$. Furthermore, we can obtain $\ell_* \sim \sqrt{\tau^2 + \tau}\,\ell$.

Next, using the above arguments, we shall derive the effective optical depth. Let us suppose again a process that a photon is emitted from a gas particle, scattered N-times in the uniform medium, and absorbed finally by another particle (Fig. 19.4). In this process, the mean free path of a photon is

$$\ell_\nu = \frac{1}{\alpha_\nu + \beta_\nu} = \frac{1}{(\kappa_\nu + \sigma_\nu)\rho}, \tag{19.26}$$

while the effective displacement ℓ_* is $\ell_* = \sqrt{N}\ell_\nu$.

In terms of the absorption and scattering coefficients, the absorption and scattering probabilities of a photon after one mean free path, ε_ν and ϖ_ν, become,

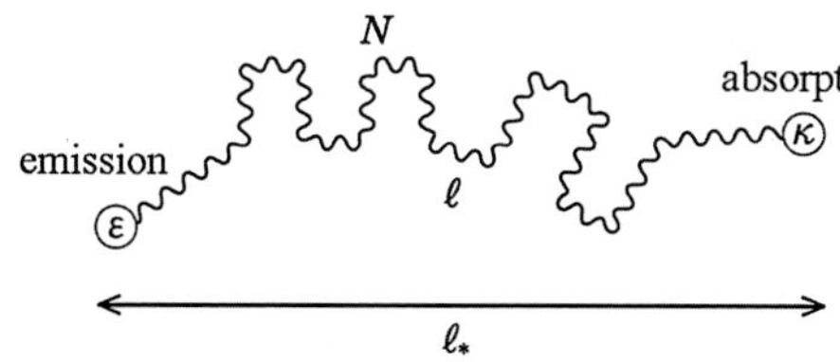

Fig. 19.4 Effective mean path. Reprinted from "Radiative transfer and radiation hydrodynamics" (in Japanese), with the permission of Nihon-Hyoronsha

respectively,

$$\varepsilon_\nu = 1 - \varpi_\nu \equiv \frac{\kappa_\nu}{\kappa_\nu + \sigma_\nu}, \tag{19.27}$$

$$1 - \varepsilon_\nu = \varpi_\nu \equiv \frac{\sigma_\nu}{\kappa_\nu + \sigma_\nu}. \tag{19.28}$$

Here, ε_ν is called a *photon destruction probability*, whereas ϖ_ν a *single-scattering albedo*.

Using these probabilities, the number of scattering becomes roughly $N = 1/\varepsilon_\nu = 1/(1 - \varpi_\nu)$. Hence, the effective displacement is expressed as

$$\ell_* = \sqrt{N}\ell_\nu = \frac{\ell_\nu}{\sqrt{\varepsilon_\nu}} = \frac{1}{\sqrt{\kappa_\nu(\kappa_\nu + \sigma_\nu)}\,\rho}. \tag{19.29}$$

This effective displacement ℓ_* means the distance between photon's birth place and its vanishing one, and called a *diffusion length*, or a *thermalization length*, or a *effective mean path*.

Furthermore, using this effective mean path, the *effective optical depth* τ_* of the whole system with size L is given by[2]

$$\tau_* \equiv \frac{L}{\ell_*} = L\sqrt{\kappa_\nu(\kappa_\nu + \sigma_\nu)}\,\rho = \sqrt{\kappa_\nu\rho L(\kappa_\nu\rho L + \sigma_\nu\rho L)}. \tag{19.30}$$

Using the absorption and scattering optical depths of the whole system,

$$\tau_{\rm abs} \equiv \kappa_\nu \rho L, \tag{19.31}$$

$$\tau_{\rm sca} \equiv \sigma_\nu \rho L, \tag{19.32}$$

2

$$d\tau = (\kappa_\nu + \sigma_\nu)\rho ds,$$

$$d\tau_* = \sqrt{\kappa_\nu(\kappa_\nu + \sigma_\nu)}\,\rho ds.$$

we can express the effective optical depth as

$$\tau_* = \sqrt{\tau_{\rm abs}(\tau_{\rm abs} + \tau_{\rm sca})}. \tag{19.33}$$

When the effective optical depth is sufficiently large ($\ell_* \ll L$; $\tau_* \sim \tau_{\rm abs} \gg 1$), the system is effectively optically thick. In the opposite case ($\ell_* \gg L$; $\tau_* \ll 1$), the system is effectively optically thin. It should be emphasized that the optical depth generally depends on frequency. For example, in the case of a hot plasma, the free-free absorption and electron scattering are important. Of these, the electron scattering opacity does not depend on frequency, but the free-free one becomes large in the low frequency regime. As a result, in the low frequency regime the hot plasma is often effectively thick, and its spectrum is blackbody-like. However, in the scattering-dominated high frequency regime, it becomes effectively thin, and its spectrum deviates from the blackbody one.

19.3 Thermodynamic Equilibrium and Blackbody Radiation

In usual gaseous objects and in many astrophysical situations, the collision timescale of particles is sufficiently shorter than the characteristic timescale of the system, and the fluid approximation holds safely (Chap. 1). In such a case, the thermodynamic equilibrium among gas particles is quickly established, and the kinetic distribution of the gas particles becomes the Maxwell–Boltzmann distribution. On the contrary, radiation itself cannot be the thermodynamic equilibrium, since radiation does not interact with one another. Under the sufficient interaction between radiation and matter, however, radiation can become the local thermodynamic equilibrium at the same temperature as matter, and the photon distribution becomes the Planck one.

19.3.1 Blackbody and Kirchhoff's Law

Let us consider an enclosure, surrounding an internal cavity (Fig. 19.5). The internal wall of the enclosure has uniform temperature and emits thermal radiation into the cavity, while it absorbs radiation from the cavity. After a sufficiently long time, radiation becomes in the thermodynamic equilibrium with the internal material of the enclosure, and the radiative spectrum I_ν of the cavity radiation[3] should be

[3]Here, I_ν is the radiation intensity at position $\boldsymbol{r}$ and time t, and at frequency ν per unit time, unit area, unit steradian, and unit frequency. The unit of I_ν is [erg s^{-1} cm^{-2} sr^{-1} Hz^{-1}]. We define I_ν in Sect. 20.1 again. Similarly, I_λ is the radiation intensity per unit wavelength.

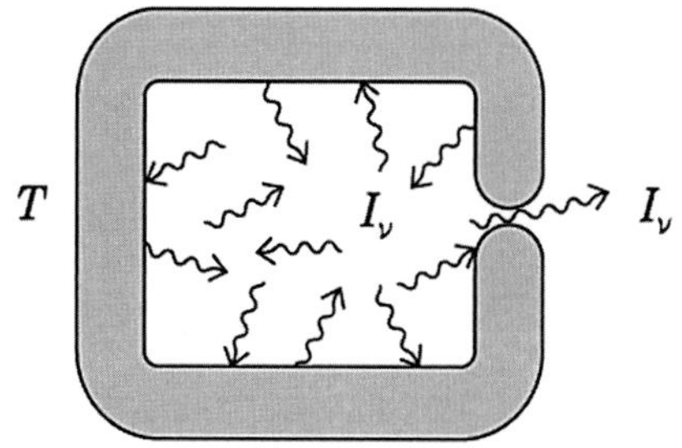

Fig. 19.5 Cavity surrounded by enclosure, and the cavity radiation. Reprinted from "Radiative transfer and radiation hydrodynamics" (in Japanese), with the permission of Nihon-Hyoronsha

expressed by some function,

$$I_\nu = B_\nu(T), \tag{19.34}$$

which is uniform and isotropic, and depends only on temperature T of the internal wall. This cavity radiation is called *blackbody*, and its spectrum is the *blackbody spectrum* $B_\nu(T)$.

Experimentally, for this cavity radiation, the following facts have been known before Planck.

Scaling Law As already stated, the cavity (blackbody) radiation only depends on temperature T and frequency ν and is independent of other quantities. The following scaling law holds:

$$I_\nu = \nu^3 F\left(\frac{\nu}{T}\right), \quad \text{and } I_\lambda = \frac{c^4}{\lambda^5} F\left(\frac{c}{\lambda T}\right). \tag{19.35}$$

Peak Wavelength The peak wavelength/frequency of the blackbody spectrum is as follows:

$$\lambda'_{\text{max}} T = 0.50995\,\text{cm K} \quad (\text{max of } I_\nu), \tag{19.36}$$

$$\lambda_{\text{max}} T = 0.28978\,\text{cm K} \quad (\text{max of } I_\lambda). \tag{19.37}$$

Stefan–Boltzmann Law The integration of the blackbody spectrum over all frequency is proportional to the fourth power of temperature T:

$$\int_0^\infty I_\nu d\nu \propto T^4. \tag{19.38}$$

Similar to this cavity radiation, in the radiative fluid, the medium emits radiation I_ν, while the radiation is absorbed by the medium. If the emission is balanced with the absorption, and the radiative fluid is under thermodynamic equilibrium,

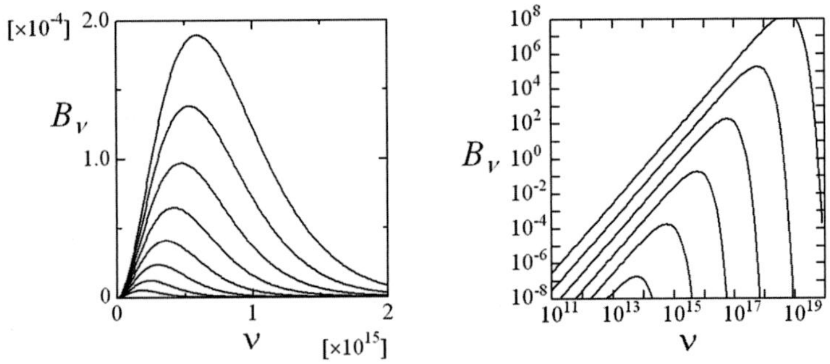

Fig. 19.6 Blackbody spectrum (Planck distribution). Left: linear scale with temperature from 3000 to 10,000 K in step of 1000 K. Right: logarithmic scale with temperature from 10^3 to 10^8 K. Scaling law is clearly seen

then the ratio of the emissivity to the absorption rate becomes the blackbody spectrum:

$$\frac{\eta_\nu}{\alpha_\nu} = \frac{j_\nu/(4\pi)}{\kappa_\nu} = I_\nu = B_\nu. \tag{19.39}$$

Here, η_ν is the volume emissivity of photons in units of [erg cm^{-3} s^{-1} Hz^{-1}sr^{-1}], and α_ν the absorption coefficient of photons in units of [cm^{-1}]. On the other hand, $j_\nu/(4\pi)$ $(= \eta_\nu/\rho)$ is the mass emissivity in units of [erg g^{-3} s^{-1} Hz^{-1}sr^{-1}], and κ $(= \alpha/\rho)$ the mass absorption coefficient (opacity) in units of [cm^2 g^{-1}].

This relation (19.39) between the emissivity and the absorption rate of photons in the medium under thermodynamic equilibrium is called *the Kirchhoff law* (Kirchhoff 1860).

19.3.2 Blackbody Radiation/Planck Distribution

Under thermodynamic equilibrium, the blackbody spectrum becomes the Planck distribution (Fig. 19.6; Planck 1900):

$$I_\nu(\boldsymbol{r}, \boldsymbol{l}, t) = B_\nu(T) = \frac{2h}{c^2} \frac{\nu^3}{\exp(h\nu/k_\mathrm{B}T) - 1}, \tag{19.40}$$

where T is the blackbody temperature, c $(= 2.9979 \times 10^{10}\,\mathrm{cm\,s^{-1}})$ the speed of light, h $(= 6.6261 \times 10^{-27}\,\mathrm{erg\,s})$ the Planck constant, and k_B $(= 1.3807 \times 10^{-16}\,\mathrm{erg\,K^{-1}})$ the Boltzmann constant.

In the low frequency regime ($h\nu \ll k_B T$), the Planck distribution is approximated by

$$I_\nu^{RJ} = \frac{2\nu^2}{c^2} k_B T \propto \nu^2 T, \tag{19.41}$$

$$I_\lambda^{RJ} = \frac{2c}{\lambda^4} k_B T \propto \frac{T}{\lambda^4}, \tag{19.42}$$

and known as the *Rayleigh-Jeans law*. In the high frequency regime ($h\nu \gg k_B T$), on the other hand, it is approximated by

$$I_\nu^{W} = \frac{2h\nu^3}{c^2} \exp\left(-\frac{h\nu}{k_B T}\right), \tag{19.43}$$

$$I_\lambda^{W} = \frac{2hc^2}{\lambda^5} \exp\left(-\frac{hc}{\lambda k_B T}\right), \tag{19.44}$$

and known as the *Wien law*.

The peak frequency/wavelength of the Planck distribution is expressed by

$$h\nu_{max} = 2.82 k_B T, \quad \frac{\nu_{max}}{T} = 5.88 \times 10^{10} \text{ [Hz deg}^{-1}], \tag{19.45}$$

$$\frac{hc}{\lambda_{max}} = 4.97 k_B T, \quad \lambda_{max} T = 0.29 \text{ [cm deg]}, \tag{19.46}$$

and known as *Wien's displacement law*.[4]

The frequency-integrated blackbody intensity (total blackbody intensity) becomes

$$B(T) = \int_0^\infty B_\nu(T) d\nu = \frac{2h}{c^2} \int_0^\infty \frac{\nu^3}{e^{h\nu/k_B T} - 1} d\nu = \frac{1}{\pi} \sigma_{SB} T^4, \tag{19.47}$$

and is proportional to T^4, as experimentally known. Here, a proportional constant,

$$\sigma_{SB} \equiv \frac{2\pi^5 k_B^4}{15 c^2 h^3} = 5.672 \times 10^{-5} \text{ erg cm}^{-2} \text{ deg}^{-4} \text{ s}^{-1}, \tag{19.48}$$

is the *Stefan–Boltzmann constant*, and the unit of $B(T)$ is [erg cm^{-2} s^{-1} sr^{-1}]. This relation (19.47) is called the *Stefan–Boltzmann law*.

[4]It should be noted that $\lambda_{max}\nu_{max} \neq c$. This is because $\nu_{max} B_{\nu_{max}}(T) = \lambda_{max} B_{\lambda_{max}}(T)$.

The radiation energy density E of the blackbody radiation (energy per unit volume) is expressed as (see also Chap. 20)

$$E(T) = \frac{4\pi}{c}B(T) = \frac{8\pi h}{c^3}\int \frac{\nu^3}{e^{h\nu/k_B T} - 1}d\nu = aT^4, \tag{19.49}$$

where

$$a \equiv \frac{8\pi^5 k_B^4}{15c^3h^3} = \frac{4\sigma_{SB}}{c} = 7.56 \times 10^{-15}\ \mathrm{erg\ cm^{-3}\ deg^{-4}} \tag{19.50}$$

is the *radiation constant*. Furthermore, the radiation pressure P of the blackbody radiation becomes

$$P(T) = \frac{1}{3}E(T) = \frac{1}{3}aT^4. \tag{19.51}$$

On the other hand, the radiation energy flux F per unit surface area of the blackbody radiator is

$$F(T) = \int B(T)\cos\theta d\Omega = \pi B(T) = \sigma_{SB}T^4. \tag{19.52}$$

In the realistic system with finite optical depth, the realization of the blackbody radiation depends on the optical depth of the system. If the effective optical depth is sufficiently large, the emergent spectrum would become a blackbody one. If, however, the optical depth is not so large, photons could escape before saturation. In such a case, the thermal radiation (e.g., free-free emission) is seen. Typical examples are shown in Fig. 19.7.

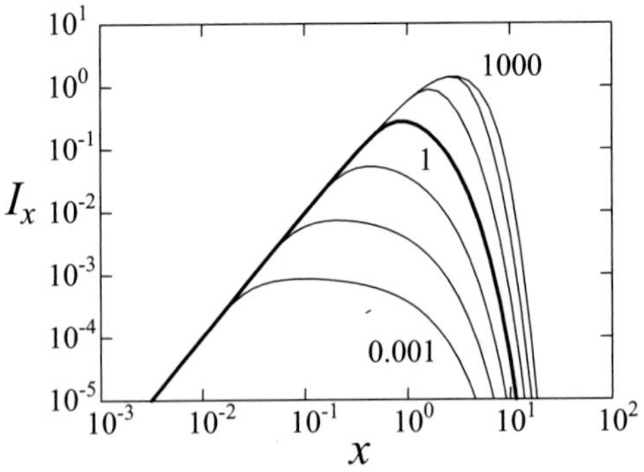

Fig. 19.7 Thermal radiation (free-free emission) from a model system with finite optical depth. The abscissae is the normalized frequency; $x = h\nu/(k_BT)$. The values of the optical depth are 0.001, 0.01, 0.1, 1, 10, 100, and 1000

19.3.3 Local Thermodynamic Equilibrium; LTE

In this subsection, we summarize several important key concepts; *thermodynamic equilibrium* (TE), *local thermodynamic equilibrium* (LTE), and *non-local thermodynamic equilibrium* (non-LTE, NLTE) (Kubát 2014).

In a solitary system, which consists of innumerable particles and photons, after a sufficiently long time longer than the relaxation time, macroscopic quantities, such as temperature or density, will not fluctuate, and microscopic distributions of constituent particles do not vary. Such a state is *thermodynamic equilibrium (TE)*. In thermodynamic equilibrium, the kinetic distribution of particles becomes the Maxwell–Boltzmann one, while photons obey the Planck one. Similarly, for atomic microscopic states, an atomic levels obeys the Boltzmann distribution, and an ionization configuration does the Saha one, although we do not treat these microscopic phenomena in this monograph.

The realistic astrophysical objects, such as stars or accretion disks, are not solitary systems, and radiative energy flows from a hot interior to a surface exposed to a cold exterior. Hence, the perfect thermodynamic equilibrium cannot be established in these realistic systems. Under the fluid approximation, however, due to the frequent collisions among particles, the particle distribution sufficiently relaxes. As a result, thermodynamic equilibrium would be *locally* established, and the particle distribution would become locally the Maxwell–Boltzmann one, the Boltzmann one, and the Saha one. This state is *local thermodynamic equilibrium (LTE)*. In such LTE state, radiation field is not always in thermodynamic equilibrium, and the photon distribution could deviate from the Planck one and should be obtained after solving the radiative transfer equation.

As already stated, one of the essential difference between radiation and matter is the interaction distance (mean free path). Since gaseous particles can travel a short distance, under the usual condition of the fluid approximation, collisions (interaction) of gaseous particles take place *locally*. On the other hand, radiation travels at the speed of light, and can interact with gaseous particles at a great distance; interaction is *non-local*. Hence, if this non-local interaction between radiation and matter becomes dominant, compared with the local collisional interaction of particles, then the local thermodynamic equilibrium does not hold in radiation, and we should solve all the detailed balance equations, instead of the Boltzmann and Saha equations. This "non-local" state is *non-local thermodynamic equilibrium (non-LTE or NLTE)*. It should be noted that even in such non-LTE state the kinetic distribution of particles is usually Maxwellian. Indeed, the word "non"is to relate not to LTE, but local; non-LTE is not non-(LTE) but (non-local)TE. In this sense, non-LTE could be called *kinetic equilibrium (KE)* (Hubeny and Mihalas 2014).

19.4 Radiation Force and Eddington Luminosity

A photon with frequency ν has energy $h\nu$, and carries momentum $h\nu/c$, where h is the Planck constant and c the speed of light. Hence, radiation of numberless photons also has energy and carries momentum. That is, the radiative flux $\boldsymbol{F}$, which is the radiation energy flow per unit area and unit time, carries momentum $\boldsymbol{F}/c$. In this radiation energy flow, a particle (e.g., electron, dust) with mass m and effective cross section S receives the *radiation force* $\boldsymbol{f}_{\rm rad}$, and the radiation force per unit mass $\boldsymbol{f}_{\rm rad}/m$,

$$\boldsymbol{f}_{\rm rad} = S\frac{\boldsymbol{F}}{c}, \tag{19.53}$$

$$\frac{1}{m}\boldsymbol{f}_{\rm rad} = \frac{S}{m}\frac{\boldsymbol{F}}{c}, \tag{19.54}$$

where the effective cross section per unit mass, S/m, is equivalent to opacity.

We now introduce a very important concept around a luminous gravitating object, the Eddington luminosity, where the radiation force is balanced by the gravitational one.

Let us consider a static gas particle (e.g., ionized hydrogen) near to a gravitating object of mass M and luminosity L. This particle is attracted inward by gravity, while it is pushed outward by radiation pressure (Fig. 19.8). When the luminosity of the object is not very large, the radiation-pressure force is smaller than the gravitational force and the particle is confined in the gravitational field of the object. As the luminosity becomes higher for a fixed mass, the radiation-pressure force also increases and eventually overcomes the gravitational force; the gas particle is ultimately blown off. Hence, gas cannot stably exist over such an extremely luminous object. Thus, for a fixed mass of the object, there exists a maximum possible luminosity for it; this is the *Eddington luminosity* $L_{\rm E}$.

At a distance r from a spherical object of luminosity L, the radiation flux (the amount of radiation energy flowing off per unit time and unit area), F, is

$$F = \frac{L}{4\pi r^2}. \tag{19.55}$$

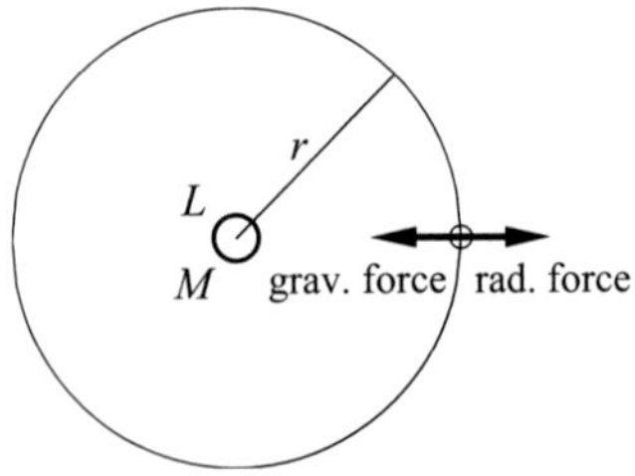

Fig. 19.8 Eddington luminosity, defined as the luminosity at which the radiation-pressure force on ionized gas is exactly balanced by the gravitational force

As already stated, this energy flux carries a momentum flux of

$$\frac{1}{c}F = \frac{1}{c}\frac{L}{4\pi r^2}. \tag{19.56}$$

Electrons in the gas receive this amount of momentum. Since the effective cross section of an electron is the Thompson scattering cross section σ_T ($= 6.65 \times 10^{-25}\,\mathrm{cm}^{-2}$), the force imposed on each electron by radiation is

$$\frac{\sigma_T}{c}F = \frac{\sigma_T}{c}\frac{L}{4\pi r^2}. \tag{19.57}$$

On the other hand, the gravitational force F_g asserted on each hydrogen atom is

$$F_g = \frac{GMm_H}{r^2}, \tag{19.58}$$

where m_H ($= m_p + m_e$) is the hydrogen mass, m_p ($= 1.67 \times 10^{-24}\,\mathrm{g}$) and m_e ($= 9.11 \times 10^{-28}\,\mathrm{g}$) being the proton mass and electron one, respectively.

We here assume fully ionized hydrogen, in which the protons and electrons are coupled with each other by the electromagnetic force. Interestingly, both the gravitational force and the radiation force have the same radial dependence in the present situation; i.e., if they are balanced at some radius, they are exactly equal everywhere. By equating these forces, Eqs. (19.57) and (19.58), we obtain the Eddington luminosity L_E,

$$L_E = \frac{4\pi cGMm_H}{\sigma_T} = \frac{4\pi cGM}{\kappa_{es}}, \tag{19.59}$$

where κ_{es} ($= \sigma_T/m_H$) is the electron scattering opacity. Note that the Eddington luminosity does not depend on r and is proportional to mass of the central object. The Eddington luminosity for typical parameters becomes

$$L_E = 1.25 \times 10^{38}\left(\frac{M}{M_\odot}\right)\,\mathrm{erg\,s^{-1}}. \tag{19.60}$$

Here, we have derived the Eddington luminosity for ionized pure hydrogen. The Eddington luminosity depends on the fluid composition. For example, the Eddington luminosity for fully ionized Helium is two times larger than that for hydrogen, while it becomes 1/1800 for electron–positron pair plasmas. For dust with typical size of 0.05 μm, the Eddington luminosity also remarkably reduces down to ~1/1000. The Eddington luminosity is modified for other absorption processes like free-free one.

A spherical static object with mass M cannot shine with luminosity greater than the Eddington luminosity L_E.[5]

19.5 Radiative Viscosity and Radiative Heat Conduction

Besides the radiative force, and the radiative heating and cooling, relating to the interaction between matter and radiation, there are several important concepts, which are higher order on the flow velocity or velocity gradient. We show several properties in the nonrelativistic regime here, and others in the relativistic regime in Sect. 24.3.

19.5.1 Radiative Viscosity

In hydrodynamics, the viscous term proportional to the velocity gradient originates from the off-diagonal components of the pressure stress tensor. In radiation hydrodynamics, the off-diagonal components of radiation stress tensor cause a similar effect called *radiative viscosity* (Jeans 1926; Milne 1929; Thomas 1930; Masaki 1971; Hsieh and Spiegel 1976; Fukue et al. 1985).

We can estimate the rough order of the radiative viscosity from physical considerations (Jeans 1926). The viscous coefficient η of the usual fluid is roughly

$$\eta \sim \frac{1}{3}\rho v \ell, \tag{19.61}$$

where ρ is the gas density, v the random velocity on the order of the sound speed c_s, and ℓ the mean free path of particles. In the case of the radiative fluid, using the radiation energy density $E\ (= aT^4)$ and opacity κ, we replace the above quantities as $\rho \to E/c^2$, $v \to c$, and $\ell \to 1/(\kappa\rho)$ to give the radiative viscosity η_{rad}:

$$\eta_{\mathrm{rad}} \sim \frac{1}{3}\frac{aT^4}{c\kappa\rho}. \tag{19.62}$$

The ratio of the radiative viscosity to the usual one is about

$$\frac{\eta_{\mathrm{rad}}}{\eta} = \frac{aT^4}{\rho c_s^2}\frac{c_s}{c}\frac{1}{\kappa\rho\ell}. \tag{19.63}$$

[5]In other words, except for a static spherical case, the luminosity can be larger than the Eddington one. For example, in neutron star winds the luminosity is slightly larger than the Eddington one. In supercritical accretion flows, which are quite aspherical, the *super-Eddington luminosity* is possible.

By the various previous studies (e.g., Hsieh and Spiegel 1976), on the order of $\mathcal{O}(v/c)^1$, the radiative stress tensor τ_{ij} is expressed as

$$\tau_{ij} = -\eta_{\rm rad}\left(\frac{\partial v_i}{\partial x_j} + \frac{\partial v_j}{\partial x_i} - \frac{2}{3}\nabla\cdot\boldsymbol{v}\delta_{ij}\right), \tag{19.64}$$

where the radiative viscous coefficient $\eta_{\rm rad}$ is

$$\eta_{\rm rad} = \frac{8aT^4}{3\rho(10\kappa + 9\sigma)c}, \tag{19.65}$$

κ and σ being absorption and scattering coefficients, respectively.

19.5.2 Radiative Heat Conduction

If there is a temperature gradient in the medium, the heat is carried from the hot region to the cold one due to conduction. The heat transfer rate per unit time and unit area is proportional to the temperature gradient, and its proportionality constant is the heat conductivity.

The heat conduction coefficient (heat conductivity) K in gaseous fluids is roughly

$$K \sim \frac{1}{3}\rho C_{\rm v} v \ell, \tag{19.66}$$

where $\rho C_{\rm v}$ is the heat capacity per unit volume, v the random velocity, and ℓ the mean free path. In the case of the radiative fluid, setting $\rho C_{\rm v} = dE/dT = 4aT^3$, and replacing $v \to c$ and $\ell \to 1/(\kappa\rho)$, we have the radiative heat conduction coefficient (radiative conductivity) $K_{\rm rad}$ [erg $\mathrm{s^{-1}\,cm^{-1}\,K^{-1}}$] as

$$K_{\rm rad} = \frac{4acT^3}{3\kappa\rho}. \tag{19.67}$$

This form is just that derived under the Rosseland approximation (Sect. 20.3.3)

In addition, the energy flux $\boldsymbol{F}$ [erg $\mathrm{s^{-1}cm^{-2}}$] due to the radiative heat conduction is

$$\boldsymbol{F} = -K_{\rm rad}\nabla\cdot T. \tag{19.68}$$

19.6 Adiabatic Exponents of Radiative Fluids

In this section, we briefly summarize the adiabatic change in an enclosure containing both matter and radiation in equilibrium, and introduce the adiabatic exponents of radiative fluids (Chandrasekhar 1967).

The matter is a perfect gas with gas pressure p, density ρ, and temperature T, while the radiation is a blackbody one with radiation pressure P and temperature T. In addition, the ratio of the specific heats is γ, and the ratio of the gas pressure to the total one is β [$\equiv p/(P+p)$].

As elegantly derived by Chandrasekhar (1967), adiabatic changes of this mixed fluid give the following relations:

$$\frac{d(P+p)}{P+p} - \Gamma_1 \frac{d\rho}{\rho} = 0, \tag{19.69}$$

$$\frac{d(P+p)}{P+p} - \frac{\Gamma_2}{\Gamma_2 - 1}\frac{dT}{T} = 0, \tag{19.70}$$

$$\frac{dT}{T} - (\Gamma_3 - 1)\frac{d\rho}{\rho} = 0, \tag{19.71}$$

where Γ_i's are generalized adiabatic exponents defined as

$$\Gamma_1 = \beta + \frac{(4-3\beta)^2(\gamma-1)}{\beta + 12(\gamma-1)(1-\beta)}, \tag{19.72}$$

$$\Gamma_2 = 1 + \frac{(4-3\beta)(\gamma-1)}{\beta^2 + 3(\gamma-1)(1-\beta)(4+\beta)}, \tag{19.73}$$

$$\Gamma_3 = 1 + \frac{(4-3\beta)(\gamma-1)}{\beta + 12(\gamma-1)(1-\beta)}. \tag{19.74}$$

In the gas-pressure dominated limit of $\beta = 1$, all of these adiabatic exponents become $\Gamma_1 = \Gamma_2 = \Gamma_3 = \gamma$, whereas they are $\Gamma_1 = \Gamma_2 = \Gamma_3 = 4/3$ in the radiation-pressure dominated limit of $\beta = 0$. For general case, these Γ_i's are shown in Fig. 19.9 as a function of β.

As an example, let us now apply the classical Jeans mass to this gas-radiation mixture (Sect. 6.4, see also Sect. 23.6). In the classical Jeans instability, the Jeans mass $M_{\rm J}$ depends on the gas sound speed $c_{\rm s}$ and density ρ as $M_{\rm J} \propto c_{\rm s}^3 \rho^{-1/2} \propto T^{3/2}\rho^{-1/2}$, where T is the gas (and radiation) temperature. In the adiabatic change, T and ρ are related by $T \propto \rho^{\Gamma_3 - 1}$, as Eq. (19.71) shows. Hence, the Jeans mass of the gas-radiation mixture becomes

$$M_{\rm J} \propto \rho^{(3\Gamma_3 - 4)/2}. \tag{19.75}$$

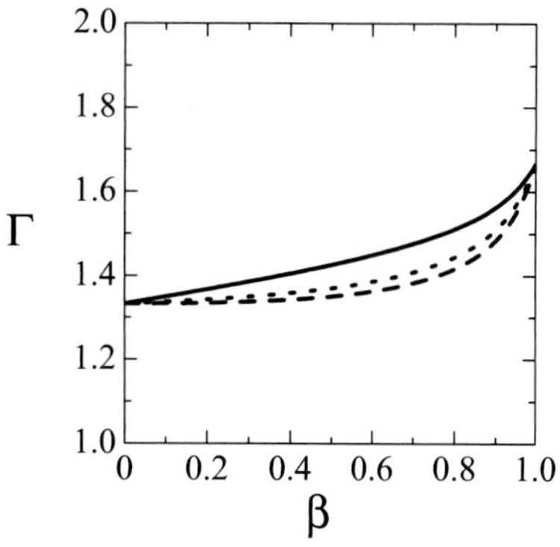

Fig. 19.9 Generalized adiabatic exponents as a function of β. A solid curve denotes Γ_1, a dashed one Γ_2, and a dotted one Γ_3. The ratio of specific heats of gaseous media is fixed as $\gamma = 5/3$

Let us consider a spherical object of a given mass, which is a mixture of gas and radiation ($4/3 < \Gamma_3 < 5/3$). If the mass of the object is larger than the Jeans mass, the object can contract by the Jeans instability. At a certain stage of the contraction, however, the Jeans mass becomes larger than the mass of the object when $\Gamma_3 > 4/3$ [see Eq. (19.75)]. This means the contraction of the object ceases at a certain contraction stage. When $\Gamma_3 = 4/3$ in the radiation-pressure dominated regime, on the other hand, the Jeans mass does not depend on density, and therefore, the radiation-pressure dominated medium of $\Gamma_3 = 4/3$ is marginally unstable for gravitational instability.

Such a radiation-pressure dominated state can be realized in, e.g., hypothetical supermassive stars (SMSs) with masses in the range 10^3–$10^8 M_\odot$ in the active galactic nuclei, or the first stars (Pop III stars) with masses in the range 10^2–$10^3 M_\odot$ in the early universe. The supermassive stars would be further general relativistically unstable.

References

Chandrasekhar, S.: An Introduction to the Study of Stellar Structure. Dover, New York (1967)
Fukue, J., Kato, S., Matsumoto, R.: Publ. Astron. Soc. Jpn. **37**, 383 (1985)
Henyey, L.G., Greenstein, J.L.: Astrophys. J. **93**, 70 (1941)
Hsieh, S.-H., Spiegel, E.A.: Astrophys. J. **207**, 244 (1976)
Hubeny, I., Mihalas, D.: Theory of Stellar Atmosphere. Princeton University Press, Princeton (2014)
Jeans, J.H.: Monthly Not. Roy. Astron. Soc. **86**, 328 (1926)
Kirchhoff, G.: Poggendorfs Annalen der Physik und Chemie **109**, 275 (1860)
Kubát, J.: (2014). arXiv:1406.3553v1
Masaki, I.: Publ. Astron. Soc. Jpn. **23**, 425 (1971)
Milne, E.A.: Monthly Not. Roy. Astron. Soc. **89**, 518 (1929)
Planck, M.: Verhandl. Dtsch. phys. Ges. **2**, 237 (1900)
Rybicki, G.B., Lightman, A.P.: Radiative Processes in Astrophysics. Wiley, New York (1979)
Thomas, L.H.: Quarterly J. Math. **1**, 239 (1930)
Wendish, M., Yang, P.: Theory of Atmospheric Radiation Transfer. Wiley, Weinheim (2012)

Chapter 20
Basic Equations for Radiative Transfer

A light-ray (a bundle of photons) travels through and interacts with gaseous materials, via emission, absorption, and scattering. The intensity of a light-ray obeys a linear integro-differential equation, the so-called *radiative transfer equation*, which is just the Boltzmann equation for photons. The distribution of gas particles is microscopically described by the Boltzmann equation in the phase space, while macroscopic quantities of fluids obey hydrodynamical equations, which are moment equations of the Boltzmann equation. Similarly, the propagation of photons is described by the radiative transfer equation, whereas moment quantities of radiation fields obey the moment equations of the radiative transfer equation. In this chapter, we first derive the fundamental equation for photon transfer and show its formal solutions in the Newtonian regime. We also derive moment equations with closure relations.

20.1 Radiation Fields

When radiation (a light-ray) propagates through matter (gas, dust, liquid), the incident radiation could be absorbed or scattered by matter, or radiation emitted from matter could append to the incident radiation. As a result, the intensity of radiation would change temporally, spatially, and directionally (Fig. 20.1). The study of the propagating way of radiation in matter is *radiative transfer*.

The theory of radiative transfer has been developed in astrophysics on the stellar atmosphere and in planetary science. In this monograph we consider fundamental concepts and basic treatments of radiative transfer. More general and detailed treatments can be found in many textbooks (e.g., Chandrasekhar 1960; Mihalas 1970; Sobolev 1975; Rybicki and Lightman 1979; Mihalas and Mihalas 1984; Shu 1991; Gray 1992, 2005; Thomas and Stamnes 1999; Peraiah 2002; Castor 2004; Wendish and Yang 2012; Hubeny and Mihalas 2014).

© Springer Nature Singapore Pte Ltd. 2020, corrected publication 2023

S. Kato, J. Fukue, *Fundamentals of Astrophysical Fluid Dynamics*,
Astronomy and Astrophysics Library,
https://doi.org/10.1007/978-981-15-4174-2_20

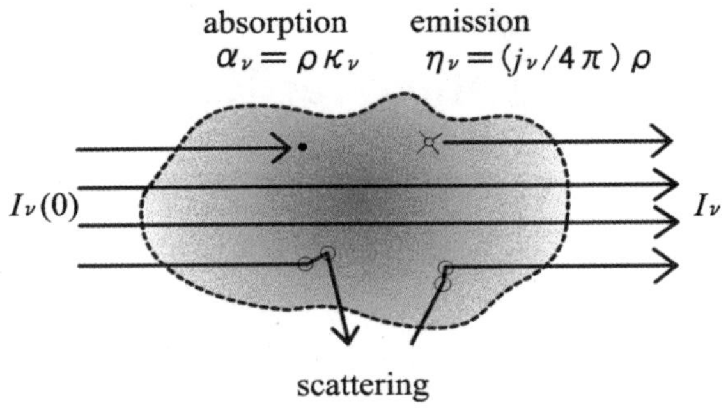

Fig. 20.1 Light-ray propagating through gaseous medium

20.1.1 Specific Intensity

We again and exactly define the intensity of radiation here. The *specific intensity*, $I_\nu(\boldsymbol{r}, \boldsymbol{l}, t)$ [erg s^{-1} cm^{-2} sr^{-1} Hz^{-1}], is the radiation energy carried off to direction $\boldsymbol{l}$ at position $\boldsymbol{r}$ and time t, by the light-rays per unit time, unit area, unit solid angle, and unit frequency (Fig. 20.2). The specific intensity is also called *brightness*. It may be defined per unit wavelength or unit energy. The specific intensity generally depends on time, position, direction, and frequency, and, therefore, I_ν is generally a function of seven independent variables.[1]

By integrating the specific intensity over frequency, we obtain the total intensity $I(\boldsymbol{r}, \boldsymbol{l}, t)$ as

$$I = \int_0^\infty I_\nu d\nu = \int_0^\infty I_\lambda d\lambda. \tag{20.1}$$

Let us consider light-rays toward various directions through a surface area dS, and suppose that some of light-rays irradiate another surface area dS', which is located in the direction with polar angle θ (Fig. 20.3). Then, the radiation energy dE flowing into a small solid angle $d\Omega$ subtended by dS' during short time interval dt is expressed as

$$dE = I_\nu dS \cos\theta d\Omega dt. \tag{20.2}$$

Since the solid angle is

$$d\Omega = \frac{dS' \cos\theta'}{r^2}, \tag{20.3}$$

[1]The time-independent case is *steady*, the position independent one is *homogenous*, the direction independent one is *isotropic*, and the frequency independent one is *gray*.

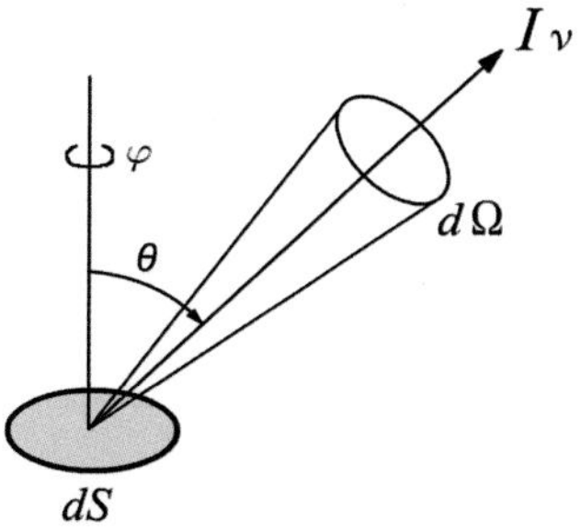

Fig. 20.2 Specific intensity of radiation (brightness)

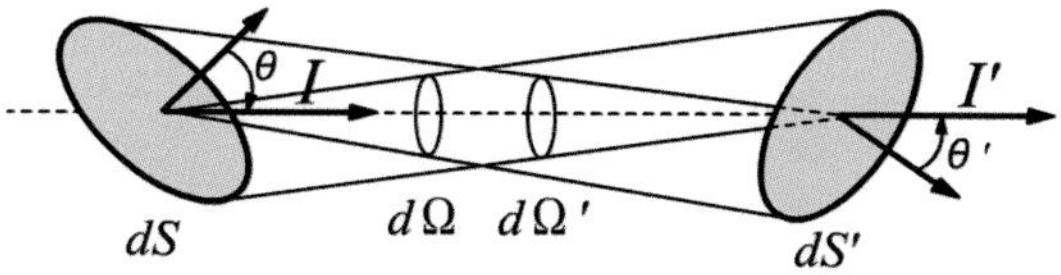

Fig. 20.3 Invariance of specific intensity

where r is the distance between dS and dS', the energy dE at dS coming from a solid angle $d\Omega$ subtended by dS' becomes

$$dE = \frac{I_\nu dS \cos\theta dS' \cos\theta' dt}{r^2}. \tag{20.4}$$

On the other hand, the radiation energy dE' at dS' coming from a solid angle $d\Omega'$ subtended by dS is

$$dE' = I'_\nu dS' \cos\theta' d\Omega' dt = \frac{I'_\nu dS' \cos\theta' dS \cos\theta dt}{r^2}. \tag{20.5}$$

If there is no absorption and emission between dS and dS', then dE is equal to dE'. Thus, we have a relation,

$$I_\nu = I'_\nu, \tag{20.6}$$

or the specific intensity is conserved along the light-ray path. This fundamental nature of the light-ray is called the *invariance of specific intensity*.

20.1.2 Moment Quantities

We here define moment quantities of the specific intensity. The radiation energy density (0th moment), the radiative flux (1st moment), and the radiation stress tensor

(2nd moment) are defined as follows[2,3] :

$$\begin{pmatrix} cE_\nu \\ F_\nu \\ cP_\nu \end{pmatrix} = \oint \begin{pmatrix} I_\nu \\ I_\nu \cos\theta \\ I_\nu \cos^2\theta \end{pmatrix} d\Omega, \quad \begin{pmatrix} J_\nu \\ H_\nu \\ K_\nu \end{pmatrix} = \frac{1}{4\pi} \oint \begin{pmatrix} I_\nu \\ I_\nu \cos\theta \\ I_\nu \cos^2\theta \end{pmatrix} d\Omega, \tag{20.7}$$

whereas in general three-dimensional space they are defined as follows:

$$\begin{pmatrix} cE_\nu \\ F_\nu^i \\ cP_\nu^{ij} \end{pmatrix} = \oint \begin{pmatrix} 1 \\ l^i \\ l^i l^j \end{pmatrix} I_\nu d\Omega, \quad l^i = \begin{pmatrix} \sin\theta\cos\varphi \\ \sin\theta\sin\varphi \\ \cos\theta \end{pmatrix}, \tag{20.8}$$

where $\boldsymbol{l}$ $(= l^i)$ is the direction cosine vector, θ the polar angle, and φ the azimuthal angle (cf. Fig. 20.2). We show the physical meanings of these quantities in turn .[4]

20.1.2.1 Radiation Energy Density

The *radiation energy density* E_ν [erg cm^{-3} Hz^{-1}], which is the energy density of radiation fields per unit volume, is defined as [5]

$$E_\nu \equiv \frac{1}{c} \oint I_\nu d\Omega, \tag{20.9}$$

while, as a mean quantity of the specific intensity I_ν, the *mean intensity* J_ν [erg s^{-1}cm^{-2} Hz^{-1}sr^{-1}] is defined as

$$J_\nu \equiv \frac{1}{4\pi} \oint I_\nu d\Omega. \tag{20.10}$$

The mean intensity J_ν is related to the radiation energy density E_ν by

$$J_\nu = \frac{c}{4\pi} E_\nu. \tag{20.11}$$

[2] The n-th moment is defined as follows: $M_\nu^n \equiv \frac{1}{4\pi} \oint I_\nu(\theta) \cos^n\theta d\Omega$.

[3] There are various expansion ways. The most generalized one may be the tensor-moment expansion (Thorne 1981).

[4] Physically, the radiation energy density E_ν, the radiative flux $\boldsymbol{F}_\nu$, and the radiation stress tensor P_ν^{ij} are important. Technically, on the other hand, the mean flux J_ν, the Eddington flux H_ν, the second moment K_ν, and the source function S_ν (defined in Sect. 20.2.2) have the same unit with the specific intensity I_ν, and are easy to handle.

[5] Since a photon travels 3×10^{10} cm per unit time, the specific intensity I_ν (a photon bundle, light-ray) contributes to the energy density of the volume which is 3×10^{10} larger than the unit volume. In other words, its contribution to the energy density of the unit volume reduces by a factor of $1/c$.

In the case of isotropic radiation fields, which extends in whole space, the isotropic intensity $\bar{I}$ is related to E as

$$E_\nu = \frac{1}{c}\oint \bar{I}_\nu d\Omega = \frac{\bar{I}_\nu}{c}\int_0^{2\pi} d\varphi \int_0^{\pi} \sin\theta d\theta = \frac{4\pi}{c}\bar{I}_\nu, \tag{20.12}$$

and $J_\nu = \bar{I}_\nu$.

The radiation energy density at a distance r from a spherical star with radius R and with uniform surface intensity $\bar{I}_\nu$ is calculated as follows. When the viewing angle of the star is θ_0 ($\sin\theta_0 = R/r$),

$$E_\nu = \frac{\bar{I}_\nu}{c}\int_0^{2\pi} d\varphi \int_0^{\theta_0} \sin\theta d\theta = \frac{2\pi \bar{I}_\nu}{c}(1 - \cos\theta_0), \tag{20.13}$$

where $\cos\theta_0 = \sqrt{1 - \sin^2\theta_0} = \sqrt{1 - (R/r)^2}$. Equation (20.13) can be written as

$$E_\nu(r) = \frac{4\pi \bar{I}_\nu}{c} W(r), \tag{20.14}$$

where

$$W(r) \equiv \frac{1}{2}\left[1 - \sqrt{1 - \left(\frac{R}{r}\right)^2}\right], \tag{20.15}$$

and is called the *dilution factor*. The dilution factor becomes $W = 1/2$ at the surface of the star, while it is $W \sim (1/4)(R/r)^2$ at a large distance ($r \gg R$).

20.1.2.2 Radiative Flux and Eddington Flux

The *radiative flux* F_ν [erg s^{-1}cm^{-2} Hz^{-1}], which is the energy flow of radiation in the direction specified by θ, per unit area, unit time, is defined as [6]

$$F_\nu \equiv \oint I_\nu \cos\theta d\Omega, \tag{20.16}$$

while, as a mean quantity, the *Eddington flux* H_ν [erg s^{-1}cm^{-2} Hz^{-1}sr^{-1}] is defined as

$$H_\nu \equiv \frac{1}{4\pi}\oint I_\nu \cos\theta d\Omega. \tag{20.17}$$

[6] The specific intensity I_ν passing through the unit area in the θ-direction times $\cos\theta$ is the energy per unit normal area and unit time carried by the light-ray. Summing up all the light-rays is the radiation energy passing through the unit normal area per unit time. The radiative flux per unit frequency is called the *monochromatic flux*, while that integrated over frequency is called the total flux. In the field of planetary science, the radiative flux is often called the *irradiance*.

The relation between the Eddington flux and radiative flux is

$$H_\nu = \frac{F_\nu}{4\pi}. \tag{20.18}$$

The radiative flux of the isotropic radiation fields, which extends in whole space, is $F = 0$.

The radiative flux F_ν at a distance r from a star with radius R and intensity $\bar{I}_\nu$ is calculated as

$$F_\nu = \bar{I}_\nu \int_0^{2\pi} d\varphi \int_0^{\theta_0} \sin\theta \cos\theta d\theta = \pi \bar{I}_\nu \sin^2\theta_0 = \pi \bar{I}_\nu \left(\frac{R}{r}\right)^2. \tag{20.19}$$

Hence, the radiative flux F_ν at the surface of the star (or on the plane uniform source) is $F_\nu = \pi \bar{I}_\nu$.

Finally, in the three-dimensional space, using the Cartesian coordinates (x, y, z) and spherical one (r, θ, φ), the radiative flux vector is expressed as

$$F_\nu^i = \begin{pmatrix} F_\nu^x \\ F_\nu^y \\ F_\nu^z \end{pmatrix} = \begin{pmatrix} \oint I_\nu \sin\theta \cos\varphi d\Omega \\ \oint I_\nu \sin\theta \sin\varphi d\Omega \\ \oint I_\nu \cos\theta d\Omega \end{pmatrix} = \begin{pmatrix} \oint I_\nu (1-\mu^2)^{1/2} \cos\varphi d\Omega \\ \oint I_\nu (1-\mu^2)^{1/2} \sin\varphi d\Omega \\ \oint I_\nu \mu d\Omega \end{pmatrix}, \tag{20.20}$$

where μ $(= \cos\theta)$ is the direction cosine.

20.1.2.3 Radiation Stress Tensor and K-Integral

The *radiation pressure* P_ν [dyn cm^{-2} Hz^{-1}] is defined as[7]

$$P_\nu \equiv \frac{1}{c} \oint I_\nu \cos^2\theta d\Omega, \tag{20.21}$$

while, as a mean quantity, the *K-integral* K_ν [erg s^{-1}cm^{-2} Hz^{-1}sr^{-1}] is defined as

$$K_\nu \equiv \frac{1}{4\pi} \oint I_\nu \cos^2\theta d\Omega. \tag{20.22}$$

[7] A light-ray hitting a unit area carries a momentum of $I_\nu \cos\theta / c$ per unit normal area. When the light-ray rebounds at the unit area, it exerts an impulse perpendicular to the unit area of a fraction of $\cos\theta$ (parallel to the unit area of $\sin\theta$). As a result, the radiation pressure perpendicular to the unit area is given by the integration of $I_\nu \cos^2\theta / c$ over solid angle, while the radiation stress parallel to the unit area is given by the integration of $I_\nu \cos\theta \sin\theta / c$ over solid angle.

The relation between the K-integral and the radiation pressure is

$$K_\nu = \frac{c}{4\pi} P_\nu. \tag{20.23}$$

In the case of isotropic radiation fields, which extends in whole space, the radiation pressure is expressed as

$$P_\nu = \frac{1}{c}\oint I_\nu \cos^2\theta d\Omega = \frac{\bar{I}_\nu}{c}\int_0^{2\pi} d\varphi \int_0^{\pi} \sin\theta\cos^2\theta d\theta = \frac{4\pi}{3c}\bar{I}_\nu = \frac{1}{3}E_\nu. \tag{20.24}$$

In the case of blackbody radiation fields at temperature T, $E = \int E_\nu d\nu = aT^4$ and $P = \int P_\nu d\nu = aT^4/3$.

The radiation pressure at a distance r from a spherical star with radius R and with uniform surface intensity $\bar{I}$ is calculated as

$$P_\nu(r) = \frac{\bar{I}_\nu}{c}\int_0^{2\pi} d\varphi \int_0^{\theta_0} \sin\theta\cos^2\theta d\theta = \frac{2\pi \bar{I}_\nu}{3c}\left(1-\cos^3\theta_0\right). \tag{20.25}$$

Hence, at the surface of the star ($\theta_0 = \pi/2$), it becomes

$$P_\nu(R) = \frac{1}{c}\oint \bar{I}_\nu \cos^2\theta d\Omega = \frac{2\pi}{3c}\bar{I}_\nu = \frac{1}{3}E_\nu(R), \tag{20.26}$$

while at far from the center ($\cos\theta_0 \sim 1$) it approaches

$$P_\nu = \frac{2\pi \bar{I}_\nu}{3c}(1-\cos\theta_0)\left(1+\cos\theta_0+\cos^2\theta_0\right) \sim \frac{2\pi \bar{I}_\nu}{c}(1-\cos\theta_0) \sim E_\nu. \tag{20.27}$$

Finally, in the three-dimensional space, the direction cosine vector is expressed as $l^i = (\sqrt{1-\mu^2}\cos\varphi, \sqrt{1-\mu^2}\sin\varphi, \mu)$, and if I_ν is axially symmetric around the azimuthal angle φ, then the radiation stress tensor is expressed as follows:

$$\begin{aligned}
P_\nu^{ij} &= \begin{pmatrix} P_\nu^{xx} & 0 & 0 \\ 0 & P_\nu^{yy} & 0 \\ 0 & 0 & P_\nu^{zz} \end{pmatrix} = \begin{pmatrix} P_\nu^{\theta\theta} & 0 & 0 \\ 0 & P_\nu^{\varphi\varphi} & 0 \\ 0 & 0 & P_\nu^{rr} \end{pmatrix} \\
&= \begin{pmatrix} P_\nu & 0 & 0 \\ 0 & P_\nu & 0 \\ 0 & 0 & P_\nu \end{pmatrix} - \frac{1}{2}\begin{pmatrix} 3P_\nu - E_\nu & 0 & 0 \\ 0 & 3P_\nu - E_\nu & 0 \\ 0 & 0 & 0 \end{pmatrix} \\
&= \frac{1}{2}\begin{pmatrix} E_\nu - P_\nu & 0 & 0 \\ 0 & E_\nu - P_\nu & 0 \\ 0 & 0 & 2P_\nu \end{pmatrix},
\end{aligned} \tag{20.28}$$

where P_ν is defined as

$$P_\nu \equiv \frac{4\pi}{c} K_\nu, \quad K_\nu \equiv \frac{1}{2}\int_{-1}^{1} I_\nu \mu^2 d\mu. \tag{20.29}$$

In addition, the mean radiation pressure $\tilde{P}_\nu$ becomes

$$\tilde{P}_\nu \equiv \frac{1}{3} P_\nu^{ii} = \frac{1}{3} E_\nu. \tag{20.30}$$

As already stated, in the case of isotropic radiation fields, $P_\nu = (1/3)E_\nu$. In the ideal fluid, the relation between gas pressure p and internal energy ρU is $p = (\gamma - 1)\rho U$, where γ is the ratio of the specific heats. This means that $\gamma = 4/3$ for the radiative fluid (relativistic fluid), while $\gamma = 5/3$ for the monoatomic gas. This is because the degree of freedom of the monoatomic gas is 3, whereas that of radiation is 2.

20.2 Radiative Transfer Equation and Formal Solution

A change in the specific intensity propagating through a medium is expressed by the *radiative transfer equation*, which corresponds to the Boltzmann equation for matter. We first show the basic equation describing the behavior of the radiation fields interacting with matter. We also write down the formal solution of the radiative transfer equation and describe the basic properties of radiation transfer.

20.2.1 Radiative Transfer Equation

A change of the specific intensity along the small path length, ds, during the propagation through medium, dI_ν, consists of the increase due to emission ($\eta_\nu ds$), the decrease by absorption ($-\alpha_\nu I_\nu ds$), scattering off ($-\beta_\nu I_\nu ds$), and scattering-in:

$$dI_\nu = \eta_\nu ds - \alpha_\nu I_\nu ds - \beta_\nu I_\nu ds + ds \cdot \beta_\nu \oint \phi_\nu(\boldsymbol{l}, \boldsymbol{l}') I_\nu(\boldsymbol{l}') d\Omega', \tag{20.31}$$

or in the differential form,

$$\begin{aligned} \frac{dI_\nu}{ds} &= \eta_\nu - \alpha_\nu I_\nu - \beta_\nu I_\nu + \beta_\nu \oint \phi_\nu(\boldsymbol{l}, \boldsymbol{l}') I_\nu(\boldsymbol{l}') d\Omega' \\ &= \frac{j_\nu}{4\pi}\rho - \kappa_\nu \rho I_\nu - \sigma_\nu \rho I_\nu + \sigma_\nu \rho \oint \phi_\nu(\boldsymbol{l}, \boldsymbol{l}') I_\nu(\boldsymbol{l}') d\Omega', \end{aligned} \tag{20.32}$$

where $\phi_\nu(\boldsymbol{l}, \boldsymbol{l}')$ is the scattering probability function ($\int \phi_\nu d\Omega = 1$). This is the basic form of the *radiative transfer equation*.

In the case of isotropic scattering, we have

$$\frac{dI_\nu}{ds} = \frac{j_\nu}{4\pi}\rho - \kappa_\nu \rho I_\nu - \sigma_\nu \rho I_\nu + \sigma_\nu \rho J_\nu, \tag{20.33}$$

where

$$J_\nu \equiv \frac{1}{4\pi}\int I_\nu(\boldsymbol{l}')d\Omega' \tag{20.34}$$

is the mean intensity.

Since a light-ray (radiation) propagates at the speed of light, the radiative transfer can be often treated as steady problems. In several cases, including waves, the time derivative of the specific intensity should be retained. The *radiative transfer equation* including time derivative is fully expressed as

$$\begin{aligned}\frac{1}{c}\frac{\partial I_\nu(\boldsymbol{l})}{\partial t} + (\boldsymbol{l}\cdot\nabla)\, I_\nu(\boldsymbol{l}) &= \frac{1}{4\pi}\rho j_\nu - \rho\kappa_\nu I_\nu(\boldsymbol{l}) \\ &\quad - \rho\sigma_\nu \int \phi_\nu(\boldsymbol{l}, \boldsymbol{l}')I_\nu(\boldsymbol{l})d\Omega' + \rho\sigma_\nu \int \phi_\nu(\boldsymbol{l}, \boldsymbol{l}')I_\nu(\boldsymbol{l}')d\Omega'.\end{aligned} \tag{20.35}$$

As already stated, this just corresponds to the Boltzmann equation for gases, but the term corresponding to the acceleration term on the left-hand side of the Boltzmann equation is absent here, since photons are massless particles.

If the scattering is isotropic, $\phi_\nu = 1/4\pi$, the transfer equation (20.35) becomes

$$\frac{1}{c}\frac{\partial I_\nu}{\partial t} + (\boldsymbol{l}\cdot\nabla)\, I_\nu = \frac{1}{4\pi}\rho j_\nu - \rho\,(\kappa_\nu + \sigma_\nu)\, I_\nu + \rho\sigma_\nu \frac{c}{4\pi}E_\nu. \tag{20.36}$$

In the weakly anisotropic case of the Thomson or Rayleigh scatterings, $\phi_\nu = [1/(4\pi)](3/4)(1+\cos^2\theta)$, we have

$$\frac{1}{c}\frac{\partial I_\nu}{\partial t} + (\boldsymbol{l}\cdot\nabla)\, I_\nu = \frac{1}{4\pi}\rho j_\nu - \rho\,(\kappa_\nu + \sigma_\nu)\, I_\nu + \frac{3}{4}\rho\sigma_\nu(J_\nu + K_\nu), \tag{20.37}$$

where K_ν is the K-integral, the second moment. When the Eddington approximation ($K_\nu = J_\nu/3$) holds, Eq. (20.37) reduces to the isotropic case (20.36).

The radiative transfer equation is an integro-differential equation on I_ν, since the scattering term includes the integral of the intensity. In addition, the specific intensity is generally a function of seven independent variables $(t, \boldsymbol{r}, \boldsymbol{l}, \nu)$. These make the radiative transfer problem too difficult to obtain precise solutions.

20.2.2 Source Function

The radiative transfer equation (20.32) can be solved in its original form. Due to the apparent simplicity and historical and technical reasons, however, the optical depth (instead of the spatial coordinates) and the source function are often used.

The radiative transfer equation (20.32) can be rearranged as

$$\frac{dI_\nu}{ds} = -(\alpha_\nu + \beta_\nu)(I_\nu - S_\nu). \tag{20.38}$$

Here, we introduce the *source function* S_ν by

$$S_\nu \equiv \varepsilon_\nu \frac{\eta_\nu}{\alpha_\nu} + (1 - \varepsilon_\nu)\Phi_\nu = \varepsilon_\nu \frac{j_\nu}{4\pi\kappa_\nu} + (1 - \varepsilon_\nu)\Phi_\nu, \tag{20.39}$$

where

$$\Phi_\nu \equiv \oint \phi_\nu(\boldsymbol{l}, \boldsymbol{l}')I_\nu(\boldsymbol{l}')d\Omega', \tag{20.40}$$

and

$$\varepsilon_\nu \equiv \frac{\alpha_\nu}{\alpha_\nu + \beta_\nu} = \frac{\kappa_\nu}{\kappa_\nu + \sigma_\nu} \tag{20.41}$$

is called the *photon destruction probability*; on the other hand, $\varpi_\nu = 1 - \varepsilon_\nu = \sigma_\nu/(\kappa_\nu + \sigma_\nu)$ is called the *single-scattering albedo*.

Finally, introducing the optical depth τ_ν along the light-ray path by

$$d\tau_\nu \equiv (\alpha_\nu + \beta_\nu)ds = (\kappa_\nu + \sigma_\nu)\rho ds, \tag{20.42}$$

we can express the radiative transfer equation as an *apparently* simple form:

$$\frac{dI_\nu}{d\tau_\nu} = -I_\nu + S_\nu. \tag{20.43}$$

In the case of isotropic scattering, $\Phi_\nu = J_\nu$, and the source function becomes

$$S_\nu = \frac{\eta_\nu + \beta_\nu J_\nu}{\alpha_\nu + \beta_\nu} = \frac{j_\nu/(4\pi) + \sigma_\nu J_\nu}{\kappa_\nu + \sigma_\nu}. \tag{20.44}$$

In addition, if LTE, then $\eta_\nu/\alpha_\nu = j_\nu/(4\pi)/\kappa_\nu = B_\nu$, and the source function becomes a symmetric form with respect to thermal and scattering parts:

$$S_\nu = \frac{\alpha_\nu B_\nu + \beta_\nu J_\nu}{\alpha_\nu + \beta_\nu} = \frac{\kappa_\nu B_\nu + \sigma_\nu J_\nu}{\kappa_\nu + \sigma_\nu} = \varepsilon_\nu B_\nu + (1 - \varepsilon_\nu)J_\nu. \tag{20.45}$$

The source function, as its name suggests, means the amount of radiation emitted at some optical depth, and the source term in the radiative transfer equation. Some part ($\varepsilon_\nu = 1 - \varpi_\nu$) comes from the thermal component emitted from material, and the other part ($1 - \varepsilon_\nu = \varpi_\nu$) originates from the scattering component scattered off from another light-ray.

20.2.3 Formal Solution

We briefly mention the formal solution of the radiative transfer equation. Multiplying e^{τ_ν} on the both sides of Eq. (20.43), we can rearrange the resultant equation in the form:

$$\frac{d}{d\tau_\nu}\left(I_\nu e^{\tau_\nu}\right) = S_\nu e^{\tau_\nu}, \tag{20.46}$$

and can integrate to yield

$$I_\nu e^{\tau_\nu} = I_\nu(0) + \int_0^{\tau_\nu} S_\nu e^{\tau_\nu'} d\tau_\nu', \tag{20.47}$$

and further multiplying $e^{-\tau_\nu}$ on both sides, we finally have a formal solution:

$$I_\nu(\tau_\nu) = I_\nu(0)e^{-\tau_\nu} + \int_0^{\tau_\nu} e^{-(\tau_\nu - \tau_\nu')} S_\nu(\tau_\nu') d\tau_\nu'. \tag{20.48}$$

The first term on the right-hand side of this formal solution (20.48) means that the incident radiation $I_\nu(0)$ is exponentially decreasing along the light-ray path by a factor of $e^{-\tau_\nu}$ (Fig. 20.4). The second term involving the source function is the summation of all the source functions $S_\nu(\tau_\nu')$ emitted at the optical depth τ_ν' along the light-ray path, which are also exponentially decreasing by a factor of $e^{-(\tau_\nu - \tau_\nu')}$.

If the source function is known as a function of the optical depth, then the integration can be performed. If, however, there is scattering, the source function

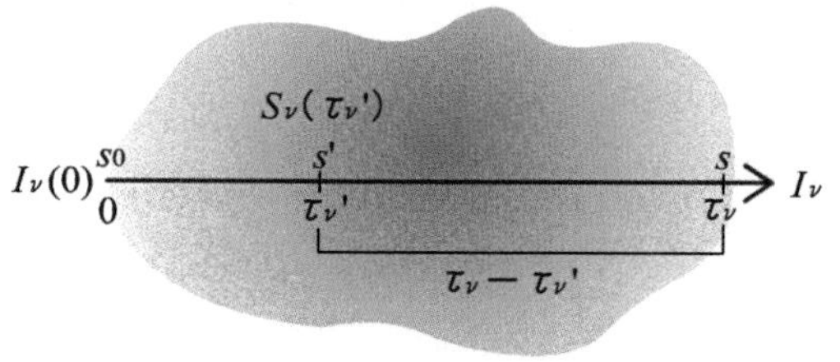

Fig. 20.4 Source function and optical depth along the light-ray path s. The emergent intensity $I_\nu(\tau_\nu)$ consists of two factors; the one is the incident radiation $I_\nu(0)$ reduced by a factor of $e^{-\tau_\nu}$, and the other is the summation of all the source function $S_\nu(\tau_\nu')$ reduced by a factor of $e^{-(\tau_\nu - \tau_\nu')}$

includes the integration of the radiative intensity, and, therefore, the formal solution is yet an integro-differential equation, in contrast to its apparently simple form.

If the source function is constant, the integration is easily performed to yield

$$I_\nu(\tau_\nu) = I_\nu(0)e^{-\tau_\nu} + S_\nu(1 - e^{-\tau_\nu}) = S_\nu + e^{-\tau_\nu}[I_\nu(0) - S_\nu]. \tag{20.49}$$

Hence, in the optically thick limit ($\tau_\nu \to \infty$), the specific intensity approaches the source function ($I_\nu \to S_\nu$), while $I_\nu \approx S_\nu \tau_\nu$ in the optically thin limit ($\tau_\nu \ll 1$) as long as the incident intensity $I_\nu(0)$ is small.

20.2.4 Radiative Transfer Equation in Various Coordinate Systems

In the previous sections, we considered the radiative transfer equation along the path of the light-ray, except for the general form in Sect. 20.2.1. As a matter of practice, we would examine the radiative transfer problem in various astrophysical situations, such as a plane-parallel atmosphere, a spherical envelope, and a three-dimensional case. Thus, in this section, we shall write down the radiative transfer equation in various coordinate systems (see, e.g., Peraiah 2002).

20.2.4.1 General Vector Form

As already shown in Sect. 20.2.2, the radiative transfer equation in the general form is

$$\begin{aligned} \frac{1}{c}\frac{\partial I_\nu}{\partial t} + \frac{\partial I_\nu}{\partial s} = \frac{1}{c}\frac{\partial I_\nu}{\partial t} + (\boldsymbol{l} \cdot \nabla)\, I_\nu &= -(\alpha_\nu + \beta_\nu)(I_\nu - S_\nu) \\ &= -(\kappa_\nu \rho + \sigma_\nu \rho)(I_\nu - S_\nu), \end{aligned} \tag{20.50}$$

where s is the coordinate along the light-ray path, α_ν ($= \rho\kappa_\nu$) the absorption coefficient, κ_ν being absorption opacity, β_ν ($= \rho\sigma_\nu$) the scattering coefficient, σ_ν being scattering opacity.

20.2.4.2 Cartesian Coordinates

In the Cartesian coordinates (x, y, z), the vector components are explicitly written as (cf. Fig. 20.5)

$$\boldsymbol{r} = (x, y, z), \tag{20.51}$$

$$\boldsymbol{l} = (\ell_x, \ell_y, \ell_z), \quad \ell_x^2 + \ell_y^2 + \ell_z^2 = 1, \tag{20.52}$$

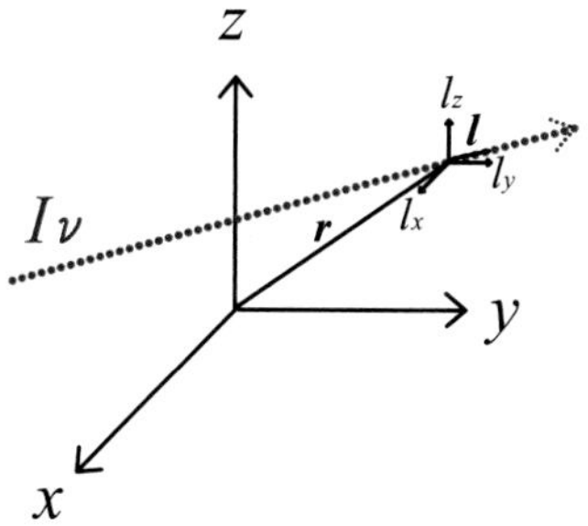

Fig. 20.5 Cartesian coordinates (x, y, z) and direction cosine vector $\boldsymbol{l}$ for the light-ray having a general direction. The polar angle θ for the light-ray direction is measured from the l_z axis, and unchanged along the path of the light-ray in the Cartesian coordinates

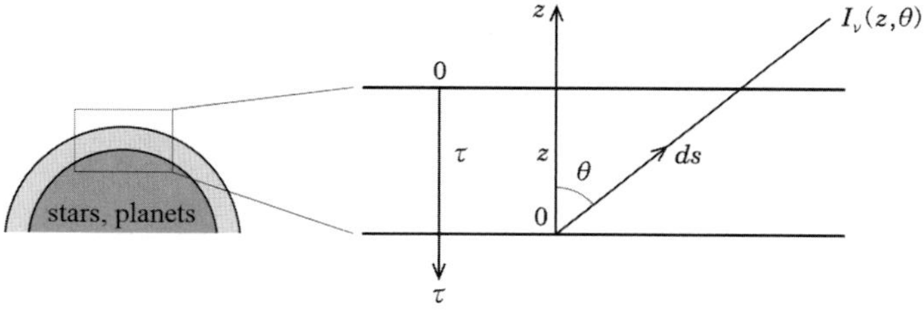

Fig. 20.6 Plane-parallel atmosphere and coordinates. The vertical axis z is measured from bottom to top against gravity, whereas the optical depth τ is measured from the surface to the interior of the atmosphere

$$ds^2 = dx^2 + dy^2 + dz^2, \tag{20.53}$$

$$I_\nu = I_\nu(x, y, z, \ell_x, \ell_y, \ell_z, t), \tag{20.54}$$

and the spatial term on the left-hand side of Eq. (20.50) becomes

$$\begin{aligned}
\frac{\partial I_\nu}{\partial s} &= \frac{\partial x}{\partial s}\frac{\partial I_\nu}{\partial x} + \frac{\partial y}{\partial s}\frac{\partial I_\nu}{\partial y} + \frac{\partial z}{\partial s}\frac{\partial I_\nu}{\partial z} \\
&= \frac{x}{s}\frac{\partial I_\nu}{\partial x} + \frac{y}{s}\frac{\partial I_\nu}{\partial y} + \frac{z}{s}\frac{\partial I_\nu}{\partial z} \\
&= \ell_x\frac{\partial I_\nu}{\partial x} + \ell_y\frac{\partial I_\nu}{\partial y} + \ell_z\frac{\partial I_\nu}{\partial z}.
\end{aligned} \tag{20.55}$$

In the steady *plane-parallel* case (Fig. 20.6), where $\partial/\partial t = \partial/\partial x = \partial/\partial y = 0$ and $I_\nu(z, \theta)$, the radiative transfer equation is expressed as

$$\mu\frac{dI_\nu}{dz} = -(\kappa_\nu\rho + \sigma_\nu\rho)(I_\nu - S_\nu), \tag{20.56}$$

where μ (= $\cos\theta$) is the direction cosine, θ being the polar angle measured from the z-axis.

Since the optical depth vanishes at the upper boundary in the usual plane-parallel atmosphere, we define the optical depth by

$$d\tau_\nu \equiv -(\kappa_\nu + \sigma_\nu)\rho dz, \tag{20.57}$$

and Eq. (20.56) becomes

$$\mu \frac{dI_\nu}{d\tau_\nu} = I_\nu - S_\nu. \tag{20.58}$$

It should be noted that the sign on the right-hand side of Eq. (20.58) is different from that of (20.43).

20.2.4.3 Cylindrical Coordinates

In the cylindrical coordinates (r, Φ, z), the vector components are expressed in terms of the coordinates and the direction cosine vector (ξ, η, μ) of the light-ray as (Fig. 20.7)

$$\boldsymbol{r} = (r, \Phi, z), \tag{20.59}$$

$$\boldsymbol{l} = (\xi, \eta, \mu) \equiv (\sin\theta\cos\varphi, \sin\theta\sin\varphi, \cos\theta), \quad \xi^2 + \eta^2 + \mu^2 = 1, \tag{20.60}$$

$$ds^2 = dr^2 + r^2 d\Phi^2 + dz^2, \tag{20.61}$$

$$I_\nu = I_\nu(r, \Phi, z, \theta, \varphi, t), \tag{20.62}$$

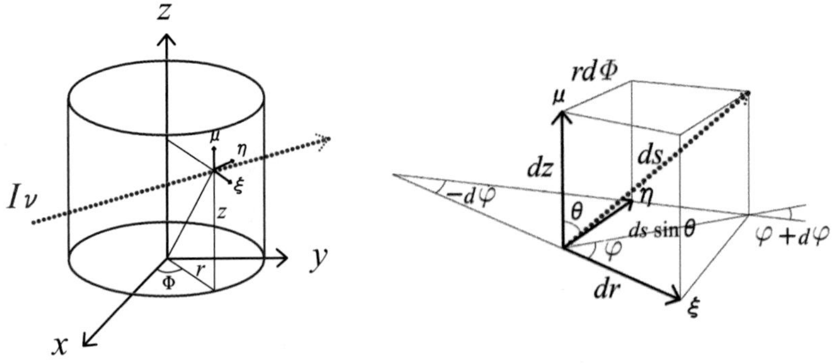

Fig. 20.7 Cylindrical coordinates (r, Φ, z) and the direction cosine vector (ξ, η, μ) of the light-ray having a general direction (left), and the enlarged view of a small volume enclosing a small path ds along the light-ray (right). The polar angle θ of the light-ray direction is measured from the μ-axis (z-axis), and does not change along the ray-path in the cylindrical coordinates, whereas the azimuthal angle φ is measured around the μ-axis and varies along the ray-path

and the spatial term on the left-hand side of Eq. (20.50) becomes

$$\frac{\partial I_\nu}{\partial s} = \frac{\partial r}{\partial s}\frac{\partial I_\nu}{\partial r} + \frac{\partial \Phi}{\partial s}\frac{\partial I_\nu}{\partial \Phi} + \frac{\partial z}{\partial s}\frac{\partial I_\nu}{\partial z} + \frac{\partial \theta}{\partial s}\frac{\partial I_\nu}{\partial \theta} + \frac{\partial \varphi}{\partial s}\frac{\partial I_\nu}{\partial \varphi}. \tag{20.63}$$

Since the lengths of each side of the small volume enclosing a small path ds along the light-ray (the right panel of Fig. 20.7) are, respectively, expressed as

$$dr = ds \sin\theta \cos\varphi, \tag{20.64}$$

$$rd\Phi = ds \sin\theta \sin\varphi, \tag{20.65}$$

$$dz = ds \cos\theta, \tag{20.66}$$

Eq. (20.63) can be written as

$$\frac{\partial I_\nu}{\partial s} = \sin\theta\cos\varphi\frac{\partial I_\nu}{\partial r} + \frac{\sin\theta\sin\varphi}{r}\frac{\partial I_\nu}{\partial \Phi} + \cos\theta\frac{\partial I_\nu}{\partial z} + \frac{\partial \theta}{\partial s}\frac{\partial I_\nu}{\partial \theta} + \frac{\partial \varphi}{\partial s}\frac{\partial I_\nu}{\partial \varphi}. \tag{20.67}$$

Since the polar angle θ of the light-ray direction measured from the μ-axis (z-axis) does not change along the ray-path ds in the cylindrical coordinates, $\partial\theta/\partial s$ in the fourth term on the right-hand side of Eq. (20.67) vanishes in this case. The azimuthal angle φ measured around the μ-axis, however, varies along the ray-path; i.e., the angle φ of the incident ray varies to the angle $\varphi + d\varphi$ (see the right panel of Figs. 20.7 and 20.8). With the help of the official of the external angle, we have the length of the side along the η-axis as

$$rd\Phi = r(-d\varphi) = ds \sin\theta \sin\varphi, \tag{20.68}$$

which determines the form of the fifth term on the right-hand side of Eq. (20.67), and we finally obtain the expression as

$$\begin{aligned}\frac{\partial I_\nu}{\partial s} &= \sin\theta\cos\varphi\frac{\partial I_\nu}{\partial r} + \frac{\sin\theta\sin\varphi}{r}\frac{\partial I_\nu}{\partial \Phi} + \cos\theta\frac{\partial I_\nu}{\partial z} - \frac{\sin\theta\sin\varphi}{r}\frac{\partial I_\nu}{\partial \varphi}\\ &= \xi\frac{\partial I_\nu}{\partial r} + \frac{\eta}{r}\frac{\partial I_\nu}{\partial \Phi} + \mu\frac{\partial I_\nu}{\partial z} - \frac{\eta}{r}\frac{\partial I_\nu}{\partial \varphi}.\end{aligned} \tag{20.69}$$

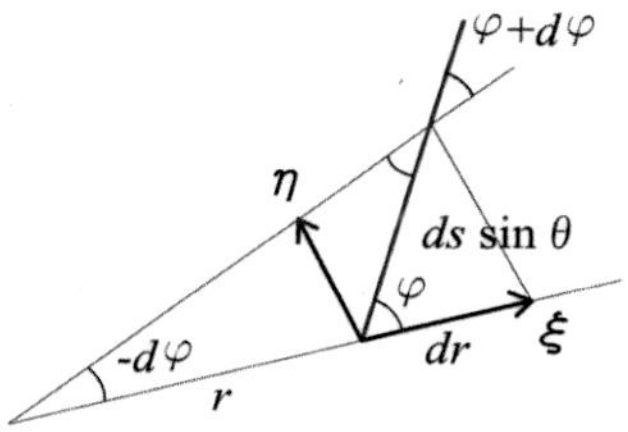

Fig. 20.8 Plane view of the enlarged figure of a small volume enclosing a projection of the light-ray, $ds \sin\theta$. The length of the side along the η-axis can be expressed by $r|-d\varphi|$ as well as $ds \sin\theta \sin\varphi$

In the steady cylindrically symmetric case, where $\partial/\partial t = \partial/\partial \Phi = \partial/\partial z = 0$, the radiative transfer equation is expressed as

$$\sin\theta\cos\varphi\frac{\partial I_\nu}{\partial r} - \frac{\sin\theta\sin\varphi}{r}\frac{\partial I_\nu}{\partial \varphi} = -(\kappa_\nu\rho + \sigma_\nu\rho)(I_\nu - S_\nu). \tag{20.70}$$

20.2.4.4 Spherical Coordinates

In the spherical coordinates (r, Θ, Φ), the vector components are expressed in terms of the coordinates and the direction cosine vector (ξ, η, μ) of the light-ray as (Fig. 20.9)

$$\boldsymbol{r} = (r, \Theta, \Phi), \tag{20.71}$$

$$\boldsymbol{l} = (\xi, \eta, \mu) \equiv (\sin\theta\cos\varphi, \sin\theta\sin\varphi, \cos\theta), \quad \xi^2 + \eta^2 + \mu^2 = 1, \tag{20.72}$$

$$ds^2 = dr^2 + r^2 d\Theta^2 + r^2\sin^2\Theta d\Phi^2, \tag{20.73}$$

$$I_\nu = I_\nu(r, \Theta, \Phi, \theta, \varphi, t) \tag{20.74}$$

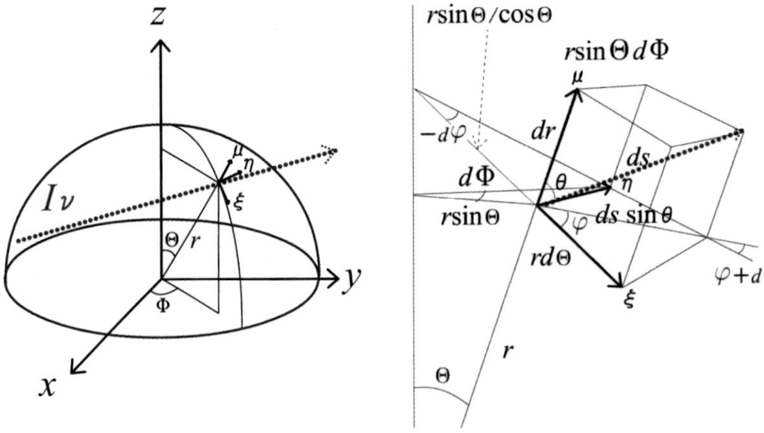

Fig. 20.9 Spherical coordinates (r, Θ, Φ) and direction cosine vector (ξ, η, μ) of the light-ray having a general direction (left), and the enlarged view of a small volume enclosing a small path ds along the light-ray (right). The polar angle θ of the light-ray direction is measured from the μ-axis (r-axis), and changes along the ray-path, whereas the azimuthal angle φ is measured around the μ axis and also varies along the ray-path

and the spatial term on the left-hand side of Eq. (20.50) becomes

$$\frac{\partial I_\nu}{\partial s} = \frac{\partial r}{\partial s}\frac{\partial I_\nu}{\partial r} + \frac{\partial \Theta}{\partial s}\frac{\partial I_\nu}{\partial \Theta} + \frac{\partial \Phi}{\partial s}\frac{\partial I_\nu}{\partial \Phi} + \frac{\partial \theta}{\partial s}\frac{\partial I_\nu}{\partial \theta} + \frac{\partial \varphi}{\partial s}\frac{\partial I_\nu}{\partial \varphi}. \tag{20.75}$$

Since the lengths of each side of the small volume enclosing a small path ds along the light-ray (the right panel of Fig. 20.9) are, respectively, expressed as

$$dr = ds\cos\theta, \tag{20.76}$$

$$rd\Theta = ds\sin\theta\cos\varphi, \tag{20.77}$$

$$r\sin\Theta d\Phi = ds\sin\theta\sin\varphi, \tag{20.78}$$

Eq. (20.75) can be written as

$$\frac{\partial I_\nu}{\partial s} = \cos\theta\frac{\partial I_\nu}{\partial r} + \frac{\sin\theta\cos\varphi}{r}\frac{\partial I_\nu}{\partial \Theta} + \frac{\sin\theta\sin\varphi}{r\sin\Theta}\frac{\partial I_\nu}{\partial \Phi} + \frac{\partial \theta}{\partial s}\frac{\partial I_\nu}{\partial \theta} + \frac{\partial \varphi}{\partial s}\frac{\partial I_\nu}{\partial \varphi}. \tag{20.79}$$

In the spherical coordinates, both the polar angle θ and azimuthal one φ of the light-ray direction vary along the ray-path ds (see the right panel of Figs. 20.9 and 20.10). Similar to the case of the cylindrical coordinates, the length of the diagonal and that of the side along μ-axis are, respectively, expressed as

$$r(-d\theta) = ds\sin\theta, \tag{20.80}$$

$$\frac{r\sin\Theta}{\cos\Theta}(-d\phi) = ds\sin\theta\sin\varphi, \tag{20.81}$$

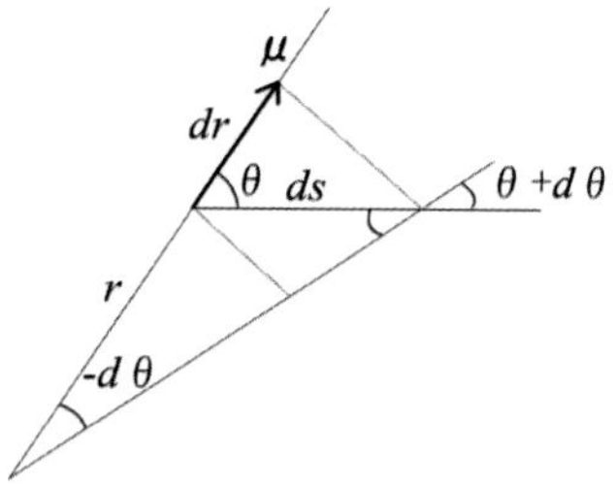

Fig. 20.10 Plane view of the enlarged figure of a small volume enclosing a small path ds along the light-ray. The length of the side perpendicular to the μ-axis can be expressed by $r|-d\theta|$ as well as $ds\sin\theta$

which determine the forms of the fourth and fifth terms on the right-hand side of Eq. (20.79), and we finally obtain the expression as

$$\begin{aligned}\frac{\partial I_\nu}{\partial s} &= \cos\theta\frac{\partial I_\nu}{\partial r} + \frac{\sin\theta\cos\varphi}{r}\frac{\partial I_\nu}{\partial\Theta} + \frac{\sin\theta\sin\varphi}{r\sin\Theta}\frac{\partial I_\nu}{\partial\Phi} \\ &\quad - \frac{\sin\theta}{r}\frac{\partial I_\nu}{\partial\theta} - \frac{\sin\theta\sin\varphi}{r}\cot\Theta\frac{\partial I_\nu}{\partial\varphi} \\ &= \mu\frac{\partial I_\nu}{\partial r} + \frac{\xi}{r}\frac{\partial I_\nu}{\partial\Theta} + \frac{\eta}{r\sin\Theta}\frac{\partial I_\nu}{\partial\Phi} + \frac{1-\mu^2}{r}\frac{\partial I_\nu}{\partial\mu} - \frac{\eta\cot\Theta}{r}\frac{\partial I_\nu}{\partial\varphi}.\end{aligned} \tag{20.82}$$

In the steady spherically symmetric case, where $\partial/\partial t = \partial/\partial\Theta = \partial/\partial\Phi = \partial/\partial\varphi = 0$, the radiative transfer equation is expressed as

$$\mu\frac{\partial I_\nu}{\partial r} + \frac{1-\mu^2}{r}\frac{\partial I_\nu}{\partial\mu} = -(\kappa_\nu\rho + \sigma_\nu\rho)(I_\nu - S_\nu). \tag{20.83}$$

20.3 Moment Equations and Closure Relation

Instead of the Boltzmann equation, we often solve the hydrodynamical equations under the fluid approximation. Similarly, instead of the radiative transfer equation, we often take moments of the transfer equation, and solve the moment equations with a closure relation.

We start from the radiative transfer equation in the form:

$$\frac{1}{c}\frac{\partial I_\nu}{\partial t} + (\boldsymbol{l}\cdot\nabla)\, I_\nu = \frac{j_\nu}{4\pi}\rho - \rho\,(\kappa_\nu + \sigma_\nu)\, I_\nu + \rho\sigma_\nu J_\nu, \tag{20.84}$$

where scattering is assumed to be isotropic, or weakly anisotropic like Thomson or Rayleigh scatterings with the Eddington approximation.

20.3.1 Moment Equations

Integrating the transfer equation (20.84) over solid angle, we obtain the zeroth moment. Integrating the transfer equation over solid angle, after multiplying it by the direction cosine, we obtain the first moment.

The zeroth and first moments of Eq. (20.84) are,[8] respectively,

$$\frac{\partial E_\nu}{\partial t} + \frac{\partial F_\nu^k}{\partial x^k} = \rho\,(j_\nu - c\kappa_\nu E_\nu)\,, \tag{20.87}$$

$$\frac{1}{c^2}\frac{\partial F_\nu^i}{\partial t} + \frac{\partial P_\nu^{ik}}{\partial x^k} = -\frac{\rho(\kappa_\nu + \sigma_\nu)}{c} F_\nu^i. \tag{20.88}$$

The former represents the energy conservation of radiation with the energy exchange with matter. That is, the first term on the right-hand side of Eq. (20.87) is the energy gain of radiation via emission from matter, while the second term is the energy loss due to absorption of radiation by matter. In the nonrelativistic limit here, the scattering process gives no net energy exchange. The latter Eq. (20.88) represents the momentum conservation of radiation with the momentum exchange with matter.

Moreover, integrating equations (20.84), (20.87), and (20.88) over the frequency, we obtain a frequency-integrated transfer equation and its moment equations[9] :

$$\frac{1}{c}\frac{\partial I}{\partial t} + (\boldsymbol{l}\cdot\nabla)\,I = \rho\left[\frac{j}{4\pi} - (\kappa+\sigma)_I I + \sigma_E \frac{c}{4\pi}E\right], \tag{20.93}$$

[8]These are equivalently written as

$$\frac{\partial J_\nu}{c\partial t} + \frac{\partial H_\nu^k}{\partial x^k} = \rho\left(\frac{j_\nu}{4\pi} - \kappa_\nu J_\nu\right), \tag{20.85}$$

$$\frac{\partial H_\nu^i}{c\partial t} + \frac{\partial K_\nu^{ik}}{\partial x^k} = -\rho(\kappa_\nu + \sigma_\nu) H_\nu^i. \tag{20.86}$$

[9]These are equivalently written as

$$\frac{1}{c}\frac{\partial I}{\partial t} + (\boldsymbol{l}\cdot\nabla)\,I = \rho\left[\frac{j}{4\pi} - (\kappa_I + \sigma_I) I + \sigma_J J\right], \tag{20.89}$$

$$\frac{\partial J}{c\partial t} + \frac{\partial H^k}{\partial x^k} = \rho\left(\frac{j}{4\pi} - \kappa_J J\right), \tag{20.90}$$

$$\frac{\partial H^i}{c\partial t} + \frac{\partial K^{ik}}{\partial x^k} = -\rho(\kappa_H + \sigma_H) H^i, \tag{20.91}$$

where

$$j \equiv \int j_\nu d\nu, \quad \kappa_I + \sigma_I \equiv \frac{1}{I}\int (\kappa_\nu + \sigma_\nu)\, I_\nu d\nu, \quad \sigma_J \equiv \frac{1}{J}\int \kappa_\nu J_\nu d\nu,$$

$$\sigma_J \equiv \frac{1}{J}\int \sigma_\nu J_\nu d\nu, \quad \kappa_H + \sigma_H \equiv \frac{1}{H^i}\int (\kappa_\nu + \sigma_\nu)\, H_\nu^i d\nu. \tag{20.92}$$

$$\frac{\partial E}{\partial t} + \frac{\partial F^k}{\partial x^k} = \rho \left(j - c\kappa_E E \right), \tag{20.94}$$

$$\frac{1}{c^2}\frac{\partial F^i}{\partial t} + \frac{\partial P^{ik}}{\partial x^k} = -\frac{1}{c}\rho(\kappa + \sigma)_F F^i, \tag{20.95}$$

where

$$j \equiv \int j_\nu d\nu, \tag{20.96}$$

$$(\kappa + \sigma)_I \equiv \frac{1}{I}\int (\kappa_\nu + \sigma_\nu)\, I_\nu d\nu, \tag{20.97}$$

$$\kappa_E \equiv \frac{1}{E}\int \kappa_\nu E_\nu d\nu, \tag{20.98}$$

$$\sigma_E \equiv \frac{1}{E}\int \sigma_\nu E_\nu d\nu, \tag{20.99}$$

$$(\kappa + \sigma)_F \equiv \frac{1}{F^i}\int (\kappa_\nu + \sigma_\nu)\, F^i_\nu d\nu. \tag{20.100}$$

It should be noted that

$$f^i_{\rm rad} = \frac{\rho}{c}\int (\kappa_\nu + \sigma_\nu)\, F^i_\nu d\nu = \frac{1}{c}\rho(\kappa + \sigma)_F F^i \tag{20.101}$$

is the radiative force per unit volume, acting on matter.

20.3.2 Source Function

We again introduce the source function S_ν by [cf. Eq. (20.45)]

$$S_\nu \equiv \frac{1}{\kappa_\nu + \sigma_\nu}\left(\frac{j_\nu}{4\pi} + \frac{c\sigma_\nu}{4\pi}E_\nu\right) = \varepsilon_\nu B_\nu + (1 - \varepsilon_\nu) J_\nu, \tag{20.102}$$

where the second equality is in the LTE case ($j_\nu/4\pi = \kappa_\nu B_\nu$).

In terms of this source function, Eqs. (20.84) and (20.87), and (20.88) are, respectively, re-expressed as[10]

$$\frac{1}{c}\frac{\partial I_\nu}{\partial t} + (\boldsymbol{l} \cdot \nabla)\, I_\nu = \rho\,(\kappa_\nu + \sigma_\nu)\,(S_\nu - I_\nu)\,, \tag{20.106}$$

$$\frac{\partial E_\nu}{\partial t} + \frac{\partial F_\nu^k}{\partial x^k} = \rho\,(\kappa_\nu + \sigma_\nu)\,(4\pi S_\nu - cE_\nu)\,, \tag{20.107}$$

$$\frac{1}{c^2}\frac{\partial F_\nu^i}{\partial t} + \frac{1}{3}\frac{\partial E_\nu}{\partial x^i} = -\frac{\rho(\kappa_\nu + \sigma_\nu)}{c} F_\nu^i. \tag{20.108}$$

20.3.3 Closure Relation

The zeroth moment Eq. (20.87) contains the radiative flux, which is determined by the first-moment Eq. (20.88). The first-moment Eq. (20.88) contains the radiation stress, which is determined by the second-moment equation. This means that in order to solve the moment equations we need some relation to cut the sequence, the so-called *closure relation*.

20.3.3.1 Eddington Approximation

As a closure relation, we often adopt the *Eddington approximation*:

$$P_\nu^{ij} = \frac{\delta^{ij}}{3} E_\nu; \quad K_\nu^{ij} = \frac{\delta^{ij}}{3} J_\nu. \tag{20.109}$$

This approximation is valid when the radiation fields are almost *isotropic*.

In the case of a flat-disk configuration, this relation holds with good accuracy in an optically thin regime as well as in an optically thick one.[11] In general cases of a two-dimensional configuration, such as geometrically thick disks, however, this

[10] These are equivalently written as

$$\frac{1}{c}\frac{\partial I_\nu}{\partial t} + (\boldsymbol{l} \cdot \nabla)\, I_\nu = \rho\,(\kappa_\nu + \sigma_\nu)\,(S_\nu - I_\nu)\,, \tag{20.103}$$

$$\frac{\partial J_\nu}{c\partial t} + \frac{\partial H_\nu^k}{\partial x^k} = \rho\,(\kappa_\nu + \sigma_\nu)\,(S_\nu - J_\nu)\,, \tag{20.104}$$

$$\frac{\partial H_\nu^i}{c\partial t} + \frac{\partial K_\nu^{ik}}{\partial x^k} = -\rho(\kappa_\nu + \sigma_\nu) H_\nu^i. \tag{20.105}$$

[11] In the spherical case $P_\nu^{rr} = E_\nu$, while all other components vanish in an optically thin regime.

relation would not be adequate in an optically thin regime, and alternative closure relation is necessary.

20.3.3.2 Variable Eddington Factor

The Eddington approximation (20.109) as a closure relation holds when the radiation field is *nearly isotropic*.[12] In the optically thin regime, or in the case where the anisotropy of radiation becomes important, we should carefully treat the Eddington approximation.[13]

Generalization of the Eddington approximation (20.109):

$$P^{ij} = f^{ij} E, \tag{20.110}$$

is useful in semi-analytical studies, where f^{ij} is the *Eddington tensor* ($\sum f^{ii} = 1$) and is generally a function of the optical depth τ. This relation (20.110) is reduced to Eq. (20.109) if we assume the radiation field is isotropic: $f^{ij} = \delta^{ij}/3$.

In the spherically symmetric case, the diagonal part of the Eddington tensor is often set as

$$\left(f(\tau), \frac{1}{2} - \frac{1}{2} f(\tau), \frac{1}{2} - \frac{1}{2} f(\tau) \right), \tag{20.111}$$

where $f(\tau)$ is the *Eddington factor*.

Suitable forms of the Eddington tensor or Eddington factor are adopted in each problem and configuration. For example, in the spherically symmetric case, the Eddington factor in the form of

$$f(\tau) = \frac{1+\tau}{1+3\tau} \tag{20.112}$$

is proposed (Tamazawa et al. 1975, Fig. 20.11).[14]

[12]The Rosseland approximation holds when the medium is sufficiently optically thick and the photon mean free path is sufficiently shorter than the typical scale of the medium, and when the velocity gradient is sufficiently small and the local diffusion is isotropic.

[13]In the relativistic regime, where the flow speed is on the order of the speed of light, we also carefully treat the closure relation.

[14]As an another example, using the results of numerical simulations, an approximate expression was proposed: $f = 1 - (2/3)e^{-1/\tau^2}$ (Hummer and Rybicki 1971).

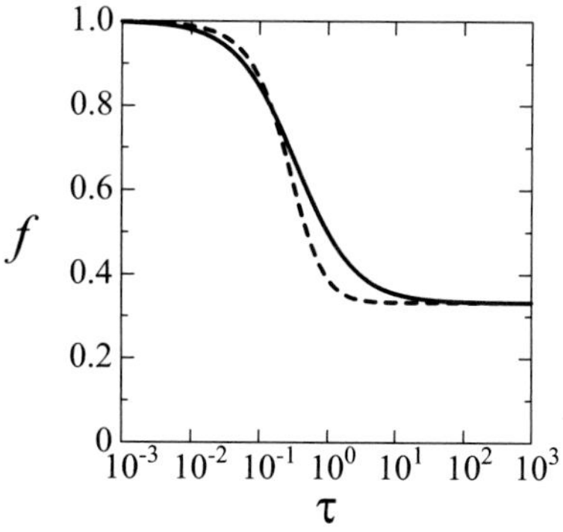

Fig. 20.11 Two models of the Eddington factors as a function of the optical depth τ. A thick solid curve denotes the variable Eddington factor (20.112), while a thick dashed one is the variable Eddington factor (20.130), where R_ν is read as τ^{-1}

20.3.3.3 Rosseland Approximation

When the medium is sufficiently *optically thick* and the velocity gradient is sufficiently small, as in the deep interior of stars, the photon mean free path is sufficiently shorter than the typical scale of the medium. In such a medium, photons diffuse out from the interior toward the outer boundary, nearly holding local thermodynamic equilibrium (LTE). Then, we may use the *diffusion approximation* (Rosseland approximation) for photon transport.

In an optically thick regime, local thermodynamic equilibrium (LTE),

$$j_\nu/(4\pi) = \kappa_\nu B_\nu(T), \tag{20.113}$$

holds, while Eq. (20.87) can be approximated as $j_\nu = c\kappa_\nu E_\nu$ as long as the radiation intensity is a slowly varying function of space and time. Hence,

$$E_\nu \sim \frac{4\pi}{c} B_\nu \quad \text{and} \quad P_\nu^{ij} \sim \frac{\delta^{ij}}{3}\frac{4\pi}{c} B_\nu. \tag{20.114}$$

In the steady state, from Eq. (20.88), we thus obtain

$$F_\nu^i = -\frac{c}{\rho(\kappa_\nu + \sigma_\nu)}\frac{\partial P^{ik}}{\partial x^k} = -\frac{4\pi}{3\rho(\kappa_\nu + \sigma_\nu)}\frac{\partial B_\nu}{\partial T}\frac{\partial T}{\partial x^i}. \tag{20.115}$$

This means that the radiation energy is transported by an isotropic diffusion of photons.

Integrating equation (20.115) over the frequency, we obtain

$$\boldsymbol{F} = \int \boldsymbol{F}_\nu d\nu = -\frac{4\pi}{3\rho}\nabla T \frac{\displaystyle\int \frac{1}{\kappa_\nu + \sigma_\nu}\frac{\partial B_\nu}{\partial T} d\nu}{\displaystyle\int \frac{\partial B_\nu}{\partial T} d\nu} \int \frac{\partial B_\nu}{\partial T} d\nu. \tag{20.116}$$

Since

$$\int \frac{\partial B_\nu}{\partial T} d\nu = \frac{d}{dT} \int B_\nu d\nu = \frac{d}{dT} \left(\frac{1}{\pi} \sigma_{\text{SB}} T^4 \right) = \frac{4}{\pi} \sigma_{\text{SB}} T^3, \tag{20.117}$$

we finally obtain a frequency-integrated radiative flux,

$$\boldsymbol{F} = -\frac{4acT^3}{3\bar{\kappa}_{\text{R}}\rho} \nabla T, \tag{20.118}$$

where $\bar{\kappa}_{\text{R}}$ is the Rosseland mean opacity[15] :

$$\frac{1}{\bar{\kappa}_{\text{R}}} \equiv \frac{\displaystyle\int_0^\infty \frac{1}{\kappa_\nu + \sigma_\nu} \frac{\partial B_\nu}{\partial T} d\nu}{\displaystyle\int_0^\infty \frac{\partial B_\nu}{\partial T} d\nu}. \tag{20.124}$$

Equation (20.118) is often called a *radiative conduction equation* of the form of Fick's law, where the effective "conductivity" is $4acT^3/3\bar{\kappa}_{\text{R}}\rho$, which is inversely proportional to the opacity.

In Eq. (20.124), $\partial B_\nu/\partial T$ is a gradually varying function in the limited frequency range, and, therefore, the frequency region of small $\kappa_\nu+\sigma_\nu$ mainly contributes to $\bar{\kappa}_{\text{R}}$. Since the small value of $\kappa_\nu + \sigma_\nu$ means that the medium is optically thin until deep

---

[15]The Rosseland mean opacities for free-free and bound-free absorptions, κ_{ff} and κ_{bf}, are approximately expressed by Kramers' law:

$$\kappa_{\text{ff}} = 3.68 \times 10^{22} g_{\text{ff}} (X+Y)(1+X)\rho T^{-7/2}\,\text{cm}^2\,\text{g}^{-1}$$
$$\sim 6.24 \times 10^{22} \rho T^{-7/2}\,\text{cm}^2\,\text{g}^{-1}, \tag{20.119}$$

$$\kappa_{\text{bf}} = 4.34 \times 10^{25} (g_{\text{bf}}/t) Z(1+X)\rho T^{-7/2}\,\text{cm}^2\,\text{g}^{-1}$$
$$\sim 1.50 \times 10^{24} \rho T^{-7/2}\,\text{cm}^2\,\text{g}^{-1}, \tag{20.120}$$

where g_{ff} and g_{bf} are the mean Gaunt factors, of order unity, for the free-free and bound-free transitions, respectively, t is the guillotine factor of order unity, and X, Y, and Z are the abundances of hydrogen, helium, and metal, respectively (Morse 1940; Schwarzschild 1958). In the low metallicity case the free-free absorption dominates the bound-free one, while the bound-free absorption will dominate the free-free one in the high metallicity case (Schwarzschild 1958).

Overall opacity is approximately expressed as follows:

$$\kappa \propto \rho^{1/2} T^4 \quad (10^4\,\text{K} \le T \le 10^5\,\text{K}), \tag{20.121}$$

$$\kappa \propto \rho T^{-7/2} \quad (10^5\,\text{K} \le T \le 10^6\,\text{K}), \tag{20.122}$$

$$\kappa = 0.2(1+X) \quad (10^6\,\text{K} \le T). \tag{20.123}$$

inside at the relevant frequency. Thus, radiation from deep inside mainly contributes to the mean opacity.

20.3.3.4 Flux-Limited Diffusion Approximation

The diffusion (Rosseland) approximation implies that in a steady state the radiation energy is transported by an isotropic diffusion of photons:

$$\boldsymbol{F} = -\frac{c}{3\bar{\kappa}_{\mathrm{R}}\rho}\nabla E \tag{20.125}$$

[see Eq. (20.118)]. This gives the correct flux in an optically thick regime, where the photon mean free path of $\sim 1/(\bar{\kappa}_{\mathrm{R}}\rho)$ is sufficiently shorter than the typical scale for the change of E. In an optically thin regime, where the mean free path diverges, this flux tends to infinity. Such a situation, however, is unphysical, since the rate at which radiation transports energy is finite even in an optically thin regime. That is, the magnitude of the flux can be no greater than the radiation energy density times the maximum transport speed.[16] Namely, the radiative flux $\boldsymbol{F}$ should be *limited* in the optically thin regime in some way, which is the *flux-limited diffusion approximation* (Levermore and Pomraning 1981; Pomraning 1983 for a relativistic correction; Melia and Zylstra 1991 for a scattering medium; Anile and Romano 1992 for a covariant form; Turner and Stone 2001 for a numerical calculation).

In the flux-limited diffusion (FLD) approximation, a Fick's type of diffusion approximation is adopted under introduction of a factor called *flux limiter* $\lambda_\nu(E_\nu)$:

$$\boldsymbol{F}_\nu = -\frac{c\lambda_\nu}{(\kappa_\nu + \sigma_\nu)\rho}\nabla E_\nu. \tag{20.126}$$

The flux limiter $\lambda_\nu(E_\nu)$ is in the range of 0 (thin) $\leq \lambda_\nu \leq 1/3$ (thick). An appropriate form of this flux limiter is presented below.

We now write the radiation pressure stress tensor P_ν^{ij} as

$$P_\nu^{ij} = f_\nu^{ij} E_\nu, \tag{20.127}$$

and express the Eddington tensor f_ν^{ij} in the form:

$$f_\nu^{ij} = \frac{1 - f_\nu}{2}\delta^{ij} + \frac{3f_\nu - 1}{2}n^i n^j. \tag{20.128}$$

[16] In a spherical case the flux is limited as $|\boldsymbol{F}| \leq cE$, while it is limited as $|\boldsymbol{F}| \leq cE/2$ in a plane-parallel case.

In this Eq. (20.128),

$$n^i \equiv \frac{\nabla E_\nu}{|\nabla E_\nu|} \tag{20.129}$$

is the unit vector in the direction of the radiation energy density gradient, i.e., the radiative flux, which is determined by the local radiation field. The Eddington factor $f_\nu(E_\nu)$ is now expressed as

$$f_\nu = \lambda_\nu + \lambda_\nu^2 R_\nu^2, \tag{20.130}$$

where

$$R_\nu \equiv \frac{|\nabla E_\nu|}{\kappa_\nu \rho E_\nu} \tag{20.131}$$

is an optical depth parameter, since $R_\nu \sim 1/\tau$, and is also determined by the local quantities.

If we know an appropriate form of λ_ν, radiation flux $\boldsymbol{F}_\nu$ can be determined by the local quantities. As a choice of λ_ν, Levermore and Pomraning (1981) proposed a relation,

$$\lambda_\nu = \frac{2 + R_\nu}{6 + 3R_\nu + R_\nu^2}, \tag{20.132}$$

although many other choices are possible, which preserve causality and are consistent with the assumption of smoothness in the radiation field.[17]

In the optically thick limit ($R_\nu \to 0$), we find $\lambda_\nu \to 1/3$ and $f_\nu \to 1/3$. In the optically thin limit ($R_\nu \to \infty$), we have $\lambda_\nu \to 1/R_\nu$ and $f_\nu \to 1 + 1/R_\nu$.

The flux-limited diffusion approximation is often used in radiation hydrodynamical simulations, since radiative quantities can be determined in terms of the local information. However, there exist various defects and problems in the FLD approximation. For example, the flux $\boldsymbol{F}$ is not generally parallel to ∇E in the multi-dimensional system. Furthermore, the FLD approximation does not reduce to the equation for the electromagnetic wave in the optically thin limit.

[17]For example, we quote two of them (Castor 2004):

$$\lambda_\nu = \frac{3}{3 + R_\nu}, \quad \lambda_\nu = \frac{1}{R_\nu}\left(\coth R_\nu - \frac{1}{R_\nu}\right).$$

20.3.3.5 M1 Closure

In M1 closure, the Eddington tensor f_ν^{ij} [cf. Eq. (20.127)] is expressed as

$$f_\nu^{ij} = \frac{1-f_\nu}{2}\delta^{ij} + \frac{3f_\nu - 1}{2}n^i n^j . \tag{20.133}$$

In contrast to the FLD approximation, however, the direction of unit vector is

$$n^i \equiv \frac{\boldsymbol{F}}{F}, \tag{20.134}$$

and the Eddington factor is expressed as follows (Levermore 1984):

$$f_\nu = \frac{3+4\xi^2}{5+2\sqrt{4-3\xi^2}}, \quad \xi \equiv \frac{|\boldsymbol{F}|}{cE}. \tag{20.135}$$

In the optically thick limit $\xi \to 0$ and $f_\nu \to 1/3$, whereas in the optically thin limit $\xi_\nu \to 1$ and $f_\nu \to 1$.

20.3.3.6 Optically Thick to Thin Regimes

Characteristics of the radiation field strongly depend on the optical depth. We briefly summarize them from optically thick to thin regimes.

Let us suppose an optically thick gaseous object in vacuum (Fig. 20.12). We do not consider any motion or flow for simplicity. From the point of view of radiative transfer, the space is divided into four regions; the diffusion region, thermalization one, isotropization one, and free streaming one, from optically thick interior to optically thin exterior.

Diffusion region: In the sufficiently optically thick deep interior, radiation and matter are in local thermodynamic equilibrium (LTE). Radiation is isotropic and blackbody, and the diffusion approximation holds. In this diffusion interior, the photon diffusion takes place almost isotropically, but globally the photon diffusively flows toward the anti-direction of the gradient of photon density.

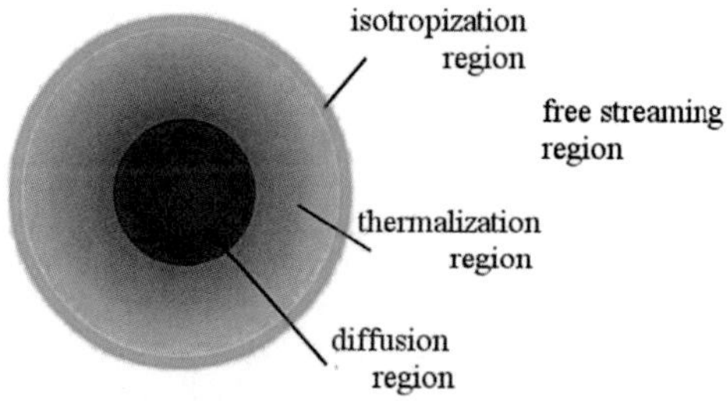

Fig. 20.12 Various regions around an optically thick gaseous cloud

Thermalization region: In the region surroundings of the diffusion region, radiation and matter are approximately in LTE, but the diffusion approximation becomes less valid. If there is no scattering, the optical depth of this thermalization region is on the order of unity. If, however, scattering works, the total optical depth of the thermalization region becomes larger than unity. In terms of the photon destruction probability ε_ν, the optical depth of the thermalization region Λ_ν, which is called *thermalization length*, is expressed as follows[18] :

$$\Lambda_\nu = \frac{1}{\sqrt{3\varepsilon_\nu}}. \tag{20.136}$$

Isotropization region: In the surface layer whose total optical depth is on the order of unity, the diffusion approximation violates, and radiation deviates from blackbody. However, the radiation field is nearly isotropic, since the thermal emission and scattering take place isotropically. Hence, in this surface *isotropization layer*, the Eddington approximation nearly holds, although the diffusion one violates. Since the spectral properties observed outside are determined at the surface layer, the radiation field near to the surface is important.

Free streaming region: Finally, in the optically thin outer exterior, the mean free path of photons is sufficiently long, and photons propagate toward the anti-direction of the source, as free streams, without absorption and scattering. In this free streaming region, $F_\nu = cE_\nu = cP_\nu$ (i.e., $f = 1$).

References

Anile, A.M., Romano, V.: Astrophys. J. **386**, 325 (1992)
Castor, J.I.: Radiation Hydrodynamics. Cambridge University Press, Cambridge (2004)
Chandrasekhar, S.: Radiative Transfer. Dover, New York (1960)
Gray, D.F.: The Observation and Analysis of Stellar Photospheres. Cambridge University Press, Cambridge (1992/2005)
Hubeny, I., Mihalas, D.: Theory of Stellar Atmosphere. Princeton University Press, Princeton (2014)
Hummer, D.G., Rybicki, G.B.: Monthly Not. Roy. Astron. Soc. **152**, 1 (1971)
Levermore, C.D.: J. Quant. Spectro. Radiat. Transfer. **31**, 149 (1984)
Levermore, C.D., Pomraning, G.C.: Astrophys. J. **248**, 321 (1981)
Melia, F., Zylstra, G.J.: Astrophys. J. **374**, 731 (1991)
Mihalas, D.: Stellar Atmospheres. W.H. Freeman, San Francisco (1970)

[18]From the definition of the optical depth and effective optical depth, roughly speaking, we have

$$\tau_\nu^* = \int \sqrt{\kappa_\nu(\kappa_\nu + \sigma_\nu)}\rho ds = \int \sqrt{\varepsilon_\nu}(\kappa_\nu + \sigma_\nu)\rho ds \sim \sqrt{\varepsilon_\nu}\tau_\nu \sim \frac{1}{\Lambda_\nu}\tau_\nu.$$

Here, a factor 3 originates from the Eddington factor. The total optical depth of the thermalization region is about $\tau_\nu \sim \Lambda_\nu$, whereas the effective optical depth is on the order of unity.

Mihalas, D., Mihalas, B.W.: Foundations of Radiation Hydrodynamics. Oxford University Press, Oxford (1984)
Morse, P.M.: Astrophys. J. **92**, 27 (1940)
Peraiah, A.: An Introduction to Radiative Transfer: Methods and applications in Astrophysics. Cambridge University Press, Cambridge (2002)
Pomraning, G.C.: Astrophys. J. **266**, 841 (1983)
Rybicki, G.B., Lightman A.P.: Radiative Processes in Astrophysics. Wiley, New York (1979)
Schwarzschild, M.: Structure and Evolution of the Stars. Dover, New York (1958)
Shu, F.H.: The Physics of Astrophysics Vol.1: Radiation. University Science Books, California (1991)
Sobolev, V.V.: Light Scattering in Planetary Atmospheres. Pergamon Press, Oxford (1975)
Tamazawa, S., Toyama, K., Kaneko, N., Ono, Y.: Astrophys. Space Sci. **32**, 403 (1975)
Thomas, G.E., Stamnes, K.: Radiative Transfer in the Atmosphere and Ocean. Cambridge University Press, Cambridge (1999)
Thorne, K.S.: Monthly Not. Roy. Astron. Soc. **194**, 439 (1981)
Turner, N.J., Stone, J.M.: Astrophys. J. Suppl. **135**, 95 (2001)
Wendish, M., Yang, P.: Theory of Atmospheric Radiative Transfer—A Comprehensive Introduction. Wiley, Weinheim (2012)

Chapter 21
Basic Equations for Radiation Hydrodynamics

Matter and radiation exchange momentum via radiative force, and exchange energy via radiative heating and cooling. The radiative force generally pushes a material to lift up or move it against, e.g., gravity, and affects the dynamics of matter. The radiative heating and cooling also change the structure and dynamics of gaseous astronomical objects. On the other hand, matter, whose dynamical and thermal states are changed, would influence on radiation. The method to simultaneously treat matter and radiation is *radiation hydrodynamics*. Radiation hydrodynamics is important in many astrophysical situations; in the cool low energy regime such as an interstellar gas, in the intermediate regime like stellar and planetary atmospheres, and in the hot high energy regime around, e.g., black-hole environments and at the early universe. In this chapter, using the results of Part I and Chap. 20, we couple the basic equations for matter and radiation step by step. In addition, we briefly discuss several stages of approximations and classification by the strength of coupling between matter and radiation.

21.1 Radiation Hydrodynamical Equations

Radiative actions on matter in the basic equations of radiation, the source terms on the right-hand side of moment equations, roughly appear as anti-actions in the basic equations of matter, and vice versa. These combinations between matter and radiation make a complicated but interesting world of *radiation hydrodynamics*.

The theory of radiation hydrodynamics has been developed together with the radiative transfer theory mainly in astrophysics. In this section we derive and summarize the basic equations of radiation hydrodynamics. More general and detailed treatments can be found in many excellent textbooks (e.g., Chandrasekhar 1960; Mihalas 1970; Sobolev 1975; Rybicki and Lightman 1979; Mihalas and Mihalas 1984; Shu 1991; Gray 1992, 2005; Thomas and Stamnes 1999; Peraiah 2002; Castor 2004; Wendish and Yang 2012; Hubeny and Mihalas 2014).

© Springer Nature Singapore Pte Ltd. 2020, corrected publication 2023

S. Kato, J. Fukue, *Fundamentals of Astrophysical Fluid Dynamics*,
Astronomy and Astrophysics Library,
https://doi.org/10.1007/978-981-15-4174-2_21

21.1.1 Single-Fluid Approximation

In Part I, we consider matter and radiation as a single fluid. Let us first start from this single-fluid approximation (Sect. 2.2). In Part I, p generally means the total pressure ($= p_{\rm gas} + p_{\rm rad}$), while the gas and radiation pressures are denoted by $p_{\rm gas}$ and $p_{\rm rad}$, respectively, if necessary. In this Part III, on the other hand, the gas and radiation pressures are denoted by p and P, respectively, since we introduce the radiation stress tensor by P^{ik}.

The continuity equation, [Eq. (2.2)], is

$$\frac{\partial \rho}{\partial t} + \nabla(\rho \boldsymbol{v}) = 0, \tag{21.1}$$

where ρ is the gas density, and $\boldsymbol{v}$ the flow velocity. The equation of motion, [Eq. (2.4)], is

$$\rho \left[\frac{\partial \boldsymbol{v}}{\partial t} + (\boldsymbol{v} \cdot \nabla)\, \boldsymbol{v} \right] = -\rho \nabla \psi - \nabla (p + P), \tag{21.2}$$

where ψ is the gravitational potential, and other general force $\boldsymbol{N}$ is omitted. The total pressure p in Part I is replaced by $p + P$ here. Finally, the energy equation, [Eq. (2.7)], is

$$\rho \left(\frac{\partial}{\partial t} + \boldsymbol{v} \cdot \nabla \right) \left(U + \frac{E}{\rho} \right) + (p + P)\, \nabla \boldsymbol{v} = q^+ - \nabla \boldsymbol{F}, \tag{21.3}$$

where q^+ the (viscous or nuclear) heating rate per unit volume, $\boldsymbol{F}$ the energy flux. The internal energy per unit mass, U, in Part I is also replaced by the sum of U and E/ρ, where E is the radiation energy per unit volume. This energy equation (21.3) is expressed in the conservative form:

$$\frac{\partial}{\partial t}(\rho U + E) + \nabla[(\rho U + E)\boldsymbol{v}] + (p + P)\nabla \boldsymbol{v} = q^+ - \nabla \boldsymbol{F}. \tag{21.4}$$

Combining the continuity equation, equation of motion, and energy equation, we have the expression in the form of the total energy conservation [Eq. (2.16)]:

$$\frac{\partial}{\partial t} \left[\rho \left(\frac{v^2}{2} + U + \psi \right) + E \right] + \mathrm{div} \left[\rho \boldsymbol{v} \left(\frac{v^2}{2} + \frac{\rho U + E}{\rho} + \frac{p + P}{\rho} + \psi \right) + \boldsymbol{F} \right] = q^+. \tag{21.5}$$

Here, we again note that p and ρU in Eq. (2.16) are replaced by $(p + P)$ and $(\rho U + E)$, respectively. In addition, the internal heating term q^+ on the right-hand side is retained.

It should be noted that when matter-radiation mixture behaves almost adiabatically as a single fluid, the relation between the total pressure $(p + P)$ and matter density ρ is described by the generalized adiabatic index Γ_i (Sect. 19.6).

21.1.2 Two-Fluid Approximation: $\mathcal{O}(v/c)^0$ Order

Next, we consider the usually adopted basic equations for radiation hydrodynamics in the non-relativistic regime.

For matter, the continuity equation is not changed due to the existence of the radiation fields:

$$\frac{\partial \rho}{\partial t} + \nabla(\rho \boldsymbol{v}) = 0. \tag{21.6}$$

Including the radiative force, under the present non-relativistic approximation, we can write the equation of motion for matter as

$$\frac{\partial \boldsymbol{v}}{\partial t} + (\boldsymbol{v} \cdot \nabla)\, \boldsymbol{v} = -\nabla \psi - \frac{1}{\rho} \nabla p + \frac{(\kappa_F + \sigma_F)}{c} \boldsymbol{F}, \tag{21.7}$$

where κ_F and σ_F are the flux-mean absorption and scattering opacities, respectively, and $\boldsymbol{F}$ the radiative flux. The third term on the right-hand side is the radiative force per unit mass. Matter gains momentum via both the absorption and scattering of radiation. Within this non-relativistic regime on the order of $\mathcal{O}(v/c)^0$, the radiation drag does not appear (cf. Chap. 23), although the radiation drag term is necessary to complete the formulation, as shown in the next subsection. The radiative viscosity (cf. Chap. 19) is usually ignored.

If the radiation field can be treated to be steady and isotropic, using the first-moment Eq. (20.95) [see also Eq. (21.16) below], we can rewrite Eq. (21.7) as

$$\frac{\partial \boldsymbol{v}}{\partial t} + (\boldsymbol{v} \cdot \nabla)\, \boldsymbol{v} = -\nabla \psi - \frac{1}{\rho} \nabla (p + P), \tag{21.8}$$

where P is the isotropic radiation pressure. This equation is just that of the single-fluid approximation, Eq. (21.2). In other words, in the single-fluid approximation we implicitly assume that the radiation field is steady and isotropic.

Including the net energy transfer rate, under the present approximation, we can write the energy equation for matter as

$$\left(\frac{\partial}{\partial t} + \boldsymbol{v} \cdot \nabla \right) U + \frac{p}{\rho} \nabla \boldsymbol{v} = \frac{1}{\rho} q^+ - (j - c\kappa_E E), \tag{21.9}$$

where U is the internal energy per unit mass, and q^+ the (viscous or nuclear) heating rate per unit volume. The second term on the right-hand side is the radiative energy exchange; $-j$ is the radiative cooling per unit mass, and $c\kappa_E E$ the radiative heating rate per unit mass via the various absorption process by matter, κ_E being the energy-mean absorption opacity. This energy equation (21.9) is expressed in the conservative form:

$$\frac{\partial}{\partial t}(\rho U) + \mathrm{div}(\rho U \boldsymbol{v}) + p \;\mathrm{div}\; \boldsymbol{v} = q^+ - \rho(j - c\kappa_E E). \tag{21.10}$$

Using the zeroth moment Eq. (20.94) [see also Eq. (21.12) below], we can rewrite Eq. (21.10) as

$$\frac{\partial}{\partial t}(\rho U + E) + \mathrm{div}(\rho U \boldsymbol{v}) + p \;\mathrm{div}\; \boldsymbol{v} = q^+ - \nabla \boldsymbol{F}. \tag{21.11}$$

This equation does not reproduce the conservative form of the energy equation (21.4) in the single-fluid approximation.

As the basic equations for radiation in radiation hydrodynamics, we often adopt the moment equations with a closure relation such as the Eddington approximation.[1] Using the results of Chap. 20 (Sect. 20.3.1), the zeroth and first moment equations, and the Eddington approximation are, respectively,

$$\frac{\partial E}{\partial t} + \nabla \boldsymbol{F} = \rho\,(j - c\kappa_E E)\,, \tag{21.12}$$

$$\frac{1}{c^2}\frac{\partial \boldsymbol{F}}{\partial t} + \frac{\partial P^{ik}}{\partial x^k} = -\frac{1}{c}\rho(\kappa + \sigma)_F \boldsymbol{F}, \tag{21.13}$$

$$P^{ik} = \frac{\delta^{ik}}{3} E, \tag{21.14}$$

where E is the radiation energy density, $\boldsymbol{F}$ the radiative flux, P^{ik} the radiation stress tensor, j the mass emissivity in all direction.[2] Substituting the Eddington approximation (21.14) into Eq. (21.13), we have

$$\frac{1}{c^2}\frac{\partial \boldsymbol{F}}{\partial t} + \frac{1}{3}\nabla E = -\frac{1}{c}\rho(\kappa + \sigma)_F \boldsymbol{F}, \tag{21.15}$$

[1] In the regime where the Eddington approximation becomes worse, we adopt another closure relations, or directly solve the radiative transfer equation:

$$\frac{1}{c}\frac{\partial I}{\partial t} + (\boldsymbol{l}\cdot\nabla)\, I = \rho\left[\frac{j}{4\pi} - (\kappa + \sigma)_I I + \sigma_E \frac{c}{4\pi} E\right].$$

[2] In the local thermodynamical equilibrium (LTE), $j = 4\pi\kappa B$, where B is the blackbody intensity from matter. The absorption opacities are usually replaced by the Rosseland mean opacity $\bar{\kappa}_{\mathrm{R}}$, and the scattering one is usually a constant electron scattering opacity κ_{es}.

whereas if we merely assume that the radiation field is isotropic, then we have

$$\frac{1}{c^2}\frac{\partial \boldsymbol{F}}{\partial t} + \nabla P = -\frac{1}{c}\rho(\kappa+\sigma)_F \boldsymbol{F}, \tag{21.16}$$

where P is the (isotropic) radiation pressure.

These Eqs. (21.6)–(21.16) are often used as the basic equations for radiation hydrodynamics in the non-relativistic regime on the order of $\mathcal{O}(v/c)^0$, where the terms on the order of $\mathcal{O}(v/c)^1$ do not considered. However, such an omission gives some contradiction, as already appeared in the energy equation above.

Similar to the single-fluid case, let us combine these equations to derive the total energy conservation, which is now in the form of

$$\frac{\partial}{\partial t}\left[\rho\left(\frac{\boldsymbol{v}^2}{2}+U+\psi\right)+E\right]$$
$$+\operatorname{div}\left[\rho\boldsymbol{v}\left(\frac{\boldsymbol{v}^2}{2}+U+\frac{p}{\rho}+\psi\right)+\boldsymbol{F}\right] = q^+ + \rho\frac{(\kappa+\sigma)_F}{c}\boldsymbol{F}\cdot\boldsymbol{v}. \tag{21.17}$$

Compared with the single-fluid expression (21.5), we easily see that the terms relating to radiation energy E and pressure P (advectively driven radiation fields) are missing in the second term on the left-hand side. Furthermore, on the right-hand side, there appears an additional term on the order of $\mathcal{O}(v/c)^1$, $\rho(\kappa+\sigma)_F \boldsymbol{F}\cdot\boldsymbol{v}/c$.

The latter additional term expresses the interaction between radiation and *moving* matter. Namely, in the zeroth moment Eq. (21.12) we consider only the radiative heating and cooling, and often overlook the dynamical effect, since we implicitly assume the radiative transfer problem in the static media. If matter moves at speed $\boldsymbol{v}$, however, radiation can work on matter or vice versa. Similarly, there exists the momentum exchange between radiation and matter due to the dynamical effect, in addition to that due to the direct flux $\boldsymbol{F}$.

Moreover, as will be shown later, the former missing terms on E and P can be reproduced, if we consider the difference between the comoving and inertial fluxes (radiation drag), which is again on the order of $\mathcal{O}(v/c)^1$.

In the current radiation hydrodynamics we usually omit the terms on the order of $\mathcal{O}(v/c)^1$ in the basic equations under the non-relativistic treatment. However, we must retain these terms on the order of $\mathcal{O}(v/c)^1$ to treat the dynamical problem between radiation and matter, even in the non-relativistic regime.

21.1.3 Two-Fluid Approximation: $\mathcal{O}(v/c)^1$ Order

Now, we shall show the complete set of basic equations for radiation hydrodynamics in the non-relativistic regime, but on the order of $\mathcal{O}(v/c)^1$. In order to do so, we must collect the terms on the order of $\mathcal{O}(v/c)^1$, or start from the basic equations for radiation hydrodynamics in the subrelativistic regime on the order of $\mathcal{O}(v/c)^1$ (Sect. 25.1.1).

For matter, the continuity equation is not changed, and we repeat it for completeness:

$$\frac{\partial \rho}{\partial t} + \nabla(\rho \boldsymbol{v}) = 0. \tag{21.18}$$

The equation of motion for matter in the subrelativistic regime is [see Eq. (25.23)]

$$\frac{\partial \boldsymbol{v}}{\partial t} + (\boldsymbol{v} \cdot \nabla)\, \boldsymbol{v} = -\nabla \psi - \frac{1}{\rho} \nabla p + \frac{(\kappa_F + \sigma_F)}{c} \left(\boldsymbol{F} - \boldsymbol{v} E - v_k P^{ik} \right). \tag{21.19}$$

The third term on the right-hand side is the radiative force per unit mass, including the radiation drag term (Chap. 25). It should be noted that $\boldsymbol{F}$ is the radiative flux measured in the inertial frame, while $\boldsymbol{F}_{\rm co}$ ($\equiv \boldsymbol{F} - \boldsymbol{v} E - v_k P^{ik}$) is that measured in the comoving frame [see Eq. (24.37)]. This means that in the frame comoving with matter there exists only the comoving flux, which affects the dynamics of matter.

Including the net energy transfer rate and a subrelativistic correction term, we can write the energy equation for matter in the conservative form as

$$\frac{\partial}{\partial t}(\rho U) + {\rm div}(\rho U \boldsymbol{v}) + p \,{\rm div}\, \boldsymbol{v} = q^+ - \rho(j_{\rm co} - c\kappa_E E) - \rho \frac{\kappa_F}{c} 2\boldsymbol{v} \cdot \boldsymbol{F}. \tag{21.20}$$

Here, $j_{\rm co}$ is the mass emissivity, the third term on the right-hand side is the subrelativistic collection one. Originally, the sum of the second and third terms on the right-hand side is the net energy exchange between radiation and matter in the *comoving* frame, $-\rho(j_{\rm co} - c\kappa_E E_{\rm co})$, where $E_{\rm co}$ is the radiation energy density measured in the comoving frame [see Eq. (24.36)]. When we transform the comoving energy density $E_{\rm co}$ to the inertial one E, due to the aberrational effect the third term on the right-hand side appears, in addition to the second term.

For an ideal gas, the internal energy is expressed as $U = [1/(\gamma - 1)](p/\rho)$, and the energy equation (21.20) is rewritten as

$$\frac{1}{\gamma - 1} \left(\frac{dp}{dt} - \gamma \frac{p}{\rho} \frac{d\rho}{dt} \right) = q^+ - \rho(j_{\rm co} - c\kappa_E E) - \rho \frac{\kappa_E}{c} 2\boldsymbol{v} \cdot \boldsymbol{F}. \tag{21.21}$$

When the radiation field is assumed to be steady, using the zeroth moment equation for radiation below [Eq. (21.23)], we can express the energy equation in terms of the radiative flux:

$$\frac{1}{\gamma - 1} \left(\frac{dp}{dt} - \gamma \frac{p}{\rho} \frac{d\rho}{dt} \right) = q^+ - \nabla \cdot \boldsymbol{F} - \rho \frac{\kappa_F + \sigma_F}{c} \boldsymbol{v} \cdot \boldsymbol{F}. \tag{21.22}$$

This shows that the third correction term on the right-hand side means the *work done by radiation on matter* when matter moves at speed $\boldsymbol{v}$.

For radiation, we again adopt the moment equations, but include the terms on the order of $\mathcal{O}(v/c)^1$ here [see Eq. (25.27)]. The zeroth moment equation is then

$$\frac{\partial E}{\partial t} + \nabla \boldsymbol{F} = \rho\, (j_{\rm co} - c\kappa_E E) + \rho \frac{2\kappa_E - \kappa_F - \sigma_F}{c} \boldsymbol{v} \cdot \boldsymbol{F}. \tag{21.23}$$

Here, the second term on the right-hand side is the subrelativistic correction one. Originally, the terms on the right-hand side are the sum of the net energy exchange in the comoving frame, $\rho(j - c\kappa_E E_{\rm co})$, and the *work done by radiation on matter* in the comoving frame, $\rho(\kappa_F + \sigma_F)\boldsymbol{v} \cdot \boldsymbol{F}_{\rm co}/c$. When we transform the comoving quantities to the inertial ones, due to the aberrational effect on $E_{\rm co}$, we have the above form. This means that the net energy exchange takes place in the comoving frame. In addition, if matter moves at speed $\boldsymbol{v}$, there appears the work done by radiation through interaction between the radiative flux $\boldsymbol{F}$ and matter at speed $\boldsymbol{v}$.

The first-moment equation [see Eq. (25.28)] is

$$\frac{1}{c^2}\frac{\partial \boldsymbol{F}}{\partial t} + \frac{\partial P^{ik}}{\partial x^k} = -\frac{1}{c}\rho(\kappa + \sigma)_F\left(\boldsymbol{F} - E\boldsymbol{v} - v_k P^{ik}\right) + \frac{\rho}{c}\left(j_{\rm co} - c\kappa_E E\right)\frac{\boldsymbol{v}}{c}. \tag{21.24}$$

Here, the radiation drag force in the first term and the second term on the right-hand side are the subrelativistic correction ones. Originally, the first term on the right-hand side is the momentum loss in the comoving frame, $\rho(\kappa_F + \sigma_F)\boldsymbol{F}_{\rm co}/c$. In other words, this means that the momentum loss takes place in the comoving frame. In addition, the second term means that there exists the momentum exchange associating with the energy gain and loss, when matter moves at speed $\boldsymbol{v}$.

Now, combining the continuity equation (21.18), equation of motion for matter (21.19), the energy equation for matter (21.20), and the zeroth moment equation for radiation (21.23), we have the expression in the form of the total energy conservation[3] :

$$\frac{\partial}{\partial t}\left[\rho\left(\frac{\boldsymbol{v}^2}{2} + U + \psi\right) + E\right] + \mathrm{div}\left[\rho\boldsymbol{v}\left(\frac{\boldsymbol{v}^2}{2} + U + \frac{p}{\rho} + \psi\right) + \boldsymbol{F}\right] = q^+. \tag{21.25}$$

[3] At the first step, multiplying equation (21.19) by $\boldsymbol{v}$, and using the continuity equation (21.18), we have the kinetic energy conservation equation in a form:

$$\frac{\partial}{\partial}\left[\rho\left(\frac{v^2}{2} + \psi\right)\right] + \nabla\left[\rho\boldsymbol{v}\left(\frac{v^2}{2} + \psi\right)\right] + \boldsymbol{v}\cdot\nabla p = \rho\frac{\kappa_F + \sigma_F}{c}\boldsymbol{F}_{\rm co}\cdot\boldsymbol{v},$$

At the next step, we sum up this kinetic energy conservation, the thermal energy one (21.20), and the radiative energy one (21.23), we finally obtain Eq. (21.26).

Furthermore, if we express the inertial flux $\boldsymbol{F}$ in terms of the comoving flux $\boldsymbol{F}_{\text{co}}$ $[\boldsymbol{F} = \boldsymbol{F}_{\text{co}} + \boldsymbol{v}(E + P)]$, we can rewrite Eq. (21.25) as

$$\frac{\partial}{\partial t}\left[\rho\left(\frac{v^2}{2} + U + \psi\right) + E\right] + \text{div}\left[\rho\boldsymbol{v}\left(\frac{v^2}{2} + \frac{\rho U + E}{\rho} + \frac{p + P}{\rho} + \psi\right) + \boldsymbol{F}_{\text{co}}\right] = q^+. \quad (21.26)$$

Ultimately, this expression (21.26) coincides with the expression (21.5) under the single-fluid approximation, except that the radiative flux is not the inertial one $\boldsymbol{F}$, but rigorously the diffusion flux $\boldsymbol{F}_{\text{co}}$ in the comoving frame.[4]

Thus, Eqs. (21.18)–(21.20), (21.23), (21.24), and some closure relation constitute the complete set of the basic equations for radiation hydrodynamics in the non-relativistic (subrelativistic) regime.[5] In many non-relativistic cases, however, we shall drop the subrelativistic correction terms for simplicity. They become important in the subrelativistic regime, and may play an important role for waves and instabilities of radiative fluids even in the non-relativistic regime.

In summary, in order to derive this total energy conservation, there are two different styles. In the (nonrelativistic) single-fluid style, as in Sect. 2.3, we consider the matter-radiation mixture, which behaves as a single fluid. In this case, as in Sect. 2.2, the internal energy per unit volume, ρU, means the total internal energy per unit volume, $(\rho U + E)$, while the pressure p should be read as the total pressure $(p + P)$. Then, the total energy conservation is straightly derived as in Eq. (21.5). Alternative is the (relativistic) two-fluids style (Sect. 24.3), where matter and radiation are not strongly coupled, and behave as two different fluids. In this case, the internal energy per unit volume, ρU, in Eq. (21.20) is just that for matter, while the pressure p is merely the gas pressure, and do not include the radiation one. In somewhat complicated manner, we collect all the subrelativistic correction terms on the order of $\mathcal{O}(v/c)^1$, combine all the relevant equations, transform the quantities between the inertial and comoving frames, we can recover the total energy conservation as in Eq. (21.26).

---

[4]On the contrary, the term $\boldsymbol{v}(E + P)$ in the second term on the left-hand side is the radiation energy carried with matter due to its motion, and called an *advection flux* (see Sects. 22.3.3, 24.3.1 later).

[5]It should be noted that these equations hold in the *equilibrium diffusion approximation*, where radiation and matter sufficiently equilibrate, and radiation temperature T_{rad} equals matter one T_{gas}; $T_{\text{rad}} = T_{\text{gas}} = T$ (see Sect. 21.2 below). When the medium is sufficiently optically thick, such as a stellar interior, this approximation is valid. In other words, when the medium is not sufficiently optically thick, or moving, or time-dependent, such as an accretion disk atmosphere, astrophysical winds, accretion flows, waves and instabilities, this approximation would be worse or violated. In such a case, we should use Eq. (21.20) with the LTE condition ($j = 4\pi\kappa B$), and treat the gas temperature T_{gas} and radiation one T_{rad}, separately.

21.2 Classification by the Coupling Strength

In this section, we briefly discuss the criterion of the equilibrium diffusion approximation, and classify several stages of approximations via the strength of coupling between matter and radiation, by comparing several timescales.

At first, we estimate the *equilibrium timescale* $t_{\rm eq}$ within which matter and radiation equilibrate and radiation temperature $T_{\rm rad}$ equals matter one $T_{\rm gas}$. Comparing $\partial E/\partial t$ ($\sim E/t_{\rm eq}$) and ρj ($= \rho 4\pi\kappa B$; LTE) in Eq. (21.23), we have

$$t_{\rm eq} \sim \frac{E}{4\pi\rho\kappa B} = \frac{1}{\rho\kappa c}\frac{J}{B} = \frac{L}{\tau c}\frac{J}{B}, \tag{21.27}$$

where L is the typical scale of the system, J the mean intensity, and τ ($\equiv \rho\kappa L$) the typical optical depth of the system. Alternatively, comparing $\partial U/\partial t$ ($\sim U/t_{\rm eq}$) and $c\kappa E$ in Eq. (21.20), we have also

$$t_{\rm eq} \sim \frac{U}{c\kappa E} = \frac{e}{c\rho\kappa E} = \frac{L}{\tau c}\frac{e}{E}, \tag{21.28}$$

where $e = \rho U$ is the internal energy of matter per unit volume. When the optical depth is sufficiently large, the equilibrium timescale is generally short, as long as $J \sim B$ and $e \sim E$.

We next estimate the *diffusion timescale* $t_{\rm diff}$ within which the thermal state of matter is changed due to the radiative diffusion. Comparing the change of the internal energy on the left-hand side of Eq. (21.22), $\rho U/t_{\rm diff}$, and the diffusion term on the right-hand side of Eq. (21.22), F/L ($\sim cE/\kappa\rho L^2$), we have

$$t_{\rm diff} \sim \frac{\rho U}{F/L} = \frac{eL^2\rho\kappa}{cE} = L\frac{\tau}{c}\frac{e}{E}. \tag{21.29}$$

When the optical depth is sufficiently large, the diffusion timescale is rather long, as long as $e \sim E$.

We finally estimate the *dynamical timescale* $t_{\rm dyn}$ within which the matter state is changed due to the fluid motion, such as accretion and wind. The dynamical timescale is given by

$$t_{\rm dyn} \sim \frac{L}{v}, \tag{21.30}$$

where v is the typical velocity of matter. The timescale of the change of the thermal state due to the adiabatic expansion or compression is also the dynamical timescale.

Now, we compare these timescales and classify several stages of approximation by the strength of coupling between matter and radiation.

In the sufficiently optically thick media with motion, when the equilibrium timescale is sufficiently shorter than the dynamical one ($t_{\rm eq} < t_{\rm dyn}$), and the diffusion timescale is sufficiently longer than the dynamical one ($t_{\rm diff} > t_{\rm dyn}$), matter and radiation strongly couple at the same temperature ($T_{\rm rad} = T_{\rm gas} = T$), and the radiative diffusion can be ignored. As a result, matter and radiation behave as a single fluid. This is a *strong equilibrium regime*. From $t_{\rm eq} < t_{\rm dyn} < t_{\rm diff}$, the criterion for the strong equilibrium regime is

$$\frac{1}{\tau}\frac{e}{E}, \frac{1}{\tau}\frac{J}{B} < \frac{c}{v} < \tau\frac{e}{E}. \tag{21.31}$$

For usual situations of $e \sim E$ and $J \sim B$, this relation reduces to

$$\frac{1}{\tau} < \frac{c}{v} < \tau. \quad \text{(strong equlibrium regime)} \tag{21.32}$$

Even if matter and radiation couple and equilibrate at the same temperature ($T_{\rm rad} = T_{\rm gas} = T$), when the diffusion timescale is shorter than the dynamical one ($t_{\rm diff} < t_{\rm dyn}$), the radiative diffusion becomes important. This is the *equilibrium diffusion regime/approximation*. From $t_{\rm eq} < t_{\rm diff} < t_{\rm dyn}$, the criterion for the equilibrium diffusion regime is

$$\frac{1}{\tau}\frac{e}{E}, \frac{1}{\tau}\frac{J}{B} < \tau\frac{e}{E} < \frac{c}{v}. \tag{21.33}$$

For usual situations of $e \sim E$ and $J \sim B$, this relation reduces to

$$\frac{1}{\tau} < \tau < \frac{c}{v}. \quad \text{(equlibrium diffusion regime)} \tag{21.34}$$

Finally, even if the system is sufficiently optically thick, when the equilibrium timescale is longer than the dynamical one ($t_{\rm eq} > t_{\rm dyn}$), there is not enough time for matter and radiation to equilibrate, and the radiation temperature is not equal to the gas one ($T_{\rm rad} \neq T_{\rm gas}$). In this regime, the radiative diffusion operates, and this is a *nonequilibrium diffusion regime*. From $t_{\rm dyn} < t_{\rm eq}$, the criterion for the nonequilibrium diffusion regime is

$$\frac{c}{v} < \frac{1}{\tau}\frac{e}{E}, \frac{1}{\tau}\frac{J}{B}. \tag{21.35}$$

For usual situations of $e \sim E$ and $J \sim B$, this relation reduces to

$$\frac{c}{v} < \frac{1}{\tau}. \quad \text{(nonequlibrium diffusion regime)} \tag{21.36}$$

When the flow is dynamically changed, and the internal energy of matter remarkably differs from that of radiation, e.g., in the shock-heated gas, matter and radiation becomes in the nonequilibrium state.

In the optically thin medium, of course, matter and radiation decouple, and are in the nonequilibrium state.

References

Castor, J.I.: Radiation Hydrodynamics. Cambridge University Press, Cambridge (2004)
Chandrasekhar, S.: Radiative Transfer. Dover, New York (1960)
Gray, D.F.: The Observation and Analysis of Stellar Photospheres. Cambridge University Press, Cambridge (1992/2005)
Hubeny, I., Mihalas, D.: Theory of Stellar Atmosphere. Princeton University Press, Princeton (2014)
Mihalas, D.: Stellar Atmospheres. W.H. Freeman, San Francisco (1970)
Mihalas, D., Mihalas, B.W.: Foundations of Radiation Hydrodynamics. Oxford University Press, Oxford (1984)
Peraiah, A.: An Introduction to Radiative Transfer: Methods and Applications in Astrophysics. Cambridge University Press, Cambridge (2002)
Rybicki, G.B., Lightman A.P.: Radiative Processes in Astrophysics. Wiley, New York (1979)
Shu, F.H.: The Physics of Astrophysics Vol.1: Radiation. University Science Books, California (1991)
Sobolev, V.V.: Light Scattering in Planetary Atmospheres. Pergamon Press, Oxford (1975)
Thomas, G.E., Stamnes, K.: Radiative Transfer in the Atmosphere and Ocean. Cambridge University Press, Cambridge (1999)
Wendish, M., Yang, P.: Theory of Atmospheric Radiative Transfer—A Comprehensive Introduction. Wiley, Weinheim (2012)

Chapter 22
Astrophysical RHD Flows

In this chapter we first consider steady radiation hydrodynamical winds in the non-relativistic regime, focusing our attention on the topological nature of critical points. These radiatively driven winds are supposed to operate in nova outbursts, neutron star winds, and black-hole ones. We next briefly show radiation-dominated accretion flows onto a compact object. We also outline a line-driven wind model, which may work in stellar winds from hot stars, in accretion disk winds in cataclysmic variables, in broad absorption line quasars, and in ultra-fast outflows from active galactic nuclei.

22.1 Radiation Hydrodynamical Winds

There exist various spherical outflows in luminous astrophysical systems, such as Wolf-Rayet stars, classical novae, supernovae, X-ray bursters, and black holes. These luminous winds are believed to originate from the very vicinity of the central gravitating object, and to be driven by strong radiation pressure.

Many researchers have examined the mechanisms of radiatively driven hydrodynamical winds under sub/special relativistic treatments (Żytkow 1972; Cassinelli and Hartmann 1975; Ruggles and Bath 1979; Kato 1983; Quinn and Paczyński 1985), or under general relativistic ones (Paczyński and Prószyński 1986; Turolla et al. 1986; Paczyński 1990; Nobili et al. 1994).

In these studies, almost all the researchers adopted the (equilibrium) diffusion approximation, except for Nobili et al. (1994), who used the nonequilibrium diffusion one with a variable Eddington factor depending on the optical depth.[1]

[1]The usage of the diffusion approximation in the moving media is physically questionable, as already stated by Thorne et al. (1981); They stated in their paper on the spherical accretion onto

© Springer Nature Singapore Pte Ltd. 2020, corrected publication 2023

S. Kato, J. Fukue, *Fundamentals of Astrophysical Fluid Dynamics*,
Astronomy and Astrophysics Library,
https://doi.org/10.1007/978-981-15-4174-2_22

In this section, we do examine the critical nature of radiation hydrodynamical winds under the equilibrium and nonequilibrium diffusion approximations (Fukue 2014). As will be stated later, the treatment using the equilibrium diffusion approximation encloses a causal problem in principle. It, however, may describe globally correct solutions, except for those around the critical points. Hence, in this section we show the treatment under the equilibrium diffusion approximation, as well as the nonequilibrium one. Under the nonequilibrium diffusion approximation with the Eddington one, where the diffusive terms do not exist in the source terms, the critical points are saddle type, and transonic solutions are not stiff. Under the diffusion approximation, however, almost all the critical points are nodal type, which are singular points making the transonic solutions stiff.

22.1.1 Basic Equations for Radiatively Driven Winds

We first describe the basic equations for the spherically symmetric, steady wind driven by radiation pressure from the central object of mass M. We restrict to the non-relativistic case with Newtonian gravity, and use the spherical coordinates (r, θ, φ).

For gas, the continuity equation is

$$4\pi r^2 \rho v = \dot{M}, \tag{22.1}$$

where ρ is the density, v the flow velocity, and $\dot{M}$ the constant mass-loss rate.

The equation of motion is

$$v\frac{dv}{dr} = -\frac{1}{\rho}\frac{dp}{dr} - \frac{GM}{r^2} + \frac{\kappa + \sigma}{c}F, \tag{22.2}$$

where p is the gas pressure, κ and σ the frequency-integrated flux-mean opacities for absorption and scattering, respectively, and F the radiative flux (in the fixed frame).

The energy equation for gas is

$$\frac{1}{\gamma - 1}\left(v\frac{dp}{dr} - \gamma\frac{p}{\rho}v\frac{d\rho}{dr}\right) = -\rho(j - c\kappa E), \tag{22.3}$$

where γ is the ratio of specific heats, j the frequency-integrated emissivity, and E the radiation energy density. We do not consider heating source. Furthermore, we do not distinguish between the flux-mean absorption opacity κ_F and the energy-mean one κ_E, and represent them by the same κ.

---

a black hole as "The diffusion approximation is notoriously acausal—it permits thermal pulses to travel faster than the speed of light, and thereby permits transfer of information out of horizon." From the viewpoint on the topological nature of critical points (critical loci), nodal-type critical points often appear in the radiative flows, including the spherical accretion onto a black hole under the diffusion approximation (Flammang 1982).

The equation of state is

$$p = \frac{\mathcal{R}}{\bar{\mu}} \rho T, \tag{22.4}$$

where $\mathcal{R}$ is the gas constant, $\bar{\mu}$ the mean molecular weight, and T the gas temperature.

For radiation, the zeroth moment equation is

$$\frac{1}{r^2}\frac{d}{dr}(r^2 F) = \rho(j - c\kappa E), \tag{22.5}$$

while the first moment equation is

$$\frac{dP}{dr} + \frac{3P - E}{r} = -\rho \frac{\kappa + \sigma}{c} F, \tag{22.6}$$

where P is the radiation pressure.

The Eddington approximation is

$$P = \frac{1}{3} E \equiv f E. \tag{22.7}$$

We use the LTE condition, if necessary,

$$\frac{j}{4\pi} = \kappa B, \tag{22.8}$$

where B is the frequency-integrated blackbody intensity ($B = \sigma_{\mathrm{SB}} T^4/\pi$), σ_{SB} being the Stephan-Boltzmann constant.

Using Eqs. (22.1), (22.2), (22.5), and (22.3), we derive the additional equation, the Bernoulli equation, for the present case:

$$\dot{M}\left(\frac{1}{2}v^2 + \frac{\gamma}{\gamma - 1}\frac{p}{\rho} - \frac{GM}{r}\right) + 4\pi r^2 F = \dot{E} \text{ (const.)}. \tag{22.9}$$

We here use the adiabatic sound speed c_{s} and isothermal one c_{T} defined, respectively, as follows:

$$c_{\mathrm{s}}^2 \equiv \gamma \frac{p}{\rho} = \gamma \frac{\mathcal{R}}{\bar{\mu}} T, \tag{22.10}$$

$$c_{\mathrm{T}}^2 \equiv \frac{p}{\rho} = \frac{\mathcal{R}}{\bar{\mu}} T. \tag{22.11}$$

Bearing in mind winds from a compact star, we introduce the non-dimensional variables,

$$x \equiv \frac{r}{r_{\mathrm{S}}}, \quad \beta \equiv \frac{v}{c}, \quad \alpha_{\mathrm{T}} \equiv \frac{c_{\mathrm{T}}}{c}, \quad \alpha_{\mathrm{s}} \equiv \frac{c_{\mathrm{s}}}{c}, \quad \ell \equiv \frac{L}{L_{\mathrm{E}}}, \tag{22.12}$$

and the non-dimensional parameters,

$$m \equiv \frac{M}{M_\odot}, \quad \dot{m} \equiv \frac{\dot{M}}{\dot{M}_\mathrm{E}}, \quad \dot{e} \equiv \frac{\dot{E}}{\dot{M}_\mathrm{E} c^2}, \tag{22.13}$$

where r_S ($\equiv 2GM/c^2$) is the Schwarzschild radius of the central object, L_E ($\equiv 4\pi cGM/\sigma$) the Eddington luminosity, and $\dot{M}_\mathrm{E}$ ($\equiv L_\mathrm{E}/c^2$) the critical mass-loss rate.[2]

If radiation and matter are strongly coupled with each other, the radiative diffusion can be ignored, and radiation and matter behave as a single fluid. In such a *strong equilibrium limit*, the solutions and structure of the radiation hydrodynamical winds are essentially the same as those of usual gaseous winds, although the adiabatic sound speed of gas is replaced by the sound speed including radiation (Chap. 3, see also Sect. 22.1.3.1). If, however, the radiative diffusion is included, the treatments and critical nature of winds differ from those of usual gaseous winds. In the subsequent sections, we examine two such cases: a *nonequilibrium case*, where radiation and matter do not equilibrate and the radiation temperature is different from the gas one, and an *equilibrium diffusion case*, where radiation and matter equilibrate and have the same temperature.

22.1.2 Nonequilibrium Diffusion Winds

We first examine the radiation hydrodynamical winds under the nonequilibrium diffusion approximation, where radiation and matter do not equilibrate and the radiation temperature is different from the gas one (*nonequilibrium diffusion winds*).

Eliminating the gas pressure gradient term from Eq. (22.2) by using Eqs. (22.3) and (22.1), we obtain the wind equation on the velocity:

$$\left(v^2 - c_\mathrm{s}^2\right)\frac{dv}{dr} = v\left(c_\mathrm{s}^2\frac{2}{r} - \frac{GM}{r^2} + \frac{\kappa+\sigma}{c}F\right) + (\gamma-1)(j - c\kappa E). \tag{22.14}$$

On the other hand, eliminating the gas pressure and density in Eq. (22.3) by using Eqs. (22.4) and (22.1), we have the wind equation on the adiabatic sound speed:

$$\left(v^2 - c_\mathrm{s}^2\right)\frac{dc_\mathrm{s}^2}{dr} = -(\gamma-1)\left[c_\mathrm{s}^2\left(v^2\frac{2}{r} - \frac{GM}{r^2} + \frac{\kappa+\sigma}{c}F\right)\right.$$
$$\left. + \left(\gamma v^2 - c_\mathrm{s}^2\right)\frac{1}{v}(j - c\kappa E)\right]. \tag{22.15}$$

[2] For the values of these non-dimensional parameters, we roughly fix as follows. In the nova wind, for example, the luminosity is almost the Eddington one $\ell_\infty \sim 1$, the wind terminal speed of 10^2–10^3 km s^{-1} roughly corresponds to $\beta_\infty \sim 0.001$–0.01, and the mass-loss rate of 10^{-5}–10^{-3} $M_\odot$ yr^{-1} are roughly $\dot{m} \sim 10^3$–10^5, and therefore, $\dot{e} = \dot{m}(\beta_\infty^2/2) + \ell_\infty \sim 1.05$–$6$. In the neutron star wind, $\ell_\infty \sim 1$, $\beta_\infty \sim 0.01$–0.1, and $\dot{m} \sim 10^3$, and therefore, $\dot{e} = \dot{m}(\beta_\infty^2/2) + \ell_\infty \sim 1.05$–$6$.

We assume LTE (22.8) in this case ($j = 4\pi\kappa B$). In addition, we use the mean intensity J ($= cE/4\pi$) instead of E, introduce the luminosity L ($= 4\pi r^2 F$), and eliminate the density by Eq. (22.1), then Eqs. (22.14), (22.15), (22.5), (22.6), and (22.9) are, respectively, rewritten as follows:

$$\left(v^2 - c_s^2\right)\frac{dv}{dr} = v\left(c_s^2\frac{2}{r} - \frac{GM}{r^2} + \frac{\kappa+\sigma}{c}\frac{L}{4\pi r^2}\right) + (\gamma-1)4\pi\kappa(B-J), \tag{22.16}$$

$$\left(v^2 - c_s^2\right)\frac{dc_s^2}{dr} = -(\gamma-1)\left[c_s^2\left(v^2\frac{2}{r} - \frac{GM}{r^2} + \frac{\kappa+\sigma}{c}\frac{L}{4\pi r^2}\right) + \frac{\gamma v^2 - c_s^2}{v}4\pi\kappa(B-J)\right], \tag{22.17}$$

$$\frac{dL}{dr} = \frac{\dot{M}}{v}4\pi\kappa(B-J), \tag{22.18}$$

$$\frac{d}{dr}\left(f\frac{4\pi}{c}J\right) = -\frac{3f-1}{r}\frac{4\pi}{c}J - \frac{\dot{M}}{4\pi r^2 v}\frac{\kappa+\sigma}{c}\frac{L}{4\pi r^2}, \tag{22.19}$$

$$\dot{E} = \dot{M}\left(\frac{1}{2}v^2 + \frac{1}{\gamma-1}c_s^2 - \frac{GM}{r}\right) + L, \tag{22.20}$$

where $f = 1/3$ for the usual Eddington approximation.

In this nonequilibrium case, the mean intensity J ($\equiv \sigma_{SB}T_{rad}^4/\pi$) generally differs from the blackbody one B ($= \sigma_{SB}T_{gas}^4/\pi$); therefore, the radiation temperature T_{rad}, which is defined in terms of the mean intensity, is also different from the gas one T_{gas}.

Next, we normalize the above set of equations in terms of the non-dimensional variables (22.12) and parameters (22.13). Then, Eqs. (22.16), (22.17), (22.18), (22.19), and (22.20) are, respectively,

$$\left(\beta^2 - \alpha_s^2\right)\frac{d\beta}{dx} = \beta\left(\frac{2\alpha_s^2}{x} - \frac{1}{2x^2} + \frac{\kappa+\sigma}{\sigma}\frac{\ell}{2x^2}\right) + (\gamma-1)\frac{\kappa}{2\sigma}(\mathcal{B}-\mathcal{J}), \tag{22.21}$$

$$\left(\beta^2 - \alpha_s^2\right)\frac{d\alpha_s^2}{dx} = -(\gamma-1)\left[\alpha_s^2\left(\frac{2\beta^2}{x} - \frac{1}{2x^2} + \frac{\kappa+\sigma}{\sigma}\frac{\ell}{2x^2}\right) + \frac{\gamma\beta^2 - \alpha_s^2}{\beta}\frac{\kappa}{2\sigma}(\mathcal{B}-\mathcal{J})\right], \tag{22.22}$$

$$\frac{d\ell}{dx} = \frac{\dot{m}}{\beta}\frac{\kappa}{2\sigma}(\mathcal{B}-\mathcal{J}), \tag{22.23}$$

$$\frac{d\mathcal{J}}{dx} = -\frac{3f-1}{f}\frac{1}{x}\mathcal{J} - \frac{\dot{m}}{2f}\frac{\kappa+\sigma}{\sigma}\frac{\ell}{x^4\beta}, \tag{22.24}$$

$$\dot{e} = \dot{m}\left(\frac{1}{2}\beta^2 + \frac{1}{\gamma-1}\alpha_s^2 - \frac{1}{2x}\right) + \ell. \tag{22.25}$$

Here, $\mathcal{J}$ and $\mathcal{B}$ are Non-dimensional mean and blackbody intensities, respectively,

$$\mathcal{J} \equiv \frac{16\pi^2 r_S^2}{L_E} J, \tag{22.26}$$

$$\mathcal{B} \equiv \frac{16\pi^2 r_S^2}{L_E} B = 1.700 \times 10^{21} \frac{m}{\gamma^4}\alpha_s^8. \tag{22.27}$$

Furthermore, for the Kramers opacity κ ($= 0.64 \times 10^{23}\rho T^{-3.5}\,\mathrm{cm^2\ s^{-1}}$) and electron scattering one σ ($= 0.4\,\mathrm{cm^2\ s^{-1}}$), their ratio is also expressed by the non-dimensional variables as

$$\frac{\kappa}{\sigma} = 1.838 \times 10^{-27}\frac{\dot{m}}{m}\frac{\gamma^{3.5}}{x^2\beta\alpha_s^7}. \tag{22.28}$$

It should be noted that $\mathcal{J}$ is related with $\mathcal{B}$ by

$$\mathcal{J} = \mathcal{B}\left(\frac{T_{\mathrm{rad}}}{T_{\mathrm{gas}}}\right)^4. \tag{22.29}$$

22.1.2.1 Purely Scattering Case of $\kappa = 0$

Here, we consider the purely scattering case of $\kappa = 0$. In this case there is no energy exchange between gas and radiation in the present non-relativistic regime, and the luminosity L (or ℓ) is constant. However, there is momentum exchange between gas and radiation, and the flow is radiatively driven.

For the constant luminosity case, the non-dimensionalized equations (22.21) and (22.25) are rewritten in the form,

$$\frac{d\beta}{dx} = \frac{\mathcal{N}}{\mathcal{D}}, \tag{22.30}$$

where

$$\mathcal{D} = \beta^2 - \alpha_s^2,$$

$$\mathcal{N} = \beta\left(\frac{2\alpha_s^2}{x} - \frac{1-\ell}{2x^2}\right), \tag{22.31}$$

$$\alpha_s^2 = (\gamma-1)\left[\dot{e} - \ell - \dot{m}\left(\frac{1}{2}\beta^2 - \frac{1}{2x}\right)\right]. \tag{22.32}$$

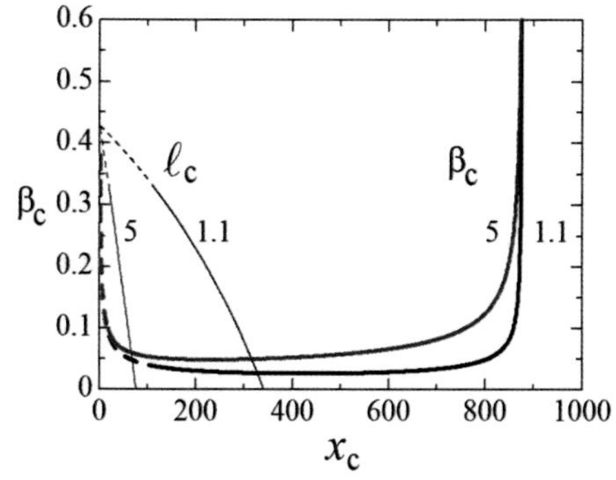

Fig. 22.1 Critical radius x_c and velocity β_c for a given luminosity ℓ under the nonequilibrium diffusion approximation of $\kappa = 0$ for the case of neutron star winds ($\gamma = 4/3$, $\dot{m} = 10^3$, $\dot{e} = 1.1, 5$). Thick curves denote β_c, while thin ones ℓ. Solid curves mean saddle type, which has both accelerated and decelerated solutions, while dashed ones also represent saddle type, but there exist only decelerated solutions

These equations reduce to the usual adiabatic gaseous wind in the limit of $\ell = 0$. In other words, this case of $\kappa = 0$ is the simplest extension of the usual wind to the case including radiation.

The critical conditions (regularity conditions)[3] now become

$$x_c = \frac{1}{2}\left[\frac{\gamma+1}{2(\gamma-1)}\frac{1-\ell}{2} - 1\right]\frac{\dot{m}}{\dot{e}-\ell}, \tag{22.33}$$

$$\beta_c^2 = \left[\frac{\gamma+1}{2(\gamma-1)}\frac{1-\ell}{2} - 1\right]^{-1}\frac{1-\ell}{2}\frac{\dot{e}-\ell}{\dot{m}}, \tag{22.34}$$

or

$$\beta_c^2 = c_{s,c}^2 = \frac{2(\dot{e}-1)x_c + \dot{m}}{2\frac{\gamma+1}{2(\gamma-1)}\dot{m}x_c - 8x_c^2}, \tag{22.35}$$

where the subscript c again denotes "critical" (Sect. 3.1.2). In this case, for given parameters $(\gamma, \dot{m}, \dot{e}, \ell)$, we obtain the critical radius x_c (see Fig. 22.1).

From analytical considerations, the critical radius x_c is limited in the range of

$$0 < x_c < \frac{\gamma+1}{8(\gamma-1)}\dot{m}, \tag{22.36}$$

[3]When the flow speed equals to the sound speed (*transonic*), the denominator of Eq. (22.30) vanishes. In order for a continuous transonic solution to exist, the numerator of Eq. (22.30) must vanish simultaneously at some radius called *critical radius*. These conditions are *critical conditions* (*regularity conditions*) (Sect. 3.1).

while the velocity β_c at the critical radius is

$$\beta_c \geq \frac{\gamma-1}{\gamma+1}\left[\sqrt{\frac{8}{\dot{m}}}+\sqrt{\frac{8}{\dot{m}}+\frac{2(\gamma+1)}{\gamma-1}\frac{\dot{e}-1}{\dot{m}}}\right]. \tag{22.37}$$

It should be emphasized that in the case of the usual adiabatic wind ($\ell = 0$), there is no critical point for $\gamma = 5/3$ (Holzer and Axford 1970). In the present case of the radiative wind ($\ell \neq 0$), in order for the critical point to exist, we have the following constraints:

$$\gamma < \frac{5-\ell}{3+\ell}, \quad \text{or} \quad \ell < \frac{5-3\gamma}{\gamma+1}. \tag{22.38}$$

The types of critical points can be determined by the eigenvalue equation, as described in Chap. 3,

$$\lambda^2 - (\lambda_{11}+\lambda_{22})\lambda + (\lambda_{11}\lambda_{22}-\lambda_{12}\lambda_{21}) = 0, \tag{22.39}$$

where

$$\lambda_{11} \equiv \left.\frac{\partial \mathcal{D}}{\partial x}\right|_c = (\gamma-1)\frac{1}{2x_c}, \tag{22.40}$$

$$\lambda_{12} \equiv \left.\frac{\partial \mathcal{D}}{\partial \beta}\right|_c = (\gamma+1)\beta_c > 0, \tag{22.41}$$

$$\lambda_{21} \equiv \left.\frac{\partial \mathcal{N}}{\partial x}\right|_c = \beta\left(\alpha_{s,c}^2\frac{2}{x_c^2} - \frac{\gamma-1}{x_c^3}\right), \tag{22.42}$$

$$\lambda_{22} \equiv \left.\frac{\partial \mathcal{N}}{\partial \beta}\right|_c = -\frac{2(\gamma-1)}{x_c}\beta_c^2 < 0. \tag{22.43}$$

After several manipulations, we can analytically show that

$$\begin{aligned} |\Lambda| &= \lambda_{11}\lambda_{22} - \lambda_{12}\lambda_{21} \\ &= -\frac{\beta_c^2}{2x_c^3}[(5-3\gamma)-(\gamma+1)\ell] < 0. \end{aligned} \tag{22.44}$$

Hence, in this case of a purely scattering flow, the solutions of the eigenvalue equation have always two real roots with opposite sign, and *the type of critical points is always a saddle* (cf. Sect. 3.1.4).

Typical examples of critical points and their types are shown in Fig. 22.1 for neutron star winds ($\dot{m} = 10^3$, $\dot{e} = 1.1,\ 5$) for a given ℓ. The ratio of specific heats is set to be 4/3 since critical points do not exist for $\gamma = 5/3$. Thick curves denote relations between x_c and β_c, whereas thin curves are relations between x_c and ℓ. The

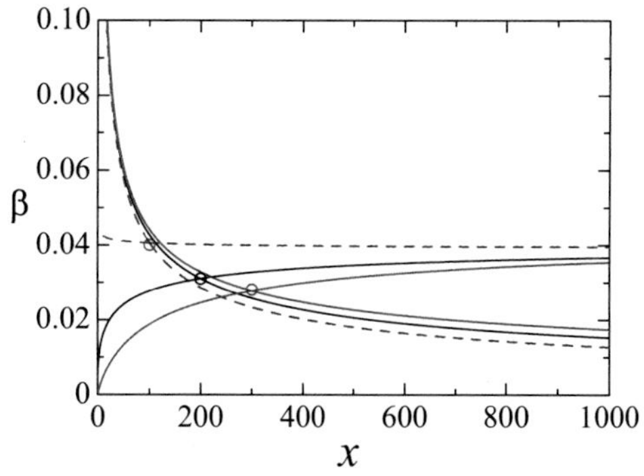

Fig. 22.2 Typical transonic solutions under the nonequilibrium diffusion approximation of $\kappa = 0$ for neutron star winds ($\gamma = 4/3$, $\dot{m} = 10^3$, $\dot{e} = 1.1$, and $x_c = 100, 200, 300$). Open circles denote the critical points

location of $\ell = 0$ corresponds to the usual adiabatic gaseous wind. As the luminosity ℓ increases, the critical radius x_c decreases, and the velocity β_c there increases.

As shown in Eqs. (22.36) and (22.37), the critical radius x_c is limited in some range, and there is the minimum value for β_c. Furthermore, for a given β_c, there exist two branch of solutions on x_c (see Fig. 22.1); a branch of smaller values of x_c and the one of larger values of x_c. In this case under the nonequilibrium diffusion approximation, the branch of large x_c is unphysical, since the luminosity ℓ is negative there (see Fig. 22.1), although ℓ is positive in the small branch.

As already stated, types of critical points are all saddles in this case. Hence, in this simple case of $\kappa = 0$ under the nonequilibrium diffusion approximation, *the transonic solution, if exists, could always pass through a saddle-type critical point.*

It should be noted that on the solid curves in Fig. 22.1 there exist both accelerated (wind) and decelerated (accretion) solutions, while on the dashed ones there exist only decelerated solutions (see also Fig. 22.2).

Examples of transonic solutions under the nonequilibrium diffusion approximation of $\kappa = 0$ for neutron star winds ($\gamma = 4/3$, $\dot{m} = 10^3$, $\dot{e} = 1.1$) are shown in Fig. 22.2. The critical radii x_c of solutions are 100 (dashed curves), 200 (solid ones), and 300 (solid ones), whereas their critical points are denoted by open circles. It should be noted that there exist only decelerated solutions in the case of $x_c = 100$. In other cases the one is an accelerated solution (wind) and the other is a decelerated one (accretion).

The behaviors of critical points and transonic solutions for the parameters of nova winds are quite similar to those for the neutron star case, although the critical radius is on the order of $x_c \sim 10^4$ and velocity there is on the order of $\beta_c \sim 10^{-3}$.

22.1.2.2 General Case of $\kappa \neq 0$

In the general case of $\kappa \neq 0$, we again obtain critical curves among x_c and β_c and ℓ_c for given parameters (γ, $\dot{m}$, $\dot{e}$). After some lengthy manipulations, we can analytically show that the eigenvalue equation for determining the type of critical points reduces to a quadratic one. Furthermore, similar to the simple case of $\kappa = 0$, *the type of critical points is always a saddle*. Hence, in the general case of $\kappa \neq 0$ under the nonequilibrium diffusion approximation, *the transonic solution, if exists, could always pass through a saddle-type critical point* (see Fukue 2014 for details).

22.1.3 Equilibrium Diffusion Winds

We next examine the radiation hydrodynamical winds under the equilibrium diffusion approximation, where radiation and matter equilibrate and have the same temperature T (*equilibrium diffusion winds*).

Eliminating the gas pressure and density from Eq. (22.2) using Eqs. (22.4) and (22.1), we obtain

$$\left(v^2 - \frac{\mathcal{R}}{\bar{\mu}}T\right)\frac{1}{v}\frac{dv}{dr} = -\frac{d}{dr}\left(\frac{\mathcal{R}}{\bar{\mu}}T\right) + \frac{\mathcal{R}}{\bar{\mu}}T\frac{2}{r} - \frac{GM}{r^2} + \frac{\kappa + \sigma}{c}F. \tag{22.45}$$

In the nonequilibrium diffusion case, energy equation (22.3) is used to eliminate the temperature gradient. Under the equilibrium diffusion approximation in a sufficiently optically thick gas, however, instead of energy equation, we use the first moment equation to eliminate the temperature gradient.

Namely, we assume that the radiation pressure is expressed by

$$P = \frac{1}{3}aT^4, \tag{22.46}$$

and we rewrite the first moment Eq. (22.6) as

$$\frac{4aT^3}{3(\mathcal{R}/\bar{\mu})}\frac{d}{dr}\left(\frac{\mathcal{R}}{\bar{\mu}}T\right) = -\rho\frac{\kappa + \sigma}{c}F, \tag{22.47}$$

as in the usual case for the internal structure of stars (the Eddington approximation is also used; $f = 1/3$). However, as already stated, this assumption has not been justified for an *optically thick flow* (Sect. 21.2).

Substituting equation (22.47) into Eq. (22.45), we obtain the wind equation under the diffusion approximation:

$$\left(v^2 - \frac{\mathcal{R}}{\bar{\mu}}T\right)\frac{1}{v}\frac{dv}{dr} = \frac{\mathcal{R}}{\bar{\mu}}T\frac{2}{r} - \frac{GM}{r^2} + \left[\frac{3(\mathcal{R}/\bar{\mu})\rho}{4aT^3} + 1\right]\frac{\kappa + \sigma}{c}F. \tag{22.48}$$

At the present stage, Eqs. (22.48), (22.47), (22.1), and (22.9) are the basic equations for ρ, v, T, and F.

Now, we use the isothermal sound speed c_{T} instead of T, introduce the luminosity L ($\equiv 4\pi r^2 F$), and eliminate the density by Eq. (22.1), then Eqs. (22.48), (22.47), and (22.9) are, respectively, rewritten as follows:

$$\left(v^2 - c_T^2\right)\frac{1}{v}\frac{dv}{dr}$$
$$= c_T^2\frac{2}{r} - \frac{GM}{r^2} + \left[\frac{3(\mathcal{R}/\bar{\mu})^4}{4a}\frac{\dot{M}}{4\pi r^2 v}\frac{1}{c_T^6} + 1\right]\frac{\kappa+\sigma}{c}\frac{L}{4\pi r^2}, \quad (22.49)$$

$$\frac{dc_T^2}{dr} = -\frac{3(\mathcal{R}/\bar{\mu})^4}{4a}\frac{\dot{M}}{4\pi r^2 v}\frac{1}{c_T^6}\frac{\kappa+\sigma}{c}\frac{L}{4\pi r^2}, \quad (22.50)$$

$$\frac{1}{2}v^2 + \frac{\gamma}{\gamma-1}c_T^2 - \frac{GM}{r} + \frac{L}{\dot{M}} = \frac{\dot{E}}{\dot{M}}. \quad (22.51)$$

Next, we represent these equations in terms of the non-dimensional variables (22.12) and parameters (22.13). Then, Eqs. (22.49), (22.50), and (22.51) are respectively reduced to

$$\left(\beta^2 - \alpha_T^2\right)\frac{1}{\beta}\frac{d\beta}{dx} = \frac{2\alpha_T^2}{x} - \frac{1}{2x^2} + \left[A\frac{\dot{m}}{m}\frac{1}{x^2\beta\alpha_T^6} + 1\right]\frac{\kappa+\sigma}{\sigma}\frac{\ell}{2x^2}, \quad (22.52)$$

$$\frac{d\alpha_T^2}{dx} = -A\frac{\dot{m}}{m}\frac{1}{x^2\beta\alpha_T^6}\frac{\kappa+\sigma}{\sigma}\frac{\ell}{2x^2}, \quad (22.53)$$

$$\ell = \dot{e} - \dot{m}\left(\frac{1}{2}\beta^2 + \frac{\gamma}{\gamma-1}\alpha_T^2 - \frac{1}{2x}\right). \quad (22.54)$$

Here, A is the dimensionless constant defined by

$$A = \frac{3(\mathcal{R}/\bar{\mu})^4}{4a}\frac{1}{4\sigma c^4 G M_\odot} = 4.4 \times 10^{-22}, \quad (22.55)$$

where we have set $\bar{\mu} = 0.5$ and $\sigma = \sigma_T = 0.4\,\mathrm{cm^2\,g^{-1}}$. Since the value of A is very small, the term including A on the right-hand side of Eq. (22.52) is usually negligible (in the radiation pressure dominated case). This in turn means that the temperature gradient term on the right-hand side of Eq. (22.45) is negligible compared with the flux term, the fourth term on the right-hand side, under the equilibrium diffusion approximation. In other words, under the present situation both the radiation pressure gradient (the radiative flux) and the gas pressure gradient drive the flow, the latter can be splitted into the density gradient and the gas temperature one. Of these three gradient terms, the gas temperature gradient is almost negligible compared with the radiation pressure gradient and the density one.[4]

Equations (22.52) and (22.53) can be rewritten in the form,

$$\frac{d\beta}{dx} = \frac{\mathcal{N}_1}{\mathcal{D}}, \tag{22.56}$$

$$\frac{d\alpha_{\mathrm{T}}^2}{dx} = \frac{\mathcal{N}_2}{\mathcal{D}}, \tag{22.57}$$

where

$$\mathcal{D} = \beta^2 - \alpha_{\mathrm{T}}^2, \tag{22.58}$$

$$\mathcal{N}_1 = \beta\left[\frac{2\alpha_{\mathrm{T}}^2}{x} - \frac{1}{2x^2} + \left(A\frac{\dot{m}}{m}\frac{1}{x^2\beta\alpha_{\mathrm{T}}^6} + 1\right)\frac{\kappa+\sigma}{\sigma}\frac{\ell}{2x^2}\right], \tag{22.59}$$

$$\mathcal{N}_2 = -(\beta^2 - \alpha_{\mathrm{T}}^2)A\frac{\dot{m}}{m}\frac{1}{x^2\beta\alpha_{\mathrm{T}}^6}\frac{\kappa+\sigma}{\sigma}\frac{\ell}{2x^2}. \tag{22.60}$$

As is well-known, both the denominator $\mathcal{D}$ and numerators $\mathcal{N}_1$ and $\mathcal{N}_2$ need to vanish simultaneously at the transonic point of $\mathcal{D} = 0$. In the present equilibrium diffusion case, for given parameters $(\gamma, \dot{m}, \dot{e})$, instead of critical points, we obtain *critical curves* between x_{c} and β_{c} and ℓ_{c}, as is shown later, where the subscript c denotes "critical" (Sect. 3.1.2).

22.1.3.1 Isothermal Case of $A = 0$

We here examine the case of $A = 0$; we drop the terms including A. In this case, from Eq. (22.53), α_{T} is constant, or in other words, the flow is isothermal-like. In addition, we assume that electron scattering dominates; we set $(\kappa + \sigma)/\sigma = 1$.

Then, Eq. (22.52) is rewritten with the help of Eq. (22.54) as

$$\frac{d\beta}{dx} = \frac{\mathcal{N}}{\mathcal{D}}, \tag{22.61}$$

[4] In the current studies, the temperature gradient is often retained, although we do not know its reason. If we eliminate F from Eqs. (22.45) and (22.47), we have

$$\left(v^2 - \frac{\mathcal{R}}{\bar{\mu}}T\right)\frac{1}{v}\frac{dv}{dr} = \frac{\mathcal{R}}{\bar{\mu}}T\frac{2}{r} - \frac{GM}{r^2} - \left(\frac{\mathcal{R}}{\bar{\mu}}T + \frac{4aT^4}{3\rho}\right)\frac{d}{dr}\ln\left(\frac{\mathcal{R}}{\bar{\mu}}T\right).$$

The fact that the value of A is very small is equivalent with that, in this expression, the first term in the parenthesis on the right-hand side is negligible compared with the second term. Or, the radiation pressure is dominant.

where

$$\mathcal{D} = \beta^2 - \alpha_T^2, \tag{22.62}$$

$$\mathcal{N} = \beta\left(\frac{2\alpha_T^2}{x} - \frac{1}{2x^2} + \frac{\ell}{2x^2}\right)$$

$$= \frac{\dot{m}\beta}{2x^2}\left(\frac{4\alpha_T^2}{\dot{m}}x + \frac{1}{2x} + \frac{\dot{e}-1}{\dot{m}} - \frac{\gamma}{\gamma-1}\alpha_T^2 - \frac{1}{2}\beta^2\right). \tag{22.63}$$

Both the denominator $\mathcal{D}$ and numerator $\mathcal{N}$ need to vanish simultaneously at the critical point. Hence, from Eqs. (22.63) and (22.54), the critical conditions become

$$\frac{2\beta_c^2}{x_c} - \frac{1-\ell_c}{2x_c^2} = 0, \tag{22.64}$$

$$1 - \ell_c = 1 - \dot{e} + \dot{m}\left[\frac{3\gamma-1}{2(\gamma-1)}\beta_c^2 - \frac{1}{2x_c}\right], \tag{22.65}$$

or

$$\beta_c^2 = \alpha_T^2 = \frac{2\dfrac{\dot{e}-1}{\dot{m}}x_c + 1}{2\dfrac{3\gamma-1}{2(\gamma-1)}x_c - \dfrac{8}{\dot{m}}x_c^2}, \tag{22.66}$$

where the subscript c again denotes "critical." In this case, for given parameters (γ, $\dot{m}$, $\dot{e}$, and $\alpha_T = \beta_c$), we again obtain the critical radius x_c and the normalized luminosity ℓ_c there (see Fig. 22.3), although the relation is somewhat different from that of the nonequilibrium diffusion case.

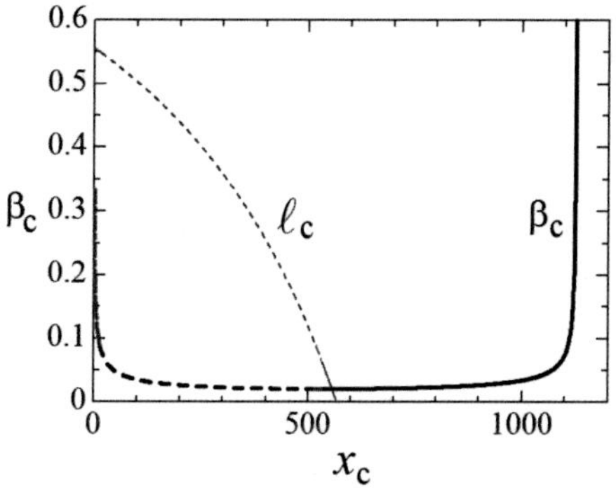

Fig. 22.3 Critical points x_c and luminosity ℓ_c for a given β_c $(= \alpha_T)$ under the equilibrium diffusion approximation of $A = 0$ for neutron star winds ($\gamma = 4/3$, $\dot{m} = 10^3$, $\dot{e} = 1.1$). Thick curves denote β_c, while thin ones ℓ_c. Solid curves mean saddle type, dashed ones a center, and chain-dotted one (at $x_c \sim 0$ and ~ 500) nodal type

From analytical considerations, the critical radius x_c is found to be limited in the range of

$$0 < x_c < \frac{3\gamma - 1}{8(\gamma - 1)}\dot{m}, \tag{22.67}$$

while the velocity β_c at the critical point is

$$\beta_c \geq \frac{\gamma - 1}{3\gamma - 1}\left[\sqrt{\frac{8}{\dot{m}}} + \sqrt{\frac{8}{\dot{m}} + \frac{2(3\gamma - 1)}{\gamma - 1}\frac{\dot{e} - 1}{\dot{m}}}\right]. \tag{22.68}$$

These constraints are also somewhat different from those under the nonequilibrium diffusion approximation.

The types of critical points can be determined similar to those for the nonequilibrium diffusion case, although the coefficients are somewhat different. That is, the coefficients in the present case are

$$\lambda_{11} \equiv \left.\frac{\partial \mathcal{D}}{\partial x}\right|_c = 0, \tag{22.69}$$

$$\lambda_{12} \equiv \left.\frac{\partial \mathcal{D}}{\partial \beta}\right|_c = 2\beta_c > 0, \tag{22.70}$$

$$\lambda_{21} \equiv \left.\frac{\partial \mathcal{N}}{\partial x}\right|_c = \frac{\dot{m}\beta}{2x_c^2}\left(\frac{4\beta_c^2}{\dot{m}} - \frac{1}{2x_c^2}\right), \tag{22.71}$$

$$\lambda_{22} \equiv \left.\frac{\partial \mathcal{N}}{\partial \beta}\right|_c = -\frac{\dot{m}\beta_c^2}{2x_c^2} < 0. \tag{22.72}$$

At large x, the sign of λ_{21} becomes positive, and $(\lambda_{11}\lambda_{22} - \lambda_{12}\lambda_{21}) = -\lambda_{12}\lambda_{21} < 0$; in this case the solutions of the eigenvalue equation have two real roots with opposite sign, and the type of critical points is a saddle. At small x, the sign of λ_{21} becomes negative, and $(\lambda_{11}\lambda_{22} - \lambda_{12}\lambda_{21}) = -\lambda_{12}\lambda_{21} > 0$; in this case the type of critical points is a node (two real roots with the same sign) or a center/spiral (complex roots) (Sect. 3.1.4).

Typical examples of critical points and their types for a given β_c ($= \alpha_T$) and for neutron star winds ($\gamma = 4/3$, $\dot{m} = 10^3$, $\dot{e} = 1.1$) are shown in Fig. 22.3. Thick curves denote relations between x_c and β_c, whereas thin curves are relations between x_c and ℓ_c. Solid, dashed, and chain-dotted curves represent that the critical point is saddle, center, and nodal types, respectively. The results are similar for different values of parameters γ or $\dot{e}$.

As shown in Eqs. (22.67) and (22.68), the critical radius x_c is limited in some range, and there is the minimum value for β_c. Furthermore, for a given β_c ($= \alpha_T$), there exist two branch of solutions on x_c; a branch of smaller values of x_c and the one of larger values of x_c. Of these, the branch of large x_c is unphysical, since the luminosity ℓ_c is negative there, although ℓ_c is positive in the branch of small x_c.

In addition, as already stated, the type of critical points of the branch of small x_c is analytically proved as a node or a center/spiral, although the type of the branch of large x_c is a saddle. Hence, in this simple case of $A = 0$ under the equilibrium diffusion approximation, *the transonic solution, if exists, should always pass through a critical point of the branch of small* x_c*, where the critical point is a nodal-type*.

22.1.3.2 General Case of $A \neq 0$

In the general case of $A \neq 0$, we obtain critical curves among x_c and β_c and ℓ_c for given parameters $(\gamma, \dot{m}, \dot{e})$. The types of critical points can be determined by the eigenvalue equation, which reduces to a quadratic one after some manipulations. Furthermore, similar to the isothermal case of $A = 0$, the type of critical points is a saddle for large x_c, where the luminosity at the critical points is negative, and therefore, such a critical locus is physically unreasonable. For small x_c the luminosity at the critical points becomes positive, but the type of the critical points is a center or node. In the case of a center-type critical point, there is no solution to pass through the critical point. In the case of a nodal-type critical point, there are numberless solutions to pass through. Hence, in this general case of $A \neq 0$ under the equilibrium diffusion approximation, *the transonic solution, if exists, would always pass through a nodal-type critical point* (see Fukue 2014 for details).

Nodal-type critical points often appear in transonic flows with the diffusive processes, such as a viscous transonic accretion flow (Kato et al. 2008). As is well-known, there exists a *causal problem* in the diffusion-type viscous transonic accretion flow around a black hole, similar to the present case of radiative transonic flow under the equilibrium diffusion approximation.

22.1.4 Topological Nature of Critical Points

As was stated and shown, under the nonequilibrium diffusion approximation, the diffusive terms do not exist in the source terms, and the basic equations for the steady, radiatively driven, optically thick winds have critical/singular points where the flow speed equals the *adiabatic* sound speed. Moreover, all the critical points with positive ℓ are saddle type, and there exists no bad behavior (Nobili et al. 1994; Fukue 2014).

On the other hand, under the equilibrium diffusion approximation, the basic equations have critical/singular points where the flow speed equals the *isothermal* sound speed. Furthermore, all the critical points (loci) for physical solutions with positive ℓ are nodal or center type, and make the solutions stiff (Turolla et al. 1986, see also Flammang 1982).

In the non-relativistic regime with small flow velocity ($0 < v \leq 0.1\,c$), the loci of critical points for both approximations are quite similar, and therefore, transonic solutions with the same parameters would be also quite similar in both approximations. But the types of critical points are completely different in each approximation. We hence carefully consider the radiatively driven winds from these viewpoints on the critical nature, physically and mathematically.

Nodal-type critical points often appear in transonic flows under the diffusive processes, such as a viscous transonic accretion flow (Kato et al. 2008). However, there exists a *causal problem* in the viscous transonic accretion flow around a black hole, similar to the present case of radiative transonic flow under the equilibrium diffusion approximation. Furthermore, in the viscous accretion flows, a saddle-type critical point is stable, while a nodal one is unstable (Kato et al. 1993). In the radiative flows, no one has derived or proved the relation between the nature and stability of critical points. Hence, as was done in the viscous accretion flows, stability analysis of critical points is required in relation to the topological nature. However, considering causality, instability, and stiff nature, we suppose that the nodal-type critical points are neither exact nor proper approximate representation of the true transonic points. The saddle type may be right and realized in nature.

We should carefully consider transonic nature in radiatively driven winds, such as nova winds, neutron star wind, black-hole winds, and radiatively driven jets.

22.1.5 Other Related Studies on Radiative Winds

In addition to open problems, such as critical nature and stability, there are several related studies on radiation hydrodynamical winds.

Massive stars in the upper part of the Hertzsprung–Russel diagram, which have luminosities close to the Eddington one, can lead to radiation-driven dense stellar winds. The highest steady mass-loss rates by stellar winds are found in the late stage of the evolution of massive stars, known as a Wolf-Rayet phenomenon. Wolf-Rayet stars are very luminous stars close to the Eddington limit, and characterized by highly supersonic dense winds. Relating to Wolf-Rayet stars, Gräfener and Vink (2013) used a semi-empirical method to determine the sonic-point conditions, and found the wind conditions for radiatively driven, optically thick winds; see also Gräfener et al. (2017). Recently, Grassitelli et al. (2018) computed the subsonic structure of optically thick winds from Wolf-Rayet stars, adopting outer boundaries at the sonic point. They found that the outflows are accelerated to supersonic velocities by the radiative force from iron or helium opacity bumps. In addition, they found that the Eddington ratio at the sonic point is very close to unity.

Observationally, there is strong evidence that massive stars undergo periods of super-Eddington phases, although the physical mechanism is not well understood. For example, in the Great Eruption in Eta Carinae from 1840 to 1860, the super-Eddington luminosity up to 5 times of the Eddington one was observed over 10 years. Super-Eddington winds driven by the continuum opacity were examined by several researchers (e.g., Shaviv 2001; Quataert et al. 2015; Owocki et al. 2017). Owocki et al. (2017) derived semi-analytical solutions for optically thick, super-Eddington stellar winds, which unify the previous several models.

For such a super-Eddington wind, the porosity wind theory has also been developed (e.g., Shaviv 1998, 2000; Owocki et al. 2004). In the usual model of radiation hydrodynamical winds, homogeneity of flows is assumed. In the porosity wind theory, the flow is not uniform, but *porous*, although the global steadiness holds.

Another type of radiation "hydrodynamical" winds, driven by continuum radiation, is a *dust driven wind*. The photon-capture cross section of a dust per unit mass is quite large, and the dust opacity is typically thousands times larger than the electron opacity. Although the dust-gas ratio is generally small (about 1%), both the gas and dust are accelerated together due to the friction between gas and dust (momentum coupling). Since the dust driven wind was first proposed by Gilman (1972) for red giants and AGB stars, there are many studies (Lamers and Cassinelli 1999 for details).

Finally, oscillations and instabilities in radiation-driven outflows are also of interest. For example, Miller and Grossman (1998) investigated dynamical behavior of radiation-driven winds, by linearizing the radiation hydrodynamical equations around steady spherical outflows. They found that radiation-driven winds are generally stable, but pointed out that radiation-driven winds act as mechanical filters that should produce quasi-periodic oscillations. When the wind is not homogeneous, the radiation-driven wind often becomes *clumpy* due to the shadowing effect. In such a case, some type of instabilities, such as the line-deshadowing instability (LDI), were found to take place (e.g., Owocki and Sundqvist 2018 and the references therein). Numerical radiation hydrodynamical simulations of the nonlinear evolution of the LDI show that the time-dependent wind develops a very inhomogeneous clumped structure (e.g., Sundqvist et al. 2018 and the references therein).

22.2 Radiation-Dominated Accretion Flows

The treatment of radiation-dominated accretion flow is almost the same as those of radiatively driven optically thick winds, although the direction of the flow is opposite. Hence, we here summarize the outline of the current studies (see also Sect. 26.2).

For the steady, spherically symmetric, optically thick accretion onto a black hole, after several pioneering works (e.g., Tamazawa et al. 1975; Schmid-Burgk 1978; Burger and Katz 1980), a complete set of moment equations for a relativistic flow was given by the projected symmetric trace-free (PSTF) formalism (Thorne 1981). Based up on Thorne's work (1981), Thorne et al. (1981) derived the basic equations of the radiation-dominated optically thick accretion flow. Furthermore, Flammang (1982) solved them under the diffusion approximation in the comoving frame, as already stated.

In the steady, spherically symmetric, adiabatic accretion, the physically meaningful solutions need to path through a critical point, where the flow speed is equal to the *adiabatic* sound speed (Bondi 1952). In the optically thick, radiation-dominated accretion flow under the equilibrium diffusion approximation, furthermore, similar to the wind case, there appears another type of critical point, where the flow speed is equal to the *isothermal* sound speed (Vitello 1978).[5] The type of this "isothermal" critical points is nodal, and solutions are stiff, similar to the wind case.

However, instead of the equilibrium diffusion approximation, if we use the nonequilibrium diffusion one with a variable Eddington factor (e.g., Nobili et al. 1991), such a bad behavior of solutions can be avoided (see also Sect. 22.1).

In the relativistic regime, however, another pathological behavior appears in the moment formalism with a closure relation. This problem of the closure relation in the relativistic regime will be discussed in Sect. 24.2.4.

22.3 Relevant Characteristic Radii

In this section we briefly summarize several characteristic radii, relating to radiation hydrodynamical flows. We use the spherical coordinates (R, θ, φ) as well as the cylindrical ones (r, φ, z).

22.3.1 Photospheric Radius and Apparent Photosphere

In contrast to dwarf stars, where the atmosphere and photosphere are sharply defined, in extended atmospheres of red giants or spherical winds, where the density gradually changes and the translucent region broadly extends, the photosphere becomes dim and ambiguous. In such a case, we define the *photospheric radius* $R_{\rm ph}$ as a radius where the optical depth $\tau(R)$ measured from infinity becomes unity[6] :

$$\tau(R) = -\int_{\infty}^{R} (\kappa + \sigma)\rho dR. \tag{22.73}$$

Here, κ and σ are the absorption and scattering opacities, respectively, and $\rho(R)$ the radial density distribution.

[5]In this case, a usual "adiabatic" critical point is diffused out, and called "subcritical point."

[6]If we consider the frequency dependence, the photospheric radius also depends on frequency, and it becomes difficult to define a simple photospheric radius.

When the electron scattering is dominant ($\kappa \ll \sigma = \kappa_{es}$), and the flow speed is constant, we can determine the density distribution from the continuity equation ($4\pi R^2 \rho v = \dot{M}$), and the photospheric radius R_{ph} explicitly becomes

$$R_{ph} = \frac{\kappa_{es}\dot{M}}{4\pi v} = \frac{\dot{m}}{2\beta} r_S, \tag{22.74}$$

where $\dot{m}$ ($\equiv \dot{M}/\dot{M}_E$) is the mass-flow rate normalized by the Eddington rate $\dot{M}_E$ ($= L_E/c^2$), β ($\equiv v/c$) the normalized flow speed, and r_S ($\equiv 2GM/c^2$) the Schwarzschild radius of the central object with mass M.

In the region outside the photospheric radius, the flow is optically thin, and the luminosity L over the photospheric radius becomes $L = 4\pi R_{ph}^2 \sigma_{SB} T_{ph}^4$, where T_{ph} is the temperature at the photospheric radius. In the region inside the photospheric radius, however, the flow luminosity L is expressed in terms of the radiative diffusion as

$$L = L_{diff} = -4\pi R^2 \frac{4acT^3}{3\kappa\rho}\frac{dT}{dR} \sim \frac{4\pi R^2 \sigma_{SB} T^4}{\tau(R)}. \tag{22.75}$$

This is called a *diffusion luminosity* (see also Sect. 26.1). The diffusion luminosity is constant, if there is no heating or cooling source.

It should be emphasized that a distant observer does *not* observe this photospheric radius, but observes a surface where the optical depth along the line-of-sight of the observer becomes unity. For example, when the electron scattering is dominant, and the flow speed is constant, the optical depth $\tau(z)$ along the line-of-sight of the observer located at infinity in the direction of z-axis is calculated as

$$\tau(z) = -\int_{\infty}^{z} \kappa_{es}\rho dz = -\frac{\kappa_{es}\dot{M}}{4\pi v}\int_{\infty}^{z}\frac{dz}{r^2+z^2} = \frac{\dot{m}}{2\beta}\frac{r_S}{r}\left(\frac{\pi}{2} - \tan^{-1}\frac{z}{r}\right). \tag{22.76}$$

Hence, from the condition that this optical depth becomes unity, we obtain the height z as a function of radius r:

$$\frac{z_{ph}}{r} = \tan\left(\frac{\pi}{2} - \frac{2\beta}{\dot{m}}\frac{r}{r_S}\right). \tag{22.77}$$

The surface expressed by this relation is not spherical but aspherical, and called a *apparent photosphere* (or a *pseudo-photosphere*) (Fig. 22.4).

This apparent photosphere coincides with the photospheric radius only at $r = 0$. As the radius r increases, the radius of the apparent photosphere becomes larger than the photospheric radius, and after passing a point of $(r, z) = (\pi\dot{m}/4\beta, 0)$ on the equator, it diverges to $z = -\infty$ at $r = \pi\dot{m}/(2\beta)$.

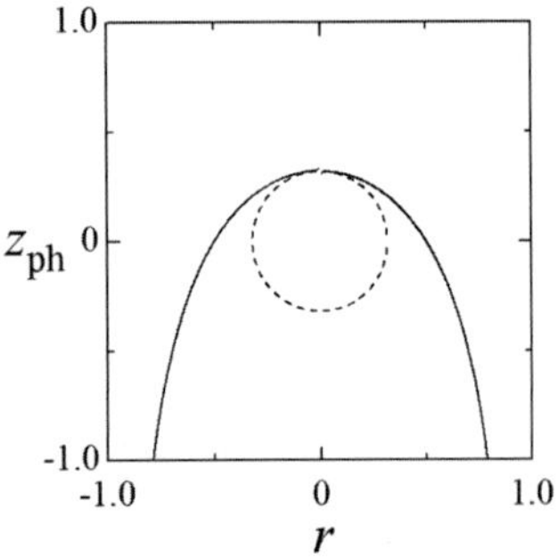

Fig. 22.4 Photospheric radius (a dashed curve) and the shape of the apparent photosphere (a solid curve). The length is normalized by the radius where the apparent photosphere diverges as $z \to -\infty$

This apparent photosphere is just the *last scattering surface*, since a distant observer receives a photon emitted from the apparent photosphere.

For accretion flows, the situation is the same. Let us suppose a spherical accretion flow, whose velocity is a freefall one, for simplicity. In this freefalling case, if we set $v = \sqrt{2GM/R}$, the optical depth measured from infinity becomes $\tau(R) = \kappa_{es}\dot{M}/(2\pi\sqrt{2GMR})$, and therefore, the photospheric radius is expressed as

$$R_{ph} = \frac{\kappa_{es}^2 \dot{M}^2}{16\pi^2 GM} = \frac{1}{2}\dot{m}^2. \tag{22.78}$$

It should be noted that this photospheric radius is proportional to $\dot{m}^2$, in contrast to the wind case, where the photospheric radius is proportional to $\dot{m}$, reflecting the difference of the density distribution.

22.3.2 *Thermalization Radius and Its Surface*

Even if the flow medium is scattering-dominated, in the sufficiently deep interior, where the effective optical depth becomes on the order of unity, radiation and matter are in the local thermodynamic equilibrium (see Sect. 20.3.3.6). We define the *thermalization radius* R_* as the radius where the effective optical depth $\tau_*(R)$ measured from infinity becomes unity:

$$\tau_*(R) = -\int_\infty^R \sqrt{\kappa(\kappa+\sigma)}\rho dR \sim -\int_\infty^R \sqrt{\frac{\kappa_{ff}}{\kappa_{es}}}\kappa_{es}\rho dR. \tag{22.79}$$

When the absorption opacity is given by Kramers' law, the flow speed is assumed to be constant, and the temperature distribution is determined by the diffusion approximation, the thermalization radius R_* is explicitly expressed as

$$R_* = 8.3 \times 10^{-5}\frac{m^{-1/18}\dot{e}^{-7/18}\dot{m}^{4/3}}{\beta^{4/3}} r_S, \tag{22.80}$$

where m $(= M/M_\odot)$ is the mass of the central object in units of the solar mass, and $\dot{e}$ $(= L/L_E)$ the diffusion luminosity normalized by the Eddington one.

We can also determine the *thermalization surface*, where the effective optical depth measured in the radial direction from the apparent photosphere becomes unity. The shape of the thermalization surface is aspherical when the scattering effect is ineffective, while it becomes spherical when the scattering effect is dominant (Ogura and Fukue 2013; Tomida et al. 2015).

22.3.3 Photon Trapping and Photon Trapping Radius

The apparent photosphere and thermalization surface are observationally important. We now introduce the concept of photon trapping and the photon trapping radius, which are physically important (e.g., slim disks in Sect. 3.3, optically thick winds and accretion, see also Sect. 24.3.1 for the relativistic case).

In the optically thick region, radiation spreads out via diffusion, and the effective speed of radiation falls down on the order of c/τ. Hence, the diffusion time $t_{\rm diff}$ passing through the typical radius R is $t_{\rm diff} \sim R/(c/\tau)$. The dynamical time $t_{\rm dyn}$ of the flow moving at speed v is $t_{\rm dyn} \sim R/v$. If the diffusion time is longer than the dynamical time, or the condition,

$$v > \frac{c}{\tau}, \tag{22.81}$$

is satisfied, radiation cannot escape from the moving medium. This phenomena is called the *photon trapping* (Begelman 1978, 1979). It should be noted that this condition is the same with that for the strong equilibrium regime given by Eq. (21.32).

In the case of a spherical wind where both the wind speed and opacity are assumed to be constant, using the continuity equation $(4\pi R^2 \rho v = \dot{M})$ we have the optical depth $\tau = \kappa_{\rm es}\rho R$, and can write the condition (22.81) in the form of $R < \kappa_{\rm es}\dot{M}/(4\pi c)$. That is, in this case, the *photon trapping radius* $R_{\rm trap}$ is obtained as

$$R_{\rm trap} = \frac{\kappa_{\rm es}\dot{M}}{4\pi c} = \frac{\dot{m}}{2} r_{\rm S}, \tag{22.82}$$

which is smaller than the photospheric radius by a factor of β.

Once the photon trapping occurs, radiation is advected with matter, and

$$L = L_{\rm adv} = 4\pi r^2 (vE + vP) \tag{22.83}$$

is the radiation energy per unit time carried with matter. This is called an *advection luminosity* (see also Sect. 24.3). At the trapping radius, this advection luminosity is equal to the diffusion one. In the advective limit, where the diffusion luminosity can be ignored, the flow becomes almost adiabatic.

For accretion flows, the photon trapping radius also exists. Let us again consider the freefall accretion. In the case of $v = \sqrt{2GM/R}$, using the continuity equation ($4\pi R^2 \rho v = \dot{M}$), we can determine the density distribution, and obtain the optical depth $\tau = \sigma \dot{M}/(4\pi c)\sqrt{16/(r_S r)}$. As a result, the condition of $v > c/\tau$ is rewritten as

$$R < \frac{\kappa_{\mathrm{es}} \dot{M}}{4\pi c} = \frac{\dot{m}}{2} r_S. \tag{22.84}$$

In this simple freefall accretion, the photon trapping radius is equal to that of a simple wind. In contrast to the photospheric radius, the photon trapping radius in wind flows is roughly equal to that in accretion flows.

22.4 Line-Force Driven Winds

Besides the radiation hydrodynamical wind driven by the continuum radiation, the *line-force driven wind* is also an important mechanism of radiatively driven winds. The line-force driven mechanism may operate in various astrophysical winds, such as a stellar wind from hot early type stars.

The line-driven mechanism can work due to the Doppler effect. Let us suppose a wind gas, which absorbs some spectral line in the continuum radiation from the photosphere at some wavelength in the comoving frame, receives the momentum of the line photons, and is accelerated outward. If the wind gas is static, or moves at a constant speed, the lower layers of winds would absorb the line photons, but the outer layers could not absorb them. If there is a velocity gradient in the wind flows, however, the outer layers can absorb the adjacent continuum radiation, which is redshifted to the line wavelength in the comoving frame of the outer layers. As a result, the continuous acceleration becomes possible (Fig. 22.5).

In this section, we introduce the CAK theory under *the Sobolev approximation*, following three steps (Castor 1970; Lucy and Solomon 1970; Castor et al. 1975; see also Lamers and Cassinelli 1999 for details).

(1) Radiative Force of a Single Absorption Line

The radiative force per unit mass is expressed as

$$g_{\mathrm{rad}} = \frac{1}{\rho} f_{\mathrm{rad}} = \frac{1}{c\rho} \int_0^\infty d\nu \kappa_\nu \rho F_\nu, \tag{22.85}$$

where f_{rad} is the radiative force per unit volume, κ_ν the absorption coefficient, and F_ν the radiative flux (Sect. 20.3.1).

In the case of a line absorption, if we assume that the line absorption profile is a rectangle with a Doppler width of $\Delta\nu_{\mathrm{D}}$, we can evaluate the line force per unit

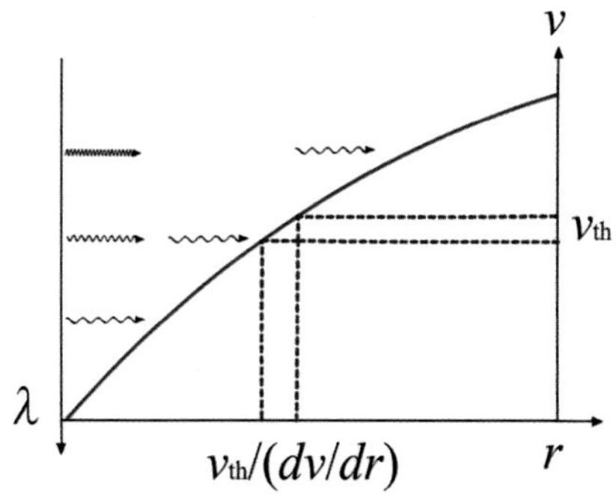

Fig. 22.5 Mechanism of the line-force driven winds. The specific light in the incident continuum radiation from the central source is redshifted to the adequate wavelength, and is line-absorbed by the moving gas. That is, the continuum radiation with a longer wavelength (a longer wave at a lower row) is slightly redshifted and absorbed by a wind gas at small distance and low speed, that with a shorter wavelength (a moderate wave at a middle row) is moderately redshifted and absorbed by a wind gas at moderately distance and speed, and that with a much shorter wavelength (a short wave at a upper row) is highly redshifted and absorbed by a wind gas at large distance and high speed. The line absorption by the wind gas with velocity gradient takes place in the finite velocity range of the thermal velocity $v_{\rm th}$ that corresponds to the radial interval of $v_{\rm th}/(dv/dr)$

mass, g_ℓ, as

$$g_\ell = \frac{1}{c\rho}\Delta\nu_{\rm D}\ell_\nu \rho F_{\rm c} = \frac{F_{\rm c}\Delta\nu_{\rm D}}{c}\ell_\nu, \tag{22.86}$$

where ℓ_ν is the line absorption coefficient, and $F_{\rm c}$ the radiative flux of the continuum.

Due to the line absorption, the incident radiation reduces by a factor of $e^{-\tau_\ell}$, where τ_ℓ is the optical depth of the spectral line. Hence, the average line force is given by

$$\langle g_\ell \rangle = g_\ell \frac{\int_0^{\tau_\ell} e^{-t}dt}{\tau_\ell} = \frac{F_{\rm c}\Delta\nu_{\rm D}}{c}\ell_\nu \frac{1-e^{-\tau_\ell}}{\tau_\ell}. \tag{22.87}$$

Furthermore, we assume that the line optical depth τ_ℓ is equal to the so-called *Sobolev optical depth*:

$$\tau_{\rm S} \sim \frac{v_{\rm th}\ell_\nu \rho}{|dv/dr|}, \tag{22.88}$$

where $v_{\rm th}$ is the thermal velocity of gas particles, and $|dv/dr|$ the velocity gradient of the wind flow. This is the *Sobolev approximation*.[7] Thus, the average line force becomes

$$\langle g_\ell \rangle = \frac{F_{\rm c} \Delta\nu_{\rm D}}{c} \frac{\ell_\nu}{\tau_{\rm S}} (1 - e^{-\tau_{\rm S}}). \tag{22.89}$$

Here, $F_{\rm c}\Delta\nu_{\rm D}/c$ means the momentum received at the line Doppler width, $\tau_{\rm S}/\ell_\nu = \rho v_{\rm th}/|dv/dr|$ the column density absorbing this momentum, and $(1 - e^{-\tau_{\rm S}})$ the absorption probability.

If we introduce the electron scattering optical depth corresponding to the Sobolev optical depth by

$$t \equiv \frac{\rho \kappa_{\rm es} v_{\rm th}}{|dv/dr|}, \tag{22.90}$$

$\tau_{\rm S} = (\ell_\nu/\kappa_{\rm es})t$, and the average line force is reexpressed as

$$\langle g_\ell \rangle = \frac{F_{\rm c} \Delta\nu_{\rm D}}{c} \frac{\kappa_{\rm es}}{t} \left[1 - e^{-(\ell_\nu/\kappa_{\rm es})t}\right] = \frac{F_{\rm c} \Delta\nu_{\rm D}}{c} \begin{cases} \frac{\kappa_{\rm es}}{t}, & (\ell_\nu/\kappa_{\rm es})t \gg 1 \\ \ell_\nu, & (\ell_\nu/\kappa_{\rm es})t \ll 1. \end{cases} \tag{22.91}$$

Hence, the line force of optically thick lines is independent of the line strength, but determined by the velocity gradient. The line force of optically thin lines does not depend on the velocity gradient, but is proportional to the line strength.

(2) Radiative Force of Multiple Absorption Lines

The radiative force of a single absorption line is expressed by Eq. (22.91), while there exist many spectral lines, and summing up all of these line forces is the total line force received by the wind gas. Using the radiative flux F $(= L/4\pi r^2)$ of the central star, the total line force per unit mass can be formally expressed as

$$g_\ell = \frac{\kappa_{\rm es} F}{c} M(t), \quad M(t) \equiv \frac{1}{F} \sum_\ell F_{\rm c}(\nu_\ell) \Delta\nu_{\rm D}(\nu_\ell) \min(1/t, \ell_\nu/\kappa_{\rm es}), \tag{22.92}$$

where $M(t)$ is a factor pushed in all the quantities on the absorption lines, and called the *radiation-force multiplier*.

If we can calculate and get this factor $M(t)$, we can obtain the total line force, but it is not easy task. For that situation, relying on the facts that each single line can be treated independently under the Sobolev approximation, and the line opacity distribution empirically has a power law form, Castor et al. (1975) assumed that the

[7] Here, we omit the details of the Sobolev approximation (Sobolev 1947, 1957). It should be noted that the Sobolev optical depth is determined only by the local quantities, such as the local density and velocity gradient.

radiation-force multiplier $M(t)$ can be fitted by the power law form of the optical depth as[8] :

$$M(t) = kt^{-\alpha}, \quad k \sim 1/30, \quad \alpha \sim 0.7. \tag{22.93}$$

This bold and basically correct model is called the *CAK theory*, in honor of the authors' name (Castor, Abbott, Klein).

In this CAK theory, using the luminosity L of the central star, the optical depth Eq. (22.90), and the continuity equation ($4\pi r^2 \rho v = \dot{M}$), we have the radiative force per unit mass as follows:

$$g_{\rm rad} = \frac{\kappa_{\rm es} L k}{4\pi c r^2} \left(\frac{1}{\kappa_{\rm es} \rho v_{\rm th}} \frac{dv}{dr} \right)^\alpha = \frac{\kappa_{\rm es} L k}{4\pi c r^2} \left(\frac{4\pi}{\kappa_{\rm es} v_{\rm th} \dot{M}} \right)^\alpha \left(r^2 v \frac{dv}{dr} \right)^\alpha . \tag{22.94}$$

Or, after the constant parts are combined into a single constant, we have the following easy-to-use form:

$$g_{\rm rad} = \frac{C}{r^2} \left(r^2 v \frac{dv}{dr} \right)^\alpha , \tag{22.95}$$

$$C = \frac{\kappa_{\rm es} L k}{4\pi c} \left(\frac{4\pi}{\kappa_{\rm es} v_{\rm th} \dot{M}} \right)^\alpha = k \left(\frac{v_{\rm th}}{c} \right)^\alpha (GM\Gamma)^{1-\alpha} \left(\frac{L}{\dot{M} c^2} \right)^\alpha , \tag{22.96}$$

where Γ is the Eddington ratio, defined by $\Gamma \equiv L/L_{\rm E}$, $L_{\rm E}$ being the Eddington luminosity of the central object (Sect. 19.4). For practical purpose, the value of C for hot stars is estimated as

$$C = \frac{(QGM\Gamma)^{1-\alpha}}{1-\alpha} \left(\frac{L}{\dot{M} c^2} \right)^\alpha , \quad Q \sim 10^5 Z \sim 2000 \ (Z = 0.02), \tag{22.97}$$

where Z is the metal abundance.

(3) Line-Force Driven Wind and Critical Points

Considering the line force, we can write the radiatively driven wind equation (22.45) as

$$\left(v^2 - c_{\rm T}^2 \right) \frac{1}{v} \frac{dv}{dr} = -\frac{dc_{\rm T}^2}{dr} + \frac{2c_{\rm T}^2}{r} - \frac{GM}{r^2} + \frac{C}{r^2} \left(r^2 v \frac{dv}{dr} \right)^\alpha . \tag{22.98}$$

Similar to (radiation) hydrodynamical winds, this wind equation also has a singular point. In contrast to those winds, however, the singular point of this equation is not a

[8]See also Lamers and Cassinelli (1999), and Hubeny and Mihalas (2014) for more detailed and refined model and discussions.

sonic point, and the treatment of the singular point is different from that of the usual hydrodynamical winds.

Let us rewrite Eq. (22.98) to set as follows:

$$\mathcal{F}(r, v, v') = r^2 v v' - \frac{c_{\rm T}^2 r^2}{v} v' + r^2 \frac{dc_{\rm T}^2}{dr} - 2c_{\rm T}^2 r + GM - C(r^2 v v')^\alpha = 0, \qquad (22.99)$$

where $v' = dv/dr$. This differential equation has a *singular locus*, where v' is not defined, in the r-v solution space. This singular locus is determined by the two conditions:

$$\mathcal{F}(r, v, v')|_{\rm c} = 0, \qquad (22.100)$$

$$\left.\frac{\partial \mathcal{F}}{\partial v'}\right|_{\rm c} = r^2 v - \frac{c_{\rm T}^2 r^2}{v} - \frac{\alpha}{v'} C\left(r^2 v v'\right)^\alpha = 0. \qquad (22.101)$$

Furthermore, in order for the solution to contact with this singular locus, the *regularity condition* is additionally imposed (Castor 1970):

$$\begin{aligned}\left.\frac{d\mathcal{F}}{dr}\right|_{\rm c} &= \frac{\partial \mathcal{F}}{\partial r} + v'\frac{\partial \mathcal{F}}{\partial v} + v''\frac{\partial \mathcal{F}}{\partial v'} = \frac{\partial \mathcal{F}}{\partial r} + v'\frac{\partial \mathcal{F}}{\partial v} \\ &= 2rvv' - \frac{2c_{\rm T}^2 r}{v} v' + 2r\frac{dc_{\rm T}^2}{dr} - 2c_{\rm T}^2 - \frac{2\alpha}{r} C\left(r^2 v v'\right)^\alpha \\ &\quad + v'\left[r^2 v' - \frac{c_{\rm T}^2}{v^2} r^2 v' - \frac{\alpha}{v} C\left(r^2 v v'\right)^\alpha\right] = 0. \end{aligned} \qquad (22.102)$$

Eliminating the term $[r^2 v v' - \alpha C(r^2 v v')^\alpha]$ from Eqs. (22.101) and (22.102), we have a relation on the singular locus:

$$\left(2 + r\frac{v'}{v}\right) r\frac{v'}{v} = 2\left(1 + r\frac{v'}{v}\right) - \left(r\frac{v'}{v}\right)^2 + \frac{2r}{c_{\rm T}^2}\frac{dc_{\rm T}^2}{dr}. \qquad (22.103)$$

If the flow is assumed to be isothermal ($c_{\rm T}$ = const.), for simplicity, Eq. (22.103) becomes a quadratic equation with respect to rv'/v, and we obtain a positive solution on the singular locus[9] :

$$\left.\frac{r}{v}\frac{dv}{dr}\right|_{\rm c} = 1. \qquad (22.104)$$

Using this solution, we can determine other quantities at the singularity.

[9]See, e.g., Lamers and Cassinelli (1999) for comprehensive general treatments.

In order to further simplify the treatment, the gas pressure is ignored in the present line-driven winds ($c_T^2 \ll v^2$). In this case, Eqs. (22.99), (22.101), (22.102) are reexpressed as the following simple equations:

$$\mathcal{F}(r, v, v') = r^2 v v' + GM - C(r^2 v v')^\alpha = 0, \tag{22.105}$$

$$\left.\frac{\partial \mathcal{F}}{\partial v'}\right|_c = r^2 v - \frac{\alpha}{v'} C(r^2 v v')^\alpha = 0, \tag{22.106}$$

$$\left.\frac{d\mathcal{F}}{dr}\right|_c = \frac{v'}{r}\left(2 + r\frac{v'}{v}\right)\left[r^2 v' - \frac{\alpha}{v} C(r^2 v v')^\alpha\right] = 0. \tag{22.107}$$

Using the first two equations, we have the solutions at the singular point as

$$r^2 v v' = \frac{\alpha}{1-\alpha} GM, \tag{22.108}$$

$$C(r^2 v v')^\alpha = \frac{1}{1-\alpha} GM, \tag{22.109}$$

which determines the constant C:

$$C = \alpha^{-\alpha}(1-\alpha)^{-(1-\alpha)}(GM)^{1-\alpha}. \tag{22.110}$$

It should be noted that Eq. (22.105) is valid on the solution curve (plane) outside the singular locus. Since a solution satisfying this Eq. (22.105) is clearly $r^2 v v' =$ const., and this general solution also satisfies equation at the singular locus, or $r^2 v v' =$ const. $= [\alpha/(1-\alpha)]GM$. Thus, Eq. (22.108) holds at the entire solution curve as well as at the singular locus. Finally, imposing the boundary condition at the stellar surface ($v = 0$ at $r = R_*$), we obtain the solution for the line-driven wind in the simple case[10] :

$$v = \left[\frac{2\alpha GM}{(1-\alpha)R_*}\left(1 - \frac{R_*}{r}\right)\right]^{1/2}. \tag{22.111}$$

Here, the effect of lines is pushed into a parameter α.

[10]This solution has the same form with a solution of continuum-driven winds in the optically thin case:

$$v = \left[\frac{2(1-\Gamma)GM}{R_*}\left(1 - \frac{R_*}{r}\right)\right]^{1/2},$$

where Γ is the Eddington parameter.

Using this solution, from the solution (22.104) at the singular point, we can determine the radius of the singular point as

$$r_{\mathrm{c}} = \frac{3}{2} R_*. \tag{22.112}$$

In addition, the mass-loss rate is also determined as

$$\dot{M} = \frac{4\pi GM}{\kappa_{\mathrm{es}} v_{\mathrm{th}}} \alpha (1-\alpha)^{(1-\alpha)/\alpha} \left(\frac{k\kappa_{\mathrm{es}} L}{4\pi GMc} \right)^{1/\alpha}. \tag{22.113}$$

Here, we briefly discuss a graphical meaning of Eq. (22.105). When we rewrite Eq. (22.105) as

$$r^2 v v' + GM = C\left(r^2 v v'\right)^{\alpha}, \tag{22.114}$$

the left-hand side means a line with the segment GM, on the graph whose abscissa is $r^2 v v'$ and ordinate is $r^2 v v' + GM$. The right-hand side, on the other hand, is a upward-convex curve that passes through the origin. Let us change the values of constant C on the graph. If C is small, there are no intersections (solutions). If C is large, there appear two intersections, and multiple solutions are also inadequate. If and only if the curve with adequate C contacts with the line expressed by the left-hand side, there exists a single solution, and the velocity and other quantities are uniquely determined. Furthermore, in this case Eq. (22.106) holds.

22.5 Self-Similar Spherical Accretion

Analytical treatments of time-dependent phenomena in astrophysical fluids are not easy, and numerical simulations are often required. In some limited situation, where the associated physical processes are quite few, we can use the *self-similar* method, where the temporal and spatial variables are combined into a similarity variable, and partial differential equations of matter are transformed to ordinary differential equations (Sect. 3.4). Since Sedov (1959), there are many investigations on self-similar flows; shock propagations in supernova explosions (Sedov solutions), self-similar flows under the central gravity (e.g., Sakashita 1974), self-similar transonic flows (e.g., Cheng 1977; Fukue 1984), self-similar spherical accretion under radiation drag (e.g., Tsuribe et al. 1995). In this section, we briefly introduce the self-similar technique to treat the time-dependent radiation hydrodynamical problem, using a simple model.

22.5.1 Basic Equations and Similarity Variables

Let us suppose a time-dependent spherically symmetric flow in a gravitational field produced by a central object of mass M and luminosity L. We use spherical coordinates (r, θ, φ), and adopt the Newtonian gravity. Furthermore, we drop the gas pressure in order to avoid the difficulty on the treatment of the transonic points, and focus our attention on the self-similar method. Moreover, we assume the radiative equilibrium (or only scattering case) for simplicity. The basic equations are described as follows.

The continuity equation is

$$\frac{\partial \rho}{\partial t} + \frac{1}{r^2}\frac{\partial}{\partial r}(r^2 \rho v) = 0, \quad (22.115)$$

where ρ is the gas density and v the radial velocity. The equation of motion in the radial direction is

$$\frac{\partial v}{\partial t} + v\frac{\partial v}{\partial r} = -\frac{GM(1-\Gamma)}{r^2}, \quad (22.116)$$

where Γ ($\equiv L/L_{\rm E}$) is the Eddington parameter, and assumed to be constant in the present simple model. Hence, the radiative flux F is independent of time. The energy equation is not used.

The zeroth moment equation is also not used, and the first moment equation becomes

$$-\frac{1}{\rho}\frac{\partial P}{\partial r} = \frac{\kappa + \sigma}{c}F = \frac{GM}{r^2}\Gamma, \quad (22.117)$$

where P is the radiation pressure, and we use the Eddington approximation. Self-similar transformations are performed for these Eqs. (22.115)–(22.117).

A similarity variable ξ is introduced as

$$\xi = (GM)^{-1/3} r t^{-\delta}, \quad \delta = 2/3. \quad (22.118)$$

Since the central mass M is assumed to be constant, δ must be $2/3$ due to the requirement of dimension (cf. Fukue 1984 for the case of variable M). Physical quantities are transformed as

$$v = -(GM)^{1/3} t^{\delta-1} V(\xi), \quad (22.119)$$

$$\rho = r^{\nu} D(\xi) = (GM)^{\nu/3} t^{\nu\delta} \xi^{\nu} D(\xi), \quad (22.120)$$

$$P = r^{2+\nu} t^{-2} Q(\xi) = (GM)^{(2+\nu)/3} t^{(2+\nu)\delta-2} \xi^{2+\nu} Q(\xi), \quad (22.121)$$

where the similarity quantities are functions of only ξ, and ν is a free parameter relating to the density distribution. In order to explicitly express *accretion*, the minus sign is attached in the velocity transformation. In addition, time t runs from infinity to zero for accretion.

In terms of the self-similar variable ξ, the partial differentials are transformed as

$$\frac{\partial}{\partial t} = \frac{\partial \xi}{\partial t}\frac{d}{d\xi} = -\delta t^{-1}\xi\frac{d}{d\xi}, \tag{22.122}$$

$$\frac{\partial}{\partial r} = \frac{\partial \xi}{\partial r}\frac{d}{d\xi} = (GM)^{-1/3}t^{-\delta}\frac{d}{d\xi}. \tag{22.123}$$

Using these similarity transformations, after some manipulations, we can transform the basic Eqs. (22.115)–(22.117) to the following set of ordinary differential equations:

$$(V+\delta\xi)\frac{1}{D}\frac{dD}{d\xi} + \frac{dV}{d\xi} = -\frac{2+\nu}{\xi}V, \tag{22.124}$$

$$(V+\delta\xi)\frac{dV}{d\xi} = -(1-\delta)V - \frac{1-\Gamma}{\xi^2}, \tag{22.125}$$

$$\frac{dQ}{d\xi} = -\frac{2+\nu}{\xi}Q - \Gamma\frac{D}{\xi^4}. \tag{22.126}$$

These transformed equations can be easily solved under the appropriate boundary conditions.

22.5.2 *Typical Solutions*

As boundary conditions for the present freefall-like accretion flow, we use asymptotic solutions at infinity, and solve Eqs. (22.124)–(22.126) inward from infinity.[11]

Considering the asymptotic behavior of Eqs. (22.124)–(22.126) at infinity, we obtain the asymptotic solutions at infinity as

$$V \to \sqrt{2(1-\Gamma)}\xi^{-1/2}, \tag{22.127}$$

$$D \to \sqrt{2(1-\Gamma)}\exp[(\nu+3/2)\xi^{-3/2}], \tag{22.128}$$

$$Q \to Q_0\xi^{-(2+\nu)}, \tag{22.129}$$

[11] We can use asymptotic solutions at center, and solve equations outward. In the present case, however, the values of quantities become very large, and numerical accuracy is needed.

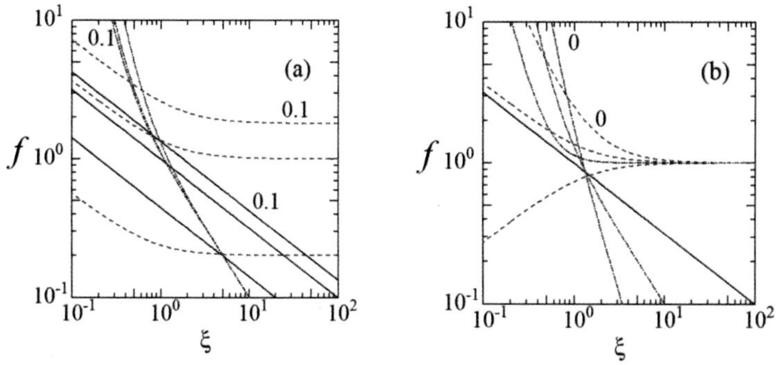

Fig. 22.6 Examples of self-similar solutions. Solid curves denote V, dashed ones express D, and chain-dotted ones represent Q. The values of parameters are (**a**) $\nu = -1$, $Q_0 = 1$, and $\Gamma = 0.1, 0.5, 0.9$, and (**b**) $\Gamma = 0.5$, $Q_0 = 1$, and $\nu = 0, -1, -2$

where Q_0 is an arbitrary constant. It should be noted that the asymptotic solution (22.127) just represents the freefall velocity, $v = -\sqrt{2(1-\Gamma)GM/r}$.

Examples of self-similar solutions are shown in Fig. 22.6, where the infall velocity V (solid curves), density D (dashed ones), and radiation pressure Q (chain-dotted ones) are plotted as a function of ξ. The values of parameters are (a) $\nu = -1$, $Q_0 = 1$, and $\Gamma = 0.1, 0.5, 0.9$, and (b) $\Gamma = 0.5$, $Q_0 = 1$, and $\nu = 0, -1, -2$.

In the present simple model, the infall velocity is freefall-like, since we assume that the central luminosity L is constant, and drop the gas pressure.

As the central luminosity increases, due to the radiation pressure, the infall speed becomes small, as expected. The density also becomes small, due to the radiative suppression of accretion. The radiation pressure does not depend on the central luminosity so much.

With regard to the dependence on parameter ν (Fig. 22.6b), the infall velocity does not depend on ν. The density in the inner region (or large time) remarkably depends on ν, although it in the outer region does not. The radiation pressure also depends on ν.

References

Begelman, M.C.: Monthly Not. Roy. Astron. Soc. **184**, 53 (1978)
Begelman, M.C.: Monthly Not. Roy. Astron. Soc. **187**, 237 (1979)
Bondi, H.: Monthly Not. Roy. Astron. Soc. **112**, 195 (1952)
Burger, H.L., Katz, J.I.: Astrophys. J. **236**, 921 (1980)
Cassinelli, J.P., Hartmann, L.: Astrophys. J. **202**, 718 (1975)
Castor, J.I.: Monthly Not. Roy. Astron. Soc. **149**, 111 (1970)
Castor, J.I., Abbott, D.C., Klein, R.I.: Astrophys. J. **195**, 157 (1975)
Cheng, A.F.: Astrophys. J. **213**, 537 (1977)
Flammang, A.N.: Monthly Not. Roy. Astron. Soc. **199**, 833 (1982)
Fukue, J.: Publ. Astron. Soc. Jpn. **36**, 87 (1984)

Fukue, J.: Publ. Astron. Soc. Jpn. **66**, 40 (2014)
Gilman, R.C.: Astrophys. J. **178**, 423 (1972)
Gräfener, G., Vink, J.S.: Astron. Astrophy. **560**, 6 (2013)
Gräfener, G., et al.: Astron. Astrophy. **608**, 34 (2017)
Grassitelli, L., et al.: Astron. Astrophy. **618**, 86 (2018)
Holzer, T.E., Axford, W.I.: ARA&A **8**, 31 (1970)
Hubeny, I., Mihalas, D.: Theory of Stellar Atmosphere. Princeton University Press, Princeton (2014)
Kato, M.: Publ. Astron. Soc. Jpn. **35**, 33 (1983)
Kato, S., Fukue, J., Mineshige, S.: Black-Hole Accretion Disks: Towards a New Paradigm. Kyoto University Press, Kyoto (2008)
Kato, S., Wu, X., Yang, L., Yang, Z.: Monthly Not. Roy. Astron. Soc. **260**, 317 (1993)
Lamers, H.J.G.L.M., Cassinelli, J.P.: Introduction to Stellar Winds. Cambridge University Press, Cambridge (1999)
Lucy, L.B., Solomon, P.M.: Astrophys. J. **159**, 879 (1970)
Miller, G.S., Grossman, S.A.: Monthly Not. Roy. Astron. Soc. **299**, 139 (1998)
Nobili, L., Turolla, R., Zampier, L.: Astrophys. J. **383**, 250 (1991)
Nobili, L., Turolla, R., Lapidus, I.: Astrophys. J. **433**, 276 (1994)
Ogura, K., Fukue, J.: Publ. Astron. Soc. Jpn. **65**, 92 (2013)
Owocki, S.P., Sundqvist, J.O.: Monthly Not. Roy. Astron. Soc. **475**, 814 (2018)
Owocki, S.P., Gayley, K.G., Shaviv, N.J.: Astrophys. J. **616**, 525 (2004)
Owocki, S.P., Townsend, R.H.D., Quataert, E.: Monthly Not. Roy. Astron. Soc. **472**, 3749 (2017)
Paczyński, B.: Astrophys. J. **363**, 218 (1990)
Paczyński, B., Prószyński, M.: Astrophys. J. **302**, 519 (1986)
Quataert, E., et al.: arXiv:1509.0637 (2015)
Quataert, E., et al.: Monthly Not. Roy. Astron. Soc. **458**, 1214 (2016)
Quinn, T., Paczyński, B.: Astrophys. J. **289**, 634 (1985)
Ruggles, C.L.N., Bath, G.T.: Astron. Astrophys. **80**, 97 (1979)
Sakashita, S.: Astrophys. Space Sci. **26**, 183 (1974)
Schmid-Burgk, J.: Astrophys. Space Sci. **56**, 191 (1978)
Sedov, L.I.: Similarity and Dimensional Methods in Mechanics. Infosearch, London (1959)
Shaviv, N.J.: Astrophys. J. **494**, L193 (1998)
Shaviv, N.J.: Astrophys. J. **532**, L137 (2000)
Shaviv, N.J.: Monthly Not. Roy. Astron. Soc. **326**, 126 (2001)
Sobolev, V.V.: Moving Atmospheres of Stas. Harvard University Press, Cambridge (1947) (English translation, 1960)
Sobolev, V.V.: Soviet Astron. AJ **1**, 678 (1957)
Sundqvist, J.O., Owocki, S.P., Puls, J.: Astron. Astrophys. **611**, A17 (2018)
Tamazawa, S., Toyama, K., Kaneko, N., Ono, Y.: Astrophys. Space Sci. **32**, 403 (1975)
Thorne, K.S.: Monthly Not. Roy. Astron. Soc. **194**, 439 (1981)
Thorne, K.S., Flammang, R.A., Zytkow, A.N.: Monthly Not. Roy. Astron. Soc. **194**, 475 (1981)
Tomida, M., et al.: Publ. Astron. Soc. Jpn. **67**, 6 (2015)
Tsuribe, T., Umemura, M., Fukue, J.: Publ. Astron. Soc. Jpn. **47**, 73 (1995)
Turolla, R., Nobili, L., Calvani, M.: Astrophys. J. **303**, 573 (1986)
Vitello, P.A.J.: Astrophys. J. **225**, 694 (1978)
Żytkow, A.N.: Acta Astron. **22**, 103 (1972)

Chapter 23
Wave and Instability in Radiative Fluids

Similar to the astrophysical and magnetohydrodynamical fluids, there exist various waves and instabilities in the radiative fluid. Some of them are usual waves modified by radiation fields, such as acoustic or gravity waves, while some are proper to the radiative fluid, such as a radiative diffusion wave. Nonlinear waves, such as shocks and ionization fronts, and various fluid instabilities are also influenced by radiation fields. In this chapter, we show typical waves and instabilities in the radiative fluid.

23.1 Acoustic Waves in the Uniform Radiative Medium

Small amplitude fluctuations and wave phenomena in a radiative fluid have been examined since 1950s, and the radiative smoothing of temperature fluctuations in a static medium has been known (e.g., Spiegel 1957; Vincenti and Baldwin 1962; Unno and Spiegel 1966; Cogley 1969; Delache and Froeschlé 1972; Bisnovatyi-Kogan and Blinnikov 1978, 1979; Schweitzer 1984). Especially, in the optically thin case the radiative damping of temperature fluctuations was examined under the Newtonian cooling approximation (e.g., Stix 1970; Souffrin 1972), whereas in the optically thick case it was examined under the diffusion or Eddington approximations (e.g., Unno and Spiegel 1966; Mihalas and Mihalas 1983). In 1990s more detailed and comprehensive investigations have been done by several researchers (e.g., Dzhalilov et al. 1992, 1994; Zhugzhda et al. 1993; Staude et al. 1994; Bogdan et al. 1996), without imposing the Eddington approximation, and using the radiative transfer equation. More thorough discussions on waves in the radiative fluid are found in, e.g., Mihalas and Mihalas (1984), and Castor (2004). In this section, we first examine the effect of radiation on acoustic waves in a homogenous medium (consult Mihalas and Mihalas 1984 for details).

© Springer Nature Singapore Pte Ltd. 2020, corrected publication 2023

S. Kato, J. Fukue, *Fundamentals of Astrophysical Fluid Dynamics*,
Astronomy and Astrophysics Library,
https://doi.org/10.1007/978-981-15-4174-2_23

23.1.1 Radiation Acoustic Wave

In order to introduce a *radiation acoustic wave*, we first consider a very simple case, where the energy exchange between radiation and matter is ignored, and the interaction between radiation and matter takes place through only the momentum exchange. We assume that the initial unperturbed state is a static and homogeneous medium, where both the matter density ρ_0 and the radiation energy density E_0 are constant, while there are no motion and no radiative flux. The absorption and scattering opacities are assumed to be constant. We further assume the Eddington approximation.

In this simple case, the basic equations of radiation hydrodynamics for the small amplitude perturbed quantities become as follows:

$$\frac{\partial}{\partial t}\left(\frac{\rho_1}{\rho_0}\right) + \nabla \boldsymbol{v}_1 = 0, \tag{23.1}$$

$$\frac{\partial \boldsymbol{v}_1}{\partial t} + \frac{p_0}{\rho_0}\nabla\frac{\rho_1}{\rho_0} + \frac{p_0}{\rho_0}\nabla\frac{T_1}{T_0} = \frac{\kappa+\sigma}{c}4\pi \boldsymbol{H}_1, \tag{23.2}$$

$$\frac{p_0}{\gamma-1}\frac{\partial}{\partial t}\left(\frac{T_1}{T_0}\right) - p_0\frac{\partial}{\partial t}\left(\frac{\rho_1}{\rho_0}\right) = 0, \tag{23.3}$$

$$\frac{1}{c}\frac{\partial J_1}{\partial t} + \nabla \boldsymbol{H}_1 = 0, \tag{23.4}$$

$$\frac{1}{c}\frac{\partial \boldsymbol{H}_1}{\partial t} + \frac{1}{3}\nabla J_1 = -\rho_0(\kappa+\sigma)\boldsymbol{H}_1, \tag{23.5}$$

$$\frac{p_1}{p_0} = \frac{\rho_1}{\rho_0} + \frac{T_1}{T_0}, \tag{23.6}$$

where and in what follows, subscripts "0" and "1" represent the unperturbed and perturbed quantities, respectively.

Looking at the above set of equations, we see easily that perturbations are separated into two modes. One is the adiabatic acoustic mode, when $J_1 = \boldsymbol{H}_1 = 0$. This mode is obviously described by

$$\left(\frac{\partial^2}{\partial t^2} - c_s^2\nabla^2\right)\rho_1 = 0, \tag{23.7}$$

where c_s ($\equiv \gamma p_0/\rho_0$) is the adiabatic sound speed.

The other is the mode expressing perturbations of radiation field, and uncoupled with fluid motions; i.e., Eqs. (23.4) and (23.5) are a closed set. This set of equations gives

$$\frac{\partial^2 J_1}{\partial t^2} - \frac{c^2}{3}\nabla^2 J_1 + c\rho_0(\kappa+\sigma)\frac{\partial J_1}{\partial t} = 0. \tag{23.8}$$

Substituting plane wave solutions in the form of $\exp(i\omega t - i\boldsymbol{k}\cdot\boldsymbol{x})$ into this equation, we have a dispersion relation for the radiation acoustic wave:

$$\omega^2 - \frac{c^2}{3}k^2 - i\omega\rho_0(\kappa+\sigma)c = 0. \tag{23.9}$$

In this dispersion relation,

$$c_{\rm rad} = \frac{c}{\sqrt{3}} \tag{23.10}$$

is the radiation "sound" speed, or the sound speed of the relativistic fluid. The factor 3 comes from the Eddington approximation, or originates from the isotropy of radiation fields. It is noticed that the perturbations associated with this oscillation are not only J_1 and $\boldsymbol{H}_1$, but fluid quantities such as v_r and ρ_1 also vary.

In the case of a temporary varying disturbance under an initial-value problem, which corresponds to complex ω and real k, the third term on the left-hand side of Eq. (23.9) represents the damping of the radiation acoustic wave. That is, if we expresses the angular frequency ω as $\omega = \omega_{\rm R} + i\omega_{\rm I}$, from Eq. (23.9) we have

$$\omega_{\rm R} = \pm\sqrt{c_{\rm rad}^2 k^2 - \omega_{\rm I}^2}, \tag{23.11}$$

$$\omega_{\rm I} = \frac{\rho_0(\kappa+\sigma)c}{2}. \tag{23.12}$$

Hence, the phase velocity decreases, and the wave damps as $\exp(-\omega_{\rm I}t)$.

On the other hand, in the case of a spatially changing disturbance under a boundary-value problem, which corresponds to real ω and complex k, the third term on the left-hand side of Eq. (23.9) represents the spatial propagation or evanescence. That is, if we expresses the wave number k as $k = k_{\rm R} - ik_{\rm I}$, from Eq. (23.9) we have

$$k_{\rm R} = \frac{1}{\sqrt{2}\ell}\sqrt{1+\sqrt{1+\tau_{\rm c}^2}}, \tag{23.13}$$

$$k_{\rm I} = \frac{1}{\sqrt{2}\ell}\frac{\tau_{\rm c}}{\sqrt{1+\sqrt{1+\tau_{\rm c}^2}}}, \tag{23.14}$$

where

$$\ell \equiv \frac{c}{\sqrt{3}\omega} \tag{23.15}$$

is the typical length associated with a disturbance of frequency ω traveling at the speed of radiation acoustic wave, while

$$\tau_{\rm c} \equiv \frac{\rho_0(\kappa+\sigma)c}{\omega} = \rho_0(\kappa+\sigma)\sqrt{3}\ell \tag{23.16}$$

is the optical thickness associated with a disturbance of frequency ω traveling at the speed of light. The phase speed v_p and the normalized damping length L/Λ ($\Lambda \equiv 2\pi/k_\mathrm{R}$ and $L \equiv 1/k_\mathrm{I}$) become, respectively, as

$$v_\mathrm{p} = \frac{c}{\sqrt{3}} \frac{\sqrt{2}}{\sqrt{1+\sqrt{1+\tau_\mathrm{c}^2}}}, \tag{23.17}$$

$$\frac{L}{\Lambda} \equiv \frac{k_\mathrm{R}}{2\pi k_\mathrm{I}} = \frac{1}{2\pi} \frac{1+\sqrt{1+\tau_\mathrm{c}^2}}{\tau_\mathrm{c}}. \tag{23.18}$$

23.1.2 Acoustic and Radiation Diffusion Waves

Next, according to Mihalas and Mihalas (1984), we consider acoustic waves propagating in a radiative fluid, taking into account the energy exchange between radiation and matter. The initial unperturbed state is assumed to be static and homogeneous, and in the radiative equilibrium (RE), while we assume the local thermodynamic equilibrium (LTE) and Eddington approximation in the perturbed state. Furthermore, scattering is ignored, and the absorption opacity is assumed to be constant. Finally, we restrict our attention only to low frequency oscillations (acoustic modes) whose frequency ω is much lower than c/λ, where λ is the characteristic length of the perturbation. In other words, we neglect the time variation of the radiation fields.

In this LTE case, the linearized equations of radiation hydrodynamics for the perturbed quantities are

$$\frac{\partial}{\partial t}\left(\frac{\rho_1}{\rho_0}\right) + \nabla \boldsymbol{v}_1 = 0, \tag{23.19}$$

$$\frac{\partial \boldsymbol{v}_1}{\partial t} + \frac{p_0}{\rho_0}\nabla\frac{\rho_1}{\rho_0} + \frac{p_0}{\rho_0}\nabla\frac{T_1}{T_0} = \frac{\kappa}{c}4\pi \boldsymbol{H}_1, \tag{23.20}$$

$$\frac{p_0}{\gamma-1}\frac{\partial}{\partial t}\left(\frac{T_1}{T_0}\right) - p_0\frac{\partial}{\partial t}\left(\frac{\rho_1}{\rho_0}\right) = 4\pi\kappa\rho_0(J_1 - B_1), \tag{23.21}$$

$$\nabla \boldsymbol{H}_1 = -\kappa\rho_0(B_1 - J_1), \tag{23.22}$$

$$\nabla(f J_1) = -\kappa\rho_0 \boldsymbol{H}_1, \tag{23.23}$$

$$\frac{p_1}{p_0} = \frac{\rho_1}{\rho_0} + \frac{T_1}{T_0}, \tag{23.24}$$

$$\frac{B_1}{B_0} = 4\frac{T_1}{T_0}, \tag{23.25}$$

where f $(= 1/3)$ is the Eddington factor, and the last equation means the LTE condition ($B = \sigma_\mathrm{SB} T^4/\pi$).

Since we concentrate our attention to longitudinal motions, a velocity potential ϕ_1 is introduced as

$$\boldsymbol{v}_1 = \nabla \phi_1. \tag{23.26}$$

Furthermore, we introduce dimensionless variables, θ_1, j_1, h_1 by

$$\frac{T_1}{T_0} = \theta_1, \quad \frac{J_1}{B_0} = j_1, \quad \frac{\boldsymbol{H}_1}{B_0} = \boldsymbol{h}_1. \tag{23.27}$$

Eliminating ρ_1 from Eqs. (23.19) and (23.20), we have

$$\frac{\partial^2 \boldsymbol{v}_1}{\partial t^2} - \frac{c_s^2}{\gamma} \nabla^2 \boldsymbol{v}_1 + \frac{c_s^2}{\gamma} \nabla \frac{\partial \theta_1}{\partial t} = 0, \tag{23.28}$$

where $\partial \boldsymbol{h}_1 / \partial t$ is dropped, since we neglect the time variation of the radiation fields here. Using the velocity potential ϕ_1, this equation is rewritten as

$$\frac{\partial^2 \phi_1}{\partial t^2} - \frac{c_s^2}{\gamma} \nabla^2 \phi_1 + \frac{c_s^2}{\gamma} \frac{\partial \theta_1}{\partial t} = 0. \tag{23.29}$$

Moreover, eliminating ρ_1 from Eqs. (23.19) and (23.21), and using ϕ_1, we have

$$\begin{aligned} \frac{\partial \theta_1}{\partial t} + (\gamma - 1)\nabla^2 \phi_1 &= \frac{\gamma - 1}{p_0} 4\pi \kappa \rho_0 B_0 (j_1 - 4\theta_1) \\ &= 4\kappa \rho_0 \frac{\gamma c_s}{\mathcal{B}o} (j_1 - 4\theta_1), \end{aligned} \tag{23.30}$$

where

$$\mathcal{B}o = \frac{\rho_0 c_p c_s}{\sigma_{SB} T^3} = \frac{1}{\pi} \frac{\gamma}{\gamma - 1} \frac{p_0 c_s}{B_0} \tag{23.31}$$

is the Boltzmann number, which is on the order of 10 in the solar photosphere, $\sim 10^{-2}$ at the center of an O-type star, and $\sim 10^{-5}$ in an X-ray source. Finally, from Eqs. (23.22) and (23.23), we have

$$\frac{1}{(\kappa \rho_0)^2} \nabla^2 f j_1 = j_1 - 4\theta_1. \tag{23.32}$$

It should be noted that combining Eqs. (23.29) and (23.30) we have an acoustic wave equation modified by radiation fields:

$$\frac{\partial^2 \phi_1}{\partial t^2} - c_s^2 \nabla^2 \phi_1 = 4\kappa \rho_0 \frac{c_s^3}{\mathcal{B}o} (j_1 - 4\theta_1). \tag{23.33}$$

For plane wave solutions in the form of $\exp(i\omega t - i\boldsymbol{k}\cdot\boldsymbol{x})$, Eqs. (23.29), (23.30), and (23.32) give

$$\begin{pmatrix} -\omega^2 + \frac{c_s^2}{\gamma}k^2 & i\omega\frac{c_s^2}{\gamma} & 0 \\ -(\gamma-1)k^2 & i\omega + 16\kappa\rho_0\frac{\gamma c_s}{\mathcal{B}o} & -4\kappa\rho_0\frac{\gamma c_s}{\mathcal{B}o} \\ 0 & -4 & 1 + f\frac{1}{(\kappa\rho_0)^2}k^2 \end{pmatrix} \begin{pmatrix} \phi_1 \\ \theta_1 \\ j_1 \end{pmatrix} = 0. \quad (23.34)$$

After several manipulations, we have a dispersion relation, equation (101.39) in Mihalas and Mihalas (1984), as

$$\left(1 - i\frac{16\tau_a}{\mathcal{B}o}\right)\left(\frac{c_s k}{\omega}\right)^4 - \left(1 - \frac{1}{f}\tau_a^2 - i\gamma\frac{16\tau_a}{\mathcal{B}o}\right)\left(\frac{c_s k}{\omega}\right)^2 - \frac{1}{f}\tau_a^2 = 0, \quad (23.35)$$

where

$$\tau_a \equiv \frac{\kappa\rho_0 c_s}{\omega} \quad (23.36)$$

is the optical thickness with a disturbance of frequency ω traveling at the adiabatic sound speed.

This dispersion relation (23.35) is quadratic in $(c_s k/\omega)^2$, and has two type solutions:

$$\left(\frac{c_s k}{\omega}\right)^2 = \frac{1}{2\left(1 - i\frac{16\tau_a}{\mathcal{B}o}\right)}\left[\left(1 - i\gamma\frac{16\tau_a}{\mathcal{B}o} - \frac{1}{f}\tau_a^2\right)\right.$$
$$\left.\pm\sqrt{\left(1 - i\gamma\frac{16\tau_a}{\mathcal{B}o} - \frac{1}{f}\tau_a^2\right)^2 + 4\left(1 - i\frac{16\tau_a}{\mathcal{B}o}\right)\frac{\tau_a^2}{f}}\,\right], \quad (23.37)$$

where the plus sign corresponds to a *radiation-modified acoustic wave*, while the minus sign a *radiation diffusion wave*, as classified by Mihalas and Mihalas (1984) (see also Bisnovatyi-Kogan and Blinnikov 1978). According to Mihalas and Mihalas (1984), we shall examine several limiting cases, bearing in mind a spatially changing disturbance under the boundary-value problem (real ω and complex k).

In the limit of the Boltzmann number $\mathcal{B}o \to \infty$, the energy exchange between radiation and matter is ignored, and the dispersion relation (23.35) reduces to the results in the previous subsection, although the radiation acoustic mode does not appear [see also Eq. (23.33)],

In the limit of $\tau_a \ll 1$ (small optical thickness and/or high frequency), from the plus-sign solution of (23.37), we have

$$\frac{c_s k}{\omega} \sim 1 - i\frac{8(\gamma-1)}{\mathcal{B}o}\tau_a, \quad (23.38)$$

which is a weakly damped acoustic wave, whose phase velocity $v_{\rm p}$ and the normalized damping length L/Λ become, respectively, as

$$v_{\rm p} = c_{\rm s}, \tag{23.39}$$

$$\frac{L}{\Lambda} = \frac{\mathcal{B}o}{16\pi(\gamma-1)\tau_{\rm a}} \gg 1, \tag{23.40}$$

where $\Lambda \equiv 2\pi/k_{\rm R}$ and $L \equiv 1/k_{\rm I}$, the wavenumber k being $k = k_{\rm R} + k_{\rm I}$. From the minus-sign solution, on the other hand, we have

$$\frac{c_{\rm s}k}{\omega} \sim \frac{8\gamma}{\sqrt{f}\mathcal{B}o}\tau_{\rm a}^2 - \frac{i}{\sqrt{f}}\tau_{\rm a}, \tag{23.41}$$

which is a strongly damped radiation diffusion wave, whose phase velocity and the normalized damping length become, respectively, as

$$v_{\rm p} = \frac{\sqrt{f}\mathcal{B}o}{8\gamma\tau_{\rm a}^2}, \tag{23.42}$$

$$\frac{L}{\Lambda} = \frac{4\gamma\tau_{\rm a}}{\pi\mathcal{B}o}. \tag{23.43}$$

It should be noted that the Eddington approximation becomes worse in the optically thin limit. Such a situation has shown by, e.g., Bogdan et al. (1996) without imposing the Eddington approximation (see also Dzhalilov et al. 1992). In their analysis, the acoustic mode is not changed so much, but the radiation diffusion mode ceases to exist in the optically thin regime (sufficiently high frequency).

In the limit of $\tau_{\rm a} \gg 1$ (large optical thickness and/or low frequency), from the plus-sign solution of (23.37), we have

$$\frac{c_{\rm s}k}{\omega} \sim 1 - i\frac{8f(\gamma-1)}{\mathcal{B}o}\frac{1}{\tau_{\rm a}}, \tag{23.44}$$

which is again a weakly damped acoustic wave, whose phase velocity and the normalized damping length become, respectively, as

$$v_{\rm p} = c_{\rm s}, \tag{23.45}$$

$$\frac{L}{\Lambda} = \frac{\mathcal{B}o}{16\pi(\gamma-1)f}\tau_{\rm a} \gg 1. \tag{23.46}$$

From the minus-sign solution, on the other hand, we have

$$\frac{c_{\rm s}k}{\omega} \sim \sqrt{\frac{\mathcal{B}o\tau_{\rm a}}{32f}}(1-i), \tag{23.47}$$

which is a strongly damped radiation diffusion wave, whose phase velocity and the normalized damping length become, respectively, as

$$v_{\mathrm{p}} = c_{\mathrm{s}}\sqrt{\frac{32f}{\mathcal{B}o\tau_{\mathrm{a}}}} \ll 1, \tag{23.48}$$

$$\frac{L}{\Lambda} = \frac{1}{2\pi}. \tag{23.49}$$

In general, Eq. (23.37) must be solved numerically. Mihalas and Mihalas (1984) showed that the acoustic mode is adiabatic at high and low frequencies (small and large τ_{a} regimes), but is isothermal at around $\tau_{\mathrm{a}} \sim 1$ over a range approximately inversely proportional to $\mathcal{B}o$. More detailed investigations on the propagation of acoustic waves in a radiative fluid are found in Mihalas and Mihalas (1984).

23.2 Gravity Waves in the Stratified Radiative Atmosphere

We here briefly comment gravity waves in a stratified radiative fluid. The propagation of gravity waves in a radiative fluid has been investigated in the optically thin case under the Newton's cooling law (Souffrin 1966, 1972; see Mihalas and Mihalas 1984), and in the optically thick case under the Eddington approximation (Spiegel and Tao 1999; Shaviv 2001; Blaes and Socrates 2001, 2003). Similar to the acoustic wave in the radiative fluid, the main effect of radiation (diffusion) is to dampen the waves. In general, we no longer obtain either pure propagating or pure standing waves separately in the diagnostic diagram, unlike in the stratified atmosphere without radiation. Radiative hydrodynamic instabilities were also investigated (Spiegel and Tao 1999; Shaviv 2001; Blaes and Socrates 2001, 2003). Particularly, Blaes and Socrates (2003) has extensively examined the radiative damping and unstable driving of short-wavelength, propagating hydrodynamic and magnetohydrodynamic waves in static, optically thick, stratified equilibria.

However, the issue of the radiative instabilities and waves in the stratified radiative atmosphere is still under a controversy in several aspects. For example, some researchers reported the radiation-hydrodynamic instabilities of acoustic waves in the purely scattering atmosphere (e.g., Spiegel and Tao 1999; Shaviv 2001), while some found no acoustic wave instabilities in the same situation (e.g., Blaes and Socrates 2003). Of these, Spiegel and Tao (1999) used the radiative quantities in the inertial frame, whereas Shaviv (2001) and Blaes and Socrates (2003) adopted the comoving quantities on the order of $\mathcal{O}(v/c)^1$.

If matter and radiation are strongly coupled, they behave as a single fluid with a modified polytropic index. If, alternatively, the unperturbed fluid is not static, but moving at speed v, advection dominates diffusion in the dynamic diffusion limit of $\tau \gg c/v$. In such cases, the radiative diffusion has negligible impact on the large scale flow, and a gas-radiation mixture can be regarded as an ideal fluid. Johnson

(2009) examined and summarized simple waves in such an ideal radiative fluids (see also Chandrasekhar 1967; Mihalas and Mihalas 1984).

23.3 Radiative Shocks

In the universe, there are various phenomena associating shock waves, such as supernova explosions, winds or accretion flows from or toward compact objects, dynamical star formation, astrophysical jets, and other various outburst phenomena. Besides astrophysical phenomena, shock waves are associated with various natural phenomena, such as volcano explosions and thunderbolts, or with various artificial origins, such as supersonic planes, blasting, and nuclear bombs, or with laboratory experiments. If the shock is sufficiently strong, then the radiation energy flux and/or radiation pressure play an essential role in the shock flows—*radiative shocks*. The radiative influence on the shock waves has been investigated in many research fields, and since Zel'dovich (1957) and Raizer (1957), there are many literatures treating radiative shocks (e.g., Zel'dovich and Raizer 1967, 2002; Weaver 1976; Mihalas and Mihalas 1984; Lacey 1988; Bouquet et al. 2000; Drake 2005, 2007; Lowrie et al. 1999; Lowrie and Rauenzahn 2007; Lowrie and Edwards 2008; Tolstov et al. 2015; Ferguson et al. 2017; Fukue 2019a).

In this section, we show the typical properties of radiative shocks in the optically thick medium in three cases of (1) the strong equilibrium shocks, (2) the equilibrium diffusion shocks, and (3) the nonequilibrium shocks.

23.3.1 Radiative Influence on Shock Waves

Before considering the radiative shocks in the optically thick regime, in this subsection, we first summarize the radiative influence on shock waves; the typical structure of the radiative shock front, and classifications of radiative shocks.

In radiative shocks, radiation emitted from the shock front generally affects both sides of the front. The radiative diffusion in the optically thick regime, or the direct radiation in the optically thin regime transports radiation energy both into the upstream medium in the pre-shock region before the shock front and the downstream medium in the post-shock region after the front.

The typical structure of radiative shocks is depicted in Fig. 23.1. In the pre-shock region ahead of the shock front, radiation emitted from the shock front heats up and/or ionizes the upstream medium to make the *radiative precursor*, whereas far from the shock there remains the undisturbed upstream medium. Even if the upstream medium is optically thick, the radiative precursor can be formed by the radiative diffusion. In the post-shock region after the shock front, the downstream medium is heated up by the shock transition, and emits radiation (*radiative zone*). The shock heats ions, and then, electrons and ions equilibrate (*relaxation zone*).

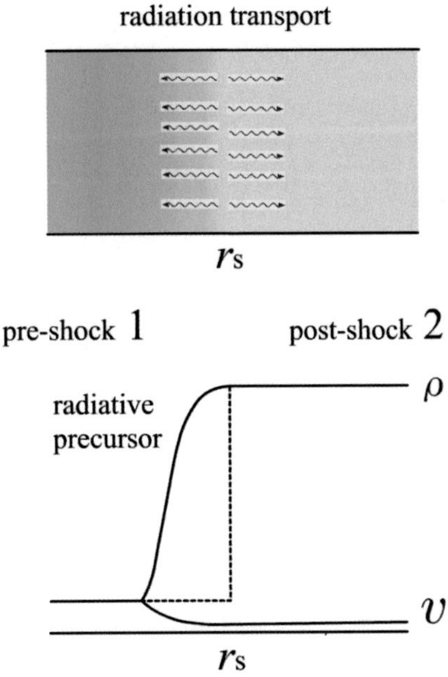

Fig. 23.1 Typical structure of radiative shocks. The radiative precursor appears in the pre-shock region, while the radiative zone and thermalization zone extends in the post-shock region

Hence, in general, one may treat two temperature plasmas just behind the shock front. However, we often assume the immediate equilibration of ions and electrons, or one often ignores the relaxation zone, where ions and electrons equilibrate, just behind the shock front. After ions and electrons equilibrate to a single temperature, in the *thermalization zone* behind the shock front, radiation emitted from the shock front (and relaxation zone) is absorbed by the medium in this region, and reemitted as blackbody, to settle down the post-shock state. Even if the shock is optically thick, and ions and electrons equilibrate, the radiation temperature generally differs from the gas one. Hence, in such a case, the relaxation zone may appear, as a region, where radiation and matter equilibrate (*nonequilbrium diffusion*). In the *equilibrium diffusion approximation*, on the other hand, it is assumed that the radiation temperature is always equal to the gas one.

The important quantities for the radiative shock behavior are the optical depth of the upstream and downstream regions. We may classify radiative shocks via the optical depth into four types (cf. Drake 2005).[1]

[1] Due to the shock strength and temperature distribution, the radiative shocks are classified into two types; the *subcritical shocks* and the *supercritical shocks* (e.g., Zel'dovich and Raizer 1967). In the radiative shocks, the radiation energy flux from the post-shock region penetrates into the upstream material and heats up to make the radiative precursor. When the temperature T_p just ahead of the shock front is smaller than the temperature T_2 in the post-shock region ($T_p < T_2$), such a shock is called a *subcritical shock*. This pre-shock temperature T_p increases rapidly with the shock strength, and eventually equals the post-shock one T_2 (a *critical shock*). For higher shock strength,

Thick-Thick shocks:

Both the upstream and downstream media are optically thick. If the both sides of the shock front are optically thick, the medium may be the local thermodynamic equilibrium (LTE) everywhere. If further both sides are sufficiently thick, radiation and matter behave as a single fluid (*strong equilibrium limit*), and in the radiation-dominated case the density compression ratio becomes up to a factor 7 (corresponding to $\gamma = 4/3$ in the nonradiative shocks), in contrast to 4 (corresponding to $\gamma = 5/3$ in the nonradiative shocks). In addition, in some situations, there is no density jump, but the density transition becomes continuous. In some cases, the equilibrium diffusion approximation does not hold, and one may consider the nonequilibrium diffusion, where radiation and matter do not equilibrate. Shocks in stellar interiors such as a blast wave within supernovae, shocks in the atmospheres of pulsating variables, and shocks in winds or accretion flows onto a gravitating body are of this type.

Thin-Thin shocks:

Both the upstream and downstream media are optically thin. If the both sides of the shock are optically thin, radiation penetrates into the upstream medium, and the entire downstream region becomes a radiative cooling layer until large enough distance. The density increase in this type is formally unlimited (corresponding to $\gamma = 1$ in the nonradiative shocks). This type of thin-thin shocks is commonly observed in astrophysics, such as type II supernovae in red supergiants, many shock-cloud interactions, and shocks in astrophysical jets.

Thin-Thick shocks:

The upstream medium is optically thin, while the downstream one is optically thick. In this type, similar to the thin-thin shocks, radiation emitted from the shock front may penetrates deep into the upstream medium to make the radiative precursor, while there exist a radiative and thermalization zone (radiative cooling layer) in the post-shock downstream region. Astrophysically, this type is seen, e.g., in the blast wave emerging from the star during the supernova explosion, and in accretion shocks during star formation.

Thick-Thin shocks:

The upstream medium is optically thick, while the downstream one is optically thin. Since the gaseous matter is usually compressed at the shock front, this type is difficult to realize in nature. The collision between the sufficiently optically thick cloud and supernova remnants may become of this type.

For the radiative shock in the optically thick medium, consult, e.g., Bouquet et al. (2000), Drake (2007), Lowrie and Rauenzahn (2007), Lowrie and Edwards (2008). For the radiative shock in the optically thin medium, see, e.g., Lacey (1988) and references therein, Drake (2005) and references therein.

---

the pre-shock temperature cannot exceed the post-shock one, and the excess energy flux enlarges the radiative precursor into the upstream region. Such a state is called a *supercritical shock*.

In the following sections, we consider the optically thick shocks, which are important in many radiation hydrodynamical flows, although the optically thin shocks are important, e.g., in the interstellar medium.

23.3.2 *Conservative Form of Radiation Hydrodynamical Equations*

In this section we summarize the basic equations of radiation hydrodynamics derived in Chap. 21 in the conservative forms. In contrast to Chap. 2 under the single-fluid style, where matter and radiation are assumed to be strongly coupled, and behave as a single fluid, we here adopt the two fluids style, where matter and radiation are not strongly coupled, and behave as two different fluids (see also Sect. 21.1.3).

Since radiation has no mass, the continuity equation does not contain radiative terms:

$$\frac{\partial \rho}{\partial t} + \nabla(\rho \boldsymbol{v}) = 0. \tag{23.50}$$

Including the radiative force, the equation of motion for matter in the nonrelativistic regime is written as

$$\frac{\partial \boldsymbol{v}}{\partial t} + (\boldsymbol{v} \cdot \nabla)\, \boldsymbol{v} = -\frac{1}{\rho} \nabla p + \frac{(\kappa_F + \sigma_F)}{c} \boldsymbol{F}, \tag{23.51}$$

where $\boldsymbol{F}$ is the radiative flux, and we drop the external gravity and radiation drag force. With the help of continuity Eq. (23.50), this equation is written as

$$\frac{\partial}{\partial t}(\rho \boldsymbol{v}) + \nabla(\rho \boldsymbol{v} \cdot \boldsymbol{v}) + \nabla p = \rho \frac{(\kappa_F + \sigma_F)}{c} \boldsymbol{F}. \tag{23.52}$$

On the other hand, the first-moment equation of radiation is expressed as

$$\frac{1}{c^2} \frac{\partial \boldsymbol{F}}{\partial t} + \nabla P = -\rho \frac{(\kappa_F + \sigma_F)}{c} \boldsymbol{F}, \tag{23.53}$$

where we assume the isotropic radiation field, P being the radiation pressure, and drop the subrelativistic correction term for simplicity. These Eqs. (23.52) and (23.53) are combined into the conservation of the net momentum:

$$\frac{\partial}{\partial t}\left(\rho \boldsymbol{v} + \frac{1}{c^2} \boldsymbol{F}\right) + \nabla\,(\rho \boldsymbol{v} \cdot \boldsymbol{v} + p + P) = 0. \tag{23.54}$$

It should be noted that if we take into account the relativistic effect up to the order of $\mathcal{O}(v/c)^1$, the radiative fluxes on the right-hand side should be read as $\boldsymbol{F}_{\rm co}$ in the comoving frame. However, the final conservative form (23.54) almost holds, since the terms on the right-hand side are canceled out.

In addition, the energy equation is expressed as

$$\frac{\partial}{\partial t}(\rho U) + \nabla(\rho U \boldsymbol{v}) + p\nabla \boldsymbol{v} = -\rho(j - c\kappa_E E), \tag{23.55}$$

where we have dropped the internal heating, and the subrelativistic correction terms for simplicity (Chap. 21). On the other hand, the zeroth-moment equation of radiation gives the net energy transfer rate to matter per unit volume:

$$\frac{\partial E}{\partial t} + \nabla \boldsymbol{F} = \rho(j - c\kappa_E E), \tag{23.56}$$

where we have again dropped the subrelativistic correction terms for simplicity. These Eqs. (23.55) and (23.56) are combined into the conservation of the net energy flux:

$$\frac{\partial}{\partial t}\left[\rho\left(\frac{\boldsymbol{v}^2}{2} + U\right) + E\right] + \nabla\left[\rho\boldsymbol{v}\left(\frac{\boldsymbol{v}^2}{2} + U + \frac{p}{\rho}\right) + \boldsymbol{F}\right] = 0. \tag{23.57}$$

It should be noted that if we take into account the relativistic effect up to the order of $\mathcal{O}(v/c)^1$, the energy density on the right-hand side should be read as $E_{\rm co}$ in the comoving frame. However, the final conservative form (23.57) does not change, although the radiation energy density and radiative flux on the left-hand side are quantities measured in the inertial frame.

In order to apply the above equations to the shock transition, the quantities should be generally evaluated in the comoving frame, since the shock front generally propagates in space. Furthermore, we consider the one-dimensional flow, bearing in mind the shock transition; i.e., we retain the normal components perpendicular to the shock front, the flow velocity v, the radiative flux F, and the radiation pressure P. Up to the order of $\mathcal{O}(v/c)^1$, the radiative quantities in the inertial frame are expressed by the quantities in the comoving (fluid) frame as (Chap. 24)

$$E = E_{\rm co} + 2\frac{v}{c^2}F_{\rm co}, \tag{23.58}$$

$$F = F_{\rm co} + v(E_{\rm co} + P_{\rm co}), \tag{23.59}$$

$$P = P_{\rm co} + 2\frac{v}{c^2}F_{\rm co}. \tag{23.60}$$

Substituting these expressions into Eqs. (23.54) and (23.57), we have the following appropriate forms for the shock transition:

$$\frac{\partial}{\partial t}\left(\rho v + \frac{1}{c^2}F_{\rm co}\right) + \frac{\partial}{\partial x}\left(\rho v^2 + p + P_{\rm co}\right) = 0, \tag{23.61}$$

$$\frac{\partial}{\partial t}\left[\rho\left(\frac{v^2}{2} + \frac{e + E_{\rm co}}{\rho}\right)\right] + \frac{\partial}{\partial x}\left[\rho v\left(\frac{v^2}{2} + \frac{e + p + E_{\rm co} + P_{\rm co}}{\rho}\right) + F_{\rm co}\right] = 0, \tag{23.62}$$

where e ($= \rho U$) is the gas internal energy per unit volume, and we have dropped the higher order terms. In the followings, we drop the subscript "co" for simplicity.

23.3.3 *Strong Equilibrium Shocks*

Thick-thick shocks are treated and discussed as well in books (Zel'dovich and Raizer 1967, 2002; Mihalas and Mihalas 1984), and in various literatures (e.g., Lowrie et al. 1999; Bouquet et al. 2000). If both the upstream and downstream media are sufficiently optically thick, radiation and matter are strongly coupled together. Furthermore, the radiative (diffusive) flux in the total energy conservation (23.62) is completely omitted, and radiation and matter are treated as a single fluid. Such a state is called the *strong equilibrium regime* or the *adiabatic limit*, since the radiative flux is omitted and the energy equation satisfies the adiabatic condition.

In such a state, the jump conditions relating the equilibrium states of the upstream and downstream flows become quite similar to those of the nonradiative shocks in Sect. 4.6:

$$[\rho u] = 0, \tag{23.63}$$

$$\left[\rho u^2 + p + P\right] = 0, \tag{23.64}$$

$$\left[\frac{1}{2}u^2 + \frac{e + p + E + P}{\rho}\right] = 0, \tag{23.65}$$

where u is the flow velocity relative to the shock front, p the gas pressure, P the radiation pressure in the fluid frame, e the internal energy per unit volume, E the radiation energy density in the fluid frame, and square brackets denote the difference across the shock front. In the case of ideal gases, furthermore, $e = p/(\gamma - 1)$, γ being the ratio of specific heats. For radiation, on the other hand, we here use the Eddington approximation; $E = 3P$. This strong equilibrium shock is a direct extension of conventional fluid shocks to situations where the radiation energy density and pressure are significant. Hence, this strong equilibrium regime occurs in the high energy flows.

Using these jump conditions, similar to the nonradiative shocks, we can derive the *Rankine–Hugoniot relation* for the present radiative shocks under a strong equilibrium regime:

$$\frac{e_2 + p_2 + E_2 + P_2}{\rho_2} - \frac{e_1 + p_1 + E_1 + P_1}{\rho_1} \\ -\frac{1}{2}\left[p_2 + P_2 - (p_1 + P_1)\right]\left(\frac{1}{\rho_1} + \frac{1}{\rho_2}\right) = 0, \tag{23.66}$$

where the subscripts "1" and "2" represent the quantities in front of the shock and after the shock, respectively. For ideal gases and isotropic radiation fields, this Rankine–Hugoniot relation is rewritten as

$$\left(\frac{\gamma}{\gamma-1}p_2 + 4P_2\right)\frac{1}{\rho_2} - \left(\frac{\gamma}{\gamma-1}p_1 + 4P_1\right)\frac{1}{\rho_1} \\ -\frac{1}{2}\left[p_2 + P_2 - (p_1 + P_1)\right]\left(\frac{1}{\rho_1} + \frac{1}{\rho_2}\right) = 0. \tag{23.67}$$

Here, we introduce the ratio of the gas pressure to the total one, β, and the generalized adiabatic exponent Γ (cf. Sect. 19.6), and rewrite the pressure part as

$$\frac{\gamma}{\gamma-1}p + 4P = \left[\frac{\gamma}{\gamma-1}\beta + 4(1-\beta)\right](p+P) = \frac{\Gamma}{\Gamma-1}(p+P), \tag{23.68}$$

where

$$\beta \equiv \frac{p}{p+P}, \tag{23.69}$$

$$\Gamma \equiv \frac{\beta + (\gamma-1)(4-3\beta)}{\beta + 3(\gamma-1)(1-\beta)}. \tag{23.70}$$

In terms of this generalized adiabatic exponent, the Rankine–Hugoniot relation (23.67) is reexpressed as

$$\frac{\Gamma_2}{\Gamma_2-1}\frac{(p+P)_2}{\rho_2} - \frac{\Gamma_1}{\Gamma_1-1}\frac{(p+P)_1}{\rho_1} \\ -\frac{1}{2}\left[(p+P)_2 - (p+P)_1\right]\left(\frac{1}{\rho_1} + \frac{1}{\rho_2}\right) = 0, \tag{23.71}$$

where the values of the generalized adiabatic index before and after the shock, Γ_1 and Γ_2, are generally different. This change can be determined by the shock condition, although we do not describe it in details.

If, however, β and therefore Γ are assumed to be constant during the shock transition, this Rankine–Hugoniot relation (23.71) is formally the same as that of the nonradiative shock (Sect. 4.6.3), and we can obtain the same form of the shock relations. For example, for the compressibility of gas density we have

$$\frac{\rho_2}{\rho_1} = \frac{(\Gamma - 1) + (\Gamma + 1)\dfrac{(p+P)_2}{(p+P)_1}}{(\Gamma + 1) + (\Gamma - 1)\dfrac{(p+P)_2}{(p+P)_1}}. \tag{23.72}$$

The figure depicting this relation is also the same as Fig. 4.15, if we replace γ and p by Γ and $(p + P)$, respectively.

In the case of nonradiative shocks ($\beta = 1$), the maximum compression is 4 for $\gamma = 5/3$. In the case of radiative shocks, on the other hand, the maximum compression becomes 7 when $\beta = 0$ and $\Gamma = 4/3$.

23.3.4 *Equilibrium Diffusion Shocks*

Except for the extremely thick case, radiation from shock fronts usually diffuses out into the pre-shock and post-shock regions, and affects the structure of shocked flows. For example, physical quantities such as temperature can continuously change in some extent, including the radiative precursor, shock front, and after-shock region. We here consider an *equilibrium diffusion approximation*, where the radiation temperature $T_{\rm rad}$ is equal to the gas temperature $T_{\rm gas}$; $T_{\rm rad} = T_{\rm gas} = T$, which is often referred as one temperature approximation or $1 - T$ limit. In addition, we focus our attention on the radiative precursor formed in the pre-shock region, and assume that the radiative diffusion does not affect the structure of the post-shock region for simplicity.

In such a radiative shock, the radiative flux in the total energy conservation (23.62) is retained, and the structure relations of the jump conditions and the radiative precursor are expressed as follows:

$$\rho u = \rho_1 u_1 = j, \tag{23.73}$$

$$\rho u^2 + p + P = \rho_1 u_1^2 + p_1 + P_1, \tag{23.74}$$

$$u\left(\frac{1}{2}\rho u^2 + e + p + E + P\right) + F = u_1\left(\frac{1}{2}\rho_1 u_1^2 + e_1 + p_1 + E_1 + P_1\right), \tag{23.75}$$

where the subscript "1" represents the quantities in the upstream region far from the shock, and no subscript means those in other regions including the radiative precursor and downstream region. The diffusive flux F is expressed in terms of the radiation pressure or temperature as a function of the coordinate x along the shock flow:

$$F = -\frac{c}{\kappa\rho}\frac{dP}{dx} = -\frac{4acT^3}{3\kappa\rho}\frac{dT}{dx}, \tag{23.76}$$

which is the first-moment equation of radiation (diffusion form). These equations are supplemented by equation of state and the opacity expression. In what follows, for simplicity, the opacity is assumed to be constant.

It should be noted again that the radiative quantities are evaluated in the fluid frame, where we can use the diffusion approximation (23.76). In addition, these generalized Rankine–Hugoniot jump conditions for radiative fluids were firstly derived and examined by Marshak (1958).

23.3.4.1 Overall Jump Conditions

Far from the shock, radiation and matter equilibrate, and the diffusive flux can be dropped. Hence, for ideal gases and isotropic radiation fields, the overall jump conditions are

$$\rho_2 u_2 = \rho_1 u_1 = j, \tag{23.77}$$

$$\rho_2 u_2^2 + p_2 + P_2 = \rho_1 u_1^2 + p_1 + P_1, \tag{23.78}$$

$$\frac{1}{2}u_2^2 + \frac{\gamma}{\gamma-1}\frac{p_2}{\rho_2} + \frac{4P_2}{\rho_2} = \frac{1}{2}u_1^2 + \frac{\gamma}{\gamma-1}\frac{p_1}{\rho_1} + \frac{4P_1}{\rho_1}, \tag{23.79}$$

where the subscripts “1” and “2” represent the quantities in the upstream and downstream regions far from the shock.

Similar to the hydrodynamical shock and that in the strong equilibrium regime, eliminating u_2, from the energy and momentum conservation equations, we have the Rankine–Hugoniot relations, respectively, as

$$\frac{\gamma}{\gamma-1}\frac{p_2}{\rho_2} + \frac{4P_2}{\rho_2} - \left(\frac{\gamma}{\gamma-1}\frac{p_1}{\rho_1} + \frac{4P_1}{\rho_1}\right)$$
$$-\frac{1}{2}\left[p_2 + P_2 - (p_1 + P_1)\right]\left(\frac{1}{\rho_1} + \frac{1}{\rho_2}\right) = 0, \tag{23.80}$$

$$\frac{p_2 + P_2 - (p_1 + P_1)}{\rho_2 - \rho_1} = \frac{\rho_1}{\rho_2}a_1^2\mathcal{M}_1^2 = \frac{1}{\rho_2}\gamma p_1\mathcal{M}_1^2, \tag{23.81}$$

where a_1 ($\equiv \sqrt{\gamma p_1/\rho_1}$) is the sound speed in the upstream region, and $\mathcal{M}_1$ ($\equiv u_1/a_1$) the Mach number of the upstream flow.

In contrast to hydrodynamical shocks, there are three unknown variables (ρ_2, p_2, P_2) in Eq. (23.80) and (23.81), when ($\rho_1, p_1, P_1, \mathcal{M}_1$) are given. After substituting equations of state, $p = (\mathcal{R}/\mu)\rho T$ and $P = aT^4/3$, we have two equations for two unknown variables (ρ_2 and T_2), and further eliminating ρ_2, we have a single ninth-order polynomial equation on T_2 (e.g., Bouquet et al. 2000; Lowrie and Rauenzahn 2007). The detailed solution procedure is described in, e.g., Lowrie and Rauenzahn (2007). Here, in order to treat the problem semi-analytically, we consider the simplest two limits, the gas-pressure and radiation-one dominant cases, in the subsequent two subsections after the next subsection.

23.3.4.2 Radiative Precursor

Using the solutions of jump conditions, Eqs. (23.80) and (23.81), as boundary conditions, the structure of the radiative precursor can be obtained. For ideal gases and isotropic radiation fields, the energy conservation (23.75) including the radiative flux is expressed as

$$\frac{4acT^3}{3\kappa\rho}\frac{dT}{dx} = \left(\frac{1}{2}\rho u^2 + \frac{\gamma}{\gamma-1}p + 4P\right)u - \left(\frac{1}{2}\rho_1 u_1^2 + \frac{\gamma}{\gamma-1}p_1 + 4P_1\right)u_1. \tag{23.82}$$

Eliminating u ($= \rho_1 u_1/\rho$), and again introducing the Mach number $\mathcal{M}_1$ ($= u_1/a_1$) in the upstream region, we can arrange the above equation as

$$\frac{4acT^3}{3\kappa\rho}\frac{dT}{dx} = \left\{\frac{1}{2}\frac{\gamma p_1}{\rho_1}\mathcal{M}_1^2\left[\left(\frac{\rho_1}{\rho}\right)^2 - 1\right] + \frac{\gamma}{\gamma-1}\left(\frac{p}{\rho} - \frac{p_1}{\rho_1}\right) + \frac{4P}{\rho} - \frac{4P_1}{\rho_1}\right\}\rho_1 u_1. \tag{23.83}$$

Similarly, the momentum conservation (23.74) is rearranged as

$$p - p_1 = \rho_1 a_1^2 \mathcal{M}_1^2\left(1 - \frac{\rho_1}{\rho}\right) - (P - P_1). \tag{23.84}$$

Substituting equations of state $p = (\mathcal{R}/\mu)\rho T$ and $P = aT^4/3$ into Eqs. (23.83) and (23.84), we have two equations on the two variables (ρ and T):

$$\frac{4acT^3}{3\kappa\rho}\frac{dT}{dx} = \left\{\frac{1}{2}\gamma\frac{\mathcal{R}}{\mu}T_1\mathcal{M}_1^2\left[\left(\frac{\rho_1}{\rho}\right)^2 - 1\right] + \frac{\gamma}{\gamma-1}\frac{\mathcal{R}}{\mu}T_1\left(\frac{T}{T_1}-1\right)\right.$$
$$\left. + \frac{1}{\rho_1}\frac{4}{3}aT_1^4\left[\frac{\rho_1}{\rho}\left(\frac{T}{T_1}\right)^4 - 1\right]\right\}\rho_1 u_1, \quad (23.85)$$

$$\frac{\mathcal{R}}{\mu}\rho_1 T_1\left(\frac{\rho}{\rho_1}\frac{T}{T_1} - 1\right) = \gamma\frac{\mathcal{R}}{\mu}\rho_1 T_1\mathcal{M}_1^2\left(1 - \frac{\rho_1}{\rho}\right) - \frac{a}{3}T_1^4\left[\left(\frac{T}{T_1}\right)^4 - 1\right]. \quad (23.86)$$

The latter Eq. (23.86) is a quadratic in ρ, and therefore, the variable ρ can be analytically eliminated from the former Eq. (23.85) to yield an equation on T.

Introducing the nondimensional quantities, $\tilde{\rho}$ ($\equiv \rho/\rho_1$), $\tilde{T}$ ($\equiv T/T_1$), and $\tilde{x}$ ($\equiv x/\ell_1$, ℓ_1 being a typical length scale), we can finally express Eqs. (23.85) and (23.86), respectively, as

$$\frac{4\alpha_1}{\beta_1\tau_1}\frac{\tilde{T}^3}{\tilde{\rho}}\frac{d\tilde{T}}{d\tilde{x}} = \frac{\gamma}{2}\mathcal{M}_1^2\left(\frac{1}{\tilde{\rho}^2} - 1\right) + \frac{\gamma}{\gamma-1}\left(\tilde{T} - 1\right) + 4\alpha_1\left(\frac{\tilde{T}^4}{\tilde{\rho}} - 1\right), \quad (23.87)$$

$$\tilde{\rho} = \frac{1 + \gamma\mathcal{M}_1^2 - \alpha_1(\tilde{T}^4 - 1) - \sqrt{[1 + \gamma\mathcal{M}_1^2 - \alpha_1(\tilde{T}^4 - 1)]^2 - 4\gamma\mathcal{M}_1^2\tilde{T}}}{2\tilde{T}}, \quad (23.88)$$

where α_1 ($\equiv P_1/p_1$), β_1 ($\equiv u_1/c$), and τ_1 ($\equiv \kappa\rho_1\ell_1$). From the physical viewpoints, we here separate several parameters, although β_1 and τ_1 can be renormalized in the coordinate scale. In addition, the root for $\tilde{\rho}$ was chosen so as to $\tilde{\rho}(\tilde{T}_1) = 1$.

As was stated, in the following subsections we shall examine the two limiting cases.

23.3.4.3 Gas-Pressure Dominant Case

In this subsection we shall examine the gas-pressure dominated radiative shock, which is often an *isothermal shock* in a sense that there is no temperature jump but the density structure is discontinuous at the shock front.

In this case, the jump conditions are the same as those for the hydrodynamical shocks (Sect. 4.5). That is,

$$\tilde{\rho}_2 \equiv \frac{\rho_2}{\rho_1} = \frac{(\gamma+1)\mathcal{M}_1^2}{(\gamma-1)\mathcal{M}_1^2+2}, \tag{23.89}$$

$$\tilde{p}_2 \equiv \frac{p_2}{p_1} = \frac{2\gamma\mathcal{M}_1^2-(\gamma-1)}{\gamma+1}, \tag{23.90}$$

$$\tilde{T}_2 \equiv \frac{T_2}{T_1} = \frac{[2\gamma\mathcal{M}_1^2-(\gamma-1)][(\gamma-1)\mathcal{M}_1^2+2]}{(\gamma+1)^2\mathcal{M}_1^2}. \tag{23.91}$$

Using the after-shock temperature $\tilde{T}_2$ as a boundary condition, we can solve the structure equations for the radiative precursor. If we drop the radiation pressure from Eqs. (23.87) and (23.88), the structure equations for the radiative precursor in this gas-pressure dominated case become

$$\frac{4\alpha_1}{\beta_1\tau_1}\frac{\tilde{T}^3}{\tilde{\rho}}\frac{d\tilde{T}}{d\tilde{x}} = \frac{\gamma}{2}\mathcal{M}_1^2\left(\frac{1}{\tilde{\rho}^2}-1\right)+\frac{\gamma}{\gamma-1}\left(\tilde{T}-1\right), \tag{23.92}$$

$$\tilde{\rho} = \frac{1+\gamma\mathcal{M}_1^2-\sqrt{(1+\gamma\mathcal{M}_1^2)^2-4\gamma\mathcal{M}_1^2\tilde{T}}}{2\tilde{T}}, \tag{23.93}$$

where $\tilde{\rho} \equiv \rho/\rho_1$ and $\tilde{T} \equiv T/T_1$, as mentioned before.

Parameters are γ, α_1 ($\equiv P_1/p_1 \ll 1$), β_1 ($\equiv u_1/c$), and τ_1 ($\equiv \kappa\rho_1\ell_1$).

An example of solutions for the isothermal shock in the gas-pressure dominated case is shown in Fig. 23.2. The abscissa is the shock coordinate $\tilde{x}$ normalized by the typical length, whereas the ordinate is the physical quantities normalized by the corresponding ones in the pre-shock region. Parameters are $\gamma = 5/3$, $\alpha_1/(\beta_1\tau_1) = 0.00001$, and $\mathcal{M}_1 = 10$.

As is seen in Fig. 23.2, the radiative energy from the hot post-shock region diffuses into the pre-shock region to make the radiative precursor. In other words, in this radiative precursor the gas temperature continuously increases from the pre-shock temperature T_1 up to the post-shock one T_2. On the other hand, the density distribution is discontinuous, since the density $\rho(T_2)$ determined by Eq. (23.93) is often smaller than ρ_2 determined by the jump conditions. As a result, the radiative shock in the gas-pressure dominated case becomes the so-called isothermal shock in a sense that there is no temperature discontinuity.

This situation mentioned in the above paragraph and in Fig. 23.2 is easily established as follows, as already stated in the footnote of Sect. 23.3.1. In the radiative shocks, the radiation energy flux from the post-shock region penetrates into the upstream material and heats up to make the radiative precursor. As a result, the pre-shock temperature $T_{\rm p}$ just ahead of the shock front increases rapidly with the shock strength, and eventually equals the post-shock one T_2 (isothermal shock).

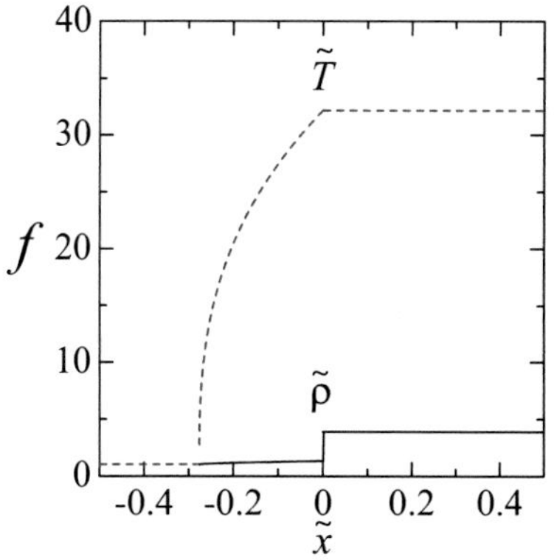

Fig. 23.2 Example of solutions of equilibrium diffusion shocks in the gas-pressure dominated case. The density distribution is discontinuous, while the temperature one is continuous. The density ρ and temperature T are normalized by ρ_1 and T_1, respectively, and x by a typical length scale. Parameters are $\gamma = 5/3$, $\alpha_1/(\beta_1\tau_1) = 0.00001$, and $\mathcal{M}_1 = 10$. In order to show the density discontinuity, we show the post-shock region, although we do not solve the post-shock structure

For higher shock strength, the pre-shock temperature cannot exceed the post-shock one, and the excess energy flux enlarges the radiative precursor into the upstream region. A rough estimation based on Eq. (23.92) shows that the extension of the radiative precursor, say $\tilde{x}_1$, is proportional to $[\alpha_1/(\beta_1\tau_1)]\tilde{T}_2^3$. Since $\tilde{T}_2$ depends on $\mathcal{M}_1$, the extension of the radiative precursor is some function of $\mathcal{M}_1$; $\tilde{x}_1 \sim [\alpha_1/(\beta_1\tau_1)]f(\mathcal{M}_1)$, although the concrete functional form of $f(\mathcal{M}_1)$ should be determined empirically (e.g., Fukue 2019b,c,d).

If the Mach number $\mathcal{M}_1$ is smaller than $\mathcal{M}_{\rm iso}$ given by

$$\mathcal{M}_{\rm iso} \sim \sqrt{\frac{3\gamma - 1}{\gamma(3-\gamma)}}, \tag{23.94}$$

the density distribution becomes continuous (Zel'dovich and Raizer 2002). In general, the radiation pressure cannot be completely ignored, and even if the amount of the radiation pressure is small, the radiative shock structure is separated into three domains for a given $\alpha_1 = P_1/p_1$ (Bouquet et al. 2000; Lowrie and Rauenzahn 2007). That is, an isothermal shock will occur over a range of $\mathcal{M}_1$:

$$\mathcal{M}_{\rm iso} < \mathcal{M}_1 < \mathcal{M}_{\rm cont}, \tag{23.95}$$

and outside of this range the shock structure, such as the density distribution, becomes continuous again. The upper boundary $\mathcal{M}_{\rm cont}$ depends on γ and α_1 in somewhat complicated way (Lowrie and Rauenzahn 2007):

$$\mathcal{M}_{\rm cont} \sim \left(\frac{1-2\eta_{\rm max}}{\alpha_1 \gamma^3 \eta_{\rm max}^8}\right)^{1/6}; \quad \eta_{\rm max} = \frac{1}{4+\sqrt{2+(\gamma+1)/(\gamma-1)}}, \tag{23.96}$$

or in the case of $\gamma = 5/3$

$$\mathcal{M}_{\rm cont} \sim 8.74\alpha_1^{-1/6}. \tag{23.97}$$

If the radiation pressure is large enough ($\alpha_1 = P_1/p_1 \sim 1.406$ for $\gamma = 5/3$), then an isothermal shock does not appear for any $\mathcal{M}_1$, and the shock structure, such as the density distribution, becomes continuous, as is shown in the next subsection.

23.3.4.4 Radiation-Pressure Dominant Case

In this subsection we shall examine the radiation-pressure dominated radiative shock, which is always a *continuous shock* in a sense that the density distribution is continuous at the shock front, in contrast to the usual hydrodynamical shock.

In this case, the Rankine–Hugoniot relations (23.80) and (23.81) become, respectively,

$$\frac{4P_2}{\rho_2} - \frac{4P_1}{\rho_1} - \frac{1}{2}(P_2 - P_1)\left(\frac{1}{\rho_1} + \frac{1}{\rho_2}\right) = 0, \tag{23.98}$$

$$\frac{P_2 - P_1}{\rho_2 - \rho_1} = \frac{\rho_1}{\rho_2} a_1^2 \mathcal{M}_1^2 = \frac{P_1}{\rho_1}\frac{\gamma}{\alpha_1}\mathcal{M}_1^2, \tag{23.99}$$

which are easily solve to yield

$$\tilde{\rho}_2 \equiv \frac{\rho_2}{\rho_1} = \frac{7\gamma\mathcal{M}_1^2/\alpha_1}{\gamma\mathcal{M}_1^2/\alpha_1 + 8}, \tag{23.100}$$

$$\tilde{P}_2 \equiv \frac{P_2}{P_1} = \frac{6\gamma\mathcal{M}_1^2/\alpha_1 - 1}{7}. \tag{23.101}$$

Furthermore, if we drop the gas pressure from Eqs. (23.87) and (23.88), the structure equations for the radiative precursor in this radiation-pressure dominated case can be expressed in terms of variables $\tilde{P}$ $(\equiv P/P_1)$ and $\tilde{\rho}$ $(\equiv \rho/\rho_1)$ without T as

$$\frac{1}{\beta_1 \tau_1} \frac{1}{\tilde{\rho}} \frac{d\tilde{P}}{d\tilde{x}} = \frac{1}{2} \frac{\gamma}{\alpha_1} \mathcal{M}_1^2 \left(\frac{1}{\tilde{\rho}^2} - 1 \right) + 4 \left(\frac{\tilde{P}}{\tilde{\rho}} - 1 \right), \tag{23.102}$$

$$\frac{1}{\tilde{\rho}} = 1 + \frac{1 - \tilde{P}}{\gamma \mathcal{M}_1^2 / \alpha_1}. \tag{23.103}$$

Parameters are γ, α_1 $(\equiv P_1/p_1 \gg 1)$, β_1 $(\equiv u_1/c)$, and τ_1 $(\equiv \kappa \rho_1 \ell_1)$.

Typical solutions for the continuous shock in the radiation-pressure dominated case are shown in Fig. 23.3. The abscissa is the shock coordinate $\tilde{x}$ normalized by the typical length, whereas the ordinate is the physical quantities normalized by the corresponding ones in the pre-shock region. Solid curves represent the density distribution, while dashed ones denote the pressure one. Parameters are (a) $\gamma = 5/3$, $\alpha_1 = 100$, $\beta_1 = 0.01$, $\tau_1 = 100$, and $\mathcal{M}_1 = 10, 10^2, 10^3, 10^4$, and (b) $\gamma = 5/3$, $\alpha_1 = 100$, $\beta_1 = 0.01$, $\mathcal{M}_1 = 100$, and $\tau_1 = 5, 10, 50, 100$.

As is seen in Fig. 23.3, similar to the gas-pressure dominated shock, the radiative energy from the hot post-shock region diffuses into the pre-shock region to make the radiative precursor. In the radiation-pressure dominated shock, however, both

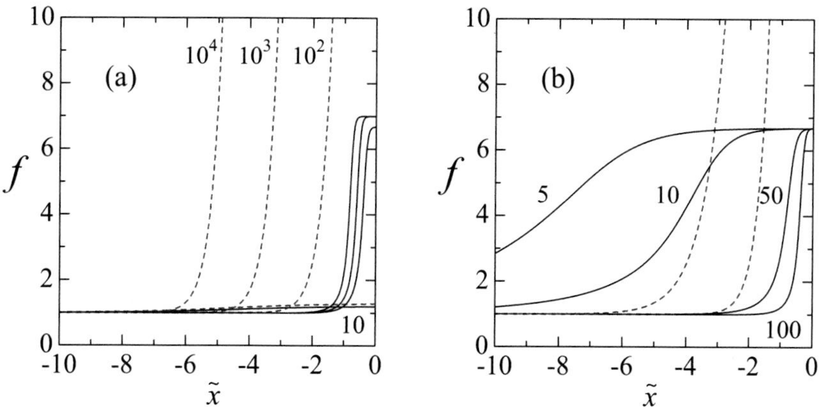

Fig. 23.3 Typical solutions of equilibrium diffusion shocks in the radiation-pressure dominated case. The density distributions are continuous. Solid curves represent the density distribution, while dashed ones denote the radiation-pressure one. The density ρ and radiation pressure are normalized by ρ_1 and P_1, respectively. Parameters are (**a**) $\gamma = 5/3$, $\alpha_1 = 100$, $\beta_1 = 0.01$, $\tau_1 = 100$, and $\mathcal{M}_1 = 10, 10^2, 10^3, 10^4$, and (**b**) $\gamma = 5/3$, $\alpha_1 = 100$, $\beta_1 = 0.01$, $\mathcal{M}_1 = 100$, and $\tau_1 = 5, 10, 50, 100$

the density and pressure distributions are continuous. As the Mach number $\mathcal{M}_1$ increases, the density compression approaches the maximum value of 7. In addition, the width of the radiation precursor depends on the typical optical depth.

It should be stressed again that several parameters are introduced from the physical viewpoints. However, parameters β_1 and τ_1 can be renormalized in the coordinate scale, and parameters γ, α_1, and $\mathcal{M}_1^2$ are combined into single parameter $\gamma \mathcal{M}_1^2/\alpha_1$. Hence, these solutions are mathematically a one-parameter family.

23.3.5 Nonequilibrium Diffusion Shocks

In the previous section, we have assumed that the radiation temperature is always equal to the gas one; $T_{\rm rad} = T_{\rm gas}$. In general, however, since the gas temperature rapidly increases at the shock, the radiation temperature cannot immediately equilibrate the gas one at the post-shock region; $T_{\rm rad} \neq T_{\rm gas}$. As a result, at the post-shock region there appears some type of the *relaxation zone*,[2] where the shock-heated gas cools down to its final post-shock equilibrium state with radiation. We here outline this general situation, called a *nonequilibrium diffusion shock*.

In such a radiative shock, we shall consider the physical changes of the total system of matter and radiation, and that of the fluid system, separately. In the former total system, radiation obeys the equilibrium diffusion equation, while it obeys the nonequilibrium diffusion equation in the latter system.

For example, the overall jump conditions for the total system of matter and radiation are the same as those for the equilibrium diffusion approximation:

$$(\rho u)_2 = (\rho u)_1 = j, \tag{23.104}$$

$$\left(\rho u^2 + p + \frac{1}{3} a T_{\rm rad}^4\right)_2 = \left(\rho u^2 + p + \frac{1}{3} a T_{\rm rad}^4\right)_1, \tag{23.105}$$

$$\left[u\left(\frac{1}{2}\rho u^2 + e + p + \frac{4}{3} a T_{\rm rad}^4\right)\right]_2 = \left[u\left(\frac{1}{2}\rho u^2 + e + p + \frac{4}{3} a T_{\rm rad}^4\right)\right]_1, \tag{23.106}$$

where the subscripts "1" and "2" represent the quantities in the upstream and downstream regions far from the shock (Fig. 23.4).

[2] At the post-shock region, there appears another type of the relaxation zone, where electrons and ions equilibrate, as already stated in Sect. 23.3.1.

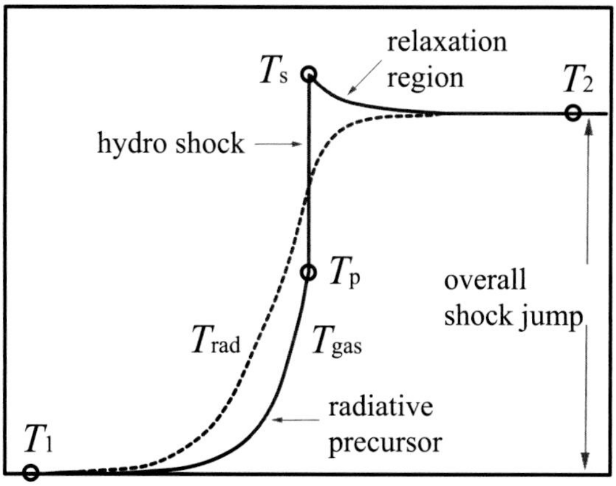

Fig. 23.4 Typical structure of the radiative shocks in the nonequilibrium diffusion case. The radiation temperature (a dashed curve) is continuous, while the gas temperature (a solid one) is discontinuous with a Zel'dovich spike

On the other hand, for the fluid part we shall use the hydrodynamical jump conditions with a radiative diffusion:

$$(\rho u)_s = (\rho u)_p = j, \tag{23.107}$$

$$\left(\rho u^2 + p\right)_s = \left(\rho u^2 + p\right)_p, \tag{23.108}$$

$$\left[u\left(\frac{1}{2}\rho u^2 + e + p\right)\right]_s = \left[u\left(\frac{1}{2}\rho u^2 + e + p\right)\right]_p, \tag{23.109}$$

$$\left[\frac{4}{3}aT_{\rm rad}^4 u - \frac{c}{3\kappa\rho}\frac{\partial}{\partial x}\left(aT_{\rm rad}^4\right)\right]_s = \left[\frac{4}{3}aT_{\rm rad}^4 u - \frac{c}{3\kappa\rho}\frac{\partial}{\partial x}\left(aT_{\rm rad}^4\right)\right]_p \tag{23.110}$$

where the subscripts "p" and "s" represent the quantities at the pre-shock and post-shock states, respectively (Fig. 23.4).

The treatment of the jump conditions and precursor in this nonequilibrium diffusion case is rather complicated, compared with that in the equilibrium diffusion shock (consult, e.g., Lowrie and Edwards 2008). We here briefly introduce the typical structure, including the so-called Zel'dovich spike (Fig. 23.4).

As is seen in Fig. 23.4, in the radiative shocks in the nonequilibrium diffusion case, the radiation temperature $T_{\rm rad}$ (a dashed curve) continuously changes from a pre-shock one to a post-shock one, due to the radiative diffusion. On the other hand, the gas temperature $T_{\rm gas}$ (a solid one) discontinuously jumps up from the pre-shock value $T_{\rm p}$ to the post-shock one $T_{\rm s}$ ($> T_2$), and gradually cools down to the post-shock equilibrium state with T_2. This gas-temperature peak is called the *Zel'dovich spike*.

As was stated in the beginning of this section, there are many papers related to the radiative shocks. In many of the current studies, however, the fluid motion is nonrelativistic, or the radiation field is approximated by the moment equations with a closure relation, or under the one-dimensional flow. Recently, Okuda et al. (2004)

performed the 1D and 2D numerical simulations of radiative shocks in rotating accretion flows around black holes. In their results, the radiative precursor and maybe the Zel'dovich spike are seen, although they did not mention these properties from the viewpoints of radiative shocks. Tolstov et al. (2015) examined the radiative shocks under the special relativistic radiation hydrodynamics by solving the radiative transfer equation. They showed quantitative differences among the Eddington approximation, M1 one, and the full radiative transfer equation. Fukue (2019a) studied the radiative shocks in disk accretion flows under the vertical hydrostatic balance, using the equilibrium diffusion and Eddington approximations. He found that the gas density in the post-shock region often decreases, but the surface density behaves like the gas density in the usual one-dimensional shock, and increases up to 7 in the radiation-pressure dominated case. As in the hydrodynamical shocks, the radiative shocks can be incorporated into the transonic accretion flows onto the central gravitating object (e.g., Fukue 1987). In contrast to the hydrodynamical shocks, which can be treated as a discontinuity with no width, the radiative shocks have generally a radiative precursor with a finite width. Hence, the radiative shocks in the transonic accretion flows onto the central object should be examined carefully, since, e.g., the gravitational field would change during shock transition.

23.4 Ionization Fronts

In addition to radiative shocks, another important radiative discontinuity is a *radiative ionization front* (Kahn 1954; Axford 1961; Goldsworthy 1961; see also Mihalas and Mihalas 1984), which is especially important in the interstellar astrophysics. Surrounding an ionizing source, such as a hot star, there appears an HII region (Strömgren sphere), where photo ionization and recombination are in equilibrium. When a hot star is born in HI regions or molecular clouds, a newly formed HII region around the star expands outward into a surrounding neutral interstellar gas. As a result, the boundary between an ionizing gas and neutral gas moves outward as an ionization front (IF), which is a kind of discontinuity governed by the incident ionizing photons. In contrast to a (radiative) shock front, which is mainly driven by gas pressure, the ionization front is driven by radiation.

23.4.1 Jump Conditions of Ionization Fronts

Let us consider an ionization front (IF) in the one-dimensional (x) direction (Fig. 23.5). Due to the ionizing photon flux J [cm^{-2} s^{-1}], neutral hydrogen atoms are ionized by J particles per unit time per unit area. In the IF frame, the initial neutral gas moves into the front at relative speed u_1, ionized at IF, and the ionized gas moves out from the front at relative speed u_2. In this case, similar to shock

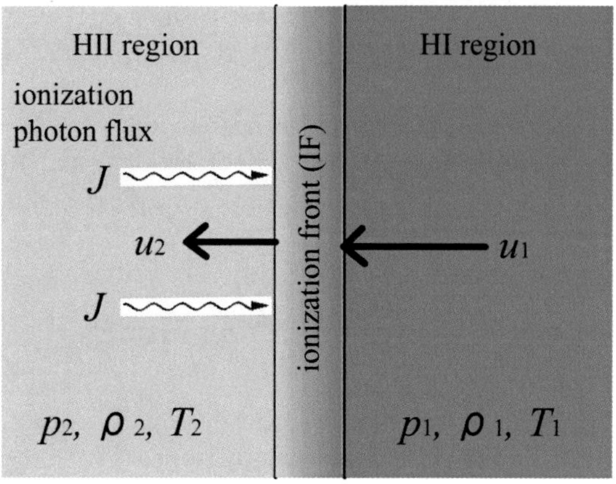

Fig. 23.5 Physical quantities at the ionization front (IF). Neutral gas with subscript "1" flows into IF, is ionized at IF, and flows out from IF as ionized gas

waves, the jump conditions for the mass and momentum fluxes are

$$\frac{\rho_2 u_2}{m_\mathrm{H}} = \frac{\rho_1 u_1}{m_\mathrm{H}} = J, \tag{23.111}$$

$$\rho_2 u_2^2 + p_2 = \rho_1 u_1^2 + p_1, \tag{23.112}$$

where m_H is the mass of the hydrogen atom, and the subscripts "1" and "2" represent the quantities in front of and after IF, respectively. Instead of the jump condition on the energy flux, both the HI and HII regions are assumed, respectively, to be isothermal, and have constant isothermal sound speeds defined as

$$c_2 = \sqrt{p_2/\rho_2}, \quad c_1 = \sqrt{p_1/\rho_1}. \tag{23.113}$$

Now, we suppose that ρ_1, c_1, c_2, and J are known. Hence, eliminating other quantities from Eqs. (23.111)–(23.113), we have the compression ratio at the ionization front as

$$\frac{\rho_2}{\rho_1} = \frac{1}{2c_2^2}\left[(c_1^2 + u_1^2) \pm \sqrt{(c_1^2 + u_1^2)^2 - 4c_2^2 u_1^2}\right]. \tag{23.114}$$

In order for the real solution to exist, the condition,

$$(c_1^2 + u_1^2)^2 - 4c_2^2 u_1^2 \geq 0 \tag{23.115}$$

must be satisfied. Or, one of the following conditions for u_1 should be satisfied:

$$u_1 \geq u_\mathrm{R} \equiv c_2 + \sqrt{c_2^2 - c_1^2} \sim 2c_2, \tag{23.116}$$

$$u_1 \leq u_\mathrm{D} \equiv c_2 - \sqrt{c_2^2 - c_1^2} \sim c_1^2/(2c_2), \tag{23.117}$$

where and in what follows we assume $c_2 \gg c_1$. In addition, the subscript "R" denotes (high-speed) rarefied gas, whereas the subscript "D" means (low-speed) dense gas.

Indeed, in the interstellar medium, the typical temperatures of HI and HII regions are, respectively, 10^2 K ($c_1 \sim 1$ km s^{-1}) and 10^4 K ($c_2 \sim 10$ km s^{-1}). Hence, $u_R \sim 2c_2 \sim 20$ km s^{-1}, and $u_D \sim c_1^2/(2c_2) \sim 0.05$ km s^{-1}.

23.4.2 Types of Ionization Fronts

According to the above mentioned boundary speeds, the properties of the ionization front are usually classified as five types (Fig. 23.6):

$$\begin{array}{ll} u_1 > u_R & \text{R (rarefied) type,} \\ u_1 = u_R & \text{R-critical type,} \\ u_D < u_1 < u_R & \text{M (intermediate) type,} \\ u_1 = u_D & \text{D-critical type,} \\ u_1 < u_D & \text{D (dense) type.} \end{array} \tag{23.118}$$

Each type we see below in turn.

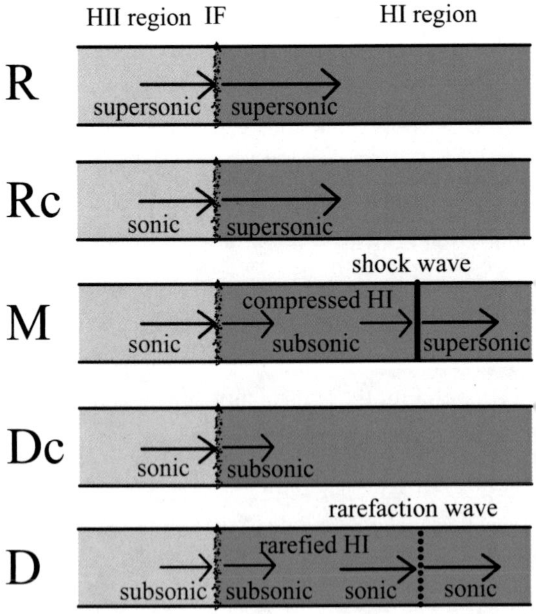

Fig. 23.6 Types and structures of the ionization front. The HII regions on the left side expand into the static HI regions on the right side passing various fronts. The arrows indicate speeds of fronts or flows in the fixed frame

(a) R (rarefied) type

In the R type, $u_1 > u_R \sim 2c_2 > 2c_1$, and the relative speeds into both sides of IF are supersonic, or the ionization front moves supersonically in the neutral HI region. Since the relative speed u_1 is high, the gas density ρ_1 is relatively low (rarefied gas).

Corresponding to the sign in the compression ratio (23.114), the R type splits into two subtypes. When the sign is minus, and $c_2 \gg c_1$, the compression ratio is roughly expressed as

$$\frac{\rho_2}{\rho_1} \sim \frac{u_1^2}{2c_2^2}\left(1 - \sqrt{1 - \frac{4c_2^2}{u_1^2}}\right). \tag{23.119}$$

Hence, the compression ratio is between 1 ($u_1 \gg u_R$) and 2 ($u_1 \sim u_R$), and the compressibility is not so large. This subtype is called a *weak R type*.

In the plus-sign case (*strong R type*), the compression ratio is roughly $\rho_2/\rho_1 \sim u_1^2/c_2^2$, and could increase indefinitely with u_1. Such a state is unrealistic, and the strong R type is not adopted as an astrophysical IF.

(b) R-critical type

The R-critical type is the case of $u_1 = u_R$. In this case, $u_1 \sim 2c_2 \gg c_1$, and the ionization front proceeds supersonically into the HI region. The compression ratio is not so large; $\rho_2/\rho_1 \sim u_1^2/(2c_2^2) \sim 2$. Hence, the relative speed to the HII region behind the front is $u_2 = (\rho_1/\rho_2)u_1 \sim c_2$, and therefore, the front moves just at the sound speed against the post front HII region.

(c) D-critical type

In contrast to the R-critical type, the D-critical type is the case of $u_1 = u_D$. In this case, $u_1 \sim c_1^2/(2c_2) \ll c_1$, and the ionization front proceeds subsonically into the HI region. The compression ratio is quite small; $\rho_2/\rho_1 \sim (c_1^2 + u_1^2)/(2c_2^2) \sim c_1^2/(2c_2^2) \sim u_1/c_2 \ll 1$. Hence, the relative speed to the HII region is $u_2 = (\rho_1/\rho_2)u_1 \sim c_2$, and therefore, the front moves just at the sound speed against the post front HII region.

(d) D (dense) type

In the D type, $u_1 < u_D \sim c_1^2/(2c_2) \ll 2c_1$, and the relative speeds in both sides of IF are subsonic. Since the relative speed u_1 is low, the gas density ρ_1 is relatively high (dense gas).

Corresponding to the sign in the compression ratio (23.114), the D type also splits into two subtypes. When the sign is plus, the compression ratio is roughly expressed as

$$\frac{\rho_2}{\rho_1} \sim \frac{c_1^2}{2c_2^2}\left(1 - \sqrt{1 - \frac{4u_1^2c_2^2}{c_1^4}}\right). \tag{23.120}$$

Hence, the gas density decreases after the front; $\rho_2 \ll \rho_1$. This subtype is called a *weak D type*. In this weak D type, the rarefaction wave proceeding at sound speed appears in front of IF.

In the minus-sign case (*strong D type*), the density ρ_2 after the front vanishes in the limit of $u_1 \to 0$. Such a state is also unrealistic, and the strong D type is not adopted as an astrophysical IF.

(e) M (intermediate) type

Finally, in the M type, $u_D < u_1 < u_R$, and there are no real solutions in Eq. (23.114). In this case, however, according to the propagation of IF, the HI region in front of IF is quickly compressed, and a radiative shock preceding IF is produced. As a result, the neutral gas flowing into this shock front is decelerated to the subsonic speed, and then flowing into the ionization front, as a D-critical type IF. Since the M type IF is accompanied by a radiative shock wave, it is important in astrophysics.

23.5 Rayleigh–Taylor and Kelvin–Helmholz Instabilities

As already described in Chap. 7, there are various important instabilities in the astrophysical fluid. Among those, in this section, we briefly summarize the current knowledge on the Rayleigh–Taylor and Kelvin–Helmholz instabilities in radiative fluids.

23.5.1 Rayleigh–Taylor Instability in Radiative Fluids

Besides the magnetic and radiation fields, the Rayleigh–Taylor instability in astrophysical fluids becomes much more complicated than that in the simple case of the two constant-density layers, since the astrophysical fluid is usually compressible,[3] and the background density is not constant under the gravitational environment. If, furthermore, there exist the magnetic or radiation fields, the construction of the equilibrium configuration is not so easy. Hence, the study of the Rayleigh–Taylor instability in radiative fluids does not enough progress, compared with, e.g., the radiative shocks. However, the radiative Rayleigh–Taylor instability becomes important in various astrophysical situations; e.g., at the boundary of a radiation-driven HII region, during massive star formation, in supernova explosions, where the inertial acceleration behaves as the gravitational field.

Radiative Rayleigh–Taylor instability has been studied by a few researchers under simple assumptions (e.g., Mathews and Blumenthal 1977; Krolik 1977). Jacquet and Krumholz (2011) have formally performed the linear stability analysis of an interface separating two stratified radiative media, and obtained the exact

[3] Although the effect of compressibility on the Rayleigh–Taylor instability has been studied by Shivamoggi (2008) (arXiv:0805.0581), the study has not been published yet.

stability conditions in several limiting cases. We show the main results by Jacquet and Krumholz (2011) below.

Let us consider two semi-infinite gaseous layers separated by an interface at $z = 0$. The upper medium 2 overlies the lower medium 1, under a uniform external gravitational acceleration $(-g)$ and a radiative flux F in the vertical direction.

In the optically thin and isothermal regime, the upward radiative flux becomes constant in both media, and it reduces the gravitational acceleration to the effective one $g_{\rm eff}$ [$= g(1-\Gamma)$], where Γ is the ratio of the radiative force to the gravitational force, and called the Eddington parameter. Although the total frequency-integrated flux is constant, the radiation spectrum may change at the interface, due to, e.g., a chemical change there. As a result, the effective gravitational acceleration may change between the media 1 and 2, say g_1 and g_2. In addition, the isothermal sound speeds of each layers are denoted as a_1 and a_2, respectively, and $\rho_{2,1} = \rho(0\pm)$. Then, the growth rate n $(= i\omega)$ is given by

$$n^2 = k\frac{\rho_2 g_2 - \rho_1 g_1}{\rho_1 + \rho_2} + \frac{(g_1+g_2)(\rho_2-\rho_1)(\rho_2 g_2 - \rho_1 g_1)}{2(a_1^2+a_2^2)(\rho_1+\rho_2)^2} + \mathcal{O}\left(\frac{1}{k}\right), \tag{23.121}$$

where k is the wavenumber of perturbations in the horizontal direction (cf. Sect. 6.2.1).

In the highly optically thick and adiabatic regime, where the perturbed radiative flux is ignored, matter and radiation behave as a single fluid, and radiation acts matter with a modified equation of state. Hence, if we restrict attention to eigenmodes localized at the interface, and assume the background quantities to be spatially constant, the results coincide with the classical Rayleigh–Taylor instability condition at the short-wavelength limit. At the long-wavelength limit, on the other hand, the frequency of perturbations converges toward a finite value, which is the Brunt–Väisälä frequency modified by compressibility, and gives a positive growth rate at (and somewhat below) the Eddington limit.

In this optically thick regime, however, the situations considered in the current studies are rather limited (local and adiabatic), and there is enough room for further analytical studies, such as the influence of the radiative diffusion on the Rayleigh–Taylor instability. For example, the radiative Rayleigh–Taylor instability of a purely scattering medium has been numerically investigated by Jiang et al. (2013). They found that the radiation field can reduce the growth rate, when the radiation pressure highly exceeds the gas pressure. Furthermore, they demonstrated that anisotropy of radiation plays an important role in the nonlinear development of the instability. Yaghoobi and Shadmehri (2018) incorporated the effect of the magnetic fields in the study by Jacquet and Krumholz (2011).

23.5.2 *Kelvin–Helmholz Instability in Radiative Fluids*

Similar to the Rayleigh–Taylor instability, the progress of the study of the Kelvin–Helmholz instability in the radiative fluid is not enough. Up to now, most of the study of the radiative Kelvin–Helmholz instability in the astrophysical fluid were done under various numerical simulations of the astrophysical jets under radiative losses (e.g., Micono et al. 2000 and references therein). Analytical studies relating to jets are still limited in the optically thin case (e.g., Massaglia et al. 1992; Hardee and Stone 1997 and references therein). Linear analyses of the Kelvin–Helmholtz instability of radiative jets show that radiative losses reduce the growth rate of perturbations (e.g., Massaglia et al. 1992; Hardee and Stone 1997). Systematic and analytical study on the radiative Kelvin–Helmholz instability are desired, including the optically thick regime.

23.6 Gravitational Instability in Radiative Fluids

In this section, we briefly mention the radiative influence on the gravitational instability (Jeans instability), the dynamical instability of self-gravitating gaseous media (cf. Sect. 6.4). As was mentioned in, e.g., introduction (Sect. 1.3.1), one of the most important forces in astrophysical fluids is self-gravitating one, which operates in the formation of planets, stars, star clusters, galaxies, and dark matter objects from small amplitude fluctuations in the background media.

The first and famous topics of self-gravitating radiative media may be the *Silk damping*, which treated the damping of acoustic waves of primordial fluctuations in the early universe by the radiative viscosity (Silk 1968; Weinberg 1971; see also Thompson 2010). Another important issue is the radiative influence on the classical Jeans instability (e.g., Vranješ and Čadež 1990; Thompson 2008).

Similar to radiative shocks in Sect. 23.3, the treatments of gravitational instabilities in optically thick radiative fluids are three step depths; (1) the strong coupling case between matter and radiation, where matter and radiation behave as a single fluid, (2) the equilibrium diffusion case, where the radiation temperature is equal to the gas one, and (3) the nonequilibrium diffusion case, where radiation no longer equilibrates matter, and the radiation temperature is not equal to the gas one.

For example, if matter and radiation are strongly coupled, and a gas-radiation mixture behaves as a single fluid, then the effective sound speed becomes

$$C_s^2 = \Gamma_1 \frac{p_0 + P_0}{\rho_0} = \frac{\Gamma_1}{\gamma}(1 + \alpha_0)c_s^2, \tag{23.122}$$

where p_0 and P_0 are the background uniform gas pressure and radiation one, respectively, ρ_0 the background uniform density, γ the ratio of the specific heats, Γ_1 the generalized adiabatic index, $\alpha_0 = P_0/p_0$, and c_s $(= \sqrt{\gamma p_0/\rho_0})$ the usual adiabatic sound speed without the radiation pressure (Sects. 4.1, 19.6, and 23.1). In such a strong equilibrium regime, according to the classical procedure (e.g., Jeans 1902; Chandrasekhar 1961), the gravitational instability of the homogeneous radiative fluid is easily examined under the "Jeans swindle," where the Poisson equation is satisfied in an ad hoc way with no background gradients in density (cf. Vranješ and Čadež 1990). The dispersion relation becomes

$$\omega^2 = f c_s^2 k^2 - 4\pi G \rho_0, \tag{23.123}$$

where

$$f \equiv \frac{\Gamma_1 (1 + \alpha_0)}{\gamma}. \tag{23.124}$$

The correction factor f is always larger than unity; $f \to 1$ at $\alpha_0 \to 0$ and $f \to [(4/3)/\gamma]\alpha_0$ at $\alpha_0 \to \infty$. Thus, the radiation pressure stabilizes the gravitational instability. For example, the critical wavelength λ_c of the gas-radiation mixture becomes greater than the classical Jeans's value λ_J by a factor of $f^{1/2}$: $\lambda_c = f^{1/2}\lambda_J$.

The effect of the radiative diffusion on the gravitational instability of radiative fluids under the equilibrium diffusion regime was also examined by, e.g., Thompson (2008). Similar to the radiative acoustic waves (Sect. 23.1), there appear two type modes. The one is a *radiato-gravity-modified acoustic wave*, whose dispersion relation is $\omega \propto \sqrt{k^2 c_s^2 - 4\pi G \rho_0}$ with correction terms, and the instability criterion is determined by the adiabatic sound speed c_s. The other is a *radiation diffusion wave*, whose dispersion relation is $\omega \propto \sqrt{k^2 c_T^2 - 4\pi G \rho_0}$ with correction terms, and the instability criterion is determined by the isothermal sound speed c_T (cf. Kato and Kumar 1960 for the gravitational instability of the conductive fluids).

Systematic and analytical studies on the radiative gravitational instability are also desired for various astrophysical situations.

References

Axford, W.I.: Phil. Trans. R. Soc. London **A253**, 301 (1961)
Bisnovatyi-Kogan, G.S., Blinnikov, S.I.: Astrophysics **14**, 563 (1978)
Bisnovatyi-Kogan, G.S., Blinnikov, S.I.: Astrophysics **15**, 165 (1979)
Blaes, O., Socrates, A.: Astrophys. J. **553**, 987 (2001)
Blaes, O., Socrates, A.: Astrophys. J. **596**, 509 (2003)
Bogdan, T.J. et al.: Astrophys. J. **456**, 879 (1996)
Bouquet, S., Teyssier, R., Chieze, J.P.: Astrophys. J. Suppl. **127**, 245 (2000)
Castor, J.I.: Radiation Hydrodynamics. Cambridge University Press, Cambridge (2004)
Chandrasekhar, S.: Hydrodynamical and Hydromagnetic Stability. Clarendon Press, Oxford (1961)
Chandrasekhar, S.: An Introduction to the Study of Stellar Structure. Dover Publ., New York (1967)
Cogley, A.C.: J. Fluid Mech. **39**, 667 (1969)
Delache, P., Froeschlé, C.: Astron. Astrophys. **16**, 348 (1972)
Drake, R.P.: Astrophys. Space Sci. **298**, 49 (2005)
Drake, R.P.: Phys. Plasma **14**, 043301 (2007)

Dzhalilov, N.S. et al.: Astron. Astrophys. **257**, 359 (1992)
Dzhalilov, N.S. et al.: Astron. Astrophys. **291**, 1001 (1994)
Ferguson, J.M., Morel, J.E., Lowrie, R.B.: J.Q.Sp.R.T. (2017) arXiv:1702.07300v1
Fukue, J.: Publ. Astron. Soc. Jpn. **39**, 309 (1987)
Fukue, J.: Publ. Astron. Soc. Jpn. **71**, 38 (2019a)
Fukue, J.: Monthly Not. Roy. Astron. Soc. **483**, 2538 (2019b)
Fukue, J.: Monthly Not. Roy. Astron. Soc. **483**, 3839 (2019c)
Fukue, J.: Publ. Astron. Soc. Jpn. **71**, 99 (2019d)
Goldsworthy, F.A.: Phil. Trans. R. Soc. London **A253**, 277 (1961)
Hardee, P.E., Stone, J.M.: Astrophys. J. **483**, 121 (1997)
Jacquet, E., Krumholz, M.: Astrophys. J. **730**, 116 (2011)
Jeans, J.H.: Phil. Trans. Roy. Soc. London **199**, 1 (1902)
Jiang, Y-F., Davis S.W., Stone, J.M.: Astrophys. J. **763**, 102 (2013)
Johnson, B.M.: Astrophys. J. **693**, 1637 (2009)
Kahn, F.D.: B.A.N. **12**, 187 (1954)
Kato, S., Kumar, S.S.: Publ. Astron. Soc. Jpn. **12**, 290 (1960)
Krolik, J.H.: Phys. Fluids **20**, 364 (1977)
Lacey, C.G.: Astrophys. J. **326**, 769 (1988)
Lowrie, R.B., Edwards, J.D.: Shock Waves **18**, 129 (2008)
Lowrie, R.B., Rauenzahn, R.M.: Shock Waves **16**, 445 (2007)
Lowrie, R.B., Morel, J.E., Hittinger, J.A.: Astrophys. J. **521**, 432 (1999)
Marshak, R.E.: Phys. Fluids **1**, 24 (1958)
Massaglia, S. et al.: Astron. Astrophys. **260**, 243 (1992)
Mathews, W.G., Blumenthal, G.R.: Astrophys. J. **214**, 10 (1977)
Micono, M., et al.: Astron. Astrophys. **360**, 795 (2000)
Mihalas, D., Mihalas, B.W.: Astrophys. J. **273**, 355 (1983)
Mihalas, D., Mihalas, B.W.: Foundations of Radiation Hydrodynamics. Oxford University Press, Oxford (1984)
Okuda, T., Teresi, V., Toscano, E., Molteni, D.: Publ. Astron. Soc. Jpn. **56**, 547 (2004)
Raizer, Y.P.: Sov. Phys. - JETP **5**, 1242 (1957)
Schweitzer, M.A.: Monthly Not. Roy. Astron. Soc. **210**, 303 (1984)
Shaviv, N.J.: Astrophys. J. **549**, 1110 (2001)
Shivamoggi, B.K.: arXiv:0805.0581 (2008)
Silk, J.: Astrophys. J. **151**, 459 (1968)
Souffrin, P.: Ann. d'Astrophys. **29**, 55 (1966)
Souffrin, P.: Astron. Astrophys. **17**, 458 (1972)
Spiegel, E.A.: Astrophys. J. **126**, 202 (1957)
Spiegel, E.A., Tao, L.: Phys. Rep. **311**, 163 (1999)
Staude, J., Dzhalilov, N.S., Zhugzhda, Y.D.: Sol. Phys. **152**, 227 (1994)
Stix, M.: Astron. Astrophys. **4**, 189 (1970)
Tolstov, A., et al.: Astrophys. J. **811**, 47 (2015)
Thompson, T.A.: Astrophys. J. **684**, 212 (2008)
Thompson, T.A.: Astrophys. J. **709**, 1119 (2010)
Unno, W., Spiegel, E.A.: Publ. Astron. Soc. Jpn. **18**, 85 (1966)
Vincenti, W.G., Baldwin, B.S., Jr.: J. Fluid Mech. **12**, 449 (1962)
Vranješ, J., Čadež, V.: Astrophys. Space Sci. **164**, 329 (1990)
Weaver, T.A.: Astrophys. J. Suppl. **32**, 233 (1976)
Weinberg, S.: Astrophys. J. **168**, 175 (1971)
Yaghoobi, A., Shadmehri, M.: Monthly Not. Roy. Astron. Soc. **477**, 412 (2018)
Zel'dovich, Y.B.: Sov. Phys. - JETP **5**, 919 (1957)
Zel'dovich, Y.B., Raizer, Y.P.: Physics of Shock Waves and High-Temperature Hydrodynamic Phenomena, vol. 1. Academic Press, New York (1967)
Zel'dovich, Y.B., Raizer, Y.P.: Physics of Shock Waves and High-Temperature Hydrodynamic Phenomena. Dover, New York (2002)
Zhugzhda, Y.D., Dzhalilov, N.S., Staude, J.: Astron. Astrophys. **278**, L9 (1993)

Chapter 24
Relativistic Radiative Transfer

In Part III, we mainly consider the radiation hydrodynamics under the nonrelativistic regime. However, the radiation field of photons is essentially relativistic, and the radiation hydrodynamical technique becomes more and more important in various relativistic astrophysical phenomena. Thus, in the last three chapters, we consider the relativistic radiation hydrodynamics. In this chapter, corresponding to Chap. 20 for the nonrelativistic case, we first derive the basic equations for relativistic radiative transfer within the framework of special relativity in somewhat details. That is, we give the metric and quantities of the radiation fields, derive the radiative transfer equation and its formal solutions, obtain moment equations under special relativity, and discuss a closure relation in the relativistic regime. In addition, we shortly show several topics on the radiative transfer in the relativistic regime.

24.1 Radiation Field Under Special Relativity

The full set of basic equations for photohydrodynamics can be found in several literature (e.g., Lindquist 1966; Anderson and Spiegel 1972; Hsieh and Spiegel 1976; Thorne 1981; Fukue et al. 1985; Mihalas and Auer 2001; Park 2006; Takahashi 2007; Kato et al. 2008). It is usually expressed in a general relativistic form. In this section and the following section, however, we derive and write explicitly the basic equations for relativistic radiative transfer within the framework of special relativity.[1]

In this section, we first summarize the relativistic quantities and transformation relating to the special relativistic radiation hydrodynamics.

---

[1]In this text book the $(+, -, -, -)$ sign system is adopted, and the Greek suffixes $\alpha, \beta, \gamma, \cdots$ take values of 0, 1, 2, and 3, while the Latin suffixes $i, j, k, \cdots$ take values of 1, 2, and 3. Finally, the quantities measured in the *comoving*/fluid frame are labeled by the suffix 0 or "co" if necessary, while the quantities in the fixed/observer/*inertial* frame are expressed without suffix.

© Springer Nature Singapore Pte Ltd. 2020, corrected publication 2023

S. Kato, J. Fukue, *Fundamentals of Astrophysical Fluid Dynamics*,
Astronomy and Astrophysics Library,
https://doi.org/10.1007/978-981-15-4174-2_24

(a) Metric

The square of the invariant line element, ds^2, is written as

$$ds^2 = c^2 d\tau^2 = g_{\mu\nu} dx^\mu dx^\nu, \tag{24.1}$$

where c is the speed of light, τ the proper time, x^μ the space-time coordinates ($x^0 = ct$), and $g_{\mu\nu}$ the space-time metric.[2] The three-dimensional part of the metric, γ_{ij}, is defined by $\gamma_{ij} = -g_{ij}$.

(b) Four-velocity of matter

The four-velocity u^μ of matter and its covariant components u_μ are, respectively, defined by

$$u^\mu \equiv \frac{dx^\mu}{ds} = \left(\gamma, \gamma \frac{v^i}{c}\right) = \gamma \left(1, \frac{\boldsymbol{v}}{c}\right), \tag{24.2}$$

$$u_\mu = g_{\mu\nu} u^\nu = \left(\gamma, -\gamma \frac{v_i}{c}\right), \tag{24.3}$$

where $\boldsymbol{v}$ $(= v^i)$ is the three-dimensional velocity defined by

$$\boldsymbol{v} = v^i \equiv \frac{dx^i}{dt}, \tag{24.4}$$

and γ the Lorentz factor,

$$\gamma = \left(1 - \frac{v^2}{c^2}\right)^{-1/2}, \tag{24.5}$$

$$v^2 = v_i v^i = \gamma_{ik} v^i v^k = -g_{ik} v^i v^k. \tag{24.6}$$

It should be noted that $u_\mu u^\mu = g_{\mu\nu} u^\mu u^\nu = 1$.

(c) Four-momentum of photon

The four-momentum k^μ of a photon and its covariant components k_μ are, respectively, defined by[3]

$$k^\mu = \left(\nu, \nu l^i\right) = \nu \left(1, \boldsymbol{l}\right), \tag{24.7}$$

$$k_\mu = \nu \left(1, -\boldsymbol{l}\right), \tag{24.8}$$

[2] In a flat space-time, the metric tensor is often expressed by $\eta_{\mu\nu}$. For example, using the cylindrical coordinates (r, φ, z), the line element (24.1) becomes $ds^2 = \eta_{\mu\nu} dx^\mu dx^\nu = c^2 dt^2 - dr^2 - r^2 d\varphi^2 - dz^2$.

[3] We drop a factor (h/c) for simplicity.

where ν is the photon frequency and $\boldsymbol{l}$ is the direction cosine vector of a photon, and expressed as

$$\boldsymbol{l} = l^i = (\sin\theta\cos\varphi, \sin\theta\sin\varphi, \cos\theta)\,. \tag{24.9}$$

Since $\boldsymbol{l}^2 = 1$, the contraction of the four-momentum is null:

$$k_\mu k^\mu = \nu^2\left(1 - \boldsymbol{l}^2\right) = 0. \tag{24.10}$$

(d) Doppler and aberration effects

The four-velocity u_μ and the four-momentum k^μ expressed in comoving frames are, respectively,

$$u_\mu = (1, 0)\,, \tag{24.11}$$

$$k^\mu = \nu_0\,(1, \boldsymbol{l}_0)\,, \tag{24.12}$$

where the subscript 0 means the values measured in the comoving (fluid) frame. Using Eqs. (24.3) and (24.7), we have

$$u_\mu k^\mu = \gamma\nu - \gamma\nu\frac{\boldsymbol{v}\cdot\boldsymbol{l}}{c} = \nu_0. \tag{24.13}$$

Thus, the transformation of the photon frequency between the inertial and comoving frames (relativistic Doppler effect) becomes

$$\nu_0 = \nu\gamma\left(1 - \frac{\boldsymbol{v}\cdot\boldsymbol{l}}{c}\right), \tag{24.14}$$

$$\nu = \nu_0\gamma\left(1 + \frac{\boldsymbol{v}\cdot\boldsymbol{l}_0}{c}\right). \tag{24.15}$$

Similarly, the transformation of the direction cosine (relativistic aberration effect)[4] becomes

$$\boldsymbol{l}_0 = \frac{\nu}{\nu_0}\left[\boldsymbol{l} + \left(\frac{\gamma - 1}{v^2/c^2}\frac{\boldsymbol{v}\cdot\boldsymbol{l}}{c} - \gamma\right)\frac{\boldsymbol{v}}{c}\right], \tag{24.16}$$

[4]In terms of the Lorentz transformation tensor between the comoving and inertial frames,

$$L_\alpha^\beta = \begin{pmatrix} \gamma & -\gamma\frac{v^i}{c} \\ -\gamma\frac{v^i}{c} & \delta_i^j + (\gamma - 1)\frac{v^i v^j}{v^2} \end{pmatrix},$$

the transformation between the comoving four-momentum k_0^β and the inertial one k^α becomes $k_0^\beta = L_\alpha^\beta k^\beta$.

$$\boldsymbol{l} = \frac{\nu_0}{\nu}\left[\boldsymbol{l}_0 + \left(\frac{\gamma - 1}{v^2/c^2}\frac{\boldsymbol{v}\cdot\boldsymbol{l}_0}{c} + \gamma\right)\frac{\boldsymbol{v}}{c}\right]. \tag{24.17}$$

Using the relativistic invariant (see below), the transformation of the solid angle is

$$d\Omega_0 = \frac{\nu}{\nu_0}\frac{d\nu}{d\nu_0}d\Omega = \left[\gamma\left(1 - \frac{\boldsymbol{v}\cdot\boldsymbol{l}}{c}\right)\right]^{-2} d\Omega, \tag{24.18}$$

$$d\Omega = \left[\gamma\left(1 + \frac{\boldsymbol{v}\cdot\boldsymbol{l}_0}{c}\right)\right]^{-2} d\Omega_0. \tag{24.19}$$

In the one-dimensional plane-parallel case, the Doppler effect [Eqs. (24.14) and (24.15)] and aberration [Eqs. (24.16) and (24.17)] become, respectively,

$$\frac{\nu_0}{\nu} = \gamma(1 - \beta\mu), \quad \frac{\nu}{\nu_0} = \gamma(1 + \beta\mu_0), \tag{24.20}$$

$$\mu_0 = \frac{\mu - \beta}{1 - \beta\mu}. \quad \mu = \frac{\mu_0 + \beta}{1 + \beta\mu_0}, \tag{24.21}$$

where β ($\equiv v/c$) is the normalized speed, and μ ($= \cos\theta$) the direction cosine.

It should be noted that, from Eq. (24.21), $\mu \to 1$ at $\beta \to 1$ irrespective of the value μ_0 in the comoving frame, or the light ray concentrates toward the forward direction due to the relativistic aberrational effect. For example, for the light ray traveling to the horizontal direction in the comoving frame ($\theta_0 = \pi/2$, $\mu_0 = 0$), the direction of the light ray in the inertial frame is $\cos\theta = \mu = \beta$ and $\sin\theta = \sqrt{1-\mu^2} = 1/\gamma$. This behavior is called *relativistic beaming* (Fig. 24.1).

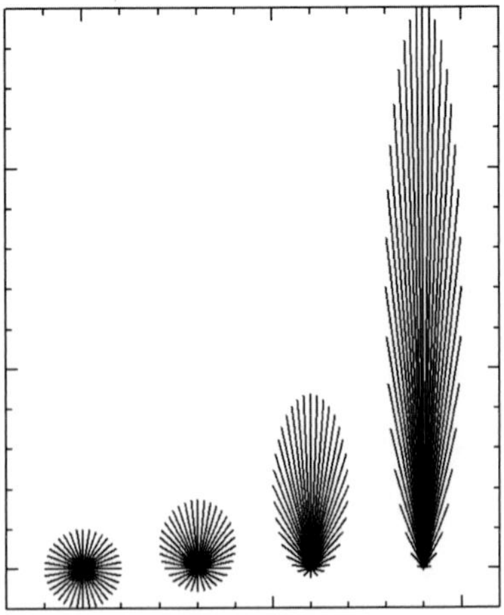

Fig. 24.1 Circular diagram of the relativistic beaming. The direction of each line represents the polar angle θ in the fixed frame, while the length of each line means the frequency ratio, ν/ν_0. The normalized velocity β is 0, 0.5, 0.9, and 0.99 from left to right

(e) Relativistic invariants

In relativity both space and time are treated as observable physical quantities, and vary by each observer, while unvaried quantities, such as the number, are called *relativistic invariants* or *Lorentz invariants; LI*. Here, we briefly derive several relativistic invariants relating to the radiation field.

Transformations of coordinate space and momentum one between inertial and comoving frames are, respectively,

$$d^3\boldsymbol{x} = \gamma^{-1} d^3\boldsymbol{x}_0, \quad d^3\boldsymbol{k} = \gamma d^3\boldsymbol{k}_0, \tag{24.22}$$

and therefore, the phase space volume is the relativistic invariant:

$$dV_{\mathrm{p}} = d^3\boldsymbol{x} d^3\boldsymbol{k} = d^3\boldsymbol{x}_0 d^3\boldsymbol{k}_0. \tag{24.23}$$

Since a particle number N is LI, the phase space density f_{p} is also LI:

$$f_{\mathrm{p}} \equiv \frac{dN}{dV_{\mathrm{p}}}. \tag{24.24}$$

If the velocity vector is in the vertical (z) direction, and the direction cosine is $\mu = \cos\theta$, then from (24.17) we have

$$\mu = \frac{\mu_0 + v/c}{1 + \boldsymbol{l}_0 \cdot \boldsymbol{v}/c}, \quad d\mu = \frac{d\mu_0}{\gamma^2 (1 + \boldsymbol{l}_0 \cdot \boldsymbol{v}/c)^2}, \tag{24.25}$$

and using Eq. (24.15), $\nu^2 d\mu$ is known to be LI:

$$\nu^2 d\mu = \nu_0^2 d\mu_0. \tag{24.26}$$

Moreover, using (24.14), we have

$$d\nu_0 = \gamma d\nu (1 - \boldsymbol{l} \cdot \boldsymbol{v}/c) = \frac{\nu_0}{\nu} d\nu, \tag{24.27}$$

and multiplying equation (24.26) by Eq. (24.27), we see that $\nu d\nu d\mu = \nu_0 d\nu_0 d\mu_0$ is LI. In addition, $\nu d\nu d\Omega$ is found to be LI, or we obtain the transformation of the solid angle:

$$\nu d\nu d\Omega = -\nu d\nu d\mu d\varphi = -\nu_0 d\nu_0 d\mu_0 d\varphi_0 = \nu_0 d\nu_0 d\Omega_0. \tag{24.28}$$

(f) Quantities of radiation field

The specific intensity I_ν is related to the photon occupation number n_ν by $I_\nu = (2h\nu^3/c^2) n_\nu$. Since n_ν is LI, I_ν/ν^3 is also LI:

$$\frac{I_\nu}{\nu^3} = \frac{I_{\nu 0}}{\nu_0^3} \equiv f. \tag{24.29}$$

For the frequency-integrated specific intensity, $I \equiv \int I_\nu d\nu$, the relativistic invariant is I/ν^4.

Using these quantities, the energy-momentum tensor of the radiation field, $R^{\mu\nu}$, is defined as

$$R^{\mu\nu} \equiv \frac{2h}{c^3}\int n_\nu l^\mu l^\nu \nu^3 d\nu d\Omega = \frac{1}{c}\int I_\nu l^\mu l^\nu d\nu d\Omega, \tag{24.30}$$

where $l^\mu = (1, l^i)$. Hence, the components of $R^{\mu\nu}$ become

$$R^{00} = \frac{1}{c}\int I_\nu d\nu d\Omega \equiv E, \tag{24.31}$$

$$R^{0i} = \frac{1}{c}\int I_\nu l^i d\nu d\Omega \equiv \frac{1}{c}F^i, \tag{24.32}$$

$$R^{ij} = \frac{1}{c}\int I_\nu l^i l^j d\nu d\Omega \equiv P^{ij}, \tag{24.33}$$

where E is the radiation energy density, F^i the radiative flux, and P^{ij} the radiation stress tensor.

Integrating over the frequency, we obtain the following frequency-integrated quantities:

$$I \equiv \int I_\nu d\nu, \quad E \equiv \int E_\nu d\nu, \quad F^i \equiv \int F^i_\nu d\nu, \quad P^{ij} \equiv \int P^{ij}_\nu d\nu. \tag{24.34}$$

(g) Transformation rules

The transformation of the frequency-integrated intensity I between the inertial and comoving frames is

$$I_0 = \left(\frac{\nu_0}{\nu}\right)^4 I = \left[\gamma\left(1 - \frac{\boldsymbol{v}\cdot\boldsymbol{l}}{c}\right)\right]^4 I. \tag{24.35}$$

Integrating equation (24.35) over solid angle, we obtain the transformation rule of E:

$$E_0 = \gamma^2\left(E - 2\frac{\boldsymbol{v}\cdot\boldsymbol{F}}{c^2} + \frac{v_i v_k}{c^2}P^{ik}\right), \tag{24.36}$$

where Eq. (24.18) is used.

Multiplying equation (24.35) by l^i_0 and integrating the resultant equation over solid angle, we have the transformation rule of F^i:

$$F^i_0 = \gamma\left\{F^i + \left[\left(\gamma + \frac{\gamma-1}{v^2/c^2}\right)\frac{\boldsymbol{v}\cdot\boldsymbol{F}}{c^2} - \gamma E - \frac{\gamma-1}{v^2/c^2}\frac{v_j v_k}{c^2}P^{jk}\right]v^i - v_k P^{ik}\right\}. \tag{24.37}$$

Multiplying equation (24.35) by $l_0^i l_0^j$ and integrating the resultant equation over solid angle, we have the transformation rule of P^{ij} [5]:

$$P_0^{ij} = P^{ij} + \frac{\gamma - 1}{v^2/c^2}\left(\frac{v^i v_k}{c^2}P^{jk} + \frac{v^j v_k}{c^2}P^{ik}\right)$$
$$+\left(\frac{\gamma - 1}{v^2/c^2}\right)^2 \frac{v^i v^j}{c^2}\frac{v_k v_m P^{km}}{c^2} + \gamma^2 \frac{v^i v^j}{c^2}E$$
$$-\gamma\left(\frac{v^i F^j}{c^2} + \frac{v^j F^i}{c^2}\right) - 2\gamma\frac{\gamma - 1}{v^2/c^2}\frac{v^i v^j}{c^2}\frac{\boldsymbol{v}\cdot\boldsymbol{F}}{c^2}. \tag{24.38}$$

In the one-dimensional plane-parallel case, these transformations are, respectively, expressed as[6]

$$cE_0 = \gamma^2\left(cE - 2\beta F + \beta^2 cP\right), \tag{24.39}$$

$$F_0 = \gamma^2\left[(1+\beta^2)F - \beta(cE + cP)\right], \tag{24.40}$$

$$cP_0 = \gamma^2\left(\beta^2 cE - 2\beta F + cP\right). \tag{24.41}$$

(h) Energy-momentum tensor

The energy-momentum tensor for an ideal gas, $T^{\mu\nu}$, is

$$T^{\mu\nu} = (\varepsilon + p)\, u^\mu u^\nu - p g^{\mu\nu}, \tag{24.42}$$

[5]Transformation rules for the frequency-dependent quantities are

$$I_{\nu 0} = \left(\frac{\nu_0}{\nu}\right)^3 I_\nu = \left[\gamma\left(1 - \frac{\boldsymbol{v}\cdot\boldsymbol{l}}{c}\right)\right]^3 I_\nu,$$

$$E_{\nu 0} = \gamma\left(E_\nu - \frac{\boldsymbol{v}\cdot\boldsymbol{F}_\nu}{c^2}\right).$$

$$F_{\nu 0}^i = F_\nu^i - \gamma\beta^i cE_\nu + \frac{\gamma - 1}{\beta^2}\beta^i\beta^k F_\nu^k.$$

[6]In terms of the mean moments,

$$J_0 = \gamma^2\left(J - 2\beta H + \beta^2 K\right),$$

$$H_0 = \gamma^2\left[(1+\beta^2)H - \beta(J + K)\right],$$

$$K_0 = \gamma^2\left(\beta^2 J - 2\beta H + K\right).$$

where ε is the internal energy per unit proper volume[7] and p is the pressure measured in the comoving frame ($\varepsilon + p$ is the enthalpy per unit proper volume) (see Sect. 9.1).

The energy-momentum tensor for radiation, $R^{\mu\nu}$, is

$$R^{\mu\nu} = \begin{pmatrix} E & \frac{1}{c}F^i \\ \frac{1}{c}F^i & P^{ij} \end{pmatrix}, \tag{24.43}$$

where E is the radiation energy density, F^i the radiative flux, and P^{ij} the radiation stress tensor.

The momentum and energy conservations are expressed, respectively, as

$$\left(T_\mu^{\ \nu} + R_\mu^{\ \nu}\right)_{;\nu} = f_\mu, \tag{24.44}$$

$$u^\mu \left(T_\mu^{\ \nu} + R_\mu^{\ \nu}\right)_{;\nu} = u^\mu f_\mu, \tag{24.45}$$

where the semicolon means the partial differentiation in the present special relativistic case. Furthermore, f^μ is the four-force defined by[8]

$$f^\mu = \left(\gamma \boldsymbol{f} \cdot \frac{\boldsymbol{v}}{c}, \gamma \boldsymbol{f}\right), \tag{24.46}$$

and $\boldsymbol{f}$ $(= f^i)$ is the three-dimensional force per unit volume expressed as

$$\boldsymbol{f} = \frac{d\boldsymbol{p}}{dt} = \frac{d(\gamma m \boldsymbol{v})}{dt}, \tag{24.47}$$

$$\boldsymbol{f} \cdot \boldsymbol{v} = \frac{dE}{dt} = \frac{d(\gamma m c^2)}{dt}, \tag{24.48}$$

where $\boldsymbol{p}$ and E are momentum and energy of particles, respectively.

[7]Here, the internal energy includes the rest mass energy, and in the case of the ideal gas it is expressed as

$$\varepsilon = \rho c^2 + \frac{1}{\Gamma - 1} p,$$

where Γ is the ratio of specific heats. It is noticed that in Parts I and II the ratio of specific heats is denoted by γ, which is assigned to the Lorentz factor in this section.

[8]This four-force includes gravity, since we consider the flat space in special relativity.

24.2 Radiative Transfer Equation Under Special Relativity

We first derive the basic equations describing the behavior of radiation interacting with matter within the framework of special relativity. As already stated, many researchers have investigated and formulated the relativistic version of radiation hydrodynamics (e.g., Lindquist 1966; Anderson and Spiegel 1972; Hsieh and Spiegel 1976; Thorne 1981; Thorne et al. 1981; Mihalas and Auer 2001; Park 2001, 2006; Takahashi 2007; Kato et al. 2008). There remain several open questions, including the closure relation in the relativistic regime.

24.2.1 Relativistic Radiative Transfer Equation

As in the case of a nonrelativistic regime (Chap. 19), a change in the specific intensity of radiation is expressed by the *radiative transfer equation*, although it should be written down in a Lorentz-invariant form.

By means of the Lorentz invariant f $(= I_\nu/\nu^3 = I_{\nu 0}/\nu_0^3)$, we can write the transfer equation of the form (Hsieh and Spiegel 1976):

$$k^i \frac{\partial f}{\partial x^i} = \rho\,(\alpha - \beta f) - \rho\sigma_{\nu 0} \int \phi_\nu(\boldsymbol{l}', \boldsymbol{l}) f(\boldsymbol{l}) \nu' d\nu' d\Omega' + \rho\sigma_{\nu 0} \int \phi_\nu(\boldsymbol{l}, \boldsymbol{l}') f(\boldsymbol{l}') \nu' d\nu' d\Omega', \quad (24.49)$$

where ρ is the proper mass density, α the invariant form of the emission coefficient, β[9] the invariant form of the absorption coefficient, $\sigma_{\nu 0}$ the scattering opacity in the comoving frame, and ϕ_ν the scattering redistribution function. It is noted that $\nu d\nu d\Omega$ $(= \nu' d\nu' d\Omega')$ is also a relativistic invariant.

Of these, α and β are related, respectively, to the mass emissivity $j_{\nu 0}$ and the mass absorption coefficient κ_ν in the comoving frame by

$$j_{\nu 0} = 4\pi \nu_0^2 \alpha \quad \text{and} \quad \kappa_{\nu 0} = \frac{\beta}{\nu_0}. \quad (24.50)$$

For Thomson scattering, the scattering redistribution function in the comoving frame is

$$\phi_\nu = \frac{3}{4}\left[1 + (\boldsymbol{l}_0 \cdot \boldsymbol{l}_0')^2\right] \delta(\nu_0 - \nu_0') \frac{1}{4\pi}. \quad (24.51)$$

It should be noted that $\int \phi_\nu \nu_0 d\nu_0 d\Omega_0 = \nu_0'$ and $\int \phi_\nu \nu_0' d\nu_0' d\Omega_0' = \nu_0$.

[9] It is noticed that the same symbol β is assigned to the normalized speed v/c.

Substituting these quantities into Eq. (24.49), we have the transfer equation in the case of Thomson scattering as

$$\nu\left[\frac{\partial f}{c\partial t}+(\boldsymbol{l}\cdot\nabla)f\right]=\rho\frac{j_{\nu 0}}{4\pi\nu_0^2}-\rho\nu_0\kappa_{\nu 0}f-\rho\nu_0\sigma_{\nu 0}f$$
$$+\frac{3}{4}\rho\sigma_{\nu 0}\nu_0\int\left[1+(\boldsymbol{l}_0\cdot\boldsymbol{l}_0')^2\right]f(\boldsymbol{l}')\frac{d\Omega_0'}{4\pi}. \tag{24.52}$$

Furthermore, replacing f by I_ν (or $I_{\nu 0}$), we finally obtain the (angle-dependent) relativistic radiative transfer equation *in the mixed frame*[10] :

$$\frac{1}{c}\frac{\partial I_\nu}{\partial t}+(\boldsymbol{l}\cdot\nabla)I_\nu=\left(\frac{\nu}{\nu_0}\right)^2\rho$$
$$\times\left[\frac{j_{\nu 0}}{4\pi}-(\kappa_{\nu 0}+\sigma_{\nu 0})I_{\nu 0}+\frac{3}{4}\sigma_{\nu 0}\frac{c}{4\pi}\left(E_{\nu 0}+l_{0i}l_{0j}P_{\nu 0}^{ij}\right)\right], \tag{24.53}$$

where we use the definitions of E and P^{ij}.

This transfer equation (24.53) seems to be similar to the nonrelativistic one, except for the ν/ν_0-term. It should be noted, however, that the left-hand side is written by the quantities evaluated in the inertial (fixed) frame, while the right-hand side by the quantities in the comoving (fluid) frame.

Integrating the transfer equation (24.53) over the frequency, with the help of the Lorentz transformation (24.15) $[d\nu=(d\nu/d\nu_0)d\nu_0=\gamma(1+\boldsymbol{v}\cdot\boldsymbol{l}_0/c)d\nu_0]$, we obtain a frequency-integrated angle-dependent transfer equation in the mixed frame:

$$\frac{1}{c}\frac{\partial I}{\partial t}+(\boldsymbol{l}\cdot\nabla)I=\rho\gamma^3\left(1+\frac{\boldsymbol{v}\cdot\boldsymbol{l}_0}{c}\right)^3$$
$$\times\left[\frac{j_0}{4\pi}-(\kappa_0+\sigma_0)I_0+\frac{3}{4}\sigma_0\frac{c}{4\pi}\left(E_0+l_{0i}l_{0j}P_0^{ij}\right)\right], \tag{24.54}$$

where

$$I\equiv\int I_\nu d\nu,\quad I_0\equiv\int I_{\nu 0}d\nu_0, \tag{24.55}$$

$$E_0\equiv\int E_{\nu 0}d\nu_0,\quad P_0^{\alpha\beta}\equiv\int P_{\nu 0}^{\alpha\beta}d\nu_0, \tag{24.56}$$

$$j_0\equiv\int j_{\nu 0}d\nu_0,\quad \kappa_0+\sigma_0\equiv\frac{1}{I_0}\int(\kappa_{\nu 0}+\sigma_{\nu 0})I_{\nu 0}d\nu_0. \tag{24.57}$$

[10]For the isotropic scattering case, $\phi_\nu=\delta(\nu_0-\nu_0')/(4\pi)$, and the third term on the right-hand side of Eq. (24.53) simply reduces $\sigma_{\nu 0}cE_{\nu 0}/(4\pi)$.

As already noted, the left-hand side of the transfer equation (24.54) is described by the quantities in the inertial (fixed) frame, while the right-hand sides by those in the comoving (fluid) frame. Thus, using the transformation rules (24.36)–(24.38), let us rewrite the right-hand side of the transfer equation. After several manipulations, we finally obtain the transfer equation expressed by the quantities in the inertial (fixed) frame:

$$
\begin{aligned}
\frac{1}{c}\frac{\partial I}{\partial t} + (\boldsymbol{l}\cdot\nabla)\,I &= \rho\gamma^{-3}\left(1-\frac{\boldsymbol{v}\cdot\boldsymbol{l}}{c}\right)^{-3} \\
&\times\left[\frac{j_0}{4\pi} - (\kappa_0+\sigma_0)\,\gamma^4\left(1-\frac{\boldsymbol{v}\cdot\boldsymbol{l}}{c}\right)^4 I + \frac{\sigma_0}{4\pi}\frac{3}{4}\gamma^{-2}\left(1-\frac{\boldsymbol{v}\cdot\boldsymbol{l}}{c}\right)^{-2}\right. \\
&\times\left\{\gamma^4\left[\left(1-\frac{\boldsymbol{v}\cdot\boldsymbol{l}}{c}\right)^2+\left(\frac{v^2}{c^2}-\frac{\boldsymbol{v}\cdot\boldsymbol{l}}{c}\right)^2\right]cE + 2\gamma^2\left(\frac{v^2}{c^2}-\frac{\boldsymbol{v}\cdot\boldsymbol{l}}{c}\right)\boldsymbol{F}\cdot\boldsymbol{l}\right. \\
&-2\gamma^4\left[\left(1-\frac{\boldsymbol{v}\cdot\boldsymbol{l}}{c}\right)^2+\left(1-\frac{\boldsymbol{v}\cdot\boldsymbol{l}}{c}\right)\left(\frac{v^2}{c^2}-\frac{\boldsymbol{v}\cdot\boldsymbol{l}}{c}\right)\right]\frac{\boldsymbol{v}\cdot\boldsymbol{F}}{c} \\
&\left.\left.+l_il_jcP^{ij} - 2\gamma^2\left(1-\frac{\boldsymbol{v}\cdot\boldsymbol{l}}{c}\right)v_il_jP^{ij} + 2\gamma^4\left(1-\frac{\boldsymbol{v}\cdot\boldsymbol{l}}{c}\right)^2\frac{v_iv_jP^{ij}}{c}\right\}\right].
\end{aligned}
\tag{24.58}
$$

For the isotropic scattering case, the transfer equations (24.54) and (24.58) reduce to a rather simple form:

$$
\begin{aligned}
\frac{1}{c}\frac{\partial I}{\partial t} + (\boldsymbol{l}\cdot\nabla)\,I &= \rho\gamma^3\left(1+\frac{\boldsymbol{v}\cdot\boldsymbol{l}_0}{c}\right)^3\left[\frac{j_0}{4\pi} - (\kappa_0+\sigma_0)\,I_0 + \sigma_0\frac{cE_0}{4\pi}\right] \\
&= \rho\gamma^{-3}\left(1-\frac{\boldsymbol{v}\cdot\boldsymbol{l}}{c}\right)^{-3}\left[\frac{j_0}{4\pi} - (\kappa_0+\sigma_0)\,\gamma^4\left(1-\frac{\boldsymbol{v}\cdot\boldsymbol{l}}{c}\right)^4 I\right. \\
&\left.+\frac{\sigma_0}{4\pi}\gamma^2\left(cE - 2\frac{\boldsymbol{v}\cdot\boldsymbol{F}}{c} + \frac{v_iv_k}{c^2}cP^{ik}\right)\right].
\end{aligned}
\tag{24.59}
$$

24.2.2 *Relativistic Formal Solutions*

Before proceeding to the moment equations, we briefly mention the relativistic formal solution of the relativistic transfer equation.

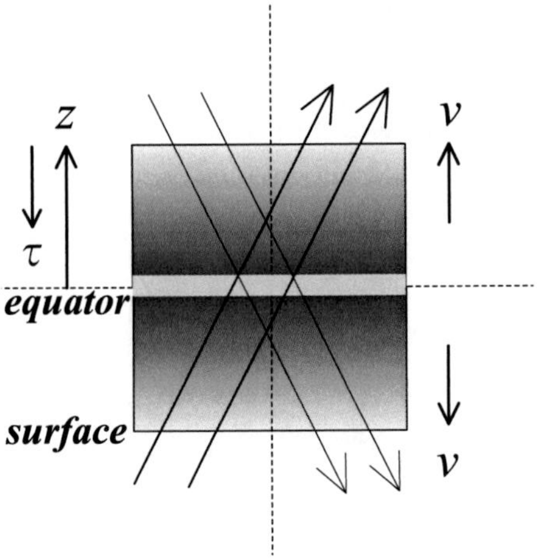

Fig. 24.2 Plane-parallel radiative flow in the vertical (z) direction

24.2.2.1 Plane-Parallel Geometry

Let us suppose a plane-parallel steady flow in the vertical (z) direction, and the flow velocity $v(z)$ is given as a function of z (Fig. 24.2). We also assume the gray approximation, where the absorption opacity κ_0 and scattering one σ_0 measured in the comoving frame do not depend on frequency. In addition, we assume scattering to be isotropic for simplicity.

Under these approximations, in the plane-parallel geometry with the vertical axis z and the direction cosine μ ($= \cos\theta$), the relativistic radiation transfer equation in the mixed frame is expressed as

$$\mu \frac{dI}{dz} = \rho_0 \frac{1}{\gamma^3(1-\beta\mu)^3} \left[\frac{j_0}{4\pi} - (\kappa_0 + \sigma_0)\, I_0 + \sigma_0 J_0 \right], \tag{24.60}$$

where I and I_0 are the frequency-integrated specific intensities in the inertial and comoving frames, respectively, ρ_0 the proper mass density, j_0 the frequency-integrated mass emissivity, and J_0 the frequency-integrated mean intensity measured in the comoving frame.

Introducing the "optical depth" defined by[11]

$$d\tau = -\,(\kappa_0 + \sigma_0)\, \rho_0 dz, \tag{24.61}$$

[11] Here, we introduce and define the "optical depth," as a convenient non-dimensional variable (see Sect. 24.3.1 below).

the transfer equation (24.60) finally becomes

$$\mu \frac{dI}{d\tau} = \frac{1}{\gamma^3(1-\beta\mu)^3}\left[\gamma^4(1-\beta\mu)^4 I - S_0\right], \tag{24.62}$$

where

$$S_0 = \frac{j_0}{4\pi}\frac{1}{\kappa_0+\sigma_0} + \frac{\sigma_0}{\kappa_0+\sigma_0}J_0 \tag{24.63}$$

is the source function in the comoving frame.

If necessary, the optical depth τ_b at the flow base of $z = 0$ is assumed to be finite,

$$\tau_b = -\int_{\infty}^{0} (\kappa_0+\sigma_0)\,\rho_0 dz. \tag{24.64}$$

Equation (24.62) is arranged as

$$\frac{dI}{d\tau} - \frac{\gamma(1-\beta\mu)}{\mu} I = -\frac{1}{\mu}\frac{1}{\gamma^3(1-\beta\mu)^3} S_0, \tag{24.65}$$

and using the integral factor:

$$X \equiv \exp\left(-\frac{1}{\mu}\int^{\tau}\gamma dt + \int^{\tau}\gamma\beta dt\right), \tag{24.66}$$

we can rearrange Eq. (24.65) as

$$\frac{d}{d\tau}(XI) = -\frac{1}{\mu}\frac{X}{\gamma^3(1-\beta\mu)^3} S_0. \tag{24.67}$$

This equation can be formally integrated as

$$e^{-G(t)/\mu+U(t)} I|^{\tau} = -\int^{\tau} \frac{e^{-G(t)/\mu+U(t)}}{\mu\gamma^3(1-\beta\mu)^3} S_0 dt, \tag{24.68}$$

where

$$G(t) \equiv \int^{t} \gamma(t')dt', \tag{24.69}$$

$$U(t) \equiv \int^{t} \gamma(t')\beta(t')dt'. \tag{24.70}$$

Integrating upward (from $\tau_{\rm b}$ to τ) or downward (from 0 to τ), we formally obtain the upward intensity $I^+(\tau, \mu > 0)$ and downward one $I^-(\tau, \mu < 0)$, respectively,

$$I^+(\tau, \mu) = \exp\left[-\frac{G(\tau_{\rm b}) - G(\tau)}{\mu} + U(\tau_{\rm b}) - U(\tau)\right] I^+(\tau_{\rm b}, \mu) + \int_\tau^{\tau_{\rm b}} \frac{\exp\left[-\frac{G(t) - G(\tau)}{\mu} + U(t) - U(\tau)\right]}{\mu\gamma^3(1 - \beta\mu)^3} S_0 dt, \quad (24.71)$$

$$I^-(\tau, \mu) = \int_0^\tau \frac{\exp\left\{-\frac{G(\tau) - G(t)}{(-\mu)} - [U(\tau) - U(t)]\right\}}{(-\mu)\gamma^3[1 + \beta(-\mu)]^3} S_0 dt. \quad (24.72)$$

These are the *relativistic formal solutions* (Fukue 2014). In the limit of $\beta = 0$, these relativistic formal solutions reduce to the usual formal solution for the static atmosphere.

24.2.2.2 Spherical Geometry

Let us next suppose a spherical flow in the radial (r) direction around a central object, e.g., an AGN wind or black-hole accretion (Fig. 24.3). The radius of the central luminous/dark core is R_* if it exists, while the outer radius of the flow region is $R_{\rm out}$. The flow quantities such as velocity v or density ρ are generally a function of radius r. According to the impact parameter method (Hummer and Rybicki 1971)/the tangent ray method (Dullemond 2015), on the other hand, the radiative intensity $I^\pm(p, z)$ is a function of the impact space coordinates (p, z); $r^2 = p^2 + z^2$. We also assume the gray approximation.

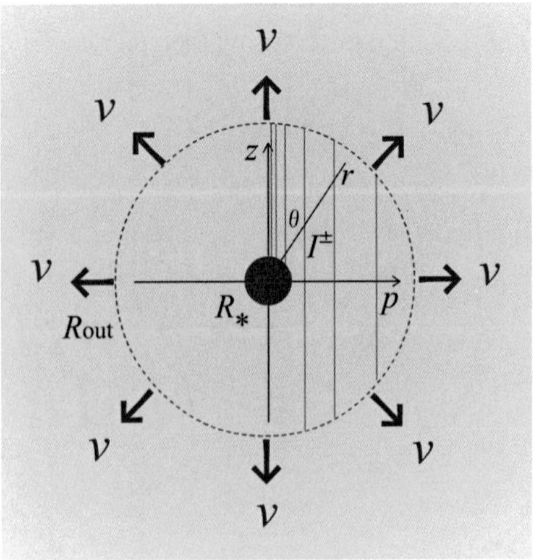

Fig. 24.3 Relativistic spherical flow in the radial (r) direction around a central object with radius R_*. The flow quantities such as velocity v or density ρ are a function of radius r, while the radiative intensity $I^\pm(p, z)$ is a function of the impact space coordinates (p, z); $r^2 = p^2 + z^2$

The relativistic transfer equation in the mixed frame for the present spherical case is expressed as

$$\pm \frac{\partial I^{\pm}(p,z)}{\partial z} = \frac{\rho_0}{\gamma^3(1-\boldsymbol{\beta}\cdot\boldsymbol{l})^3}\left[\frac{j_0}{4\pi} - (\kappa_0+\sigma_0)\, I_0 + \sigma_0 J_0\right], \quad (24.73)$$

where $\gamma = 1/\sqrt{1-\beta^2}$ is the Lorentz factor, $\beta(r)$ $(= v/c)$ being the normalized radial speed, and $\boldsymbol{\beta}\cdot\boldsymbol{l}$ the scalar product between the flow velocity $\boldsymbol{\beta}$ and the ray direction $\boldsymbol{l}$ in the inertial frame;

$$\boldsymbol{\beta}\cdot\boldsymbol{l} = \pm\beta(r)\cos\theta = \pm\beta(r)\frac{z}{r} \quad (24.74)$$

in the present coordinate.

Using the source function S_0, Eq. (24.63), the relativistic transfer equation (24.73) becomes

$$\pm \frac{\partial I^{\pm}(p,z)}{\partial z} = -\frac{(\kappa_0+\sigma_0)\rho_0}{\gamma^3(1-\boldsymbol{\beta}\cdot\boldsymbol{l})^3}\left[\gamma^4(1-\boldsymbol{\beta}\cdot\boldsymbol{l})^4 I^{\pm} - S_0\right]. \quad (24.75)$$

In a manner similar to the plane-parallel case, Eq. (24.75) is arranged as

$$\pm \frac{\partial I^{\pm}(p,z)}{\partial z} = -(\kappa_0+\sigma_0)\rho_0\gamma(1-\boldsymbol{\beta}\cdot\boldsymbol{l})I^{\pm}(p,z) + \frac{(\kappa_0+\sigma_0)\rho_0}{\gamma^3(1-\boldsymbol{\beta}\cdot\boldsymbol{l})^3}S_0(r). \quad (24.76)$$

In the present case, using the integral factor:

$$X \equiv \exp\left[+\int^z (\kappa_0+\sigma_0)\rho_0\gamma\, d\zeta - \int^z (\kappa_0+\sigma_0)\rho_0\gamma\,\boldsymbol{\beta}\cdot\boldsymbol{l}\, d\zeta\right], \quad (24.77)$$

where the integration is done along the $z(\zeta)$-axis, we further rearranged Eq. (24.76) as

$$\pm \frac{\partial}{\partial z}\left(X I^{\pm}\right) = X\frac{(\kappa_0+\sigma_0)\rho_0}{\gamma^3\left(1-\boldsymbol{\beta}\cdot\boldsymbol{l}\right)^3}S_0, \quad (24.78)$$

or integrate as

$$e^{G(p,\zeta)-U(p,\zeta)} I^{\pm}|^z = \int^z X\frac{(\kappa_0+\sigma_0)\rho_0}{\gamma^3\left(1-\boldsymbol{\beta}\cdot\boldsymbol{l}\right)^3}S_0 d\zeta, \quad (24.79)$$

where

$$G(p,z) \equiv \int^{z} (\kappa_0 + \sigma_0)\rho_0 \gamma d\zeta, \tag{24.80}$$

$$U(p,z) \equiv \int^{z} (\kappa_0 + \sigma_0)\rho_0 \gamma \boldsymbol{\beta} \cdot \boldsymbol{l} d\zeta. \tag{24.81}$$

Integrating from 0 to z, we formally have the upward intensity $I^+(p,z)$ as

$$\begin{aligned} I^+(p,z) &= \exp\left[G(p,z_*) - G(p,z) - U(p,z_*) + U(p,z)\right] I^*(p,z_*) \\ &+ \int_0^z \frac{\exp\left[G(p,\zeta) - G(p,z) - U(p,\zeta) + U(p,z)\right]}{\gamma^3 \left(1 - \beta \frac{\zeta}{r}\right)^3} \\ &\times (\kappa_0 + \sigma_0)\rho_0 S_0 d\zeta, \end{aligned} \tag{24.82}$$

where $z_* \equiv \sqrt{R_*^2 - p^2}$ if the luminous core exists. On the other hand, integrating from z_{out} $(= \sqrt{R_{\text{out}}^2 - p^2})$ to z, we have the downward intensity $I^-(p,z)$ as

$$\begin{aligned} I^-(p,z) = -\int_{z_{\text{out}}}^{z} & \frac{\exp\{[G(p,z) - G(p,\zeta)] + [U(p,z) - U(p,\zeta)]\}}{\gamma^3 \left(1 + \beta \frac{\zeta}{r}\right)^3} \\ & \times (\kappa_0 + \sigma_0)\rho_0 S_0 d\zeta, \end{aligned} \tag{24.83}$$

where we assume that $I^-(p,\infty) = 0$ as a boundary condition. These are the *relativistic formal solutions for the relativistic spherical flows* (Fukue 2017).

24.2.3 Relativistic Moment Equations

Next, we derive the (frequency-integrated) moment equations. After a long time since Eddington, who first introduced a moment expansion to radiation transfer in the early twentieth century, moment equations for relativistic radiation transfer have been studied for special relativistic cases (Thomas 1930; Hazlehurst and Sargent 1959; Castor 1972; Mihalas and Mihalas 1984; Mihalas and Auer 2001) and for cases of a curved space-time (Lindquist 1966; Anderson and Spiegel 1972; Thorne 1981; Udey and Israel 1982; Nobili et al. 1993; Park 2003, 2006; Takahashi 2007). A complete set of moment equations for a relativistic flow is given by the projected symmetric trace-free (PSTF) formalism (Thorne 1981).

Integrating the transfer equation (24.54) over solid angle, with the help of the transformation of the solid angle (24.19), we obtain the zeroth moment of

Eq. (24.54):

$$\frac{\partial E}{\partial t} + \frac{\partial F^k}{\partial x^k} = \rho\gamma\left(j_0 - c\kappa_0 E_0\right) - \rho\gamma\left(\kappa_0 + \sigma_0\right)\frac{\boldsymbol{v}\cdot\boldsymbol{F}_0}{c}. \tag{24.84}$$

Integrating the transfer equation (24.54) over solid angle, after being multiplied by the direction cosine, with the help of transformations (24.17) and (24.19), we obtain the first-moment of Eq. (24.54):

$$\frac{1}{c^2}\frac{\partial F^i}{\partial t} + \frac{\partial P^{ik}}{\partial x^k} = -\frac{1}{c}\rho\left(\kappa_0 + \sigma_0\right)F_0^\alpha + \rho\gamma\frac{v^i}{c^2}\left(j_0 - c\kappa_0 E_0\right)$$
$$-\rho\left(\kappa_0 + \sigma_0\right)\frac{\gamma - 1}{v^2}\frac{v^i}{c}\left(\boldsymbol{v}\cdot\boldsymbol{F}_0\right). \tag{24.85}$$

In general, the method of a moment expansion gives an infinite set of equations. In order to make the transfer problem tractable, one must truncate the expansion at a finite order by adopting a suitable closure assumption. For example, we here truncate the equations at the second order, and we introduce some closure relation among E, F^i, and P^{ik}, as given in the next subsection.

As already noted, the left-hand sides of moment equations (24.84) and (24.85) are described by the quantities in the inertial (fixed) frame, while the right-hand sides by those in the comoving (fluid) frame. Thus, using the transformation rules (24.36)–(24.38), let us rewrite the right-hand side of these moment equations by quantities in the inertial frame. After several manipulations, we finally obtain the moment equations expressed by the quantities in the inertial (fixed) frame:

$$\frac{\partial E}{\partial t} + \frac{\partial F^k}{\partial x^k} = \rho\gamma\left(j_0 - c\kappa_0 E + \kappa_0\frac{\boldsymbol{v}\cdot\boldsymbol{F}}{c}\right)$$
$$+\rho\gamma^3\sigma_0\left[\frac{v^2}{c}E + \frac{v_i v_j}{c}P^{ij} - \left(1 + \frac{v^2}{c^2}\right)\frac{\boldsymbol{v}\cdot\boldsymbol{F}}{c}\right], \tag{24.86}$$

$$\frac{1}{c^2}\frac{\partial F^i}{\partial t} + \frac{\partial P^{ik}}{\partial x^k} = \frac{\rho\gamma}{c}\left(\frac{v^i}{c}j_0 - \kappa_0 F^i + \kappa_0 v_k P^{ik}\right)$$
$$-\frac{\rho\gamma}{c}\sigma_0\left[F^i - \gamma^2 E v^i - v_k P^{ik} + \gamma^2 v^i\left(\frac{2\boldsymbol{v}\cdot\boldsymbol{F}}{c^2} - \frac{v_j v_k}{c^2}P^{jk}\right)\right]. \tag{24.87}$$

24.2.4 Closure Relation in the Relativistic Regime

As in the nonrelativistic regime, we need some closure relation for the moment equations. However, a well-behaved closure relation has not been known in the relativistic regime.

One of choices for the closure relation, we may adopt the diffusion approximation in the comoving frame, as used in many literatures referred in Chap. 21, or the flux-limited diffusion (FLD) approximation in the comoving frame, similar to the nonrelativistic case (Chap. 20), as used in various radiation hydrodynamical simulations. As already stated in Sect. 21.1, however, even in the nonrelativistic regime, the diffusion approximation brings a bad behavior at the critical points. Hence, we hesitate in using the diffusion approximations in the relativistic regime.

As an alternative choice for the closure relation, we may adopt the Eddington approximation *in the comoving frame*:

$$P_0^{ij} = \frac{\delta^{ij}}{3} E_0. \tag{24.88}$$

Substituting the transformation rules (24.36)–(24.38) into this relation (24.88), we obtain the closure relation in the inertial frame[12] :

$$\begin{aligned} P^{ij} &- \frac{\delta^{ij}}{3}\gamma^2 \frac{v_k v_m}{c^2} P^{km} + \frac{\gamma^2}{\gamma+1}\left(\frac{v^i v_k}{c^2} P^{jk} + \frac{v^j v_k}{c^2} P^{ik}\right) \\ &+ \left(\frac{\gamma^2}{\gamma+1}\right)^2 \frac{v^i v^j}{c^2} \frac{v_k v_m}{c^2} P^{km} = \frac{\delta^{ij}}{3}\gamma^2 \left(E - 2\frac{\boldsymbol{v}\cdot\boldsymbol{F}}{c^2}\right) - \gamma^2 \frac{v^i v^j}{c^2} E \\ &+ \gamma\left(\frac{v^i F^j}{c^2} + \frac{v^j F^i}{c^2}\right) + 2\gamma \frac{\gamma^2}{\gamma+1} \frac{v^i v^j}{c^2} \frac{\boldsymbol{v}\cdot\boldsymbol{F}}{c^2}. \end{aligned} \tag{24.90}$$

In order to improve the simple Eddington approximation, where the radiation field is assumed to be isotropic in the comoving frame, we can adopt a *variable Eddington factor* which depends on the flow velocity β $(= v/c)$ and the velocity gradient $d\beta/d\tau$ as well as the optical depth τ (Fukue 2006; Akizuki and Fukue 2008; Fukue 2008a,b, 2009). In one-dimensional flows, for example, the variable Eddington factor $f(\tau, \beta)$ is generally defined as

$$P_0 = f(\tau, \beta) E_0, \tag{24.91}$$

---

[12] To the first order in $\boldsymbol{v}/c$, this closure relation becomes (Hsieh and Spiegel 1976)

$$P^{ij} = \frac{\delta^{ij}}{3} E + \frac{v^i F^j}{c^2} + \frac{v^j F^i}{c^2} - \frac{2}{3}\delta^{ij} \frac{\boldsymbol{v}\cdot\boldsymbol{F}}{c^2}. \tag{24.89}$$

Yin and Miller (1995) pointed out, however, that if we solve the relativistic problem within the approximation of $\mathcal{O}(v/c)^1$, the solution would be quite inaccurate in some cases.

where E_0 and P_0 are the radiation energy density and the radiation pressure in the comoving frame, respectively. The closure relation in the inertial frame for one-dimensional flows then becomes

$$cP(1 - f\beta^2) = cE(f - \beta^2) + 2F\beta(1 - f), \tag{24.92}$$

where E, F, and P are the radiation energy density, the radiative flux, and the radiation pressure in the inertial frame, respectively.

The function $f(\tau, \beta)$ must reduce to 1/3 or an appropriate value in the nonrelativistic limit of $\beta = 0$, whereas it would approach unity in the extremely relativistic limit of $\beta = 1$. Furthermore, in the sufficiently optically thick regime f approaches 1/3 except for the case of $\beta = 1$, while in the optically thin limit it reduces to an appropriate form determined by the geometry under the considerations.[13]

The radiation moment formalism is quite convenient and essential, and it is a powerful tool for tackling problems of relativistic radiation hydrodynamics (e.g., Thorne et al. 1981; Flammang 1982, 1984; Nobili et al. 1991, 1993; Park 2001, 2006 for spherically symmetric problems; Takahashi 2007 for the Kerr metric). However, its validity is never verified unless a fully angle-dependent radiation transfer equation is solved. Thus, the relativistic moment equations with a closure relation must be carefully treated and applied to black-hole accretion flows, relativistic jets and winds, and relativistic explosions such as gamma-ray bursts.

Actually, a pathological behavior in relativistic radiation moment equations has been pointed out and examined (Turolla and Nobili 1988; Nobili et al. 1991; Turolla et al. 1995; Dullemond 1999). Namely, the moment equations for radiation transfer in relativistically moving media can generally have singular (critical) points. As a result, solutions behave pathologically in a relativistic regime. The appearance of singularities is related to the approximation of the full transfer equations with a finite number of moments, and purely mathematical artifacts of the moment expansion (Dullemond 1999).

For example, in one-dimensional relativistic flows using the closure relation (24.88), where the moment equations are truncated at the second order, a singularity appears when the flow velocity v becomes $\pm c/\sqrt{3}$, which corresponds to the relativistic sound speed (Turolla and Nobili 1988; Turolla et al. 1995). Hence, we cannot obtain solutions accelerated beyond $c/\sqrt{3}$, although there exists a decelerating solution (Fukue 2005).

The invalidity of the Eddington approximation in such a relativistic flow can be understood as follows. In adopting the Eddington approximation (24.88), we

[13]In the plane-parallel case, for instance, the variable Eddington factor in the optically thin limit is analytically derived as [Kato et al. (2008)]

$$f = \frac{1 - 3\beta + 3\beta^2}{3 - 3\beta + \beta^2}.$$

assume that within the photon mean free path the radiation field is *isotropic* in the comoving frame. However, in the relativistic regime, where the velocity gradient becomes large and there exist the Doppler and aberration effects of photons, the isotropy of the radiation field may break down even in the comoving frame.

For example, the photon mean free path ℓ in the comoving frame is $\ell \sim 1/(\kappa\rho)$, where κ is the opacity measured in the comoving frame and ρ is the proper density. When there exists a (strong) velocity gradient, say dv/ds, the velocity increase at the distance of ℓ is estimated as

$$\Delta v = \ell \frac{dv}{ds} = \frac{dv}{\kappa\rho ds} = \frac{dv}{d\tau}, \tag{24.93}$$

where τ $(= \kappa\rho s)$ is the optical depth. In order for the radiation fields to be isotropic in the comoving frame, this velocity increase should be sufficiently smaller than the speed of light. Otherwise, we should modify the closure relation in the case of subrelativistic to relativistic regimes, as in the case of optically thick to thin regimes.

24.3 Several Topics on the Relativistic Radiative Transfer

Similar to the hydrodynamics and magnetohydrodynamics, many properties of radiation hydrodynamics in the nonrelativistic regime are qualitatively extended in the relativistic regime, although they are quantitatively modified. However, there exist several properties which show up only in the relativistic regime. In this section, we shall introduce several characteristic properties that appeared in the relativistic radiative transfer. These relativistic properties become important in the relativistic astrophysical phenomena, such as astrophysical jets and outflows from a compact object, relativistic explosions, accretion flows onto accretion disks around a black hole, and so on.

24.3.1 Optical Depth and Apparent Photosphere

Since the optical depth is a relativistic invariant, which is the same both for the inertial and comoving frames, the correspondent lengths in both frames are different, due to the Lorentz–Fitzgerald contraction. Abramowicz et al. (1991) found that, in the relativistically moving media, the apparent optical depth becomes small in the downstream direction, while it is large in the upstream direction. As a result, the apparent photosphere is modified, compared with the nonrelativistic case (Sect. 22.3.1); the shape of the photosphere appears convex in the nonrelativistic case, but concave for relativistic regimes. Hence, in order to calculate spectra and luminosities of *relativistic* winds, we should determine the location of the photosphere and the temperature distribution there under the viewpoint of the *relativistic radiation transfer* (cf. Sumitomo et al. 2007; Fukue and Sumitomo 2009). Similar effects would take place in the relativistic jets, accretion disks, and relativistically moving objects.

24.3.1.1 Invariant Optical Depth

Let us consider a small distance ds along the light path. Due to the relativistic effect (Lorentz–Fitzgerald contraction), the mean free path of photons in the fixed frame, ℓ, is related to that in the comoving frame, ℓ_0, by

$$\ell = \frac{1}{\gamma(1-\beta\cos\theta)}\ell_0, \tag{24.94}$$

where β $(= v/c)$ is the flow speed v normalized by the speed of light, γ $(= 1/\sqrt{1-\beta^2})$ the Lorentz factor, and θ the viewing angle measured from the z-axis (observer's line-of-sight; see Fig. 24.4 below). Here and hereafter, quantities with subscript "0" refer to physical quantities measured in the comoving frame. Then, the optical depth in the fixed frame is given by

$$d\tau = d\tau_0 = \kappa_0\rho_0 ds_0 = \gamma(1-\beta\cos\theta)\kappa_0\rho_0 ds, \tag{24.95}$$

where the opacity κ_0 is assumed to be electron scattering. As is seen in Eq. (24.95), the optical depth strongly depends on the viewing angle θ as well as the flow speed v in the relativistic flow. It is obvious that the optical depth is the largest in the upstream direction at $\theta = \pi$, where the photons move in the opposite direction to the fluid, while it becomes the smallest in the downstream direction at $\theta = 0$, where the photons move in the same direction with the fluid.

An "apparent photosphere" of the flow is defined as the surface where the optical depth τ measured from an observer at infinity becomes unity (see Fig. 24.4).

The integrated optical depth $\tau_{\rm ph}$ from an observer at infinity is calculated, from Eq. (24.95), as

$$\tau_{\rm ph} = \int_{z_{\rm ph}}^{\infty} \gamma(1-\beta\cos\theta)\kappa_0\rho_0 ds = 1, \tag{24.96}$$

where $z_{\rm ph}$ is the height of the apparent photosphere from the equatorial plane. Although the nonrelativistic and moderately relativistic winds have convex photospheres, the photospheres of relativistic winds become concave for $\beta > 2/3$ (see Fig. 24.5 later).

24.3.1.2 Apparent Photosphere of Relativistic Spherical Wind

According to Abramowicz et al. (1991), let us suppose a spherically symmetric, relativistic wind, which blows off from a central object with mass M at a constant speed v. The Schwarzschild radius of the central object is defined by $r_S = 2GM/c^2$. We use the spherical coordinates (R, θ, φ) and the cylindrical coordinate (r, φ, z), whose z-axis is along the line-of-sight (see Fig. 24.4).

From continuity equation, for the spherically symmetric stationary wind, the rest mass density ρ_0 measured in the comoving frame varies as

$$\rho_0 = \frac{\dot{M}}{4\pi\gamma v}\frac{1}{R^2}, \tag{24.97}$$

where $\dot{M}$ is the constant mass-outflow rate and $R = \sqrt{r^2 + z^2}$ is a distance from the central object. Hence, if the wind velocity v is assumed to be constant, the density varies as $\rho_0 \propto 1/R^2$.

In this simple model, we first note that the photospheric radius R_{ph} (Sect. 22.3.1), the height $z_{ph}(0)$ of the apparent photosphere at $r = 0$ $(\theta = 0)$, becomes

$$R_{ph} = z_{ph}(0) = \frac{1-\beta}{\beta}\dot{m} r_S, \tag{24.98}$$

where $\dot{m}$ $(= \dot{M}/\dot{M}_{crit})$ is the normalized mass-loss rate, $\dot{M}_{crit}$ $(= L_E/c^2)$ being the critical mass-loss rate. Hence, the size of the apparent photosphere of a black-hole wind roughly expands in proportion to the mass-outflow rate. In addition, for the

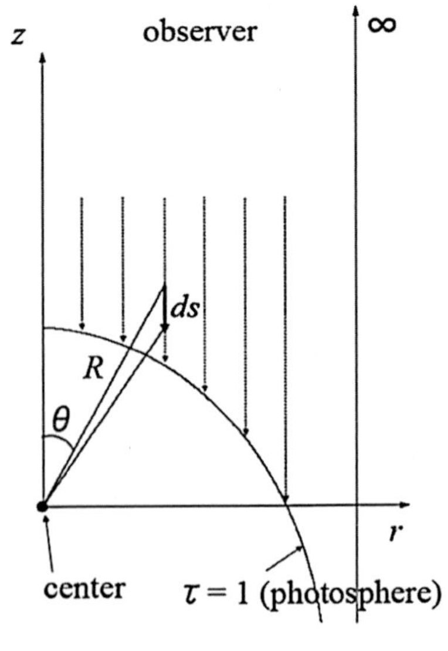

Fig. 24.4 Schematic picture of an apparent "photosphere" in the spherical wind blown off from the origin at a relativistic speed. The observer is located at infinity in the z-direction. An apparent "photosphere" is defined so that the optical depth measured from the infinity becomes unity

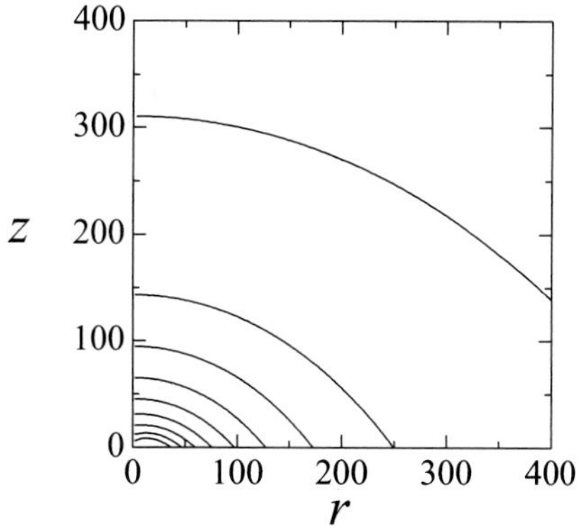

Fig. 24.5 Location of the apparent photosphere for various wind velocity β in the case of $\dot{m} = 100$. The wind velocity is varied from 0.1 (outer region) to 0.9 (central region) in steps of 0.1. The units of the r- and z-axis is the Schwarzschild radius r_S

extremely relativistic case of $\beta \sim 1$, the size is proportional to $1/\gamma^2$, as Abramowicz et al. (1991) showed.

In this simple model, furthermore, Eq. (24.96) can be analytically integrated to give the location of the "apparent" photosphere (cf. Sect. 22.3.1 for the nonrelativistic case):

$$\tau(z) = -\int_{\infty}^{z} \gamma(1 - \beta\cos\theta)\kappa_{es}\rho_0 dz = \frac{\dot{m}}{2\beta}\frac{r_S}{r}\left[\frac{\pi}{2} - \tan^{-1}\frac{z}{r} - \beta\left(1 + \frac{z^2}{r^2}\right)^{-1/2}\right]. \tag{24.99}$$

The location of the apparent photosphere seen by the observer at infinity in the z-direction is shown in Fig. 24.5 for various wind velocities. When the wind speed is low, the shape of the photosphere is the same as that of the nonrelativistic case. When the wind speed becomes high, the shape of the apparent photosphere changes, because the optical depth depends on the angle θ and the wind velocity v. In particular, when the wind blows off at highly relativistic speed ($\beta \geq 2/3$), the photosphere looks like a concave shape.

24.3.1.3 Photon Trapping Radius in the Relativistic Regime

Due to the optical depth in the relativistic regime, the photon trapping radius (Sect. 22.3.3) is also modified. In the case of a simple optically thick spherical wind at constant speed v, instead of the nonrelativistic case of $\dot{m} r_S/2$, the photon trapping radius is evaluated as

$$R_{trap} = \frac{\dot{m}(1 - \beta)}{2} r_S. \tag{24.100}$$

References

Abramowicz, M.A., Novikov, I.D., Paczyński, B.: Astrophys. J. **369**, 175 (1991)

Akizuki, C., Fukue, J.: Publ. Astron. Soc. Jpn. **60**, 337 (2008)

Anderson, J.L., Spiegel, E.A.: Astrophys. J. **171**, 127 (1972)

Castor, J.I.: Astrophys. J. **178**, 779 (1972)

Dullemond, C.P.: Astron. Astrophys. **343**, 1030 (1999)

Dullemond, C.P.: Radiative Transfer in Astrophysics: Theory, Numerical Methods and Applications (e-book) (2015)

Flammang, R.A.: Monthly Not. Roy. Astron. Soc. **199**, 833 (1982)

Flammang, R.A.: Monthly Not. Roy. Astron. Soc. **206**, 589 (1984)

Fukue, J.: Publ. Astron. Soc. Jpn. **57**, 1023 (2005)

Fukue, J.: Publ. Astron. Soc. Jpn. **58**, 461 (2006)

Fukue, J.: Publ. Astron. Soc. Jpn. **60**, 377 (2008a)

Fukue, J.: Publ. Astron. Soc. Jpn. **60**, 1209 (2008b)

Fukue, J.: Publ. Astron. Soc. Jpn. **61**, 367 (2009)

Fukue, J.: Publ. Astron. Soc. Jpn. **66**, 73 (2014)

Fukue, J.: Publ. Astron. Soc. Jpn. **69**, 8 (2017)

Fukue, J., Sumitomo, N.: Publ. Astron. Soc. Jpn. **61**, 615 (2009)

Fukue, J., Kato S., Matsumoto, R.: Publ. Astron. Soc. Jpn. **37**, 383 (1985)

Hazlehurst, T., Sargent, W.L.W.: Astrophys. J. **130**, 276 (1959)

Hsieh, S.-H., Spiegel E.A.: Astrophys. J. **207**, 244 (1976)

Hummer, D.G., Rybicki, G.B.: Monthly Not. Roy. Astron. Soc. **152**, 1 (1971)

Kato, S., Fukue, J., Mineshige, S.: Black-Hole Accretion disks – Toward a New Paradigm. Kyoto University Press, Kyoto (2008)

Lindquist, R.W.: Ann. Phys. **37**, 487 (1966)

Mihalas, D., Auer, L.H.: JQSTR **71**, 61 (2001)

Mihalas, D., Mihalas, B.W.: Foundations of Radiation Hydrodynamics. Oxford University Press, Oxford (1984)

Nobili, L., Turolla, R., Zampieri, L.: Astrophys. J. **383**, 250 (1991)

Nobili, L., Turolla, R., Zampieri, L.: Astrophys. J. **404**, 686 (1993)

Park, M.-G.: JKAS **34**, 305 (2001)

Park, M.-G.: Astron. Astrophys. **274**, 642 (2003)

Park, M.-G.: Monthly Not. Roy. Astron. Soc. **367**, 1739 (2006)

Sumitomo, N., Nishiyama, S., Akizuki, C., Watarai, K., Fukue, J.: Publ. Astron. Soc. Jpn. **59**, 1043 (2007)

Takahashi, R.: Monthly Not. Roy. Astron. Soc. **382**, 1041 (2007)

Thomas, L.H.: Quart J. Math **1**, 239 (1930)

Thorne, K.S.: Monthly Not. Roy. Astron. Soc. **194**, 439 (1981)

Thorne, K.S., Flammang, R.A., Żytkow, A.N.: Monthly Not. Roy. Astron. Soc. **194**, 475 (1981)

Turolla, R., Nobili, L.: Monthly Not. Roy. Astron. Soc. **235**, 1273 (1988)

Turolla, R., Zampieri, L., Nobili, L.: Monthly Not. Roy. Astron. Soc. **272**, 625 (1995)

Udey, N., Israel, W.: Monthly Not. Roy. Astron. Soc. **199**, 1137 (1982)

Yin, W.-W., Miller, G.S.: Astrophys. J. **449**, 826 (1995)

Chapter 25
Relativistic Radiation Hydrodynamics

Continuing to the previous chapter (Chap. 24), in this chapter, corresponding to Chap. 21 for the nonrelativistic case, we derive the basic equations for relativistic radiation hydrodynamics within the framework of special relativity. In addition, we briefly show several topics on the radiation hydrodynamics in the relativistic regime. We also examine relativistic radiative shocks in some details.

25.1 Radiation Hydrodynamical Equations Under Special Relativity

We can now write the basic equations for radiation hydrodynamics under special relativity (Hsieh and Spiegel 1976; Fukue et al. 1985; Mihalas and Auer 2001; Park 2006 for the Schwarzschild metric; Takahashi 2007 for the Kerr metric). In the literatures the basic equations are usually expressed in a general relativistic form. In this section, however, we derive and write explicitly the basic equations for relativistic radiative hydrodynamics within the framework of special relativity

25.1.1 Radiation Hydrodynamical Equations in the Fixed Frame

We first derive the radiation hydrodynamical equations in the fixed (inertial) frame. In terms of the momentum and energy conservations (24.44) and (24.45), we can evaluate the radiative force exerting on matter and net energy transfer rate to matter, similar to the nonrelativistic case in Sect. 20.1.

© Springer Nature Singapore Pte Ltd. 2020, corrected publication 2023

S. Kato, J. Fukue, *Fundamentals of Astrophysical Fluid Dynamics*,
Astronomy and Astrophysics Library,
https://doi.org/10.1007/978-981-15-4174-2_25

(a) Continuity equation

The particle number conservation is

$$(nu^\mu)_{;\mu} = \frac{1}{\sqrt{-g}}\frac{\partial}{\partial x^\mu}\left(\sqrt{-g}nu^\mu\right) = 0, \tag{25.1}$$

where x^μ is the space-time coordinates, u^μ the four velocity, and n the proper number density.

In the three-dimensional form, the continuity equation becomes

$$\frac{\partial}{\partial t}(\rho\gamma) + \mathrm{div}(\rho\gamma\boldsymbol{v}) = 0, \tag{25.2}$$

where ρ $(= nmc^2)$ is the proper density, and γ $(= 1/\sqrt{1 - v^2/c^2})$ the Lorentz factor.

(b) Hydrodynamic equations

The relativistic equations of motion (24.44), $\left(T_i^{\ \mu} + R_i^{\ \mu}\right)_{;\mu} = f_i$, are written as[1]

$$\begin{aligned}(\varepsilon + p)\left(u^\mu \tfrac{\partial u^i}{\partial x^\mu} + \Gamma^i_{\mu\nu}u^\mu u^\nu\right) - \left(g^{i\mu} - u^i u^\mu\right)\tfrac{\partial p}{\partial x^\mu} \\ = -(g^{i\mu} - u^i u^\mu)R_{\mu\ ;\nu}^{\ \nu} + (g^{i\mu} - u^i u^\mu)f_\mu,\end{aligned} \tag{25.3}$$

where ε is the internal energy per unit proper volume, p the pressure measured in the comoving frame, and $R^{\mu\nu}$ the stress-energy tensor of radiation. For equation of motion in the three-dimensional space, the Latin suffix $i = 1, 2, 3$.

The radiative part on the right-hand side of Eq. (25.3) are, from Eqs. (24.43), (24.84), (24.85), (24.86), and (24.87),

$$\begin{aligned}&-\left(g^{i\mu} - u^i u^\mu\right)R_{\mu\ ;\nu}^{\ \nu} \\ &= -\left(\frac{1}{c^2}\frac{\partial F^i}{\partial t} + \frac{\partial P^{ik}}{\partial x^k}\right) \\ &\quad -\frac{\gamma^2}{c^2}v^i\left[-\left(\frac{\partial E}{\partial t} + \frac{\partial F^k}{\partial x^k}\right) + v_j\left(\frac{1}{c^2}\frac{\partial F^j}{\partial t} + \frac{\partial P^{jk}}{\partial x^k}\right)\right] \\ &= \frac{\rho}{c}(\kappa_0 + \sigma_0)\left[F_0^i + \frac{\gamma - 1}{v^2}v^i(\boldsymbol{v}\cdot\boldsymbol{F}_0)\right] \\ &= \frac{\rho\gamma}{c}(\kappa_0 + \sigma_0) \\ &\quad \times\left[F^i - \gamma^2 E v^i - v_k P^{ik} + \gamma^2 v^i\left(\frac{2\boldsymbol{v}\cdot\boldsymbol{F}}{c^2} - \frac{v_j v_k}{c^2}P^{jk}\right)\right],\end{aligned} \tag{25.4}$$

---

[1] $(T^{i\mu} + R^{i\mu})_{;\mu} - u^i u_\alpha(T^{\alpha\beta} + R^{\alpha\beta})_{;\beta} = (g^{i\mu} - u^i u^\mu)f_\mu$.

whereas the pressure-gradient term and force one become as follows ($u^\mu f_\mu = 0$):

$$\left(g^{i\mu} - u^i u^\mu\right) \frac{\partial p}{\partial x^\mu} = -\nabla p - \gamma^2 \frac{\boldsymbol{v}}{c^2} \left(\frac{\partial p}{\partial t} + \boldsymbol{v} \cdot \nabla p\right), \tag{25.5}$$

$$(g^{i\mu} - u^i u^\mu) f_\mu = f^i = \gamma \boldsymbol{f}. \tag{25.6}$$

Thus, the relativistic equations of motion are finally expressed as

$$\begin{aligned} c^2 \left(u^\mu \frac{\partial u^i}{\partial x^\mu} + \Gamma^i_{\mu\nu} u^\mu u^\nu\right) &= \frac{c^2}{\varepsilon + p} \gamma \boldsymbol{f} - \frac{c^2}{\varepsilon + p} \nabla p - \frac{\gamma^2 \boldsymbol{v}}{\varepsilon + p} \left(\frac{\partial p}{\partial t} + \boldsymbol{v} \cdot \nabla p\right) \\ &\quad + \frac{\rho c^2}{\varepsilon + p} \frac{1}{c} (\kappa_0 + \sigma_0) \left[F_0^i + \frac{\gamma - 1}{v^2} v^i (\boldsymbol{v} \cdot \boldsymbol{F}_0)\right] \\ &= \frac{c^2}{\varepsilon + p} \gamma \boldsymbol{f} - \frac{c^2}{\varepsilon + p} \nabla p - \frac{\gamma^2 \boldsymbol{v}}{\varepsilon + p} \left(\frac{\partial p}{\partial t} + \boldsymbol{v} \cdot \nabla p\right) \\ &\quad + \frac{\rho c^2}{\varepsilon + p} \frac{\gamma}{c} (\kappa_0 + \sigma_0) \\ &\quad \times \left[F^i - \gamma^2 E v^i - v_k P^{ik} + \gamma^2 v^i \right. \\ &\quad \times \left. \left(\frac{2\boldsymbol{v} \cdot \boldsymbol{F}}{c^2} - \frac{v_j v_k}{c^2} P^{jk}\right)\right]. \end{aligned} \tag{25.7}$$

On the right-hand side of equation of motion (25.7), the first term is the external force, the second and third terms denote, respectively, the pressure-gradient force and its relativistic correction, and the fourth term represents the radiative interaction, including the radiative force ($\boldsymbol{F}$) and the radiative drag force ($\boldsymbol{v}E$ and $v_k P^{ik}$). It should be noted that both the radiative force and radiative drag one exist in the fixed (inertial) frame, the latter drag force does not appear in the comoving (fluid) frame.

(c) Energy equation

The relativistic energy equation (24.45), $u^\mu (T_\mu^{\ \nu} + R_\mu^{\ \nu})_{;\nu} = u^\mu f_\mu = 0$, is written as ($u^\mu f_\mu = 0$)

$$\frac{1}{\sqrt{-g}} \frac{\partial}{\partial x^\mu} \left(\sqrt{-g} \varepsilon u^\mu\right) + \frac{p}{\sqrt{-g}} \frac{\partial}{\partial x^\mu} \left(\sqrt{-g} u^\mu\right) = -u^\mu R_{\mu\ ;\nu}^{\ \nu}. \tag{25.8}$$

The right-hand side of Eq. (25.8) is, from (24.43), (24.84), (24.85), (24.86), and (24.87),

$$-u^\mu R_{\mu\ ;\nu}^{\ \nu} = -\frac{\gamma}{c} \left(\frac{\partial E}{\partial t} + \frac{\partial F^k}{\partial x^k}\right) + \frac{\gamma v_i}{c} \left(\frac{1}{c^2} \frac{\partial F^i}{\partial t} + \frac{\partial P^{ik}}{\partial x^k}\right)$$

$$= -\frac{\rho}{c}\left(j_0 - c\kappa_0 E_0\right)$$
$$= \frac{\gamma^2\rho}{c}\left(-\frac{j_0}{\gamma^2} + c\kappa_0 E - \kappa_0\frac{2\boldsymbol{v}\cdot\boldsymbol{F}}{c} + \kappa_0\frac{v_i v_k}{c}P^{ik}\right). \tag{25.9}$$

Thus, the energy equation is finally expressed as

$$\frac{c}{\sqrt{-g}}\frac{\partial}{\partial x^\mu}\left[\sqrt{-g}\left(\varepsilon - \rho c^2\right)u^\mu\right] + c\frac{p}{\sqrt{-g}}\frac{\partial}{\partial x^\mu}\left(\sqrt{-g}u^\mu\right)$$
$$= -\rho\left(j_0 - c\kappa_0 E_0\right)$$
$$= \gamma^2\rho\left(-\frac{j_0}{\gamma^2} + c\kappa_0 E - \kappa_0\frac{2\boldsymbol{v}\cdot\boldsymbol{F}}{c} + \kappa_0\frac{v_i v_k}{c}P^{ik}\right). \tag{25.10}$$

Here, we have used the continuity equation (25.1).

In the case of an ideal gas, $\varepsilon - \rho c^2 = p/(\Gamma - 1)$, where Γ is the ratio of the specific heats, and therefore, the left-hand side of the energy equation (25.10) is reexpressed as

$$\frac{p}{\Gamma - 1}\left(cu^\mu\frac{\partial}{\partial x^\mu}\ln p - cu^\mu\frac{\partial}{\partial x^\mu}\ln\rho^\Gamma\right) = \frac{p}{\Gamma - 1}cu^\mu\frac{\partial}{\partial x^\mu}\left(\ln\frac{p}{\rho^\Gamma}\right). \tag{25.11}$$

On the right-hand side of energy equation (25.10), the first term in the parentheses is the radiative cooling, the second denotes the radiative heating, and other terms represent the relativistic correction.

Now, Eqs. (25.2), (25.7), and (25.10) for matter, and Eqs. (24.86), (24.87), and (24.90) for radiation constitute the basic equations for radiation hydrodynamics under special relativity.

(d) Total energy conservation and relativistic Bernoulli equation

The zeroth component of Eq. (24.44) gives the equation of the total energy conservation and the relativistic Bernoulli equation as follows. That is, when there is no general four-force, $f_0 = 0$, we have

$$\left(T_0^\nu + R_0^\nu\right)_{;\nu} = \frac{\partial}{c\partial t}\left(T_0^0 + R_0^0\right) + \frac{1}{\sqrt{-g}}\frac{\partial}{\partial x^k}\left[\sqrt{-g}\left(T_0^k + R_0^k\right)\right] = 0, \tag{25.12}$$

or explicitly,

$$\frac{\partial}{\partial t}\left[\gamma^2(\varepsilon + \beta^2 p) + E\right] + \frac{1}{\sqrt{-g}}\frac{\partial}{\partial x^k}\left\{\sqrt{-g}\left[(\varepsilon + p)u_0 cu^k + F^k\right]\right\} = 0. \tag{25.13}$$

This equation is the total energy conservation of matter and radiation.

In the case of the steady one-dimensional flow, from the second term of Eq. (25.13), we have the relativistic Bernoulli equation along the flow:

$$\sqrt{-g}\left[(\varepsilon + p)u_0cu^k + F^k\right] = \text{constant}. \tag{25.14}$$

(e) Diffusion and advection luminosities

In the spherically symmetric case, the relativistic Bernoulli equation becomes

$$4\pi r^2(\varepsilon + p)u_0cu + 4\pi r^2 F = \dot{E}\ (\text{constant}), \tag{25.15}$$

while the continuity equation is

$$4\pi r^2 \rho cu = \dot{M}\ (\text{constant}). \tag{25.16}$$

Substituting the continuity equation into the Bernoulli equation, we have

$$\dot{M}\frac{\varepsilon + p}{\rho}\gamma + 4\pi r^2 F = \dot{E}. \tag{25.17}$$

From the transformation between the inertial quantities and comoving ones (Sect. 23.1), the radiative flux F in the inertial frame is expressed by the quantities in the comoving frame as

$$F = \gamma^2\left[(1+\beta^2)F_0 + \beta(cE_0 + cP_0)\right]. \tag{25.18}$$

Here, the first term in the brackets on the right-hand side is the energy flux flowing in the comoving frame, and called the *diffusion flux*. The second term, on the other hand, is the energy flux carried with the flow, and called the *advection flux*. Furthermore, in the comoving frame, we define the *diffusion luminosity* L^{diff} and *advection one* L^{adv}, respectively, as[2] :

$$L^{\text{diff}} = 4\pi r^2 F_0, \tag{25.20}$$

$$L^{\text{adv}} = 4\pi r^2\beta(cE_0 + cP_0). \tag{25.21}$$

[2] Several luminosities are distinguished as follows:

$$\begin{aligned}
L &= 4\pi r^2 F && \cdots \text{ luminosity in the inertial frame}\\
L_0^{\text{diff}} &= 4\pi r^2 F_0 && \cdots \text{ diffusion luminosity in the comoving frame}\\
L_\infty^{\text{diff}} &= \gamma^2(1+\beta^2)L_{\text{diff}} && \cdots \text{ diffusion luminosity observed at infinity}\\
L_0^{\text{adv}} &= 4\pi r^2\beta(cE_0 + cP_0) && \cdots \text{ advection luminosity in the comoving frame}\\
L_\infty^{\text{adv}} &= \gamma^2 L_{\text{adv}} && \cdots \text{ advection luminosity observed at infinity.}
\end{aligned} \tag{25.19}$$

Finally, substituting Eq. (25.18) into Eq. (25.17), we have another expression:

$$\dot{M}\frac{\varepsilon + p + E_0 + P_0}{\rho}\gamma + 4\pi r^2\gamma^2(1+\beta^2)F_0 = \dot{E}. \tag{25.22}$$

In this expression, the advection flux is combined with the material part in the first term on the left-hand side, and the second term is merely the diffusion luminosity observed at infinity.

25.1.1.1 Subrelativistic Regime on the Order of $\mathcal{O}(v/c)^1$

There are various subrelativistic astrophysical phenomena, such as outflows and jets from compact objects, where the full relativistic treatments are not necessary. Hence, in this subsection we show the basic equations in the subrelativistic regime on the order of $\mathcal{O}(v/c)^1$. As was discussed and shown in Sect. 21.1, in order to complete the set of the basic equations for radiation hydrodynamics, the subrelativistic correction terms are necessary, even in the nonrelativistic regime.

The continuity equation is the same as the nonrelativistic one within the subrelativistic approximation.

To the first order of $(\boldsymbol{v}/c)$, the equation of motion is

$$\frac{\partial \boldsymbol{v}}{\partial t} + (\boldsymbol{v}\cdot\boldsymbol{\nabla})\,\boldsymbol{v} = -\boldsymbol{\nabla}\psi - \frac{1}{\rho}\boldsymbol{\nabla}p + \frac{\kappa_F + \sigma_F}{c}\left(\boldsymbol{F} - E\boldsymbol{v} - v_k P^{ik}\right), \tag{25.23}$$

where $\boldsymbol{v}$ is the velocity, ψ the gravitational potential, and p the gas pressure. In the third term on the right-hand side of this Eq. (25.23), $\boldsymbol{F}$ is the radiative flux in the *inertial/observer frame* (the coordinate of the observer at rest), while the term $(Ev^i + v_k P^{ik})$ is the *radiation drag* received by the moving material, where E and P^{ik} are quantities measured in the inertial frame (see Sect. 23.4). It should be noted that the term $(\boldsymbol{F} - E\boldsymbol{v} - v_\beta P^{\alpha\beta})$ is the radiative flux $\boldsymbol{F}_0$ measured in the *comoving/fluid frame* (the coordinate of observer comoving with fluid).

The energy equation on the order of $\mathcal{O}(v/c)^1$ is expressed as

$$\left(\frac{\partial}{\partial t} + \boldsymbol{v}\cdot\boldsymbol{\nabla}\right)e + \frac{p}{\rho}\boldsymbol{\nabla}\boldsymbol{v} = \frac{1}{\rho}q^+ - j_0 + c\kappa_E E - \kappa_E\frac{2\boldsymbol{v}\cdot\boldsymbol{F}}{c}, \tag{25.24}$$

or in the case of an ideal gas,

$$\frac{1}{\Gamma - 1}\left(\frac{dp}{dt} - \Gamma\frac{p}{\rho}\frac{d\rho}{dt}\right) = q^+ - \rho\left(j_0 - c\kappa_E E + \kappa_E\frac{2\boldsymbol{v}\cdot\boldsymbol{F}}{c}\right), \tag{25.25}$$

where e is the internal energy per unit mass, and q^+ the (viscous) heating rate per unit volume. On the right-hand side, there appears an additional term on the order of $\mathcal{O}(v/c)^1$, which means the work done on the fluid by the radiative flux.

The radiative transfer equation (24.58), moment equations (24.86) and (24.87), and the closure relation (24.90) become, respectively, on the order of $\mathcal{O}(v/c)^1$:

$$\frac{1}{c}\frac{\partial I}{\partial t} + (\boldsymbol{l}\cdot\nabla)\, I = \rho\left[\frac{j}{4\pi} - (\kappa_I + \sigma_I)I + \frac{\sigma_E}{4\pi}\left(cE - \frac{2\boldsymbol{v}\cdot\boldsymbol{F}}{c}\right)\right]$$
$$+\rho\frac{\boldsymbol{v}\cdot\boldsymbol{l}}{c}\left[3\frac{j}{4\pi} + (\kappa_I + \sigma_I)I + 3\frac{\sigma_E}{4\pi}\left(cE - \frac{2\boldsymbol{v}\cdot\boldsymbol{F}}{c}\right)\right] \quad (25.26)$$

$$\frac{\partial E}{\partial t} + \nabla\cdot\boldsymbol{F} = \rho\left[j - c\kappa_E E + (2\kappa_E - \kappa_F - \sigma_F)\frac{\boldsymbol{v}\cdot\boldsymbol{F}}{c}\right], \quad (25.27)$$

$$\frac{1}{c^2}\frac{\partial \boldsymbol{F}}{\partial t} + \frac{\partial P^{ik}}{\partial x^k} = -\rho\frac{\kappa_F + \sigma_F}{c}\left(\boldsymbol{F} - E\boldsymbol{v} - v_k P^{ik}\right)$$
$$+\frac{\rho}{c}\left(j - c\kappa_E E\right)\frac{\boldsymbol{v}}{c}. \quad (25.28)$$

$$P^{ij} = \frac{\delta^{ij}}{3}E + \frac{v^i F^j}{c^2} + \frac{v^j F^i}{c^2} - \frac{2}{3}\frac{\boldsymbol{v}\cdot\boldsymbol{F}}{c^2}\delta^{ij}. \quad (25.29)$$

25.1.1.2 Plane-Parallel Geometry

For a relativistically moving atmosphere in the plane-parallel geometry (z), i.e., a plane-parallel vertical flow, the hydrodynamic equations and transfer equations become as follows.

For matter, the continuity equation is

$$\frac{\partial}{\partial t}(\rho\gamma) + \frac{\partial}{\partial z}(\rho c u) = 0, \quad (25.30)$$

where ρ is the proper gas density, u the vertical component of four velocity, and c the speed of light. The vertical component of four velocity u is related to the proper three velocity v by $u = \gamma\beta = \gamma v/c$, where γ is the Lorentz factor, $\gamma = \sqrt{1+u^2} = 1/\sqrt{1-(v/c)^2}$. In addition, in the steady case this continuity equation is integrated to yield

$$\rho c u = \rho\gamma\beta c = \dot{J}\ (= \text{const.}), \quad (25.31)$$

where $\dot{J}$ is the mass-loss rate per unit area.

The equation of motion is

$$c\gamma\frac{\partial u}{\partial t}+c^2u\frac{\partial u}{\partial z}=c\gamma^4\frac{\partial\beta}{\partial t}+c^2\gamma^4\beta\frac{\partial\beta}{\partial z}$$
$$=-\frac{\partial\psi}{\partial z}-\gamma^2\frac{c^2}{\varepsilon+p}\frac{\partial p}{\partial z}$$
$$+\frac{\rho c^2}{\varepsilon+p}\frac{\kappa+\sigma}{c}\gamma^3\Big[F(1+\beta^2)-(cE+cP)\beta\Big], \quad (25.32)$$

where ψ is the gravitational potential, p the gas pressure, κ and σ are the absorption and scattering opacities (gray), defined in the comoving frame, E the radiation energy density, F the radiative flux, and P the radiation pressure observed in the inertial frame. Here and hereafter, the subscripts on opacities (i.e., $_E$, $_F$, $_0$) are dropped for simplicity. In other words, we consider the frequency independent gray case.

The energy equation is

$$\frac{1}{\Gamma-1}\left(\gamma\frac{\partial p}{\partial t}+\Gamma p\frac{\partial\gamma}{\partial t}+cu\frac{\partial p}{\partial z}+\Gamma cp\frac{\partial u}{\partial z}\right)$$
$$=q^+-\rho\Big(j_0-\kappa cE\gamma^2-\kappa cPu^2+2\kappa F\gamma u\Big), \quad (25.33)$$

where q^+ is the internal heating and j_0 is the emissivity defined in the comoving frame.

For radiation, the frequency-integrated transfer equation (24.58), the zeroth-moment equation (24.86), and the first-moment equation (24.87) become, respectively:

$$\frac{1}{c}\frac{\partial I}{\partial t}+\mu\frac{\partial I}{\partial z}=\rho\frac{1}{\gamma^3(1-\beta\mu)^3}\left[\frac{j_0}{4\pi}-(\kappa+\sigma)\gamma^4(1-\beta\mu)^4 I\right.$$
$$+\frac{\sigma}{4\pi}\frac{3}{4}\gamma^2\left\{\left[1+\frac{(\mu-\beta)^2}{(1-\beta\mu)^2}\beta^2+\frac{(1-\beta^2)^2}{(1-\beta\mu)^2}\frac{1-\mu^2}{2}\right]cE\right.$$
$$-\left[1+\frac{(\mu-\beta)^2}{(1-\beta\mu)^2}\right]2F\beta$$
$$\left.\left.+\left[\beta^2+\frac{(\mu-\beta)^2}{(1-\beta\mu)^2}-\frac{(1-\beta^2)^2}{(1-\beta\mu)^2}\frac{1-\mu^2}{2}\right]cP\right\}\right], \quad (25.34)$$

$$\frac{\partial E}{\partial t}+\frac{\partial F}{\partial z}=\rho\gamma\Big[j_0-\kappa cE+\sigma(cE+cP)\gamma^2\beta^2$$
$$+\kappa F\beta-\sigma F(1+\beta^2)\gamma^2\beta\Big]. \quad (25.35)$$

$$\frac{1}{c^2}\frac{\partial F}{\partial t} + \frac{\partial P}{\partial z} = \frac{\rho\gamma}{c}\Big[j_0\beta - \kappa F + \kappa c P\beta$$
$$-\sigma F\gamma^2(1+\beta^2) + \sigma(cE + cP)\gamma^2\beta \Big], \tag{25.36}$$

where $\mu = \cos\theta$.

Finally, a closure relation is

$$cP(1 - f\beta^2) = cE(f - \beta^2) + 2F\beta(1 - f), \tag{25.37}$$

where $f(\tau, \beta)$ is the variable Eddington factor depending on the velocity as well as the optical depth.

25.1.1.3 Spherical Geometry

In the case of the spherical flow in the radial (r) direction, the hydrodynamic equations and transfer equations are explicitly written as follows.

For matter, the continuity equation is

$$\frac{\partial}{\partial t}(\rho\gamma) + \frac{1}{r^2}\frac{\partial}{\partial r}(r^2\rho c u) = 0, \tag{25.38}$$

where u is the radial component of four velocity. In the steady case this continuity equation is integrated to yield

$$4\pi r^2\rho c u = 4\pi r^2\rho\gamma\beta c = \dot{M}\ (= \text{const.}), \tag{25.39}$$

where $\dot{M}$ is the constant mass-flow.

The equation of motion is

$$c\gamma\frac{\partial u}{\partial t} + c^2 u\frac{\partial u}{\partial r} = c\gamma^4\frac{\partial\beta}{\partial t} + c^2\gamma^4\beta\frac{\partial\beta}{\partial r}$$
$$= -\frac{\partial\psi}{\partial r} - \gamma^2\frac{c^2}{\varepsilon + p}\frac{\partial p}{\partial r} + \frac{\rho c^2}{\varepsilon + p}\frac{\kappa + \sigma}{c}\gamma F_0, \tag{25.40}$$

where F_0 the radiative flux in the comoving frame.

The energy equation is

$$\frac{1}{\Gamma - 1}\left[\gamma\frac{\partial p}{\partial t} + \Gamma p\frac{\partial\gamma}{\partial t} + cu\frac{\partial p}{\partial r} + \Gamma c p\frac{1}{r^2}\frac{\partial}{\partial r}(r^2 u)\right]$$
$$= q^+ - \rho\left(j_0 - \kappa_0 c E_0\right), \tag{25.41}$$

where E_0 is the radiation energy density in the comoving frame.

For radiation, the frequency-integrated transfer equation (24.58), the zeroth-moment equation (24.86), and the first-moment equation (24.87) become, respectively:

$$\frac{1}{c}\frac{\partial I}{\partial t} + \mu\frac{\partial I}{\partial r} + \frac{1-\mu^2}{r}\frac{\partial I}{\partial \mu}$$
$$= \rho\frac{1}{\gamma^3(1-\beta\mu)^3}\left[\frac{j_0}{4\pi} - (\kappa+\sigma)\,\gamma^4\,(1-\beta\mu)^4\,I\right.$$
$$+\frac{\sigma}{4\pi}\frac{3}{4}\gamma^2\left\{\left[1+\frac{(\mu-\beta)^2}{(1-\beta\mu)^2}\beta^2+\frac{(1-\beta^2)^2}{(1-\beta\mu)^2}\frac{1-\mu^2}{2}\right]cE\right.$$
$$-\left[1+\frac{(\mu-\beta)^2}{(1-\beta\mu)^2}\right]2F\beta$$
$$\left.\left.+\left[\beta^2+\frac{(\mu-\beta)^2}{(1-\beta\mu)^2}-\frac{(1-\beta^2)^2}{(1-\beta\mu)^2}\frac{1-\mu^2}{2}\right]cP\right\}\right], \tag{25.42}$$

$$\frac{\partial E}{\partial t} + \frac{1}{r^2}\frac{\partial}{\partial r}(r^2F) = \rho\gamma\left[\,j_0 - \kappa cE + \sigma(cE+cP)\gamma^2\beta^2\right.$$
$$\left.+\kappa F\beta - \sigma F(1+\beta^2)\gamma^2\beta\,\right]. \tag{25.43}$$

$$\frac{1}{c^2}\frac{\partial F}{\partial t} + \frac{\partial P}{\partial r} + \frac{3P-E}{r} = \frac{\rho\gamma}{c}\left[\,j_0\beta - \kappa F + \kappa cP\beta\right.$$
$$\left.-\sigma F\gamma^2(1+\beta^2) + \sigma(cE+cP)\gamma^2\beta\,\right], \tag{25.44}$$

where $\mu = \cos\theta$.

Finally, a closure relation is the same as Eq. (25.37).

25.1.2 Radiation Hydrodynamical Equations Using the Comoving Frame Quantities

By transforming the left-hand sides of the basic equations in the mixed frame, Eq. (24.54), we can derive the basic equations expressed by the comoving quantities (Mihalas 1980; Mihalas and Mihalas 1984). We show here the plane-parallel and spherical cases.

25.1.2.1 Plane-Parallel Geometry

The relativistic radiative transfer equation in the vertical (z) direction is described in the mixed frame as

$$\frac{1}{c}\frac{\partial I}{\partial t} + \mu\frac{\partial I}{\partial z} = \left(\frac{\nu}{\nu_0}\right)^3 \rho\left[\frac{j_0}{4\pi} - (\kappa_0 + \sigma_0)\, I_0 + \sigma_0\frac{cE_0}{4\pi}\right], \tag{25.45}$$

where the isotropic scattering is assumed for simplicity. Substituting the transformation rule (24.35) into the left-hand side, we can write this Eq. (25.45) as

$$\frac{\nu}{\nu_0}\left(\frac{1}{c}\frac{\partial I_0}{\partial t} + \mu\frac{\partial I_0}{\partial z}\right) - 4\frac{\nu}{\nu_0^2} I_0\left(\frac{1}{c}\frac{\partial \nu_0}{\partial t} + \mu\frac{\partial \nu_0}{\partial z}\right)$$
$$= \rho\left[\frac{j_0}{4\pi} - (\kappa_0 + \sigma_0)\, I_0 + \sigma_0\frac{cE_0}{4\pi}\right]. \tag{25.46}$$

To calculate the derivatives of I_0, we apply the chain rules, and after some manipulations we have

$$\left.\frac{\partial}{\partial t}\right|_{z\mu\nu} = \left.\frac{\partial}{\partial t}\right|_{z\mu_0\nu_0} + \left.\frac{\partial \mu_0}{\partial t}\right|_{z\mu_0\nu_0}\frac{\partial}{\partial \mu_0} + \left.\frac{\partial \nu_0}{\partial t}\right|_{z\mu_0\nu_0}\frac{\partial}{\partial \nu_0}$$
$$= \left.\frac{\partial}{\partial t}\right|_{z\mu_0\nu_0} - \gamma^2(1-\mu_0^2)\frac{\partial \beta}{\partial t}\frac{\partial}{\partial \mu_0} - \gamma^2\mu_0\nu_0\frac{\partial \beta}{\partial t}\frac{\partial}{\partial \nu_0}, \tag{25.47}$$

$$\left.\frac{\partial}{\partial z}\right|_{t\mu\nu} = \left.\frac{\partial}{\partial z}\right|_{t\mu_0\nu_0} + \left.\frac{\partial \mu_0}{\partial z}\right|_{t\mu_0\nu_0}\frac{\partial}{\partial \mu_0} + \left.\frac{\partial \nu_0}{\partial z}\right|_{t\mu_0\nu_0}\frac{\partial}{\partial \nu_0}$$
$$= \left.\frac{\partial}{\partial z}\right|_{t\mu_0\nu_0} - \gamma^2(1-\mu_0^2)\frac{\partial \beta}{\partial z}\frac{\partial}{\partial \mu_0} - \gamma^2\mu_0\nu_0\frac{\partial \beta}{\partial z}\frac{\partial}{\partial \nu_0}, \tag{25.48}$$

where μ_0 is the direction cosine in the comoving frame. In addition, in the present vertical case, the Doppler shift and aberration are, respectively, expressed by Eqs. (24.20) and (24.21).

Using these expressions, after some manipulations, we have the radiative transfer equation expressed by the comoving quantities for the plane-parallel flow:

$$\gamma(1+\beta\mu_0)\frac{1}{c}\frac{\partial I_0}{\partial t} + \gamma(\mu_0+\beta)\frac{\partial I_0}{\partial z}$$
$$-\gamma^3(1+\beta\mu_0)\left[(1-\mu_0^2)\frac{\partial I_0}{\partial \mu_0} - 4\mu_0 I_0\right]\frac{1}{c}\frac{\partial \beta}{\partial t}$$
$$-\gamma^3(\mu_0+\beta)\left[(1-\mu_0^2)\frac{\partial I_0}{\partial \mu_0} - 4\mu_0 I_0\right]\frac{\partial \beta}{\partial z}$$
$$= \rho\left[\frac{j_0}{4\pi} - (\kappa_0 + \sigma_0)\, I_0 + \sigma_0\frac{cE_0}{4\pi}\right]. \tag{25.49}$$

Integrating this transfer equation (25.49) over a solid angle in the comoving frame, we have the zeroth- and first-moment equations expressed by the comoving quantities:

$$\gamma\frac{\partial cE_0}{c\partial t}+\gamma\frac{\partial F_0}{\partial z}+\gamma\beta\frac{\partial F_0}{c\partial t}+\gamma\beta\frac{\partial cE_0}{\partial z}$$
$$+\gamma^3\left[2F_0+\beta(cE_0+cP_0)\right]\frac{\partial\beta}{c\partial t}+\gamma^3\left[2\beta F_0+(cE_0+cP_0)\right]\frac{\partial\beta}{\partial z}$$
$$=\rho\left(j_0-\kappa_0 cE_0\right), \tag{25.50}$$

$$\gamma\frac{\partial F_0}{c\partial t}+\gamma\frac{\partial cP_0}{\partial z}+\gamma\beta\frac{\partial cP_0}{c\partial t}+\gamma\beta\frac{\partial F_0}{\partial z}$$
$$+\gamma^3\left[2\beta F_0+(cE_0+cP_0)\right]\frac{\partial\beta}{c\partial t}+\gamma^3\left[2F_0+\beta(cE_0+cP_0)\right]\frac{\partial\beta}{\partial z}$$
$$=-\rho\left(\kappa_0+\sigma_0\right)F_0. \tag{25.51}$$

25.1.2.2 Spherical Geometry

The relativistic radiative transfer equation in the radial (r) direction is described in the mixed frame as

$$\frac{1}{c}\frac{\partial I}{\partial t}+\mu\frac{\partial I}{\partial r}+\frac{1-\mu^2}{r}\frac{\partial I}{\partial\mu}=\left(\frac{\nu}{\nu_0}\right)^3\rho\left[\frac{j_0}{4\pi}-(\kappa_0+\sigma_0)I_0+\sigma_0\frac{cE_0}{4\pi}\right], \tag{25.52}$$

where the isotropic scattering is assumed for simplicity. Substituting the transformation rule (24.35) into the left-hand side, this Eq. (25.52) is expressed as

$$\frac{\nu}{\nu_0}\left(\frac{1}{c}\frac{\partial I_0}{\partial t}+\mu\frac{\partial I_0}{\partial r}+\frac{1-\mu^2}{r}\frac{\partial I_0}{\partial\mu}\right)-4\frac{\nu}{\nu_0^2}I_0\left(\frac{1}{c}\frac{\partial\nu_0}{\partial t}+\mu\frac{\partial\nu_0}{\partial z}+\frac{1-\mu^2}{r}\frac{\partial\nu_0}{\partial\mu}\right)$$
$$=\rho\left[\frac{j_0}{4\pi}-(\kappa_0+\sigma_0)I_0+\sigma_0\frac{cE_0}{4\pi}\right]. \tag{25.53}$$

We apply the chain rules to the derivatives on the left-hand side, and we have

$$\left.\frac{\partial}{\partial t}\right|_{r\mu\nu}=\left.\frac{\partial}{\partial t}\right|_{r\mu_0\nu_0}+\left.\frac{\partial\mu_0}{\partial t}\right|_{r\mu_0\nu_0}\frac{\partial}{\partial\mu_0}+\left.\frac{\partial\nu_0}{\partial t}\right|_{r\mu_0\nu_0}\frac{\partial}{\partial\nu_0}$$
$$=\left.\frac{\partial}{\partial t}\right|_{r\mu_0\nu_0}-\gamma^2(1-\mu_0^2)\frac{\partial\beta}{\partial t}\frac{\partial}{\partial\mu_0}-\gamma^2\mu_0\nu_0\frac{\partial\beta}{\partial t}\frac{\partial}{\partial\nu_0}, \tag{25.54}$$

$$\left.\frac{\partial}{\partial r}\right|_{t\mu\nu} = \left.\frac{\partial}{\partial r}\right|_{t\mu_0\nu_0} + \left.\frac{\partial \mu_0}{\partial r}\right|_{t\mu_0\nu_0} \frac{\partial}{\partial \mu_0} + \left.\frac{\partial \nu_0}{\partial r}\right|_{t\mu_0\nu_0} \frac{\partial}{\partial \nu_0}$$
$$= \left.\frac{\partial}{\partial r}\right|_{t\mu_0\nu_0} - \gamma^2(1-\mu_0^2)\frac{\partial \beta}{\partial r}\frac{\partial}{\partial \mu_0} - \gamma^2 \mu_0 \nu_0 \frac{\partial \beta}{\partial r}\frac{\partial}{\partial \nu_0}, \quad (25.55)$$

$$\left.\frac{\partial}{\partial \mu}\right|_{rt\nu} = \left.\frac{\partial \mu_0}{\partial \mu}\right|_{rt\nu_0} \frac{\partial}{\partial \mu_0} + \left.\frac{\partial \nu_0}{\partial \mu_0}\right|_{rt\nu_0} \frac{\partial}{\partial \nu_0}$$
$$= \gamma^2(1+\beta\mu_0)^2 \frac{\partial}{\partial \mu_0} - \gamma^2 \beta (1+\beta\mu_0)\nu_0 \frac{\partial}{\partial \nu_0}. \quad (25.56)$$

Using these expressions, after some manipulations, we have the radiative transfer equation expressed by the comoving quantities for the spherical flow:

$$\gamma(1+\beta\mu_0)\frac{1}{c}\frac{\partial I_0}{\partial t} + \gamma(\mu_0+\beta)\frac{\partial I_0}{\partial r} + \gamma(1+\beta\mu_0)\frac{1-\mu_0}{r}\frac{\partial I_0}{\partial \mu_0} + 4\gamma\beta\frac{1-\mu_0^2}{r}I_0$$
$$-\gamma^3(1+\beta\mu_0)\left[(1-\mu_0^2)\frac{\partial I_0}{\partial \mu_0} - 4\mu_0 I_0\right]\frac{1}{c}\frac{\partial \beta}{\partial t}$$
$$-\gamma^3(\mu_0+\beta)\left[(1-\mu_0^2)\frac{\partial I_0}{\partial \mu_0} - 4\mu_0 I_0\right]\frac{\partial \beta}{\partial r}$$
$$= \rho\left[\frac{j_0}{4\pi} - (\kappa_0+\sigma_0)\, I_0 + \sigma_0 \frac{cE_0}{4\pi}\right]. \quad (25.57)$$

Integrating this transfer equation (25.57) over a solid angle in the comoving frame, we have the zeroth- and first-moment equations expressed by the comoving quantities:

$$\gamma\frac{\partial cE_0}{c\partial t} + \gamma\frac{\partial F_0}{\partial r} + \gamma\beta\frac{\partial F_0}{c\partial t} + \gamma\beta\frac{\partial cE_0}{\partial r} + \frac{\gamma}{r}\left[2F_0 + \beta(3cE_0 - cP_0)\right]$$
$$+\gamma^3\left[2F_0 + \beta(cE_0 + cP_0)\right]\frac{\partial \beta}{c\partial t} + \gamma^3\left[2\beta F_0 + (cE_0 + cP_0)\right]\frac{\partial \beta}{\partial r}$$
$$= \rho\left(j_0 - \kappa_0 cE_0\right), \quad (25.58)$$

$$\gamma\frac{\partial F_0}{c\partial t} + \gamma\frac{\partial cP_0}{\partial r} + \gamma\beta\frac{\partial cP_0}{c\partial t} + \gamma\beta\frac{\partial F_0}{\partial r} + \frac{\gamma}{r}\left[2\beta F_0 - cE_0 + 3cP_0\right]$$
$$+\gamma^3\left[2\beta F_0 + (cE_0 + cP_0)\right]\frac{\partial \beta}{c\partial t} + \gamma^3\left[2F_0 + \beta(cE_0 + cP_0)\right]\frac{\partial \beta}{\partial r}$$
$$= -\rho\left(\kappa_0 + \sigma_0\right) F_0. \quad (25.59)$$

25.2 Several Topics on the Relativistic Radiation Hydrodynamics

As was stated at the beginning of Sect. 24.3, there exist several properties which show up only in the relativistic regime. In this section, we shall introduce several characteristic properties appeared in the relativistic radiation hydrodynamics. These relativistic properties become important in the relativistic astrophysical phenomena, such as astrophysical jets and outflows from a compact object, relativistic explosions, accretion flows onto accretion disks around a black hole, and so on.

25.2.1 *Radiation Drag and Terminal Speed*

One of the important properties appeared in the radiation hydrodynamical relativistic flows is the *radiation drag force* (e.g., Phinney 1987). The radiative force accelerates matter, while the radiation drag decelerates it. As a result, the radiatively-driven flow would have a finite terminal speed, where the radiative force is balanced with the radiation drag force (Icke 1989).

25.2.1.1 Radiation Drag

In the steady one-dimensional flow, from Eq. (25.23), the radiative force $f_{\rm rad}$ on the order of $\mathcal{O}(v/c)^1$ in the inertial frame is expressed as

$$f_{\rm rad} = \frac{\kappa_F + \sigma_F}{c} F_0 = \frac{\kappa_F + \sigma_F}{c} [F - (E + P)v], \tag{25.60}$$

where F_0 is the radiative flux in the comoving frame, while E, F, and P are quantities measured in the inertial frame.

Let us first consider the forces in the inertial frame (rest frame). In the radiative environments, in addition to the radiative flux F, which accelerates fluid particles, there exist radiation energy E and pressure P. Since energy is equivalent with mass and has inertia, the moving particles in the radiation field receive a force against the motion, which is proportional to the fluid velocity. This is called *radiation drag*.

Let us consider the above situation in the comoving frame (fluid frame). In the comoving frame, photons come to the matter from the top-side in average by aberration, even in the case where photon distribution is isotropic in the inertial frame. Hence, photons exert the momentum to the fluid particles in the backward direction.[3] This acts as the deceleration force in the inertial frame.

Radiation drag is important and operative in various radiative environments. For example, in the early universe, where the cosmic background radiation is dense, the cosmic background radiation can remove the angular momentum from the collapsing gas cloud to assist the evolution of massive black holes (e.g., Umemura et al. 1993). Under the starburst environments, mass accretion may be driven by angular momentum extraction due to radiation drag exerted by stellar radiation from circumnuclear starburst regions (e.g., Umemura et al. 1997). Surrounding an black-hole accretion disk, which is an intense radiation source, the radiatively-driven astrophysical jets and winds would be affected by the intense radiation from the accretion disk (e.g., Icke 1989; Tajima and Fukue 1998). Even around such low energy objects as red giants and protoplanetary disks, where dust receives the radiation drag force as well as radiative force, the effect of radiation drag works in the nonrelativistic regime (Poynting–Robertson effect).

Relating to radiation drag, we briefly comment on the diffusion flux and advection one. Replacing v by $-v$, we can express the radiative flux in the inertial frame by the quantities in the comoving one:

$$F = F_0 + (E_0 + P_0)v \tag{25.61}$$

on the order of $\mathcal{O}(v/c)^1$. Of the components on the right-hand side, F_0 means the energy flux in the comoving frame, and called the *diffusion flux*. On the other hand, $(E_0 + P_0)v$ represents the radiation energy carried with the moving fluid, and called the *advection flux*.

25.2.1.2 Magic Terminal Speed

Under an intense radiative environment, gaseous particles are accelerated by the radiative flux, while decelerated by the radiation drag, which is proportional to the velocity v of the particle, and includes the radiation energy density and stress. Icke (1989) examined the *photon surfing* of gaseous particles in such an intense radiation environment for various situations and configurations of sources. He found that the particle reaches a terminal speed, where in the fixed frame the radiative flux force becomes equal to the radiation drag force, or in the comoving frame the net flux

[3] In the one-dimensional flow, aberration is expressed as

$$\mu_0 = \frac{\mu - \beta}{1 - \beta\mu},$$

where μ_0 is the direction cosine of photons in the comoving frame, and μ that in the inertial frame. Hence, for example, photons from the horizontal direction in the inertial frame ($\mu = 0$) are observed in the comoving frame, as those coming from the forward direction of $\mu_0 = -\beta = -v/c$.

from the top-side and bottom-side vanishes. In the case of an infinite flat uniform source, this *magic terminal speed* is

$$\beta_{\mathrm{m}} = \frac{v_{\mathrm{m}}}{c} = \frac{4-\sqrt{7}}{3} \sim 0.45. \tag{25.62}$$

In the radiatively-driven jets and accretion disk winds, the effect of radiation drag becomes very important, and many researchers investigate the relativistic flow with radiation drag.

Let us derive this magic terminal speed both in the inertial and comoving viewpoints (Fukue 2014a).

At first, in order to consider the force balance for a moving stratus in the inertial frame, we start from the equation of motion (25.32) for the stratus (or particle) with mass m and surface area S, moving at speed v in the vertical (z) direction perpendicular to the infinite flat source of uniform intensity I^* under special relativity:

$$\begin{aligned} m\frac{d}{dt}(\gamma v) &= -m\frac{d\psi}{dz} + \frac{S}{c}\gamma F_0 \\ &= -m\frac{d\psi}{dz} + \frac{S}{c}\gamma^3\left[\left(1+\beta^2\right)F - (cE + cP)\beta\right], \end{aligned} \tag{25.63}$$

where ψ is the gravitational potential, which we do not consider here, F_0 the vertical flux in the comoving frame, E the radiation energy density, F the (vertical) radiative flux, and P the zz-component of the radiation stress tensor in the inertial frame, respectively.

In the inertial frame these components of radiation field (E, F, and P) above an infinite flat source of uniform intensity I^* become

$$cE = 2\pi I^*, \tag{25.64}$$

$$F = \pi I^*, \tag{25.65}$$

$$cP = 2\pi I^*/3. \tag{25.66}$$

Substituting these expressions for E, F, and P into the right-hand side of (25.63), we have the radiative flux in the comoving frame as

$$F_0 = \gamma^2 \pi I^* \left(1 - \frac{8}{3}\beta + \beta^2\right). \tag{25.67}$$

Hence, at $\beta = \beta_m$, the net radiative force in the comoving frame vanishes, and the moving cloud reaches a terminal speed of β_m, if the gravity or other forces are ignored.[4]

Next, in the comoving frame the distributions of the direction and intensity of incident radiation are changed from those in the inertial frame due to aberration and Doppler effects (24.20) and (24.21). The incident radiation I_0^* in the comoving frame is expressed as

$$I_0^* = \left(\frac{\nu_0}{\nu}\right)^4 I^* = [\gamma\,(1-\beta\cos\theta)]^4 I^* = \frac{1}{[\gamma\,(1+\beta\cos\theta_0)]^4} I^*. \tag{25.68}$$

In the inertial frame the incident radiation comes from $0 \le \theta \le \pi/2$, while in the comoving frame the intensity I_0^* exists only for $0 \le \theta_0 \le \theta_{max}$, satisfying $\cos\theta_{max} = -\beta$, due to aberration.

Hence, in the comoving frame, the comoving radiative flux F_0^+ from the bottom-side, which accelerates the stratus, is calculated as

$$\begin{aligned} F_0^+ &= \int_0^{\pi/2} I_0^* \cos\theta_0 d\Omega_0 = \frac{2\pi I^*}{\gamma^4} \int_0^{\pi/2} \frac{\sin\theta_0 \cos\theta_0 d\theta_0}{(1+\beta\cos\theta_0)^4} \\ &= \frac{2\pi I^*}{\gamma^4} \frac{3+\beta}{6\,(1+\beta)^3} = \pi I^* \gamma^2 \frac{(3+\beta)(1-\beta)^3}{3}. \end{aligned} \tag{25.69}$$

On the other hand, the comoving radiative flux F_0^- from the top-side, which decelerates the stratus, is

$$\begin{aligned} F_0^- &= \int_{\pi/2}^{\cos^{-1}(-\beta)} I_0^* \cos\theta_0 d\Omega_0 \\ &= \frac{2\pi I^*}{\gamma^4} \frac{3\beta^2-\beta^4}{6\left(1-\beta^2\right)^3} = \pi I^* \gamma^2 \frac{3\beta^2-\beta^4}{3}. \end{aligned} \tag{25.70}$$

[4] In this simple case, equation of motion without gravity is expressed as

$$\frac{d\beta}{dt} = \frac{S}{mc^2}\pi I^* \left(1 - \frac{8}{3}\beta + \beta^2\right).$$

This equation can be solved analytically to yield a solution:

$$\frac{\beta_+ - \beta}{\beta_- - \beta}\frac{\beta_-}{\beta_+} = \exp\left(\frac{2\sqrt{7}}{3}\frac{S}{mc^2}\pi I^* t\right),$$

where $\beta_\pm \equiv (4 \pm \sqrt{7})/3$. Hence, a particle asymptotically approaches to $\beta_- = \beta_m$ as time goes by.

Thus, the net radiative force $(S/c)\gamma F_0$ is

$$\frac{S}{c}\gamma F_0 = \frac{S}{c}\pi I^* \gamma^3 \left(1 - \frac{8}{3}\beta + \beta^2\right), \tag{25.71}$$

and we can obtain the magic terminal speed again.

It should be noted that this magic terminal speed β_m is for the particles over an infinite flat uniform source. It is modified when the source size is finite or the source has an intensity gradient (Icke 1989). Moreover, in the case of a gaseous stratus with finite optical depth over an infinite source, the terminal speed slightly decreases in the optically thin regime (Masuda and Fukue 2016).

25.3 Relativistic Radiative Shocks

Relativistic (radiative) shocks are also remarkably important in various energetic astrophysical situations, such as accretion flows onto a compact object, astrophysical jets, gamma-ray bursts, and early stages of the universe.

For the relativistic hydrodynamical shocks, the jump conditions (shock adiabat) were firstly derived by Taub (1948), and examined by many researchers (e.g., Landau and Lifshitz 1959; Johnson and McKee 1971; Thorne 1973). The relativistic Rankine–Hugoniot relations have been also applied to various astrophysical phenomena; supernova explosions (e.g., Colgate and White 1966), disk accretion onto a black hole (e.g., Fukue 1987), fireballs in the gamma-ray bursts (e.g., Mészáros 2002), relativistic combustion waves (e.g., Gao and Law 2012).

The relativistic radiative shocks have been also examined by several researchers (e.g., Cissoko 1997; Farris et al. 2008; Budnik et al. 2010) and applied to various astrophysical relativistic flows; supernova shock breakouts (e.g., Katz et al. 2010), gamma-ray bursts (e.g., Budnik et al. 2010; Beloborodov 2017), Bondi–Hoyle accretion (e.g., Zanotti et al. 2011), spherical accretion onto a black hole (e.g., Fragile et al. 2012).

The fundamental studies on the relativistic radiative shocks, however, are not sufficient yet in semi-analytical levels. For example, Farris et al. (2008) assumed the strong shock, and neglected the radiation pressure and energy density in their semi-analytical model in the appendix. The gas pressure was also neglected in some studies (Budnik et al. 2010; Beloborodov (2017). Although the shock adiabats were well examined and displayed, other useful relations have not been shown explicitly.

Some researchers have numerically examined relativistic radiation-mediated shocks, by using the relativistic moment equations with a closure relation (e.g., Takahashi et al. 2013; Sądowski et al. 2013), or by solving the full relativistic radiative transfer (e.g., Budnik et al. 2010; Tolstov et al. 2015). Semi-analytical studies would be useful to check and interpret numerical studies.

Thus, in this section we shall analytically examine and summarize characteristics of relativistic radiative shock waves in the one-dimensional flow under the equilibrium diffusion approximation: e.g., we derive the overall jump conditions, and solve the relativistic radiative precursor (see Fukue 2019b for details).

25.3.1 Equilibrium Diffusion Shock

Except for the extremely thick case, radiation from shock fronts usually diffuses out into the pre-shock and post-shock regions, and affects the structure of shocked flows (Sect. 23.3 for the nonrelativistic case). For example, physical quantities such as temperature can continuously change in some extent, including the radiative precursor, shock front, and after-shock region.

In this section, we consider the case where an *equilibrium diffusion approximation* can be adopted, and we assume that the radiation temperature $T_{\rm rad}$ is equal to the gas temperature $T_{\rm gas}$ (one temperature approximation); $T_{\rm rad} = T_{\rm gas} = T$. We assume the following situations: (1) We treat the radiative shock within the framework of special relativity, and the effect of the curved spacetime is ignored. (2) The shock frame (comoving frame) is one dimensional in the x direction. (3) The gas flow is optically thick, and the diffusion approximation holds. (4) The Eddington approximation is adopted for radiation fields. (5) We only consider the radiative diffusion into the upstream region, and do not consider that into the downstream region; we examine the structure of the radiative precursor before the shock front. Under these assumptions, the basic conservation equations are described in the following subsections.

25.3.1.1 Conservation Equations

In the radiative shocks under the equilibrium diffusion approximation, the radiative flux in the total energy conservation is retained, and the jump conditions for radiation and matter are derived (see, e.g., Lowrie and Rauenzahn 2007 for the nonrelativistic radiative case; Thorne 1973 for the relativistic hydrodynamical case).

We can give the mass flux, momentum flux, and energy flux conservations for the comoving quantities such as gas density ρ, gas internal energy density ε, gas pressure p, radiation energy density E, radiation pressure P, and four velocity u ($\equiv \gamma\beta$; $\gamma = \sqrt{1-\beta^2}$ and $\beta = v/c$) relative to the shock front as follows [Thorne 1973; Sect. 23.3.4, equations (23.73)–(23.75)]:

$$\rho c u = \rho_1 c u_1 = j \text{ (constant)}, \tag{25.72}$$

$$(\varepsilon + p)u^2 + p + P = (\varepsilon_1 + p_1)u_1^2 + p_1 + P_1, \tag{25.73}$$

$$\rho c u \frac{\varepsilon + p + E + P}{\rho}\gamma + \gamma^2(1+\beta^2)F = \rho_1 c u_1 \frac{\varepsilon_1 + p_1 + E_1 + P_1}{\rho_1}\gamma_1, \tag{25.74}$$

where the subscript “1” on the right-hand side represents the quantities in the upstream region far from the shock front, and no subscript means those in other regions. When we consider quantities in the downstream region far from the shock front, the subscript “2” will be attached. Here, we consider the structure of the radiative precursor, and ignore the effect of the radiative diffusion into the downstream region, whose optical depth is assumed to be sufficiently large. Hence, no subscript means the quantities in the radiative precursor, and the subscript “2” actually denotes those just after the shock front.

For the gas, we adopt the polytropic equation of state (EoS), Eq. (9.1), and the gas energy density is expressed as

$$\varepsilon = \rho c^2 + \frac{1}{\Gamma - 1} p, \tag{25.75}$$

where Γ is the polytropic index. For the radiation field, under the Eddington approximation in the comoving frame, we can set the radiation energy density E as $E = 3P$. The diffusive flux F in the comoving frame is expressed in terms of the radiation pressure or temperature as a function of the coordinate x along the shock flow:

$$F = -\frac{c}{\kappa\rho}\frac{dP}{dx} = -\frac{4acT^3}{3\kappa\rho}\frac{dT}{dx}, \tag{25.76}$$

which is the first-moment equation (diffusion form). Here, κ is opacity and assumed to be constant.

It should be again stressed that all the quantities are measured in the comoving (shock) frame (The suffix “0” is ignored for simplicity.).

It is further noted that the *relativistic* has two meanings; the flow speed relative to the shock front is relativistic and/or the gas internal energy is relativistic.

25.3.1.2 Overall Jump Conditions

Far from the shock, radiation and matter equilibrate, and the diffusive flux can be ignored. Hence, for ideal gases and isotropic radiation fields, the overall jump conditions in the present case are [cf. Eqs. (23.77)–(23.79) for the nonrelativistic case]

$$\rho_2 c u_2 = \rho_1 c u_1 = j, \tag{25.77}$$

$$(\varepsilon_2 + p_2)u_2^2 + p_2 + P_2 = (\varepsilon_1 + p_1)u_1^2 + p_1 + P_1, \tag{25.78}$$

$$\frac{\varepsilon_2 + p_2 + E_2 + P_2}{\rho_2}\gamma_2 = \frac{\varepsilon_1 + p_1 + E_1 + P_1}{\rho_1}\gamma_1, \tag{25.79}$$

where the subscripts "1" and "2" represent the quantities in the upstream and down stream regions far from the shock.

Similar to the hydrodynamical case (e.g., Thorne 1973), substituting $u_2 = j/(\rho_2 c)$ and $u_1 = j/(\rho_1 c)$ into Eq. (25.78), we can express the mass flux as

$$j^2 = \frac{p_2 + P_2 - (p_1 + P_1)}{\dfrac{\varepsilon_1 + p_1}{\rho_1 c^2}\dfrac{1}{\rho_1} - \dfrac{\varepsilon_2 + p_2}{\rho_2 c^2}\dfrac{1}{\rho_2}}. \tag{25.80}$$

Since $\gamma_2^2 = 1 + u_2^2 = 1 + j^2/(\rho_2 c)^2$, the square of Eq. (25.79) is expressed as [cf. equation (23.80) for the nonrelativistic case]

$$\left(\frac{\varepsilon_2 + p_2}{\rho_2} + \frac{E_2 + P_2}{\rho_2}\right)^2 \left[1 + \frac{1}{\rho_2^2 c^2}\frac{p_2 + P_2 - (p_1 + P_1)}{\dfrac{\varepsilon_1 + p_1}{\rho_1 c^2}\dfrac{1}{\rho_1} - \dfrac{\varepsilon_2 + p_2}{\rho_2 c^2}\dfrac{1}{\rho_2}}\right]$$
$$= \left(\frac{\varepsilon_1 + p_1}{\rho_1} + \frac{E_1 + P_1}{\rho_1}\right)^2 \left[1 + \frac{1}{\rho_1^2 c^2}\frac{p_2 + P_2 - (p_1 + P_1)}{\dfrac{\varepsilon_1 + p_1}{\rho_1 c^2}\dfrac{1}{\rho_1} - \dfrac{\varepsilon_2 + p_2}{\rho_2 c^2}\dfrac{1}{\rho_2}}\right]. \tag{25.81}$$

This is the generalized Rankine–Hugoniot relation for the relativistic radiative shocks, and reduces to Taub's (1948) condition (Thorne 1973) when we drop the radiation pressure and energy density.

Similar to the nonrelativistic case, Eq. (25.78) can be deformed as follows. Since the relativistic sound speed a of the present polytropic case is

$$a^2 \equiv c^2 \frac{\partial p}{\partial \varepsilon} = c^2 \frac{\Gamma p}{\varepsilon + p}, \tag{25.82}$$

Eq. (25.78) is rewritten as

$$\begin{aligned} p_2 + P_2 - (p_1 + P_1) &= (\varepsilon_1 + p_1)u_1^2 - (\varepsilon_2 + p_2)u_2^2 \\ &= \frac{\Gamma p_1}{a_1^2}\gamma_1^2 v_1^2 - (\varepsilon_2 + p_2)\frac{j^2}{\rho_2^2 c^2}. \end{aligned} \tag{25.83}$$

Introducing the Mach number $\mathcal{M}_1$ ($\equiv v_1/a_1$), and substituting Eq. (25.80) into γ_1^2 and j^2, we have another Rankine–Hugoniot relation [cf. Eq. (23.81) for the nonrelativistic case] :

$$[p_2 + P_2 - (p_1 + P_1)]\frac{\dfrac{\varepsilon_1 + p_1}{\rho_1 c^2} - \dfrac{\Gamma p_1 \mathcal{M}_1^2}{\rho_1 c^2}}{\dfrac{\varepsilon_1 + p_1}{\rho_1 c^2} - \dfrac{\varepsilon_2 + p_2}{\rho_2 c^2}\dfrac{\rho_1}{\rho_2}} = \Gamma p_1 \mathcal{M}_1^2. \tag{25.84}$$

In the nonrelativistic limit, this relation reduces to the usual Rankine–Hugoniot relation.

25.3.1.3 Radiative Precursor Region

Using the solutions of jump conditions as boundary conditions, the structure of the radiative precursor can be solved. For ideal gases and isotropic radiation fields, the energy conservation (25.74) including the radiative flux (25.76) is expressed as [cf. Eq. (23.82) for the nonrelativistic case]

$$\begin{aligned}\gamma^2(1+\beta^2)\frac{c}{\kappa\rho}\frac{dP}{dx} &= \gamma^2(1+\beta^2)\frac{4acT^3}{3\kappa\rho}\frac{dT}{dx}\\ &= \rho c u\frac{\varepsilon+p+E+P}{\rho}\gamma - \rho_1 c u_1\frac{\varepsilon_1+p_1+E_1+P_1}{\rho_1}\gamma_1\\ &= \left(\frac{\varepsilon+p+E+P}{\rho}\gamma - \frac{\varepsilon_1+p_1+E_1+P_1}{\rho_1}\gamma_1\right)\rho_1 c u_1. \end{aligned} \tag{25.85}$$

Introducing the typical optical depth τ_1 ($\equiv \kappa\rho_1\ell_1$), ℓ_1 being the relevant scalelength, we can rearrange Eq. (25.85) as

$$\begin{aligned}&\frac{\gamma^2}{\gamma_1^2}(1+\beta^2)\frac{1}{\tau_1\beta_1}\frac{\rho_1}{\rho}\frac{d}{d\tilde{x}}\frac{P}{P_1}\\ &\quad = \frac{\varepsilon+p+E+P}{P_1}\frac{\rho_1}{\rho}\frac{\gamma}{\gamma_1} - \frac{\varepsilon_1+p_1+E_1+P_1}{P_1}, \end{aligned} \tag{25.86}$$

where $\tilde{x}$ ($\equiv x/\ell_1$) is the normalized coordinate, and from Eq. (25.72) the Lorentz factor is expressed as

$$\frac{\gamma^2}{\gamma_1^2} = \frac{\gamma_1^2-1}{\gamma_1^2}\frac{\rho_1^2}{\rho^2} + \frac{1}{\gamma_1^2}. \tag{25.87}$$

Similarly, the momentum conservation (25.73) is rearranged as [cf. Eq. (23.84) for the nonrelativistic case]

$$p + P - (p_1 + P_1) = \Gamma p_1\gamma_1^2\mathcal{M}_1^2\left(1 - \frac{\varepsilon+p}{\rho c^2}\frac{\rho_1 c^2}{\varepsilon+p}\frac{\rho_1}{\rho}\right). \tag{25.88}$$

Parameters are Γ, τ_1, α_1 ($\equiv P_1/p_1$), β_1 ($\equiv v_1/c$), γ_1, and $\mathcal{M}_1^2$. From the physical viewpoints, we specify and separate these individual parameters. However, since τ_1 and β_1 are combined into a single parameter ($\tau_1\beta_1$), and furthermore, they are renormalized into the coordinate, the independent parameters are, e.g., Γ, α_1, γ_1, and $\mathcal{M}_1$.

In order to treat the problem semi-analytically, we consider the two limiting cases, the gas-pressure/radiation-pressure dominant cases.

25.3.2 *Gas-Pressure Dominated Shocks*

In this subsection we shall examine the *gas-gas* shocks, where the gas pressure is dominant in both sides of the shock. In the nonrelativistic radiative shocks, the gas-pressure dominated case is often an *isothermal shock* with discontinous density structure (cf. Sect. 23.3).

25.3.2.1 Overall Jump Conditions

In the gas-pressure dominated case, we drop the radiation-pressure terms, and the Rankine–Hugoniot relation (25.81) becomes

$$\left(\frac{\varepsilon_2+p_2}{\rho_2 c^2}\right)^2\left[1+\frac{1}{\rho_2^2 c^2}\frac{p_2-p_1}{\frac{\varepsilon_1+p_1}{\rho_1 c^2}\frac{1}{\rho_1}-\frac{\varepsilon_2+p_2}{\rho_2 c^2}\frac{1}{\rho_2}}\right]$$

$$=\left(\frac{\varepsilon_1+p_1}{\rho_1 c^2}\right)^2\left[1+\frac{1}{\rho_1^2 c^2}\frac{p_2-p_1}{\frac{\varepsilon_1+p_1}{\rho_1 c^2}\frac{1}{\rho_1}-\frac{\varepsilon_2+p_2}{\rho_2 c^2}\frac{1}{\rho_2}}\right]. \tag{25.89}$$

Introducing the enthalpy per unit rest mass energy, μ, by

$$\mu \equiv \frac{\varepsilon+p}{\rho c^2}, \tag{25.90}$$

we can be easily rewritten Eq. (25.89) in the form of Taub's (1948) jump condition (Thorne 1973):

$$\mu_2^2-\mu_1^2=(p_2-p_1)\left(\frac{\mu_2}{\rho_2 c^2}+\frac{\mu_1}{\rho_1 c^2}\right). \tag{25.91}$$

In the present case adopting the polytropic equation of state (25.75), instead of μ, it is convenient to use the gas pressure per unit rest mass energy, ν:

$$\nu \equiv \frac{p}{\rho c^2}; \quad \mu = 1+\frac{\Gamma}{\Gamma-1}\nu. \tag{25.92}$$

Substituting the polytropic equation of state (25.75) into Eq. (25.91), and using ν instead of μ, after several manipulations, we have a relation between p_2 and ρ_2:

$$\frac{\Gamma\nu_1}{\Gamma-1}(\tilde{p}_2+\Gamma-1)\,\tilde{p}_2\frac{1}{\tilde{\rho}_2^2}+\left[(\Gamma+1)\tilde{p}_2+\Gamma-1\right]\frac{1}{\tilde{\rho}_2}$$
$$-\left[(\Gamma\nu_1+\Gamma-1)\tilde{p}_2+\Gamma+1+\frac{\Gamma\nu_1}{\Gamma-1}\right]=0. \qquad (25.93)$$

where $\tilde{p}_2 = p_2/p_1$ and $\tilde{\rho}_2 = \rho_2/\rho_1$. This equation is a quadratic on $1/\tilde{\rho}_2$, and analytically solved for a given parameter Γ and a pre-shock quantity ν_1. Other post-shock quantities are given as, e.g., $\tilde{T}_2 = T_2/T_1 = \nu_2/\nu_1$, $\mu_1 = 1+\Gamma/(\Gamma-1)\nu_1$, $\mu_2 = 1+\Gamma/(\Gamma-1)\nu_1\tilde{p}_2/\tilde{\rho}_2$, $u_1^2 = \nu_1(\tilde{p}_2-1)/(\mu_1-\mu_2/\tilde{\rho}_2)$, and so on.

In Fig. 25.1 several of the post-shock quantities are depicted as a function of $\tilde{p}_2$ $(\equiv p_2/p_1)$ in the gas-pressure dominated case ($\Gamma = 5/3$). Solid curves represent $\tilde{\rho}_2$ $(\equiv \rho_2/\rho_1)$, where $\nu_1 = 0.00001, 0.001, 0.1$ from bottom to top, whereas dashed curves are corresponding $\tilde{\rho}_2/\tilde{\mu}_2$, where $\nu_1 = 0.00001, 0.001, 0.1$ from top to bottom. Chain-dotted ones mean $\tilde{T}_2$ $(\equiv T_2/T_1)$, where $\nu_1 = 0.00001$, 0.001, 0.1 from right to left. Finally, dashed ones denote the pre-shock u_1, where $\nu_1 = 0.001, 0.1$ from bottom to top.

As is seen in Fig. 25.1, similar to the nonrelativistic case, the post-shock gas density increases with $\tilde{p}_2$. When the pre-shock gas internal energy is sufficiently nonrelativistic ($\nu_1 = 0.00001$), the post-shock gas density increases and approaches 4 as $\tilde{p}_2$ becomes large in the gas-pressure dominated case ($\Gamma = 5/3$), as in the nonrelativistic case. For sufficiently large $\tilde{p}_2$, however, the gas density becomes slightly larger than 4. When the gas internal energy is mildly relativistic ($\nu_1 =$

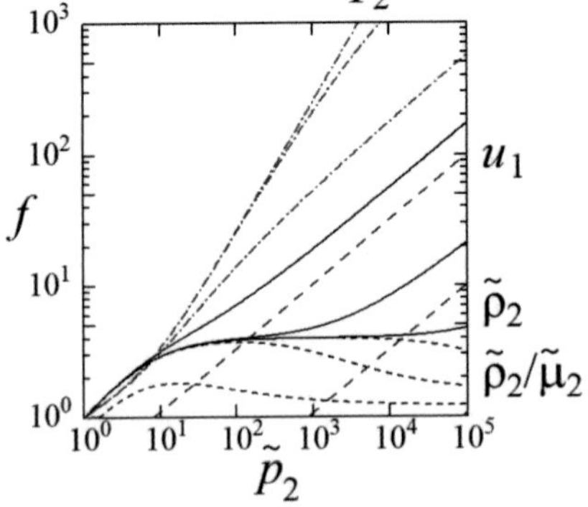

Fig. 25.1 Post-shock quantities as a function of $\tilde{p}_2$ $(\equiv p_2/p_1)$ in the gas-pressure dominated case ($\Gamma = 5/3$). Solid curves represent $\tilde{\rho}_2$ $(\equiv \rho_2/\rho_1)$, where $\nu_1 = 0.00001, 0.001, 0.1$ from bottom to top, whereas dashed curves are corresponding $\tilde{\rho}_2/\tilde{\mu}_2$, where $\nu_1 = 0.00001, 0.001, 0.1$ from top to bottom. Chain-dotted ones mean $\tilde{T}_2$ $(\equiv T_2/T_1)$, where $\nu_1 = 0.00001, 0.001, 0.1$ from right to left. Finally, dashed ones denote the pre-shock u_1, where $\nu_1 = 0.001, 0.1$ from bottom to top

0.001, 0.1), the post-shock gas density no longer approaches 4, but monotonically increases with $\tilde{p}_2$. In such a case, the pre-shock flow speed u_1 must be also relativistic, as shown in Fig. 25.1. It should be noted that in the extremely relativistic regime $\tilde{p}_2 \sim \tilde{\rho}_2^2$.

On the other hand, another Rankine–Hugoniot relation (25.84) becomes

$$(p_2 - p_1)\frac{\mu_1 - \Gamma \nu_1 \mathcal{M}_1^2}{\mu_1 - \mu_2\rho_1/\rho_2} = \Gamma p_1 \mathcal{M}_1^2 \tag{25.94}$$

or in the nondimensional form:

$$\mathcal{M}_1^2 = \frac{1}{\Gamma}\frac{\mu_1(\tilde{p}_2 - 1)}{\nu_1(\tilde{p}_2 - 1) + \mu_1 - \mu_2/\tilde{\rho}_2}. \tag{25.95}$$

Now, we can implicitly express the post-shock quantities by the pre-shock Mach number $\mathcal{M}_1$.

In Fig. 25.2 several of the post-shock quantities are depicted as a function of $\mathcal{M}_1$ in the gas-pressure dominated case ($\Gamma = 5/3$). Solid curves represent $\tilde{\rho}_2$, where $\nu_1 = 0.00001, 0.001, 0.1$ from right to left. Chain-dotted ones mean $\tilde{T}_2$, where $\nu_1 = 0.00001, 0.001, 0.1$ from right to left, while two-dot chain ones are $\tilde{p}_2$, where $\nu_1 = 0.00001, 0.001, 0.1$ from right to left. Finally, dashed ones denote the pre-shock u_1, where $\nu_1 = 0.00001, 0.001, 0.1$ from bottom to top.

As is seen in Fig. 25.2, similar to the nonrelativistic case, the gas density and other quantities generally increases with $\mathcal{M}_1$. When the pre-shock gas internal energy is sufficiently nonrelativistic ($\nu_1 = 0.00001$), the post-shock gas density increases and approaches 4 as $\mathcal{M}_1$ becomes large in this parameter range in the gas-

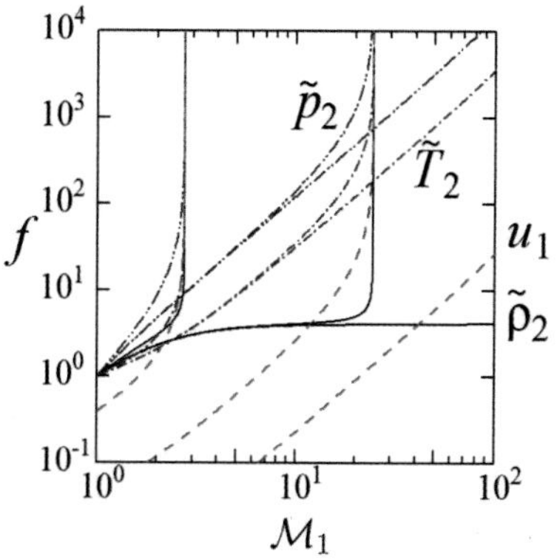

Fig. 25.2 Post-shock quantities as a function of $\mathcal{M}_1$ in the gas-pressure dominated case ($\Gamma = 5/3$). Solid curves represent $\tilde{\rho}_2$, where $\nu_1 = 0.00001, 0.001, 0.1$ from right to left. Chain-dotted ones mean $\tilde{T}_2$, where $\nu_1 = 0.00001, 0.001, 0.1$ from right to left, while two-dot chain ones are $\tilde{p}_2$, where $\nu_1 = 0.00001, 0.001, 0.1$ from right to left. Finally, dashed ones denote the pre-shock u_1, where $\nu_1 = 0.00001, 0.001, 0.1$ from bottom to top

pressure dominated case ($\Gamma = 5/3$), as in the nonrelativistic case. When the gas internal energy is mildly relativistic ($\nu_1 = 0.001, 0.1$), the post-shock gas density becomes larger than 4.

It should be noted that in the relativistic case the pre-shock Mach number $\mathcal{M}_1$ cannot increase without limit, but has some limiting value, as is seen in Fig. 25.2. This is understood as follows. If the pre-shock gas internal energy becomes relativistic, the corresponding sound speed a_1 is also relativistic, and becomes on the order of unity (but less than $c/\sqrt{3}$). On the other hand, the pre-shock flow speed v_1 cannot exceed the speed of light. As a result, the pre-shock Mach number $\mathcal{M}_1$ ($\equiv v_1/a_1$) cannot increase and approach infinity, but has some limiting value on the order of unity at the sufficiently relativistic case. Indeed, from Eq. (25.95), in the limit of strong shocks, we have

$$\mathcal{M}_1^2 = \frac{1}{\Gamma}\frac{1+\Gamma/(\Gamma-1)\nu_1}{\nu_1}, \tag{25.96}$$

which is the maximum value of $\mathcal{M}_1$ in Fig. 25.2.

In Fig. 25.3 the so-called shock adiabats (Rankine–Hugoniot curves) are plotted, although they are essentially the same as Fig. 25.1. Solid curves denote the usual relations between $\tilde{p}_2$ and $1/\tilde{\rho}_2$, where $\nu_1 = 0.00001, 0.001, 0.1$ from right to left. Dashed curves represent the relations between $\tilde{p}_2$ and $\tilde{\mu}_2/\tilde{\rho}_2$, where $\nu_1 = 0.00001, 0.001, 0.1$ from left to right.

If we adopt $\tilde{\mu}_2/\tilde{\rho}_2$ as an independent variable, instead of $1/\tilde{\rho}_2$, as recommended by, e.g., Thorne (1973), the shock adiabats become rather different from the usual cases.

It should be noted that Gao and Law (2012) has shown and discussed the Rankine–Hugoniot curves of this type for various parameters in details, although their plots were limited within the range of $p_2/p_1 < 10$.

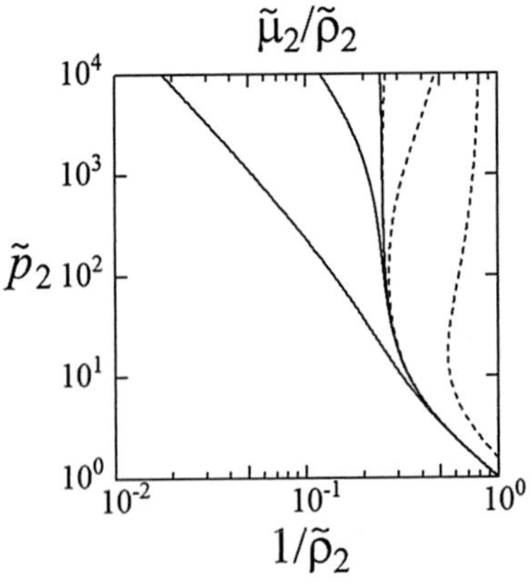

Fig. 25.3 Shock adiabats in the gas-pressure dominated case ($\Gamma = 5/3$). Solid curves represent the relations between $\tilde{p}_2$ and $1/\tilde{\rho}_2$, where $\nu_1 = 0.00001, 0.001, 0.1$ from right to left. Dashed ones denote the relations between $\tilde{p}_2$ and $\tilde{\mu}_2/\tilde{\rho}_2$, where $\nu_1 = 0.00001, 0.001, 0.1$ from left to right

25.3.2.2 Radiative Precursor Structure

In the gas-pressure dominated case, the structure equations for the radiative precursor (25.86)–(25.88) can be expressed in terms of variables $\tilde{T}$ ($\equiv T/T_1$), $\tilde{\rho}$ ($\equiv \rho/\rho_1$), and γ as

$$\frac{\gamma^2}{\gamma_1^2}(1+\beta^2)\frac{4\alpha_1}{\tau_1\beta_1}\frac{\tilde{T}^3}{\tilde{\rho}}\frac{d\tilde{T}}{d\tilde{x}} = \frac{1}{\nu_1}\left(\frac{\gamma}{\gamma_1}-1\right)+\frac{\Gamma}{\Gamma-1}\left(\tilde{T}\frac{\gamma}{\gamma_1}-1\right), \quad (25.97)$$

$$\frac{\gamma^2}{\gamma_1^2} = \frac{\gamma_1^2-1}{\gamma_1^2}\frac{1}{\tilde{\rho}^2}+\frac{1}{\gamma_1^2}. \quad (25.98)$$

$$\tilde{T}\tilde{\rho}^2-(\Gamma\gamma_1^2\mathcal{M}_1^2+1)\tilde{\rho}+\Gamma\gamma_1^2\mathcal{M}_1^2\frac{\Gamma-1+\Gamma\nu_1\tilde{T}}{\Gamma-1+\Gamma\nu_1} = 0, \quad (25.99)$$

where $\tau_1 = \kappa\rho_1\ell_1$, $\beta_1 = v_1/c$, and $\alpha_1 = P_1/p_1$. Equation (25.99) is a quadratic on $\tilde{\rho}$, and analytically solved for given parameters. We choose the minus-sign solution, which approaches unity at $\tilde{T} \to 1$.

Using the overall jump relations as boundary conditions, we can solve these Eqs. (25.97)–(25.99) to obtain the radiative precursor structure. Since τ_1, β_1, and α_1 are renormalized into the coordinate scale, the independent parameters are Γ, ν_1, and $\mathcal{M}_1$ (then a_1, u_1, γ_1 are automatically determined.).

Typical solutions for the radiative precursor region of the relativistic radiative shocks are shown in Fig. 25.4 as a function of the normalized coordinate. Thick solid curves represent $\tilde{\rho}$, thin chain-dotted ones the gas temperature $\tilde{T}$, thin two-dot chain ones the gas pressure $\tilde{p}$, thin dashed ones the flow speed v/c (upper) and sound one a/c, thick dashed ones the Mach number $\mathcal{M}$, and thin solid ones the comoving radiative flux $\tilde{F}$. The parameters are $\Gamma = 5/3$ and (a) $\nu_1 = 0.00001$ and $\mathcal{M}_1 = 2.46$ ($\tilde{\rho}_2 = 7.3$, $\beta_1 = 0.01$), and (b) $\nu_1 = 0.1$ and $\mathcal{M}_1 = 2.20$ ($\tilde{\rho}_2 = 9.3$, $\beta_1 = 0.74$). That is, Fig. 25.4a corresponds to a nonrelativistic case, where both the gas internal energy and flow speed are nonrelativistic ($\nu_1 = 0.00001$ and $\beta_1 = 0.01$), and both flow speed v/c and sound speed a/c are on the order of 0.01, while Fig. 25.4b is a relativistic case, where both the gas internal energy and flow speed are relativistic ($\nu_1 = 0.1$ and $\beta_1 = 0.74$), and both flow speed v/c and sound speed a/c are on the order of 1.

Similar to the nonrelativistic case, the radiative energy from the hot post-shock region diffuses into the pre-shock region to make the radiative precursor. Furthermore, similar to the nonrelativistic case, the density distribution is discontinuous, although the temperature one is continuous; i.e., the so-called *isothermal shocks*. Namely, in this gas-pressure dominated case the density (25.99) in the radiative precursor just before the shock front is usually smaller than the density in the post-shock region determined by the jump conditions.

Comparing the nonrelativistic case (Fig. 25.4a; $v/c \sim a/c \sim 0.01$), with the relativistic case (Fig. 25.4b; $v/c \sim a/c \sim 1$), we see that the relative values of the other quantities such as ρ/ρ_1, p/p_1, T/T_1, and $\mathcal{M}$ are apparently not different so

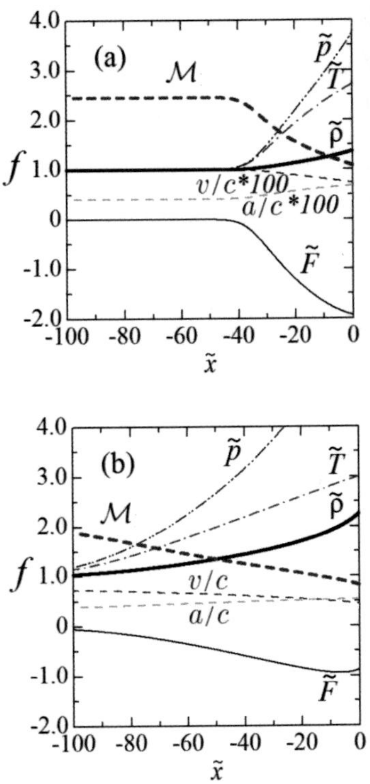

Fig. 25.4 Typical solutions for the radiative precursor region of the relativistic radiative shocks in the gas-pressure dominated case as a function of the normalized coordinate. Thick solid curves represent $\tilde{\rho}$, thin chain-dotted ones $\tilde{T}$, thin two-dot chain ones $\tilde{p}$, thin dashed ones v/c (upper) and a/c, thick dashed ones $\mathcal{M}$, and thin solid ones $\tilde{F}$. The parameters are $\Gamma = 5/3$ and (**a**) $\nu_1 = 0.00001$ and $\mathcal{M}_1 = 2.46$ ($\tilde{p}_2 = 7.3$, $\beta_1 = 0.01$), and (**b**) $\nu_1 = 0.1$ and $\mathcal{M}_1 = 2.20$ ($\tilde{p}_2 = 9.3$, $\beta_1 = 0.74$). The density distribution is discontinuous, while the temperature one is continuous

much. However, the absolute values are of course different. In addition, the physical quantities in the relativistic precursor are gradually changed, compared with those in the nonrelativistic precursor.

When the gas internal energy is nonrelativistic, but the flow speed is relativistic (e.g., $\nu_1 \sim 0.00001$ and $\beta_1 \sim 0.5$), the precursor quantities become almost constant, and extend to the upstream region. Let us suppose the situation where we keep $\nu_1 \sim 0.00001$, but increase β_1 (or $\mathcal{M}_1$), the shock strength. In the radiative shocks, even in the gas-pressure dominated case, the radiation energy flux from the post-shock region penetrates into the upstream material and heats up to make the radiative precursor. When the temperature T_p just ahead of the shock front is smaller than the temperature T_2 in the post-shock region ($T_p < T_2$), such a shock is called *subcritical shock*. This pre-shock temperature T_p increases rapidly with the shock strength, and eventually equals the post-shock one T_2. For the higher shock strength, the pre-shock temperature cannot exceed the post-shock one, and the excess energy flux enlarges the radiative precursor into the upstream region (*supercritical shock*). Indeed, from Eq. (25.97), if we set γ and ρ on the order of unity, the extension $\tilde{x}$ of the radiative precursor is roughly estimated as $\tilde{x} \propto \tilde{T}^3$, which quickly enlarges with the temperature increase.

25.3.3 Radiation-Pressure Dominated Shocks

In this subsection we shall examine the *radiation-radiation* shocks, where the radiation pressure is dominant in both sides of the shock. In the nonrelativistic radiative shocks, the radiation-pressure dominated case is usually a *continuous shock* (cf. Sect. 23.3).

25.3.3.1 Overall Jump Conditions

In the radiation-pressure dominated case, we drop the gas-pressure terms, although retaining the rest mass energy, and the Rankine–Hugoniot relation (25.81) becomes

$$\left(1+\frac{4P_2}{\rho_2 c^2}\right)^2\left[1+\frac{1}{\rho_2^2 c^2}\frac{P_2-P_1}{\frac{1}{\rho_1}-\frac{1}{\rho_2}}\right] = \left(1+\frac{4P_1}{\rho_1 c^2}\right)^2\left[1+\frac{1}{\rho_1^2 c^2}\frac{P_2-P_1}{\frac{1}{\rho_1}-\frac{1}{\rho_2}}\right], \tag{25.100}$$

where we use the Eddington approximation: $E = 3P$.

Introducing the variables:

$$N \equiv \frac{P}{\rho c^2} \quad \text{and} \quad \tilde{\eta}_2 \equiv \frac{\rho_1}{\rho_2} = \frac{1}{\tilde{\rho}_2}, \tag{25.101}$$

after several manipulations, Eq. (25.100) can be rewritten as

$$\begin{aligned} &N_1(1+4N_2)^2\eta_2^3 \\ &+[(1+4N_2)^2-(1+4N_1)^2-N_2(1+4N_2)^2]\eta_2^2 \\ &-[(1+4N_2)^2-(1+4N_1)^2+N_1(1+4N_1)^2]\eta_2 \\ &+N_2(1+4N_1)^2=0. \end{aligned} \tag{25.102}$$

By simple analysis, we can show that this equation has one root between $0 < \eta_2 < 1$. By deriving the solution numerically for a given pre-shock quantity N_1, we obtain post-shock quantities such as, e.g., $P_2/P_1 = (N_2/N_1)\tilde{\eta}_2$, $T_2/T_1 = (P_2/P_1)^{1/4}$, $u_1^2 = (N_2/\tilde{\eta}_2 - N_1)/(1-\tilde{\eta}_2)$, $u_2^2 = (N_2 - N_1\tilde{\eta}_2)/(1/\tilde{\eta}_2 - 1)$, and so on.

In Fig. 25.5 several of the post-shock quantities are depicted as a function of $\tilde{P}_2$ in the radiation-pressure dominated case. Solid curves represent $\tilde{\rho}_2$, where $N_1 = 0.00001, 0.001, 0.1$ from bottom to top. Chain-dotted ones mean $\tilde{T}_2$, which do not depend on N_1, and is proportional to $\tilde{P}_2^{1/4}$. Finally, dashed ones denote the pre-

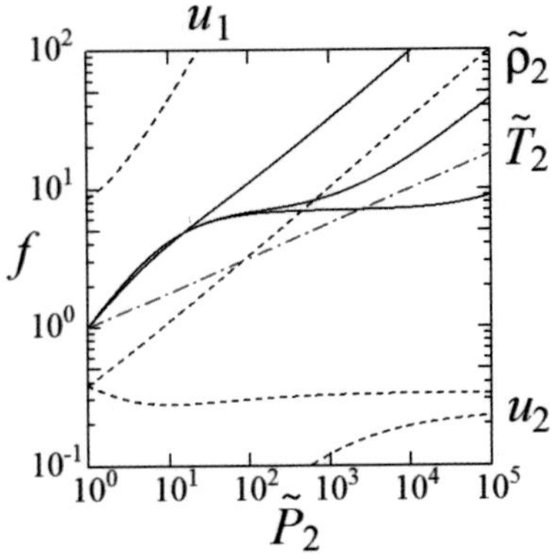

Fig. 25.5 Post-shock quantities as a function of P_2 in the radiation-pressure dominated case. Solid curves represent ρ_2, where $N_1 = 0.00001, 0.001, 0.1$ from bottom to top. Chain-dotted ones mean T_2, which do not depend on N_1, and is proportional to $P_2^{1/4}$. Finally, dashed ones denote the pre-shock u_1, where $N_1 = 0.001, 0.1$ from top to bottom, and the post-shock u_2, where $N_1 = 0.001, 0.1$ from bottom to top

shock u_1, where $N_1 = 0.001, 0.1$ from top to bottom, and the post-shock u_2, where $N_1 = 0.001, 0.1$ from bottom to top.

As is seen in Fig. 25.5, the post-shock gas density increases with $\tilde{P}_2$. When the pre-shock state is sufficiently nonrelativistic ($N_1 = 0.00001$), the post-shock gas density increases and approaches 7 as $\tilde{P}_2$, as is known in the nonrelativistic case. For sufficiently large $\tilde{P}_2$, however, the gas density becomes slightly larger than 7. When the pre-shock state is mildly relativistic ($N_1 = 0.001, 0.1$), the post-shock gas density no longer approaches 7, but monotonically increases with $\tilde{P}_2$. It should be noted that in the extremely relativistic regime $\tilde{P}_2 \sim \tilde{\rho}_2^2$.

In radiation-pressure dominated case, Eq. (25.84) becomes

$$\frac{P_2 - P_1}{1 - \tilde{\eta}_2} = \Gamma p_1 \mathcal{M}_1^2, \tag{25.103}$$

or for the Mach number,

$$\mathcal{M}_1^2 = \frac{\alpha_1}{\Gamma} \frac{P_2/P_1 - 1}{1 - \tilde{\eta}_2}, \tag{25.104}$$

where $\alpha_1 = P_1/p_1$. Using this relation, we can express the post-shock quantities by the pre-shock Mach number $\mathcal{M}_1$, instead of $\tilde{P}_2$.

In Fig. 25.6 several of the post-shock quantities are depicted as a function of $\mathcal{M}_1$ in the radiation-pressure dominated case. Solid curves represent $\tilde{\rho}_2$, where $\nu_1 = 0.00001, 0.001, 0.1$ from right to left. Chain-dotted ones mean $\tilde{T}_2$, while two-dot chain ones are $\tilde{P}_2$; both do not depend on N_2. Finally, dashed ones denote N_2, where

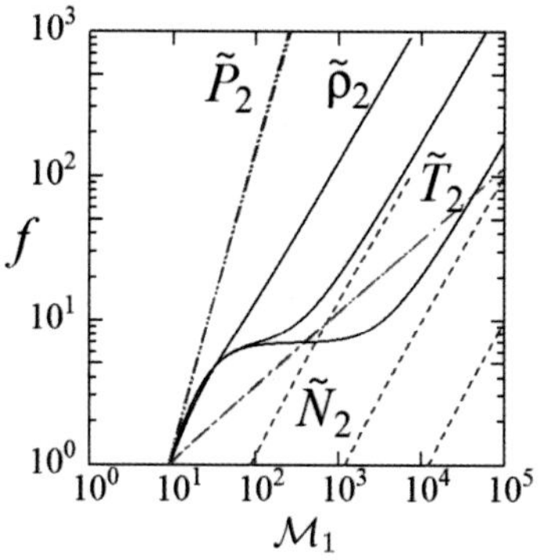

Fig. 25.6 Post-shock quantities as a function of $\mathcal{M}_1$ in the radiation-pressure dominated case. Solid curves represent $\tilde{\rho}_2$, where $\nu_1 = 0.00001, 0.001, 0.1$ from right to left. Chain-dotted ones mean $\tilde{T}_2$, while two-dot chain ones are $\tilde{P}_2$; both do not depend on N_2. Finally, dashed ones denote N_2, where $N_1 = 0.00001, 0.001, 0.1$ from right to left. The parameters are $\Gamma = 5/3$ and $\alpha_1 = 100$

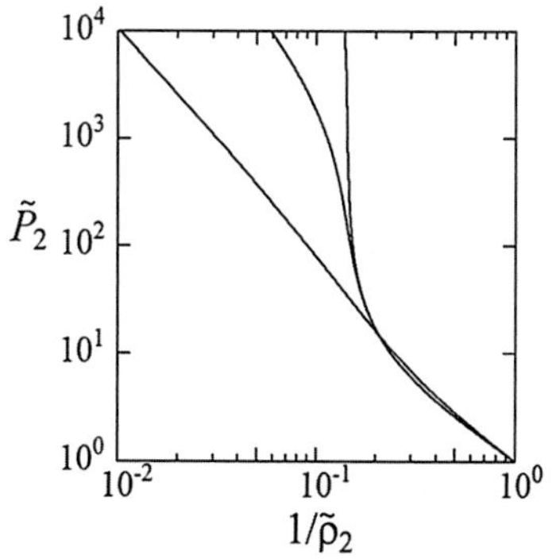

Fig. 25.7 Shock adiabats in the radiation-pressure dominated case. Solid curves represent the relations between $\tilde{P}_2$ and $1/\tilde{\rho}_2$, where $N_1 = 0.00001, 0.001, 0.1$ from right to left

$N_1 = 0.00001, 0.001, 0.1$ from right to left. The parameters are $\Gamma = 5/3$ and $\alpha_1 = 100$.

As is seen in Fig. 25.6, the gas density and other quantities generally increase with $\mathcal{M}_1$. When the pre-shock state is sufficiently nonrelativistic ($\nu_1 = 0.00001$), as $\mathcal{M}_1$ becomes large, the post-shock gas density increases and approaches 7 at once, and again increases. When the pre-shock state is mildly relativistic ($\nu_1 = 0.001, 0.1$), the post-shock gas density no longer approaches 7, but monotonically increases with $\mathcal{M}_1$.

In contrast to the gas-pressure dominated case, there is no limit on the value of the pre-shock Mach number $\mathcal{M}_1$.

In Fig. 25.7 the so-called shock adiabats (Rankine–Hugoniot curves) are plotted, although they are essentially the same as Fig. 25.5. Solid curves represent the relations between $\tilde{P}_2$ and $1/\tilde{\rho}_2$, where $N_1 = 0.00001, 0.001, 0.1$ from right to left.

25.3.3.2 Radiative Precursor Structure

In the radiation-pressure dominated case, the structure equations for the radiative precursor (25.86)–(25.88) can be expressed in terms of variables $\tilde{P}$ ($\equiv P/P_1$), $\tilde{\rho}$ ($\equiv \rho/\rho_1$), and γ without T as

$$\frac{\gamma^2}{\gamma_1^2}(1+\beta^2)\frac{1}{\tau_1\beta_1}\frac{1}{\tilde{\rho}}\frac{d\tilde{P}}{d\tilde{x}} = \frac{4\tilde{P}}{\tilde{\rho}}\frac{\gamma}{\gamma_1} - 4, \tag{25.105}$$

$$\frac{\gamma^2}{\gamma_1^2} = \frac{\gamma_1^2-1}{\gamma_1^2}\frac{1}{\tilde{\rho}^2} + \frac{1}{\gamma_1^2}. \tag{25.106}$$

$$\frac{1}{\tilde{\rho}} = 1 - \frac{\tilde{P}-1}{(\Gamma/\alpha_1)\gamma_1^2\mathcal{M}_1^2}, \tag{25.107}$$

where $\tau_1 = \kappa\rho_1\ell_1$, $\beta_1 = v_1/c$, and $\alpha_1 = P_1/p_1$ ($\gg 1$ now).

Using the overall jump relations as boundary conditions, we can solve these Eqs. (25.105) and (25.107) to obtain the radiative precursor structure. Since τ_1 and β_1 are renormalized into the coordinate scale, the independent parameters are Γ, α_1, N_1, and $\mathcal{M}_1$ (then a_1, u_1, γ_1 are automatically determined.).

Typical solutions for the radiative precursor region of the relativistic radiative shocks are shown in Fig. 25.8 as a function of the normalized coordinate. Thick solid curves represent $\tilde{\rho}$, thin chain-dotted ones $\tilde{T}$, thin two-dot chain ones $\tilde{P}$, and thin solid ones $\tilde{F}$. The parameters are $\Gamma = 5/3$ and $\alpha_1 = 100$ (a) $N_1 = 0.00001$ and $\mathcal{M}_1 = 24.8$ ($N_2 = 0.000022$, $\beta_1 = 0.01$), and (b) $N_1 = 0.1$ and $\mathcal{M}_1 = 52.85$ ($N_2 = 0.55$, $\beta_1 = 0.91$). That is, Fig. 25.8a corresponds to the nonrelativistic case, where both the gas initial state and flow speed are nonrelativistic ($N_1 = 0.00001$ and $\beta_1 = 0.01$), while Fig. 25.8b is the relativistic case, where both the gas initial state and flow speed are relativistic ($N_1 = 0.1$ and $\beta_1 = 0.91$).

Similar to the gas-pressure dominated case, the radiative energy from the hot post-shock region diffuses into the pre-shock region to make the radiative precursor. Furthermore, in Fig. 25.8a, similar to the nonrelativistic case, both the density and temperature distributions are continuous. However, in the relativistic case (Fig. 25.8b), the density distribution is discontinuous, although the temperature one is continuous. That is, in the relativistic case the density (25.107) in the radiative precursor just before the shock front is smaller than the density in the post-shock region determined by the jump conditions.

These solutions for relativistic radiative shocks are simple typical cases, and there are various developments and several aspects should be improved. For example, although we adopt the equilibrium diffusion approximation in the optically thick regime, in general we must solve the nonequilibrium diffusion shocks in order to obtain more accurate solutions. In addition, although we adopt the Eddington approximation to close the radiative moment equations, in order to treat the radiative shock more accurately, we should solve the radiative transfer equation numerically (cf. Tolstov et al. 2015). Furthermore, as in the hydrodynamical shocks, the radiative

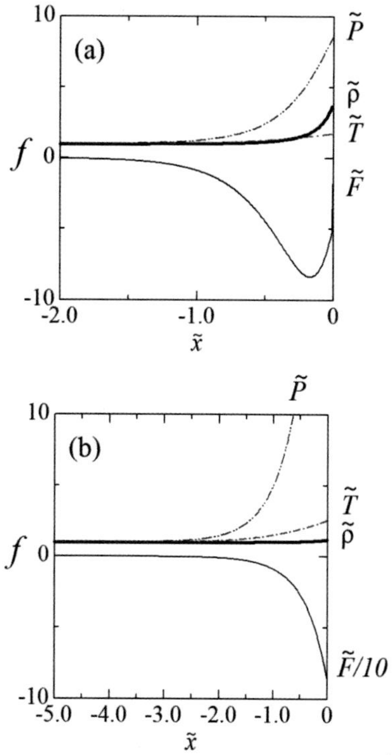

Fig. 25.8 Typical solutions for the radiative precursor region of the relativistic radiative shocks in the radiation-pressure dominated case as a function of the normalized coordinate. Thick solid curves represent $\tilde{\rho}$, thin chain-dotted ones $\tilde{T}$, thin two-dot chain ones $\tilde{P}$, and thin solid ones $\tilde{F}$. The parameters are $\Gamma = 5/3$ and $\alpha_1 = 100$ (**a**) $N_1 = 0.00001$ and $\mathcal{M}_1 = 24.8$ ($N_2 = 0.000022$, $\beta_1 = 0.01$), and (**b**) $N_1 = 0.1$ and $\mathcal{M}_1 = 52.85$ ($N_2 = 0.55$, $\beta_1 = 0.91$). The temperature distribution is continuous

shocks can be incorporated into the transonic accretion flows onto the central gravitating object (e.g., Fukue 1987). Relativistic radiative shocks in disk accretion flows under hydrostatic equilibrium in the vertical direction was recently studied (Fukue 2019c; see also Fukue 2019a). However, in contrast to the hydrodynamical shocks, which can be treated as a discontinuity with no width, the radiative shocks have generally a radiative precursor with a finite width. Hence, the radiative shocks in the transonic accretion flows onto the central object should be examined carefully, since, e.g., the gravitational field would change during shock transition (e.g., Fukue 2019d).

References

Belobolodov, A.M.: Astrophys. J. **838**, 125 (2017)
Budnik, R., Katz, B., Sagiv, A., Waxman, E.: Astrophys. J. **725**, 63 (2010)
Cissoko, M.: Phys. Rev. D. **55**, 4555 (1997)
Colgate, S.A., White, R.H.: Astrophys. J. **143**, 626 (1966)
Farris, B.D., Li, T.K., Liu, Y.T., Shapiro, S.L.: Phys. Rev. D. **78**, 024023 (2008)
Fragile, P.C., Gillespie, A., Monahan, T., Rodriguez, M., Anninos, P.: Astrophys. J.S. **201**, 9 (2012)
Fukue, J.: Publ. Astron. Soc. Jpn. **39**, 309 (1987)

Fukue, J.: Publ. Astron. Soc. Jpn. **66**, 13 (2014a)
Fukue, J.: Publ. Astron. Soc. Jpn. **71**, 38 (2019a)
Fukue, J.: Monthly Not. Roy. Astron. Soc. **483**, 2538 (2019b)
Fukue, J.: Monthly Not. Roy. Astron. Soc. **483**, 3839 (2019c)
Fukue, J.: Publ. Astron. Soc. Jpn. **71**, 99 (2019d)
Fukue, J., Kato, S., Matsumoto, R.: Publ. Astron. Soc. Jpn. **37**, 383 (1985)
Gao, Y., Law, C.K.: Astrophys. J. **760**, 122 (2012)
Hsieh, S.-H., Spiegel, E.A.: Astrophys. J. **207**, 244 (1976)
Icke, V.: Astron. Astrophys. **216**, 294 (1989)
Johnson, M.H., McKee, C.F.: Phys. Rev. D. **3**, 858 (1971)
Katz, B., Budnik, R., Waxman, E.: Astrophys. J. **716**, 781 (2010)
Landau, L.D., Lifshitz, E.M.: Fluid Mechanics. Pergamon Press, London (1959)
Lowrie, R.B., Rauenzahn, R.M.: Shock Waves **16**, 445 (2007)
Masuda, T., Fukue, J.: Publ. Astron. Soc. Jpn. **68**, 42 (2016)
Mészáros, P.: ARAstron. Astrophys. **40**, 137 (2002)
Mihalas, D.: Astrophys. J. **237**, 574 (1980)
Mihalas, D., Auer, L.H.: JQSTR **71**, 61 (2001)
Mihalas, D., Mihalas, B.W.: Foundations of Radiation Hydrodynamics. Oxford University Press, Oxford (1984)
Park, M.-G.: Monthly Not. Roy. Astron. Soc. **367**, 1739 (2006)
Phinney, E.S.: Superluminal Radio Sources. In: Zensuz, J.A., Pearson, T.J., p. 301. Cambridge University Press, Cambridge (1987)
Sądowski, A., Narayan, R., Tchekhovskoy, A., Zhu, Y.: Monthly Not. Roy. Astron. Soc. **429**, 3533 (2013)
Tajima, Y., Fukue, J.: Publ. Astron. Soc. Jpn. **50**, 483 (1998)
Takahashi, R.: Monthly Not. Roy. Astron. Soc. **382**, 1041 (2007)
Takahashi, H.R., Ohsuga, K., Sekiguchi, Y., Inoue, T., Tomida, K.: Astrophys. J. **764**, 122 (2013)
Taub, A.: Phys. Rev. **74**, 328 (1948)
Thorne, K.S.: Astrophys. J. **179**, 897 (1973)
Tolstov, A., Blinnikov, S., Nagataki, S., Nomoto, K.: Astrophys. J. **811**, 47 (2015)
Umemura, M., Loeb, A., Turner, E.L.: Astrophys. J. **419**, 459 (1993)
Umemura, M., Fukue, J., Mineshige, S.: Astrophys. J. **479**, L97 (1997)
Zanotti, O., Roedig, C., Rezzolla, L., Del Zanna, L.: Monthly Not. Roy. Astron. Soc. **417**, 2899 (2011)

Chapter 26
General Relativistic Radiation Hydrodynamics

In the previous two chapters, we consider radiative fluids within the framework of special relativity. At the last, in this chapter, we summarize the basic equations for radiation hydrodynamics under general relativity. In addition, we briefly review the general relativistic radiation hydrodynamical flows.

26.1 General Relativistic Radiation Hydrodynamics

The basic equations for radiative transfer and radiation hydrodynamics under general relativity have been derived by several researchers (e.g., Lindquist 1966; Anderson and Spiegel 1972; Schmid-Burgk 1978; Thorne 1981; Udey and Israel 1982; Turolla and Nobili 1988; Nobili et al. 1993; Park 2003, 2006; Takahashi 2007). In particular, all components of relativistic moment equations were written down in the Schwarzschild metric (Park 2006) and in the Kerr case (Takahashi 2007).

In this section we briefly summarize the basic equations for general relativistic radiation hydrodynamics, bearing in mind the spherically symmetric Schwarzschild space-time.[1]

[1] As already stated in the beginning of Chap. 24, in this book the $(+, -, -, -)$ signature is adopted, and the Greek suffixes $\alpha, \beta, \gamma, \cdots$ take values of 0, 1, 2, and 3, while the Latin suffixes $i, j, k, \cdots$ take values of 1, 2, and 3. Finally, the quantities measured in the *comoving*/fluid frame are labeled by the suffix 0 if necessary, while the quantities in the fixed/observer/*inertial* frame are expressed without suffix.

© Springer Nature Singapore Pte Ltd. 2020, corrected publication 2023

S. Kato, J. Fukue, *Fundamentals of Astrophysical Fluid Dynamics*,
Astronomy and Astrophysics Library,
https://doi.org/10.1007/978-981-15-4174-2_26

26.1.1 Radiation Field Under General Relativity

In this section, we first summarize the relativistic quantities relating to the general relativistic radiation hydrodynamics.

(a) Metric tensor

In the case of the spherically symmetric Schwarzschild space-time, the space-time metric is given by

$$
\begin{aligned}
ds^2 &= c^2 d\tau^2 = g_{\mu\nu} dx^\mu dx^\nu \\
&= \left(1 - \frac{r_S}{r}\right) c^2 dt^2 - \left(1 - \frac{r_S}{r}\right)^{-1} dr^2 - r^2 (d\theta^2 + \sin^2\theta d\varphi^2),
\end{aligned}
\tag{26.1}
$$

where c is the speed of light, τ the proper time, x^μ the space-time coordinates ($x^0 = ct$), and $g_{\mu\nu}$ the space-time metric.

The nonzero components of the space-time metric are

$$g_{00} = (g^{00})^{-1} = 1 - \frac{r_S}{r}, \tag{26.2}$$

$$g_{11} = (g^{11})^{-1} = -\left(1 - \frac{r_S}{r}\right)^{-1}, \tag{26.3}$$

$$g_{22} = (g^{22})^{-1} = -r^2, \tag{26.4}$$

$$g_{33} = (g^{33})^{-1} = -r^2 \sin^2\theta, \tag{26.5}$$

and

$$\sqrt{-g} = \sqrt{-|g_{\rm ij}|} = r^2 \sin\theta. \tag{26.6}$$

In addition, the nonzero components of the Riemann–Christoffel tensor $\Gamma^\sigma_{\mu\nu}$ are

$$\Gamma^0_{10} = \Gamma^0_{01} = \frac{r_S}{2r^2}\left(1 - \frac{r_S}{r}\right)^{-1}, \tag{26.7}$$

$$\Gamma^1_{00} = \frac{r_S}{2r^2}\left(1 - \frac{r_S}{r}\right), \tag{26.8}$$

$$\Gamma^1_{11} = -\frac{r_S}{2r^2}\left(1 - \frac{r_S}{r}\right)^{-1}, \tag{26.9}$$

$$\Gamma^1_{22} = -r\left(1 - \frac{r_S}{r}\right), \tag{26.10}$$

$$\Gamma^1_{33} = -r\sin^2\theta\left(1 - \frac{r_S}{r}\right), \tag{26.11}$$

$$\Gamma^2_{12} = \Gamma^2_{21} = \frac{1}{r}, \tag{26.12}$$

$$\Gamma^2_{33} = -\sin\theta\cos\theta, \tag{26.13}$$

$$\Gamma^3_{13} = \Gamma^3_{31} = \frac{1}{r}, \tag{26.14}$$

$$\Gamma^3_{23} = \Gamma^3_{32} = \cot\theta. \tag{26.15}$$

(b) Tetrad or natural frame

A fixed tetrad (natural frame), which means a four-part structure in Greek, is a set of four four-vectors attached to each point x^μ of space-time, and consists of a set of axes. We usually choose an orthogonal basis. A tetrad (contravariant) four-vector, which is numbered by subscript tetrad indices, is often expressed as

$$e^\mu_{\hat{\alpha}}, \quad e^\mu_{(a)}, \quad \boldsymbol{e}_m, \tag{26.16}$$

where Latin indices μ label coordinate frames (tensor indices), while hats $\hat{\alpha}$ or Latin indices (a) label tetrad frames (tetrad indices). Furthermore, in the case of an orthonormal tetrad $e^\mu_{\hat{\alpha}}$, the axes form a locally inertial frame at each point, so that the scalar products of the tetrad vector constitute the Minkowski metric $\eta_{\alpha\beta}$:

$$e^\mu_{\hat{\alpha}} e_{\hat{\beta}\mu} = \eta_{\alpha\beta}, \quad \boldsymbol{e}_{\hat{\alpha}} \cdot \boldsymbol{e}_{\hat{\beta}} = \eta_{\alpha\beta}. \tag{26.17}$$

A dual vector corresponding to a tetrad vector, which is numbered by superscript tetrad indices, such as

$$e^{\hat{\alpha}\mu}, \quad e^{(a)\mu}, \quad \boldsymbol{e}^m, \tag{26.18}$$

is defined by the orthogonal condition:

$$e^{\hat{\alpha}}_\mu e^\mu_{\hat{\beta}} = \delta^\alpha_\beta, \tag{26.19}$$

$$e^{\hat{\alpha}}_\mu e^\nu_{\hat{\alpha}} = \delta^\nu_\mu. \tag{26.20}$$

The subscript raised and lowered are done by the Minkowski metric as

$$e^{\hat{\beta}}_\mu = \eta^{\beta\gamma} e_{\hat{\gamma}\mu}, \quad e_{\hat{\beta}\mu} = \eta_{\beta\gamma} e^{\hat{\gamma}}_\mu. \tag{26.21}$$

One of the meanings to introduce such a tetrad is that we can constitute the metric tensor from the tetrad. That is, using these tetrad vectors, the metric tensor can be expressed as

$$g_{\mu\nu} = e^{\hat{\alpha}}_\mu e_{\hat{\alpha}\nu}, \tag{26.22}$$

$$ds^2 = g_{\mu\nu} dx^\mu dx^\nu = \eta_{\alpha\beta} e^{\hat{\alpha}}_\mu dx^\mu e^{\hat{\beta}}_\nu dx^\nu, \tag{26.23}$$

and therefore,

$$g_{\mu\nu} = \eta_{\alpha\beta} e^{\hat{\alpha}}_{\mu} e^{\hat{\beta}}_{\nu}, \tag{26.24}$$

$$\eta_{\alpha\beta} = e^{\mu}_{\hat{\alpha}} e^{\nu}_{\hat{\beta}} g_{\mu\nu}. \tag{26.25}$$

Furthermore, if we choose the orthonormal tetrad $e^{\mu}_{\hat{\alpha}}$, the axes form a locally inertial frame, where the physics is more transparent, as opposed to the coordinate frame. For example, we can use the transformation in the flat Minkowski space-time on the tetrad frame.

In the case of the Schwarzschild metric, the tetrad vector is explicitly written as

$$e_{\hat{t}\mu} = \left(\sqrt{g_{00}}, 0, 0, 0\right), \tag{26.26}$$

$$e_{\hat{r}\mu} = \left(0, -1/\sqrt{g_{00}}, 0, 0\right), \tag{26.27}$$

$$e_{\hat{\theta}\mu} = (0, 0, -r, 0), \tag{26.28}$$

$$e_{\hat{\varphi}\mu} = (0, 0, 0, -r\sin\theta), \tag{26.29}$$

and

$$e^{\mu}_{\hat{t}} = \left(1/\sqrt{g_{00}}, 0, 0, 0\right), \tag{26.30}$$

$$e^{\mu}_{\hat{r}} = \left(0, \sqrt{g_{00}}, 0, 0\right), \tag{26.31}$$

$$e^{\mu}_{\hat{\theta}} = (0, 0, 1/r, 0), \tag{26.32}$$

$$e^{\mu}_{\hat{\varphi}} = (0, 0, 0, 1/r\sin\theta), \tag{26.33}$$

where $\sqrt{g_{00}} = \sqrt{1 - r_S/r}$.

The dual vector, on the other hand, is also explicitly expressed as

$$e^{\hat{t}}_{\mu} = \left(\sqrt{g_{00}}, 0, 0, 0\right), \tag{26.34}$$

$$e^{\hat{r}}_{\mu} = \left(0, 1/\sqrt{g_{00}}, 0, 0\right), \tag{26.35}$$

$$e^{\hat{\theta}}_{\mu} = (0, 0, r, 0), \tag{26.36}$$

$$e^{\hat{\varphi}}_{\mu} = (0, 0, 0, r\sin\theta), \tag{26.37}$$

and

$$e^{\hat{t}\mu} = \left(1/\sqrt{g_{00}}, 0, 0, 0\right), \tag{26.38}$$

$$e^{\hat{r}\mu} = \left(0, -\sqrt{g_{00}}, 0, 0\right), \tag{26.39}$$

$$e^{\hat{\theta}\mu} = (0, 0, -1/r, 0), \tag{26.40}$$

$$e^{\hat{\varphi}\mu} = (0, 0, 0, -1/r\sin\theta). \tag{26.41}$$

In addition, the transformation between the tetrad and coordinate frames becomes

$$\begin{pmatrix} d\hat{t} \\ d\hat{r} \\ d\hat{\theta} \\ d\hat{\varphi} \end{pmatrix} = \begin{pmatrix} \sqrt{g_{00}} & 0 & 0 & 0 \\ 0 & 1/\sqrt{g_{00}} & 0 & 0 \\ 0 & 0 & r & 0 \\ 0 & 0 & 0 & r\sin\theta \end{pmatrix} \begin{pmatrix} dt \\ dr \\ d\theta \\ d\varphi \end{pmatrix}, \tag{26.42}$$

and

$$\begin{pmatrix} \partial/\partial\hat{t} \\ \partial/\partial\hat{r} \\ \partial/\partial\hat{\theta} \\ \partial/\partial\hat{\varphi} \end{pmatrix} = \begin{pmatrix} 1/\sqrt{g_{00}} & 0 & 0 & 0 \\ 0 & \sqrt{g_{00}} & 0 & 0 \\ 0 & 0 & 1/r & 0 \\ 0 & 0 & 0 & 1/r\sin\theta \end{pmatrix} \begin{pmatrix} \partial/\partial t \\ \partial/\partial r \\ \partial/\partial\theta \\ \partial/\partial\varphi \end{pmatrix}. \tag{26.43}$$

(c) Four-velocity of matter

In the curved space, the four-velocity u^μ of matter and its covariant components u_μ are, respectively, defined by

$$u^\mu \equiv \frac{dx^\mu}{ds} = \left(\frac{\gamma}{\sqrt{g_{00}}}, \gamma\frac{v^i}{c}\right), \tag{26.44}$$

$$u_\mu = g_{\mu\nu}u^\nu = \left(\gamma\sqrt{g_{00}}, -\gamma\frac{v_i}{c}\right), \tag{26.45}$$

where γ is the Lorentz factor,

$$\gamma = \left(1 - \frac{v^2}{c^2}\right)^{-1/2}, \tag{26.46}$$

$$v^2 = v_i v^i = \gamma_{ik}v^i v^k = -g_{ik}v^i v^k, \tag{26.47}$$

and $u_\mu u^\mu = g_{\mu\nu}u^\mu u^\nu = 1$.

In the spherically symmetric space-time, the contravariant and covariant components of the four-velocity are, respectively, expressed as

$$u^0 \equiv \frac{dt}{d\tau} = \frac{\gamma}{\sqrt{g_{00}}}, \tag{26.48}$$

$$u^1 \equiv \frac{1}{c}\frac{dr}{d\tau} = u^r = \frac{1}{c}\gamma v^r, \tag{26.49}$$

$$u^2 \equiv \frac{1}{c}\frac{d\theta}{d\tau} = \frac{u^\theta}{r} = \frac{1}{c}\frac{\gamma v^\theta}{r}, \tag{26.50}$$

$$u^3 \equiv \frac{1}{c}\frac{d\varphi}{d\tau} = \frac{u^\varphi}{r\sin\theta} = \frac{1}{c}\frac{\gamma v^\varphi}{r\sin\theta}, \tag{26.51}$$

$$u_0 = g_{00}u^0 = \gamma\sqrt{g_{00}} \equiv y \text{ (energy parameter)}, \tag{26.52}$$

$$u_1 = g_{11}u^1 = -(g_{00})^{-1}u^1, \tag{26.53}$$

$$u_2 = g_{22}u^2 = -r^2u^2, \tag{26.54}$$

$$u_3 = g_{33}u^3 = -r^2 \sin^2\theta\, u^3, \tag{26.55}$$

where v^i is the three velocity.

The energy parameter y and the Lorentz factor γ are further expressed as

$$\begin{aligned} y = u_0 &= g_{00}u^0 \\ &= \left[g_{00} + (u^1)^2 + g_{00}(ru^2)^2 + g_{00}(r\sin\theta u^3)^2\right]^{1/2} \\ &= \left[g_{00} + (u^r)^2 + g_{00}(u^\theta)^2 + g_{00}(u^\varphi)^2\right]^{1/2} = \gamma\sqrt{g_{00}}, \end{aligned} \tag{26.56}$$

$$\begin{aligned} \gamma &= \left[1 + \frac{1}{g_{00}}(u^r)^2 + (u^\theta)^2 + (u^\varphi)^2\right]^{1/2} \\ &= \left[1 - \frac{1}{g_{00}}\left(\frac{v^r}{c}\right)^2 - \left(\frac{v^\theta}{c}\right)^2 - \left(\frac{v^\varphi}{c}\right)^2\right]^{-1/2}. \end{aligned} \tag{26.57}$$

In addition, a three vector $v_{\hat{i}}$ in the fixed tetrad is related to the four-velocity in the fixed coordinate frame as

$$v_{\hat{r}} = v^{\hat{r}} = \frac{1}{\gamma\sqrt{g_{00}}}cu^r, \tag{26.58}$$

$$v_{\hat{\theta}} = v^{\hat{\theta}} = \frac{1}{\gamma}cu^\theta, \tag{26.59}$$

$$v_{\hat{\varphi}} = v^{\hat{\varphi}} = \frac{1}{\gamma}cu^\varphi. \tag{26.60}$$

(d) Quantities of radiation field

The expressions of radiation field in each tetrad are the same as those under special relativity (Sect. 23.1); e.g.,

$$E_0 \equiv \frac{1}{c}\int I_{\nu 0}d\nu_0 d\Omega_0,\ F_0^i \equiv \int I_{\nu 0}l_0^i d\nu_0 d\Omega_0,\ P_0^{ij} \equiv \frac{1}{c}\int I_{\nu 0}l_0^i l_0^j d\nu_0 d\Omega_0, \tag{26.61}$$

$$E \equiv \frac{1}{c}\int I_\nu d\nu d\Omega,\ F^i \equiv \int I_\nu l^i d\nu d\Omega,\ P^{ij} \equiv \frac{1}{c}\int I_\nu l^i l^j d\nu d\Omega \tag{26.62}$$

in the comoving and fixed frames, respectively.

Similarly, the transformation rules are the same as those under special relativity (Sect. 23.1), although the velocity is the tetrad component. In other words, if the transformation is done among tetrad quantities, it is valid in the curved space-time as well, as long as the proper velocity in the tetrad is used.

(e) Energy-momentum tensor

The energy-momentum tensor for an ideal gas is

$$T^{\mu\nu} = (\varepsilon + p)\, u^\mu u^\nu - p g^{\mu\nu}, \tag{26.63}$$

where ε is the gas internal energy per unit proper volume[2] and p the gas pressure measured in the comoving frame ($\varepsilon + p$ is the gas enthalpy per unit proper volume).

The energy-momentum tensor for radiation is as follows. The tetrad components for the comoving and fixed tetrads are the same as those under special relativity (Sect. 24.1):

$$R_0^{\hat{\mu}\hat{\nu}} = \begin{pmatrix} E_0 & \frac{1}{c}F_0^i \\ \frac{1}{c}F_0^i & P_0^{ij} \end{pmatrix}, \quad R^{\hat{\mu}\hat{\nu}} = \begin{pmatrix} E & \frac{1}{c}F^i \\ \frac{1}{c}F^i & P^{ij} \end{pmatrix}, \tag{26.64}$$

where E is the radiation energy density, F^i the radiative flux, and P^{ij} the radiation stress tensor.

On the other hand, the contravariant components of the energy-momentum tensor of radiation field in the spherically symmetric Schwarzschild space-time are

$$R^{\mu\nu} = \begin{pmatrix} \frac{1}{g_{00}}E & \frac{F^r}{c} & \frac{1}{\sqrt{g_{00}}}\frac{F^\theta}{cr} & \frac{1}{\sqrt{g_{00}}}\frac{F^\varphi}{cr\sin\theta} \\ \frac{F^r}{c} & g_{00}P^{rr} & \sqrt{g_{00}}\frac{P^{r\theta}}{r} & \sqrt{g_{00}}\frac{P^{r\varphi}}{r\sin\theta} \\ \frac{1}{\sqrt{g_{00}}}\frac{F^\theta}{cr} & \sqrt{g_{00}}\frac{P^{r\theta}}{r} & \frac{P^{\theta\theta}}{r^2} & \frac{P^{\theta\varphi}}{r^2\sin\theta} \\ \frac{1}{\sqrt{g_{00}}}\frac{F^\varphi}{cr\sin\theta} & \sqrt{g_{00}}\frac{P^{r\varphi}}{r\sin\theta} & \frac{P^{\theta\varphi}}{r^2\sin\theta} & \frac{P^{\varphi\varphi}}{r^2\sin^2\theta} \end{pmatrix}. \tag{26.65}$$

The mass, momentum, and energy conservations are expressed, respectively, as

$$\left(nu^\mu\right)_{;\nu} = 0, \tag{26.66}$$

$$\left(T_\mu^{\ \nu} + R_\mu^{\ \nu}\right)_{;\nu} = 0, \tag{26.67}$$

[2]Here, the internal energy includes the rest mass energy, and in the case of the ideal gas it is expressed as

$$\varepsilon = \rho c^2 + \frac{1}{\Gamma - 1}p,$$

where Γ is the ratio of specific heats.

$$u^{\mu}\left(T_{\mu}^{\ \nu}+R_{\mu}^{\ \nu}\right)_{;\nu}=0. \tag{26.68}$$

Finally, the radiation four-force density is

$$G^{\mu}=-R^{\mu\nu}_{\ \ ;\nu}. \tag{26.69}$$

The tetrad expressions of the radiation four-force density are also the same as those under special relativity. That is, the tetrad components in the comoving frame are

$$G_0^{\hat{t}}=\frac{1}{c}\left(-\rho j_0+\rho\kappa_0 c E_0\right), \tag{26.70}$$

$$G_0^{\hat{i}}=\frac{1}{c}\rho\left(\kappa_0+\sigma_0\right)F_0^i, \tag{26.71}$$

where ρj_0 is a usual cooling function per unit proper volume, $\rho\kappa_0 c E_0$ the usual heating function per unit proper volume, and opacities are energy or flux mean opacities.[3]

On the other hand, the contravariant components of radiation four-force density in the spherically symmetric Schwarzschild space-time are

$$\begin{aligned} G^t &= \frac{\gamma}{\sqrt{g_{00}}}\left(G_0^{\hat{t}}+\frac{v_{\hat{i}}}{c}G_0^{\hat{i}}\right)\\ &= \frac{y}{c g_{00}}\left[-\rho j_0+\rho\kappa_0 c E_0+\rho(\kappa_0+\sigma_0)\frac{v_{\hat{i}}}{c}F_0^i\right], \end{aligned} \tag{26.72}$$

$$\begin{aligned} G^r &= \sqrt{g_{00}}\left(G_0^{\hat{r}}+\gamma\frac{v_{\hat{r}}}{c}G_0^{\hat{t}}+\frac{\gamma-1}{v^2}v_{\hat{r}}v_{\hat{i}}G_0^{\hat{i}}\right)\\ &= \sqrt{g_{00}}\frac{\rho(\kappa_0+\sigma_0)}{c}F_0^r+\sqrt{g_{00}}\frac{\gamma v_{\hat{r}}}{c^2}\left(-\rho j_0+\rho\kappa_0 c E_0\right)\\ &\quad+\sqrt{g_{00}}\frac{\gamma-1}{v^2}v_{\hat{r}}v_{\hat{i}}\frac{\rho(\kappa_0+\sigma_0)}{c}F_0^i, \end{aligned} \tag{26.73}$$

$$\begin{aligned} G^\theta &= \frac{1}{r}\left(G_0^{\hat{\theta}}+\gamma\frac{v_{\hat{\theta}}}{c}G_0^{\hat{t}}+\frac{\gamma-1}{v^2}v_{\hat{\theta}}v_{\hat{i}}G_0^{\hat{i}}\right)\\ &= \frac{1}{r}\left[\frac{\rho(\kappa_0+\sigma_0)}{c}F_0^\theta+\frac{\gamma v_{\hat{\theta}}}{c^2}\left(-\rho j_0+\rho\kappa_0 c E_0\right)\right.\\ &\quad\left.+\frac{\gamma-1}{v^2}v_{\hat{\theta}}v_{\hat{i}}\frac{\rho(\kappa_0+\sigma_0)}{c}F_0^i\right], \end{aligned} \tag{26.74}$$

[3]Rigorously speaking, κ_0 attached on E_0 is energy mean, while $(\kappa_0+\sigma_0)$ attached on F_0^i are flux mean, as was defined in Sect. 20.3.1 for the nonrelativistic case. We here drop the suffix E or F for simplicity, but retain the suffix 0 denoting the comoving quantities.

$$G^{\varphi} = \frac{1}{r\sin\theta}\left(G_0^{\hat{\varphi}} + \gamma\frac{v_{\hat{\varphi}}}{c}G_0^{\hat{t}} + \frac{\gamma-1}{v^2}v_{\hat{\varphi}}v_{\hat{i}}G_0^{\hat{i}}\right)$$
$$= \frac{1}{r\sin\theta}\left[\frac{\rho(\kappa_0+\sigma_0)}{c}F_0^{\varphi} + \frac{\gamma v_{\hat{\varphi}}}{c^2}(-\rho j_0 + \rho\kappa_0 c E_0)\right.$$
$$\left. + \frac{\gamma-1}{v^2}v_{\hat{\varphi}}v_{\hat{i}}\frac{\rho(\kappa_0+\sigma_0)}{c}F_0^i\right], \tag{26.75}$$

where $v_{\hat{i}}$ is a three vector in the fixed tetrad, as already defined.

26.1.2 *Radiation Hydrodynamical Equations Under General Relativity*

We here write down the basic equations for radiation hydrodynamics under general relativity (e.g., Park 2006 for the Schwarzschild metric; Takahashi 2007 for the Kerr metric). We bear in mind the spherically symmetric Schwarzschild space-time (see Park 2006).

(a) Continuity equation

The particle number conservation (26.66) is

$$(nu^{\mu})_{;\mu} = \frac{1}{\sqrt{-g}}\frac{\partial}{\partial x^{\mu}}\left(\sqrt{-g}nu^{\mu}\right) = 0, \tag{26.76}$$

where x^{μ} is the space-time coordinates, u^{μ} the four-velocity, and n the proper number density.

Using the energy parameter y ($\equiv u_0 = g_{00}u^0$), the continuity equation in the spherical coordinate becomes

$$\frac{1}{g_{00}}\frac{\partial}{\partial t}(y\rho) + \frac{1}{r^2}\frac{\partial}{\partial r}(r^2\rho c u^r) + \frac{1}{r\sin\theta}\frac{\partial}{\partial\theta}(\sin\theta\rho c u^{\theta}) + \frac{1}{r\sin\theta}\frac{\partial}{\partial\varphi}(\rho c u^{\varphi}) = 0, \tag{26.77}$$

where ρ ($= nmc^2$) is the proper density.

(b) Hydrodynamic equations

The relativistic equations of motion (26.67) are written as[4]

$$(\varepsilon + p)\left(u^{\mu}\frac{\partial u^i}{\partial x^{\mu}} + \Gamma^i_{\mu\nu}u^{\mu}u^{\nu}\right) - \left(g^{i\mu} - u^i u^{\mu}\right)\frac{\partial p}{\partial x^{\mu}}$$
$$= -(g^{i\mu} - u^i u^{\mu})R_{\mu\ ;\nu}^{\ \nu} = G^i - u^i u_{\mu}G^{\mu}, \tag{26.78}$$

[4]The contravariant form of Eq. (26.67) is $(T^{i\mu} + R^{i\mu})_{;\mu} = 0$. We here use the relation: $(T^{i\mu} + R^{i\mu})_{;\mu} - u^i u_{\mu}(T^{\mu\nu} + R^{\mu\nu})_{;\nu} = 0$.

where ε is the internal energy per unit proper volume, p the pressure measured in the comoving frame, $R^{\mu\nu}$ the stress-energy tensor of radiation, and G^{μ} the radiation four-force density.

The components of equation of motion (26.78) in the spherically symmetric Schwarzschild space-time are explicitly written as follows.

The radial (r) component is

$$
\begin{aligned}
&(\varepsilon+p)\left\{u^0\frac{\partial u^r}{c\partial t}+u^i\frac{\partial u^r}{\partial x^i}+\frac{r_S}{2r^2}\left[g_{00}(u^0)^2-\frac{1}{g_{00}}(u^r)^2\right]-\frac{g_{00}}{r}\left[(u^\theta)^2+(u^\varphi)^2\right]\right\}\\
&+u^r u^0\frac{\partial p}{c\partial t}+g_{00}\frac{\partial p}{\partial r}+u^r u^i\frac{\partial p}{\partial x^i}\\
&=-yu^r G^t+\left[1+\frac{1}{g_{00}}(u^r)^2\right]G^r+u^r u^\theta r G^\theta+u^r u^\varphi r\sin\theta\, G^\varphi.
\end{aligned}
\tag{26.79}
$$

The third term in the curly braces on the left-hand side means the gravitational acceleration, and the fourth term represents the centrifugal acceleration. The right-hand side is the radiative force.

The tangential (θ) component is

$$
\begin{aligned}
&(\varepsilon+p)\left[u^0\frac{\partial u^\theta}{c\partial t}+ru^i\frac{\partial}{\partial x^i}\left(\frac{u^\theta}{r}\right)+\frac{2}{r}u^r u^\theta-\frac{\cot\theta}{r}(u^\varphi)^2\right]\\
&+u^\theta u^0\frac{\partial p}{c\partial t}+\frac{1}{r}\frac{\partial p}{\partial\theta}+u^\theta u^i\frac{\partial p}{\partial x^i}\\
&=-yu^\theta G^t+\frac{1}{g_{00}}u^r u^\theta G^r+\left[1+(u^\theta)^2\right]rG^\theta+u^\theta u^\varphi r\sin\theta\, G^\varphi.
\end{aligned}
\tag{26.80}
$$

On the left-hand side, there is no gravitational acceleration, but there exists the centrifugal acceleration, as in the nonrelativistic case. The right-hand side is the radiative force.

Finally, the azimuthal (φ) component is

$$
\begin{aligned}
&(\varepsilon+p)\left[u^0\frac{\partial u^\varphi}{c\partial t}+r\sin\theta u^i\frac{\partial}{\partial x^i}\left(\frac{u^\varphi}{r\sin\theta}\right)+\frac{2}{r}u^r u^\varphi+\frac{2\cot\theta}{r}u^\theta u^\varphi\right]\\
&+u^\varphi u^0\frac{\partial p}{c\partial t}+\frac{1}{r\sin\theta}\frac{\partial p}{\partial\varphi}+u^\varphi u^i\frac{\partial p}{\partial x^i}\\
&=-yu^\varphi G^t+\frac{1}{g_{00}}u^r u^\varphi G^r+u^\theta u^\varphi rG^\theta+\left[1+(u^\varphi)^2\right]r\sin\theta\, G^\varphi.
\end{aligned}
\tag{26.81}
$$

On the left-hand side, there is no gravitational acceleration, but there exists the centrifugal acceleration, as in the nonrelativistic case. The right-hand side is the radiative force.

(c) Energy equation

The relativistic energy equation (26.68), $u^\mu(T_\mu^{\ \nu} + R_\mu^{\ \nu})_{;\nu} = 0$, is written as

$$\frac{1}{\sqrt{-g}}\frac{\partial}{\partial x^\mu}\left(\sqrt{-g}\varepsilon u^\mu\right) + \frac{p}{\sqrt{-g}}\frac{\partial}{\partial x^\mu}\left(\sqrt{-g}u^\mu\right) = -u^\mu R_{\mu\ ;\nu}^{\ \nu}. \tag{26.82}$$

The left-hand side of energy equation (26.82) is, using continuity equation (26.76), also written as

$$u^\mu T_{\mu\ ;\nu}^{\ \nu} = \frac{1}{\sqrt{-g}}\frac{\partial}{\partial x^\mu}\left[\sqrt{-g}(\varepsilon - \rho c^2)u^\mu\right] + \frac{p}{\sqrt{-g}}\frac{\partial}{\partial x^\mu}\left(\sqrt{-g}u^\mu\right), \tag{26.83}$$

whereas the right-hand side of Eq. (26.82) is expressed as

$$\begin{aligned}
-u^\mu R_{\mu\ ;\nu}^{\ \nu} &= -u_\mu R^{\mu\nu}_{\ \ ;\nu} = u_\mu G^\mu \\
&= yG^t - \frac{1}{g_{00}}u^r G^r - u^\theta r G^\theta - u^\varphi r \sin\theta G^\varphi, \\
&= G_0^{\hat{t}} = \frac{1}{c}\left(-\rho j_0 + \rho\kappa_0 c E_0\right).
\end{aligned} \tag{26.84}$$

Thus, the energy equation is finally expressed as

$$\begin{aligned}
&\frac{c}{\sqrt{-g}}\frac{\partial}{\partial x^\mu}\left[\sqrt{-g}\left(\varepsilon - \rho c^2\right)u^\mu\right] + c\frac{p}{\sqrt{-g}}\frac{\partial}{\partial x^\mu}\left(\sqrt{-g}u^\mu\right) \\
&= -\rho\left(j_0 - c\kappa_0 E_0\right),
\end{aligned} \tag{26.85}$$

or

$$\frac{c}{\Gamma - 1}\left(u^\mu\frac{\partial p}{\partial x^\mu} - \frac{\Gamma p}{\rho}u^\mu\frac{\partial \rho}{\partial x^\mu}\right) = -\rho\left(j_0 - c\kappa_0 E_0\right) \tag{26.86}$$

in the case of an ideal gas, where $\varepsilon - \rho c^2 = p/(\Gamma - 1)$, Γ being the ratio of the specific heats. In the right-hand side, the exchange of energy between matter and radiation is expressed in terms of quantities in the comoving frame.

(d) Radiation moment equations

The radiation four-force density (26.69) is just the relativistic moment equation:

$$R^{\mu\nu}_{\ \ ;\nu} = -G^\mu, \tag{26.87}$$

where G^μ is the radiation four-force density, and

$$R^{\mu\nu}_{;\nu} = \frac{1}{\sqrt{-g}}\frac{\partial}{\partial x^\nu}\left(\sqrt{-g}R^{\mu\nu}\right) + \Gamma^\mu_{\nu\sigma}R^{\nu\sigma}, \tag{26.88}$$

since $R^{\mu\nu}$ is a symmetric tensor.

The components of moment equation (26.87) in the spherically symmetric Schwarzschild space-time are explicitly written as follows.

The zeroth moment equation ($\mu = 0 = t$) is

$$R^{0\nu}_{;\nu} = \frac{1}{\sqrt{-g}}\frac{\partial}{\partial x^\nu}\left(\sqrt{-g}R^{0\nu}\right) + 2\Gamma^0_{01}R^{01} = -G^t, \tag{26.89}$$

or explicitly,

$$\frac{1}{g_{00}}\frac{\partial E}{\partial t} + \frac{1}{g_{00}r^2}\frac{\partial}{\partial r}\left(r^2 g_{00}F^r\right) + \frac{1}{\sqrt{g_{00}}}\frac{1}{r\sin\theta}\frac{\partial}{\partial\theta}\left(\sin\theta F^\theta\right) + \frac{1}{\sqrt{g_{00}}}\frac{1}{r\sin\theta}\frac{\partial F^\varphi}{\partial\varphi}$$
$$= -cG^t = \frac{\gamma}{g_{00}}\left[\rho(j_0 - c\kappa_0 E_0) - \frac{\rho(\kappa_0+\sigma_0)}{c}v_{\hat{i}}F_0^i\right], \tag{26.90}$$

where $v_{\hat{i}}$ is a three vector in the fixed tetrad, Eqs. (26.58)–(26.60). It should be noted that this is the mixed-frame description, similar to the special relativistic case. That is, both the temporal and spatial derivatives on the left-hand side are applied to the radiation moments in the fixed frame, while the interaction between matter and radiation on the right-hand side is described by the comoving quantities.

The radial first moment equation ($\mu = 1 = r$) is

$$R^{1\nu}_{;\nu} = \frac{1}{\sqrt{-g}}\frac{\partial}{\partial x^\nu}\left(\sqrt{-g}R^{1\nu}\right) + \Gamma^1_{00}R^{00} + \Gamma^1_{11}R^{11} + \Gamma^1_{22}R^{22} + \Gamma^1_{33}R^{33} = -G^r, \tag{26.91}$$

or explicitly,

$$\frac{1}{c^2}\frac{\partial F^r}{\partial t} + g_{00}\frac{\partial P^{rr}}{\partial r} + \frac{\sqrt{g_{00}}}{r\sin\theta}\frac{\partial}{\partial\theta}\left(\sin\theta P^{r\theta}\right) + \frac{\sqrt{g_{00}}}{r\sin\theta}\frac{\partial P^{r\varphi}}{\partial\varphi}$$
$$+ \frac{r_S}{2r^2}(E + P^{rr}) + \frac{g_{00}}{r}(2P^{rr} - P^{\theta\theta} - P^{\varphi\varphi})$$
$$= -G^r = \sqrt{g_{00}}\gamma\frac{v^{\hat{r}}}{c^2}\rho(j_0 - c\kappa_0 E_0)$$
$$- \sqrt{g_{00}}\frac{\rho(\kappa_0+\sigma_0)}{c}\frac{\gamma-1}{v^2}v^{\hat{r}}v_{\hat{i}}F_0^i - \sqrt{g_{00}}\frac{\rho(\kappa_0+\sigma_0)}{c}F_0^r. \tag{26.92}$$

The tangential first moment equation ($\mu = 2 = \theta$) is

$$R^{2\nu}_{;\nu} = \frac{1}{\sqrt{-g}}\frac{\partial}{\partial x^\nu}\left(\sqrt{-g}R^{2\nu}\right) + 2\Gamma^2_{12}R^{12} + \Gamma^2_{33}R^{33} = -G^\theta, \tag{26.93}$$

or explicitly,

$$\frac{1}{\sqrt{g_{00}}c^2}\frac{\partial F^\theta}{\partial t}+\frac{1}{r}\frac{\partial}{\partial r}\left(\sqrt{g_{00}}r P^{r\theta}\right)+\frac{1}{r\sin\theta}\frac{\partial}{\partial\theta}\left(\sin\theta P^{\theta\theta}\right)+\frac{1}{r\sin\theta}\frac{\partial P^{\theta\varphi}}{\partial\varphi}$$
$$+\frac{2g_{00}}{r}P^{r\theta}-\frac{1}{r\tan\theta}P^{\varphi\varphi}=-rG^\theta=\gamma\frac{v^{\hat{\theta}}}{c^2}\rho(j_0-c\kappa_0 E_0)$$
$$-\frac{\rho(\kappa_0+\sigma_0)}{c}\frac{\gamma-1}{v^2}v^{\hat{\theta}}v_{\hat{i}}F_0^i-\frac{\rho(\kappa_0+\sigma_0)}{c}F_0^\theta. \tag{26.94}$$

Finally, the azimuthal first moment equation ($\mu=3=\varphi$) is

$$R^{3\nu}_{;\nu}=\frac{1}{\sqrt{-g}}\frac{\partial}{\partial x^\nu}\left(\sqrt{-g}R^{3\nu}\right)+2\Gamma^3_{13}R^{13}+\Gamma^3_{23}R^{23}=-G^\varphi, \tag{26.95}$$

or explicitly,

$$\frac{1}{\sqrt{g_{00}}c^2}\frac{\partial F^\varphi}{\partial t}+\frac{1}{r}\frac{\partial}{\partial r}\left(\sqrt{g_{00}}r P^{r\varphi}\right)+\frac{1}{r}\frac{\partial P^{\theta\varphi}}{\partial\theta}+\frac{1}{r\sin\theta}\frac{\partial P^{\varphi\varphi}}{\partial\varphi}$$
$$+\frac{2\sqrt{g_{00}}}{r}P^{r\varphi}+\frac{2}{r\tan\theta}P^{\theta\varphi}=-r\sin\theta G^\varphi=\gamma\frac{v^{\hat{\varphi}}}{c^2}\rho(j_0-c\kappa_0 E_0)$$
$$-\frac{\rho(\kappa_0+\sigma_0)}{c}\frac{\gamma-1}{v^2}v^{\hat{\varphi}}v_{\hat{i}}F_0^i-\frac{\rho(\kappa_0+\sigma_0)}{c}F_0^\varphi. \tag{26.96}$$

Similar to the nonrelativistic and special relativistic cases, we need some closure relations in order to close the moment equations. One of them is the Eddington approximation in the comoving frame:

$$P_0^{ij}=f^{ij}E_0, \tag{26.97}$$

where f^{ij} is the Eddington tensor, which generally depends on the flow speed as well as the optical depth (see Sect. 24.2.4).

(e) Total energy conservation

The zeroth component of Eq. (26.67), $(T^{0\nu}+R^{0\nu})_{;\nu}=0$, or

$$T^{0\nu}_{;\nu}=-R^{0\nu}_{;\nu} \tag{26.98}$$

gives the equation of the total energy conservation as follows.

The left-hand side of Eq. (26.98) is expressed as

$$\begin{aligned}
T^{0\nu}_{;\nu} &= \frac{1}{\sqrt{-g}}\frac{\partial}{c\partial t}\left(\sqrt{-g}T^{00}\right) + \frac{1}{\sqrt{-g}}\frac{\partial}{\partial x^i}\left(\sqrt{-g}T^{0i}\right) + 2\Gamma^0_{01}T^{01} \\
&= \frac{1}{g_{00}}\frac{\partial}{c\partial t}\left[(\varepsilon + p)\gamma^2 - p\right] + \frac{1}{g_{00}}\frac{1}{r^2}\frac{\partial}{\partial r}\left[r^2(\varepsilon + p)u^0 u^r g_{00}\right] \\
&\quad + \frac{1}{g_{00}}\frac{1}{r\sin\theta}\frac{\partial}{\partial\theta}\left[\sin\theta(\varepsilon + p)u_0 u^\theta\right] \\
&\quad + \frac{1}{g_{00}}\frac{1}{r\sin\theta}\frac{\partial}{\partial\varphi}\left[(\varepsilon + p)u_0 u^\varphi\right],
\end{aligned} \tag{26.99}$$

whereas the right-hand side of Eq. (26.98) is expressed as

$$\begin{aligned}
-R^{0\nu}_{;\nu} &= \frac{1}{\sqrt{-g}}\frac{\partial}{c\partial t}\left(\sqrt{-g}R^{00}\right) + \frac{1}{\sqrt{-g}}\frac{\partial}{\partial x^i}\left(\sqrt{-g}R^{0i}\right) + 2\Gamma^0_{01}R^{01} \\
&= -\left[\frac{1}{g_{00}}\frac{\partial E}{c\partial t} + \frac{1}{g_{00}}\frac{1}{r^2}\frac{\partial}{\partial r}\left(r^2 g_{00}\frac{F^r}{c}\right)\right. \\
&\quad + \frac{1}{\sqrt{g_{00}}}\frac{1}{r\sin\theta}\frac{\partial}{\partial\theta}\left(\sin\theta\frac{F^\theta}{c}\right) \\
&\quad \left. + \frac{1}{\sqrt{g_{00}}}\frac{1}{r\sin\theta}\frac{\partial}{\partial\varphi}\left(\frac{F^\varphi}{c}\right)\right].
\end{aligned} \tag{26.100}$$

Thus, Eq. (26.98) is explicitly written in the spherically symmetric Schwarzschild space-time as

$$\begin{aligned}
&\frac{\partial}{\partial t}\left[(\varepsilon + p)\gamma^2 - p + E\right] + \frac{1}{4\pi r^2}\frac{\partial}{\partial r}\left[4\pi r^2(\varepsilon + p)u_0 c u^r + 4\pi r^2 g_{00}F^r\right] \\
&\quad + \frac{1}{r\sin\theta}\frac{\partial}{\partial\theta}\left\{\sin\theta\left[(\varepsilon + p)u_0 c u^\theta + \sqrt{g_{00}}F^\theta\right]\right\} \\
&\quad + \frac{1}{r\sin\theta}\frac{\partial}{\partial\varphi}\left[(\varepsilon + p)u_0 c u^\varphi + \sqrt{g_{00}}F^\varphi\right] = 0.
\end{aligned} \tag{26.101}$$

As is expected, the first term of the left-hand side of Eq. (26.101) means the temporal change of the total internal energy of matter and radiation, while the other terms represent the spatial divergence of the total energy flux of matter and radiation in the spherical coordinate.

26.1.2.1 Steady Spherical Flows

In the case of steady spherically symmetric flows, radiation hydrodynamical equations under general relativity become in somewhat simple forms as follows (cf. Park 2006).

For matter, the continuity equation becomes

$$4\pi r^2 \rho c u^r = \dot{M}, \tag{26.102}$$

where $\dot{M}$ is a constant mass-flow rate. The four-velocity u^r is related to the proper three velocity v by

$$u^r = \beta\gamma\sqrt{g_{00}}, \tag{26.103}$$

where $\beta = v/c$, $\gamma = 1/\sqrt{1-\beta^2}$, and $g_{00} = (1 - r_S/r)$, r_S ($\equiv 2GM/c^2$) being the Schwarzschild radius.

The radial component of the equation of motion [see Eq. (26.79)] is

$$\begin{aligned}(\varepsilon + p)\left(u^r \frac{du^r}{dr} + \frac{r_S}{2r^2}\right) + y^2 \frac{dp}{dr} \\ = -yu^r G^t + yu^0 G^r = yG_0^{\hat{r}} = y\frac{\rho(\kappa_0 + \sigma_0)}{c} F_0^r,\end{aligned} \tag{26.104}$$

which is further rearranged as

$$\begin{aligned}c^2 u^r \frac{du^r}{dr} &= -\frac{r_S c^2}{2r^2} - y^2 \frac{c^2}{\varepsilon + p}\frac{dp}{dr} + y\frac{\rho c^2}{\varepsilon + p}\frac{(\kappa_0 + \sigma_0)}{c} F_0^r \\ &= -\frac{r_S c^2}{2r^2} - y^2 \frac{c^2}{\varepsilon + p}\frac{dp}{dr} \\ &\quad + y\frac{\rho c^2}{\varepsilon + p}\frac{(\kappa_0 + \sigma_0)}{c}\gamma^2\left[(1+\beta^2)F^r - \beta(cE + cP^{rr})\right] \\ &= -\frac{r_S c^2}{2r^2} - y^2 \frac{c^2}{\varepsilon + p}\frac{dp}{dr} + y\frac{\rho c^2}{\varepsilon + p}\frac{(\kappa_0 + \sigma_0)}{c} \\ &\quad \times \left[\frac{g_{00} + 2(u^r)^2}{g_{00}} F^r - \frac{\gamma u^r}{\sqrt{g_{00}}}(cE + cP^{rr})\right],\end{aligned} \tag{26.105}$$

where $(\kappa_0 + \sigma_0)$ is the flux mean opacity, E is the radiation energy density, F^r is the radial component of the radiative flux, and P^{rr} is the rr component of the radiation stress tensor in the inertial frame.

The energy equation in the case of an ideal gas [see Eq. (26.86)] is

$$\frac{c}{\Gamma-1}\left(u^r\frac{dp}{dr}-\Gamma\frac{p}{\rho}u^r\frac{d\rho}{dr}\right)=-\rho\left(j_0-c\kappa_0 E_0\right)$$
$$=-\rho\left[j_0+\gamma^2\kappa_0\left(cE-\beta F^r+\beta^2 cP^{rr}\right)\right], \quad (26.106)$$

where κ_0 is the energy density mean opacity.

For radiation field, the radial component of the zeroth moment equation [see Eq. (26.90)] becomes

$$\frac{d}{dr}\left(4\pi r^2 g_{00}F^r\right)=4\pi r^2\rho y[j_0-\kappa_0 cE_0-(\kappa_0+\sigma_0)\beta F_0^r]$$
$$=4\pi r^2\rho y\left\{j_0-\gamma^2\kappa_0\left(cE-2\beta F^r+\beta^2 cP^{rr}\right)\right.$$
$$\left.-\gamma^2\beta(\kappa_0+\sigma_0)\left[(1+\beta^2)F^r-\beta(cE+cP^{rr})\right]\right\}. \quad (26.107)$$

The radial component of the first moment equation [see Eq. (26.92)] is

$$\frac{d}{dr}\left(4\pi r^2 g_{00}P^{rr}\right)=4\pi r\left(1-\frac{3r_S}{2r}\right)\left(E-P^{rr}\right)$$
$$-4\pi r^2\rho y\left[\frac{(\kappa_0+\sigma_0)}{c}F_0^r-\frac{\beta}{c}(j_0-\kappa_0 cE_0)\right]. \quad (26.108)$$

In order to close moment equations for radiation field, we adopt the Eddington approximation,

$$P_0^{rr}=f(\tau,\beta)E_0, \quad (26.109)$$

as the closure relation, which is transformed in the inertial frame as

$$\left(1-f\beta^2\right)cP^{rr}=\left(f-\beta^2\right)cE+(1-f)2\beta F^r. \quad (26.110)$$

One of the trial form of the Eddington factor for the spherically symmetric flow in the relativistic regime is

$$f(\tau,\beta)=\frac{\gamma(1+\beta)+\tau}{\gamma(1+\beta)+3\tau}, \quad (26.111)$$

which depends both on the optical depth and the flow speed (cf. Tamazawa et al. 1975; Abramowicz et al. 1991; Akizuki and Fukue 2008). It should be noted that a closure relation (26.111) is correct only in the limiting cases; $f\to 1/3$ at $\beta\to 0$

and $\tau \to \infty$, $f \to 1$ at $\tau \to 0$ or $\beta \to 1$. In the limit of $\beta = 0$ this form reduces to that of Tamazawa et al. (1975). In contrast to the nonrelativistic case (see, e.g., Bisnovatyi-Kogan and Dorodnitsyn 1999), at the present knowledge, we do not know the precise or suitable form of a relativistic Eddington factor in the spherical geometry.

Finally, in the case of the steady one-dimensional flow, the energy conservation for matter and radiation [see Eq. (26.101)], the relativistic Bernoulli equation, is expressed as

$$4\pi r^2(\varepsilon + p)u_0 c u^r + 4\pi r^2 g_{00} F^r = \dot{E}\ (= \text{const}), \tag{26.112}$$

or using the continuity equation (26.102), we have

$$\dot{M}\frac{\varepsilon + p}{\rho}\gamma\sqrt{g_{00}} + 4\pi r^2 g_{00} F^r = \dot{E}\ (= \text{const}). \tag{26.113}$$

Using the comoving quantities, the total energy conservation is also expressed as

$$\dot{M}\frac{\varepsilon + p + E_0 + P_0^{rr}}{\rho}\gamma\sqrt{g_{00}} + 4\pi r^2\gamma^2(1+\beta^2)g_{00}F_0^r = \dot{E}\ (= \text{const}). \tag{26.114}$$

Here, it should be noted that

$$L_0^{\text{diff}} \equiv 4\pi r^2 F_0, \tag{26.115}$$

$$L_\infty^{\text{diff}} \equiv \gamma^2(1 + \beta^2)g_{00}L_0^{\text{diff}}, \tag{26.116}$$

are the diffusion luminosity measured by a comoving observer, and that measured by an observer at infinity, respectively. On the other hand,

$$L_0^{\text{adv}} \equiv 4\pi r^2 v(E_0 + P_0^{rr}), \tag{26.117}$$

$$L_\infty^{\text{adv}} \equiv \gamma^2 g_{00} L_0^{\text{adv}}, \tag{26.118}$$

are the advection luminosity in the comoving frame, and that measured at infinity, respectively.

26.2 General Relativistic RHD Flows

As was mentioned in Sect. 22.1, there exist various astronomical outflows from luminous objects, and some of them from luminous compact objects are mildly or highly relativistic, e.g., neutron star winds in X-ray bursters, black-hole winds and jets in microquasars and active galactic nuclei. These relativistic outflows are believed to be driven by radiation pressure or magnetic force. In the case of the relativistic outflow from the luminous objects, which highly exceed the Eddington luminosity, the outflow seems to be driven by radiation pressure. There also exist various astronomical accretion flows onto such compact objects as neutron stars and black holes. These accretion flows must be highly relativistic in the vicinity of the central compact objects. Moreover, when the mass-accretion rate is on the order of or exceeds the Eddington rate, the accretion flow must be radiation-dominated one. In such a case, most of the accreting material cannot be settled down or swallowed into the central object, but may be turned to radiatively driven outflows, observed in microquasars, active galactic nuclei, and gamma-ray bursts.

Theoretically, spherically symmetric relativistic winds driven by radiation pressure have been studied by several researchers (Castor 1972; Cassinelli and Hartmann 1975; Ruggles and Bath 1979; Mihalas 1980; Quinn and Paczyński 1985; Paczyński 1986; Prószyński and Prószyński 1986; Turolla et al. 1986; Paczyński 1990; Nobili et al. 1994; Akizuki and Fukue 2009). Some of them were treated under general relativity (Prószyński and Prószyński 1986; Turolla et al. 1986; Paczyński 1990; Nobili et al. 1994; Akizuki and Fukue 2009). As was already mentioned in Sect. 22.2, on the other hand, radiation-dominated relativistic accretion onto compact objects have been also investigated by many researchers (Tamazawa et al. 1975; Schmid-Burgk 1978; Burger and Katz 1980; Thorne 1981; Thorne et al. 1981; Flammang 1982, 1984; Nobili et al. 1991).

In order to solve the relativistic radiation hydrodynamical problems, as in the nonrelativistic case, we need some closure relations for moment equations. In the relativistic regime, however, a satisfactory closure relation is not known.

In the current studies, well-used is the diffusion approximation in the comoving frame, or the flux-limited diffusion approximation (e.g., Ruggles and Bath 1979; Prószyński and Prószyński 1986; Turolla et al. 1986; Paczyński 1990). The diffusion approximation is convenient, and has been extensively adopted by many researchers. However, it is well-known that the diffusion approximation easily breaks down in the optically thin regime. In addition, it has causal problems as already stated (Sect. 24.2.4), and lacks accuracy in the relativistic regime. Hence, there is no assurance whether the diffusion approximation can be valid in the highly relativistic regime, even in the comoving frame.

Alternative closure relation is the Eddington approximation in the comoving frame, where the Eddington factor depends on the flow speed as well as the optical depth (Nobili et al. 1994; Akizuki and Fukue 2009). Indeed, in the nonrelativistic case, the Eddington factor for the spherical geometry depends on the optical depth; i.e., it varies from 1/3 in the optically thick regime to unity in the optically thin regime, and its suitable form can be determined with the fully self-consistent solutions (e.g., Bisnovatyi-Kogan and Dorodnitsyn 1999, 2001). As already stated (Sect. 24.2.4), the usual Eddington approximation is easily violated in the relativistic regime, due to the strong anisotropy of radiation field, and to the finite truncation of moment equations (Turolla et al. 1995; Dullemond 1999). If we give a suitable form of the Eddington factor in the comoving frame, we could solve moment equations in the relativistic regime, while the accuracy is not guaranteed, since we do not know the correct form of the Eddington factor. In summary, for spherical geometry the

Eddington factor in the comoving frame would depend both on the optical depth and the flow velocity, and the traditional Eddington factor would be generally inadequate for the relativistic momentum formalism.

The third way is to solve both the relativistic radiative transfer equation and relativistic hydrodynamical equations simultaneously without introducing moment procedures. If we use the transfer equation itself, instead of moment equations with a closure relation, various troubles on the closure relation do not appear. As is well-known, however, to solve the transfer equation needs a lot of numerical costs. In the relativistic regime, moreover, since the light-ray concentrates in the forward direction due to aberration, a great deal of numerical accuracy is necessary in the angular direction in addition to the optical depth. Hence, there is no fully satisfactory study in such a full transfer calculation (see, however, Knop et al. 2007; Takahashi and Umemura 2017 for general relativistic radiative transfer).

References

Abramowicz, M.A., Novikov, I.D., Paczyński, B.: Astrophys. J. **369**, 175 (1991)
Akizuki, C., Fukue, J.: Publ. Astron. Soc. Jpn. **60**, 337 (2008)
Akizuki, C., Fukue, J.: Publ. Astron. Soc. Jpn. **61**, 543 (2009)
Anderson, J.L., Spiegel, E.A.: Astrophys. J. **171**, 127 (1972)
Bisnovatyi-Kogan, G.S., Dorodnitsyn, A.V.: Astron. Astrophys. **344**, 647 (1999)
Bisnovatyi-Kogan, G.S., Dorodnitsyn, A.V.: Astron. Rep. **45**, 995 (2001)
Burger, H.L., Katz, J.I.: Astrophys. J. **236**, 921 (1980)
Cassinelli, J.P., Hartmann, L.: Astrophys. J. **202**, 718 (1975)
Castor, J.I.: Astrophys. J. **178**, 779 (1972)
Dullemond, C.P.: Astron. Astrophys. **343**, 1030 (1999)
Flammang, R.A.: Monthly Not. Roy. Astron. Soc. **199**, 833 (1982)
Flammang, R.A.: Monthly Not. Roy. Astron. Soc. **206**, 589 (1984)
Knop, S., Hauschildt, P.H., Baron, E.: Astron. Astrophys. **315**, 320 (2007)
Lindquist, R.W.: Ann. Phys. **37**, 487 (1966)
Mihalas, D.: Astrophys. J. **237**, 574 (1980)
Nobili, L., Turolla, R., Lapidus, I.: Astrophys. J. **433**, 276 (1994)
Nobili, L., Turolla, R., Zampieri, L.: Astrophys. J. **383**, 250 (1991)
Nobili, L., Turolla, R., Zampieri, L.: Astrophys. J. **404**, 686 (1993)
Paczyński, B.: Astrophys. J. **308**, L43 (1986)
Paczyński, B.: Astrophys. J. **363**, 218 (1990)
Paczyński, B., Prószyński, M.: Astrophys. J. **302**, 519 (1986)
Park, M.-G.: Astron. Astrophys. **274**, 642 (2003)
Park, M.-G.: Monthly Not. Roy. Astron. Soc. **367**, 1739 (2006)
Quinn, T., Paczyński, B.: Astrophys. J. **289**, 634 (1985)
Ruggles, C.L.N., Bath, G.T.: Astron. Astrophys. **80**, 97 (1979)
Schmid-Burgk, J.: ApSS **56**, 191 (1978)
Takahashi, R.: Monthly Not. Roy. Astron. Soc. **382**, 1041 (2007)
Takahashi, R., Umemura, M.: Monthly Not. Roy. Astron. Soc. **464**, 456 (2017)
Tamazawa, S., Toyama, K., Kaneko, N., Ôno, Y.: Astrophys. Space Sci. **32**, 403 (1975)
Thorne, K.S.: Monthly Not. Roy. Astron. Soc. **194**, 439 (1981)

Thorne, K.S., Flammang R.A., Żytkow A.N.: Monthly Not. Roy. Astron. Soc. **194**, 475 (1981)
Turolla, R., Nobili, L.: Monthly Not. Roy. Astron. Soc. **235**, 1273 (1988)
Turolla, R., Nobili, L., Calvani, M.: Astrophys. J. **303**, 573 (1986)
Turolla, R., Zampieri, L., Nobili, L.: Monthly Not. Roy. Astron. Soc. **272**, 625 (1995)
Udey, N., Israel, W.: Monthly Not. Roy. Astron. Soc. **199**, 1137 (1982)

Correction to: Fundamentals of Astrophysical Fluid Dynamics

Correction to:
S. Kato and J. Fukue, *Fundamentals of Astrophysical Fluid Dynamics*, Astronomy and Astrophysics Library, https://doi.org/10.1007/978-981-15-4174-2

The original version of this book was inadvertently published with multiple typographical errors and the same has been updated now.

The updated original version for this book can be found at https://doi.org/10.1007/978-981-15-4174-2

© Springer Nature Singapore Pte Ltd. 2023
S. Kato, J. Fukue, *Fundamentals of Astrophysical Fluid Dynamics*, Astronomy and Astrophysics Library,
https://doi.org/10.1007/978-981-15-4174-2_27

Appendix A
Quasi-Nonlinear Terms in Equation of Motions in Lagrangian Expression

Expressions for quasi-nonlinear terms of equation of motions are important in studying (i) turbulent excitation of oscillations in astrophysical objects (helioseismology, astroseismology, and discoseismology) (Sect. 5.3) and (ii) resonant excitation of oscillations in astrophysical objects (Sect. 6.7). In these studies the Lagrangian expression of quasi-nonlinear terms is better than that of the Eulerian ones. In particular, in studying the issue (ii), the stability criterion can be derived in a quite concise form by using the Lagrangian expression of the quasi-nonlinear terms. The presence of this concise expression for stability criterion comes from a kind of commutative relation related to quasi-nonlinear terms. In this appendix we derive an expression of quasi-nonlinear terms by Lagrangian formalism and present the commutative relation. It is noted that the characteristics of the commutability still hold in MHD fluids, and thus the stability criterion of the wave–wave resonant instability derived in Sect. 6.7 can be formally extended to MHD systems.

A.1 Quasi-Nonlinear Expression of Equation of Motion

The equation of motions describing adiabatic hydrodynamical motions is (Sect. 2.2)

$$\frac{d\boldsymbol{v}}{dt} = -\nabla\psi - \frac{1}{\rho}\nabla p, \tag{A.1}$$

where ψ is gravitational potential and other notations have their usual meanings. If the Lagrangian variation is expressed by δ, from the above equation we have

$$\delta\left(\frac{d\boldsymbol{v}}{dt}\right) = \delta\left(-\nabla\psi - \frac{1}{\rho}\nabla p\right). \tag{A.2}$$

© Springer Nature Singapore Pte Ltd. 2020, corrected publication 2023

S. Kato, J. Fukue, *Fundamentals of Astrophysical Fluid Dynamics*,
Astronomy and Astrophysics Library,
https://doi.org/10.1007/978-981-15-4174-2

An important characteristic of the Lagrangian operator δ is that δ and the Lagrangian time variation along perturbed flow, d/dt, are commuted in the following sense (Lynden-Bell and Ostriker 1967):

$$\delta\left(\frac{dQ}{dt}\right) = \frac{d_0}{dt}(\delta Q), \tag{A.3}$$

where $Q(\boldsymbol{r}, t)$ is an arbitrary variable of $\boldsymbol{r}$ and t, and d_0/dt is the Lagrangian time variation along the unperturbed flow. Hence, from Eq. (A.2) we have

$$\frac{d_0^2\boldsymbol{\xi}}{dt^2} = \delta\left(- \nabla\psi - \frac{1}{\rho}\nabla p\right), \tag{A.4}$$

since

$$\delta\left(\frac{d\boldsymbol{v}}{dt}\right) = \frac{d_0}{dt}(\delta\boldsymbol{v}) = \frac{d_0}{dt}\left(\frac{d_0\boldsymbol{\xi}}{dt}\right), \quad \delta\boldsymbol{v} = \frac{d_0\boldsymbol{\xi}}{dt}. \tag{A.5}$$

An important point to be emphasized here is that Eq. (A.4) is valid for arbitrary amplitude of displacement vector $\boldsymbol{\xi}$, not restricted to small amplitude perturbations.

The linearized part of Eq. (A.4) with respect to $\boldsymbol{\xi}$ is (Lynden-Bell and Ostriker 1967, see also Sect. 2.5)

$$\rho_0\frac{\partial^2\boldsymbol{\xi}}{\partial t^2} + 2\rho_0(\boldsymbol{v}_0\cdot\nabla)\frac{\partial\boldsymbol{\xi}}{\partial t} + \mathcal{L}(\boldsymbol{\xi}) = 0, \tag{A.6}$$

where an explicit form of operator $\mathcal{L}(\boldsymbol{\xi})$ in nonself-gravitating systems is (see Sect. 2.5)

$$\mathcal{L}(\boldsymbol{\xi}) = \rho_0(\boldsymbol{v}_0\cdot\nabla)(\boldsymbol{v}_0\cdot\nabla)\boldsymbol{\xi} + \rho_0(\boldsymbol{\xi}\cdot\nabla)(\nabla\psi_0) + \boldsymbol{P}_{\rm linear}(\boldsymbol{\xi}) \tag{A.7}$$

with

$$\boldsymbol{P}_{\rm linear} = \nabla\left[(1-\Gamma_1)p_0\frac{\partial\xi_i}{\partial r_i}\right] - p_0\nabla\left(\frac{\partial\xi_i}{\partial r_i}\right) - \nabla\left(\xi_j\frac{\partial p_0}{\partial r_j}\right) + \xi_j\frac{\partial}{\partial r_j}\nabla p_0. \tag{A.8}$$

Our purpose here is to explicitly write down Eq. (A.4) up to the second-order quantities with respect to $\boldsymbol{\xi}$, assuming that $\boldsymbol{\xi}$ is a small quantity. Following Kato (2004, 2008), we present the procedures and the results.

The Lagrangian change of a quantity, $Q(\boldsymbol{r},t)$, i.e., δQ, is defined by

$$\delta Q = Q[\boldsymbol{r} + \boldsymbol{\xi}(\boldsymbol{r}, t), t] - Q_0(\boldsymbol{r}, t), \tag{A.9}$$

where the subscript 0 to Q represents unperturbed quantity. The Eulerian change of Q, i.e., Q_1, on the other hand, is defined by

$$Q_1 = Q(\boldsymbol{r}, t) - Q_0(\boldsymbol{r}, t). \tag{A.10}$$

If $Q[\boldsymbol{r} + \boldsymbol{\xi}(\boldsymbol{r}, t), t]$ is Taylor-expanded around $Q(\boldsymbol{r}, t)$ up to the second-order terms with respect to perturbations, we have from Eqs. (A.9) and (A.10),

$$\delta Q = Q_1 + \xi_j \frac{\partial Q_0}{\partial r_j} + \xi_j \frac{\partial Q_1}{\partial r_j} + \frac{1}{2}\xi_i \xi_j \frac{\partial^2 Q_0}{\partial r_i \partial r_j}. \tag{A.11}$$

This is a relation between δQ and Q_1, up to the second-order small quantities with respect to perturbations. Here and hereafter, the summation abbreviation is adopted, using Cartesian coordinates.

Since the self-gravity of disks is neglected here, i.e., $\psi_1 = 0$, we have easily

$$\delta(\nabla\psi) = \xi_j \frac{\partial}{\partial r_j}(\nabla\psi_0) + \frac{1}{2}\xi_i \xi_j \frac{\partial^2}{\partial r_i \partial_j}(\nabla\psi_0). \tag{A.12}$$

The second term on the right-hand side of Eq. (A.12) represents quasi-nonlinear terms.

Expressing $\delta(\nabla p/\rho)$ in terms of Lagrangian displacements is somewhat complicated. Using Eq. (A.11) and the definition of Lagrangian change, we have

$$\delta\left(\frac{1}{\rho}\nabla p\right) = \frac{1}{\rho_0 + \delta\rho}\Big[\nabla p_0 + \nabla p_1 + \xi_j \frac{\partial}{\partial r_j}\nabla(p_0 + p_1) \\ + \frac{1}{2}\xi_i \xi_j \frac{\partial^2}{\partial r_i \partial r_j}\nabla p_0 + \dots\Big] - \frac{1}{\rho_0}\nabla p_0. \tag{A.13}$$

The Eulerian pressure variation, p_1, which appears on the right-hand side of Eq. (A.13), is expressed in terms of δp and $\boldsymbol{\xi}$ by using

$$\delta p = p_1 + \xi_j \frac{\partial}{\partial r_j}(p_0 + p_1) + \frac{1}{2}\xi_i \xi_j \frac{\partial^2 p_0}{\partial r_i \partial r_j}. \tag{A.14}$$

We have then

$$\rho_0 \delta\left(\frac{1}{\rho}\nabla p\right) = \nabla\left[\delta p - \xi_j \frac{\partial p_0}{\partial r_j} - \xi_j \frac{\partial}{\partial r_j}\left(\delta p - \xi_i \frac{\partial p_0}{\partial r_i}\right) - \frac{1}{2}\xi_i \xi_j \frac{\partial^2 p_0}{\partial r_i \partial r_j}\right] \\ + \xi_j \frac{\partial}{\partial r_j}\nabla\left[p_0 + \delta p - \xi_i \frac{\partial p_0}{\partial r_i}\right] + \frac{1}{2}\xi_i \xi_j \frac{\partial^2}{\partial r_i \partial r_j}\nabla p_0 \\ - \frac{\delta\rho}{\rho_0}\left[\nabla p_0 + \nabla(\delta p) - (\nabla\xi_j)\frac{\partial p_0}{\partial r_j}\right] + \left(\frac{\delta\rho}{\rho_0}\right)^2 \nabla p_0. \tag{A.15}$$

This explicitly expresses $\rho_0\delta(\nabla p/\rho)$ up to the second-order quantities in terms of Lagrangian quantities, δp, $\delta\rho$, and $\boldsymbol{\xi}$.

Next, δp and $\delta\rho$ on the right-hand side of Eq. (A.15) are explicitly expressed in terms of $\boldsymbol{\xi}$. To do so we use the equation of continuity and adiabatic relation. The equation of continuity is expressed as (e.g., Kato and Unno 1967)

$$\delta\rho + \rho_0\frac{\partial\xi_i}{\partial r_i} = \rho_0 N_{\rm c}, \tag{A.16}$$

where

$$N_{\rm c} = \frac{1}{2}\left[\left(\frac{\partial\xi_j}{\partial\xi_j}\right)^2 + \frac{\partial\xi_i}{\partial\xi_j}\frac{\partial\xi_j}{\partial r_i}\right]. \tag{A.17}$$

The adiabatic relation is written as (e.g., Kato and Unno 1967)

$$\delta p - \Gamma_1\frac{p_0}{\rho_0}\delta\rho = p_0 N_p, \tag{A.18}$$

or

$$\delta p + \Gamma_1 p_0\frac{\partial\xi_j}{\partial r_j} = p_0(\Gamma_1 N_{\rm c} + N_{\rm p}), \tag{A.19}$$

where

$$N_{\rm p} = \frac{1}{2}\Gamma_1(\Gamma_1 - 1)\left(\frac{\delta\rho}{\rho_0}\right)^2. \tag{A.20}$$

Substitution of Eqs. (A.16) and (A.18) into Eq. (A.15) finally gives an expression for $\rho_0\delta(\nabla p/\rho)$, explicitly expressed in terms of $\boldsymbol{\xi}$ alone. This is summarized as

$$\rho_0\left(\frac{1}{\rho}\nabla p\right) = \boldsymbol{P}_{\rm linear} + \boldsymbol{P}_{\rm nonlinear}, \tag{A.21}$$

where $\boldsymbol{P}_{\rm linear}$ and $\boldsymbol{P}_{\rm nonlinear}$ are the linear and quasi-nonlinear parts, respectively, and the former is already given by Eq. (A.8). After some manipulations we can express the quasi-nonlinear part of $\rho_0\delta[(1/\rho)\nabla p]$, namely $\boldsymbol{P}_{\rm nonlinear}$, as (Kato 2008)[1] :

$$\begin{aligned}-(P_{\rm nonlinear})_k = &-\frac{\partial}{\partial r_j}\left(p_0\frac{\partial\xi_i}{\partial r_k}\frac{\partial\xi_j}{\partial r_i}\right) + \frac{\partial}{\partial r_j}\left[(\Gamma_1 - 1)p_0\frac{\partial\xi_j}{\partial r_k}\frac{\partial\xi_i}{\partial r_i}\right]\\ &+\frac{1}{2}\frac{\partial}{\partial r_k}\left[(\Gamma_1 - 1)p_0\frac{\partial\xi_i}{\partial r_j}\frac{\partial\xi_j}{\partial r_i}\right] + \frac{1}{2}\frac{\partial}{\partial r_k}\left[(\Gamma_1 - 1)^2 p_0\frac{\partial\xi_i}{\partial r_i}\frac{\partial\xi_j}{\partial r_j}\right],\end{aligned} \tag{A.22}$$

[1] In expression for $\boldsymbol{P}_{\rm nonlinear}$ given by equation (82) by Kato (2008) there are typographical errors.

where $(P_{\mathrm{nonlinear}})_k$ represents the k-component of $\boldsymbol{P}_{\mathrm{nonlinear}}$.

Substituting Eqs. (A.12) and (A.21)–(A.22) into Eq. (A.4), we find that the linear wave equation (A.6) is now extended into a quasi-nonlinear equation as

$$\rho_0 \frac{\partial^2 \boldsymbol{\xi}}{\partial t^2} + 2\rho_0(\boldsymbol{v}_0 \cdot \nabla)\frac{\partial \boldsymbol{\xi}}{\partial t} + \mathcal{L}(\boldsymbol{\xi}) = \boldsymbol{C}(\boldsymbol{\xi}), \tag{A.23}$$

where $\boldsymbol{C}(\boldsymbol{\xi})$ is the quasi-nonlinear term and its Cartesian component is written as

$$C_k(\boldsymbol{\xi}) = -\frac{1}{2}\rho_0 \xi_i \xi_j \frac{\partial^2 \psi_0}{\partial r_i \partial r_j \partial r_k} - (P_{\mathrm{nonlinear}})_k. \tag{A.24}$$

A.2 Commutative Relations

The quasi-nonlinear term $\boldsymbol{C}$ has a characteristic which is important when a wave–wave resonant interaction is considered (Sect. 6.7). This characteristic is mentioned here briefly.

Let us consider two independent, small-amplitude normal mode oscillations, say, $\boldsymbol{\xi}(\boldsymbol{r}, t)$ and $\boldsymbol{\eta}(\boldsymbol{r}, t)$. When we consider a quasi-nonlinear coupling between two oscillations $\boldsymbol{\xi}(\boldsymbol{r}, t)$ and $\boldsymbol{\eta}(\boldsymbol{r}, t)$, we need to examine coupling terms, $\boldsymbol{D}(\boldsymbol{\xi}, \boldsymbol{\eta})$. This coupling term is obtained by changing one of ξ_n $(n = i, j, k, \ldots)$ in all binomial terms of $\boldsymbol{\xi}$ in $\boldsymbol{C}_k$ [Eq. (A.24)] to η_n. That is, the coupling term $\boldsymbol{D}(\boldsymbol{\xi}, \boldsymbol{\eta})$ we need is given by

$$\begin{aligned} D_k(\boldsymbol{\xi}, \boldsymbol{\eta}) &= -\rho_0 \xi_i \eta_j \frac{\partial^2 \psi_0}{\partial r_i \partial r_j \partial r_k} \\ &- \frac{\partial}{\partial r_j}\left[p_0\left(\frac{\partial \xi_i}{\partial r_k}\frac{\partial \eta_j}{\partial r_i} + \frac{\partial \eta_i}{\partial r_k}\frac{\partial \xi_j}{\partial r_i}\right)\right] + \frac{\partial}{\partial r_j}\left[(\Gamma_1 - 1)p_0\left(\frac{\partial \xi_j}{\partial r_k}\frac{\partial \eta_i}{\partial r_i} + \frac{\partial \eta_j}{\partial r_k}\frac{\partial \xi_i}{\partial r_i}\right)\right] \\ &+ \frac{\partial}{\partial r_k}\left[(\Gamma_1 - 1)p_0 \frac{\partial \xi_i}{\partial r_j}\frac{\partial \eta_j}{\partial r_i}\right] + \frac{\partial}{\partial r_k}\left[(\Gamma_1 - 1)^2 p_0 \frac{\partial \xi_i}{\partial r_i}\frac{\partial \eta_j}{\partial r_j}\right]. \end{aligned} \tag{A.25}$$

It is noted that $\boldsymbol{D}(\boldsymbol{\xi}, \boldsymbol{\eta})$ is symmetric with respect to exchange of $\boldsymbol{\xi}$ and $\boldsymbol{\eta}$, i.e., $\boldsymbol{D}(\boldsymbol{\xi}, \boldsymbol{\eta}) = \boldsymbol{D}(\boldsymbol{\eta}, \boldsymbol{\xi})$.

Let us introduce an another perturbation which is expressed as $\boldsymbol{\zeta}(\boldsymbol{r}, t)$, and consider an integration of $\boldsymbol{\zeta} \cdot \boldsymbol{D}(\boldsymbol{\xi}, \boldsymbol{\eta})$ over the whole volume V where $\boldsymbol{\xi}$, $\boldsymbol{\eta}$, and $\boldsymbol{\zeta}$ exist:

$$\int \boldsymbol{\zeta} \cdot \boldsymbol{D}(\boldsymbol{\xi}, \boldsymbol{\eta}) dV. \tag{A.26}$$

If density ρ_0 vanishes on the surface of the system, we have the following commutative relations by performing the integration by part:

$$\int \boldsymbol{\xi} \cdot \boldsymbol{D}(\boldsymbol{\eta}, \boldsymbol{\zeta}) dV = \int \boldsymbol{\eta} \cdot \boldsymbol{D}(\boldsymbol{\xi}, \boldsymbol{\zeta}) dV = \int \boldsymbol{\zeta} \cdot \boldsymbol{D}(\boldsymbol{\xi}, \boldsymbol{\eta}) dV. \tag{A.27}$$

The presence of these commutative relations allows us to derive a simple form of instability criterion for wave–wave resonant processes (see Sect. 6.7).

Finally, it is noted that the above commutative relations can be extended to cases of hydromagnetic perturbations (Kato 2014). The presence of wave–wave resonant instability resulting from the above commutative relations might be related to pattern formation in deformed systems.

Appendix B
Useful Formulae of Vector Analyses

In studying astrophysical fluid dynamics, we treat many vector quantities such as velocity fields $\boldsymbol{v}$ and magnetic fields $\boldsymbol{B}$. Thus, formulae of vector analyses as well as expressions for some scalar and vector quantities in orthogonal curvilinear coordinate systems are helpful. In this appendix some of them are summarized for convenience of readers.

B.1 Some Formulae of Vector Analyses

Some useful formulae of vector analyses are summarized here. Let φ and ψ be arbitrary scalar functions, and $\boldsymbol{A}$ and $\boldsymbol{B}$ be arbitrary vector functions. Then, we have the following formulae:

$$\mathrm{grad}(\varphi\psi) = \nabla(\varphi\psi) = \psi\nabla\varphi + \varphi\nabla\psi, \tag{B.1}$$

$$\mathrm{div}(\varphi\boldsymbol{A}) = \nabla(\varphi\boldsymbol{A}) = \nabla\varphi\cdot\boldsymbol{A} + \varphi\,\nabla\boldsymbol{A} = \nabla\,\varphi\cdot\boldsymbol{A} + \varphi\,\mathrm{div}\,\boldsymbol{A}, \tag{B.2}$$

$$\begin{aligned}\mathrm{curl}(\varphi\boldsymbol{A}) &= \nabla\times(\varphi\boldsymbol{A}) = \nabla\varphi\times\boldsymbol{A} + \varphi\,\nabla\times\boldsymbol{A}\\ &= \nabla\varphi\times\boldsymbol{A} + \varphi\,\mathrm{curl}\,\boldsymbol{A},\end{aligned} \tag{B.3}$$

$$\begin{aligned}\mathrm{div}(\boldsymbol{A}\times\boldsymbol{B}) &= \nabla(\boldsymbol{A}\times\boldsymbol{B}) = \boldsymbol{B}(\nabla\times\boldsymbol{A}) - \boldsymbol{A}(\nabla\times\boldsymbol{B})\\ &= \boldsymbol{B}\,\mathrm{curl}\,\boldsymbol{A} - \boldsymbol{A}\,\mathrm{curl}\,\boldsymbol{B},\end{aligned} \tag{B.4}$$

$$\begin{aligned}\mathrm{curl}(\boldsymbol{A}\times\boldsymbol{B}) &= \nabla\times(\boldsymbol{A}\times\boldsymbol{B})\\ &= (\boldsymbol{B}\nabla)\boldsymbol{A} - (\boldsymbol{A}\nabla)\boldsymbol{B} + \boldsymbol{A}\,\mathrm{div}\boldsymbol{B} - \boldsymbol{B}\,\mathrm{div}\boldsymbol{A},\end{aligned} \tag{B.5}$$

$$\begin{aligned}\mathrm{grad}(\boldsymbol{A}\boldsymbol{B}) &= \nabla(\boldsymbol{A}\boldsymbol{B})\\ &= (\boldsymbol{A}\nabla)\boldsymbol{B} + (\boldsymbol{B}\nabla)\boldsymbol{A} + \boldsymbol{A}\times\mathrm{curl}\boldsymbol{B} + \boldsymbol{B}\times\mathrm{curl}\boldsymbol{A}.\end{aligned} \tag{B.6}$$

© Springer Nature Singapore Pte Ltd. 2020, corrected publication 2023
S. Kato, J. Fukue, *Fundamentals of Astrophysical Fluid Dynamics*,
Astronomy and Astrophysics Library,
https://doi.org/10.1007/978-981-15-4174-2

Other important relations are

$$\operatorname{curl}\operatorname{grad}\varphi = \nabla \times \nabla\varphi = 0, \tag{B.7}$$

$$\operatorname{div}\operatorname{curl}\boldsymbol{A} = \nabla(\nabla \times \boldsymbol{A}) = 0, \tag{B.8}$$

$$\operatorname{curl}\operatorname{curl}\boldsymbol{A} = \nabla \times (\nabla \times \boldsymbol{A}) = \nabla(\nabla\boldsymbol{A}) - \nabla^2\boldsymbol{A}. \tag{B.9}$$

Next, a formula of vector triple product is presented. Let $\boldsymbol{A}$, $\boldsymbol{B}$, and $\boldsymbol{C}$ be arbitrary vector functions. Then, we have

$$\boldsymbol{A} \times (\boldsymbol{B} \times \boldsymbol{C}) = \boldsymbol{B}(\boldsymbol{A}\boldsymbol{C}) - \boldsymbol{C}(\boldsymbol{A}\boldsymbol{B}). \tag{B.10}$$

B.2 Expressions of Some Scalar and Vector Quantities in Curved Coordinate Systems

We consider cylindrical and spherical coordinates.

B.2.1 Cylindrical Coordinates

Let us consider cylindrical coordinates (r, φ, z), whose z-axis is in the z direction of Cartesian ones (x, y, z) and r and φ are the polar coordinates in the x–y plane: That is, $x = r\cos\varphi$, $y = r\sin\varphi$, and $z = z$.

Then, we have

$$\frac{\partial}{\partial x} = \cos\varphi\frac{\partial}{\partial r} - \sin\varphi\frac{\partial}{r\partial\varphi}, \tag{B.11}$$

$$\frac{\partial}{\partial y} = \sin\varphi\frac{\partial}{\partial r} + \cos\varphi\frac{\partial}{r\partial\varphi}, \tag{B.12}$$

$$\frac{\partial}{\partial z} = \frac{\partial}{\partial z}. \tag{B.13}$$

Using the above relations, we have, for an arbitrary vector $\boldsymbol{A}$,

$$\operatorname{div}\boldsymbol{A} = \frac{\partial(rA_r)}{r\partial r} + \frac{\partial A_\varphi}{r\partial\varphi} + \frac{\partial A_z}{\partial z}. \tag{B.14}$$

$$\nabla^2\psi = \frac{\partial}{r\partial r}\left(r\frac{\partial\psi}{\partial r}\right) + \frac{\partial^2\psi}{r^2\partial\varphi^2} + \frac{\partial^2\psi}{\partial z^2}. \tag{B.15}$$

Furthermore, we have

$$\mathrm{curl}_r \boldsymbol{A} = \frac{\partial A_z}{r\partial\varphi} - \frac{\partial A_\varphi}{\partial z}, \tag{B.16}$$

$$\mathrm{curl}_\varphi \boldsymbol{A} = \frac{\partial A_r}{\partial z} - \frac{\partial A_z}{\partial r}, \tag{B.17}$$

$$\mathrm{curl}_z \boldsymbol{A} = \frac{\partial (rA_\varphi)}{r\partial r} - \frac{\partial A_r}{r\partial\varphi}. \tag{B.18}$$

Finally, a useful expression in cylindrical coordinates is

$$(\boldsymbol{B}\cdot\nabla)\boldsymbol{A} = \left(B_r\frac{\partial}{\partial r} + B_\varphi\frac{\partial}{r\partial\varphi} + B_z\frac{\partial}{\partial z}\right)\boldsymbol{A} + \frac{B_\varphi}{r}\boldsymbol{i}_z \times \boldsymbol{A}, \tag{B.19}$$

where $\boldsymbol{B}$ is also an arbitrary vector and $\boldsymbol{i}_z$ is the unit vector in the z-direction.

B.2.2 Spherical Coordinates

Let us introduce the spherical coordinates (r, θ, φ), where the angle between z-axis of the Cartesian coordinates and the direction of r is taken to be θ. That is, the relation among coordinates is

$$x = r\sin\theta\cos\varphi, \tag{B.20}$$

$$y = r\sin\theta\sin\varphi, \tag{B.21}$$

$$z = r\cos\theta. \tag{B.22}$$

The above relations give

$$\frac{\partial}{\partial x} = \sin\theta\cos\varphi\frac{\partial}{\partial r} + \cos\theta\cos\varphi\frac{\partial}{r\partial\theta} - \sin\varphi\frac{\partial}{r\sin\theta\partial\varphi}, \tag{B.23}$$

$$\frac{\partial}{\partial y} = \sin\theta\sin\varphi\frac{\partial}{\partial r} + \cos\theta\sin\varphi\frac{\partial}{r\partial\theta} + \cos\varphi\frac{\partial}{r\sin\theta\partial\varphi}, \tag{B.24}$$

$$\frac{\partial}{\partial z} = \cos\frac{\partial}{\partial r} - \sin\theta\frac{\partial}{r\partial\theta}. \tag{B.25}$$

Using the above relations we have

$$\mathrm{div}\boldsymbol{A} = \frac{\partial}{r^2\partial r}(r^2A_r) + \frac{1}{r\sin\theta}\frac{\partial}{\partial\theta}(\sin\theta\, A_\theta) + \frac{1}{r\sin\theta}\frac{\partial A_\varphi}{\partial\varphi} \tag{B.26}$$

and

$$\nabla^2\psi = \frac{1}{r^2}\frac{\partial}{\partial r}\left(r^2\frac{\partial\psi}{\partial r}\right) + \frac{1}{r^2\sin\theta}\frac{\partial}{\partial\theta}\left(\sin\theta\frac{\partial\psi}{\partial\theta}\right) + \frac{1}{r^2\sin^2\theta}\frac{\partial^2\psi}{\partial\varphi^2}. \tag{B.27}$$

Furthermore, we have

$$\mathrm{curl}_r\boldsymbol{A} = \frac{1}{r\sin\theta}\left\{\frac{\partial}{\partial\theta}(\sin\theta\, A_\varphi) - \frac{\partial A_\theta}{\partial\varphi}\right\}, \tag{B.28}$$

$$\mathrm{curl}_\theta\boldsymbol{A} = \frac{1}{r}\left\{\frac{1}{\sin\theta}\frac{\partial A_r}{\partial\varphi} - \frac{\partial}{\partial r}(rA_\varphi)\right\}, \tag{B.29}$$

$$\mathrm{curl}_\varphi\boldsymbol{A} = \frac{1}{r}\left\{\frac{\partial}{\partial r}(rA_\theta) - \frac{\partial A_r}{\partial\theta}\right\}. \tag{B.30}$$

Appendix C
Brief Summary of Hydrodynamical Equations

We summarize the basic equations for a viscous fluid, the *Navier–Stokes equations*, in the nonrelativistic regime. The basic equations include the continuity equation, the equation of motion, and the energy equation, supplemented by the equation of state. We first present these basic equations in vector forms, and then express them in cylindrical and spherical coordinate frames.

C.1 General Form

(a) **Continuity equation**

The conservation of mass (continuity equation) is written as

$$\frac{\partial \rho}{\partial t} + \mathrm{div}(\rho \boldsymbol{v}) = 0, \tag{C.1}$$

where ρ is the density and $\boldsymbol{v}$ the velocity vector. In terms of the *Lagrange time derivative*:

$$\frac{d}{dt} = \frac{\partial}{\partial t} + (\boldsymbol{v} \cdot \nabla), \tag{C.2}$$

the continuity equation (C.1) is expressed as

$$\frac{d\rho}{dt} + \rho\ \mathrm{div}\boldsymbol{v} = 0. \tag{C.3}$$

© Springer Nature Singapore Pte Ltd. 2020, corrected publication 2023

S. Kato, J. Fukue, *Fundamentals of Astrophysical Fluid Dynamics*, Astronomy and Astrophysics Library,
https://doi.org/10.1007/978-981-15-4174-2

(b) **Equation of motion**

The equation of motion is described as

$$\rho \frac{d\boldsymbol{v}}{dt} = \rho \left[\frac{\partial \boldsymbol{v}}{\partial t} + (\boldsymbol{v} \cdot \nabla)\boldsymbol{v} \right] = -\rho \nabla \psi - \nabla p + \rho \boldsymbol{N} + \rho \boldsymbol{K}, \tag{C.4}$$

where ψ is the gravitational potential, p the pressure, $\boldsymbol{N}$ the viscous force per unit mass, and $\boldsymbol{K}$ the external force per unit mass.

The viscous force $\boldsymbol{N}$ is expressed as

$$\rho N_i = \frac{\partial t_{ik}}{\partial x_k}, \tag{C.5}$$

where t_{ik} is the viscous stress tensor:

$$t_{ik} \equiv \eta \left(\frac{\partial v_i}{\partial x_k} + \frac{\partial v_k}{\partial x_i} - \frac{2}{3} \delta_{ik} \frac{\partial v_j}{\partial x_j} \right) + \zeta \delta_{ik} \frac{\partial v_j}{\partial x_j}, \tag{C.6}$$

where η is the dynamical viscosity (first viscosity) and ζ the bulk viscosity (second viscosity) (the latter is usually ignored). By introducing a symmetric tensor σ_{ik} defined by

$$\sigma_{ik} \equiv \frac{\partial v_i}{\partial x_k} + \frac{\partial v_k}{\partial x_i}, \tag{C.7}$$

we can rewrite t_{ik} in the form:

$$t_{ik} = \eta \sigma_{ik} + \left(\zeta - \frac{2}{3} \eta \right) \delta_{ik} \frac{\partial v_j}{\partial x_j}. \tag{C.8}$$

The Navier–Stokes equation is thus finally expressed in the form:

$$\begin{aligned} \rho \frac{dv_i}{dt} &= -\rho \frac{\partial \psi}{\partial x_i} - \frac{\partial p}{\partial x_i} + \rho K_i \\ &\quad + \frac{\partial}{\partial x_k} \left[\eta \left(\frac{\partial v_i}{\partial x_k} + \frac{\partial v_k}{\partial x_i} - \frac{2}{3} \delta_{ik} \frac{\partial v_j}{\partial x_j} \right) + \zeta \delta_{ik} \frac{\partial v_j}{\partial x_j} \right]. \end{aligned} \tag{C.9}$$

(c) **Energy equation**

The conservation of thermal energy is described as

$$\rho T \frac{ds}{dt} = \rho T \left[\frac{\partial s}{\partial t} + (\boldsymbol{v} \cdot \nabla) s \right] = \Phi + \rho \epsilon - \mathrm{div} \boldsymbol{F}, \tag{C.10}$$

where T is the temperature, s the specific entropy, Φ the viscous dissipative function, ϵ the other heating rate (such as nuclear energy) per unit mass (i.e., $\rho\epsilon$ per unit volume), and $\boldsymbol{F}$ the energy flux (which includes, e.g., the radiative flux $\boldsymbol{F}_{\mathrm{rad}}$, the convective one $\boldsymbol{F}_{\mathrm{conv}}$, and the conductive one $\boldsymbol{F}_{\mathrm{cond}}$)

Using the first law of thermodynamics: $Tds = dU + pdV$, where U the internal energy per unit mass, we can rewrite the left-hand side of the energy equation (C.10) as $\rho T ds/dt = \rho dU/dt + p\,\mathrm{div}\boldsymbol{v}$. For an ideal gas, moreover, the internal energy U is expressed as $U = [1/(\gamma-1)](p/\rho)$, where $\gamma = C_p/C_v$ is the ratio of the specific heat. Hence, the left-hand side of the energy equation (C.10) becomes, with the help of the continuity equation (C.1),

$$\rho T\frac{ds}{dt} = \frac{1}{\gamma-1}\left(\frac{dp}{dt} - \gamma\frac{p}{\rho}\frac{d\rho}{dt}\right). \tag{C.11}$$

The *viscous dissipative function* Φ, which expresses the viscous heating rate per unit volume, is expressed as

$$\begin{aligned}\Phi = t_{ik}\frac{\partial v_i}{\partial x_k} &= \frac{1}{2}\eta\left(\frac{\partial v_i}{\partial x_k} + \frac{\partial v_k}{\partial x_i}\right)^2 + \left(\zeta - \frac{2}{3}\eta\right)\left(\frac{\partial v_j}{\partial x_j}\right)^2 \\ &= \frac{1}{2}\eta\sigma_{ij}^2 + \left(\zeta - \frac{2}{3}\eta\right)\left(\frac{\partial v_j}{\partial x_j}\right)^2 .\end{aligned} \tag{C.12}$$

The energy flux $\boldsymbol{F}$ is generally written in the form:

$$\boldsymbol{F} = -K\ \mathrm{grad}T, \tag{C.13}$$

where K is the "conductivity." For example, in the optically thick regime, the radiative flux $\boldsymbol{F}_{\mathrm{rad}}$ is

$$\boldsymbol{F}_{\mathrm{rad}} = -\frac{4acT^3}{3k_{\mathrm{tot}}\rho}\ \mathrm{grad}T, \tag{C.14}$$

where a the radiation constant and k_{tot} the (Rosseland-mean) total opacity.

Thus, the energy equation is finally expressed as

$$\frac{1}{\gamma-1}\left(\frac{dp}{dt} - \gamma\frac{p}{\rho}\frac{d\rho}{dt}\right) = \Phi + \rho\epsilon + \mathrm{div}\,(K\ \mathrm{grad}T)\,. \tag{C.15}$$

If the pressure consists of gas and radiation ones as shown in Eq. (C.17), the energy equation is written as

$$\frac{1}{\Gamma_3-1}\left(\frac{dp}{dt} - \Gamma_1\frac{p}{\rho}\frac{d\rho}{dt}\right) = \Phi + \rho\epsilon + \mathrm{div}\,(K\ \mathrm{grad}T)\,, \tag{C.16}$$

where Γ_1 and Γ_3 are the generalized adiabatic exponents (see Sect. 19.6).

(d) **Equation of state**

Finally, the equation of state is

$$p = \frac{\mathcal{R}}{\bar{\mu}}\rho T + \frac{1}{3}aT^4, \tag{C.17}$$

where $\mathcal{R}$ is the gas constant and $\bar{\mu}$ the mean molecular weight.

C.2 Cylindrical Coordinate Expression

For the convenience of readers, we explicitly write down the basic equations for a viscous fluid in cylindrical coordinates (r, φ, z), where the z-axis is coincident with the axis of symmetry.

(a) **Continuity equation**

$$\frac{\partial \rho}{\partial t} + \frac{\partial}{r\partial r}(r\rho v_r) + \frac{\partial}{r\partial \varphi}(\rho v_\varphi) + \frac{\partial}{\partial z}(\rho v_z) = 0, \tag{C.18}$$

where (v_r, v_φ, v_z) are the components of velocity in the cylindrical coordinates.

(b) **Equation of motion**

$$\frac{\partial v_r}{\partial t} + v_r\frac{\partial v_r}{\partial r} + v_\varphi\frac{\partial v_r}{r\partial \varphi} + v_z\frac{\partial v_r}{\partial z} - \frac{v_\varphi^2}{r} = -\frac{\partial \psi}{\partial r} - \frac{1}{\rho}\frac{\partial p}{\partial r} + N_r + K_r, \tag{C.19}$$

$$\frac{\partial v_\varphi}{\partial t} + v_r\frac{\partial v_\varphi}{\partial r} + v_\varphi\frac{\partial v_\varphi}{r\partial \varphi} + v_z\frac{\partial v_\varphi}{\partial z} + \frac{v_r v_\varphi}{r} = -\frac{\partial \psi}{r\partial \varphi} - \frac{1}{\rho}\frac{\partial p}{r\partial \varphi} + N_\varphi + K_\varphi, \tag{C.20}$$

$$\frac{\partial v_z}{\partial t} + v_r\frac{\partial v_z}{\partial r} + v_\varphi\frac{\partial v_z}{r\partial \varphi} + v_z\frac{\partial v_z}{\partial z} = -\frac{\partial \psi}{\partial z} - \frac{1}{\rho}\frac{\partial p}{\partial z} + N_z + K_z, \tag{C.21}$$

where (N_r, N_φ, N_z) are the components of the viscous force $\boldsymbol{N}$ and (K_r, K_φ, K_z) are those of external force $\boldsymbol{K}$.

In cylindrical coordinates the components of the viscous force are expressed as

$$\rho N_r = \frac{1}{r}\frac{\partial}{\partial r}(r t_{rr}) + \frac{1}{r}\frac{\partial t_{r\varphi}}{\partial \varphi} - \frac{t_{\varphi\varphi}}{r} + \frac{\partial t_{rz}}{\partial z}, \tag{C.22}$$

$$\rho N_\varphi = \frac{1}{r^2}\frac{\partial}{\partial r}\left(r^2 t_{r\varphi}\right) + \frac{1}{r}\frac{\partial t_{\varphi\varphi}}{\partial \varphi} + \frac{\partial t_{z\varphi}}{\partial z}, \tag{C.23}$$

$$\rho N_z = \frac{1}{r}\frac{\partial}{\partial r}(r t_{rz}) + \frac{1}{r}\frac{\partial t_{\varphi z}}{\partial \varphi} + \frac{\partial t_{zz}}{\partial z}, \tag{C.24}$$

where the viscous stress tensor t_{ik} is given by

$$t_{rr} = \eta\,\sigma_{rr} + \left(\zeta - \frac{2}{3}\eta\right) \mathrm{div}\,\boldsymbol{v}$$

$$= 2\eta\frac{\partial v_r}{\partial r} + \left(\zeta - \frac{2}{3}\eta\right) \mathrm{div}\,\boldsymbol{v}, \tag{C.25}$$

$$t_{\varphi\varphi} = \eta\,\sigma_{\varphi\varphi} + \left(\zeta - \frac{2}{3}\eta\right) \mathrm{div}\,\boldsymbol{v}$$

$$= 2\eta\left(\frac{1}{r}\frac{\partial v_\varphi}{\partial \varphi} + \frac{v_r}{r}\right) + \left(\zeta - \frac{2}{3}\eta\right) \mathrm{div}\,\boldsymbol{v}, \tag{C.26}$$

$$t_{zz} = \eta\,\sigma_{zz} + \left(\zeta - \frac{2}{3}\eta\right) \mathrm{div}\,\boldsymbol{v}$$

$$= 2\eta\frac{\partial v_z}{\partial z} + \left(\zeta - \frac{2}{3}\eta\right) \mathrm{div}\,\boldsymbol{v}, \tag{C.27}$$

$$t_{r\varphi} = t_{\varphi r} = \eta\,\sigma_{\varphi r} = \eta\left[r\frac{\partial}{\partial r}\left(\frac{v_\varphi}{r}\right) + \frac{1}{r}\frac{\partial v_r}{\partial \varphi}\right], \tag{C.28}$$

$$t_{rz} = t_{zr} = \eta\,\sigma_{rz} = \eta\left(\frac{\partial v_z}{\partial r} + \frac{\partial v_r}{\partial z}\right), \tag{C.29}$$

$$t_{\varphi z} = t_{z\varphi} = \eta\,\sigma_{\varphi z} = \eta\left(\frac{\partial v_\varphi}{\partial z} + \frac{1}{r}\frac{\partial v_z}{\partial \varphi}\right), \tag{C.30}$$

$$\mathrm{div}\,\boldsymbol{v} = \frac{1}{r}\frac{\partial}{\partial r}(r v_r) + \frac{1}{r}\frac{\partial v_\varphi}{\partial \varphi} + \frac{\partial v_z}{\partial z}. \tag{C.31}$$

In viscous accretion flows (e.g., geometrically this accretion disks), only the $r\varphi$-component of the viscous stress tensor, $t_{r\varphi}$, is dominant, and the main component of $\boldsymbol{N}$ is N_φ. In this case we have

$$\rho N_\varphi = \frac{1}{r^2}\frac{\partial}{\partial r}\left(r^2 t_{r\varphi}\right) \quad \text{and} \quad t_{r\varphi} = \eta r\frac{\partial}{\partial r}\left(\frac{v_\varphi}{r}\right) = \eta r\frac{\partial \Omega}{\partial r}; \tag{C.32}$$

$$\rho N_\varphi = \frac{1}{r^2}\frac{\partial}{\partial r}\left(\eta r^3\frac{d\Omega}{dr}\right), \tag{C.33}$$

where Ω is the angular velocity of rotation: $\Omega = v_\varphi/r$.

(c) **Energy equation**

$$\frac{1}{\Gamma_3 - 1}\left[\left(\frac{\partial}{\partial t} + v_r\frac{\partial}{\partial r} + v_\varphi\frac{\partial}{r\partial\varphi} + v_z\frac{\partial}{\partial z}\right)p \right.$$
$$\left. -\Gamma_1\frac{p}{\rho}\left(\frac{\partial}{\partial t} + v_r\frac{\partial}{\partial r} + v_\varphi\frac{\partial}{r\partial\varphi} + v_z\frac{\partial}{\partial z}\right)\rho\right]$$
$$= \Phi + \rho\epsilon + \frac{\partial}{r\partial r}\left(rK\frac{\partial T}{\partial r}\right)$$
$$+\frac{\partial}{r\partial\varphi}\left(K\frac{\partial T}{r\partial\varphi}\right) + \frac{\partial}{\partial z}\left(K\frac{\partial T}{\partial z}\right). \quad \text{(C.34)}$$

In cylindrical coordinates the viscous dissipative function Φ is expressed as

$$\Phi = \frac{1}{2}\eta\,(\sigma_{rr}^2 + \sigma_{\varphi\varphi}^2 + \sigma_{zz}^2 + 2\sigma_{r\varphi}^2 + 2\sigma_{rz}^2 + 2\sigma_{\varphi z}^2) + \left(\zeta - \frac{2}{3}\eta\right)(\mathrm{div}\,\boldsymbol{v})^2,$$

or more explicitly as

$$\Phi = \eta\left[2\left(\frac{\partial v_r}{\partial r}\right)^2 + 2\left(\frac{\partial v_\varphi}{r\partial\varphi} + \frac{v_r}{r}\right)^2 + 2\left(\frac{\partial v_z}{\partial z}\right)^2\right.$$
$$\left. +\left(\frac{\partial v_\varphi}{r\partial\varphi} + \frac{\partial v_\varphi}{\partial r} - \frac{v_\varphi}{r}\right)^2 + \left(\frac{\partial v_\varphi}{\partial z} + \frac{\partial v_z}{r\partial\varphi}\right)^2 + \left(\frac{\partial v_z}{\partial r} + \frac{\partial v_r}{\partial z}\right)^2\right]$$
$$+\left(\zeta - \frac{2}{3}\eta\right)\left[\frac{\partial(rv_r)}{r\partial r} + \frac{\partial v_\varphi}{r\partial\varphi} + \frac{\partial v_z}{\partial z}\right]^2. \quad \text{(C.35)}$$

In the case of accretion disks, where the shear due to the rotational motion is important, we retain the dominant term and have

$$\Phi = t_{r\varphi}\left(\frac{\partial v_\varphi}{\partial r} - \frac{v_\varphi}{r}\right) = \eta\left(\frac{\partial v_\varphi}{\partial r} - \frac{v_\varphi}{r}\right)^2 = \eta r^2\left(\frac{d\Omega}{dr}\right)^2. \quad \text{(C.36)}$$

C.3 Spherical Coordinate Expression

Basic equations expressed in terms of spherical coordinates (r, θ, φ) are summarized here.

(a) **Continuity equation**

$$\frac{\partial \rho}{\partial t} + \frac{1}{r^2}\frac{\partial}{\partial r}(r^2 \rho v_r) + \frac{1}{r\sin\theta}\frac{\partial}{\partial \theta}(\sin\theta\, \rho v_\theta) + \frac{1}{r\sin\theta}\frac{\partial \rho v_\varphi}{\partial \varphi} = 0, \tag{C.37}$$

where $(v_r, v_\theta, v_\varphi)$ are the components of velocity in the spherical coordinates.

(b) **Equation of motion**

$$\frac{\partial v_r}{\partial t} + v_r\frac{\partial v_r}{\partial r} + \frac{v_\theta}{r}\frac{\partial v_r}{\partial \theta} + \frac{v_\varphi}{r\sin\theta}\frac{\partial v_r}{\partial \varphi} - \frac{v_\theta^2 + v_\varphi^2}{r} = -\frac{\partial \psi}{\partial r} - \frac{1}{\rho}\frac{\partial p}{\partial r} + N_r + K_r, \tag{C.38}$$

$$\frac{\partial v_\theta}{\partial t} + v_r\frac{\partial v_\theta}{\partial r} + \frac{v_\theta}{r}\frac{\partial v_\theta}{\partial \theta} + \frac{v_\varphi}{r\sin\theta}\frac{\partial v_\theta}{\partial \varphi} + \frac{v_r v_\theta}{r} - \frac{v_\varphi^2 \cot\theta}{r} = -\frac{\partial \psi}{r\partial \theta} - \frac{1}{\rho r}\frac{\partial p}{\partial \theta} + N_\theta + K_\theta, \tag{C.39}$$

$$\frac{\partial v_\varphi}{\partial t} + v_r\frac{\partial v_\varphi}{\partial r} + \frac{v_\theta}{r}\frac{\partial v_\varphi}{\partial \theta} + \frac{v_\varphi}{r\sin\theta}\frac{\partial v_\varphi}{\partial \varphi} + \frac{v_r v_\varphi}{r} + \frac{v_\theta v_\varphi \cot\theta}{r} = -\frac{1}{r\sin\theta}\frac{\partial \psi}{\partial \varphi} - \frac{1}{\rho r\sin\theta}\frac{\partial p}{\partial \varphi} + N_\varphi + K_\varphi, \tag{C.40}$$

where $(N_r, N_\theta, N_\varphi)$ are the components of the viscous force $\boldsymbol{N}$ and $(K_r, K_\theta, K_\varphi)$ are those of external force.

In spherical coordinates the components of the viscous force per unit volume can be expressed in terms of stress tensor as

$$\rho N_r = \frac{1}{r^2}\frac{\partial}{\partial r}(r^2 t_{rr}) + \frac{1}{r\sin\theta}\frac{\partial}{\partial \theta}(\sin\theta\, t_{r\theta}) + \frac{1}{r\sin\theta}\frac{\partial}{\partial \varphi}(t_{r\varphi}) - \frac{1}{r}(t_{\theta\theta} + t_{\varphi\varphi}), \tag{C.41}$$

$$\rho N_\theta = \frac{1}{r^3}\frac{\partial}{\partial r}(r^3 t_{r\theta}) + \frac{1}{r\sin\theta}\frac{\partial}{\partial \theta}(t_{\theta\theta}\sin\theta) + \frac{1}{r\sin\theta}\frac{\partial t_{\theta\varphi}}{\partial \varphi} - \frac{\cot\theta}{r}t_{\varphi\varphi}, \tag{C.42}$$

$$\rho N_\varphi = \frac{1}{r^3}\frac{\partial}{\partial r}(r^3 t_{\varphi r}) + \frac{1}{r\sin\theta}\frac{\partial}{\partial \theta}(\sin\theta\, t_{\varphi\theta}) + \frac{1}{r\sin\theta}\frac{\partial t_{\varphi\varphi}}{\partial \varphi} + \frac{\cot\theta}{r}t_{\varphi\theta}. \tag{C.43}$$

Explicit expressions for components of the viscous stress tensor in spherical coordinates are

$$t_{rr} = \eta\,\sigma_{rr} + \left(\zeta - \frac{2}{3}\eta\right)\mathrm{div}\,\boldsymbol{v} = 2\eta\frac{\partial v_r}{\partial r} + \left(\zeta - \frac{2}{3}\eta\right)\mathrm{div}\,\boldsymbol{v}, \tag{C.44}$$

$$\begin{aligned} t_{\theta\theta} &= \eta\,\sigma_{\theta\theta} + \left(\zeta - \frac{2}{3}\eta\right)\mathrm{div}\,\boldsymbol{v} \\ &= 2\eta\left(\frac{1}{r}\frac{\partial v_\theta}{\partial \theta} + \frac{v_r}{r}\right) + \left(\zeta - \frac{2}{3}\eta\right)\mathrm{div}\,\boldsymbol{v}, \end{aligned} \tag{C.45}$$

$$\begin{aligned} t_{\varphi\varphi} &= \eta\,\sigma_{\varphi\varphi} + \left(\zeta - \frac{2}{3}\eta\right)\mathrm{div}\,\boldsymbol{v} \\ &= 2\eta\left(\frac{1}{r\sin\theta}\frac{\partial v_\varphi}{\partial \varphi} + \frac{v_r}{r} + \frac{v_\theta\cot\theta}{r}\right) + \left(\zeta - \frac{2}{3}\eta\right)\mathrm{div}\,\boldsymbol{v}, \end{aligned} \tag{C.46}$$

$$t_{r\theta} = t_{\theta r} = \eta\,\sigma_{r\theta} = \eta\left(\frac{1}{r}\frac{\partial v_r}{\partial \theta} + \frac{\partial v_\theta}{\partial r} - \frac{v_\theta}{r}\right), \tag{C.47}$$

$$t_{\theta\varphi} = t_{\varphi\theta} = \eta\,\sigma_{\theta\varphi} = \eta\left(\frac{1}{r\sin\theta}\frac{\partial v_\theta}{\partial \varphi} + \frac{1}{r}\frac{\partial v_\varphi}{\partial \theta} - \frac{v_\varphi\cot\theta}{r}\right), \tag{C.48}$$

$$t_{\varphi r} = t_{r\varphi} = \eta\,\sigma_{\varphi r} = \eta\left(\frac{\partial v_\varphi}{\partial r} + \frac{1}{r\sin\theta}\frac{\partial v_r}{\partial \varphi} - \frac{v_\varphi}{r}\right), \tag{C.49}$$

$$\mathrm{div}\,\boldsymbol{v} = \frac{1}{r^2}\frac{\partial}{\partial r}(r^2 v_r) + \frac{1}{r\sin\theta}\frac{\partial}{\partial \theta}(\sin\theta\, v_\theta) + \frac{1}{r\sin\theta}\frac{\partial v_\varphi}{\partial \varphi}. \tag{C.50}$$

(c) **Energy equation**

$$\begin{aligned} &\frac{1}{\Gamma_3 - 1}\left[\left(\frac{\partial}{\partial t} + v_r\frac{\partial}{\partial r} + \frac{v_\theta}{r}\frac{\partial}{\partial \theta} + \frac{v_\varphi}{r\sin\theta}\frac{\partial}{\partial \varphi}\right)p \right. \\ &\left. -\Gamma_1\frac{p}{\rho}\left(\frac{\partial}{\partial t} + v_r\frac{\partial}{\partial r} + \frac{v_\theta}{r}\frac{\partial}{\partial \theta} + \frac{v_\varphi}{r\sin\theta}\frac{\partial}{\partial \varphi}\right)\rho\right] \\ &= \Phi + \rho\epsilon + \frac{1}{r^2\partial r}\left(r^2\frac{\partial T}{\partial r}\right) + \frac{1}{r\sin\theta}\frac{\partial}{\partial \theta}\left(\frac{1}{r}\frac{\partial T}{\partial \theta}\right) \\ &+ \frac{1}{r\sin\theta}\frac{\partial}{\partial \varphi}\left(\frac{1}{r\sin\theta}\frac{\partial T}{\partial \varphi}\right), \end{aligned} \tag{C.51}$$

In spherical coordinates the viscous dissipative function Φ is expressed as

$$\Phi = \frac{1}{2}\eta\left(\sigma_{rr}^2 + \sigma_{\theta\theta}^2 + \sigma_{\varphi\varphi}^2 + 2\sigma_{r\theta}^2 + 2\sigma_{r\varphi}^2 + 2\sigma_{\theta\varphi}^2\right) + \left(\zeta - \frac{2}{3}\eta\right)(\mathrm{div}\,\boldsymbol{v})^2, \tag{C.52}$$

or more explicitly as

$$
\begin{aligned}
\Phi = \eta \Bigg[& 2\left(\frac{\partial v_r}{\partial r}\right)^2 + 2\left(\frac{\partial v_\theta}{r\partial\theta} + \frac{v_r}{r}\right)^2 + 2\left(\frac{1}{r\sin\theta}\frac{\partial v_\varphi}{\partial\varphi} + \frac{v_r}{r} + \frac{v_\theta\cot\theta}{r}\right)^2 \\
& + \left(\frac{\partial v_r}{r\partial\theta} + \frac{\partial v_\theta}{\partial r} - \frac{v_\theta}{r}\right)^2 + \left(\frac{1}{r\sin\theta}\frac{\partial v_\theta}{\partial\varphi} + \frac{\partial v_\varphi}{r\partial\theta} - \frac{v_\varphi\cot\theta}{r}\right)^2 \\
& + \left(\frac{\partial v_\varphi}{\partial r} + \frac{1}{r\sin\theta}\frac{\partial v_r}{\partial\varphi} - \frac{v_\varphi}{r}\right)^2\Bigg] \\
& + \left(\zeta - \frac{2}{3}\eta\right)\left[\frac{1}{r^2}\frac{\partial}{\partial r}(r^2 v_r) + \frac{1}{r\sin\theta}\frac{\partial}{\partial\theta}(\sin\theta\, v_\theta) + \frac{1}{r\sin\theta}\frac{\partial v_\varphi}{\partial\varphi}\right]^2 . \qquad \text{(C.53)}
\end{aligned}
$$

Appendix D
Brief Summary of Magnetohydrodynamical Equations

In this appendix some of basic equations for ionized fluids subject to the magnetic fields under magnetohydrodynamical approximations, the *magnetohydrodynamical equations*, are described in the nonrelativistic regime. The basic equations include the continuity equation, the equation of motion, the energy equation, and the induction equation, supplemented by the equation of state. In order to avoid repetition, we describe here only (a) the equation of motion and (b) the induction equation. We first show them in vector forms, and then we give them in cylindrical and spherical coordinates.

D.1 General Form

The equation of motion and the induction equation are described here.

In ionized fluids the term of the Lorentz force is added to the equation of motion:

$$\rho\frac{d\boldsymbol{v}}{dt} = \rho\left[\frac{\partial \boldsymbol{v}}{\partial t} + (\boldsymbol{v}\cdot\nabla)\boldsymbol{v}\right] = -\rho\nabla\psi - \nabla p + \frac{1}{c}\boldsymbol{j}\times\boldsymbol{B} + \rho\boldsymbol{N} + \rho\boldsymbol{K}, \qquad \text{(D.1)}$$

where $\boldsymbol{j}$ is the electric current and $\boldsymbol{B}$ the strength of magnetic fields, and the Lorentz force $\boldsymbol{j}\times\boldsymbol{B}/c$ can be also expressed in terms of magnetic fields as

$$\frac{1}{c}\boldsymbol{j}\times\boldsymbol{B} = \frac{1}{4\pi}\mathrm{curl}\,\boldsymbol{B}\times\boldsymbol{B} = -\nabla\left(\frac{1}{8\pi}\boldsymbol{B}^2\right) + \frac{1}{4\pi}(\boldsymbol{B}\cdot\nabla)\boldsymbol{B}. \qquad \text{(D.2)}$$

© Springer Nature Singapore Pte Ltd. 2020, corrected publication 2023

S. Kato, J. Fukue, *Fundamentals of Astrophysical Fluid Dynamics*,
Astronomy and Astrophysics Library,
https://doi.org/10.1007/978-981-15-4174-2

The induction equation is an equation describing the time evolution of the magnetic fields, and is

$$\frac{\partial \boldsymbol{B}}{\partial t} = \mathrm{curl}\,(\boldsymbol{v} \times \boldsymbol{B}) - \mathrm{curl}(\eta\,\mathrm{curl}\,\boldsymbol{B})$$
$$= \mathrm{curl}\,(\boldsymbol{v} \times \boldsymbol{B}) - \eta\,\mathrm{curl}\,\mathrm{curl}\,\boldsymbol{B} - \nabla\eta \times \mathrm{curl}\,\boldsymbol{B}, \tag{D.3}$$

where the magnetic diffusivity η is given by $\eta = c^2/4\pi\sigma$ (σ is electric conductively) and should not be confused with the dynamical viscosity used in Appendix C. If $\eta =$ const., the above induction equation is written as

$$\frac{\partial \boldsymbol{B}}{\partial t} = \mathrm{curl}\,(\boldsymbol{v} \times \boldsymbol{B}) + \eta\nabla^2\boldsymbol{B}, \tag{D.4}$$

where we should be careful when $\nabla^2\boldsymbol{B}$ is expressed in terms of coordinate systems other than the Cartesian ones.

D.2 Cylindrical Coordinate Expression

Let us express the Lorenz force and the induction equation in cylindrical coordinates (r, φ, z).

(a) **Lorentz force**

The Lorentz force is explicitly written in cylindrical coordinates as

$$\frac{1}{c}[\boldsymbol{j} \times \boldsymbol{B}]_r = \frac{1}{4\pi}[\mathrm{curl}\boldsymbol{B} \times \boldsymbol{B}]_r$$
$$= \frac{1}{4\pi}\left[\left(\frac{\partial B_r}{\partial z} - \frac{\partial B_z}{\partial r}\right)B_z - \left(\frac{\partial}{r\partial r}rB_\varphi - \frac{\partial B_r}{r\partial\varphi}\right)B_\varphi\right], \tag{D.5}$$

$$\frac{1}{c}[\boldsymbol{j} \times \boldsymbol{B}]_\varphi = \frac{1}{4\pi}[\mathrm{curl}\boldsymbol{B} \times \boldsymbol{B}]_\varphi$$
$$= \frac{1}{4\pi}\left[\left(\frac{\partial}{r\partial r}rB_\varphi - \frac{\partial B_r}{r\partial\varphi}\right)B_r - \left(\frac{\partial B_z}{r\partial\varphi} - \frac{\partial B_\varphi}{\partial z}\right)B_z\right], \tag{D.6}$$

$$\frac{1}{c}[\boldsymbol{j} \times \boldsymbol{B}]_z = \frac{1}{4\pi}[\mathrm{curl}\boldsymbol{B} \times \boldsymbol{B}]_z$$
$$= \frac{1}{4\pi}\left[\left(\frac{\partial B_z}{r\partial\varphi} - \frac{\partial B_\varphi}{\partial z}\right)B_\varphi - \left(\frac{\partial B_r}{\partial z} - \frac{\partial B_z}{\partial r}\right)B_r\right]. \tag{D.7}$$

(b) **Induction equation**

In the limit of the frozen-in approximation ($\eta = 0$), we have

$$\frac{\partial B_r}{\partial t} = \frac{\partial}{r\partial\varphi}\left(v_r B_\varphi - v_\varphi B_r\right) - \frac{\partial}{\partial z}\left(v_z B_r - v_r B_z\right), \tag{D.8}$$

$$\frac{\partial B_\varphi}{\partial t} = \frac{\partial}{\partial z}\left(v_\varphi B_z - v_z B_\varphi\right) - \frac{\partial}{\partial r}\left(v_r B_\varphi - v_\varphi B_r\right), \tag{D.9}$$

$$\frac{\partial B_z}{\partial t} = \frac{\partial}{r\partial r}\left[r\left(v_z B_r - v_r B_z\right)\right] - \frac{\partial}{r\partial\varphi}\left(v_\varphi B_z - v_z B_\varphi\right). \tag{D.10}$$

If η is not constant, the last term on the right-hand side of Eq. (D.3) should be added to the above equations. It is easy to add the term in the form expressed in terms of cylindrical coordinates [see Eqs. (B.11)–(B.13)].

D.3 Spherical Coordinate Expression

Expressing the Lorentz force and the induction equation in terms of spherical coordinates (r, θ, φ) is also straightforward. We summarize them as follows.

(a) **Lorentz force**

The components of the Lorentz force in spherical coordinates are

$$\begin{aligned}\frac{1}{c}[\boldsymbol{j} \times \boldsymbol{B}]_r &= \frac{1}{4\pi}[\mathrm{curl}\boldsymbol{B} \times \boldsymbol{B}]_r \\ &= \frac{1}{4\pi}\left[\left(\frac{1}{r\sin\theta}\frac{\partial B_r}{\partial\varphi} - \frac{1}{r}\frac{\partial}{\partial r}rB_\varphi\right)B_\varphi - \left(\frac{\partial}{r\partial r}rB_\theta - \frac{\partial B_r}{r\partial\theta}\right)B_\theta\right],\end{aligned} \tag{D.11}$$

$$\begin{aligned}\frac{1}{c}[\boldsymbol{j} \times \boldsymbol{B}]_\theta &= \frac{1}{4\pi}[\mathrm{curl}\boldsymbol{B} \times \boldsymbol{B}]_\theta \\ &= \frac{1}{4\pi}\left[\left(\frac{\partial}{r\partial r}rB_\theta - \frac{\partial B_r}{r\partial\theta}\right)B_r - \frac{1}{r\sin\theta}\left(\frac{\partial}{\partial\theta}\sin\theta B_\varphi - \frac{\partial B_\theta}{\partial\varphi}\right)B_\varphi\right],\end{aligned} \tag{D.12}$$

$$\begin{aligned}\frac{1}{c}[\boldsymbol{j} \times \boldsymbol{B}]_\varphi &= \frac{1}{4\pi}[\mathrm{curl}\boldsymbol{B} \times \boldsymbol{B}]_\varphi \\ &= \frac{1}{4\pi}\left[\frac{1}{r\sin\theta}\left(\frac{\partial}{\partial\theta}\sin\theta B_\varphi - \frac{\partial B_\theta}{\partial\varphi}\right)B_\theta - \left(\frac{1}{r\sin\theta}\frac{\partial B_r}{\partial\varphi} - \frac{\partial}{r\partial r}rB_\varphi\right)B_r\right].\end{aligned} \tag{D.13}$$

(b) Induction equation

In the case of magnetic diffusivity η is zero, we have

$$\frac{\partial B_r}{\partial t} = \frac{1}{r\sin\theta}\frac{\partial}{\partial\theta}[\sin\theta(v_r B_\theta - v_\theta B_r)] - \frac{1}{r\sin\theta}\frac{\partial}{\partial\varphi}(v_\varphi B_r - v_r B_\varphi), \quad \text{(D.14)}$$

$$\frac{\partial B_\theta}{\partial t} = \frac{1}{r\sin\theta}\frac{\partial}{\partial\varphi}(v_\theta B_\varphi - v_\varphi B_\theta) - \frac{1}{r}\frac{\partial}{\partial r}[r(v_r B_\theta - v_\theta B_r)], \quad \text{(D.15)}$$

$$\frac{\partial B_\varphi}{\partial t} = \frac{1}{r}\frac{\partial}{\partial r}[r(v_\varphi B_r - v_r B_\varphi)] - \frac{1}{r}\frac{\partial}{\partial\theta}(v_\theta B_\varphi - v_\varphi B_\theta). \quad \text{(D.16)}$$

Appendix E
Brief Summary of Radiation Hydrodynamical Equations

In this appendix we summarize the basic equations for a radiative fluid, the *radiation hydrodynamical equations*, in the nonrelativistic regime. The basic equations include the continuity equation, the equation of motion, the energy equation, the zeroth moment equation, the first moment equation, the closure relation such as the Eddington approximation, supplemented by the equation of state. We first present these basic equations in vector forms on the order of $\mathcal{O}(v/c)^0$ (cf. Chap. 21), and then express them in several coordinate systems.

E.1 General Form

(a) **Continuity equation**

For matter, the continuity equation is written as

$$\frac{\partial \rho}{\partial t} + \nabla(\rho \boldsymbol{v}) = 0, \tag{E.1}$$

where ρ is the density and $\boldsymbol{v}$ the flow velocity.

(b) **Equation of motion**

Including the radiative force, the equation of motion for matter is

$$\frac{\partial \boldsymbol{v}}{\partial t} + (\boldsymbol{v} \cdot \nabla)\, \boldsymbol{v} = -\nabla \psi - \frac{1}{\rho} \nabla p + \frac{(\kappa_F + \sigma_F)}{c} \boldsymbol{F}, \tag{E.2}$$

where ψ is the gravitational potential, p the gas pressure ($p = \mathcal{R} \rho T / \bar{\mu}$), κ_F and σ_F are the flux-mean absorption and scattering opacities, respectively, and $\boldsymbol{F}$ the radiative flux. We here omit the viscous force and other ones.

© Springer Nature Singapore Pte Ltd. 2020, corrected publication 2023

S. Kato, J. Fukue, *Fundamentals of Astrophysical Fluid Dynamics*,
Astronomy and Astrophysics Library,
https://doi.org/10.1007/978-981-15-4174-2

(c) **Energy equation**

Including the net energy transfer rate, under the present approximation, we can write the energy equation for matter as

$$\left(\frac{\partial}{\partial t} + \boldsymbol{v} \cdot \nabla\right) U + \frac{p}{\rho} \nabla \boldsymbol{v} = \frac{1}{\rho} q^{+} - (j - c\kappa_E E), \tag{E.3}$$

where U is the internal energy per unit mass, and q^{+} the (viscous or nuclear) heating rate per unit volume, j the emissivity per unit mass,[1] E the radiation energy density, and κ_E the energy-mean absorption opacity.

(d) **Zeroth moment equation**

The zeroth moment equation is

$$\frac{\partial E}{\partial t} + \nabla \boldsymbol{F} = \rho\, (j - c\kappa_E E)\,, \tag{E.4}$$

where the right-hand side is the net energy transfer rate per unit volume.

(e) **First moment equation**

The first moment equation is

$$\frac{1}{c^2} \frac{\partial \boldsymbol{F}}{\partial t} + \frac{\partial P^{ik}}{\partial x^k} = -\frac{1}{c} \rho (\kappa + \sigma)_F \boldsymbol{F}, \tag{E.5}$$

where P^{ik} is the radiation stress tensor.

(f) **Eddington approximation**

The Eddington approximation[2] is

$$P^{ik} = \frac{\delta^{ik}}{3} E. \tag{E.6}$$

---

[1]In the local thermodynamical equilibrium (LTE), $j = 4\pi\kappa B$, where B is the blackbody intensity from matter. The absorption opacities are usually replaced by the Rosseland mean opacity $\bar{\kappa}_{\mathrm{R}}$, and the scattering one is usually an electron scattering opacity κ_{es}.

[2]In the regime, where the Eddington approximation becomes worse, we adopt another closure relations, or directly solve the radiative transfer equation:

$$\frac{1}{c} \frac{\partial I}{\partial t} + (\boldsymbol{l} \cdot \nabla)\, I = \rho \left[\frac{j}{4\pi} - (\kappa + \sigma)_I I + \sigma_E \frac{c}{4\pi} E \right].$$

Substituting the Eddington approximation (E.6) into Eq. (E.5), we have

$$\frac{1}{c^2}\frac{\partial \boldsymbol{F}}{\partial t} + \frac{1}{3}\nabla E = -\frac{1}{c}\rho(\kappa+\sigma)_F \boldsymbol{F}, \tag{E.7}$$

whereas if we merely assume that the radiation field is isotropic, then we have

$$\frac{1}{c^2}\frac{\partial \boldsymbol{F}}{\partial t} + \nabla P = -\frac{1}{c}\rho(\kappa+\sigma)_F \boldsymbol{F}, \tag{E.8}$$

where P is the (isotropic) radiation pressure.

E.2 Cartesian Coordinate Expression

For the convenience of readers, we explicitly write down the basic equations for a radiative fluid in Cartesian coordinates (x, y, z).

In the Cartesian coordinates (x, y, z), the continuity equation, equation of motion, and energy equation are explicitly written as

$$\frac{\partial \rho}{\partial t} + \frac{\partial}{\partial x}(\rho v_x) + \frac{\partial}{\partial y}(\rho v_y) + \frac{\partial}{\partial z}(\rho v_z) = 0, \tag{E.9}$$

$$\frac{\partial v_x}{\partial t} + v_x\frac{\partial v_x}{\partial x} + v_y\frac{\partial v_x}{\partial y} + v_z\frac{\partial v_x}{\partial z} = -\frac{1}{\rho}\frac{\partial p}{\partial x} - \frac{\partial \psi}{\partial x} + \frac{\kappa_F+\sigma_F}{c}F_x, \tag{E.10}$$

$$\frac{\partial v_y}{\partial t} + v_x\frac{\partial v_y}{\partial x} + v_y\frac{\partial v_y}{\partial y} + v_z\frac{\partial v_y}{\partial z} = -\frac{1}{\rho}\frac{\partial p}{\partial y} - \frac{\partial \psi}{\partial y} + \frac{\kappa_F+\sigma_F}{c}F_y, \tag{E.11}$$

$$\frac{\partial v_z}{\partial t} + v_x\frac{\partial v_z}{\partial x} + v_y\frac{\partial v_z}{\partial y} + v_z\frac{\partial v_z}{\partial z} = -\frac{1}{\rho}\frac{\partial p}{\partial z} - \frac{\partial \psi}{\partial z} + \frac{\kappa_F+\sigma_F}{c}F_z, \tag{E.12}$$

$$\frac{p}{\gamma-1}\left(\frac{\partial}{\partial t} + v_x\frac{\partial}{\partial x} + v_y\frac{\partial}{\partial y} + v_z\frac{\partial}{\partial z}\right)\ln\left(\frac{p}{\rho^\gamma}\right) = q^+ - \rho(j - c\kappa_E E). \tag{E.13}$$

On the other hand, the moment equations for radiation are written as

$$\frac{\partial E}{\partial t} + \frac{\partial F_x}{\partial x} + \frac{\partial F_y}{\partial y} + \frac{\partial F_z}{\partial z} = \rho(j - c\kappa_E E), \tag{E.14}$$

$$\frac{1}{c^2}\frac{\partial F_x}{\partial t} + \frac{\partial P_{xx}}{\partial x} + \frac{\partial P_{xy}}{\partial y} + \frac{\partial P_{xz}}{\partial z} = -\rho\frac{\kappa_F+\sigma_F}{c}F_x, \tag{E.15}$$

$$\frac{1}{c^2}\frac{\partial F_y}{\partial t} + \frac{\partial P_{yx}}{\partial x} + \frac{\partial P_{yy}}{\partial y} + \frac{\partial P_{yz}}{\partial z} = -\rho\frac{\kappa_F+\sigma_F}{c}F_y, \tag{E.16}$$

$$\frac{1}{c^2}\frac{\partial F_z}{\partial t} + \frac{\partial P_{zx}}{\partial x} + \frac{\partial P_{zy}}{\partial y} + \frac{\partial P_{zz}}{\partial z} = -\rho\frac{\kappa_F+\sigma_F}{c}F_z. \tag{E.17}$$

In a steady plane-parallel geometry (z), where $\partial/\partial t = \partial/\partial x = \partial/\partial y = 0$, $F_x = F_y = 0$, the continuity equation, the vertical momentum balance, and energy equation are, respectively,

$$\frac{d}{dz}(\rho v_z) = 0, \tag{E.18}$$

$$v_z \frac{dv_z}{dz} = -\frac{d\psi}{dz} - \frac{1}{\rho}\frac{dp}{dz} + \frac{(\kappa+\sigma)_F}{c}F, \tag{E.19}$$

$$\frac{p v_z}{\gamma - 1}\frac{d}{dz}\ln\left(\frac{p}{\rho^\gamma}\right) = q^+_{\rm vis} - \rho\,(j - c\kappa_E E)\,, \tag{E.20}$$

The frequency-integrated transfer equation, and the zeroth and first moment equations become, respectively,

$$\cos\theta\frac{dI}{dz} = \rho\left(\frac{j}{4\pi} - (\kappa+\sigma)_I I + \kappa_E\frac{cE}{4\pi}\right), \tag{E.21}$$

$$\frac{dF}{dz} = \rho\,(j - c\kappa_E E)\,, \tag{E.22}$$

$$\frac{dP}{dz} = -\frac{1}{c}\rho(\kappa+\sigma)_F F, \tag{E.23}$$

where I is the frequency-integrated specific intensity, E the radiation energy density, F the vertical component of the radiative flux, P the zz-component of the radiation stress tensor, and θ the polar angle.

E.3 Cylindrical Coordinate Expression

In the cylindrical coordinates (r, φ, z), the continuity equation, equation of motion, and energy equation are explicitly written as

$$\frac{\partial\rho}{\partial t} + \frac{1}{r}\frac{\partial}{\partial r}(r\rho v_r) + \frac{1}{r}\frac{\partial}{\partial\varphi}(\rho v_\varphi) + \frac{\partial}{\partial z}(\rho v_z) = 0, \tag{E.24}$$

$$\frac{\partial v_r}{\partial t} + v_r\frac{\partial v_r}{\partial r} + \frac{v_\varphi}{r}\frac{\partial v_r}{\partial\varphi} + v_z\frac{\partial v_r}{\partial z} - \frac{v_\varphi^2}{r}$$
$$= -\frac{1}{\rho}\frac{\partial p}{\partial r} - \frac{\partial\psi}{\partial r} + \frac{\kappa_F+\sigma_F}{c}F_r, \tag{E.25}$$

$$\frac{\partial v_\varphi}{\partial t} + v_r\frac{\partial v_\varphi}{\partial r} + \frac{v_\varphi}{r}\frac{\partial v_\varphi}{\partial\varphi} + v_z\frac{\partial v_\varphi}{\partial z} + \frac{v_r v_\varphi}{r}$$

$$= -\frac{1}{\rho r}\frac{\partial p}{\partial \varphi} - \frac{1}{r}\frac{\partial \psi}{\partial \varphi} + \frac{\kappa_F + \sigma_F}{c} F_\varphi, \tag{E.26}$$

$$\frac{\partial v_z}{\partial t} + v_r \frac{\partial v_z}{\partial r} + \frac{v_\varphi}{r}\frac{\partial v_z}{\partial \varphi} + v_z \frac{\partial v_z}{\partial z}$$

$$= -\frac{1}{\rho}\frac{\partial p}{\partial z} - \frac{\partial \psi}{\partial z} + \frac{\kappa_F + \sigma_F}{c} F_z, \tag{E.27}$$

$$\frac{p}{\gamma - 1}\left(\frac{\partial}{\partial t} + v_r \frac{\partial}{\partial r} + \frac{v_\varphi}{r}\frac{\partial}{\partial \varphi} + v_z \frac{\partial}{\partial z}\right) \ln\left(\frac{p}{\rho^\gamma}\right)$$

$$= q^+ - \rho(j - c\kappa_E E). \tag{E.28}$$

On the other hand, the moment equations for radiation are written as

$$\frac{\partial E}{\partial t} + \frac{1}{r}\frac{\partial}{\partial r}(r F_r) + \frac{1}{r}\frac{\partial F_\varphi}{\partial \varphi} + \frac{\partial F_z}{\partial z} = \rho(j - c\kappa_E E), \tag{E.29}$$

$$\frac{1}{c^2}\frac{\partial F_r}{\partial t} + \frac{1}{r}\frac{\partial}{\partial r}(r P_{rr}) + \frac{1}{r}\frac{\partial P_{r\varphi}}{\partial \varphi} - \frac{P_{\varphi\varphi}}{r} + \frac{\partial P_{rz}}{\partial z} = -\rho \frac{\kappa_F + \sigma_F}{c} F_r, \tag{E.30}$$

$$\frac{1}{c^2}\frac{\partial F_\varphi}{\partial t} + \frac{1}{r^2}\frac{\partial}{\partial r}(r^2 P_{\varphi r}) + \frac{1}{r}\frac{\partial P_{\varphi\varphi}}{\partial \varphi} + \frac{\partial P_{\varphi z}}{\partial z} = -\rho \frac{\kappa_F + \sigma_F}{c} F_\varphi, \tag{E.31}$$

$$\frac{1}{c^2}\frac{\partial F_z}{\partial t} + \frac{1}{r}\frac{\partial}{\partial r}(r P_{zr}) + \frac{1}{r}\frac{\partial P_{z\varphi}}{\partial \varphi} + \frac{\partial P_{zz}}{\partial z} = -\rho \frac{\kappa_F + \sigma_F}{c} F_z. \tag{E.32}$$

E.4 Spherical Coordinate Expression

In the spherical coordinates (r, θ, φ), the continuity equation, equation of motion, and energy equation are explicitly written as

$$\frac{\partial \rho}{\partial t} + \frac{1}{r^2}\frac{\partial}{\partial r}(r^2 \rho v_r) + \frac{1}{r\sin\theta}\frac{\partial}{\partial \theta}(\sin\theta \rho v_\theta) + \frac{1}{r\sin\theta}\frac{\partial}{\partial \varphi}(\rho v_\varphi) = 0, \tag{E.33}$$

$$\frac{\partial v_r}{\partial t} + v_r \frac{\partial v_r}{\partial r} + \frac{v_\theta}{r}\frac{\partial v_r}{\partial \theta} + \frac{v_\varphi}{r\sin\theta}\frac{\partial v_r}{\partial \varphi} - \frac{v_\theta^2 + v_\varphi^2}{r}$$

$$= -\frac{1}{\rho}\frac{\partial p}{\partial r} - \frac{\partial \psi}{\partial r} + \frac{\kappa_F + \sigma_F}{c} F_r, \tag{E.34}$$

$$\frac{\partial v_\theta}{\partial t} + v_r \frac{\partial v_\theta}{\partial r} + \frac{v_\theta}{r}\frac{\partial v_\theta}{\partial \theta} + \frac{v_\varphi}{r\sin\theta}\frac{\partial v_\theta}{\partial \varphi} + \frac{v_r v_\theta}{r} - \frac{\cot\theta\, v_\varphi^2}{r}$$

$$= -\frac{1}{\rho r}\frac{\partial p}{\partial \theta} - \frac{1}{r}\frac{\partial \psi}{\partial \theta} + \frac{\kappa_F + \sigma_F}{c} F_\theta, \tag{E.35}$$

$$\frac{\partial v_\varphi}{\partial t} + v_r \frac{\partial v_\varphi}{\partial r} + \frac{v_\theta}{r}\frac{\partial v_\varphi}{\partial \theta} + \frac{v_\varphi}{r \sin\theta}\frac{\partial v_\varphi}{\partial \varphi} + \frac{v_r v_\varphi}{r} + \frac{\cot\theta\, v_\theta v_\varphi}{r}$$

$$= -\frac{1}{\rho r \sin\theta}\frac{\partial p}{\partial \varphi} - \frac{1}{r\sin\theta}\frac{\partial \psi}{\partial \varphi} + \frac{\kappa_F + \sigma_F}{c} F_\varphi, \tag{E.36}$$

$$\frac{p}{\gamma - 1}\left(\frac{\partial}{\partial t} + v_r \frac{\partial}{\partial r} + \frac{v_\theta}{r}\frac{\partial}{\partial \theta} + \frac{v_\varphi}{r\sin\theta}\frac{\partial}{\partial \varphi}\right) \ln\left(\frac{p}{\rho^\gamma}\right)$$

$$= q^+ - \rho(j - c\kappa_E E). \tag{E.37}$$

On the other hand, the moment equations for radiation are written as

$$\frac{\partial E}{\partial t} + \frac{1}{r^2}\frac{\partial}{\partial r}(r^2 F_r) + \frac{1}{r\sin\theta}\frac{\partial}{\partial \theta}(\sin\theta F_\theta) + \frac{1}{r\sin\theta}\frac{\partial F_\varphi}{\partial \varphi}$$

$$= \rho(j - c\kappa_E E), \tag{E.38}$$

$$\frac{1}{c^2}\frac{\partial F_r}{\partial t} + \frac{1}{r^2}\frac{\partial}{\partial r}(r^2 P_{rr}) + \frac{1}{r\sin\theta}\frac{\partial}{\partial \theta}(\sin\theta P_{r\theta}) + \frac{1}{r\sin\theta}\frac{\partial P_{r\varphi}}{\partial \varphi}$$

$$-\frac{P_{\theta\theta}}{r} - \frac{P_{\varphi\varphi}}{r} = -\rho\frac{\kappa_F + \sigma_F}{c} F_r, \tag{E.39}$$

$$\frac{1}{c^2}\frac{\partial F_\theta}{\partial t} + \frac{1}{r^2}\frac{\partial}{\partial r}(r^2 P_{\theta r}) + \frac{1}{r\sin\theta}\frac{\partial}{\partial \theta}(\sin\theta P_{\theta\theta}) + \frac{1}{r\sin\theta}\frac{\partial P_{\theta\varphi}}{\partial \varphi}$$

$$+\frac{P_{\theta r}}{r} - \frac{\cot\theta\, P_{\varphi\varphi}}{r} = -\rho\frac{\kappa_F + \sigma_F}{c} F_\theta, \tag{E.40}$$

$$\frac{1}{c^2}\frac{\partial F_\varphi}{\partial t} + \frac{1}{r^2}\frac{\partial}{\partial r}(r^2 P_{\varphi r}) + \frac{1}{r\sin\theta}\frac{\partial}{\partial \theta}(\sin\theta P_{\varphi\theta}) + \frac{1}{r\sin\theta}\frac{\partial P_{\varphi\varphi}}{\partial \varphi}$$

$$+\frac{P_{\varphi r}}{r} + \frac{\cot\theta\, P_{\varphi\theta}}{r} = -\rho\frac{\kappa_F + \sigma_F}{c} F_\varphi. \tag{E.41}$$

In the steady spherically symmetric case, where $\partial/\partial t = \partial/\partial\theta = \partial/\partial\varphi = 0$, the continuity equation, equation of motion, and energy equation are, respectively,

$$\frac{1}{r^2}\frac{d}{dr}(r^2 \rho v_r) = 0, \tag{E.42}$$

$$v_r \frac{dv_r}{dr} = -\frac{1}{\rho}\frac{dp}{dr} - \frac{d\psi}{dr} + \frac{\kappa_F + \sigma_F}{c} F, \tag{E.43}$$

$$\frac{p}{\gamma - 1} v_r \frac{d}{dr} \ln\left(\frac{p}{\rho^\gamma}\right) = q^+ - \rho(j - c\kappa_E E). \tag{E.44}$$

The radiative transfer equation, the zeroth and first moment equations are expressed as

$$\mu\frac{\partial I}{\partial r}+\frac{1-\mu^2}{r}\frac{\partial I}{\partial \mu}=-(\kappa+\sigma)\rho(I-S), \tag{E.45}$$

$$\frac{\partial F}{\partial r}+\frac{2}{r}F=\frac{1}{r^2}\frac{\partial}{\partial r}\left(r^2F\right)=-(\kappa+\sigma)\rho(cE-4\pi S), \tag{E.46}$$

$$\frac{\partial P}{\partial r}+\frac{3P-E}{r}=-\frac{(\kappa+\sigma)\rho}{c}F, \tag{E.47}$$

where F is the radial component of the radiative flux, and P the rr-component of the radiation stress tensor.

References

Kato, S.: Publ. Astron. Soc. Jpn. **56**, 905 (2004)
Kato, S.: Publ. Astron. Soc. Jpn. **60**, 111 (2008)
Kato, S.: Publ. Astron. Soc. Jpn. **66**, 24 (2014)
Kato, S., Unno, W.: Publ. Astron. Soc. Jpn. **19**, 1 (1967)
Lynden-Bell, D., Ostriker, J.P.: Mon. Not. R. Astron. Soc. **136**, 293 (1967)

Index

© Springer Nature Singapore Pte Ltd. 2020, corrected publication 2023

S. Kato, J. Fukue, *Fundamentals of Astrophysical Fluid Dynamics*,
Astronomy and Astrophysics Library,
https://doi.org/10.1007/978-981-15-4174-2

L

M

N

O

P

R

Printed in the United States
by Baker & Taylor Publisher Services